Wilderness Management

Fifth Edition

Wilderness Management

Fifth Edition

Stewardship and Protection of Resources and Values

Chad P. Dawson

State University of New York–College of Environmental Science and Forestry

John C. Hendee

late of University of Idaho

WAVELAND PRESS, INC.

Long Grove, Illinois

For information about this book, contact:
Waveland Press, Inc.
4180 IL Route 83, Suite 101
Long Grove, IL 60047-9580
(847) 634-0081
info@waveland.com
www.waveland.com

10-digit ISBN 1-4786-5273-X
13-digit ISBN 978-1-4786-5273-1

Printed in the United States of America

7 6 5 4 3 2 1

In Memory of John C. Hendee (1938–2016)

Lead coauthor of the first three editions of this book; coauthor of the fourth edition; coauthor of this fifth edition by virtue of his leadership and contributions to previous editions and his suggested revisions to this fifth edition prior to his death; and beloved husband, father, grandfather, author, teacher, researcher, mentor, colleague, and friend.

Dedication

This fifth edition is dedicated to Robert C. Baron and Charlotte Baron for their generous support of this book in its second, third, and fourth editions and long-term support for wilderness nationally and internationally.

Acknowledgement

The cover photo was donated by Ken Rimany © 2009. "In the Middle of Everything Wild: Elk Lake the first Adirondack Conservation Easement. Elk Lake Lodge, North Hudson, New York.

Contents

IV. WILDERNESS RESOURCES, VALUES, AND THREATS TO THEM

V. WILDERNESS USES AND THEIR MANAGEMENT

Preface

This book was first conceived in 1971 at a U.S. Forest Service–sponsored symposium, "Management Implications of Wilderness Research," where managers, researchers, and environmentalists discussed mutual concerns for wilderness management and the need for new and relevant information. The first edition of *Wilderness Management*, coauthored by John C. Hendee, George H. Stankey, and Robert C. Lucas, with many collaborating chapter coauthors and reviewers, was published in 1978; a revised second edition by the same coauthors was published in 1990. A third edition, by John C. Hendee and Chad P. Dawson in 2002, retained relevant material from the first and second editions but was rewritten to incorporate new changes and materials appropriate under the now broader title *Wilderness Management: Stewardship and Protection of Resources and Values*. The fourth (2009) and fifth editions, with Dawson assuming senior authorship, continues the evolution of content in response to expanded wilderness knowledge, challenges, and its application as a land-use designation in the United States and abroad.

We thank and recognize our colleagues George Stankey and Bob Lucas for their major coauthorship and contributions to the first two editions and to Vance Martin and The WILD Foundation for their support and encouragement for the continued publication of this book. The International Wilderness Leadership (WILD) Foundation retains all net proceeds from book sales of this edition for its wilderness leadership programs.

We thank the dozens of agency managers, scientists, university and environmental colleagues, and contributing chapter authors, coauthors, and reviewers for their contributions to this and earlier editions. The book, the entire field, and all of us in wilderness work have changed, broadened, and evolved through the collective efforts of many.

Like the first four editions, this fifth edition is directed toward wilderness stewards everywhere, especially in U.S. federal land management agencies—the U.S. Forest Service, National Park Service, U.S. Fish and Wildlife Service, and Bureau of Land Management—and toward teachers, students, resource managers, citizens, conservationists, scientists, and wilderness users. Because wilderness literature has expanded severalfold since previous editions, we've tried to be less encyclopedic, citing the most important new studies and classic ones, but presenting now accepted practices and concepts without extensive citations. Each chapter has been revised, reviewed, and edited, new material added, and other material deleted.

Our case examples and background material represent the expertise of all the U.S. agencies and, although we recognize specific policy differences among the agencies, we have sought a level of presentation that transcends agency differences. This is not a book constrained by any agency policy—our goal is to inspire and guide creative stewardship of the National Wilderness Preservation System independent of differences in current organizational perspectives, policies, or practices. Although each agency is directed by internal policies and federal guidelines, all operate under the Wilderness Act and all have high aspirations for preserving designated wilderness under their management. The expanded and updated coverage of international wilderness seeks to inspire protective designation, management, and stewardship of wilderness values where possible worldwide.

We have included authorship and coauthorship of chapters or sections of chapters by colleagues who are experts in these areas. Chapter coauthors include: Vance G. Martin, Alan Watson, Christopher A. Armatas, and Amy Lewis on "International Wilderness in the Twenty-First Century" (Chapter 3); Robert G. Dvorak on "Managing for Appropriate Wilderness Conditions" (Chapter 9); Gregory H. Aplet, Jerry F. Franklin, and Andrew J. Larson on "Wilderness Ecosystems" (Chapter 10); Sean A. Parks on "Fire in Wilderness Ecosystems" (Chapter 11); Jeffrey L. Marion on "Ecological Impacts of Wilderness Recreation and Their Management" (Chapter 15); and Chad P. Dawson and Paul A. Dawson on "Wilderness Visitor Management: Stewardship for Quality Experiences" (Chapter 16).

All parts of the book bear the imprint of the two book coauthors, either writing new material or editing contributed material as necessary for continuity with the rest of the book. Wherever the word *we* appears in the text, it refers to a consensus opinion held by us.

We are grateful to our many colleagues in universities, the U.S. wilderness management agencies, international colleagues, and the conservation community who provided encouragement and constructive review for the previous editions of this book. At the end of each chapter, we do acknowledge colleagues who contributed information or provided review for that chapter in this or previous editions.

To our supporters and critics, we offer sincere gratitude—their review and input were invaluable, but they share no responsibility for our interpretations or statements. We thank the staff at Fulcrum Publishing for their work to help us make past editions of this book as good as we can, especially Patty Maher, and to Bob Baron, president of Fulcrum Publishing, for accepting the challenge of publishing the second, third, and fourth editions of this textbook. We thank Don Rosso, Lily Capstick, and the staff at Waveland Press for taking on the continued evolution of the book contents and publishing this fifth edition.

The book's seventeen chapters bring together important information, established concepts, and examples pertaining to wilderness designation and stewardship. The book's objectives are:

1. To describe the evolution of the U.S. National Wilderness Preservation System: its philosophical, historical, and legal origins; its current size, number of areas, and distribution; and its probable future. We also describe

state wilderness in the United States and some wilderness systems in other countries.
2. To provide a common reference for managers, students, scientists, educators, and citizens, who must work together to steward the U.S. National Wilderness Preservation System, state wilderness systems, and wilderness in other countries to protect their resources and values.
3. To propose principles and concepts from which management policy and actions to preserve wilderness might be derived and to describe current wilderness management approaches, policies, procedures, and techniques.
4. To introduce readers to pressing wilderness management issues, impacts, the implications of alternative methods of dealing with them, current problems, solutions, and research.
5. To describe differences and interrelationships between wilderness designation and wilderness management and between management of wilderness and contiguous non-wilderness lands.

We recognize that readers harbor diverse views about wilderness designation and management, and we do not expect universal agreement with our treatment of this value-laden topic. We have tried to avoid polarity of opinion about wilderness and have attempted to maintain a broad, conceptual perspective. Rather than offer stock solutions, we have tried to identify alternatives and their implications, and to cite examples where possible. Where we do advocate specific action, we have tried to clearly state our position and our reasons. Although public and private agencies and many individuals continue to hold various notions about wilderness and its management, we are gratified by an increasing convergence of views on wilderness stewardship. We hope this book will stimulate even more discussion and encourage consensus to meet the challenges of wilderness designation, stewardship, and management that face governmental agencies, managers, and citizens in the decades ahead.

This edition follows the same mega-organization as earlier editions. This framework is designed for a logical flow through the subject matter, but each section and chapter stands alone to invite reading and study independent of the rest of the book. *Wilderness Management: Stewardship and Protection of Resources and Values* contains seventeen chapters divided into six sections:

I. The Setting (Chapters 1–3)
II. U.S. Legal Authority and Process for Wilderness (Chapters 4–6).
III. Wilderness Management and Planning Concepts (Chapters 7–9)
IV. Wilderness Resources, Values, and Threats to Them (Chapters 10–13)
V. Wilderness Uses and Their Management (Chapters 14–16)
VI. The Future (Chapter 17)

Throughout the book, unfamiliar terms, or words used in a special way, are explained where they first occur in the text, and subsequent uses are without additional explanation. Before 1964, the term *wilderness* was used to describe humanity's changing perception of unknown areas or lands modified primarily by natural forces. But after 1964, although retaining this historic and familiar meaning, the term has acquired a new, more precise legal definition. In the United States, the word *wilderness* now refers to areas designated by Congress as units of the National Wilderness Preservation System. This legal designation sometimes prompts reference to "Capital 'W' Wilderness" for legally designated wilderness and "little 'w' wilderness" for roadless land that is not yet, but could someday be, designated. For readability and clarity, we have elected to capitalize the word *Wilderness* only when it is used as a proper noun, for example, the Bob Marshall Wilderness, the National Wilderness Preservation System, or the Wilderness Act. All other references to wilderness as a general term or in the legal sense use a lowercase *w*, and this generally applies to discussions of international wilderness.

The term *act* used alone might occasionally raise questions. As it is used in this book, if the context is to a period before January 1975, the term refers only to the Wilderness Act of 1964. After January 1975, the term refers to the general intent or provisions of the Wilderness Act of 1964 *and* the so-called Eastern Wilderness Act of 1975. The term *wilderness designation act* refers to legislation designating specific areas as wilderness or as wilderness study areas to be evaluated for their suitability to become units of the National Wilderness Preservation System. The term *wilderness acts* refers collectively to all the just mentioned terms.

Statistical sources for information concerning the National Wilderness Preservation System used in this book include information published by the Aldo Leopold Wilderness Research Institute and at www.wilderness.net, a website supported by the Aldo Leopold Wilderness Center, the Arthur Carhart National Wilderness Training Center, and the University of Montana's College of Forestry and Conservation's Wilderness Institute. Authoritative sources differ slightly on several statistical characteristics of the National Wilderness Preservation System, including the number of units, their acreage, and the agencies that manage them. Several factors affect these figures and account for the discrepancies: units can be in more than one state; land acquisitions might have changed the acreage; boundaries might have been adjusted; measurements might have been refined; and official mapping might have been completed after publication of the statistics. But these discrepancies, especially concerning unit acreage, are insignificant in assessing total area of the National Wilderness Preservation System, which includes more than 800 units and more than 111 million acres.

We acknowledge and honor all the ancestral stewards of these lands and their descendants living in over 1,000 Tribes, Nations, Bands, Pueblos, Communities, or Native Villages.

—Chad P. Dawson,
March 2024

I. The Setting

CHAPTER 1

Wilderness Management: Philosophical Direction

INTRODUCTION

Since 1964, important progress in wilderness preservation has been made in the United States as a result of the combined efforts of conservation organizations, concerned citizens, federal agencies, wilderness users, and the U.S. Congress. *The Wilderness Act of 1964 (U.S. Public Law 88-577) and additional wilderness designation laws that have classified more land as wilderness seek to maintain a portion of the nation in a natural and undeveloped condition for future generations. The U.S. National Wilderness Preservation System (NWPS) includes separate wilderness areas in the national forests, national parks, fish and wildlife refuges, and public lands administered by the Bureau of Land Management (BLM). By 2023, more than 111.9 million acres in more than 806 areas were protected in the NWPS.* States protect another 3 million acres of lands as wilderness under state laws, and several nations besides the United States also protect wilderness—all inspired to some degree by the U.S. Wilderness Act of 1964 (Fig. 1.1).

Public interest in wilderness has been strong and consistent over many decades now (Cordell et al. 2003; Cordell et al. 2005; Scott 2004; Schuster et al. 2005). Wilderness is often treated as an icon of nature and naturalness and portrayed in the visual arts as a primitive and natural landscape with striking physical features like glaciers, mountains, deserts, shorelines, rivers and waterfalls, wildlife, and primeval forests and grasslands. Our interest in nature has been discussed in popular books and reports in recent years including the positive linkages between the natural environment, biodiversity, and mental health. The Biophilia Hypothesis of E. O. Wilson has caught the imagination of many with his idea that human development through the ages has resulted in a "human dependence on nature that extends far beyond the simple issues of material and physical sustenance to encompass as well the human craving for aesthetic, intellectual, cognitive, and even spiritual meaning and satisfaction" (Kellert and Wilson 1993, 20). Scientists and educators have expressed concerns that future generations are growing up without that strong connection with nature (Louv 2006).

FIGURE 1.1 The view southeast from Silver Pass in the John Muir Wilderness, on the Sierra National Forest, California. Once an area is legally designated as wilderness, it joins the National Wilderness Preservation System. Preservation of its wilderness qualities then depends on its management. Photo courtesy U.S. Forest Service.

Our final achievements in wilderness stewardship and preservation are as yet unknown because designating areas as part of the NWPS is only part of the task. As George Marshall, former president of the Sierra Club and brother of wilderness advocate Robert Marshall, noted, "At the same time that wilderness boundaries are being established and protected by Acts of Congress, attention must be given to the quality of wilderness within these boundaries, or we may be preserving empty shells" (Marshall 1969, 14). This is the challenge of wilderness management to formulate and implement programs of stewardship and protection that will achieve the objectives of the NWPS.

Wilderness management has evolved since the early 1960s and passage of the Wilderness Act. Wilderness management is increasingly important because of growing direct pressures from all kinds of wilderness users and indirect human impacts on all lands. In brief, *the goals of wilderness management are to protect a designated wilderness area's naturalness and solitude, that is, its wilderness character.* This is not an easy task while also providing for its use and enjoyment as wilderness by increasing numbers of people engaging in primitive forms of recreation, hunting and fishing in some areas, and by traditional uses such as grazing, mining, and air access in place before wilderness designation. Further, management must accommodate special provisions and exceptions to the Wilderness Act that may be in the wilderness designation legislation for some areas. For example, there are special provisions for wilderness in Alaska, such as snowmobile access and subsistence hunting, fishing, and gathering under the Alaska National Interest Lands Conservation Act, or ANILCA (U.S. Public Law 96-487). Increasing wilderness use compromises solitude, and naturalness may be compromised by fire protection, fish stocking, grazing, heavy visitor use, and climate change. *Simply designating an area as wilderness does not ensure its preservation. Science-based wilderness management and stewardship are needed.*

Probably most of the areas and acreage to be set aside as wilderness in the United States have already been designated, although another 10 to 20 million acres are possible. Therefore, protecting wilderness values in the United States rests primarily on the careful stewardship of areas already in the NWPS and to a lesser extent on designating more areas. This is a sobering responsibility for natural-resource professionals and wilderness advocates, and future generations will be our critics. Resource professionals are challenged to develop and apply effective and proactive approaches to wilderness management now, rather than when problems are out of control and our efforts can, at best, be reactions to pressing needs. Gaining the support and involvement of wilderness advocates is an important part of the challenge.

WILDERNESS AND OTHER LAND USES

This book focuses primarily on the management of *legally designated wilderness* areas in the United States on federal lands formally protected by the 1964 Wilderness Act and its extension to lands in the eastern United States by the so-called Eastern Wilderness Act of 1975 (U.S. Public Law 93-622); to public lands administered by the Bureau of Land Management by the Federal Land Policy and Management Act of 1976, or FLPMA (U.S. Public Law 94-579); and, in 1980, to Alaska by the ANILCA (U.S. Public Law 96-487). The NWPS consists of more than 111.9 million acres of wilderness administered by four federal agencies: one under the Department of Agriculture, the U.S. Forest Service (USFS), and three under the Department of the Interior, the National Park Service (NPS), the U.S. Fish and Wildlife Service (FWS) and the Bureau of Land Management (BLM).

Our goal is to further the protection and appropriate use of legally designated wilderness areas through their proper stewardship and management. We have not addressed the issues of how much wilderness should be set aside, where it should be located, how large an area should be, and so forth. We believe that such issues are primarily political and will be decided largely through the political process. Resource managers can best facilitate that process by helping to define and assess potential land-use alternatives for areas proposed as wilderness. Managers' recommendations are useful when they include their expert knowledge, data, and analyses. Providing factual analysis of the consequences and outcomes of proposed alternatives is a key responsibility for natural-resource managers.

We are fortunate in the United States to have a variety of environments, ranging from paved and densely populated urban areas to uninhabited primeval forests and deserts. That range can be described as being along an environmental modification spectrum (EMS). Providing a full spectrum of natural environments is a desirable response to the broad range of tastes and interests in any society. At one end of the spectrum are cities, the most modified of environments; at the opposite end are the least modified environments and most primeval, wildernesses. Considering landscapes for recreational purposes leads to the parallel concept of a recreation opportunity spectrum (ROS) that includes urban and suburban recreation areas, rural countryside, highly developed campgrounds, intensively managed private lands, publicly owned multiple-use forests, national parks, recreation and scenic areas, roadless wildlands, and wilderness. Through the Wilderness Act, U.S. society has chosen to preserve, unimpaired for future generations, as wilderness, a selection of our least modified environments and landscapes at the primeval end of the spectrum.

Wilderness, as the least modified extreme on the spectrum of land uses, is sensitive to a variety of uses, including those on adjacent public and private lands. Thus, wilderness managers must consider the compatibility of uses within a given wilderness and also respond to influences and interrelationships among uses and management on adjacent lands.

Moreover, because U.S. federal agencies that manage wilderness—the NPS, FWS, BLM, and USFS—also manage lands devoted to many other purposes and activities, wilderness management must be carefully coordinated with other agency and interagency land-management plans and the laws that direct them.

While most of this book focuses on wilderness managed by federal agencies in the United States, the concepts are relevant to wildernesses designated by states (see Chapter 6), and by law, regulation, or policy in other countries (see Chapter 3). Thus, by outlining wilderness stewardship and management philosophies; approaches, ideas, and concepts; and some policies, procedures, actions, and likely consequences based on experience and research, *we anticipate that the book will be useful not only to federal and state wilderness managers in the United States, but also in other countries where wilderness is protected or contemplated.*

WHAT IS WILDERNESS?

What is wilderness? This crucial question affects all wilderness designation and management decisions. *Legally*, wilderness can be defined narrowly as *an area possessing qualities outlined in section 2(c) of the Wilderness Act of 1964. Socially*, it is whatever people think it is, potentially the entire universe, the terra incognita of people's minds. We can call these two definitions *legal wilderness* and *sociological wilderness*. Deriving a universally accepted definition of sociological wilderness seems unlikely because perceptions of wilderness vary so widely. For example, for some urbanites with scant outdoor experience, wilderness might be perceived as any relatively undeveloped wildland, uncut forest, or woodlot. In addition, as illustrated in Chapter 3, people from different cultures and countries may have widely different perceptions of what qualifies as wilderness.

On the other hand, legal wilderness as defined by the Wilderness Act, section 2(c) is much more precise: *"A wilderness, in contrast with those areas where man and his own works dominate the landscape, is hereby recognized as an area where the earth and its community of life are untrammeled by man, where man himself is a visitor who does not remain."* This legal definition places wilderness on the "untrammeled" or "primeval" portion of the EMS. Furthermore, it is sanctioned by the traditions of land use in America and rests on ideas espoused decades ago. For example, Aldo Leopold (1921, 719) envisioned wilderness as "a continuous stretch of country preserved in its natural state, open to lawful hunting and fishing, devoid of roads, artificial trails, cottages, or other works of man." Robert Marshall (1930, 141) offered a similar definition:

> I shall use the word wilderness to denote a region which contains no permanent inhabitants, possesses no possibility of conveyance by any mechanical means and is sufficiently spacious that a person in crossing it must have the experience of sleeping out. The dominant attributes of such an area are: First, that it requires any one who exists in it to depend exclusively on his own effort for survival; and second, that it preserves as nearly as possible the primitive environment. This means that all roads, power transportation and settlements are barred. But trails and temporary shelters, which were common long before the advent of the white race, are entirely permissible.

Because this book is primarily about managing legal wilderness on federal lands in the United States, our definition of wilderness mirrors that outlined in the Wilderness Act. This definition prescribes conditions for areas included in the NWPS and indicates the purposes that management programs for these areas are designed to achieve. The Wilderness Act lends both quantitative and qualitative substance to the traditionally elusive question, What is wilderness?

Although recognizing that wilderness has taken on added precision in its legal definition, we should not forget the evolution of the concept. *Wilderness is still largely a social and political phenomenon of twentieth-century North America. Recently, however, its recognition has spread internationally.* Chapter 3 describes in greater detail the status of such lands elsewhere in the world and the cultural uniqueness of wilderness in any country. Even in the United States, the perception of wilderness has evolved from that of a forbidding landscape to that of a valued cultural, economic, social, and environmental resource (Cordell et al. 2003; Cordell et al. 2005). For example, substantial economic and social values are often tied to the protection of wilderness in particular vicinities that make nearby locations attractive places to live or to visit as a recreationist or tourist, to conserve biodiversity, and as sources of ecological services such as clean air and water, climate modification, and more (Power 1996a, 1996b; Rudzitis 1996; Holmes and Hecox 2004; Green et al. 2005).

Our intent is not to trace the origins of the word *wilderness* and its cultural evolution, or to annotate the extensive literature about the many values and philosophies of wilderness. Many other books and articles do this (Scott 2004); even a complete listing is beyond the scope of this book. However, *readers should understand the origin of the term wilderness and the diversity of human values associated with it. This is necessary to understand the Wilderness Act and to be able to relate it to ways in which these values can be depreciated or enhanced through wilderness management or stewardship.* Thus, the remainder of this chapter briefly reviews the origins of the wilderness concept and basic values that underlie its legal definition, and management or stewardship, because we use these terms interchangeably.

HISTORICAL ORIGINS OF THE WILDERNESS CONCEPT

A strong religious flavor influenced the early origins of the word *wilderness.* The word appeared in the fourteenth-century English translation of the Bible from Latin and was used as a synonym for uninhabited and arid lands of the Near East (Nash 2001). Lands described as wilderness were virtually uninhabited, desolate and arid, and vast (Outdoor Recreation Resources Review Commission 1962). In such lands, people could not long survive *Eventually, with the rise of the physical and natural sciences, people began to associate wilderness with inspiration, not terror, and to explain wilderness and other natural phenomena based on science rather than theology.*

This new appreciation of nature was reflected in many ways. For instance, the symmetrical, formal gardens commonly found in the late 1700s, such as the Garden of Versailles, gradually gave way to more informal, pastoral, and natural settings. In art and literature, the wild, turbulent panoramas of the Alps became favorite scenes. Some literary heroes were the persons who knew how to live in harmony with nature (McCloskey 1966). *This gradual evolution of thinking about relationships of humans to their environment represented an important precondition to the recognition of wilderness as a source of human values and to the eventual development of programs for its preservation.*

WILDERNESS AND THE EARLY AMERICAN SCENE

When the first European explorers reached what is now the conterminous forty-eight United States, they found a continent that was inhabited by Native Americans, whose lands, in contrast to the more dense European level of human habitation and cultivation, were considered undeveloped wilderness by the arriving Europeans. We begin this history at the arrival of European immigration and do not address Indigenous knowledge of these wildlands. We do acknowledge and honor the ancestral stewards of these lands and their descendants; however, the complexity of that important story is not included herein. This book focuses on the management of these lands and their utilization or protection following colonization. Much can be learned from Indigenous knowledge and, hopefully, will be added in the future to wildland and wilderness management for the mutual benefit of all peoples.

The wilderness was considered a barrier and a threat to sixteenth- and seventeenth-century settlers. It hindered movement and it frequently possessed little that could directly help settlers prosper. However, as settlements by European immigrants rolled across the landscape, a movement was begun to retain some unmodified lands for perpetuity. *Nash (2001) has argued that interest in maintaining these wildlands was motivated in part by the desire to lend a distinctive quality to American culture. Although literary and artistic accomplishments were almost nonexistent in the young nation, it did possess one thing for which there was no European counterpart: wilderness.* Thus, even while strong motivations existed to conquer the wilderness, there were also stirrings of opinions that valued its retention.

One early observer who foresaw the need for long-term protection of the natural environment was George Catlin (1796–1892), a nineteenth-century lawyer, painter, and student of American Indians. Following a series of trips through the northern Great Plains, Catlin called for establishment of *"a nation's park,* containing man and bear, in all the wild and freshness of the nature's beauty!" (Nash 1982, 101).

Catlin's remarks received little attention, but the seed was planted. In 1858, writing in *Atlantic Monthly,* Henry David

Thoreau (1817–1862) asked, "Why should not we have our national preserves in which the bear and panther, and some even of the hunter race, may still exist" (as cited in Nash 1982, 102). Thoreau spoke primarily for the view that credited wilderness with the values most important in the molding of humankind, a perspective summarized in the oft-cited statement attributed to him, "in wildness is the preservation of the world." (Nash 1976). He provided a philosophical framework for literary and artistic expressions of feelings about wilderness as a uniquely American asset, the one attribute of the new nation that made it superior to the tired, settled lands of the Old World (Evans 1981). Gradually, this idea became a great force among the leaders of American thought and culture.

More than a century passed between the warnings of Catlin and Thoreau and passage of the Wilderness Act. During that time, numerous advocates argued for the preservation of some of the remaining wilderness landscape, and some important governmental laws indicated increasing recognition of the importance of reserving lands for public purposes. Yellowstone National Park was established in 1872, the first in a long series of important reserves. Earlier, in 1864, the federal government had granted Yosemite Valley to the state of California "to hold inalienable for all time" (Ise 1961, n.p.). Although it is almost certain that neither area was intended to remain wilderness, these actions did set precedents for the federal government to allocate lands for nonconsumptive purposes. Additionally, public policy debates over such issues as roadless areas in national parks transformed into later support for the wilderness movement (Sutter 2002). Moreover, these parks represented official recognition of values and philosophies expressed by people like Thoreau.

HISTORICAL WILDERNESS THEMES AND VALUES

The Wilderness Act of 1964 was the product of more than eight years of debate in Congress and multiple legislative proposals (Scott 2004). To fully understand the meaning of that legislation, however, one must reach back into history. The Wilderness Act and the movement leading to it reflect a synthesis of diverse philosophical values, which evolved over many decades (Callicott and Nelson 1998). As McCloskey (1966, 295) notes: "The evolution has blended many political, religious, and cultural meanings into deeply felt personal convictions. Those who administer that law must look to these convictions to understand why the law exists."

The diversity of motives and values among individuals and groups supporting the wilderness movement has been instrumental in its success. We need to understand the appeal of the wilderness idea that inspires such broad endorsement by a majority of Americans and intense commitment by a fervent minority. *The success of wilderness management efforts depends on how clearly we understand these wilderness values and how effectively we protect them.*

The contributions of many wilderness proponents suggest distinct themes around which the wilderness cause has been argued. *Although wilderness may mean something different to each person, four central themes have consistently emerged: experiential, the direct value of the wilderness experience; the value of wilderness as a scientific resource and environmental baseline; the symbolic and spiritual values of wilderness to the nation and the world; and the value of wilderness as a commodity or place that generates direct and indirect economic benefits.*

EXPERIENTIAL

The wilderness experience is seen as valuable in its own right, and as a factor in forming our national character, part of what makes Americans unique. *Historically, American writers have extolled the closeness to nature, education, freedom, solitude, and simplicity, as well as spiritual, aesthetic, and mystical dimensions of the wilderness experience.*

John Muir, who founded the Sierra Club in 1892, was an articulate and influential early proponent of experiential values of wilderness. Muir, a Scottish immigrant, was raised on a Wisconsin farm. Although a talented inventor, Muir was more intrigued by the ideas he found in science and literature. He was heavily influenced by the writings and philosophies of Thoreau and Ralph Waldo Emerson (Davis 1966). Nevertheless, he found them both wanting in some respects. Muir thought Emerson had failed to express an appropriate amount of excitement after hiking in the mountains at Yosemite, and he was amused that Thoreau, who repeatedly proclaimed, "In wildness is the preservation of the world," could refer to orchards as forests (as cited in Nash 1982).

To Muir, the essence of wilderness was the freedom, solitude, and beauty of the mountains. These qualities, he felt, could satisfy all human needs. The wilderness experience to Muir was spiritual, the forests were temples, and the trees sang psalms. In the Sierra wilderness, "everything seems equally divine, one smooth, pure wild glow of heaven's love" (Muir 1938, n.p.). The wilderness also offered personal insight; during a raging windstorm in the Sierra, Muir climbed a tree and lashed himself to it to experience nature more closely.

The experiential theme was also reflected in many of the writings of Robert Marshall. Marshall was an extraordinary individual. In a brief but fruitful life, he accomplished much for wilderness. In 1935, along with Aldo Leopold, he helped found The Wilderness Society, and in 1939 as director of Recreation for the U.S. Forest Service, he formulated the U regulations; both actions strengthened protection of roadless areas before passage of the Wilderness Act in 1964. As an extremely enthusiastic hiker, Marshall routinely hiked thirty-five to forty miles a day, and set a goal of walking thirty miles a day in every state (Edwards 1985). To Marshall, wild scenery was similar to great works of art. In a major paper outlining the future of wilderness, he wrote, "Wilderness furnishes perhaps the best opportunity for pure aesthetic rapture" (Marshall 1930, 145). *Marshall believed the restorative powers of wilderness could help prevent moral deterioration.*

Like Marshall, Aldo Leopold was a USFS employee who helped form that agency's wilderness concept. As a young forester in New Mexico following World War I, he became apprehensive about the expansion of USFS road systems into the backcountry he loved. His prompting led to administrative protection of the Gila River country as a wilderness, the first in the nation. Over the years, his wilderness ideas developed and matured, and his *Sand County Almanac and Sketches Here and There,* published after his death in 1948, is a classic in American conservation literature (Leopold 1949).

In the early 1920s, Leopold argued especially hard for wilderness designation because he wanted to preserve a particular kind of recreational experience, the pack trip. Unless steps were taken to preserve large tracts of land, he stated, the day would come "when a pack train must wind its way up a graveled highway and turn its bell-mare in the pasture of a summer hotel. When that day comes the pack train will be dead, the diamond hitch will be merely rope, and Kit Carson and Jim Bridger will be names in a history lesson" (Leopold 1925, 403).

Leopold pressed for preservation of areas that would provide recreational experiences he thought developed both individual and national character (Leopold 1921). Another young forester of the period, Arthur Carhart, agreed with Leopold's philosophy and pressed for protection of an area surrounding Trappers Lake in Colorado. *To Carhart, recreation was not merely an incidental use of forests; it ranked among the highest of all possible uses because of the moral benefits associated with it*: "Recreation in the open is of the finest grade. The moral benefits are all positive. The individual with any soul cannot live long in the presence of towering mountains or sweeping plains without getting a little of the high moral standard of Nature infused into his being" (Carhart 1920, 268).

Carhart's immediate superior in Colorado was Carl J. Stahl, a firm supporter of the Trappers Lake proposal. Stahl (1921, 529) also saw forests (and wilderness) as a source for strengthening moral values:

> An appreciation of nature, a stimulation of vigor of the mind and body, and the contentment of soul contributed by association with the forests, go far toward making a useful and contented citizenry. If the American population can be made to feel contented and its effort directed to useful channels, enlistment in the Red organizations of this critical period of unrest can be averted. I can conceive of no more useful purpose the forests can be made to serve.

In the 1920s, contact between humans and nature was seen as a character-building activity.

Today, the experiential theme of wilderness includes personal growth and healing. Now hundreds of wilderness experience programs use wilderness for environmental education, personal growth, and therapy, such as Outward Bound and the National Outdoor Leadership School (NOLS) (Friese et al. 1998; Russell et al. 2000). These programs form an important and growing wilderness industry; they aim at growth in desirable personal qualities (e.g., development of self) such as self-esteem and confidence, independence, outdoor skills, improved group skills (e.g., development of community), and enhanced performance through leadership and team building, spiritual development, and environmental education. Some programs are aimed at disadvantaged groups and individuals and help people in crisis or transition find inspiration and new goals. They provide healing for those dealing with the trauma of domestic instability and chaos or abuse, those adjusting to emotional losses such as death and broken relationships, and those fighting chemical dependency and delinquency.

All of these wilderness experience programs derive from a belief that, in the natural environment (ideally in wilderness), away from social pressures, excessive stimuli, and diversions, we can confront our true and deeper selves, identify our values and priorities, and recover a sense of wholeness. This is the *primal hypothesis*, that the experience of wilderness naturalness and solitude yields important benefits to people, including desirable personal qualities, group skills and cohesion, and spiritual development (White and Hendee 2000). Such programs reflect one of the central beliefs of the founders of our wilderness system: that the character-building values of wilderness are vital to our society (Scott 2004).

Many recreational experiences take place in designated wilderness, and the kinds of physical challenge, naturalness, and solitude individuals seek are increasingly found only in wilderness. Such programs fulfill Aldo Leopold's dream; he believed that wilderness areas should be places where primitive travel and subsistence skills could be perpetuated (Leopold 1921). Today, such programs challenge wilderness management in its dual role of providing opportunity for experiences aimed at human growth, enrichment, and healing, while also protecting the wilderness resources of naturalness and solitude (Gager et al. 1988).

SCIENTIFIC

Numerous values arise from the ecological and environmental services provided by wilderness (Cordell et al. 2005). For example, due to its generally undisturbed environmental setting, wilderness is an important benchmark source of scientific information about the world around us, how it evolved, how the effects of civilization have altered natural systems, and what the unmodified environment holds for us. *Wilderness is a baseline or control area with which to compare change in other world environments.* Areas where natural processes remain intact are increasingly scarce. Wilderness areas are thus valuable assets: as natural baselines that reveal the extent of impacts elsewhere, as sites where scientists can study natural processes, as gene pools maintaining and reflecting the diversity of nature and providing a gene reservoir we are only now learning to use, and as refuges for certain flora and fauna that cannot survive outside wilderness conditions.

The intricate interrelationships among all organisms have been important to the scientific theme. Muir, for example, saw in wilderness a place where people could feel "part of wild nature, kin to everything" (as cited in Nash 1982, 129). Moreover, it was Muir who expressed the fundamental principle of ecology: whenever you pluck up something, you find everything in the universe attached to it.

Leopold saw scientific value in wilderness as a laboratory for the study of land health. *In 1935, Leopold, Marshall, and others founded The Wilderness Society to promote the protection of lands retaining such naturalness* (Leopold 1941). Although Leopold had long recognized historical and recreational values of wilderness, in 1935 his justifications for such areas predominantly turned to ecological and ethical reasons (Flader 1974).

Because wilderness areas have remained relatively undisturbed over long periods, they are reservoirs of genetic materials that have evolved over eons. Such gene pools hold answers to questions yet unasked; once lost, they are impossible to replicate. Similarly, it is important to retain species whose chemical and biological makeup might be useful in the future as the sources of important drugs, for example. Large, undisturbed tracts of wilderness are important sources of genetic diversity and havens of stability in animal and plant populations so essential to retaining genetic pools.

Sometimes the relationship of humans to the world around them can be understood only by analyzing biological systems that have escaped human impact. Wilderness offers an important opportunity to examine ecosystems as they have evolved outside human influence. Understanding such evolution can help prevent errors, at best careless and at worst catastrophic, as we shape and modify the rest of the earth to our purposes. As Leopold once noted, the first principle of intelligent tinkering is to save all the parts, ecologically speaking; a sufficiently large and ecologically representative wilderness system will help that effort.

However, the 1964 Wilderness Act did not directly specify biological diversity or ecological integrity criteria as part of the wilderness designation process; thus, some important physical-biological systems are underrepresented or not included in the current NWPS. The USFS's second Roadless Area Review and Evaluation (RAREII) from 1977 to 1979 specified that the nation's ecosystems, landforms, and wilderness wildlife be important criteria for identifying candidate areas for wilderness designation. An analysis revealed that of the nation's 261 ecotypes, only 157 were included in the NWPS by the late 1980s (Davis 1989). By 2005, Cordell et al. (2005) reported that only nine of the fifty-two categories in Bailey's Province-level Ecosystem classification were not represented somewhere in the NWPS; however, they observed that some of those categories were only represented in a few NWPS areas or the areas were small in size.

Does it really matter if a natural system disappears? In a plea for an end to wilderness destruction, E. O. Wilson (1984, 15) gave this example:

> Natural products have been called the sleeping giants of the pharmaceutical industry. One in every ten plant species contains compounds with some anticancer activity. Among the leading successes from the screening conducted thus far is the rosy periwinkle, a native of the West Indies. It is . . . the kind of inconspicuous flowering plant that might otherwise have been unknowingly consigned to extinction by the growth of sugarcane plantations and parking lots. But it also happens to produce two alkaloids, vincristine and vinblastine, that achieve 80 percent remission from Hodgkin's disease, a cancer of the lymphatic system, as well as 99 percent remission from acute lymphocytic leukemia.

Wilderness also provides large tracts of unmodified habitat that some threatened species, for example, the grizzly bear and the timber wolf, *need to survive. Such wilderness offers the opportunity to study these species to ensure their maintenance.* Wilderness tracts have served as laboratories for greatly increasing our knowledge of the biota around us, and studies of wildlife in wilderness settings have substantially enlarged our understanding of these animals. Similarly, the presence of extensive tracts of undeveloped (wilderness) land has made possible important baseline research on vegetative communities, fire history, and studies of other natural biological systems that simply could not have occurred without such tracts.

Wilderness also provides an important laboratory for scientists concerned with human behavior. How individuals relate to one another, how they react to stress and challenge, and how natural environments affect behavior are important topics for wilderness research. One series of studies over a ten-year period concluded that restored functioning was a key benefit of wilderness experience (Kaplan and Kaplan 1989), and numerous studies document the values of wilderness experience for personal growth and healing dysfunctional behavior (Russell et al. 2000). Further, studies of wilderness recreation users can provide important insights on the experiential values espoused by Muir, Marshall, and others.

SYMBOLIC AND SPIRITUAL

In a world characterized by rapid change and complexity, both exciting and frightening, wilderness symbolizes comforting stability and simplicity. The existence of wilderness reflects self-imposed limits on the technological imperative that we must subdue all the earth just because we can. *Wilderness is a symbol of respect for the naturalness and solitude that have been displaced by civilization on most of the earth.*

In 1962, The Outdoor Recreation Resources Review Commission (ORRRC) called on Wallace Stegner, head of the Creative Writing Center at Stanford University, to comment on the significance of wilderness as "an intangible which has altered the American consciousness" (Outdoor Recreation Resouces Review Commission 1962, 34). Stegner's reply argued forcefully for the maintenance of wilderness for the sake of survival:

> Something will have gone out of us as a people if we ever let the remaining wilderness be destroyed; if we permit the last virgin forests to be turned into comic books and plastic cigarette cases; if we drive the few remaining members of the wild species into zoos or to extinction; if we pollute the last clear air and dirty the last clean streams and push our paved roads through the last of the silence, so that never again will Americans be free in their own country from the noise, the exhausts, the stinks of human automotive waste.

Wilderness is needed, Stegner concluded, because it is "a means of reassuring ourselves of our sanity as creatures, a part of the geography of hope" (ORRRC 1962, 34).

The extensive writings of Sigurd Olson reveal many important human values derived from wilderness. In a series of books published over two decades, Olson describes wilderness as a source of inspiration, insight, and personal peace: "The singing wilderness has to do with the calling of the loons. It is concerned with the simple joys, the timelessness and perspective found in a way of life that is close to the past" (Olson 1957, n.p.). While exploring the Knife River in the Quetico-Superior country, Olson found himself nearly overwhelmed by the environment around him: "I was aware of a fusion with the country, an overwhelming sense of completion in which all my hopes and experiences seemed crystallized into one shining vision" (Olson 1963, n.p.).

Spiritual development from wilderness experience is defined in more recent writings and studies as a deep sense of connection to all things, such as the larger universe, a higher power, nature, a feeling of oneness, what is referred to as "connection to other" as opposed to "connection to self" (McDonald and Schreyer 1991; White and Hendee 2000; McDonald 2003). Spiritual development as a wilderness benefit has received little attention and study, in part because spiritual experiences are intensely personal and often inexpressible, and because of the varied personal meanings of spirituality that make it difficult to define them operationally. However, some studies do report findings on spirituality in wilderness. For example, Riley and Hendee (2001), in a study of participants in a commercial wilderness vision-quest program over a ten-year period, found that to "go on a spiritual journey" was a leading motive for participating, and "spirituality connectedness" and "connection to nature" accounted for 26 percent of the reported benefits.

The accounts of many prominent wilderness writers in addition to Muir, Stegner, and Olson extol spiritual benefits from wilderness experience, and spiritual benefits are beginning to be documented in studies of wilderness experience. The *spiritual values* of wilderness will be a challenge for scholars and a desire of users in the future as wilderness stands in greater and greater contrast to societies that are ever more impersonal.

ECONOMICS

Wilderness, wildlands, and natural protected areas provide valuable ecological services and direct and indirect income from recreation and nature-based tourism, and thus economic benefits for local, regional, and national economies (Bowker et al. 2005; Rosenberger and English 2005). *The diversity of economic values of wilderness is a social and political force for protecting and managing wilderness.* Public opinion survey studies consistently report that the majority of the public supports setting aside wilderness areas for present and future generations (Cordell et al. 2003; Scott 2004; Schuster et al. 2005). Most of these people will not set foot in a wilderness, but they value wilderness from several perspectives: *for bequest value—protection of natural ecosystems and national heritage for future generations; existence value—protection of natural processes and conditions just to know that there are places that are relatively unaffected by humans; and option values—maintaining the option to either protect or develop an area in the future. These are often referred to as the "nonuse" values of wilderness.*

Recreationists traveling to and from wilderness areas use retail lodging and service businesses, and buy food and beverages to support their trips and wilderness experiences. Similarly, proximity to wilderness as a setting for travel and tourism can bring new income from providing goods and services to tourists who enjoy the wilderness from a distance but may not actually hike there. Such revenues may be considered in economic development strategies for communities near wilderness or states with numerous wilderness opportunities (Moisey and Yuan 1992; Yuan and Christensen 1992; Bowker et al. 2005; Rosenberger and English 2005).

Local residents can also benefit from wilderness by enjoying a high-quality local environment with natural amenities. Because of the possibility of a higher quality of life, pristine conditions and natural environments raise property values and attract people and businesses to areas that may have few other economic opportunities (Power 1996a).

The designation of wilderness has been perceived by some opponents as having negative economic consequences based on the misconception that wilderness designation prohibits maintenance or expansion of local, extractive industries. Power (1996b, 9) outlines these concerns, reviews relevant economic information, and concludes:

> Wilderness protection does not impoverish communities by locking up resources. Rather, it protects the economic future of those communities by preserving high quality natural environments that are increasing in demand across the nation. Wilderness protection does not threaten the ongoing development of non-metropolitan economies in any significant way. Rather, it lays part of the long-run basis for their ongoing development by providing attractive places to live, work, and do business. Because of this, the economic problem posed by protecting landscapes is not how to maintain local economic health, but almost the opposite: How to keep the economic activity attracted to areas adjacent to wilderness from undermining the environmental quality that wilderness protection seeks to insure in the first place.

Economic analysis of wilderness benefits is evolving as new approaches are applied and better information becomes available (Holmes and Hecox 2004; Russell et al. 1998; Loomis and Richardson 2001). *Clearly, wilderness is extremely valuable, generating benefits we are beginning to better document and understand. Wilderness values will only increase in the future as improved methods of estimating benefits are applied and as the supply of wilderness diminishes and demand for its services increases.*

RELATING WILDERNESS THEMES AND VALUES TO MANAGEMENT

The preceding categorized some of the basic wilderness values identified by early and contemporary authors under four broad themes: experiential, scientific, spiritual, and economic.

McCloskey's statement (1966), that *to understand the Wilderness Act, we must understand its historical and philosophical origin*, is the key to this discussion. The philosophies and perspectives discussed earlier increase our understanding of what wilderness is, why we have it, and what values it should provide society. These issues are important in applying the Wilderness Act, both in designating wilderness and in its subsequent management. *Once wilderness designation decisions have been made, the extent to which the values espoused by philosophers like Muir, Marshall, and Leopold are realized will depend on wilderness management. It is essential that managers, educators, and citizens be guided in their efforts by a personal philosophy of wilderness that recognizes the values set forth by both early and contemporary philosophers.*

Bergstrom et al. (2005) suggested a connection between wilderness attributes and human values by using a sequential framework based on four levels: (1) wilderness attributes (e.g., naturalness, wildness, geographic); (2) wilderness functions (e.g., biodiversity preserve, recreation setting); (3) wilderness services (e.g., personal health and growth, animal habitat, scientific discovery); and (4) wilderness values (e.g., social). Their research organizes wilderness values into four categories: social, ecological, economic, and ethical and intrinsic. For example, the social values relate to the noneconomic benefits that accrue to society and individuals from direct use (e.g., recreational activities, experiential use, grazing) and indirect use of wilderness (e.g., existence, bequest to future generations, option for future use, human health and well-being).

WILDERNESS MANAGEMENT AND WILDERNESS DESIGNATION

There is a fundamental difference between wilderness designation and wilderness management. Wilderness designation *includes all processes and activities of government agencies and interested publics to identify areas for potential protection as wilderness and for their designation by Congress under the Wilderness Act.* Wilderness management *includes government and citizen activity, within the constraints of the Wilderness Act, to identify goals and objectives for designated wilderness areas and the planning, implementation, and administration of policies and management actions to achieve them.* Wilderness management *applies concepts, criteria, guidelines, standards, and procedures derived from the physical, biological, social, and management sciences to preserve naturalness, and outstanding opportunities for solitude or primitive, unconfined recreation, in the stewardship and protection of resources and values in designated wilderness areas.*

In general, the wilderness designation process has typically been that: the managing agency of a national forest, national park, wildlife refuge, or BLM-administered land unit reviews its roadless areas, and with public involvement, identifies lands to be studied more intensively for potential wilderness designation. Often intense public interest in a particular area precipitates the process. Then the agency studies these selected areas, again with public involvement, to determine their suitability, the demand and need for their legal designation as wilderness, and the values and developments such as timber, ski resorts, and mining that would have to be given up. Based on this study, a wilderness proposal is submitted by the agency to Congress for review, evaluation, and possible legislation, to designate the area as wilderness.

The congressional wilderness study and review process includes formal public hearings and extensive deliberation in committees as alternative bills are proposed, debated, revised, and perhaps finally enacted to legally designate a specific area as wilderness. During this process, interested groups may submit alternative wilderness proposals. For example, a wilderness study on the Mount Baker–Snoqualmie and Okanogan-Wenatchee National Forests in Washington state led to a USFS wilderness proposal to Congress for a 292,000-acre Alpine Lakes Wilderness, with the addition of 82,000 acres if money were to be allocated to purchase private land. However, two other special interest groups proposed smaller and larger areas. Congress considered all three proposals and, after public hearings and congressional debate, finally passed compromise legislation in 1976 (U.S. Public Law 94-357) establishing a 303,508-acre Alpine Lakes Wilderness. An additional 86,000 acres of intermingled private and public land was defined as "intended wilderness," and Congress authorized funds for acquisition of private land needed to complete the intended Alpine Lakes Wilderness. The Alpine Lakes Wilderness was completed in 1980, following years of public involvement and USFS efforts (Knibb 1982), and included 393,000 acres. The Alpine Lakes example illustrates the political nature of wilderness designation and the competing demands for use of an area.

Wilderness management involves applying guidelines and principles to achieve established goals and objectives. The guidelines are first to protect the area as wilderness and implement any special legislative provisions and purposes established for the area in its wilderness designation legislation. Management generally is not as political as wilderness designation, although it has become more so because the process of establishing management objectives, policies, and actions must address the desires of competing users. In some wilderness designation acts, Congress gives special direction for handling a specific management issue, for example, grazing guidelines in the 1980 Colorado Wilderness Act (U.S. Public Law 96-560) and its accompanying committee report. In addition, management challenges facing the responsible agency are a direct function of the attributes of the designated areas. The agency must manage "as wilderness" what it inherits through the designation process (Fig. 1.2). If that process yields heavily impacted areas, with high levels of established or incompatible uses, providing "naturalness and solitude" or other wilderness-dependent values will be more difficult and require more intensive management actions.

FIGURE 1.2 Desert tortoise habitat is on sixty-nine BLM Wilderness Areas designated by the California Desert Protection Act of 1994. The tortoise, however, is at risk on desert roads near wilderness boundaries. Photos courtesy Marilyn and John Hendee.

THE EFFECTS OF WILDERNESS DESIGNATION DECISIONS ON MANAGEMENT

Wilderness designation, including its often-lengthy process, affects wilderness management in several ways: (1) there must be *interim management* before designation and after designation but before development of a management plan; (2) designation can establish precedents, such as *purity,* and *nondegradation*; (3) designation of areas can include *unusual management problems*; and (4) the designation legislation can prescribe special provisions and *limitations on uses and management.*

INTERIM MANAGEMENT

Areas under study or consideration for wilderness designation must be managed during the review or study period in a way that does not preclude their being designated by Congress as units of the NWPS. In other words, the decision to *study* an area for possible wilderness designation is a decision to temporarily *manage* that area as a wilderness. The management of roadless lands to maintain their wilderness character until Congress decides is based on Wilderness Act directives that only Congress has the right to permit or deny any area's designation as wilderness. If interim management (or lack of management) allows the wilderness character of a proposed area to deteriorate before congressional review, it has effectively reduced the options for wilderness designation by Congress.

While wilderness study and the designation process go on, nonconforming uses or inappropriate or excessive levels of use can become well established and protected by their own advocates. Irreversible damage can result from unmanaged use: concentrations of visitors can compact and erode soils and threatened wildlife dependent on wilderness can be further jeopardized. Correction is always far more difficult, expensive, and controversial than prevention, so *an important interim management task is to prevent damage to areas for which wilderness designation is likely, possible, or pending.*

Thus, roadless areas identified as candidates for possible addition to the NWPS are generally managed and used essentially as wilderness, pending designation decisions. This principle is so well established that a major point of conflict in wilderness designation legislation concerns the *release* of roadless areas from any further consideration for wilderness designation: that is, whether roadless areas, having been considered by Congress but not designated, should be *released* from further consideration, thus allowing them to be used for purposes such as logging or developed recreation that could forever preclude their being designated as wilderness. Following wilderness designation there is a period in which interim management continues before the formulation, acceptance, and implementation of an approved wilderness management plan. Unfortunately, this period can exceed a decade or two as state and federal agencies attempt the management-planning process with limited staff, budgets, other heavy workload demands, numerous other plans to write, diverse user-group demands, and potential litigation from special-interest groups. In 1989, Reed et al. (1989) reported that more than 60 percent of federally designated wilderness areas lacked a completed and approved management plan, and management capacities are scarcely better today. The longer a wilderness or study area continues under lax or inappropriate interim management, the more users who experience it might conclude that what they experience is what was intended; hence, the more difficult it will be later to protect the resources and provide wilderness experiences more rigorously.

PRECEDENTS, PURITY, AND NONDEGRADATION

There are concerns that designating wilderness areas with substandard qualities of naturalness and solitude, which thus require special management attention, could establish precedents affecting the designation and management of other areas. This is the so-called purity issue. The fear is that if areas with less pristine qualities are designated as wilderness, the standards used in admitting such areas will set a precedent for designating other substandard wilderness or will be used as a rationale for lowering standards in the more pristine areas already admitted (Allin 1982). Likewise, if an area designated as wilderness requires special management for some existing feature or established use, purists argue that this could establish a precedent requiring similar management in other wilderness areas. Purity has been a major wilderness designation issue because Congress has increasingly considered and designated wilderness areas with impacts and features that some argued should disqualify them for consideration (such as abandoned roads and railroads, cabins, and homesteads). Now, as new wilderness designation proposals emerge for more impacted areas, the concerns about precedents could focus on past management. If one wilderness contains undesirable conditions, such as intensively developed campsites, or allows aerial fish stocking, relocation of nonnative wildlife, or limited use of motorboats and airplanes, will those lower standards apply to other wilderness areas?

The issue of purity in wilderness designation and management will remain, but *fears of setting precedents have been largely quieted by adopting the principle of nondegradation*—a concept widely used in managing air, water, and noise (Mihaley 1972)—to wilderness.

Under nondegradation, management's obligation is to prevent further environmental degradation of individual areas that meet wilderness standards, while managing to upgrade areas that are below minimum wilderness standards. Thus, no one area sets a standard for another, because each in effect is managed to maintain its own baseline conditions of naturalness, solitude, and other wilderness characteristics, as long as they are above some minimum standard. For example, the presence of high use levels requiring tight management controls in the San Gorgonio Wilderness in California does not justify permitting similar conditions to evolve in the relatively pristine Selway-Bitterroot Wilderness in Idaho and Montana. Likewise, practices essential for managing the typically smaller wilderness areas in the East do not establish standards for western areas that are typically larger. Nor do conditions in heavily used areas near large population centers set precedents for more lightly used, remote areas.

DESIGNATION OF AREAS WITH UNUSUAL MANAGEMENT PROBLEMS

Wilderness designation decisions can affect management if an area is designated as wilderness that presents problems for which technical solutions are limited, unavailable, or infeasible. For example, locating boundaries in vulnerable or unwieldy places (e.g., roadsides and lake shorelines) can make wilderness management difficult, as can the inclusion of private land inholdings or popular locations already affected by high recreational use.

The Wilderness Act explicitly recognized that designation criteria can affect management; in section 2(c), the act prescribes the minimum size of a wilderness area: "at least five thousand acres of land or of sufficient size as to make practicable its preservation and use in an unimpaired condition." Below some minimum size, the act is saying, management to preserve an area's wilderness qualities becomes extremely difficult. Maintaining naturalness and solitude in smaller areas is now a challenge in many wilderness areas in the East where, under the so-called Eastern Wilderness Act (U.S. Public Law 93-622), areas of less than 5,000 acres have been considered "of sufficient size" to be designated as wilderness. In addition, highly irregularly shaped boundaries, boundaries that do not follow easily recognized terrain features, and boundaries that adjoin areas of commodity production or urban development could all accentuate management problems. Agencies have generally attempted to keep wilderness boundaries along ridges or other easily recognized topographical features, arguing that such boundaries make management easier.

Some areas cannot be managed easily as wilderness for other reasons. Marion Lake in Oregon's Mount Jefferson Wilderness illustrates the difficulties that can occur when an accessible and popular area is included in a wilderness. Marion Lake, an easy one-mile hike from the road, had a long history of heavy recreation use before its inclusion in the wilderness system. Nonetheless, environmentalists strongly supported its inclusion, even though a permanent USFS cabin shared space at the lake with scores of private boats stored along the shoreline. The USFS contended it would be extremely difficult to scale back this established use to a level consistent with that required by the Wilderness Act. Congress disagreed, and the area was included in the Mount Jefferson Wilderness. Despite a substantial reduction in boat storage, Marion Lake continued to attract heavy use, heavier use than managers considered appropriate in wilderness. To deal with the undesirable impacts and resulting intensive management required at the lake, Oregon senator Mark Hatfield sponsored legislation to remove it from the Mount Jefferson Wilderness. His proposal, however, failed to attract the support necessary for enactment. Similar situations continue to arise where easy access to popular locations in proposed wilderness poses an interim and subsequent challenge to wilderness management.

The potential for mining in many USFS-managed and BLM-managed wilderness areas is another complication with which wilderness managers must deal. The Wilderness Act provided for continued prospecting (mineral exploration) until 1983, and operation of valid claims after that date, subject to regulations to protect the wilderness. Some wilderness designation laws withdrew areas from mineral exploration or claim operation. In 1982 and 1983, Congress added legislation that blocked the Secretaries of Agriculture and the Interior from any mineral exploration in wilderness. That legislation remained in effect in later years, serving as imminent protection for wilderness study areas (WSAs) and areas recommended for wilderness should they be designated. Mining claims can be staked on BLM study areas until the areas are designated as wilderness. Valid claims with valuable deposits discovered before the cutoff can be activated in the future. Although BLM and USFS regulations attempt to ensure that mining activities conform to the mandates of the act, balancing the legal rights of miners with the preservation of wilderness values presents managers with yet another complicated challenge.

DESIGNATION LIMITING USES AND MANAGEMENT

Designating an area as wilderness limits the range of uses and management alternatives and techniques. Areas managed as wilderness will offer only part of the ROS the area might otherwise provide. For example, motorized recreational activities such as trail bikes and snowmobiles generally are not allowed. Certain management activities are likewise generally restricted by wilderness designation, such as the use of motorized equipment and mechanical transport, although some exceptions are provided in certain wilderness designation laws.

The issue of what management actions, facilities, and equipment are allowed in wilderness was an important point of debate in the Endangered American Wilderness Act of 1978 (U.S. Public Law 95-237), which sought to designate more than twenty new areas as wilderness and was a forerunner of the Forest Service's second Roadless Area Review and Evaluation (RARE II). Opponents claimed that wilderness designation could mean wildfires could not be suppressed, trails could be closed, and hunting and fishing could be drastically reduced. To clarify the issue of what is permissible in wilderness, Weaver and Cutler (1977) outlined some important statements of wilderness management latitude. It exposed several misconceptions about the extreme purity of USFS wilderness management policy, which was much less restrictive than opponents of wilderness claimed. Subsequently the Endangered American Wilderness Act passed, and the RARE II was initiated (see Chapters 2 and 5 for more information about RARE I and RARE II).

The restrictions on recreational and management activity in wilderness make it all the more important for land managers to provide a spectrum of opportunities and activities in areas that are roadless, but *not* designated wilderness: use of motorized vehicles, trail bikes, concentrated recreation use requiring intense management and development, and so on. A broad range of opportunities in the middle of the EMS, with relatively clear standards for recreational and other uses and management, will help meet the public's desire for diverse kinds of off-road recreation and ease the demands on wilderness to accommodate all kinds of roadless recreation activities. Provision for roadless but non-wilderness land-use categories will provide alternatives for areas that cannot be successfully managed as wilderness.

THE NEED FOR WILDERNESS MANAGEMENT

Managing wilderness has been a common practice since 1964. The early leaders of the wilderness movement—John Muir, Aldo Leopold, Arthur Carhart, and Robert Marshall—were primarily concerned with saving wilderness from development. They assumed that designating lands as wilderness and prohibiting road construction, logging, and similar uses would ensure the preservation of wilderness, at least for the time being.

"Draw a line around it and leave it alone" well described the prevailing opinion. Today, most people concerned about wilderness have concluded that a "no-management" approach will not work with the many pressures and threats confronting wilderness. Thus, *some of the public concern about wilderness is directed at how wildernesses are to be managed to protect their wilderness characteristics and values that led to their designation.*

THE MANAGEMENT PARADOX

The term wilderness management *is a paradox (Nash 1982). Wilderness is conceptualized to be an area where the influence of modern people is absent (or at least minimized), but the word* management *suggests humans controlling nature.* In most kinds of intensive resource management, humans do alter and control natural processes. Forest management changes the number, size, distribution, and even species of trees, along with its habitat qualities for different fauna.

Many people react negatively when wilderness management is mentioned, because they envision resource and environmental manipulation. Part of the emerging appeal of the term *wilderness stewardship* in lieu of *management* is to soften the implications of the latter. However, *wilderness management does not necessarily require manipulative activity within the designated area. Wilderness management (stewardship) is essentially the management of human use and internal and external influences to preserve an area's naturalness and solitude. It includes everything done to administer an area, from planning in an office to public education and information programs near the wilderness and to management actions and enforcement activities in the field.* Management includes planning and implementing visitor education and, when necessary, rules, regulations, and visitor management to control overuse; minimal facilities such as trails and signs; decisions about access roads to the wilderness boundary; regulations for recreation stock and livestock grazing; wilderness patrols to monitor conditions and use; and public education, including information in maps, brochures, and guidebooks for visitors by wilderness rangers in the field and in schools and other public settings.

Wilderness management does not need to be heavy-handed on people or on resources. In recent decades, the emphasis on educating and informing wilderness users has emerged and grown. *Wilderness managers are guardians, not gardeners or guards.* Managers should not mold nature to suit people; they should manage human use and influences to not impact natural conditions and processes. *The guiding rule is: wilderness managers should only do what is necessary to meet wilderness objectives using the minimum tools, regulation, or force required to achieve those objectives.*

MORE INTERNAL AND EXTERNAL THREATS TO WILDERNESS

Why is management so important now if the founders of the wilderness movement apparently were not that concerned about it? *The main reason for management is the overwhelming increase in the number of threats to wilderness compared to the 1930s, in the time of Marshall and Leopold.* For example, soils and vegetation are impacted from increased recreational use far beyond anything experienced a generation or more ago. Some features of the wilderness experience, the quiet and solitude that were once ensured, are now difficult to find in many places and impossible to experience in others. The founders of the wilderness movement did not ignore management because they provided direction for it in the Wilderness Act. But, it was certainly of lower priority than getting wilderness designated. Today, wilderness management and stewardship is increasingly a concern of managers and the environmental community alike (Fig. 1.3).

More than twenty types of internal and external threats to wilderness conditions and values have been identified (see Chapter 13). For example, population growth and residential development near wilderness areas stem from many factors: rising incomes, greater mobility, and growing interest in the environment and outdoor recreation; and increased interest in health and physical activity. Heightened interest and increased use of wilderness are logical outgrowths of the intense controversies and publicity over the designation of particular areas. These battles tend to be lengthy, sustained by dedicated local environmentalists and accompanied by continued publicity (Allin 1982; Evans 1981; Frome 1997; Scott 2004). They have occurred throughout the nation. Public recitation of the virtues of and threats to particular areas contributes to the growth of wilderness visitation and appreciation. Inevitably, this interest has been accompanied by a growing realization of the necessity for wilderness management.

FIGURE 1.3 U.S. Forest Service wilderness manager Greg Hansen (now retired) is an example of federal agency professionals in the U.S. Forest Service, Bureau of Land Management, National Park Service, and U.S. Fish and Wildlife Service who have made a career in wilderness management. Wilderness management includes planning and implementing policies and plans to achieve objectives of the Wilderness Act. Photo courtesy Marilyn Hendee.

Paralleling increased threats to wilderness has been a decline in acreage of undeveloped land that might qualify as wilderness. The amount of land that has wilderness potential is fixed; more cannot be created within any reasonable time span. Designation as wilderness ensures only that development will not occur; it does not create additional acreage or raise its capacity for use. In other words, the wilderness was there, and it was being used as such before it was designated. So one benefit of designation is better management and protection of the existing supply of wilderness.

PAST DISRUPTION OF NATURE

Wilderness management is needed to restore the dynamics that have been altered by human interference with natural processes. For example, fire prevention and suppression have limited fire as a natural ecological force in some ecosystems for more than fifty years. This constitutes control of nature just as surely as does a dam across a wild river. Before modern fire prevention and suppression, fires occurred much more frequently and over much larger areas, and thereby prevented, for example, unnatural accumulations of fuel that can produce fires more intense than those that might otherwise occur. Fire management to restore natural conditions is needed in many areas, and wilderness management agencies are implementing fire policies to allow fire a more natural role (see Chapter 11). Insect infestations and disease are allowed to run their course in wilderness where they do not threaten resources outside the protected area, and such infestations provide natural demonstrations of ecological processes that are unlikely to be tolerated elsewhere.

Increasingly, as more comprehensive stewardship strategies are adopted for all public wildlands under the label of ecosystem and landscape management, wilderness provides the natural model against which results can be compared and monitored.

GIVING MEANING TO WILDERNESS DESIGNATION

Wilderness designation loses much of its meaning if subsequent management policies do not define what it will actually accomplish. *Without management, wilderness designation verges on being an empty symbol, a mere name and a boundary line on a map.*

Some people who recognize the need for management want it to wait until all the wilderness designation decisions have been made: that is, "We'll worry about management when all the wilderness designation is all done." We think this is an indefensible stand. Even though a majority of wilderness designation decisions may have been made, many areas are still under consideration. Threats to wilderness (e.g., global climate change, fragmentation of ecosystems, nonnative plant and animal introductions) have greatly increased in the past fifty years and more are expected. The impacts of such threats will be substantial, and in some places have already affected areas and threaten to change the quality and character of available wilderness experiences.

ALTERNATIVES TO MANAGEMENT?

Is managing wilderness the only alternative? Two policies could obviate that necessity. First, all use could be prohibited. Some problems, like fire management, could still require action, but essentially no use could perhaps justify no management. However, the Wilderness Act and wilderness philosophy make clear that wilderness is to be used and enjoyed by people. Public use is legally mandated. Second, we could just designate areas and forget about management. Any kind of use not clearly illegal under the Wilderness Act could be allowed in an unlimited amount, and environmental damage or changes in the ecosystem could simply be accepted. Under this option, wilderness could vanish from most places. This result would violate the Wilderness Act, which requires the protection and perpetuation, the preservation, of wilderness.

The middle ground between these two extremes requires management. We see no other course of action. This course was endorsed early in the history of designated wilderness by natural-resource leaders reflecting on the new challenge and who largely concluded that managed wilderness is the only possible course (Spurr 1966; Zivnuska 1973). *The real question is not whether to manage but how to manage wilderness.* That is the topic of this book, and the subject of an expanding compendium of experience and research in agency manuals, scientific and popular literature, conference proceedings, and Internet sources. This expanding knowledge continues to greatly strengthen wilderness stewardship. However, the knowledge is scattered; we have synthesized and summarized it for ease of teaching and application.

WILDERNESS PHILOSOPHY AND WILDERNESS MANAGEMENT

Each federal agency having responsibilities under the Wilderness Act has developed policies and guidelines to apply it to lands under their jurisdiction. Such direction is important so that management does not significantly deviate from area to area. However, policies cannot and should not be so detailed as to cover all contingencies because managers need to retain some flexibility to respond to unique conditions occurring in individual areas. There is always a gap between specific policy direction and unanticipated contingencies, the need for broad guidelines on one hand and flexibility on the other, and this gap must be filled by a manager's philosophical perspective. *A deep philosophical understanding of wilderness, grounded in knowledge of its legislated direction and inherent ecological and social values, is a wilderness manager's most important tool.*

Before considering philosophical perspectives, we call attention to a principle discussed in more detail in Chapter 7: *that wilderness is preserved and managed for the benefits and values it provides people.* Many laws, among them the Endangered Species Act, the Wild and Scenic Rivers Act, and the Wilderness Act, have been enacted to protect and preserve the natural environment for its own sake, but under the assumption that retaining natural features provides important human benefits. Even the statement that a feature is unique and worthy of protection is a human judgment based on the belief that such things are valuable for our pleasure, survival, and well-being (Fig. 1.4).

The idea that wilderness is for use and enjoyment by people has clear statutory support. The Wilderness Act specifically notes that wildernesses will be administered "for the use and enjoyment of the American people." But what kinds of use? The act rules out some uses for instance, motorized equipment, with some minor exceptions included in wilderness designation acts, is not permitted. Still, the act permits considerable diversity in styles of use and in the accompanying developments. Simply arguing that wilderness is for public use only states the obvious. It does little to resolve the issues of what kinds of use, how much, where, and when. Understanding different philosophical perspectives about wilderness and its values is important in addressing these questions regarding the act.

FIGURE 1.4 Visitors aboard a cruise ship observe tidewater glaciers in 2.6-million-acre Glacier Bay Wilderness, in Glacier Bay National Park, Alaska. The need for wilderness management is growing, partly because of steadily increasing use, the impacts of that use, and the special permitted uses such as motorized access in Alaskan wilderness areas. Photo courtesy Carol Dawson.

ANTHROPOCENTRIC AND BIOCENTRIC PERSPECTIVES

Consider two alternative philosophical notions about wilderness and the basis for its value: *anthropocentric* and *biocentric* (Hendee and Stankey 1973). *The anthropocentric position takes the "use and enjoyment" phrase of the Wilderness Act quite literally.* Under this philosophy, wilderness is viewed primarily from a sociological or human-oriented perspective; *the naturalness of the wilderness is less important than maximizing direct human use.* Programs to alter the physical and biological environment to produce desired settings are encouraged: big trees, open vistas, lots of fish and wildlife. Developed facilities to increase recreational use of wilderness are appropriate. In fact, increasing direct human use could increase some types of human values and benefits under this philosophy. The concept of wilderness carrying capacity would not exist, because people's ever-changing adaptation to their environment (and vice versa) continually changes (lowers) standards of crowding and naturalness. The character of wilderness could change to reflect human desires and contemporary standards of naturalness. Wilderness could still be one extreme on the environmental spectrum, but a shifting extreme, not grounded in any definite standards. "Let's open up our wilderness areas and do away with restrictions" could be a slogan for believers in pure anthropocentrism.

In contrast to the anthropocentric perspective, the *biocentric perspective emphasizes maintaining natural systems, if necessary, at the expense of recreational and other human uses* (Hendee and Stankey 1973). The goal of the *biocentric* philosophy is *to permit natural ecological processes to operate as freely as possible, because wilderness values for society ultimately depend on retaining naturalness.* To the extent that naturalness is lost, the experiential, spiritual, and scientific values of wilderness are lessened.

Labeling these perspectives as *biocentric* versus *anthropocentric* might create a false distinction between "wilderness for people's sake" and "wilderness for wilderness's sake." But, as discussed earlier, wilderness is for people. *The important distinction between these philosophies is the extent to which the human benefits of wilderness are seen as dependent on the natural integrity (naturalness and solitude) of the wilderness setting.*

Their respective proponents have argued these alternative perspectives persuasively. Before turning to a more detailed look at each, two points should be emphasized. First, these philosophies represent extreme, polarized concepts about wilderness management and stewardship, and it is unlikely either could be completely followed. However, they do highlight alternative orientations toward wilderness stewardship. Second, it is difficult, if not impossible, to say that either idea is wrong or right. It seems more important to examine the long-range implications of each and to judge the appropriateness of each considering society's objectives as reflected in the Wilderness Act. With that in mind, let us look at each perspective more closely. *Readers interested in testing their own anthropocentric versus biocentric instincts can find a diagnostic quiz to take in the "Reader Exercise" at the end of this chapter.*

THE ANTHROPOCENTRIC PHILOSOPHY

Advocates of the anthropocentric approach would have us facilitate direct human use of wilderness. Wilderness managers could emphasize recreation and comfort. They might develop high-standard trail systems; expand stocking fish to most wilderness lakes; increase and upgrade campsite facilities, shelters, toilets, and similar features; and generally, increase recreational carrying capacity, aesthetic features, and user comfort and convenience. The perception of wilderness held by the largest number of users could be the most important guideline for managers.

This orientation emphasizes society's demands on wilderness, not its natural condition, and has important implications for users and the environment. Initially the emphasis on the aesthetic and recreational qualities of wilderness settings could lead to substantial alteration of the environment, particularly vegetation. Given current knowledge and technology, we could engineer the wilderness scene to produce specific environmental conditions. For example, some have argued that a wilderness should represent an early point in our history, perhaps the land as it was at the time the continent was first settled by Europeans. To create such a setting would require sharp interference with natural processes to steer ecological succession in the desired direction. Fire, chemicals, or machines might be needed. Some desirable results, from an anthropocentric point of view, might be increased scenic views from cleared viewpoints, additional forage for stock from intensively managed (perhaps irrigated and fertilized) range, alpine meadows enlarged and maintained by uprooting invading tree seedlings, and more observations of wildlife stimulated by strategically placed salt licks. Traditional forestry, silviculture, and habitat management might be in order. Wilderness managers would be more like gardeners rather than guardians.

However, the anthropocentric approach would mean losing an essential wilderness quality: naturalness. Furthermore, such an approach (at least after a certain point) would be illegal in designated wilderness. The Wilderness Act says quite clearly that wilderness should be a setting where the forces of nature operate free from human influence. Human influence on ecosystems is already pervasive worldwide, ranging from the introduction of atmospheric pollutants to direct recreational impacts. But minimal influence with natural evolution and processes seems to be the clear intent of the Wilderness Act (sec. 2[a]): "In order to insure that an expanding settlement and growing mechanization does not occupy and modify all areas within the United States, [and] to secure for the American people of present and future generations the benefits of an enduring resource of wilderness." Meeting this goal would not be possible under the anthropocentric approach.

An anthropocentric approach might be particularly detrimental to the scientific values of wilderness. The notion of wilderness as a genetic pool, an environmental baseline, and a refuge for the survival of species especially sensitive to human influence would be lost in a wilderness manipulated and altered to fit changing human tastes. Because the loss of naturalness could in many cases be irreversible, there could be incalculable costs in terms of forgone scientific values and benefits.

Styles of recreation tuned to anthropocentric management philosophy might be convenience oriented. Recreational experiences would be a primary goal and would mean increasing access, reducing difficulty and danger, and facilitating use. Conversely, programs hindering or restricting use might be rejected. For example, if wilderness exists primarily for use and enjoyment, programs limiting use (e.g., wilderness permit systems, rationing, rough trails, minimum party size) are bureaucratic hindrances that would be eliminated. Under an anthropocentric philosophy, then, if problems of environmental impact or excessive congestion did arise, the managerial response might be to harden sites so they could tolerate impacts, revegetating, installing more facilities, upgrading trails, and otherwise intensively managing the wilderness to handle increased levels of use. Opportunities for solitude would decline.

Proponents of the anthropocentric philosophy argue that biocentric wilderness regulations discriminate against people who might want to visit wilderness but cannot because of lack of physical ability and, in effect, close off public lands (wilderness) to a large segment of the public. Such proponents might point to the European Alps, where large numbers of people are accommodated through extensive road systems, cog railways, groomed trails, and mountain huts.

THE BIOCENTRIC PHILOSOPHY

The biocentric perspective emphasizes preserving natural processes and encouraging nearly natural conditions and processes within wilderness ecosystems as they existed in the absence of human influence. This requires controlling human influences in the ecosystems from sources such as excess recreational use and eliminating restrictions on normal processes caused by such policies as fire prevention and suppression. Recreational use in wilderness might be consistent with this perspective only to the point that it does not unduly alter natural conditions and processes. *Like the anthropocentric philosophy, a biocentric approach focuses on human benefits, but to a lesser extent. The important distinction we make between them is that, biocentrically, these benefits are viewed as being dependent on the naturalness of wilderness ecosystems. Thus, protecting naturalness is essential to providing human benefits. It is the experience of wild (untrammeled) nature, and the ecosystem services that flow from it, on which human benefits depend.*

The biocentric approach to wilderness management has specific implications for the environment and users. Over an extended period of time, we might expect to see environmental conditions evolve that reflect historical patterns of ecological succession. The natural processes such as wind, erosion, insects, disease, and fire that have shaped and altered the landscape might continue to operate much as they always have. One consequence could be a wilderness that is aesthetically unattractive. Insect infestation, erosion, fire, forest disease, and similar processes might be allowed to run their course without human interference; as a result, wilderness landscapes would reflect these natural perturbations. This approach could mean that particularly desirable recreational features, such as high mountain meadows as important stock forage areas or areas of high wildlife visibility, might gradually, or dramatically, shift with advancing ecological succession or following natural disturbances.

These management challenges would present little opportunity to apply intensive natural-resource skills such as silviculture and habitat management. Rather, much more emphasis could be placed on monitoring conditions and controlling visitor behavior to preserve dynamic natural processes. Nature would roll the dice to determine ecological outcomes.

Under a biocentric management philosophy, recreational use of a wilderness might be secondary to maintaining the natural order. Management actions to increase and facilitate use, such as high-standard trails and campsite facilities, might not be appropriate. Where recreational use would cause significant impacts, management could modify visitor behavior and limit or disperse use rather than institute methods to absorb greater impact. Moreover, management programs could promote opportunities emphasizing the primitive environment, challenge, and solitude, activities with fewer environmental impacts than those available on non-wilderness lands or emphasized under an anthropocentric philosophy. *A biocentric philosophy requires that visitors take wilderness on its own terms.*

These two wilderness management philosophies, anthropocentric and biocentric, represent polar viewpoints that are especially useful for the sake of discussion and planning. A more moderate view between these two extremes has been called ecocentric (Clark and Kozacek 1997), but we call it midcentric, thinking that ecocentric might be construed as an extreme beyond biocentric.

IN SUPPORT OF BIOCENTRIC MANAGEMENT

Which is the most appropriate management philosophy: biocentric or anthropocentric? As noted earlier, the question of which approach is right and which is wrong has no absolute answer. However, an answer must relate to long-term implications and to the legal mandates set forth in the Wilderness Act.

Facilitating maximum use within wilderness will gradually diminish its naturalness and solitude, alter ecological regimes in subtle or drastic ways, and diminish opportunities for experiences dependent on wild and unaltered settings. The result might be to eliminate one extreme on the environmental modification spectrum (EMS) and recreation opportunity spectrum (ROS) and, thus, lose diversity among wildland settings.

We believe that to achieve the legal goals of the NWPS, management should emphasize the natural integrity of wilderness ecosystems as much as possible, and this reflects a more biocentric management philosophy. This position and its implications are most consistent with the mandate of the Wilderness Act and with the intent of the legislative debate that fashioned the act, and with its historical and philosophical foundations that evolved over the past century. *Consider the following arguments supporting a biocentric perspective.*

First, providing for the diversity of recreationists' tastes could ensure that people who prefer a wild and pristine setting would not be displaced in favor of users whose tastes can be met in non-wilderness settings. Some users depend on a pristine wilderness to satisfy their desires for a wilderness experience. As the supply of pristine settings diminishes, management philosophies should strive to maintain an ROS of diverse opportunities to protect all of them and, from our perspective, especially pristine wilderness.

Second, a human ecology view of outdoor recreation suggests that as people gain outdoor experience through activities such as car camping, they may begin to seek more demanding kinds of experiences. For example, young people may choose to expand the boundaries of childhood car camping into wilderness backpacking experience as young adults. Many campers using developed facilities might in the future opt for wilderness. Increased demands on wilderness are virtually certain. Biocentric management will help maintain opportunities at the primitive end of the EMS and ROS to meet this increasing demand.

Third, management responses to increasing use can lead to unanticipated shifts in the kinds of recreational opportunity an area offers. Developments to protect a site can attract a different clientele. For example, campgrounds developed to protect natural qualities of a site such as tent pads, tables, fireplaces, and toilets may attract a new clientele drawn not by nature but by the comfort and convenience of the facilities and socialization with other users. The biocentric approach is challenged not just by recreational use, but also by wilderness designation proposals enacted to include various exceptions and special provisions for mechanized access and use that negatively impact wilderness. Applying the biocentric criteria of naturalness and solitude might minimize such changes in wilderness and limit the growth of inappropriate kinds of use and impacts.

Fourth, a biocentric approach would preserve the greatest range of future options. Management decisions that facilitate increased use through development are often irreversible, and they narrow the range of environmental opportunities by eliminating unmodified areas that are already in short supply. This leads to loss of diversity and loss of important biological and scientific values.

WILDERNESS NATURALNESS AND SOLITUDE

The Wilderness Act, in section 2(c), defines wilderness, in part, as a place that "has outstanding opportunities for solitude *or* a primitive and unconfined type of recreation" (italics added).

The view of this phrasing was generally understood to mean the "or" was intended to clarify the nature of the experiences wilderness should offer and that the descriptive phrase "solitude or a primitive and unconfined type of recreation" is required in the definition of wilderness (Worf 2001; Worf et al. 1972). Proponents of this interpretation claim that it is supported by the legislative history of the Act and ensuing legislation (Saylor 1962). In the so-called Eastern Wilderness Act of 1975 (U.S. Public Law 93-622), the statement of policy notes that wildernesses designated by the Act become a part of the NWPS and that management shall "promote and perpetuate the wilderness character of the land and its specific values of solitude, physical and mental challenge, scientific study, inspiration and primitive recreation." Thus, management to provide these experiences, *including solitude,* would seem required.

Another interpretation of section 2(c) of the Act has emerged: that opportunities for either solitude *or* a primitive kind of recreation are required (that is, either one could qualify an area). This interpretation raises new management possibilities, as the following example illustrates.

At the turn of the millennium, following decades of increased wilderness use now concentrating in high-use locations, the USFS proposed a new wilderness recreation strategy that would accept higher levels of use at popular wilderness locations to keep from diverting excessive use and impacts to more pristine, low-use wilderness (Oye 2001). Thus, under this strategy, the interpretation of section 2(c) of the Wilderness Act would seem to be that solitude *or* (rather than *and*) a primitive and unconfined type of recreation is what is required. This interpretation was very controversial. Opponents of the proposed new wilderness recreation strategy claim that a nondegradation mandate is being violated by accepting lower standards of naturalness and solitude at the high-use sites, thereby creating two classes of wilderness (Worf 2001). However, supporters argue that this is needed; they also point to the dramatically increased use at popular wilderness sites and defeat by environmental groups of wilderness management plans to greatly reduce use. "If we lock people out who will fight to save the wilderness?" asked Spring in his statement supporting the proposed new policy (2001, 17). In many ways, the proposed new policy is a fait accompli because use levels at popular wilderness locations are already beyond levels providing solitude. Rolling back use now might not be politically possible, and use will inevitably divert to more lightly used but pristine sites. A statement of the proposed USFS wilderness recreation policy (Oye 2001), preceded by a statement of the situation requiring action (Cole 2001), followed by arguments against the policy (Worf 2001), and supportive of the policy (Spring 2001), appear in the April 2001 issue of the *International Journal of Wilderness. The policy has not been implemented fully across USFS wilderness areas, but it is cited in some USFS plans, such as a management plan for the White Mountain National Forest in New Hampshire and Vermont* (U.S. Department of Agriculture, Forest Service 2005).

We believe that section 2(c) of the Wilderness Act does require that both solitude and a primitive and unconfined type of recreation be provided. Some flexibility seems possible by approaching the requirement with an NWPS-wide view, that is, every area cannot provide equal amounts of each requirement and some variation is expected.

Obviously, the levels of solitude found within and between areas will vary, perhaps substantially. The wilderness acts do not define a single standard for solitude that all areas must meet. The existing pattern of trails and campsites in a candidate or designated wilderness precludes such a standard anyway. Visitor use intensities will be higher near trailheads and at popular locations, and it seems neither necessary nor possible to manage for some uniform level of interparty contact. The 1964 act calls for provision of "*outstanding opportunities* for solitude" (italics added), and we interpret that to mean there should be places and times within the NWPS and within individual wildernesses where visitors find little or no contact with others. However, this does not mean on every wilderness acre all the time.

We recognize that the wilderness experience is a product of human perception and cannot be precisely described and packaged. However, managers do need some guidelines. To us, the Wilderness Act, in section 2(c) (1), provides them. In answering the question "What distinguishes wilderness from other settings?" we have sought an answer that would be true to congressional intent and offer specific criteria for managers. We think wilderness is an area (1) featuring substantially natural ecological conditions and (2) offering visitors outstanding opportunities for solitude or a primitive and unconfined type of recreation. There will be naturalness everywhere in the wilderness all the time, but at some places in the wilderness, at some times, there will also be high-use sites where the simultaneous provision of naturalness *and* solitude will not be possible. Further, it may not be possible in such situations to roll back use to a level where naturalness and solitude can be provided without diverting excessive use to more sensitive pristine areas or prompting political repercussions that might be destructive to the wilderness—such as removing the wilderness designation from popular sites. So judgment is needed. And ironically, it may be by allowing more use at high-use sites (with appropriate environmental safeguards) that a biocentric approach featuring solitude can be applied to the remaining majority of a wilderness area.

TWO QUALIFICATIONS TO BIOCENTRISM

The issue of what philosophy will underlie wilderness management is crucial to the future of the NWPS. The wilderness system we have in the future will be a direct product of the philosophy that guides policy and management decisions today. We judge that a predominantly biocentric philosophy is appropriate, necessary, and defensible. The policies and practices of federal agencies tend to reflect such a perspective over most of their wildernesses (some high-use sites excepted), and we hope this direction continues.

We qualify our support for the biocentric approach in two important respects: First, it must be practical, because the biocentric philosophy, as we describe it, represents an ideal, and its implementation will be inhibited by practical constraints. We noted earlier the proposed Forest Service wilderness recreation strategy to cope with high-use sites and arguments for and against such a strategy. Other examples also apply: ideally, biocentric management would allow wilderness fires to burn. However, this is not practical. In many areas, because of several decades of fire suppression, unnaturally large supplies of highly combustible fuels have accumulated. Fires in these areas might become extraordinarily intense, causing catastrophic damage to wilderness resources and/or to resources outside the wilderness boundary endangering people and property (e.g., the Yellowstone forest fires of 1988).

Because the NWPS includes many relatively small tracts of land scattered throughout areas managed for other purposes, non-wilderness considerations will always influence what happens inside the wilderness boundary and vice versa. *Thus, we endorse the biocentric philosophy, while recognizing that its extreme application might be unrealistic and impractical, and create a backlash against it. A rule of reason must temper its application.* Nevertheless, we think that a management and policy orientation that judiciously strives toward the *intent* of biocentricity is proper and feasible.

Our second qualification is that biocentricity and the entire wilderness preservation movement in a broader sense are viable philosophies only if they are accompanied by equitable provision for alternative outdoor recreation opportunities (ROS), and comprehensive efforts to humanize the places where we work and live. The elitist overtones of biocentricity and wilderness concern us. Are we endorsing a philosophy that offers access only to a privileged few at the expense of the majority? That is only true if alternative outdoor recreation opportunities—a full ROS of them—are not provided. In a democratic society, we see little chance that support for wilderness, much less a biocentric philosophy of wilderness management, can survive unless an equitable range of outdoor recreation opportunities are provided.

As Sebastian de Grazia noted, "Only if you give the city a pleasant and healthful outdoor environment, can you slacken the . . . drive for the wilderness. Only the city can save the wilderness" (de Grazia 1970, 96). Obviously, citizens who press for wilderness designation and the agencies that manage wilderness have only a limited capacity to change the poverty that plagues our inner cities, the social inequities that divide our people, the haphazard land-use patterns and transportation systems that blight our landscapes, or the pollution that clouds our land, air, and water. However, even though wilderness is our primary focus, we should not lose sight of these broader issues. This wider perspective links wilderness to issues such as energy use, social equity, land-use planning, and federal programs to provide outdoor opportunities for people of the inner city.

A COMMONSENSE POLICY IS NEEDED

While we lean toward a biocentric wilderness management philosophy, we are not extremists. We call for management with a biocentric emphasis but that it be applied with common sense and sensitivity to local conditions; perhaps it might be more appropriate to call it stewardship (Fig. 1.5). We reiterate the idea that *wilderness management should not mold nature to suit people. Rather, it should manage human use and influences so as not to alter natural conditions and processes. Managers should do only what is necessary to meet wilderness objectives and use only the minimum tools, regulation, and enforcement required to achieve those objectives.* In wilderness, people adapt to nature, to naturalness, and to solitude, and that is the source of human benefits from wilderness experience as well as the ecological and nonuse benefits.

The details of management may vary among wildernesses, but we support our broad notions by citing the late Senator Frank Church who in a 1977 Wilderness Resource Distinguished Lecture at the University of Idaho (Church 1977, 13) argued:

> It was *not* the intent of Congress that wilderness be administered in so pure a fashion as to needlessly restrict their customary public use and enjoyment. Quite to the contrary, Congress fully intended that wilderness should be managed to allow its use by a wide spectrum of Americans. There is a need for a rule of reason in interpreting the Act, of course, because wilderness values are to be protected. As I stated in 1972, while chairing an oversight hearing of the Subcommittee on Public Lands: . . . The Wilderness Act was not deliberately contrived to hamstring reasonable and necessary management activities. We intend to permit the managing agencies . . . latitude . . . where the purpose is to protect the visitors within the area . . . [including, for example] minimum sanitation facilities . . . fire protection necessities . . . [and] the development of potable water supplies. . . . The issue is not whether necessary management facilities are prohibited; they are not. The test is whether they are necessary.

Thus, *the wilderness management framework intended by Congress was for the agencies to do only what is necessary.* The facilities just mentioned may be required, such as at high-use sites. But, restrictions on use may sometimes be needed to protect especially fragile locations such as pristine, low-use wilderness zones. However, in adopting regulations, common sense is required.

FIGURE 1.5 On Public Lands Day, September 23, 2000, the BLM Needles, California, field office organized a project in which more than forty volunteers (top) gathered to pick up trash from an old mining and ranching homestead in the Old Woman Mountains Wilderness. Such public involvement in wilderness management provides much-needed help to busy managers. Volunteers (middle) take trash to a staging area (bottom) where a horse-drawn wagon is loaded with old wire, irrigation pipe, fencing, cans, and tin and hauled to the trailhead a mile away. Photos courtesy Marilyn Hendee.

SUMMARY

This chapter's objective has been to set the stage for systematic and progressively more detailed discussions related to management of the NWPS in the United States. So far, we have focused on the broadest direction—wilderness management philosophy.

More specifically, this chapter explored the meanings and definition of wilderness in light of some basic, albeit overlapping, themes and values espoused by historical wilderness spokespeople in the United States. We briefly discussed the difference between wilderness designation and wilderness management. We argued the need for wilderness management from several vantage points. Finally, we described two alternative wilderness management philosophies, anthropocentric and biocentric—acknowledging a middle ground as midcentric—and evaluated their applicability in light of objectives of the Wilderness Act. We concluded that the judicious application of a biocentric philosophy of management (stewardship) could result in the most appropriate protection of the NWPS. Stewardship should strive to maintain the historical natural processes that formed the great American wildernesses experienced by our American forebears.

STUDY QUESTIONS

1. How did early European settlers of America view wilderness? How has that view changed in recent history and why?
2. Name four categories of wilderness values or themes recognized by the wilderness movement in this nation. Briefly explain each.
3. Why do managers need to understand the values people hold for wilderness?
4. How does wilderness designation differ from wilderness management?
5. Who decides which lands shall be designated as wilderness?
6. Why is wilderness management more necessary today than in the past?
7. Contrast the *anthropocentric* and *biocentric* philosophies of wilderness management.
8. Which of the two philosophies do you favor for managing wilderness in the United States. Why?
9. What does de Grazia mean by the statement "Only the city can save the wilderness"? Do you agree or disagree? Why?
10. Do the exercise at the end of the chapter, "How Do Your Wilderness Values Rate?" Where does your wilderness management philosophy fall on the anthropocentric versus biocentric spectrum? How do you think it will change by reading this book?

READER EXERCISE: HOW DO YOUR PERSONAL WILDERNESS VALUES RATE?

A person's personal orientation and philosophy about wilderness and its management can be an important factor in his or her perception of what are appropriate policies and actions for wilderness stewardship. Just recognizing that wilderness visitors and one's colleagues might have a slightly different philosophy contributes to understanding that wilderness stewardship is not a cut-and-dried business; rather, there is a lot of room for interpretation in making the right decision.

PUT YOUR VALUES TO THE TEST!

How do your personal wilderness values rate? Take the test, adapted from Clark and Kozacek (1997), and see. It's easy—and you might be interested in seeing where you fit compared to some others who have taken the test. Remember, answer these questions based on your personal values—not what you think the Wilderness Act or an agency's policy requires.

Stop here and answer the thirty-five questions that follow either "yes" or "no." When you have finished the test, count your number of "yes" answers and then continue reading. There you will find the rest of the story and see how your score might compare. We suggest you take the test before seeing how your score might compare with others.

WILDERNESS VALUES QUESTIONS

1. Do you feel hunting is an appropriate activity in wilderness?
2. Do you feel it is okay to stock native fish in lakes that historically have not had fish?
3. In an area that has established wildlife watering devices (e.g., guzzlers), do you feel it is appropriate to maintain these and leave them in wilderness?
4. Do you feel it is appropriate to control predators in wilderness that are killing a substantial number of livestock?
5. Are low-level aerial-game surveys in wilderness acceptable to you?
6. Do you feel we should be protecting known threatened and endangered species habitat from Prescribed Natural Fires (PNF)?
7. Is it acceptable to you to have Management Ignited Fires (MIF) in a wilderness area?
8. Do you feel it is appropriate to have technologically advanced data-collecting stations in wilderness to monitor temperature, moisture content, wind, and other factors that might allow better information for PNF and MIF?
9. Do you feel we should be suppressing any fires in wilderness?
10. In your opinion, is it okay to maintain historic cabins in wilderness?
11. Do you feel that there is a point when air quality is more important than allowing extended periods of PNF?
12. Is it okay to interpret in a publicly available book historic structures and cultural resources that are in wilderness?
13. Do you feel that cattle or sheep grazing is an appropriate use for wilderness?
14. Do you feel grazing permittees should be allowed to use motorized equipment for maintaining water developments in wilderness where this has been a historical method of maintenance (for example, using a dozer to clean out a dirt stock tank in wilderness)?
15. Do you feel a hazardous tree along a well-used trail should be cut to protect public safety?
16. Do you feel that cutting logs in trails to facilitate passage by pack strings is appropriate in wilderness?
17. Do you feel we should be placing signs by natural caves in wilderness that pose safety hazards?
18. Do you feel it is appropriate for a visitor center to be giving users more information about hazards in wilderness so we can lessen the potential of search-and-rescue operations?
19. Do you feel that signs should be placed at historic structures to warn people of the potential for Hantavirus?
20. Do you feel we should rescue a person with a broken leg (but not in a life-threatening situation) in wilderness with a helicopter?
21. Do you feel it is okay to use llamas or pack goats in wilderness?
22. Do you feel that it is appropriate to leave some established rock-bolt routes for climbers in wilderness areas?
23. Does the value of having the number of users controlled by a permit system outweigh the value of unregulated use and freedom in wilderness (i.e., do you believe permit systems should be used in wilderness)?
24. Do you feel it is okay to allow people to collect crystals in wilderness?
25. Do you feel it is okay to allow people to collect antlers in wilderness?
26. Do you feel that recreation opportunities are the dominant value of wilderness?
27. Do you feel it is okay to have trail signs in wilderness?
28. Do you feel it is okay to put mileage on signs in the wilderness?
29. If a free one were available to you, would you take a cellular phone into wilderness with the intention that it would only be used to help in an emergency situation?
30. Do you feel okay about burying decomposable garbage in wilderness?
31. If you had a well-behaved dog, would you feel okay about taking it with you to the wilderness?
32. Do you think it is appropriate for outfitters to have business operations dependent on wilderness?
33. Do you feel it is okay to film in wilderness a movie about wilderness values?
34. Do you feel it is appropriate to allow a one- or two-week window for chainsaw use to open trails after an intense blow-down event?
35. Do you feel it is okay to apply a mandatory party size or limited permits to promote solitude in wilderness?

Now count your number of "yes" answers and read the following.

ANTHROPOCENTRIC AND BIOCENTRIC PHILOSOPHIES

Two contrasting orientations are often used to characterize philosophies of wilderness stewardship: anthropocentric and biocentric. An *anthropocentric* philosophy sees wilderness primarily from a human-oriented perspective. The naturalness of wilderness is less important than facilitating human use and convenience. Programs that might alter the physical and biological environment to produce desired settings are encouraged. A *biocentric* philosophy emphasizes maintaining natural systems at the expense of recreational and other human uses, if necessary, because wilderness values depend on naturalness and solitude. The goal of this philosophy is to permit natural ecological processes to operate as freely as possible.

The wilderness values test is based on thirty-five questions (Clark and Kozacek 1997) that could be answered "yes" or "no," such as question 9: "Do you feel we should be suppressing any fires in wilderness?" In general, a "yes" answer would place a person on the anthropocentric end of the wilderness values scale, and a "no" answer would reflect a biocentric philosophy. The test is scored by tabulating the number of "yes" answers recorded after all questions have been answered. Of course, the questions present choices that are oversimplified compared to the real world, so you have to respond in a generalized way. In addition, you must keep in mind that it is your "personal" wilderness values that are being measured—not the Wilderness Act or an interpretation of agency policy.

The authors of the test reported that most of the managers they tested respond with between fifteen and twenty-five "yes" answers, and they might characterize them as being in the middle of the anthropocentric–biocentric continuum (we call this midcentric). A few responded with fewer than fifteen "yes" answers, which puts them on the biocentric side of the continuum. Moreover, there are some who have more than twenty-five "yes" answers, reflecting an anthropocentric view.

When the article describing this test was written, Kendall Clark was district ranger on the Eagle Cap Ranger District of the Wallowa-Whitman National Forest in Enterprise, Oregon, and Susan Kozacek was district ranger on the Wilderness Ranger District of the Gila National Forest in Mimbres, New Mexico.

CASE DISCUSSION QUESTIONS: ANTHROPOCENTRIC AND BIOCENTRIC VALUES

1. How do your personal test results compare to those reported by Clark and Kozacek or to the combined results of others in your discussion group? Are you in the anthropocentric, midcentric, or biocentric portions of the range? Does this accurately reflect what you believe are your values about wilderness?
2. How do your personal wilderness values test results compare to your interpretation of the Wilderness Act? Do you perceive the Wilderness Act to be in the anthropocentric, midcentric, or biocentric portions of the range? Defend with specific supporting references to various sections or sentences of the act.
3. Select some readings of a wilderness visionary (Leopold, Marshall, Olson, or others) and try to determine where he or she might be placed along the anthropocentric-to-biocentric continuum. How do you compare to that visionary? Why?
4. Assume the role of a wilderness visitor or user, such as a big-game hunter, long-distance backpacker, local sheep rancher, floatplane pilot and outfitter, or local resident day hiking, and take the test as you imagine he or she would respond. Compare his or her total test result to what you reported in the three previous questions.
5. How do you expect that your wilderness values and philosophy affect your interpretation of the Wilderness Act compared to others? Do others in your discussion group interpret the Wilderness Act the same or differently from you?

REFERENCES

Allin, Craig W. 1982. *The Politics of Wilderness Preservation.* Westport, CT: Greenwood Press.

Bergstrom, John C.; Bowker, J. M.; Cordell, H. Ken. 2005. An organizing framework for wilderness values. In: Cordell, H. Ken; Bergstrom, John C.; Bowker, J. Michael. eds. *The Multiple Values of Wilderness.* State College, PA: Venture Publishing, pp. 47–55.

Bowker, J. Michael; Harvard, J. E., III; Beregstrom, John C.; Cordell, H. Ken; English, Donald B. K.; Loomis, John B. 2005. The net economic value of wilderness. In: Cordell, H. Ken; Bergstrom, John C.; Bowker, J. Michael, eds. *The Multiple Values of Wilderness.* State College, PA: Venture Publishing, pp. 161–180.

Callicott, J. Baird; Nelson, Michael P. 1998. *The Great New Wilderness Debate.* Athens: University of Georgia Press.

Carhart, Arthur H. 1920. Recreation in the forests. *American Forests.* 26: 268–272.

Church, Frank. 1977. Wilderness in a balanced land use framework. *First Annual Wilderness Resource Distinguished Lecture.* University of Idaho Wilderness Research Center, Moscow, Idaho. March 21. [Reprinted as Whither wilderness. *American Forests.* 83(7): 10–12, 38–41.]

Clark, Kendall; Kozacek, Susan. 1997. How do your wilderness values rate? *International Journal of Wilderness.* 3(1): 12–13.

Cole, David. 2001. Wilderness recreation management strategies: Balancing freedom and protection. *International Journal of Wilderness.* 7(1): 12–13.

Cordell, H. Ken; Bergstrom, John C.; Bowker, J. Michael. 2005. *The Multiple Values of Wilderness.* State College, PA: Venture Publishing.

Cordell, H. Ken; Tarrant, Michael A.; Green, Gary T. 2003. Is the public viewpoint of wilderness shifting? *International Journal of Wilderness.* 9(2): 27–32.

Davis, George D. 1989. Preservation of natural diversity: The role of ecosystem representation within wilderness. In: Freilich, Helen R., comp. *Proceedings of the National Wilderness Colloquium*; January 13–14, 1988; Tampa, FL. General Technical Report SE-51. Athens, GA: U.S. Department of Agriculture, Forest Service, Southeastern Forest Experiment Station, pp. 76–82.

Davis, Millard C. 1966. The influence of Emerson, Thoreau, and Whitman on the early American naturalists, John Muir and John Barrows. *Living Wilderness.* 39(95): 19–23.

de Grazia, Sebastian. 1970. Some reflections on the history of outdoor recreation. In: Driver, Beverly L., ed. *Elements of Outdoor Recreation Planning.* Ann Arbor: University of Michigan Press, pp. 89–97.

Edwards, Mike. 1985. A short hike with Bob Marshall. *National Geographic.* 167(5): 664–689.

Evans, Brock. 1981. The wilderness idea as a moving force in American cultural and political history. In: *Congressional Record*; April 27, 1981. Washington, DC: U.S. Senate: S4010–S4014.

Flader, Susan L. 1974. *Thinking Like a Mountain: Aldo Leopold and the Evolution of an Ecological Attitude Toward Deer, Wolves, and Forests.* Columbia: University of Missouri Press.

Friese, Gregory T.; Hendee, John C.; Kinziger, Mike. 1998. The Wilderness Experience Program (WEP) industry in the United States: Characteristics and dynamics. *Journal of Experiential Education.* 2(1): 40–45.

Frome, Michael. 1997. *Battle for the Wilderness.* Salt Lake City: University of Utah Press.

Gager, Dan; Hendee, John C.; Kinziger, Mike; Krumpe, Ed. 1988. What managers are saying—and doing—about wilderness experience programs. *Journal of Forestry.* 96(8): 33–37.

Green, Gary T.; Tarrant, Michael A.; Raychaudhuri, Uttiyo; Zhang, Yangjian. 2005. Wilderness in whose backyard? *International Journal of Wilderness.* 11(3): 31–38.

Hendee, John C.; Stankey, George H. 1973. Biocentricity in wilderness management. *BioScience.* 23(9): 535–538.

Holmes, F. Patrick; Hecox, Walter E. 2004. Does wilderness impoverish rural regions? *International Journal of Wilderness.* 10(3): 34–39.

Ise, John. 1961. *Our National Park Policy: A Critical History.* Baltimore: Johns Hopkins University Press.

Kaplan, Rachel; Kaplan, Stephen. 1989. *The Experience of Nature: A Psychological Perspective.* Cambridge, England: Cambridge University Press.

Kellert, Stephen R.; Wilson, Edward O., eds. 1993. *The Biophilia Hypothesis.* Washington, DC: Island Press.

Knibb, David. 1982. *Backyard Wilderness: The Alpine Lakes Story.* Seattle, WA: Mountaineers Books.

Leopold, Aldo. 1921. The wilderness and its place in forest recreational policy. *Journal of Forestry.* 19(7): 718–721.

———. 1925. Wildernesses as a form of land use. *Journal of Land and Public Utility Economics.* 1(4): 398–404.

———. 1941. Wilderness as a land laboratory. *Living Wilderness.* 6(6): 3.

———. 1949. *A Sand County Almanac and Sketches Here and There.* New York: Oxford University Press.

Loomis, John B.; Richardson, Robert. 2001. Economic values of the U.S. wilderness system: Research evidence to date and questions for the future. *International Journal of Wilderness.* 7(1): 31–34.

Louv, Richard. 2006. *Last Child in the Woods: Saving Our Children from Nature-Deficit Disorder.* Chapel Hill, NC: Algonquin Books.

Marshall, George. 1969. Introduction. In: McCloskey, Maxine E.; Gilligan, James P., eds. *Wilderness and the Quality of Life.* San Francisco: Sierra Club, pp. 13–15.

Marshall, Robert. 1930. The problem of the wilderness. *Scientific Monthly.* 30: 141–148.

McCloskey, Michael. 1966. The Wilderness Act: Its background and meaning. *Oregon Law Review.* 45(4): 288–321.

McDonald, Barbara. 2003. The soul of environmental activists. *International Journal of Wilderness.* 9(2): 14–17.

McDonald, Barbara; Schreyer, Richard. 1991. Spiritual benefits of leisure participation and leisure settings. In: Driver, Beverly et al.; Brown, Perry J.; Peterson, George L., eds. *Benefits of Leisure.* State College, PA: Venture Publishing, pp. 179–193.

Mihaley, Marc B. 1972. The Clean Air Act and the concept of non-degradation: *Sierra Club v. Ruckelhaus. Ecology Law Review.* 2(4): 801–836.

Moisey, Neil; Yuan, Michael S. 1992. Economic significance and characteristics of select wild land-attracted visitors to Montana. In: Payne, Claire; Bowker, J. Michael; Reed, Patrick C., eds. *The Economic Value of Wilderness: Proceedings of the National Wilderness Conference;* May 8–11, 1991; Jackson, WY. General Technical Report SE-78. Asheville, NC: U.S. Department of Agriculture, Forest Service, Southeastern Forest Experiment Station, pp. 181–189.

Muir, John. 1938. *John of the Mountains: The Unpublished Journals of John Muir.* Wolfe, Linnie Marsh, ed. Boston: Houghton Mifflin.

Nash, Roderick, ed. 1976. *The American Environment: Readings in the History of Conservation,* 2d ed. Reading, MA: Addison-Wesley.

———. 1982. *Wilderness and the American Mind,* 3rd ed. New Haven, CT: Yale University Press.

———. 2001. *Wilderness and the American Mind,* 4th ed. New Haven, CT: Yale University Press.

Oelschlaeger, Max. 1991. *The Idea of Wilderness: From Prehistory to the Age of Ecology.* New Haven: Yale University Press.

Olson, Sigurd F. 1957. *The Singing Wilderness.* New York: Alfred A. Knopf.

———. 1963. *Runes of the North.* New York: Alfred A. Knopf.

Outdoor Recreation Resources Review Commission [ORRRC]. 1962. *Wilderness and Recreation, a Report on Resources, Values, and Problems.* Study Report 3. Washington, DC: U.S. Government Printing Office. 352p.

Oye, Garry. 2001. A new recreation strategy for national forest wilderness. *International Journal of Wilderness.* 7(1): 13–14.

Power, Thomas M. 1996a. *Environmental Protection and Local Economic Well-Being: The Economic Pursuit of Quality.* New York: M. E. Sharpe.

———. 1996b. Wilderness economics must look through the windshield, not the rearview mirror. *International Journal of Wilderness.* 2(1): 5–9.

Reed, Patrick; Haas, Glenn; Beum, Frank; Sherrick, Lois. 1989. Non-recreational uses of national wilderness preservation systems: A 1988 telephone survey. In: Freilich, Helen R., comp. *Wilderness benchmark 1988: Proceedings of the National Wilderness Colloquium:* Freilich, Helen R., comp.; January 13–14, 1988; Tampa, FL. General Technical Report SE-51. Ashville, NC: U.S. Department of Agriculture, Forest Service, pp. 220–228.

Riley, Marilyn; Hendee, John C. 2001. Wilderness vision quest clients, motivations and reported benefits from an urban based program 1988–1997. In: Martin, Vance G.; Sarathy, Partha, eds. *Wilderness and Humanity: The Global Issue: 6th World Wilderness Congress Proceedings;* October 24–29, 1998; Bangalore, India. Golden, CO: Fulcrum, pp. 267–276.

Rosenberger, R. S.; English, Donald B. K. 2005. Impacts of wilderness on local economic development. In: Cordell, H. Ken; Bergstrom, John C.; Bowker, J. Michael, eds. *The Multiple Values of Wilderness.* State College, PA: Venture Publishing, pp. 181–204.

Rudzitis, Gundars. 1996. *Wilderness and the Changing American West.* New York: John Wiley.

Russell, Keith C.; Hendee, John C.; Cooke, Steve. 1998. Social and economic benefits of a U.S. wilderness experience program for youth at risk in the Federal Job Corps. *International Journal of Wilderness.* 4(3): 32–38.

Russell, Keith; Hendee, John C.; Phillips-Miller, Dianne. 2000. How wilderness therapy works: An examination of the wilderness therapy process to treat adolescents with behavioral problems and addictions. In: Cole, David N.; McCool, Stephen F.; Borrie, William T.; O'Loughlin, Jennifer, comps. *Proceedings: Wilderness Science in a Time of Change Conference;* May 23–28, 1999; Missoula, MT. RMRS-P-15 Vol. 3. Fort Collins, CO: U.S. Department of Agriculture, Forest Service, Rocky Mountain Research Station, pp. 207–217.

Saylor, John. 1962. A report on wilderness. In: *Congressional Record;* May–June 1962. Washington, DC: House of Representatives.

Schroeder, H. W. 2007. Symbolism, experience, and the value of wilderness. *International Journal of Wilderness.* 13(1): 13–18.

Schuster, Rudy M.; Cordell, H. Ken; Phillips, Brad. 2005. Understanding the cultural, existence, and bequest values of wilderness. *International Journal of Wilderness.* 11(3): 22–25.

Schuster, Rudolph M.; Tarrant, Michael; Watson, Alan. 2005. The social values of wilderness. In: Cordell, H. Ken; Bergstrom, John C.; Bowker, J. Michael, eds. *The Multiple Values of Wilderness.* State College, PA: Venture Publishing, pp. 113–142.

Scott, Doug. 2004. *The Enduring Wilderness.* Golden, CO: Fulcrum Publishing.

Spring, Ira. 2001. If we lock people out, who will fight to save the wilderness? *International Journal of Wilderness.* 7(1): 17–19.

Spurr, Stephen H. 1966. Wilderness management. *The Horace M. Albright Conservation Lectureship,* Vol. 6. Berkeley: University of California, School of Forestry and Conservation.

Stahl, Carl J. 1921. Where forestry and recreation meet. *Journal of Forestry.* 19(5): 526–529.

Sutter, Paul S. 2002. *Driven Wild: How the Fight Against Automobiles Launched the Modern Wilderness Movement.* Seattle: University of Washington Press.

U.S. Department of Agriculture, U.S. Forest Service. 2005. White Mountain National Forest Land and Resources Management Plan. Laconia, NH: USFS Eastern Region.

U.S. Public Law 88-577. The Wilderness Act of September 3, 1964. 78 Stat. 890.

U.S. Public Law 93-622. (Eastern Wilderness) Act of January 3, 1975. 88 Stat. 2096.

U.S. Public Law 94-579. Federal Land Policy and Management Act (FLPMA) of 1976. 90 Stat. 2743.

U.S. Public Law 95-237. Endangered American Wilderness Act of February 24, 1978. 92 Stat. 40.

U.S. Public Law 96-487. Alaska National Interest Lands Conservation Act (ANILCA) of December 2, 1980. 94 Stat. 2371.

U.S. Public Law 96-560. (Colorado Wilderness) Act of December 22, 1980. 94 Stat. 3266.

Weaver, James W.; Cutler, Rupert. 1977. Wilderness policy: A colloquy between Congressman Weaver and Assistant Secretary Cutler. *Journal of Forestry*. 75(7): 392–394.

White, David; Hendee, John C. 2000. Primal hypotheses: The relationships between naturalness, solitude and the wilderness experience benefits of development of self (DOS), development of community (DOC) and spiritual development (SD). In: McCool, Stephen F.; Cole, David N.; Borrie, William T.; O'Loughlin, Jennifer, comps. *2000 Proceedings: Wilderness Science in a Time of Change Conference,* Vol. 3; May 23–27, 1999; Missoula, MT. RMRS-P-15. Fort Collins, CO: U.S. Department of Agriculture, Forest Service, pp. 223–228.

White, Lynn. 1967. The historical roots of our ecological crisis. *Science.* 155(3767): 1203–1207.

Wilson, Edward O. 1984. Million-year histories: Species diversity as an ethical goal. *Wilderness.* 48(165): 12–17.

Worf, William A. 2001. The Forest Service recreation strategy spells doom to the National Wilderness Preservation System. *International Journal of Wilderness.* 7(1): 15–16.

Worf, William A.; Jorgenson, Glen; Lucas, Robert. 1972. *Wilderness Policy Review.* Washington, DC: U.S. Department of Agriculture, Forest Service.

Yuan, Michael S.; Christensen, N. A. 1992. Wilderness-influenced economic impacts on portal communities: the case of Missoula, Montana. In: Payne, Claire; Bowker, J. Michael; Reed, Patrick C., eds. *The Economic Value of Wilderness: Proceedings of the National Wilderness Conference*; May 8–11, 1991; Jackson, WY. General Technical Report SE-78. Asheville, NC: U.S. Department of Agriculture, Forest Service, Southeastern Forest Experiment Station, pp. 191–199.

Zivnuska, John A. 1973. The managed wilderness. *American Forests.* 79(8): 16–19.

CHAPTER

2

Historical Roots of Wilderness Management

INTRODUCTION

Management of a designated wilderness can be thought of as a contradiction in terms. It could even be said that any area that is proclaimed wilderness and managed as such is not wilderness by these very acts. The problem is that *wilderness* traditionally means an environment that humans do *not* influence, a place they do *not* control.

Before herding and agriculture began, there was no distinction between wilderness and civilization. As hunters and gatherers, people did not control their environment; they simply lived in it as part of the ecosystem. However, with the beginning of herding, and later of agriculture, *Homo sapiens* began to experiment with environmental modification by domesticating animals and managing plants, soil, and water. In time, totally humanized environments, called towns and cities, developed. In the process, people indirectly created wilderness by drawing a physical and, even more importantly, a mental distinction between the places they controlled and the places they did not control (Nash 1975).

Etymologically, the word *wilderness* is derived from the Old English wild-deor-ness, the place of untamed beasts (Nash 1970b). Some scholars believe the origin is the concept of "self-willed" land. *Civilization*, conversely, is an environment under human control. Understandably, since the advent of the civilization that created the word *wilderness*, it has stood for the dark, the chaotic, the unknown and fearful, and the back of beyond. It was defined by the absence of the controlling structures of institutions and technologies. Outlaws and brigands of times past, like today's revolutionary guerrillas and marijuana growers, sought wild country for the same reason it attracts some of today's backpackers who seek to escape civilization's nearly omnipresent cloak of control.

THE INTELLECTUAL DILEMMA

The only wilderness true to the etymological roots of the word is that which humans do not influence in any way whatsoever. The more people learn about wilderness, the more they visit it, map it, manage it, and write about it, the less wild it becomes. From such a perspective, even knowledge about a region disqualifies it as wilderness in the true etymological sense. Management of any kind compromises wildness.

Even maps, trails, and signs are civilizing, steps toward ordering the environment in the interest of people, toward lessening the amount of the unknown. The association of rangers, wardens, "adopt a wilderness" groups, "save the whatever" organizations, and search-and-rescue teams for a given area obviously detracts from its wildness.

More subtle, but just as foreign to pure wilderness, are sophisticated management techniques. The notions of carrying capacities, use permits and quotas, regulations on behavior, prescribed fire, and fire control gradually erode the "wild" from "wilderness." For many, the wilderness experience is affected when recreational demand makes it necessary to wait years or trust to a lottery for an entrance permit (e.g., private users seeking trips on the Colorado River in Grand Canyon National Park, Arizona, or wilderness hikers to Mount Whitney on the Inyo National Forest, California) or, at the peril of arrest, or citation and fine, to maintain a rigid backcountry travel itinerary so other parties, following a day behind, have places to camp (for example, in the Boundary Waters Canoe Area Wilderness on the Superior National Forest, Minnesota). For many users, this kind of intensive management transforms wilderness into an open-air motel with advance registration and checkout times. The resulting dissatisfaction underscores the need for less restrictive, *light-handed management techniques* that emphasize visitor education, and voluntary rather than enforced restraint. Indeed, as this book describes, such techniques are evolving from experience and research, along with a management ethic to use only the minimum tools to achieve desired results.

The intellectual dilemma posed by a managed wilderness is compounded by the fact that, *in the last analysis, wilderness is a state of mind. Like beauty, wilderness is defined by human perception.* For some individuals, regulations will not be distracting. However, for others, just the *knowledge* that they visit an area by the grace of, and under conditions established by, a government agency detracts from a wilderness experience. Ironically, the success of management in protecting the wilderness experience may decline in proportion to its effectiveness. *Wilderness management in its first few decades has been largely people management. Now more and more focus is being placed on wilderness ecosystems.* The test may be whether we can resist tinkering with them, or as Glover (2000, 4) asks, "Can we stop trying to control Nature?"

It is hard to deny the principle that management is essential today if wilderness is to have any meaning at all. The pure definition of wilderness (no maps, no knowledge, and a totally blank space on the map) is, at least between the 60th parallels, a thing of the past. No one in the United States can ever again have the experience of a Lewis or a Clark, a Jim Bridger, or a John Wesley Powell. It is even wishful thinking to suppose one might today duplicate David Brower's 1930s experience of making a recorded first ascent almost every time he climbed a Sierra peak. The best that can be hoped for, in the American West for instance, is a chance to be in beautiful and comparatively natural country, away from roads, relatively alone, and dependent, in the short run, on one's own resources for comfort and survival (Fig. 2.1).

Thus, a key factor compelling acceptance of managed wilderness is awareness that *without management, contradictory as it may seem, wilderness would surely be impaired or lost* (Nash 1978, 1982, 2001).

FIGURE 2.1 In the twentieth century, increasing numbers of Americans began to consider wilderness a resource; they no longer regarded it as an obstacle to be conquered. Early wilderness visitors, such as this man, did not have lightweight equipment, portable stoves, or dehydrated food. Photo courtesy U.S. Forest Service.

WILDERNESS RECREATION DEMAND

The public flocks in increasing numbers to wilderness for recreation activities and experiences. Mount Whitney, the highest peak in the United States outside Alaska, is a good example. Mount Whitney is part of the Sierra Nevada and lies in the John Muir Wilderness on the Inyo National Forest and in Sequoia Wilderness in national park wilderness. The peak's first recorded ascent was in 1873; in 1973, approximately 14,000 persons made the climb. A dramatic illustration of the change popularity has brought to Mount Whitney comes from a man who, on August 4, 1949, climbed the peak with his father. Proudly, they signed the summit register, the sixth and seventh individuals to have done so *that year*. On August 11, 1972, this same man climbed Mount Whitney with his son. Upon signing the register, they noted with some shock that they were the 259th and 260th persons on record *that day*! Presumably, there was less wilderness solitude in the experience.

Today's visitation to Mount Whitney requires a permit year-round and from May 1 to October 3 there is a limited daily entry. According to Brian Spitek (2007), wilderness manager on the Mount Whitney Ranger District of the Inyo National Forest, in 2007, a total of 23,197 visitors had legal permits to climb Mount Whitney, plus an unknown number of illegal visitors who dodged the permit system.

Popularity can affect the experience and resource as noted for the Grand Canyon, where the more-than-200-mile float trip of the Colorado River is perhaps the most intensively supervised wilderness activity in the United States today. Close control by national park officers is facilitated by severely limited access to the river and the expedition-level difficulty of the trip. The number of visitors floating the Colorado River through the Grand Canyon had risen to several thousand per year by the mid-1960s and then to 20,000 or more by the mid-1980s.

In 2007, a new Colorado River Management Plan (U.S. Department of the Interior, National Park Service 2006) went into effect with a strategy to implement a "launch-based" system that is intended to move toward a more even distribution of use through the year and to achieve more equity between commercial and noncommercial visitor use. More than 114,006 commercial user days were recorded in 2007, which converts to 18,417 commercial visitors. The number of visitor days for private, noncommercial use was increased in 2007 under the new permit system to 105,438 visitor days, which converts to 6,318 visitors. Starting in 2007, the old waiting list system for permits was replaced by a "weighted" lottery system that is based on length of time the applicant has been applying for a permit. Also, any trip cancellations can be used by a replacement applicant and are not lost from use (Jalbert 2008).

The quality of wilderness as experienced by the early Grand Canyon river runners has declined. Some argue that, enjoyable as it is, the locale can no longer be considered wilderness. Some visitors turn to the few remaining wild rivers, perhaps in Alaska and the Canadian northland. However, many others comply, albeit reluctantly, with the strict management policies currently in effect for the Grand Canyon. The logic that persuades them might be illustrated by comparing access to the Grand Canyon and other popular wildernesses with access to playing time on tennis courts.

Tennis players would obviously prefer to play when they wish, for as long as they wish. However, the popularity of the game does not permit this luxury except on private courts, which can be compared to the game reserves of medieval nobility. On public, tax-supported courts (as in publicly supported wilderness areas) demand frequently exceeds available space. Hence management devices are instituted, such as sign-up sheets, time and frequency limitation rules for waiting players, and encouraging doubles games. Players recognize that tennis is a game that is played by two or four persons. Out of respect for the integrity of the game, and with their own self-interest in mind, players support management. They sign up, wait their turn, and vacate the court at the appointed hour.

Like tennis, wilderness recreation is also an activity that cannot be played at any one time and place by more than a few persons. Moreover, it is an activity that depends on the existence of a relatively unmodified natural and physical environment. These realizations prompt many to accept management. The resulting regulations may be distasteful, and are violations of the traditional sense of wilderness, but they are the best hope of salvaging an approximation of the wilderness experience from the pressures of popularity.

The basis for any historical discussion of wilderness management recognizes that management is a necessary newcomer to the wilderness movement. The concept of preservation existed long before there was management except for fire control and associated lookouts, trails, and guard cabins. For decades, few even thought about managing wilderness. Perhaps this tendency sprang from the contradiction between the concepts of wilderness and management. Wilderness was not *supposed* to be managed. It was the region that began where management stopped. However, certainly equal in importance was the fact that for years there were few significant overuse or visitor management problems. As late as 1949, when only a dozen persons a season were climbing Mount Whitney and about the same number were running the Colorado River through the Grand Canyon, their control could hardly be regarded as a pressing issue. For a transitory, enchanted moment in American environmental history, wilderness preservation could exist without wilderness management.

ROOTS OF WILDERNESS APPRECIATION

The first accounts of wilderness pleasure trips make us realize just how spectacularly empty the country was at that time. Consider, for example, the 625-mile hike that Joseph N. LeConte and three companions made in the southern section of the Sierra in 1890 and recorded by LeConte in his journal (LeConte 1972). The trip is interesting for its parallels and its contrasts with contemporary patterns of wilderness recreation. The four men, all in their early twenties, were students at the University of California, Berkeley. The trip was their summer vacation; they went into the mountains for fun. One is struck, immediately, by the total lack of regulation. There were no permits in 1890, no regulations, no fish and game laws, not even clear maps. The students simply packed up—using burros rather than backpacks—and headed out. Much of the time they had only a general idea of where they were, and a considerable part of their adventure was knowing no one was poised to bail them out of trouble. In 1890, the wilderness was noteworthy for its emptiness. Except for a few miners on the lower Kings River and one of their geology professors conducting experiments above Yosemite Valley, they saw no people.

The Sierra was even emptier twenty years earlier when John Muir ranged through the mountains with a pocketful of bread and a little tea. Even the Sierra Club's organized outings, begun under Muir's leadership in 1901, hardly compromised the Sierra's isolation.

"Off the beaten track" has been transformed by contemporary backpacking, in part due to the evolution of backpacking equipment. Old-timers are astonished at what the outdoor-equipment industry has wrought. In an earlier era, huge bedrolls, heavy tents, and the weight of canned foods sharply limited the places one could visit and the amount of time a party could spend in wilderness. So did the lack of portable, efficient gear for winter camping and rock climbing. Today's light, streamlined equipment has opened the most remote places in all seasons (Fig. 2.2). Then, too, in earlier years, the population of the United States was much smaller. Moreover, compared to contemporary Americans, most of our grandparents had limited mobility (fewer cars, poor roads), less leisure, and greater dedication to the work ethic.

However, the principal reason wild places were empty was that very few Americans cared to visit them. Even as late as World War II, wilderness appreciation was still in its infancy. Wilderness was not easy to appreciate, as it was something fought since the dawn of civilization. It was a matter of being too close to wilderness, of having too much of it. For appreciation to flourish, wilderness had to become more scarce, and this, in turn, depended on the rise of an urbanized, industrialized society. The United States was at the brink of the transition from a developing to a developed nation in the late

FIGURE 2.2 Better access, light and efficient recreational equipment, population growth, and pressures of urban life are a few reasons for increased recreational use of wilderness. Here visitors contemplate Green Lake in North Cascades National Park, Washington, after a tough cross-country hike. Photo courtesy John C. Hendee.

nineteenth century. The frontier ended, according to the U.S. census, in 1890, and then large numbers of Americans began to consider wilderness a resource to be enjoyed rather than an adversary to be conquered.

EVOLUTION OF WILDERNESS IN THE NATIONAL PARKS

Understandably, the first interest in wilderness for recreation was tentative, and only flirted with the wilderness experience. People wanted wilderness, but not too much. They preferred to be on its edge, to look at it, and to have the security and comforts of civilization. This was the context that generated the first wilderness management decisions. For example, the preservation of the world's first national park, Yellowstone, in 1872, had little to do with providing a true wilderness experience for vacationing Americans. The intent of Congress, as stated in the act creating Yellowstone National Park in 1872, was to create a "public park or pleasuring ground for the benefit and enjoyment of the people." Study of the intent of Yellowstone's proponents indicates that they expected the "enjoyment" to be derived from viewing scenic wonders such as geysers, hot springs, and waterfalls from the civilized vantage point afforded by luxurious lodges. Even Nathaniel P. Langford, a leading explorer and publicist of the first national park, enthusiastically predicted that it would not be long "before the march of civil improvements will reclaim this delightful solitude and garnish it with all the attractions of cultivated taste and refinement" (Langford 1972, 97). This was entirely consistent with the established pattern of nature tourism of the nineteenth century, which emphasized the edge of wildness, convenient transportation (usually railroads), and lavish hotel accommodations (Runte 1979).

What did visitors to the early national parks expect? The brochures and literature distributed by park promoters invariably featured the attributes of civilization: comfortable coaches, grand lodges, elegantly dressed tourists. Far from enticing the visitor with visions of wilderness camping, the advertisements tried to convince the tourist that there was no need to rough it. Wildness was to be enjoyed, but at a distance. Too wild a park, it was rightly assumed in the late nineteenth and early twentieth centuries, could be a deterrent to tourism. Interest in wilderness was growing, but it had not yet affected recreational desires enough for wilderness management to exist, even as a concept.

As for the roads and hotels that "opened up" Yellowstone and the other early parks and determined their dominant use, park personnel or Congress *did not decide* to feature wilderness as an attraction. It wasn't even an issue; people took conveniences for granted. The language used in the Yellowstone Act made such an interpretation easy. As long as the timber, mineral deposits, natural curiosities, or wonders were preserved in their natural condition, there was no problem with developing the park for mass tourism. Specifically, Old Faithful and Yellowstone Falls were the objects of concern, not the wild backcountry of the park. As long as these "wonders" were kept in public ownership and free from vandalism, the nineteenth-century purposes of the park were fulfilled.

Even the most ardent wilderness preservationists of the time, people like Sierra Club president John Muir, accepted this premise of mass tourism development and its management implications. In 1913, Muir *supported* admitting the first private automobiles into Yosemite Valley (Lillard 1968). His reasoning centered on the need to bring people into the parks to build citizen support for the park idea. Along with cars came civilized lifestyles. Hotels, like the posh Ahwahnee completed in 1926 in the scenic heart of Yosemite Valley, contributed to the people's outdoor pleasure as it was defined in the first decades of the twentieth century. In addition, because the national parks were established by law to be "pleasuring grounds," who could object?

Muir bitterly fought economic development of park wilderness. He opposed grazing, mining, logging, and, unsuccessfully, the 1913 decision in favor of hydropower development that inundated Hetch Hetchy Valley in northern Yosemite National Park. However, Muir did not recognize development for recreation, for the public's pleasure, as a comparable threat to wilderness. He died in 1914, long before wilderness advocates understood that management was an unpleasant necessity.

The passage of the National Park Service Act on August 25, 1916, did nothing to change earlier conceptions of the meaning, purpose, and appropriate uses of national parks. Although the legislation stipulated that anything done in the parks must leave the scenery and wildlife "unimpaired," the whole reason for their existence was indisputably public enjoyment. However, pleasure-seeking people could impair nature. The ambiguity inherent in the National Park Service Act has been the source of extensive commentary and still more extensive agony for subsequent park managers (Sax 1980). However, for Americans in 1916, there was considerably less inconsistency in the act. Because wilderness protection and the provision of a wilderness experience were not recognized goals of park management, few questioned developments such as roads and lodges that eroded wildness. Why should they have questions, after all? Hardly anyone went into the park backcountry at this time. Dramatic assertions (probably true) that more than 90 percent of these early visitors saw only 3 percent of the park were not the result of any conscious management policy. They reflected quite accurately the tastes of recreation-minded Americans in the early twentieth century. Most people did not *want* to experience park wilderness. The practical implications for management of these conceptions of park means and purposes can be found in a letter of management direction, May 13, 1918, from Secretary of the Interior Franklin K. Lane to Stephen T. Mather, the first director of the National Park Service (U.S. Department of the Interior, National Park Service 1970).

The letter opens with the standard insistence that the parks be kept in "absolutely unimpaired form," but quickly makes compromises on behalf of public enjoyment. From the standpoint of wilderness preservation, the most damaging aspect of Lane's letter is the assumption that the public should be encouraged to enjoy the parks "in the manner that best satisfies the individual taste." There is no attempt to define what kind of enjoyment is appropriate in a national park, and no effort to distinguish uses that are consistent with the mandate to leave parkland unimpaired. In effect, Lane is saying that the citizen will bring individual preferences to the parks and the parks will fulfill them. The 1918 letter makes clear that "automobiles and motorcycles will be permitted in all of the national parks; in fact, the parks will be kept accessible by any means practicable." The implications of this statement are extraordinary and clearly work against wilderness. So does the secretary's directive to encourage a full range of accommodations from "luxurious hotels" to "free campsites." Nothing is said about low-density, off-road wilderness uses of the parks. Again, the point is that in 1918, wilderness recreation was not considered part of the statutory purpose of national parks. Americans of this period gave little evidence of being disappointed with such a definition and the resulting management policy.

Stephen T. Mather was the ideal director of the National Park Service (NPS) under the explicit and implicit mandates of the early twentieth century. His talent was public relations, and he recognized that national park survival and growth depended on skillful playing of the "number of visitors game" in the political arena. Immediately after passage of the 1916 legislation, Mather launched a vigorous program to boost visitation to national parks. It included a series of publications, the work of Robert Sterling Yard, and initiating or continuing management policies designed to attract and please visitors. Mather and Yard knew that wilderness might not "sell" to their contemporaries. Instead, they cultivated a resort or circus image of parks. Drive-through sequoia trees, cut initially in the 1880s, continued to be a tourist attraction at Yosemite. At Yellowstone, soap was regularly dumped into the geysers to break their surface tension and cause eruptions at times convenient to tourists. At Old Faithful, the symbol of America's national park in this period, colored spotlights from adjacent hotels illuminated night eruptions. During the hour between eruptions, tourists were entertained by radio music.

Yellowstone's famous roadside bears shared top billing with Old Faithful. By explicit direction of Director Mather and his assistant and subsequent director, Horace M. Albright, the bears were regularly fed hotel garbage before grandstands of camera-wielding tourists. In the 1920s, it must be remembered, bear feedings and caged wildlife around hotels did not violate national park purpose; rather, they expressed it. Public enjoyment could easily be stretched to cover such activities.

At Yosemite National Park in the 1870s, the "firefall" replaced the "chicken fall," in which live chickens were tossed over the cliffs. It continued, under Mather and Albright, to dominate the park experience for most tourists. The firefall involved constructing a huge wood fire on the lip of Glacier Point 3,000 feet above the floor of Yosemite Valley. As dusk fell, the crowds gathered. Music played, and at a voice signal, "Let the fire fall!", the burning logs and embers were pushed over the cliff. The potential for forest fire was fully recognized and carefully avoided, but for decades no one even questioned whether the firefall was an appropriate activity for management to sponsor in a national park. No one asked if this was the *kind* of "enjoyment" parks were created to provide. It was not until the late 1960s that changing interpretations of the meaning and purpose of national parks led to limiting road access and replacing the *resort concept* with more wilderness orientation in national parks (Sutter 2002).

There had, however, been earlier indications of wilderness consciousness in national park circles. In 1929, the phrase "original wilderness character" was used in certain versions of the bill establishing Grand Teton National Park, Wyoming. No hotels or new roads were to be permitted in the park. Although stricken from the final text of the bill (Grand Teton National Park Act of 1929), the omitted phrase clearly indicated a desire to emphasize wilderness in Grand Teton National Park. The first explicit recognition of wilderness in national park legislation appeared five years later in the act establishing Everglades National Park. Section 4 specified that the Florida wetlands would be "permanently preserved as a wilderness." With an eye toward management, the bill went on to say that "no development of the project or plan for the entertainment of visitors shall be undertaken which will interfere with the preservation of the . . . essential primitive natural conditions now prevailing in this area" (Everglades National Park Act of 1934). Another milestone in the development of the wilderness idea was the 1940 establishment of Sequoia–Kings Canyon National Park, California, south of Yosemite, as a roadless and hotelless park. Providing a wilderness experience was, necessarily, its main objective. Indeed, a preliminary version of the establishing legislation even used the name "Kings Canyon Wilderness National Park." The Sierra Club regarded Kings Canyon as compensation, in a sense, for the heavily developed character of Yosemite. Together the two parks would serve a wide spectrum of recreational interests.

Other evidence shows some early awakening of the need to be concerned about management. In 1928 and 1929, George M. Wright of the NPS saw the need for an organizational unit to monitor impacts on wildlife in national park ecosystems. Wright organized a small group of individuals to begin a nationwide systematic survey of the status of wildlife in the parks, with development of a well-defined wildlife policy as its goal. The work of this group (Wright et al. 1932; Wright and Thompson 1934) provided a historic data baseline concerning wildlife in the parks, including such wilderness-dependent species as the wolf and grizzly bear. Wright's work led to the establishment of the Division of Wildlife Research in the NPS, with Wright as its first head, and ultimately the George Wright Society, which today provides for an annual conference and proceedings of research papers by national park scientists.

The expansion of thinking to include park and wilderness management as well as designation was beginning. One of the first written examples of this transformation as it concerned national parks was a 1936 report by Lowell Sumner, a regional wildlife technician. In his policy recommendations for Sierra parks, Sumner (1936) wondered "how large a crowd can be turned loose in a wilderness without destroying its essential qualities" (n.p.). He realized that for wilderness to exist in the parks, the areas "cannot hope to accommodate unlimited numbers of people" (n.p.). Construction of tourist facilities would have to be restricted. Finally, Sumner's insights extended to the understanding that wilderness managers could also pose a threat to wilderness values. He urged that only "the very simplest maintenance activity" be undertaken in wilderness (Sumner 1936, n.p.). Heavy management, in other words, could be a liability in dealing with an experience featuring solitude, self-reliance, and freedom from the controls normally present in civilization.

Sumner's thinking on these points matured so that six years later he could discuss the adverse effects of pack-stock grazing, fishing, and sheer numbers of visitors on the biological balances of wilderness areas. Then, in one of the first uses of the term, Sumner (1942, 21) urged that use of wilderness be kept "within the carrying capacity or 'recreational saturation point.'" His 1942 definition described carrying capacity as "the maximum degree of the highest type of recreational use which a wilderness can receive, consistent with its long-term preservation." Wilderness managers should "determine in advance the probable maximum permissible use, short of impairment, of all wilderness areas" (Sumner 1942, 22). Here, in 1942, was the basic logic of modern wilderness recreation management.

The evolution of wilderness in the national parks is not complete. While 84 percent of the NPS is protected, most of it is still waiting to be legislatively designated by Congress (Potts 2007).

EVOLUTION OF WILDERNESS IN THE NATIONAL FORESTS

Although the national parks of the early twentieth century were playing to crowds who had little interest beyond visiting "pleasuring grounds," the USFS took the first steps toward explicitly identifying wilderness as a specific recreational resource and developing appropriate management techniques. While Gifford Pinchot headed the U.S. Division of Forestry (after 1905, called the U.S. Forest Service), the emphasis was on producing commodities, consistent with the 1897 Organic Act, which mentioned only timber, grazing, and watershed protection as uses of the forest reserves (Roth 1984). The forests were to be used, albeit carefully, as a constant source of valuable products. After Pinchot's departure from office in 1910, the meaning of "forest products" was expanded. Some people began to understand that forests were valuable for more than just commodities. Henry Graves, the new chief of the USFS, began to conceive of the national forests as valuable for recreation. Of course, in these early years, *recreation* meant almost every imaginable outdoor activity, but wilderness had a small and growing significance. For instance, in 1910 Graves asked Treadwell Cleveland, Jr., to write an essay on public-recreation facilities for the American Academy of Political and Social Science. The resulting discussion of the use of logging roads, bridges, and trails by the hunter, angler, and picnicker was unprecedented in the history of American forestry. Moreover, Cleveland (1910, 245) made a significant prediction:

> So great is the value of national forest area for recreation, and so certain is this value to increase with the growth of the country and the shrinkage of the wilderness, that even if the forest resources of food and water were not to be required by the civilization of the future, many of the forests ought certainly to be preserved . . . for recreation use alone.

However, the USFS, like the NPS, was constrained by the anti-wilderness bias of public opinion in this era. Few people wanted to rough it. Recreational development, therefore, consisted of extending forest roads and leasing sites for summer home and hotel construction. Chief Forester Graves was enthusiastic about progress in these areas in his 1912 report; three years later, he obtained permission from the secretary of agriculture to extend summer home leases to thirty years. The result? More permanent structures were built. Wilderness suffered, but at the time few Americans really cared.

In 1918, landscape architect Frank A. Waugh prepared a report for the USFS titled "Recreation Uses on the National Forests" (Waugh 1918). It marked the emergence of full awareness that recreation was an established rationale for national forests. William B. Greeley, who became chief forester in 1920, and his associate forester L. F. Kneipp gave increasing emphasis to this use and even secured budgetary appropriations for recreation beginning in 1922. Greeley, in particular, valued forest scenery, and on several occasions in the early 1920s vetoed tourist-development plans. The most important decision of the era affecting wilderness concerned the spectacular Trappers Lake in Colorado on national forestland. In 1919, a young USFS landscape architect named Arthur H. Carhart was assigned to survey the area for road access and several hundred vacation homes. The plan was entirely in keeping with USFS definitions of recreation, but Carhart was troubled. The beaver of Trappers Lake had been exploited in the 1850s, but otherwise the lake was untouched and reachable only by a tough five-mile trail. Realizing the rarity of such wilderness in the American West, Carhart had misgivings about developing Trappers Lake even for recreational purposes. He was encouraged in this line of thinking by several wealthy hunters who wanted the country to remain pristine. So, after a summer spent not only surveying, but also developing a conviction that the area should be preserved in its pristine condition, Carhart had the courage to recommend doing nothing at all to Trappers Lake. Probably to his surprise, the Denver District Office of the USFS approved the idea. Trappers Lake was left without roads or summer homes (Allin 1982; Baldwin 1972).

Arthur Carhart followed his pro-wilderness recommendation in Colorado with a similar one for the Superior National Forest in Minnesota. Moreover, late in 1919, he met with the young, nontraditional forester Aldo Leopold. The disappearance of large roadless areas in Arizona and New Mexico was evoking in Leopold misgivings similar to Carhart's; Leopold's efforts to retain large sections of country devoid of human influence included a call for a wilderness of at least 500,000 acres for each of the eleven states west of the Great Plains (Leopold 1921; Roth 1984). In 1924, Leopold had the satisfaction of seeing the USFS designate 574,000 acres of the Gila National Forest, New Mexico, as a reserve for wilderness recreation. The efforts of Carhart and Leopold produced the first designation of public lands specifically for wilderness values in America and, indeed, in the world. (Fig. 2.3).

The management consequences of establishing the Gila Wilderness Reserve were minimal. A laissez-faire approach prevailed—prohibit building roads and hotels and then leave it alone. In the 1920s, no attempt was made to manage

FIGURE 2.3 Arthur Carhart (left) and Aldo Leopold (above) were instrumental in setting aside 574,000 acres of the Gila National Forest, New Mexico (right), in 1924 as the first designated wilderness in the United States. Photos courtesy U.S. Forest Service.

positively for wilderness values, recreational or otherwise. Wilderness was simply set aside.

Chief Forester William B. Greeley exemplified this philosophy in action in the 1920s. He was enthusiastic about creating wilderness reserves on national forests, largely because he feared that the aggressive leadership of Stephen T. Mather, head of the National Park Service, was threatening his own empire (Gilligan 1953). If the USFS did not move to protect its spectacular scenery and develop its recreational resources, there was a good chance that some of its land would be turned over to the NPS. Such considerations unquestionably supported the intentions of some foresters to preserve wilderness simply because it was a good thing to do.

In 1926, Greeley formulated a policy for wilderness. Commercial uses such as grazing and even logging could continue, but campsites, meadows for pack-stock forage, and special scenic spots, as they were called (Gilligan 1953), would be protected. Greeley also instructed Associate Forester L. F. Kneipp to inventory national forest wilderness, the first roadless area review. The result showed 74 areas, each at least 360 square miles, in the 48 states. The chief forester's ideas of management stopped at this point.

In a 1926 communication to his several districts, Greeley explicitly disavowed any intention to regulate the numbers or the behavior of recreational users of wilderness. "I have no sympathy," he declared, "for the viewpoint that people should be kept out of wilderness areas in any large numbers because the presence of human beings destroys the wilderness aspect. . . . Public use and enjoyment are the only justification for having wilderness reserves at all" (as cited in Gilligan 1953, 104). As for the numbers of visitors, "the only limitation should be the natural one set up by the modes of travel possible" (as cited in Gilligan 1953, 104). Clearly, Greeley did not foresee the time when such limitations would be insufficient to keep wilderness from being degraded, ironically, by those who loved it. In the 1920s and 1930s, there was little reason to worry about overuse because the backcountry was still relatively empty.

At a 1926 session of the National Conference on Outdoor Recreation, Aldo Leopold made a strong plea for more systematic planning to protect wilderness (National Conference on Outdoor Recreation 1926). In Associate Forester Kneipp, Leopold found a supporter close to the center of power in Washington. Three years later, on July 12, 1929, Kneipp wrote Forest Service Regulation L-20, to order and consolidate what had until then been piecemeal preservation. This directive, while not law, was only an expression of agency policy, and standardized the term *primitive area* for a decade. Interestingly, the term *wilderness* was not used in L-20 because Kneipp and his colleagues thought the public might be repelled by its connotations (Pomeroy 1957). Kneipp admitted that the term *wilderness* did not apply to regions that had been, and still were being, commercially used.

L-20, with the amendments and instructions that followed it, required field staff to submit definite management plans for each primitive area. (Further discussion of the L-20 Regulation is found in Chapter 4.) These first, extremely vague management instructions amounted to little more than a list of prohibited and permitted activities. Among those permitted activities were virtually the full range of commercial endeavors customarily pursued in national forests. A notable exception was a section of L-20 that established research reserves (after 1930, called experimental forests). These areas, usually small, embraced natural forest ecosystems of scientific importance. Commercial use of the research reserves was prohibited; even recreational use was discouraged. Here, at least, was implicit recognition that recreation could have an impact on the biological integrity of an area.

L-20 raised important management questions about recreational developments in primitive areas. Some USFS officials, with every good intention, responded to the regulation with an aggressive program of trail and shelter construction to compete with the civilized style of developments common in the national parks and attractive to the majority of vacationers in the 1930s.

Kneipp had a different idea of wilderness management. In 1930, he wrote with some impatience to the field staff: "There should be no need for developing these areas to take care of the large numbers of people who are not capable of exploring wild country without considerable aid" (as cited in Gilligan 1953, 147). Kneipp went on to direct his forest supervisors to stop plans for trail signs, latrines, corrals, and shelters in the wilderness. He recommended that "primitive simplicity" be used as a criterion for development decisions and concluded: "These primitive areas are for the class who seek almost absolute detachment from the evidences of civilization" (as cited in Gilligan 1953, 147).

Such sentiments must have cheered Robert Marshall. The New York–born son of a well-to-do constitutional lawyer, Marshall devoted his short but intense life to wilderness. Professionally, he trained as a forester and plant pathologist. For recreation, he explored the nation's wildest remaining corners, including the Brooks Range in northern Alaska. A prodigious hiker, Marshall resented any kind of convenience in wilderness, and his management ideas reflected this viewpoint. From his position after 1933 as director of the Forestry Division of the U.S. Office of Indian Affairs, Marshall crusaded to curtail road building in wild places. So-called fire roads particularly offended him. Easy to build, especially when the Great Depression brought thousands of job-hungry men under federal care in work programs like the Civilian Conservation Corps, dirt roads threatened to divide and conquer the last really large wildernesses in the West. Marshall's greatest achievement, really a memorial because he died two months later at thirty-eight, was promulgating the U regulations by the USFS on September 19, 1939. Superseding the L-20 Regulation with respect to more than 14 million acres of wilderness on the national forests, the U regulations tightened protection. (See Chapter 4 for additional details on the U regulations.) In administratively designated *wilderness* and *wild areas* (the term *primitive* was no longer to be used in classifying areas), there would be no roads or other provision for motorized transportation, no lumbering, and no hotels, lodges, or permanent camps (Baldwin 1972). Very little was said about management, either in the U regulations or in the subsequent instructions for their implementation. Wilderness preservation was a caretaker function with an emphasis on guarding against outside influences. Wilderness inventory and designation, not management, were the preoccupations of the 1930s.

The USFS, through efforts by Carhart, Leopold, and Marshall, was in the forefront of the wilderness struggles (Fig. 2.4). This was possible in large part because of the discretion accorded the USFS in deciding uses of the national forests. This would change under the 1964 Wilderness Act, which provided only Congress would have the power to designate a wilderness and took land-use discretion concerning wilderness away from the agency. As fights over what qualified as wilderness and the designation of many wilderness areas attest, the USFS, interest groups, and Congress have often disagreed over what areas, and what kinds of areas, should be designated as wilderness.

Still, pride of authorship and leadership stemming from its early efforts to establish wilderness runs deep in the USFS. This legacy has complicated the agency's response to expanded proposals and liberalized definitions of wilderness thrust on it by the environmental movement of recent years.

THE WILDERNESS MANAGEMENT IDEA BEGINS

The same emphasis on "drawing a circle around it and letting it alone" characterized citizen conservation groups in the 1930s. The Wilderness Society had its origins in 1934 and 1935 among a small group of men, Marshall and Leopold included, whose declared objective was "holding wild areas *soundproof* as well as *sightproof* from our increasingly mechanized life" (as quoted in Nash 1982, 206). The whole thrust of their effort was to keep adverse influences *out* of wilderness, and not to try to understand and control what was happening *within* its borders.

The first recognition of a management dimension to wilderness preservation began in the 1930s. Marshall's contribution to the "National Plan for American Forestry" (1933), the so-called Copeland Report, contained sections on the overuse of backcountry campsites and the need to educate recreationists in outdoor etiquette, today called minimum-impact or Leave No Trace (LNT) camping.

Further recognition that wilderness required management came in the summer of 1937 when Marshall, the new chief of the Division of Recreation and Lands in the USFS, toured the Sierra with members of the Sierra Club. On the trip, the party visited high country severely damaged by pack-stock grazing and by campers. Discussions begun on the trip led to Marshall's requesting Professor Joel H. Hildebrand, president

FIGURE 2.4 Many people contributed to the wilderness movement in the United States. Robert Marshall (left), chief of the Division of Recreation and Lands in the USFS. Photo courtesy The Wilderness Society. Lowell Sumner (right) of the NPS helped inventory wilderness conditions in the Sierra Nevada of California in the 1930s. Photo courtesy National Park Service.

of the Sierra Club, to organize a committee to advise the USFS about wilderness management. Marshall provided the committee with key questions that revealed the direction of his thinking about wilderness. One question, for example, concerned the feasibility of distributing use by zoning wilderness, in effect, to achieve certain ends. Specifically, Marshall was anxious that "certain areas may still be preserved in what might be termed a super wilderness condition, or, in other words, kept entirely free even from trails, in order that a traveler can have the feeling of being where no one has been before" (Hildebrand 1938, 90).

The Hildebrand Committee replied with a list of trails currently in the Sierra and a recommendation that construction of new trails be sharply limited and, if necessary, kept at a low (that is, primitive) standard. Responding to other questions from Marshall, the Sierra Club advised restricting trail signs, limiting the use and grazing of pack stock, and prohibiting cutting pine boughs for beds. To manage the wilderness and enforce such regulations, the club suggested appointing high-country rangers or guards. Finally, in a significant forecast, both Marshall and the Sierra Club expressed concern that wilderness be made available to the entire public by encouraging use by younger and poorer people.

The Marshall–Sierra Club interchange in 1937 began a new era in wilderness management. It recognized that recreation was only one value associated with wilderness and that, to maintain wild conditions, recreation should be regulated and restricted. Subsequently, a November 1940 article entitled "Certified Outdoorsmen" by J. V. K. Wagar observed that "nature once certified outdoorsmen." According to him, the weak, foolish, and careless just did not return from the wilderness they entered. "But now," he continued, "there is such ease of transportation and so much improvement in equipment that anyone can become a wilderness traveler" (Wagar 1940, 490).

Wagar's point was that many people in the wilderness did not know how to care for themselves or for the country. His suggested remedy was a program conducted by rangers from the NPS and the USFS to certify outdoorsmen. Those attaining the rank of Expert Outdoorsman were safe to live in the woods. Included in their knowledge was the ability to respect and live gently on the land.

In 1940, Wagar did not go so far as to suggest that *only* certified outdoorsmen be admitted to designated wilderness areas, but the implication was clearly present. More than forty years later Roderick Nash proposed a mandatory "wilderness license" in *Backpacker* magazine (Nash 1981). Critics of the idea point out that such a license might further compromise the freedom of the wilderness. The counterargument is that educated (licensed) visitors could be permitted *more* freedom than visitors unfamiliar with minimum-impact wilderness skills. The same logic prevails in the current effort by the agencies to "educate visitors" (but without licensing) as a major wilderness management tool, including certifying instructors and practitioners in LNT practices, so that they can teach other wilderness managers and users (Swain 1996).

Wilderness enthusiasts have long recognized that too many people, even too many qualified outdoor enthusiasts, can spoil a particular place. As early as August 15, 1926, the *New York Herald Tribune* featured a before-and-after cartoon of a mountain lake. In the first frame, a lone horseman approaches the lake, which is surrounded with pines and full of leaping trout; in the second, a solid rank of fishermen surround the lake, and their camps obliterate the scenery. In this case, the extension of a road to the lake was represented as the cause of the change. However, by the 1930s, some Americans understood that, even without roads, wilderness values could be threatened by overuse. If that solitary horseman were joined by fifty other riders and a hundred backpackers, the problem remained much the same.

As a prime practitioner of wilderness recreation, the Sierra Club continued to take keen interest in developing techniques of wilderness management. In 1947, the *Sierra Club Bulletin* featured another article on the recreational impact on wilderness. It was coauthored by Lowell Sumner of the NPS and Richard M. Leonard, the chair of the Sierra Club's Outing Committee, which was by then coordinating a number of large high-country trips annually.

Sumner and Leonard focused particularly on the mountain meadows of the Sierra, and their article included a photographic sequence depicting stages in the transformation of a lush meadow grassland into a dustbowl. The cause was excessive recreational use. Discussing the problem under the heading "Saturation of the Wilderness," the authors declared, "We need more than just a concept. . . . We need a comprehensive technique of use that will prevent oversaturation of wilderness and still enable people, in reasonable numbers, to enjoy wilderness" (Leonard and Sumner 1947, 60).

Among the suggested management tools were rotating camping and grazing sites, limiting the length of a permissible stay by one party in one area, and transporting oats rather than relying on natural grasses for pack stock. According to Leonard and Sumner, twenty-four-hour camping limits existed in 1947 in some meadows—the earliest such rules in wilderness management history.

In 1949, the Sierra Club sponsored a High Sierra Wilderness Conference, which grew into a biennial event that flourished for a quarter of a century. At the initial conference, about a hundred federal and state administrators, outing-club representatives, and professional outfitters and guides met to discuss a common concern: wilderness preservation and management. Attendees conceded that the designation and permanent protection of wilderness from outside influences such as roads and commercial development addressed only part of the problem. The other part was the impact of recreation users on wild country. The attendees had the courage to recognize that they were part of the problem.

By the fourth Wilderness Conference in 1955, a full range of wilderness management concerns was being discussed. So was the idea, still a decade away from fruition, for an NWPS; but most commentators recognized that without proper management the designation of wilderness could be meaningless.

In the 1950s and 1960s, the related concerns of designation and management continued to dominate the American discussion of wilderness. Inventory and designation of wild places progressed, as did the protection of established reserves. Notable here was the Echo Park Dam controversy involving Dinosaur National Monument in Colorado and Utah and, many felt, the integrity of all national parks (Nash 1982, 2001). Part of the price of a 1956 decision not to build a dam in Dinosaur was approval of one in Glen Canyon on the Colorado River.

Completion of Glen Canyon Dam in 1963 intensified efforts of dam builders and wilderness protectors when the Grand Canyon itself became the subject of controversy three years later (Nash 1970a). The success of wilderness advocates in defending the Grand Canyon from dams, coming on top of the passage of the Wilderness Act and its establishment of the NWPS in 1964, constituted dramatic evidence of the new political muscle of the preservation idea. However, these successes were limited to the external dimension of wilderness designation. The internal one—how a designated wilderness would be managed—continued to generate problems. The fact was that the NPS had not substantially departed from the tourism development and management assumptions of the Mather-Albright era.

Park management continued to emphasize visitor numbers, conveniences, and viewing spectacular scenic splendors rather than the wilderness experience. This became clear in 1956, when the NPS launched Mission 66. The program responded to rapidly increasing park visitation, but some feared it was the wrong response because its major thrust was further development. More than a billion dollars were poured into it, mostly for constructing roads, visitor centers, and motel-type accommodations. The entire emphasis of Mission 66 was to improve a park's capability for handling *more* tourists, and little was said about wilderness values and wilderness management. The prevailing management philosophy was more appropriate to an amusement park or resort than to a wilderness. The NPS was not alone in this posture. Operation Outdoors, the USFS counterpart to Mission 66, similarly emphasized tourism facilities and conveniences.

The facility and convenience orientation also appeared in backcountry-management practices of the mid-1960s. Trail standards were improved in the interest of easier access. Picnic tables, bulletin boards, fireplace grates, latrines, and corrals for pack stock were often placed at wilderness campsites. In some California areas, wilderness visitors even found rakes for use in tidying up their campsites (Snyder 1966).

However, the tide was beginning to turn toward visitor education, self-sufficiency, and light-handed styles of management. A visitor to the John Muir Wilderness in California in 1973 found stoves, latrines, and fences gone. Instead of concentrating visitors at designated campsites, management had opted to encourage dispersed camping. People were not as visible due to the lack of centralized, formal camp improvement.

The year 1973 marked the first use of permits to limit the number of visitors to some of the more heavily used wildernesses, for example the San Gorgonio Wilderness in California managed by the USFS and the Colorado River in Grand Canyon National Park. Granted, wilderness permits had existed in Minnesota's Boundary Waters Canoe Area managed by the USFS since 1966 and were required in many wildernesses for record keeping, but they were freely given to every applicant. New to the early 1970s was the idea of using permits to limit use to a predetermined recreational carrying capacity. Use of other wilderness rules and regulations, such as banning wood fires and a pack-it-out policy for litter, also increased. Grand Canyon National Park rules went so far as to require river runners to pack out solid human waste. At a few heavily used wildernesses and backcountry areas, personnel were assigned to coordinate and disperse visitors.

WILDERNESS MANAGEMENT EVOLVES

Wilderness management direction has evolved from the initial heavy-handed visitor-control perspectives of the late 1960s and early 1970s to a light-handed, visitor-education approach that emerged in the 1980s and continues today as the first choice of wilderness managers. Many trends and events facilitated this evolution.

During the late 1970s and early 1980s, volunteer worker programs expanded considerably in national parks and forests, while at the same time recreation management budgets and personnel ceilings were being reduced. Wilderness management research and field experience were revealing the power and wisdom of education and appeals for cooperation as effective methods of managing visitor behavior.

The professionalism of wilderness managers increased. Managers eagerly exchanged views and experiences in regional and national wilderness conferences and workshops sponsored by universities, the USFS, and other agencies that included wilderness users, scientists, and university professors. The Arthur Carhart National Wilderness Training Center, Missoula, Montana, was established in the early 1990s to provide specific training for field managers and agency executives concerned with wilderness. The Aldo Leopold Wilderness Research Institute was established in the early 1990s to further research supporting wilderness management.

Increasing public involvement, a trend spurred by the Wilderness Act, has become a way of life for federal agencies; in addition, agencies are bound in partnership with pro-wilderness clients (Fig. 2.5). Most recently, the Internet (for example, www.wilderness.net) has become a major forum for citizen concerns and input to management proposals and environmental impact statements.

With these new initiatives and conditions, light-handed and educational approaches to wilderness management have evolved, with some tentative success. For example, in 1983 the Eagle Cap Wilderness on the Wallowa-Whitman National Forest, Oregon, was the first unit of the NWPS to drop a permit system once instituted (Scholz 1983). This move to a light-handed management strategy was based, in part, on the view that wilderness camping ethics and skills were improving, due to an educational emphasis featuring public and wilderness-user contact by volunteer rangers. Indeed, a survey

FIGURE 2.5 Arizona State University students conduct campsite inventory work under the supervision of their instructors, U.S. Forest Service and Bureau of Land Management personnel. Photos courtesy Greg Hansen and Jim Mahoney.

of Eagle Cap visitors in 1993, repeating many questions asked in a 1965 survey, documented a positive change in visitors' wilderness-behavior norms (Watson et al. 1996).

Wilderness users are concerned about protecting wilderness resources and values and respecting the interests of other visitors in having a wilderness experience. The hope was that the Eagle Cap Wilderness and other wilderness areas, in response to light-handed management and user education, are healing, and that restrictive permits are no longer needed. However, in 1994, a mandatory-permit system was restored to the Eagle Cap Wilderness, but this time mainly to establish use levels, educate visitors, and see that visitors are apprised of wilderness regulations and LNT principles. After three years with the renewed permit system, use levels were determined but the permit system continued through 2007—without restricting the numbers of visitors—because of its value as a wilderness education technique (Carlson 2000).

Literature that confirms the presence of new wilderness attitudes and influence abounds, as do wilderness education programs by organizations such as Outward Bound, the National Outdoor Leadership School, many environmental organizations, youth groups like the Boy Scouts and 4-H, in addition to college and university courses and conferences and agency workshops. Well-trained wilderness users may need fewer rules and less policing (Nash 1981). The emergence of a new generation of visitors who are sensitive to and respectful of the meaning and value of wilderness and who know how to minimize their impacts on wilderness—to "leave no trace"—increase the possibility that light-handed and educational approaches to wilderness management will be successful (Swain 1996) (Fig. 2.6).

Wilderness experience is extremely delicate, and wilderness management must be correspondingly deft. In the new millennium, wilderness managers, working on behalf of the public, must make judgments on several difficult questions. One is access to wilderness by commercially guided groups versus private or self-reliant user groups. Should outfitters and their need to show a profit be favored in allocating wilderness recreational opportunity, or do guides actually diminish the essence of the wilderness experience for visitors? What about the increasingly popular use of wilderness for personal growth and wilderness therapy for troubled adolescents? Should such use of wilderness be favored over other commercial or recreational use? To what extent should the Wilderness Act be bent to accommodate, for example, semipermanent outfitter camps and permanently fixed rock-climbing protection? Are certain kinds of nonmotorized recreation (geocaching, for example) less compatible with wilderness preservation than other activities? To what extent should the absence of solitude at popular wilderness locations be tolerated in order to provide more *visitors* a wilderness experience, or to concentrate rather than disperse visitor impacts? How and where can electronic technologies be appropriately used for education of wilderness managers and visitors and for communication between them in wilderness (Nash 1996; Van Horn 2007; Dawson 2007)? Such issues are at the heart of present and future wilderness management.

FIGURE 2.6 Wilderness management has moved into schools and the community. Rich Hanson, wilderness manager from the BLM's Phoenix Field Office, explains the visitor entry permit system to educators in the North Maricopa Mountains Wilderness. Photo courtesy Chad P. Dawson.

STUDY QUESTIONS

1. What is the "intellectual dilemma" of wilderness management?
2. Compare an expedition through the Grand Canyon today with what one might have experienced a century ago.
3. What accounts for the huge increases in wilderness recreation participation since the early twentieth century?
4. Compare and contrast the purposes for establishment of national parks and national forests. How did those purposes influence early notions of wilderness management in the two agencies?
5. How did Robert Marshall's contributions change ideas about wilderness use and management?
6. Who were Arthur Carhart and Aldo Leopold, and how did they contribute to early recognition of wilderness values by the USFS?
7. The Sierra Club is one of the oldest and most influential environmental groups interested in wilderness. How has the Club's position on wilderness management changed over time? How do you think the Sierra Club and other wilderness user groups will further evolve?
8. How has wilderness management evolved since the 1960s and 1970s, and what challenges does it face now?

ACKNOWLEDGMENTS

Roderick Frazier Nash, professor emeritus of History and Environmental Studies, University of California, Santa Barbara, originally wrote this chapter. Dr. Nash and John C. Hendee revised it for the second, third, and fourth editions.

CASE DISCUSSION: MODERN-DAY AMERICAN VIEWS ON WILDERNESS

While the wilderness values of early wilderness leaders and visionaries have been the subject of much literature on the history of the movement, it is the current generation who must manage and sustain the protection of the National Wilderness Preservation System (NWPS). What Americans see as valuable about protecting wilderness areas is important for legislators and policy makers to understand and to take into account regarding both new wilderness area designations and supporting the existing NWPS.

In 2008, Cordell et al. reported on some of the results of a National Survey on Recreation and the Environment (NSRE) that asked Americans sixteen years of age and older what they valued about wilderness protection. During the fall of 2006 through the fall of 2007, the NSRE asked about the relative importance of thirteen wilderness values on a five-point scale, from "not important" to "extremely important." The percentages of respondents who reported that a value was very or extremely important are as follows:

Wilderness Value	Percentage
Protecting air quality	93.0
Protecting water quality	90.3
Protecting wildlife habitat	87.7
Knowing future generations will have wilderness areas	85.6
Preserving unique wild plants and animals	83.0
Protecting rare and endangered species	82.7
Providing scenic beauty	76.5
Having an option to visit wilderness areas in the future	76.2
Knowing that wilderness areas exist	75.7
Providing recreation opportunities	71.3
Preserving natural areas for science	64.3
Providing spiritual inspiration	57.8
Providing income for tourist industry	39.6

Clearly, modern-day Americans have strong and positive feelings about twelve of these thirteen wilderness values. The study found no statistically significant differences between the responses of residents of metropolitan and rural areas.

CASE DISCUSSION QUESTIONS

1. How do these modern-day values compare to the implied and expressed values inherent in the roots of the wilderness movement?
2. Would wilderness visionaries like Arthur Carhart, Aldo Leopold, and Bob Marshall agree with these modern-day wilderness values? Why or why not?
3. Do you expect the percentages to increase or decrease on any of these wilderness values in the coming decades as the NSRE is repeated? What do you expect to change and why?

REFERENCES

Allin, Craig W. 1982. *The Politics of Wilderness Preservation.* Westport, CT: Greenwood Press.

Baldwin, Donald N. 1972. *The Quiet Revolution: The Grass Roots of Today's Wilderness Preservation Movement.* Boulder, CO: Pruett.

Carlson, Tom. 2000. The Eagle Cap Wilderness Permit System: A visitor education tool. *International Journal of Wilderness.* 6(2): 27–28.

Cleveland, Treadwell, Jr. 1910. National Forests as recreation grounds. *Annals of the American Academy of Political and Social Science.* 35(3): 241–247.

Cordell, H. Ken; Betz, Carter J.; Stephens, Becky; Mou, Shela; Green, Gary T. 2008. How do Americans view wilderness: Part I. http://warnell.forestry.uga.edu/nrrt/nsre/IrisWild1.html; accessed 2/14/08.

Dawson, Chad P. 2007. New opportunities for educating future wilderness and wildland managers in a changing technological world. *International Journal of Wilderness.* 13(3): 36–39.

Gilligan, James P. 1953. *The development of policy and administration of Forest Service primitive and wilderness areas in the Western United States.* Ann Arbor: Dissertation. University of Michigan.

Glover, James M. 2000. Spirit of the wilderness: Can we stop trying to control nature? *International Journal of Wilderness.* 6(1): 4–8.

Hildebrand, Joel H. 1938. Maintenance of recreation values in the High Sierra: A report to the United States Forest Service. *Sierra Club Bulletin.* 23(5): 85–96.

Jalbert, Linda. 2008. Correspondence between Linda Jalbert, wilderness planner and coordinator, Grand Canyon National Park and Chad P. Dawson, January 21, 2008.

Langford, Nathaniel P. 1972. *Discovery of Yellowstone Park: Journal of the Washburn Expedition to the Yellowstone and Firehole Rivers in the Year 1870.* Haines, Aubrey L., ed. Lincoln: University of Nebraska Press. 125p.

LeConte, Joseph N. 1972. *A Summer of Travel in the High Sierra.* Sargent, Shirley, ed. Ashland, OR: Lewis Osborne. 144p.

Leonard, Richard; Sumner, E. Lowell. 1947. Protecting mountain meadows. *Sierra Club Bulletin.* 32(5): 53–62.

Leopold, Aldo. 1921. Wilderness and its place in forest recreational policy. *Journal of Forestry.* 19(7): 718–721.

Lillard, Richard G. 1968. The siege and conquest of a national park. *American West.* 5(1): 28–31, 67, 69–71.

Nash, Roderick. 1970a. *Grand Canyon of the Living Colorado.* Ballantine, NY: Sierra Club.

———. 1970b. Wild-deor-ness, the place of wild beasts. In: McCloskey, Maxine E., ed. *Wilderness: The Edge of Knowledge.* San Francisco: Sierra Club, pp. 34–37.

———. 1975. The creation of wilderness by herding and agriculture. In: *Program/Journal, 14th Biennial Wilderness Conference;* June 5–8, 1975; New York. Audubon, NY: Sierra Club, pp. 51–55.

———. 1978. Wilderness management: A contradiction in terms? Wilderness Resource Distinguished Lecture. Moscow: University of Idaho Wilderness Research Center. [Also published as Wilderness is all in your mind. *Backpacker.* 1979: 7(Feb.–Mar.): 39–41, 70, 72–75.]

———. 1981. Worldview: Protecting the wilderness from its friends. *Backpacker.* 9(Feb.–Mar.): 15–16.

———. 1982. *Wilderness and the American Mind,* rev. ed. New Haven, CT: Yale University Press.

———. 1996. A wilderness ethic for the age of cyberspace. *International Journal of Wilderness.* 2(3): 4–5.

———. 2001. *Wilderness and the American Mind,* 4th ed. New Haven, CT: Yale University Press.

National Conference on Outdoor Recreation. 1926. S. Doc. 117, 69th Congress, 1st Session.

Pomeroy, Earl. 1957. *In Search of the Golden West: The Tourist in Western America.* New York: Knopf.

Potts, R. 2007. Changing human relationships with wilderness and wildlands. *International Journal of Wilderness.* 13(3): 4–6, 11.

Roth, Dennis. 1984. The national forests and the campaign for wilderness legislation. *Journal of Forest History.* 28(3): 112–125.

Runte, Alfred. 1979. *National Parks: The American Experience.* Lincoln: University of Nebraska Press.

Sax, Joseph. 1980. *Mountains Without Handrails: Reflections on the National Parks.* Ann Arbor: University of Michigan Press.

Scholz, Sue. 1983. The human approach: Wilderness permits not required. *Wallowa County* [OR] *Chieftain.* July 7: 3 (col. 3).

Snyder, Arnold P. 1966. Wilderness management: A growing challenge. *Journal of Forestry.* 64(7): 441–446.

Spitek, Brian. 2007. Correspondence between Brian Spitek, wilderness manager, Mount Whitney Ranger District, Inyo National Forest, and Chad P. Dawson, December 13, 2007.

Sumner, E. Lowell. 1936. Special report on a wildlife study of the High Sierra in Sequoia and Yosemite National Parks and adjacent territory. In-service report. Washington, DC: National Park Service Archives.

———. 1942. The biology of wilderness protection. *Sierra Club Bulletin.* 27(8): 14–22.

Sutter, Paul S. 2002. *Driven Wild: How the Fight Against Automobiles Launched the Modern Wilderness Movement.* Seattle: University of Washington Press.

Swain, Ralph. 1996. Leave No Trace (LNT)—outdoor skills and ethics program. *International Journal of Wilderness.* 2(3): 24–26.

U.S. Department of the Interior, National Park Service. 1970. Administrative policies for natural areas of the National Park System. Washington, DC: National Park Service.

———. 2006. Colorado River Management Plan.

U.S. Public Law. Yellowstone National Park Establishment Act of March 1, 1872.

U.S. Public Law. National Park Service Act of August 25, 1916. 39 Stat. 535.

U.S. Public Law. Grand Teton National Park Act of February 26, 1929. 70 Stat. 1314.

U.S. Public Law. Everglades National Park Act of May 30, 1934. 48 Stat. B16.

U.S. Public Law. Kings Canyon National Park Act of March 4, 1940.

U.S. Public Law. Grand Teton National Park Act of 1950; Repeal of Grand Teton National Park Act of 1929. 64 Stat. 849.

Van Horn, Joe. 2007. GPS. *International Journal of Wilderness.* 13(3): 7–11.

Wagar, J. V. K. 1940. Certified outdoorsmen. *American Forests.* 46(11): 490–492, 524–525.

Watson, Alan; Hendee, John C.; Zaglauer, Hans P. 1996. Human values and codes of behavior: Changes in Oregon's Eagle Cap Wilderness visitors and their attitudes. *National Areas Journal.* 16(2): 89–93.

Waugh, Frank A. 1918. *Recreation Uses on the National Forests.* Washington, DC: Government Printing Office.

Wright, George M.; Dixon, Joseph S.; Thompson, Ben H. 1932. Fauna of the national parks of the United States. *Fauna Series 1.* Washington, DC: U.S. Department of the Interior, National Park Service.

Wright, George M.; Thompson, Ben H. 1934. Fauna of the national parks of the United States. *Fauna Series 2.* Washington, DC: U.S. Department of the Interior, National Park Service.

CHAPTER

3

International Wilderness in the Twenty-First Century: A Global Review

by Vance G. Martin, Alan Watson, Christopher A. Armatas, and Amy Lewis

INTRODUCTION

Today it is recognized that wilderness has many dimensions, at once touching upon history, ecology, culture, mystery, risk, management, and personal experiences. The evolution of the wilderness concept, and how it is perceived, is influenced by diverse cultural perspectives, a continually expanding understanding of how humans are supported by nature, and existential threats to humanity, both socioeconomic (e.g., income inequality, access to nature and natural resources) and environmental (e.g., climate change, biodiversity loss). The primary purpose of this chapter is to review and summarize global contributions to the protection of wilderness land-and seascapes ("wild nature") thus illustrating how the many dimensions of wilderness are imagined and applied across institutions and cultures.

In our global review, wilderness is primarily conceptualized through the lens of the U.S. Wilderness Act of 1964, in large part because the idea outlined in the Act is reflected in a variety of intranational "wilderness" protection systems across various global contexts, and international adaptive applications. Indeed, the concept of "wilderness" lands and waters may be commonly considered as: "In contrast with those areas where man and his works dominate the landscape, is hereby recognized as an area where the earth and its community of life are untrammeled by man, where man himself is a visitor who does not remain" (U.S. Public Law 88-577). However, it is worth highlighting that this conceptualization of wilderness is both recent and not universally embraced. That is, the earliest human relationships with the natural world were so intimately entwined that the "idea of wilderness as a distinctive environment, let alone an area deserving special attention, was beyond comprehension" (Stankey 1989, 9). Numerous factors contributed to this. One view is that with the rise of monotheism and Judaism across the Middle East, people were increasingly viewed as separate from wild nature, which led, in part, to the creation of a wilderness idea where wastelands and evil forces impeded human progress and development (Nash 1982; Stankey 1989). Another view looks simply at the constantly improved and more efficient methods of human development that simply began to incentivize increased exploitation of natural resources. Consequently, these wilderness lands, from the perspective of the dominant cultural and political institutions (i.e., those formed and perpetuated by imperial concepts manifesting in multiple cultures throughout the world and carried forward by American settlers, for example), were seen as areas to be conquered, dominated, and exploited (Stankey 1989). These concepts were inextricably tied to widely embraced philosophies and movements such as the Doctrine of Discovery and Manifest Destiny and, eventually, the Industrial Revolution. As a result, over several hundred years, colonization and rapid development of significant portions of the Earth led to an efficient reduction in wilderness lands and waters. In response to the rapidly decreasing amount of wilderness and changing perspectives about the benefits of wild spaces, the U.S. Congress passed the Wilderness Act and the President signed it into effect in 1964.

From the passage of the Act to today, in America and across various global contexts, protection of wilderness largely serves as an antidote to the view that nature should be dominated and developed. However, to some, domination over nature was never embraced in the first place, nor is it embraced today. For instance, while Indigenous cultures and perspectives are not monolithic, it is certainly true that nature reverence, philosophy, and practice among Indigenous cultures is a common core concept in the survival of their cultures. For example, Chief Luther Standing Bear of the Oglala Lakota (as quoted in McLuhan 1971, 45) famously commented in the nineteenth century on the difference between his culture and that of the European settlers who were replacing it: "We did not think of the great open plains, the beautiful rolling hills, and the winding streams with their tangled growth as 'wild.' Only to the white man was nature a 'wilderness' and only to him was the land 'infested' with 'wild' animals and 'savage' people. To us it was tame."

Further, the outcomes of protecting wilderness in a world where wild spaces are less accessible to an increasingly urban population and are in higher demand may be perceived as the creation of elite spaces for high-income people to recreate in, and as a global resource to absorb ecological crises created primarily by colonial ideals and forces. These two outcomes benefit a relatively small and homogenous group and appear inequitable to developing nations and Indigenous peoples. These perceptions have merit.

Wilderness protection per the U.S. Wilderness Act of 1964 provides benefits that may be widely recognized across cultures—even if the original motivations and meanings are not shared by all, and if the pressing need to preserve wilderness was driven by a Eurocentric domination of nature that manifested in colonialization and global capitalism. More generally, as global society pursues "sustainable development" (i.e., realization of a good life within planetary limits) (World Commission on Environment and Development 1987), and works to address social-ecological crises such as human rights for all, climate change, and widespread biodiversity loss, wilderness protection is one integral tool. For example, in the financial sector, there are wide-ranging discussions around innovative financial instruments such as debt-for-nature swaps, "green bonds," impact-oriented philanthropy, and elimination of perverse subsidies; all are directly related to protecting additional land and seas, with a focus on equitable compensation and distribution of benefits (European Commission 2018). In other words, the retention of wild nature can be a major source of income and economic progress, while also bolstering national heritage and health (Shah 2014).

As we review and summarize global contributions to the protection of wilderness land-and seascapes, with a focus on expansion since 2008, we recognize that the meaning of the word wilderness, and how it is regarded, differs around the world. One of the many objectives of this chapter is to introduce some of the challenges in understanding wilderness in the global context, as well as the importance of appreciating and working for it because of its key role in addressing many of the existential threats faced by our planet today.

For example, it is important to note that attitudes towards wilderness are not always positive and are often generic, as when some people use the term simply to refer to all natural areas that are undeveloped, sometimes with a connotation of the land being abandoned and in need of development. However, to many other people, a wilderness area means a place that you can go to and express identity, its protection shows responsibility to future populations and is a place of spiritual inspiration; a place that is protected so that it remains ecologically whole. Nelson (1998) cited some thirty different reasons for wilderness conservation, including intrinsic value; for dependent indigenous cultures whose livelihood and cultural identity are inextricably linked to wildlands; for biodiversity protection, national identity, and ecotourism; and many more. Since 2015 there has been a new awareness of a broadened approach to sustainability, one that includes less-technological and more cost-efficient "nature-based" solutions (International Union for Conservation of Nature 2016). Protecting the remaining wilderness is beginning to be recognized as a key element of this new approach to sustainability, offering multiple benefits and being essential to simultaneously address both climate change and species extinction.

There is increasing acceptance of the term "wilderness area," also, to mean those areas legislated or zoned for protection in their natural condition but often accommodating a wider spectrum of human activity than the U.S. definition might allow (Fig. 3.1). Additionally, an increasing number of areas have been protected in many countries without wilderness protection systems, but that meet international standards established by the International Union for Conservation of Nature (IUCN). Globalization of the wilderness concept and protection of areas in countries with rigorous protection systems have continued to evolve in both developed and developing nations. This evolution is driven, in part, by mutual learning through dialogue and proceedings of the World Wilderness Congresses (WWC), the *International Journal of Wilderness* (IJW), a growing network of wilderness training exchanges and a growing worldwide community of activists committed to protecting the wildest remaining places or restoring wildness where it has disappeared. Further, new partnerships continue to expand support for wilderness science, to share information on methods and benefits of wilderness protection, and to carry out wilderness stewardship training.

In this chapter, by focusing on the expansion of the wilderness concept in global contexts, we aim to highlight the various ways that wilderness protection was achieved, as well as the nuance of what wilderness means as different cultures operationalize the concept. At the same time, we stress that wilderness and wild nature have never been more imperiled than they are today, and for this reason the need for better public understanding, applied science, enhanced management, and political willpower has never been greater.

FIGURE 3.1 Around the globe many nations are following the American idea of maintaining certain lands in a wild state, as wilderness or in some protected area category appropriate to their culture. In Africa, wildlife, such as the elephant, are the international and enduring symbol of wildness and wilderness. These charismatic megafauna are part of the natural system and the history and culture that are Africa. Photo courtesy Ulf Doerner.

FORCES OF CHANGE FOR WILDERNESS PROTECTION IN THE TWENTY-FIRST CENTURY

There are wide ranging initiatives to both restore and protect wilderness across the globe. Specifically, the fifteenth meeting of the parties to the UN Convention on Biodiversity (CBD) (COP15) approved four "goals" and twenty-three specific "targets" for 2030, of which two targets are particularly relevant to our discussion:

Target 2 addresses restoration: "By 2030 . . . at least 30 percent of areas of degraded terrestrial, inland water, and coastal and marine ecosystems are under effective restoration, in order to enhance biodiversity and ecosystem functions and services, ecological integrity and connectivity."

Target 3 addresses protected areas: "At least 30 percent of terrestrial, inland water, and coastal and marine areas, especially areas of particular importance for biodiversity and ecosystem functions and services, are effectively conserved and managed through ecologically representative, well-connected and equitably governed systems of protected areas and other effective area-based conservation measures, recognizing indigenous and traditional territories." Rights and full inclusion of Indigenous Peoples are especially emphasized. Though popularized as "30x30," this protected area target still begs the question of how much is actually needed, which we address later in this chapter.

A prerequisite to realizing these targets is understanding how much wild nature exists on Earth. Numerous, ongoing efforts have tried to assess how much wilderness is left in the world, as the first step in deciding protection priorities. The answer often differs depending on the wilderness criteria used. There have been ongoing inventories since 1987, with estimates until recently of remaining terrestrial wilderness, as between 25 and 55 percent of our planet. The specific criteria vary in each inventory, so we encourage you to review each one. The first inventory—referred to by its authors as "reconnaissance-level" was pre-digital and, using Jet Navigation Charts and Operational Navigation Charts, produced in 1987 as part of the 4th World Wilderness Congress (WWC). It was undertaken by the Sierra Club, with participation by the United Nations Environment Program, the World Bank, and the World Resources Institute (McCloskey and Spalding 1988). This attempt took almost two years and was admittedly a crude but necessary first estimate, limited to areas 1 million acres (405,000 ha) or greater, and concluding that 33 percent of the world was still in wilderness condition.

Later, Hannah et al. (1994) produced a GIS-based map of human disturbance of natural ecosystems using a habitat index with three categories: undisturbed, partially disturbed, and human-dominated. Their analysis suggested that approximately 52 percent of the planet's land was undisturbed.

The Wildlife Conservation Society's assessment (Sanderson et al. 2002) used four criteria to assess wilderness: population density, land transformation, human access points, and electrical power infrastructure. These factors were combined to generate a Human Influence Index rating, and the resulting map showed that 17 percent of the planet's land surface remained wild in 2002, though that assessment did not include Antarctica (Fig. 3.2).

Next, in 2003, Conservation International identified all areas over 2.47 million acres (1 million ha) having: at least 70 percent intact habitat, intact faunal assemblages of mammals, birds, and large predators, and a population density of fewer than five people per square kilometer (0.38 sq mi) (Mittermeier et al. 2003). Under these criteria, 38 to 46 percent of the planet was still wilderness.

Most recently, an international team of scientists mapped the world's remaining terrestrial wilderness and produced a similar map for intact ocean ecosystems (Watson et al. 2018). To map Earth's remaining terrestrial wilderness, instead of any universal definition of wilderness they used the best available data on eight indicators of human pressures at a resolution of 1 square kilometer (0.38 sq mi). These eight indicators were built environments, crop lands, pasture lands, population density, night-time lights, railways, major roadways, and navigable waterways (data from 2009). For mapping of intact ocean ecosystems, Watson et al. (2018) used 2013 data on fishing, industrial shipping and fertilizer run-off, along with sixteen other indicators. They identified wilderness land or ocean areas as those that were free of human pressures, with a contiguous area of more than 10,000 sq km (3,861 sq mi). They concluded that 77 percent of land (excluding Antarctica) and 87 percent of the ocean has been modified by the direct effects of human activities. They also concluded that more than 70 percent of the world's remaining terrestrial wilderness are contained in five countries (Russia, Canada, Australia, United States and Brazil—in descending order). These varying inventories use datasets that do not include relatively small areas of high biodiversity wilderness, often very important for protection and linking through natural corridors.

FIGURE 3.2 Wild weather and sudden flooding on rivers are major obstacles to New Zealand wilderness travel. Wilderness visitors are subject to the changes of nature, and travel schedules can be changed for days without warning. Photo courtesy Les Molloy.

Finally, and to add an extra layer of complexity, it is important to note that recognition or inventories of wilderness are most generally related to measuring human impact, and this should not be confused with "ecological intactness." A very recent study focused purely on "faunal assemblages" as a criterion for intactness, and it concluded that only 2.9 percent of terrestrial areas remain intact as per these criteria (Plumptre et al. 2021). In this chapter we will report other efforts to measure and map wilderness in specific countries or continents (e.g., China and Europe). The "forces of change" summarized below can be seen as a response to the recognized threat to the declining amount of wilderness remaining in the world.

WILD NATURE RE-EMERGES IN EUROPE

Despite the remaining undeveloped, wild, and remote areas, in many countries there is no wilderness left to protect. In Switzerland, for example, wilderness has long since vanished, despite its rugged mountain landscapes. From border to border, this small nation has been transformed by human activity into a thoroughly modified landscape.

All over Europe, even the wildest places have been converted for agriculture and industry for thousands of years. Fisher et al. (2010) reported that no more than 1.5 percent of Europe can be classified as wilderness, which generally occurs in boreal zones, mountains, and rural areas. Most German national parks (the most strictly protected areas in Germany), for example, do not meet IUCN international criteria for even the national park category (World Commission on Protected Areas [WCPA] and IUCN Category 2). The first national park in Germany was established in 1969—the Bavarian Forest National Park, which receives about 1.5 million visitors annually (Brüggemann 1997). Logging and development for tourism access were among the initial plans for the park; however, conservation interests prevailed. Today, the principal objective of German park management is non-intervention in natural processes. This represents a change from the highly managed landscape and intensive forestry traditionally practiced in Germany and most of Europe. The name of this new management approach is translated as "let nature be."

Křenova and Kiener (2012) gave an update on a wilderness experiment on the border of Germany and the Czech Republic. Two national parks, representing the two largest terrestrial Natura 2000 sites in both countries, Šumava National Park in the Czech Republic and the neighboring Bavarian Forest NP were described as Europe's Wild Heart for several years. The main goal of Europe's Wild Heart, which started in 2008, was to designate a transboundary wilderness area in the core zones of the two national parks. The project area was 13,060 ha and was intended to be the first and largest transboundary wilderness area in Central Europe. However, a bark beetle outbreak followed the Kyrill hurricane in 2008 and 2009 and escalated the discussion about appropriate forest management in Šumava National Park. Opponents of the national park principles of non-intervention and wilderness became more vocal. The situation escalated after the election in 2010 when the Green Party was replaced by conservatives (ODS, Civic Democratic Party) at the Czech Ministry of Environment. Clearcutting was started in some former non-intervention parts of Šumava National

Park, and hunting was again allowed in the core zone. Since then, both the management strategies and practical management measures in Šumava National Park and Bavarian Forest NP have become increasingly divergent. Currently, Europe's Wild Heart project is on hold in both countries.

There is, however, still another "wild heart" of Europe. The Carpathian Mountains extend from southern Poland southward almost to the Balkans, including a small area of Ukraine, but are largely contained within Romania. According to reports, there are a minimum of 864,838 acres (350,000 ha) of original forest, up to 400 years old, in a landscape mosaic that includes meadows lightly managed as hay fields for hundreds of years by subsistence farmers. The landscape is rich in biodiversity, and it features charismatic megafauna such as bears, wolves, lynx, and stags. Since 2001, the Romanian government has undergone a process of land restitution, giving national parks and forestry lands back to the families of their original owners (Rosu 2014). Despite prohibitions on logging, clearcuts are evident on many of the restituted parcels. Nonetheless, the Carpathian Mountains present a major opportunity for wilderness protection in Europe.

In southern Europe, land abandonment started early in the twenty-first century; the rural exodus was the result of several socioeconomic drivers, including declining ability to compete in the globalized food markets, lowering EU Common Agricultural Policy (CAP) subsidy, and tenure insecurity after the collapse of the socialist era in Eastern Europe. The amount of land taken out of production between 2002 and 2012 has been estimated at 114 million acres (46.1 mha or 461,000 sq km) of permanently fallowed farmland (Ustaoglu and Collier 2018).

Simultaneously, many wildlife species have made a return across Europe over the last four decades, highlighting the potential resilience of wildlife. The most recent study (Ledger et al. 2022) covers mammals such as the Eurasian beaver, gray seal, and European bison, which all show a strong comeback. It also includes birds such as the barnacle goose, griffon vulture, great white egret, Dalmatian pelican, and black stork which show the best comeback among some 25 species of birds which are also recovering. Factors such as increased enforced legal protection, the creation of corridors between protected areas, recovery of prey species, rewilding, reintroductions, and an ever-greater willingness and desire by Europeans to live alongside and enjoy the presence of wild animals have all contributed to this recovery.

Great Britain provides additional evidence of why the American model of wilderness preservation cannot be followed in many places, even if the culture would welcome islands of wildness in its midst. The British landscape has been settled continuously for more than 5,000 years with long-term impact on landscapes and biodiversity. British national parks serve primarily to preserve a rural lifestyle, complete with traditional agricultural practices. The British situation further demonstrates that, by the time old and intensely developed nations decide to establish a system of parks and reserves, there may be little alternative but to include substantial amounts of civilization. There is simply no opportunity to preserve an extensive wilderness system like that in North America or Australia.

Nevertheless, the quest for protection of the wildest areas in the UK continues. The wilderness concept emerged in the last 50 years in Europe, with a turning point perhaps when the 3rd World Wilderness Congress convened in Scotland (1983), at which the Secretary of State for Scotland announced UK ratification of the World Heritage Convention, and numerous wilderness-specific NGOs were created. Until that point, the general opinion was that wilderness had disappeared in Europe during the centuries of intensive growth and development of human society. To the contrary, the presence and appreciation of wild nature is accelerating.

For example, areas within the Highlands of Scotland that appear inhospitable still generate feelings of wildness. However, many of these seemingly inhospitable areas were once inhabited, but were cleared of local people by the English in the early 1800s to establish sporting (hunting) estates and sheep grazing. These areas have often experienced declines in biodiversity, but many people nonetheless perceive them as wild. Restoration of the original forest cover would be needed to restore ecosystems. Thus, numerous "rewilding" projects are advancing throughout the UK, with perhaps one of the oldest and most respected being the Trees for Life project, started in the mid-1980s in the NW Highlands to restore the original Caledonian Forest (see www.treesforlife.org.uk). As of 2021, Trees for Life is an integral part of a larger program under Rewilding Europe, now called the "Affric Highlands," (McDonnell 2021) with an ambitious 30-year vision.

Europe's attention to wilderness is changing in several ways. Most recently this has been empowered by the European Parliament Resolution on Wilderness in Europe (European Parliament 2009). This sweeping resolution has many directives and calls for action by Member States, including: definition and mapping; enhanced protection; creation of wilderness zones within the European Union's Natura 2000 system of protected areas (the largest such network in the world); and sharing of best practices. It has also led to significantly enhanced NGO-driven initiatives such as that of Wild Europe and the criteria and standards for wilderness quality promoted by the European Wilderness Society.

European Wilderness Network

The European Wilderness Network (EWN) is an initiative of the European Wilderness Society (EWS), working in accord with the European Parliament Resolution on Wilderness in Europe (European Parliament 2009). EWS aims to strengthen the wilderness concept in Europe through creation of standard criteria for wilderness areas and providing NGO-based accreditation for those areas that qualify.

EWS has codified these standards under the European Wilderness Quality and Audit System (EWQAS), consisting of a hierarchy of nine principles, fifty-four criteria and more than 300 indicators. The structure of these standards is similar to other systems, such as that used by the Forest Stewardship Council and the Marine Stewardship Council. The European Wilderness Network currently includes forty-one areas across Europe, representing various biomes, including rivers, forests, coasts, and islands (Table 3.1).

There is potential to enlarge the EWN in the coming years, for example through many areas that are currently under a wilderness-focused restoration process. In this process, EWQAS provides an important, guiding protocol and framework for unified standards of quality across these new potential wilderness areas.

TABLE 3.1 European Wilderness Network Accredited Areas (Source: European Wilderness Society in 2022)

Name of Protected Area	Country	Total Area (ha)	Total Area (acres)
WILDERNESS			
1. Borjomi-Kharagauli	Georgia	50,325	124,356
2. Central Balkan	Bulgaria	20,019	49,469
3. Čepkeliai	Lithuania	13,004	32,134
4. Chornyy Dil	Ukraine	1,580	3,904
5. Darvaika	Ukraine	1,450	3,583
6. Fulufjället	Sweden	19,000	46,950
7. Gorgany	Ukraine	4,800	11,861
8. Hohe Tauern	Austria	6,728	16,625
9. Hoverla	Ukraine	4,477	11,063
10. Kalkalpen	Austria	13,034	32,208
11. Konigsbrücker Heide	Germany	5,600	13,838
12. Küre	Turkey	26,162	64,648
13. Majella	Italy	25,895	63,988
14. Maramarosh	Ukraine	3,162	7,813
15. Oulanka	Finland	12,924	31,936
16. Paanajärvi	Russia	30,000	74,132
17. Peneda-Geres	Portugal	5,000	12,355
18. Podyjí	Czech Republic	254	628
19. Retezat	Romania	10,440	25,798
20. Rila	Bulgaria	16,222	40,085
21. Soomaa	Estonia	11,530	28,491
22. Teiču	Latvia	4,570	11,293
23. Uholka-Shyrokyy Luh	Ukraine	7,117	17,586
24. Zacharovanyy Kray	Ukraine	1,332	3,291
WILDCOAST			
25. Archipelago	Finland	10,600	26,193
26. Jasmund	Germany	806	1,992
WILDFOREST			
27. Hainich	Germany	1,570	3,880
28. Kalkalpen	Austria	5,251	12,976
29. Kuziy-Trybushany	Ukraine	5,251	12,976
30. Thayatal	Austria	250	618
31. Uholka-Shyrokyy Luh	Ukraine	7,117	17,586
32. Zacharovanyy Kray	Ukraine	1,258	3,109

(continued)

TABLE 3.1 European Wilderness Network Accredited Areas (Source: European Wilderness Society in 2022) (*continued*)

Name of Protected Area	Country	Total Area (ha)	Total Area (acres)
WILDISLAND			
33. Vilm	Germany	171	423
WILDRIVER		Total Length (km)	Total Length (miles)
34. Dyje	Czech Republic	2.7	1.7
35. Kitka	Finland	16	9.9
36. Mala Uholka	Ukraine	7	4.3
37. Olanga	Russia	36	22.4
38. Oulanka	Finland	26	16.2
39. Shyrokyy Luh	Ukraine	9	5.6
40. Thaya	Austria	2	1.2
41. Velyka Uholka	Ukraine	8	5.0

THE INTERNATIONAL UNION FOR CONSERVATION OF NATURE (IUCN)

The IUCN (once known for a short period as the World Conservation Union) was created in 1948 and as of 2022 has 1,700 members comprised of sovereign states, governmental agencies, and nongovernmental organizations (NGOs), and affiliates.

IUCN's purpose is to influence, encourage, and help societies throughout the world to conserve the integrity and diversity of nature and to ensure that any use of natural resources is equitable and ecologically sustainable. IUCN works with agencies in individual nations as well as various international bodies, such as the United Nations Environmental Program (UNEP), Food and Agricultural Organization (FAO), and United Nations Educational, Scientific & Cultural Organization (UNESCO).

The IUCN is administered through its secretariat in Switzerland, has numerous regional offices, and performs its conservation work primarily through seven commissions. The commission with the most relevance to wilderness is the World Commission on Protected Areas (WCPA), which undertakes a variety of responsibilities in aiding nations, particularly developing nations, in planning of protected areas. WCPA provides strategic advice to policy makers, guidance to protected area managers, and advice to the public, corporate donors, and governments on the value of protected areas, to increase investment.

Members and the Steering Committee of the WCPA support the compilation of global protected area data, and the following institutions or sources of information can provide a comprehensive, frequently updated, global view of protected nature. WCPA and its members co-founded the World Conservation Monitoring Centre (WCMC), administered by the United Nations Environment Programme (UNEP) and located in Cambridge (UK), which provides detailed computer-based analyses of the status of protected areas around the world through its World Database on Protected Areas that is updated monthly. More recently, WCPA launched the Protected Planet website, which provides information on protected nature under eight themes: protected areas (the World Database on Protected Areas); management effectiveness; other effective conservation-based measures (OECMs); territories and areas conserved by Indigenous peoples and local communities (ICCAs); the IUCN Green List of Protected and Conserved Areas; connectivity; equitable governance, and marine protected areas. The WCPA also assists with the UN List of Protected Areas. Finally, the WCPA takes primary responsibility for organizing the World Parks Congress held about every ten years in various locations around the world.

Wilderness was not always recognized as a conservation category by the IUCN. Following a resolution adopted at the 3rd World Wilderness Congress in Scotland in 1983, a recommendation was passed at the IUCN General Assembly meeting in Madrid in 1984, requesting that a specific category—wilderness—be added to the IUCN protected area category list. This request assumed that formal international recognition of wilderness as a conservation category would lend important credibility to wilderness protection around the world. Wilderness was included as a Category 1 protected area when the revised IUCN conservation categories were finally approved in 1992. The six categories are presented under the Guidelines for Applying Protected Area Categories, in which Category 1 includes two types of protected areas: strict nature reserves, largely for science purposes only (Category 1a), and wilderness areas (Category 1b).

IUCN protected area management Category 1b areas are generally large-scale sites in which ecological processes can function with minimal human disturbance. These sites are defined as, "Protected areas that are usually large, unmodified or slightly modified areas, retaining their natural character and influence, without permanent or significant human habitation, which are protected and managed so as to preserve their natural condition" (Dudley 2013, 14).

In 2002, in a move to reinforce wilderness in the WCPA's Guidelines for Protected Areas, a temporary Wilderness Task Force (WTF) was officially established within the WCPA.

This Task Force was quickly elevated to a permanent Specialist Group (WSG) within WCPA, founded by the WILD Foundation, thus creating a permanent institutional voice for wilderness within the WCPA and the global environmental institution framework. The WSG initially led a series of workshops at the World Parks Congress in Durban, South Africa, in 2003; created a web portal for global wilderness professionals connected to IUCN; successfully maintained the integrity of Category 1b during the 2007–2008 process of revising the Guidelines for Protected Areas; and, in 2016, produced management criteria for the wilderness category (Casson et al. 2016).

WILDERNESS PROTECTED AREAS: GUIDELINES FOR MANAGEMENT OF IUCN CATEGORY 1B WILDERNESS

Casson et al. (2016) established the first, and to date, only global guidelines for understanding and managing formally designated wilderness areas. The Guidelines begin by acknowledging that the term "wilderness" is used in a variety of ways. It is a biological descriptor, referring to places that are mainly ecologically intact. It is a type of protected-area classification, referring to a category of protected areas that seeks to maintain wilderness quality over time, while still allowing for human uses that are compatible with those wilderness qualities. It is also used to describe an essential dimension of human culture, which is that humans, like all other species, originated in the wilderness: they evolved for millions of years in caves, trees, and open savannahs (Martin and Robles Gil 2009).

The Guidelines also address management principles; governance and authority; management tools and issues; and evaluating effectiveness. In addition, the Guidelines—produced by forty volunteer experts from around the world—emphasize that *wilderness areas do not exclude humans . . . they exclude certain human uses*. In addition, *the Guidelines emphasize that wilderness areas recognize, promote, and insist upon the inclusion of Indigenous Peoples and Local Communities (IPLC) as full partners (not just stakeholders) in wilderness designation and management as codified under the UN Declaration of the Rights of Indigenous Peoples (UNDRIP).* Overall, the Guidelines present tools and techniques for wilderness management, while also providing insights into why and how wilderness areas are an important part of the international protected area estate. They increasingly serve as a priority solution to climate change, species extinction, and supporting/partnering with IPLC that live in or adjacent to the wilderness.

Further, and to amplify the above points about arguments used by opponents of wilderness, the authors of the Guidelines offer two specific wilderness critiques early on, of which wilderness advocates and managers need to be aware. They also provide information on how to respond when faced with these critiques. First, IPLC resist using the term "wilderness" because of the divide between their nature-based cultures and those of westernized, time-driven cultures, and because the concept of wilderness drives out Indigenous inhabitants. The Guidelines demonstrate that IPLC are not a holistic bloc, and there are clear examples in all regions of the world of use of the wilderness concept by IPLC. Second, the "neo-greens" argue against using the term "wilderness," rejecting it as postmodern and deconstructionist because they believe "there really is no such thing as wilderness." This argument insists that wilderness ignores local people and that human impacts such as climate change and pollution are globally pervasive and therefore there are no "pristine" areas left on earth. Such an argument is unsupported by conservation science, and one also has simply to note that the word "pristine" is absent from formal definitions of wilderness, especially that of the IUCN.

Finally, the Guidelines also demonstrate that, in the absence of corruption and illegal uses, wilderness areas can support human-rights-based approaches to creating a world in which humans and nature coexist and are mutually reinforcing and reciprocal.

INTERNATIONAL INSTITUTION SUPPORT FOR WILDERNESS

In the U.S., wilderness is well defined in the Wilderness Act of 1964 and has a rich history of research and policy decisions to guide management by the four federal land management agencies (Forest Service, National Park Service, Bureau of Land Management, and the U.S. Fish & Wildlife Service). However, "wilderness" in the U.S. and internationally can also have a broader official definition. As noted earlier, the IUCN protected area Category 1b (wilderness area) is defined as "usually large unmodified or slightly modified areas, retaining their natural character and influence, without permanent or significant human habitation, protected and managed to preserve their natural condition." Wild nature (or "wildlands"), while not being formally designated wilderness, is still supported by international institutions.

The World Bank conveyed international recognition for wilderness in a 1986 policy statement explaining how certain pristine areas yield more benefits to present and future generations if maintained in their natural state (Goodland 1987). To protect these long-term values, the World Bank announced that it would usually decline to finance development projects in such areas. If development in wildlands is to occur, careful justification is required and compensation (in the form of financing) to preserve an ecologically similar area or some other mitigating measure must be undertaken (World Bank 1986).

The establishment by the World Bank in 1990 of the Global Environmental Facility (prompted by the work of and a resolution approved at the 4th World Wilderness Congress) created one of the largest sources of intergovernmental funding to address biodiversity protection, often in wildlands and seas. Further progress is the trend among global financial experts and think tanks to monetize the values of ecosystem services provided by wild nature, and therefore determine the financial cost of their destruction (Johnson et al. 2021). When funding for such destruction is provided by governments, this genre of finance is now regarded as "perverse subsidies" (Myers 1998). Most recently, the UK government commissioned a two-year study specifically focused on the urgent need to change financial models to address the state of nature, with special interest in reallocating subsidies to instead protect nature see, for instance, (Sylven and Martin 2020). The Dasgupta Review (Dasgupta 2021) directly addressed the futility and danger of measuring economic growth and prosperity without including the value and destruction of the essential natural processes produced by wild nature, upon which all life depends.

Given this progress, it is important to recognize that adherence to these new policies and trends in the mainstream financial sectors is, however, uneven and challenging. According to the Forest and Finance Policy Assessment (2022), which researched 200 financial institutions, since the Paris Agreement was signed in 2015, banks have provided USD 267 billion in credit (from 2016 to September 2022) to 300 forest-risk commodity companies operating in the world's three largest high-functioning, tropical forest regions: Southeast Asia, Central and West Africa, and parts of South America. Referring to destructive land-transformation projects, the report summarizes the situation as "the finance sector is pouring money in at an alarming rate."

General, international, natural-area designations exist and are widely used. While not based per se on designated wilderness, most of them interact by existing adjacent to or overlapping with wilderness. These types of designations exist as Conventions under the auspices of United Nations Educational, Scientific and Cultural Organization (UNESCO): Biosphere Reserves (multi-use sites including a core zone much akin to wilderness); Ramsar Sites (wetlands); and World Heritage Areas (specifically "natural" sites that have implications for wilderness) (Kormos et al. 2016). These areas are not under any formal legal statute, but they are rather prestigious global designations adopted by nations after strict review and approval by the relevant Convention's governing body. These designations, especially the Natural World Heritage one, can be used effectively in wilderness campaigns by helping build public understanding of their local and global importance (Fig. 3.3). Perhaps the most famous such matter is that of the South West Wilderness in Tasmania (declared a World Heritage Area in 1982), one of the largest temperate wilderness areas in the world and one of the very first wilderness campaigns that assumed global status when it was presented in plenary at the 2nd World Wilderness Congress in Australia, in 1980 (Brown 1982). The South West Wilderness has been an ongoing issue, and as recently as 2016 the Australian and Tasmanian governments tried to remove official status from this wilderness to allow logging. However, rejection of this action by the World Heritage Convention was a deciding factor in making the government abandon its plan (Glenday 2014).

FIGURE 3.3 World heritage site status is given to areas possessing outstanding natural qualities, exceptional natural beauty, or threatened species of plants or animals. The Daintree area of northern Queensland, Australia, typifies a rain forest setting. Photo courtesy Ralph Lindsey.

In the United States, several biosphere reserves overlap with designated wilderness areas, such as the Noatak Wilderness in the Noatak National Preserve and the Aleutian Islands Wildernesses in Alaska, Everglades National Park Wilderness in Florida, and Three Sisters Wilderness in Oregon. The wilderness cores of these biosphere reserves provide a critical baseline for helping determine the extent of human influence on the surrounding biotic environment and are a source for understanding how natural processes can accommodate human impacts. As public awareness grows about the critical role wilderness plays in improving understanding of our environment, the security of such areas should increase.

DIFFERENT TYPES OF RECOGNITION AND PROTECTION OF WILDERNESS: A GLOBAL REVIEW

It is important to recognize some inherent confusion and overlap in the many ways that the term "wilderness" is used internationally. The status field in the World Database on Protected Areas (WDPA) provides information on whether a protected area has been established or designated, at the time the data were submitted. According to UNEP-WCMC (2017), designation means it is recognized or dedicated through legal means and implies a specific binding commitment to conservation in the long term. In this chapter we focus on some of those areas designated since 2008. Actual management plans were usually found for designated areas, but not all. While it is not entirely clear, the likely case is that many Category 1b areas are "recognized" as wilderness but are without a formal management plan. First, we consider an overview of Category 1b, and then look at another suggested type of general classification.

IUCN PROTECTED AREA CATEGORY 1B—WILDERNESS (AN INTERNATIONAL PERSPECTIVE)

Since 2008 and the most recent, former edition of this book, a total of 1,083 designated entries for Category 1b areas have been added to the WDPA (UCN and UNEP-WCMC 2022), with area, ownership, year designated, and type of protection indicated. There were also ten "established" (nine in Canada and one in Antigua and Barbuda). For the 1b areas added since 2008, there are eight administered by collaborative governments (all in Canada), one indicated to be in multiple ownership but federally or ministry controlled in Austria, 878 federal areas, twelve government-delegated areas, two Indigenous areas in Canada (Wehexlaxodıale Land Use Exclusion Zone through an Indigenous Land Use Plan, and Thaidene Nëné Territorial Protected Area designated in 2019 in the Northwest Territories with collaborative governance), one with individual landowners (Tierra Silvestre Canon del Diablo in Mexico), five in joint governance (Australia, Costa Rica, and Finland [Private Nature Reserve]), one local community in the Marshall Islands, six non-profits, three where jurisdiction is not reported (in Brazil and the Dominican Republic), and 175 sub-national areas (states or provinces, mostly in Canada, but also in Australia and Malaysia). In the U.S., since 2008, there were 163 new designation entries (101 BLM, 34 FS, 5 NPS, and 23 FWS). The number of "new" Category 1b classifications is a topic deserving of additional study. We do not fully understand how many are simply additions to existing 1b areas, how many are re-classification of areas within other protection designations, who submits these areas for addition to the WDPA for each country, or whether there is consistency in reporting, the screening process, and the implications for management or future designations. But we do see a trend of recognizing the wilderness qualities of these places and growing designations across the world. We also do not know how much the IUCN Category 1b guidelines are in use to guide designation and protection of these areas.

Since 2008, many countries have reported new designations of protected areas meeting Category 1b criteria. These include Argentina, Australia, Austria, Burundi, Bosnia and Herzegovina, Brazil, Canada, China, Costa Rica, Denmark, Dominican Republic, Spain, Estonia, Finland, France, Haiti, Japan, Lebanon, Sri Lanka, Luxembourg, Mexico, Marshall Islands, Mongolia, Malaysia, Niger, Norway, Oman, Philippines, Poland, Portugal, Russia, Slovenia, Sweden, Tanzania, USA, and Uzbekistan. While many of these countries do not have official intranational systems of wilderness recognition, or official processes to designate areas (both of which do exist in several countries), they are protecting areas that they report do meet the criteria of IUCN Category 1b and their specific information have been added to the WDPA.

THREE OBSERVED CLASSES OF WILDERNESS PROTECTION (AN INTRANATIONAL PERSPECTIVE)

Following is a summary of some of the de jure (existing in law) and de facto (existing in fact) protection of wilderness areas and values in countries where such protection exists. A useful way to view this spectrum would be in three classes (as suggested by Martin 1994):

Class I wilderness includes statutory or legal protection of wilderness areas, usually with management plans. Eight nations including the United States have legislatively designated areas that protect wilderness attributes: Australia, Canada, Finland, South Africa, Sri Lanka, the Flathead Indian Reservation (a sovereign nation with the authority to self-govern within the United States), and Russia (technically, protected nature reserves with protection of wilderness attributes).

Class II wilderness includes recognition and protection of wilderness through administrative zoning as found in several countries, such as New Zealand, Zimbabwe, and Italy.

Class III wilderness includes wilderness recognition and protection in resource-management programs in which use of the concept *wilderness* is a key component in the overall conservation effort, such as in Namibia and the Philippines.

The following sections provide information obtained from some of the countries that demonstrate the characteristics of each class of the three classes listed above and some details about the history of their recognition system.

Class I—Statutory (Legal) Protection of Wilderness

Finland

Commercial forestry in large continuous areas of wildlands and the need to protect traditional means of livelihood in northern Finland stimulated a wilderness movement that began in the 1950s (Sippola 2000; Kajala 2008). Finland's Legislative Act on Wilderness Reserves, however, did not pass until 1991, providing statutory protection for about 3.75 million acres (1.5 million ha) in twelve units, about 5 percent of the area of Finland and all north of the Arctic Circle (Kajala 2008). No additional areas have been designated under this law.

The Finnish word that is translated as wilderness (*erämaa*) is defined culturally and historically different from wilderness in the English language. In Finland, some of the areas protected as wilderness have always been part of some people's everyday life for hunting and fishing (Saarinen 1998; Sippola 2000).

These areas in Finland were legislatively established to preserve their wilderness character, to protect the native Sami culture, maintain traditional subsistence uses of the areas, and enhance possibilities for multiple use of nature. All Finnish wilderness areas belong to IUCN protected area Category 1b: Wilderness. To be considered for inclusion, areas had to be a minimum of 37,000 acres (15,000 ha), over six miles across, and mostly roadless and uninhabited. The smallest Finnish wilderness area, Tsarmitunturi, is 37,000 acres (15,000 ha), and the largest, Kaldoaivi, is 726,500 acres (294,000 ha) (Kajala 2008). In addition, there are, in total, 334 Category 1b areas in Finland, with 112 designated since 2008. Of those designated since 2008, all are either in Private Nature Reserves or State Nature Reserves (either designated by Metsähallitus or Regional Environmental Centers).

Additional designation of wilderness lands under *Finland's Legislative Act on Wilderness Reserves* is not anticipated, although the Finnish wilderness legislation does state that if the national government acquires private inholdings within a designated wilderness reserve, that land area shall become part of the wilderness. In addition to national wilderness areas, Finnish Lapland has about 2.5 million acres (1 million ha) of protected natural areas that are valued for their wilderness character. Human activities are even more restricted in these areas than they are in wilderness areas, including within national parks, where a common management zone was once called a "wilderness zone," but recently this term has been changed to "remote zone." These are typically low-use recreation zones with little regulation in effect (Kajala 2008).

Protecting traditional relationships with nature and Indigenous Sami culture is an important aspect of wilderness in Finland (Pietikäinen 1995). Roads, mining, and large-scale tourism are forbidden, but small-scale forestry was allowed, with important restrictions, in some parts of five of the twelve wilderness units. However, Metsähallitus decided by the year 2003 that no forestry will be practiced in wilderness areas and before that, logging had only occurred in one wilderness area, Hammastunturi.

The wilderness areas are administered by Metsähallitus, which governs most state-owned lands. Recreation and tourism associated with the wilderness areas and national parks of northern Finland have become increasingly important. Saarinen (1998) found that the most important reasons growing numbers of tourists visit some northern Finland sites include opportunities to see natural landscapes and wilderness and enjoy peace and quiet. Wilderness areas provide open, reservable and rentable huts, marked hiking trails, and snowmobile routes. However, to protect their pristine character, the largest parts of the wilderness areas do not have such facilities and services.

The Finnish people have protected many areas under a variety of designations, including wilderness (Table 3.2).

Australia

Having one of the lowest population densities of any settled country, Australia is also highly urbanized. Hence, a large part of the country is sparsely populated, although extensive areas show the influence of rangeland agriculture, logging and mining.

In 1944, the 3.2-million-acre (1.3-million-ha) Kosciuszko National Park in New South Wales was designated with the provision that up to 10 percent of the park be set aside as a primitive area. In 1962 this designation prevented the construction of hydroelectric power works in the higher elevations of the Snowy Mountains. From the late 1960s, wilderness issues were prominently featured in conservation campaigns opposing hydroelectric schemes in Western Tasmania and native forest logging in several states.

A series of inventories of Australia's wilderness resources began in the mid-1970s (Helman et al. 1976), covering parts of NSW, Queensland, Victoria, and Tasmania. Lesslie and Taylor (1985) developed a procedure to inventory wilderness quality as a continuum, characterized by remoteness and primitiveness. The survey was extended to nearly all of terrestrial Australia as the National Wilderness Inventory (NWI) (Lesslie and Maslen 1995). Little progress has been made since then, probably because the wilderness concept represented an ongoing constraint on development, was not acceptable to some sections of the Indigenous community, and there was a shift in emphasis to biodiversity conservation (Sawyer 2015). The NWI was not subsequently updated, and the Commonwealth Wilderness Programme under which it operated finished in 1999. However, variants of the NWI method have been used to assess wilderness qualities in several other parts of the world (Hawes et al. 2018) and a revised version of the NWI was prepared and remains in use in Tasmania (Hawes et al. 2015).

TABLE 3.2 Protected Areas and Wilderness in Finland (Metsähallitus, Parks & Wildlife Finland, December 2022)

Protected Area	Number	Total Area (ha)	Total Area (acres)
National Parks	41	1,014,619	2,507,178
Strict Nature Reserves	19	153,998	380,537
Other Nature Reserves	1,186	921,897	2,278,057
Privately Owned Nature Reserves	13,142	407,624	1,007,260
Wilderness Areas	12	1,489,310	3,680,165

Each of Australia's eight states and mainland territories holds principal authority over the management of its natural resources, including park and conservation management. Thus, the spread of the national parks idea and, later, that of wilderness was driven by regional rather than national issues and interests. In Australia today, two states, New South Wales and South Australia, have wilderness legislation. Most other states and territories have wilderness provisions within their general nature conservation legislation or within their parks and reserves legislation, although definitions of what constitutes wilderness varies.

Since the 1990s, the Australian federal government has taken a predominantly antipathetical approach to wilderness zoning on reserves partly or entirely under the control of its agencies. For example, the large wilderness zone in Kakadu National Park was removed, as was the marine wilderness zone of the northernmost part of the Great Barrier Reef, although both remain conservation reserves.

The *Environment Protection and Biodiversity Conservation Act 1999* is the main piece of national legislation that deals with environmental conservation, but it makes no mention of wilderness. Regulations under this Act include management provisions for IUCN Category 1b Wilderness, but in the absence of any such designated reserves under federal government control they remain unapplied. The World Heritage Properties Conservation Act 1983 recognizes Australia's international treaty obligations to protect World Heritage-listed areas, and federal guidelines recognize wilderness among the natural heritage values of World Heritage areas. However, the actual identification of wilderness is determined on a case-by-case basis in the reservation and planning processes conducted largely by the state governments.

Since much of Australia's unprotected wilderness is located on Aboriginal land, its protection depends on harmonizing non-Indigenous and Indigenous perspectives (Muir 2007). Some members of the Australian Indigenous community are supportive of the wilderness concept. Others are antagonistic, maintaining a contested view that wilderness implies a lack of past human activity and precludes traditional rights. The Malimup Communiqué (Commonwealth of Australia 1998) included principles for Indigenous use of wilderness but has not seen wilderness protection extended across northern Australia. Most Australian states are developing a co-management approach to protected area management.

The Collaborative Australian Protected Area Database [CAPAD] compiled by the Commonwealth of Australia (2021) (Table 3.3) indicates the country has 66 IUCN Category 1b protected areas, totalling 9.5 million acres (3.85 million ha), although categorization as 1b does not automatically confer protection of wilderness values. Twenty-five areas have been designated since 2008 (UCN and UNEP-WCMC 2022). However, Australia lacks a nationally consistent approach to assigning IUCN protected area categories, which are compiled by various state-based agencies. This information is then provided uncritically for the World Database on Protected Areas (WDPA). Hence, the information listed in the

TABLE 3.3 IUCN Protected Area Category 1b Areas in Australia, by State and Territory (CAPAD 2021)

State or Territory	Areas with Category 1b Classification	Total Area (ha)	Total Area (acres)
Australian Capital Territory	1 wilderness zone	28,793	71,149
New South Wales	27 wilderness areas	1,762,383	4,354,937
Northern Territory	0	0	0
Queensland	1	4	10
South Australia	14 wilderness areas	1,850,067	4,571,608
Tasmania	5	4,884	12,069
Victoria	16 wilderness parks or zones	200,233	494,786
Western Australia	2 wilderness zones	1,281	3,165
Commonwealth of Australia	0	0	0

WDPA for Australia, historically and now, at least for Category 1b areas, gives a poor and inconsistent indication of the extent of protected wilderness in the country.

Despite high public support for protecting Australia's remaining wilderness (Fig. 3.4) (Hawes and Dixon 2021), the achievement of this continues to be hampered by limited political support, a lack of wilderness legislation at a national level, and a lack of current data on the extent and quality of the country's wilderness. Ongoing physical threats to wilderness include land clearance, logging, fracking, dam construction, and tourism development. Additionally, climate change poses a major threat to the ecological integrity and biodiversity of many wilderness areas.

Canada

A comprehensive appraisal of wilderness protection in Canada (Hummel 1989) called for a coordinated effort at provincial, territorial, and federal levels to secure legal protection of at least 12 percent of the nation's landscape by the turn of the century. Canada now intends to protect 25 percent of the terrestrial environment by 2025 and 30 percent by 2030 (Environment and Climate Change Canada [ECCC] 2022). The protected areas will include national/provincial/territorial parks, Indigenous protected and conserved areas, national wildlife areas, migratory bird sanctuaries, and other effective area-based conservation measures. Canada is the second largest country in the world and as of 2021 had about 12.6 percent of its land and freshwater in protected areas (ECCC 2022).

FIGURE 3.4 The Valley of the Monoliths, part of the 70,000-acre (28,000-ha) wilderness located within the Budawang Range, in Morton National Park, New South Wales, Australia. Although recreational use is permitted here, overnight camping is prohibited because of the highly erodible terrain. Photo courtesy Rob Jung.

Table 3.4 summarizes the areal extent of Canadian IUCN Category 1b Wilderness that is protected under national, provincial, territorial, and Indigenous designations. Table 3.5 represents knowledge about wilderness designation statuses published in 2008. These numbers are not directly comparable, but they do indicate accelerated interest in designation of areas (likely already protected at some level) protected for wilderness attributes, most beyond the intranational wilderness designation system. Since 2008, WDPA entries suggest that 157 new areas have been designated in Category 1b in Canada, mostly at the provincial level, with a few in collaboration with non-profit organizations and two Indigenous community protected areas (UCN and UNEP-WCMC 2022).

In public opinion polls, Canadians have expressed strong support for the protection of wilderness lands (Rutledge and Vold 1995, Hummel 1989, Parks Canada Agency 2000). There was almost unanimous support for the government of Canada to take actions to protect wilderness, and 83 percent believed it was important for Canada to be seen as an international leader in protecting wilderness (Parks Canada Agency 2000). While wild landscapes remain vital, the protection of wilderness (without an integral human component) as a primary objective for protected areas has been eclipsed by objectives for biodiversity, climate change mitigation, and contributing to Indigenous Peoples conservation priorities (Maxwell et al. 2020, Fletcher et al. 2021, No'kmaq et al. 2021, Hebblewhite et al. 2022). Public support for protected areas remains strong (Fig. 3.5), as is the support for achieving Canada's protected areas targets (Wright et al. 2019).

TABLE 3.4 IUCN Protected Area Category 1b (Wilderness) Areas in Canada (ECCC 2021)

Province/Territory	Authority	Total Area (ha)	Total Area (acres)
Alberta	Alberta	3,771,492	9,319,560
	Canadian Wildlife Service	10,798	26,682
	Parks Canada	1,784,900	4,410,584
British Columbia	British Columbia	3,624,801	8,957,078
	Parks Canada	261,400	645,933
Manitoba	Manitoba	3,438,513	8,496,753
New Brunswick	New Brunswick	17,570	43,416
	Parks Canada	18,400	45,467
Newfoundland & Labrador	Newfoundland & Labrador	398,268	984,142
Nova Scotia	Nova Scotia	424,664	1,049,368
Nunavut	Nunavut	3,494,459	8,634,996
	Canadian Wildlife Service	6,808,419	16,823,970
Northwest Territories	Northwest Territories	2,127,009	5,255,954
	Indigenous Protected & Conserved Area	2,183,988	5,396,752
	Canadian Wildlife Service	3,509,817	8,672,947
	Parks Canada	30,026,000	74,195,862
Ontario	Ontario	4,801,690	11,865,234
	Canadian Wildlife Service	21,277	52,577
	Parks Canada	217,308	536,980
Quebec	Quebec	781,374	1,930,817
Saskatchewan	Saskatchewan	2,054,220	5,076,088
Yukon	Yukon	520,320	1,285,739
	Parks Canada	325,900	805,316
Total		70,622,587	174,512,213

TABLE 3.5 Wilderness Lands by Province in Canada (McNamee 2008)

Province or Authority	Number of Units	Total Area (ha)	Total Area (acres)
British Columbia	1 Wilderness Area	19,000	46,930
Alberta	3 Wilderness Areas	100,800	248,976
Newfoundland/Labrador	2 Wilderness Areas	396,500	979,355
Ontario	9 Wilderness Parks	4,823,745	11,914,650
Nova Scotia	31 Wilderness Areas	285,000	703,950
Parks Canada	2 Wilderness Areas	85,000	209,950
Total	48 Wilderness Units	5,710,045	14,103,811

FIGURE 3.5 Much of Canada's far north holds outstanding opportunities for wilderness recreation. Here a kayaker enjoys the solitude and natural landscape of Pelly Bay. Photo courtesy Department of Economic Development and Tourism, Northwest Territories, Canada.

Indigenous Peoples are a growing force for establishing protected areas in Canada. Indigenous Protected and Conserved Areas (IPCA) are lands and waters where Indigenous Peoples have the primary role in identifying and designating their location and purpose. Some IPCAs are co-designated with federal, provincial, or territorial governments and legislation, while others are designated under Indigenous laws, governance, and knowledge systems. IPCA establishment, governance, and management objectives share three elements: being Indigenous-led, having a long-term commitment to conservation, and elevating of Indigenous rights and responsibilities (Indigenous Circle of Experts 2018). Ivvavik National Park (the majority of which is zoned as wilderness) was Canada's first IPCA and created through the Inuvialuit Final Agreement in 1984. Many IPCAs have been established through Indigenous land claim agreements, national park establishment processes, and in the past five years, five large IPCAs were established in the Northwest Territories, British Columbia, and Alberta.

Transboundary protection of wilderness areas is in place between provinces, provinces and territories, territories, and Canada and the U.S. Some examples include the protection of Manitoba-Ontario Interprovincial Wilderness Area, Tatshenshini-Alsek Provincial Park, Kluane National Park, and Glacier Bay and Wrangell-St. Elias National Park and Preserves; Torngat Mountains National Park, Kuururjuaq Quebec National Park, and Waterton-Glacier International Peace Park; and Ivvavik National Park, Vuntut National Park, and Arctic National Wildlife Area.

Seven of the ten Canadian provinces have legislation to designate wilderness protected areas and/or to support wilderness management zones within protected areas: Alberta, British Columbia, Manitoba, Newfoundland and Labrador, Nova Scotia, Ontario, and Saskatchewan. The three territories, Quebec, and New Brunswick do not have wilderness legislation, but each has IUCN Category 1b areas identified (Table 3.4).

At the federal level, the Canada's National Parks Act was amended in 1988 to include provision for wilderness designation (McNamee 2008). In 2000, Parks Canada declared its first wilderness areas in Banff, Jasper, Kootenay, and Yoho (McNamee 2008) using the National Parks Wilderness Area Declaration Regulation. As of 2022, eight national parks have legislated wilderness areas. In addition, through each national park management plan a large proportion of land and water are zoned wilderness. Some migratory bird sanctuaries and national wildlife areas have been identified as Category 1b Wilderness by the Canadian Wildlife Service because they are large unmodified areas that retain their natural character, lack permanent habitation, and are managed to preserve their natural condition, but the legislation for these protected areas does not identify wilderness as a management objective.

Currently Alberta manages wilderness areas established under three legislations: the Willmore Wilderness Park Act (1962), the Provincial Parks Act (1964), and the Wilderness Areas, Ecological Reserves, Natural Areas and Heritage Rangelands Act (2000). Historically, other legislation was used: the Wilderness Provincial Park Act (1959), Forest Reserves Act (1961), Public Lands Act (1965), and Wilderness Areas Act (1971).

British Columbia now establishes wilderness parks/areas through the Protected Areas of British Columbia Act (2022). This Act consolidates legislation still used for designating and managing protected areas: the Parks Act and Ecological Reserve Act. British Columbia has been a leader by establishing large, protected areas since the early 1900s, with logging, mining, and hydro dams considered threats to wild spaces. Public opinion has supported protecting large wilderness areas as a top priority for the province.

In Manitoba, the Provincial Parks Act (1993) includes the designation of a wilderness park with the purpose of preserving representative areas of a natural region and provides for a wilderness land use category within the management plan for any type of park.

Newfoundland and Labrador created its Wilderness and Ecological Reserves Act in 1980 that has an emphasis on ensuring that traditional relationships (hunting, trapping, fishing, and recreational travel) with the land are supported (McNamee 2008).

In Nova Scotia, the Wilderness Areas Protection Act of 1998 provided immediate protection of almost 5 percent of the province's total land base (McNamee 2008). This legislation is unique in a province that is 70 percent privately owned.

Ontario passed a Wilderness Areas Act in 1959 (McNamee 2008). In 1978, the Ontario government recognized wilderness parks as one of six official categories of Ontario provincial parks, and in 2006, a new Provincial Parks and Conservation Reserves Act repealed the old Wilderness Areas Act and made wilderness parks an official category in the province's protected areas system.

In Saskatchewan, the Parks Act (1986) includes a category for the designation of wilderness parks that are to be used primarily for preserving large remote natural landscapes and for low intensity, non-mechanized recreation. The Saskatchewan Protected and Conserved Areas Roadmap (2020) describes the overall goal of conserving and protecting 12 percent of land and water but does not include a specific goal for wilderness.

South Africa

The modern concept of wilderness was introduced to South Africa (SA) through a group of rangers working in remote conserved areas in KwaZulu-Natal during the 1950s. The manner in which they lived and worked had a deep effect on their worldview and this common perception created enduring bonds. It is noteworthy that these bonds sometimes circumvented the legally enforced racial segregation that had plagued the country. One ranger, Jim Feely, circulated an article from the American Wilderness Society, "11 Fundamentals of the Wilderness Concept" (Trippensee 1948). These concepts had a profound effect on fellow ranger Ian Player who, against considerable political opposition, initiated two innovations (Draper 1998).

To firstly prepare himself, Player contacted Howard Zahniser of The Wilderness Society (U.S.)—the unassuming, modest "architect" of the U.S. Wilderness Act—and "Zahnie" mailed him transcripts of the U.S. Senate hearings on the proposed legislation for wilderness. Player read these transcripts around the campfire while on patrol in the iMfolozi, and consigned to memory all the arguments used against wilderness so that he could be prepared for the policy struggle he knew would come in his own country as the wilderness concept advanced (Martin 1983).

Then came the innovations. Under the rationale that the value of wildness would not be appreciated while it remained unknown, Player initiated multiple-day hikes for the public where participants would live simply: camping under the stars or a basic canvas, and traversing the wilderness on trails made by rhinoceros and hippopotamus. Such trips commonly terminated with a group meeting (*indaba* in Zulu; Mutwa 1999) with participants articulating appreciation of the intrinsic and societal values of these "wild places" and their personal experiences while there. Player's Zulu elder and mentor/colleague/friend, Magqubu Ntombela (Fig. 3.6) suggested that a larger *indaba* should be held to expand this understanding, leading directly to what Player then named the first World Wilderness Congress (in Johannesburg, 1977).

The second innovation entailed establishing portions of two conservation areas as wilderness, a designation never-before formally used in South Africa or anywhere else in Africa. These areas have been expanded and continue to this day as the iMfolozi and the Tewati Wilderness Areas (see Table 3.6). Determined to expand wilderness awareness, Player formed the *Wilderness Leadership School*, prioritising selected participants from all walks of life, races, and social levels to promote advocacy for environmental care and wilderness awareness. The symbol of this organisation, selected by Ntombela, was the trifoliate coral tree leaf (signifying the interconnectedness of humanity, spirit, and nature) later adopted by the sister organizations—the WILD Foundation, Wilderness Foundation UK, and the Wilderness Foundation Africa. The advocacy by these groups and the *Wilderness Action Group*, increased the number of SA wilderness areas. Table 3.6 below shows the extent and nature of wilderness areas in SA.

Independent of this introduction of wilderness, the forester Danie Ackerman (then the director) and W.R. "Bill" Bainbridge (then a planner) of the national Forestry Department led the way in amending the 1971 Forest Act to protect forest wilderness, with successive amendments retaining this within the National Forest Act No. 84 of 1998. Interestingly, this Act does not define wilderness, relying, rather, on either common usage of the term or subsequent legal definitions appearing elsewhere in SA law. Bainbridge was primarily instrumental in protecting huge swathes of the Drakensberg mountains as catchment areas. Later much of this was formally designated as wilderness.

FIGURE 3.6 Dr. Ian Player, noted South African and international wilderness advocate, and his friend, Magqubu Ntombela, famous game guard and Zulu tribal chief. Player was instrumental in establishing part of the Umfolozi Game Reserve as wilderness in 1958 and in founding the Wilderness Leadership School (WLS), a nonprofit group dedicated to introducing leaders, especially young people, to wilderness. Photo courtesy Ian Player.

The legal definition of wilderness occurred with the National Environmental Management: Protected Areas Act No. 57 of 2003 (NEMPAA), which denotes their intrinsically wild character, devoid of roads or human habitation. It limits access to unmechanised means and prioritises biodiversity and ecosystem services including opportunities for solitude (Bainbridge and Lax 2008). It describes how state, communal, or private wilderness can be designated directly or zoned within management plans approved by the government minister. To ensure its efficacy, zoned wilderness must have two inherent requisites: (1) the management plan must be current and (2) the management system must withstand scientific rigour. In other words, the process of zoning wilderness must be consultatively lawful and adhere to a defendable peer-reviewed methodology (Cryer 2009).

Commitment to wilderness protection is varied: the increased number of wilderness areas shows promise, but both the preference for zoning wilderness rather than officially designating it, and the tendency for the areas to be smaller in extent, indicates less or diluted commitment to the wilderness concept. SA's environmental policy, especially since the late 1990's, focuses more on social benefit. This, and increasing law-enforcement needs, result in core conservation (including wilderness) being de-prioritized. The mixed enthusiasm is rooted in the historic perception of wilderness as a post-colonial romantic relic (Zealand 2007), resulting in the concept narrowly surviving SA's democratic transition in the 1990s. Early drafts of NEMPAA excluded wilderness as a protected area category. Instead, it was retrofitted into subsequent drafts as an attribute to be applied to protected areas. The survival of the wilderness concept in SA in the face of

TABLE 3.6 Properties of Wilderness Areas in South Africa (Source: Paul Cryer, South Africa)

Name	Reserve/Park or Area in Which It Is Embedded	Management Authority	Total Area (ha)	Total Area (acres)
Kruger National Park Wilderness Complex (15)	Kruger National Park	South African National Parks	814,999	2,013,906
Richtersveld Wilderness Complex (2)	Richtersveld National Park	South African National Parks	67,777	167,481
Wolkberg Wilderness Area	Wolksberg Nature Reserve	Department of Environment, Forestry & Fisheries	22,000	54,363
Imfolozi Wilderness Area	Hluhluwe-iMfolozi Park	Ezemvelo KZN Wildlife	31,711	78,360
Tewati Wilderness Area	Isimangaliso	Greater St Lucia Wetland Park Authority	42,486	104,985
Ithala Wilderness Area	IGRIthala Game Reserve	Ezemvelo KZN Wildlife	6,338	15,662
uKhahlamba Drakensberg Wilderness Complex (4)	uKhahlamba Drakensberg Park	Ezemvelo KZN Wildlife	157,421	388,996
Ntendeka Wilderness Area	Ngome State Forest	Department of Environment, Forestry & Fisheries	5,250	12,973
Groot Swartberg Wilderness Complex (2)	Swartberg Complex	Western Cape Nature Conservation Board	32,437	80,154
East Swartberg Wilderness Complex (2)	Swartberg Complex	Western Cape Nature Conservation Board	16,382	40,481
Kammanassie Wilderness Area	Swartberg Complex	Western Cape Nature Conservation Board	8,136	20,104
Rooiolifantskloof Wilderness Area	Swartberg Complex	Western Cape Nature Conservation Board	5,550	13,714
Paardenberg Wilderness Area	Swartberg Complex	Western Cape Nature Conservation Board	1,522	3,761
Rooiberg Wilderness Area	Swartberg Complex	Western Cape Nature Conservation Board	12,844	31,738
Cederberg Wilderness Complex (3)	Cederberg Complex	Western Cape Nature Conservation Board	49,773	122,992
Groot Winterhoek Wilderness Area	Groot Winterhoek Complex	Western Cape Nature Conservation Board	22,251	54,983
Boosmanbos Wilderness Area	Langeberg Complex	Western Cape Nature Conservation Board	14,000	34,595
Marlroth Nature Reserve Wilderness Area	Langeberg Complex	Western Cape Nature Conservation Board	5,821	14,384
Padderberg Nature Reserve Wilderness Area	Langeberg Complex	Western Cape Nature Conservation Board	560	1,384
Kogelberg Wilderness Area	Kogelberg Nature Reserve	Western Cape Nature Conservation Board	12,858	31,773
Fonteinjtiesberg Wilderness Area	Fonteinjtiesberg Nature Reserve	Western Cape Nature Conservation Board	2,522	6,232

(*continued*)

TABLE 3.6 Properties of Wilderness Areas in South Africa (Source: Paul Cryer, South Africa) (*continued*)

Name	Reserve/Park or Area in Which It Is Embedded	Management Authority	Total Area (ha)	Total Area (acres)
Anysberg Wilderness Complex (2)	Anysberg Nature Reserve	Western Cape Nature Conservation Board	18,621	46,013
De Hoop Wilderness Area	De Hoop Nature Reserve Complex	Western Cape Nature Conservation Board	1,519	3,754
Pilanesberg Wilderness Area	Pilanesberg National Park	Department of Environment, Forestry & Fisheries	11,000	27,182
Baviaanskloof Wilderness Complex (3)	Baviaanskloof World Heritage Site	Eastern Cape Parks & Tourism	84,117	207,858
Total			1,447,895	3,577,828

criticisms (including its exclusion of human habitation) rests with two trends in global and local environmental thinking:

- The acknowledgement of sacred sites to Indigenous/local peoples has increased significance in SA, with many protected areas (and embedded wilderness areas) now being owned by local communities through the land restitution process, which reversed the loss of tenure during apartheid. The idea that certain sacred sites could be protected within a wilderness designation offers new opportunities (Wild and McLeod 2008). The overlap between sacred sites and areas of remote wildness is well understood and could further justify limited access to the community that holds the area as sacred. Such thinking may help convert local critics of wilderness into ardent advocates (Beltrán 2000).
- Within an accentuated awareness of planetary boundaries, averting a biodiversity crisis involves revaluing the complexity inherent in large intact ecosystems. This long-known value of wilderness may induce invigorated support within the country's protected area network.

As a result, current SA wilderness designation prioritizes ecological and Indigenous knowledge systems over simple recreation (a word sometimes associated in SA with elitism) (Fig. 3.7). Even to fatigued urbanites, the value of wilderness is then perceived as benefitting the human spirit rather than just providing fun. This offers challenges in that wilderness management methodologies and language remain structured around recreation. Broader acceptance of wilderness in SA is dependent on incorporating within wilderness management practice this linguistic and conceptual evolution of a society vastly changed since the election of President Mandela in 1994.

Sri Lanka

Sri Lanka is the only South or East Asian country with Class I (legally designated) wilderness. Sri Lanka's 1988 Natural Heritage Wilderness Act provides for the conservation and protection of unique ecosystems, genetic resources, and outstanding natural features by regulating the entry to, and residence within, these areas and the use of their natural resources by local communities. Natural heritage wilderness is created by order of the minister of Lands and Development, but only after consulting the ministers of the Environment, Wildlife Conservation, Fisheries, Agriculture, Cultural Affairs, and Indigenous Medicine. The recommendation must then be approved by the president and confirmed by Parliament. Management of protected areas falls under the Forest Department and Department of Wildlife Conservation.

Two main rationales form the historical basis for the Act. First is the historical need to protect wild landscapes for watershed values; this practice originated with the ancient kings to protect status as an agricultural power some 800 years ago. Singhalese kings declared specific areas as *thahansi kale*, or forbidden forests. These were typically dense, evergreen tropical forests with rivers and streams. Between 1600 and 1800, colonial rulers destroyed much of these areas for tropical hardwood exports. The few that remain, such as the Sinharaja Rainforest, Peak Wilderness, and Horton Plains, are dramatically reduced in size and now protected under the 1998 Natural Heritage Wilderness Act.

The second rationale for the Act is cultural and religious. Buddhist thought and practice has influenced ideas of nature and conservation for more than 2,300 years. Thus, modern Sri Lanka is steeped in the tenets of compassion, protection, and sanctuary for all life. One ancient teaching explained that humanity and all living beings were equal: "Oh great King, the birds of the air and the animals have an equal right to live in this land as thou, the land belongs to the people and all living beings and thou art only the guardian of it." The concept of sanctuary for animals dates back more than 800 years. However, conservation historians remind us that Sri Lanka's several modern protected areas (e.g., Yala, Horton Plains) were originally established by colonial administrators to protect "game" for sportsmen to kill.

Today Sri Lanka's large network of protected areas covers a diverse range of habitats and covers a significant portion of the island. These areas protect a wide variety of terrestrial, aquatic, and marine species and contribute to the thriving ecotourism industry. In fact, Sri Lanka serves as a case study illustrating how long-held cultural beliefs about nature and humans can demonstrate the wide variety of ecological, social, and economic benefits that arise from protecting nature.

FIGURE 3.7 Wilderness in Africa includes some charismatic megafauna—elephant, lion, and rhinoceros—that are considered icons of wilderness. Photos courtesy Ulf Doerner.

Flathead Indian Nation (Confederated Salish and Kootenai Tribes of Montana, U.S.)

There is a link between Indigenous people and nature that is not described or acknowledged in the U.S. Wilderness Act. The Wilderness Act suggests that these protected areas should be "untrammeled," or unmanipulated, unfettered, when in fact it is common knowledge that, for most areas in North America, Indigenous people have intervened, with respect, for generations. The Mission Mountains Tribal Wilderness in Montana, though not part of the National Wilderness Preservation System, was designated to protect many of these same values but also extended to protect important cultural meanings assigned to this wild landscape. The meanings that members of the Confederated Salish and Kootenai Tribes of Montana attach to their (formally designated under Tribal law) Mission Mountains Tribal Wilderness were found to include: (1) protecting nature and culture; (2) functional, emotional, and symbolic attachments; (3) wildlife and watershed protection; and (4) access, beauty, privacy, and recreation (Watson et al. 2011a). Protecting the relationship between Indigenous people and relatively intact, complex systems, which we commonly refer to as wilderness in North America, can be an important contributor to sustainability of the landscape and cultural heritage (Watson et al. 2011a).

The Confederated Salish and Kootenai Tribes live on 1.2 million acres (518,000 ha) in northern Montana, United States, and are a quasi-autonomous jurisdiction under tribal governance. The Mission Mountain Tribal Wilderness Area on the Flathead Indian Reservation in northwest Montana is the only wilderness protected under tribal law found on the more than 300 federally recognized Indian reservations in the United States, and only one of three Category 1b areas designated in the world in the WDPA (UCN and UNEP-WCMC 2022) protected by an Indigenous community.

The Confederated Salish and Kootenai Tribes are the descendants of Salish, Pend d'Oreille, and Kootenai Indians who traditionally occupied a 20-million-acre (8.3-million-ha) area stretching from central Montana to eastern Washington and north into Canada. After the Hellgate Treaty of 1855, most of these ancestral lands were ceded to the U.S. government, with 1.2 million acres (518,000 ha) remaining and now known as the Flathead Indian Reservation (McDonald 1995).

Starting in 1936, there were several attempts to protect the Mission Mountain Range, which is central to the Indians' physical and spiritual geography, but the main effort to protect it came in the mid-1970s in response to a proposal by the Bureau of Indian Affairs to log the mountains. Three greatly respected grandmothers (yayas) raised the initial protest and led the way for other community leaders to organize a "Save the Mission Mountains Committee." About this time, Thurman Trosper (a tribal member, retired U.S. Forest Service and National Park Service employee, and past president of The Wilderness Society) returned home to the Flathead Reservation and proposed the idea of a tribal wilderness. Following extended community action, the Tribal Council in 1982 approved Tribal Ordinance 79A and the Mission Mountains Tribal Wilderness management plan, designating 92,000 acres (38,333 ha) as wilderness. It was the first time that an Indian tribe had taken action to designate wilderness on its own land and to establish policy direction and provide personnel for its protection and management. CSKT Tribal Ordinance 79A states that "Wilderness . . . is the essence of traditional religion and has served Indian people of these Tribes as a place to hunt, as a place to gather medicinal herbs and roots, as a vision seeking ground, as a sanctuary, and in countless ways for thousands of years" (CSKT 2005).

Salish Elder Clarence Woodock (CSKT 2005) described these mountains as lands where the people walked and lived. They are important to elders and for the perpetuation of Indian culture because of stories created and told there. A significant difference between the Tribal Ordinance and the federal Wilderness Act, however, is that Ordinance 79A states unambiguously that a primary purpose of the Mission Mountains Tribal Wilderness is the preservation of Tribal culture, and it acknowledges the importance of wilderness to the perpetuation of traditional Indian religion.

Further management plans ensued for protecting grizzly habitat and for fire management, and a Wilderness Buffer Zone of 23,000 acres (9,600 ha) was later established. The importance of the Mission Mountain Wilderness Buffer Zone (a policy or designation not applied to federal Wilderness in the U.S.) to achieving Wilderness protection purposes has been highlighted in recent research, particularly its role in restoration of fire in the larger ecosystem through buffering the non-wilderness from the wilderness (Watson et al. 2013, Watson et al. 2009).

There are also important differences between managing wilderness on the Flathead Nation and managing federal lands under the U.S. Wilderness Act. The needs and values of tribal members take precedence over those of nontribal members resident on the reservation or visiting. Research by Watson et al. (2011a) demonstrates that protecting this tribal wilderness showcases the wisdom of tribal communities in safeguarding their cultural landscape. This protection extends to off-site benefits such as water resources, wildlife habitats, aesthetic values, and other essential resources.

Russia

Although a concept exists in the Russian language that is analogous to wilderness—*dikaya mestnost* (wild place)—a different word is used for some of Russia's wildest protected places: *zapovedniks*—literally "forbidden areas" (Table 3.7). The Soviet Union established these areas, withdrawn from economic use, to preserve representative natural ecosystems and typical and rare landscapes; to protect gene pools for wildlife; and for research and educational purposes (Pryde 1987).

The first zapovednik, Barguzin Zapovednik, was established as a wildlife preserve to protect a sable population by a regional government in 1916 (Ostergren and Shvarts 2000). In the 1920s, the Ilmen'ski Zapovednik was established to protect its unique geological and mineralogical features, making it the first area in the world protected primarily for scientific purposes (Weiner 1988a). The number of areas expanded until 1951, when 128 areas totaled more than 31 million acres (12 million ha). Then, in 1951, about seven-eighths of the reserves were abruptly eliminated, and it was further reduced by one-third in 1961 (Weiner 1988b). Since 1961, the number of zapovedniks has slowly rebuilt and in 2022 there are 1,019 areas totaling more than 84 million acres (34.5 million ha) (Table 3.8).

Despite the privatization of industries and homes following the fall of the Soviet Union in 1991, the vast majority of land has remained under federal control. Total public control creates, in theory, a promising political framework for all kinds of conservation, including that of protected areas. Within the Russian Federation there are also 64 national parks, 60 national nature refuges, and 17 national nature monuments (Table 3.8).

There are important differences between the Russian zapovednik and Class 1b Wilderness: absence of parliamentary or other public legislative control; lack of consistent, enforced management plans; some use of motorized vehicles and, originally, the lack of people as part of the environment. However, there is enough regard for maintaining "wild" values present in the Russian zapovedniks that they are included here as legally protected wild places.

Despite the original dominant role of nature protection in zapovedniks, their value for human uses has also become acknowledged. In the 1960s and 1970s, growing pressures on several nature reserves forced authorities to consider that the areas may have demand for recreation. The recreational demands contributed in part to the creation of several Soviet National Parks in the 1970s. Some parklands were zoned to include wilderness conditions (e.g., natural conditions, access by foot or horse only, no permanent structures), but unlike zapovedniks the parks also include economic zones for tourism infrastructure and may have had limited logging.

Probably the most significant post-Soviet event for protected areas was the passage of an "organic act" for all agencies in 1995 and the establishment of the Department of Zapovednik Management within the State Committee of Environmental Protection. Historically, zapovedniks were protected by various administrative authorities, even at local and provincial levels. This legislation in 1995 (which emphasized biodiversity protection and long-term ecological monitoring) provided a new sense of direction and legal strength to

TABLE 3.7 Types of Protected Areas in Russia (Ministry of Natural Resources of Russia 2021)

Type of PA (Russian and English)	Meaning	IUCN Equivalent	IUCN Category
Заповедник sounds: zapovednik	strictly protected preserve of federal (national) level only	National Nature Reserve or Zapovednik	1a
Национальный парк sounds: natsionalny park	national park, sometimes closer to U.S. National Forest of recreational areas; federal level only	National Park	2 (but some parks closer to 6 because of natural resources used)
Заказник sounds: zakaznik	nature refuge or sanctuary mostly for ecosystem/species protection and management; can be of federal, regional, municipal or local level	(National) Nature Refuge or (National) Nature Sanctuary	4
Памятник природы sounds: pam'yatnik prirody	nature monument—"the least important" PA with quite "soft" regime usually; can be of federal, regional, municipal or local level	(National) Nature Monument	3

TABLE 3.8 Protected Areas (PA) System in Russia (2020) (Ministry of Natural Resources of Russia 2021)

	Federal (National) Level		Regional Level		Municipal and Local PAs	
Category of Protected Area	**Number of PAs**	**Total Area (million ha)**	**Number of PAs**	**Total area (million ha)**	**Number of PAs**	**Total Area (million ha)**
National Nature Reserves (Zapovedniks)	109	34.5	na	na	na	na
National Parks	64	26.9	na	na	na	na
Nature Park	na	na	106	-	-	-
Nature Refuges	60	13.6	2,422	55.9	-	-
Nature Monuments	17	0.024	7,390	-	-	-
Dendrological parks and Botanic gardens	46	0.005	27	0.002	-	-
others	-	-	494	-	-	-
Total:	296	75	10,439	117.7	1,090	47.5

na = not applicable
- = no data

protect these critical areas in Russia. In addition to clarifying responsibilities and strengthening law enforcement authority, the 1995 law expanded the purpose of zapovedniks to include conservation training for professionals, participation in regional environmental impact assessments, and, very controversially, environmental education that may include ecotourism. After eighty years of traditional exclusion, it became legal to bring people into the territory of zapovedniks for purposes other than ecological research (Ostergren 2001).

Discussions about the possibility or necessity of tourism in the zapovedniks went on for many years, with some managers supporting this idea, while some believed that it could destroy wildlife. In 2011, amendments were adopted to the basic law on protected areas, where the development of tourism has become one of the main tasks of the zapovedniks. In fact, tourism has already become established in and around some particularly attractive protected areas, and since 2011 has become legal.

The impetus behind this change in the law is the popularization of nature conservation in Russia, the extensive system of protected areas, and the formation of very positive public attitudes and support for nature conservation associated with the Protected Areas system. However, this decision was not followed by specific policies and direction, leading many managers to begin to develop recreation services with some impact on natural resources, which caused conflicts with nature conservation interests and the loss of value for some territories. In Russia, there is the need to develop ecological tourism in zapovedniks and use it as a tool for nature conservation, without negative impact to the natural resources of protected areas.

According to Watson et al. (2018), Russia contains more remaining "wilderness" than any other country. Because Russia covers eleven time zones and a wide array of ecosystems (e.g., tundra, taiga, bogs, deciduous forest, steppe, mountains, deserts), there is potential for additional protection of wild lands. Endangered wildlife species are most frequently located in these "wilderness" places, but inadequate funding and staffing, coupled with pressures from competing economic interests (e.g., poaching, food gathering), are continuing problems.

Interestingly, the twelve Russian protected areas listed in the WDPA as Category 1b: Wilderness (six since 2008), are not in the strict nature reserves (zapovednik). Eight are in national parks and the other four are located in State Natural Zakaznik areas (UCN and UNEP-WCMC 2022).

Class II—Administrative Zoning of Wilderness

Some nations protect wilderness by administrative zoning. This protection is not as secure as Class I because administrative rules and regulations are easier to change than statutory protection.

New Zealand

New Zealand protects a wilderness system through administrative designation or zoning under provisions provided in its Reserve Act of 1977, the National Parks Act of 1980, and the Conservation Act of 1987 (Cessford and Dingwall 1997; Somers 2008). Physical conditions, cultural values, and vigorous wilderness advocacy combine to make New Zealand's wilderness have extremely high standards, supported by strict management guidelines.

In 2023, there were eleven designated wilderness areas and two recommended wilderness proposals that totaled around 1.80 million acres (732,000 ha) in New Zealand (Table 3.9).

TABLE 3.9 Wilderness Areas in New Zealand in 2022 (Cessford 2008 and Barnett et al. 2015)

Wilderness Areas	Total Area (ha)	Total Area (acres)	Gazetted (Legislated)
Raukumara	40,000	99,000	1988. Forested highlands, Raukumara Conservation Park.
Ruakituri	24,000	59,500	2006. Remote forested highlands, Te Urewera
Te Tatau-Pounamu	6,500	13,529	1962. Forested subalpine landscapes, Tongariro N.P.
Hauhungatahi	8,500	20,999	1966. Forested subalpine landscapes, Tongariro N.P.
Tasman	87,000	215,000	1988. Forested highlands, Kahurangi N.P.
Paparoa	31,000	76,500	2002. Forested highlands/mountains, Paparoa N.P.
Adams	47,000	116,000	2002. Mountains & valleys, Southern Alps
Hooker-Landsborough*	41,000	101,000	1990. Mountain/valleys, Southern Alps, by Mt Cook N.P.
Olivine*	83,000	205,000	1997. Mountains, valleys, and glaciers, Mt Aspiring N.P.
Pembroke*	18,000	44,500	1974. Forested/alpine mountain lands, Fiordland N.P.
Glaisnock*	125,000	309,000	1974. Forested/alpine mountain lands, Fiordland N.P.
Gazetted Area Subtotal	**511,000**	**1,263,000**	**1.9 percent of New Zealand land**
Proposed Wilderness	**ha**	**acres**	**Area Proposals**
South-West/Cameron*	182,000	450,000	Lowland forest and lakes, Fiordland N.P.
Pegasus	40,000	99,000	Lowland Stewart Island forests/coast, Rakiura N.P.
Proposed Subtotal	**222,000**	**549,000**	**0.8 percent of New Zealand land area**
Wilderness Area Total	**733,000**	**1,811,000**	**2.7 percent of New Zealand land area**

These designated and proposed wilderness areas total about 3 percent of New Zealand's land area. These New Zealand "wildernesses" reflect standards of physical and social conditions that are among the most rigorous in the world and exceed the Category 1b criteria specified by IUCN (1998) for the most protected class of wilderness areas, because IUCN criteria make more allowance for minor human modification and habitation. Indeed, over 30 percent of New Zealand's total area is protected natural public conservation land, with over 20,288,760 acres (8,2000,000 ha) managed by the Department of Conservation for biodiversity conservation and wilderness-type values (Fig. 3.8). This is one of the highest proportions of protected areas of any country in the world (Green and Paine 1997; IUCN 1998). Most of these lands are remote and natural, with high wilderness quality by any standards.

Further, possibly 25 percent of New Zealand's conservation lands outside the designated wilderness areas might

FIGURE 3.8 Most high alpine areas have significant wilderness values in New Zealand and internationally. Some have access by aircraft, the only major intrusion, other than people, to otherwise pristine places. Photo courtesy Gordon Cessford.

qualify as the most remote kind of wilderness under IUCN criteria. But these lands also have a legislative basis for conservation management in New Zealand, and potential pressures on them for development and recreational use are not great. Eighty-five percent of New Zealand's 5 million residents are urbanized coastal dwellers, most concentrated in the upper half of the North Island. Very few live in remote areas and, unlike most other countries of the world, very few have any occupancy options in lands that are managed as protected natural areas.

New Zealand created one of the world's first national parks in 1894 when the volcanoes and lava fields of the central North Island became Tongariro National Park. These lands had been given to the nation in 1887 by its Indigenous Maori guardians, who wanted these sacred mountains enshrined in public protected-area status (Cessford and Reedy 2000).

The arrival of humans introduced new ecological impacts against which the native biota had few natural defenses. In less than one thousand years, it is estimated that humans and their accompanying animals removed 70 percent of the forest cover and drove 32 percent of the Indigenous terrestrial bird species to extinction. Moreover, around a thousand species of Indigenous plants and animals are now threatened, including thirty-seven of the fifty remaining native terrestrial bird species. Thus, retaining biological uniqueness by maintaining natural biodiversity is a high priority for the management of New Zealand's conservation lands.

Wilderness has a very specific meaning in New Zealand conservation management, relating most often to specifically designated "Wilderness Areas" and based most often on recreational experience criteria. Most wilderness advocacy from the earliest days to today has been generated by the Federated Mountain Clubs of New Zealand (FMC), although the earliest concepts were heavily influenced by those developing in the United States (Barnett et al. 2015). In 1939, Lance McCaskill, a prominent conservationist, visited the United States and met Aldo Leopold and, through his influence, introduced the concept of wilderness into New Zealand National Park circles (Cumberland 1981). In 1949, the American conservationist Olaus Murie visited New Zealand and supported the idea of introducing wilderness protection provisions into the pending 1952 national park legislation. This legislation outlined a procedure whereby distinct areas of wilderness could be established in national park settings, and that such areas should be kept and maintained in a state of nature, without any buildings, roads, or tracks, and only foot access permitted.

Although there was general agreement about what wilderness was, there was no consensus or management philosophy. Because the mechanisms for wilderness protection were spread across three different statutes, three different governmental agencies, and two governmental ministries, it was difficult to promote the assessment and protection of wilderness. In 1976, public conservation groups, led by the FMC, began to consider options for wilderness protection (Molloy 1983). Working with three government management agencies, the FMC helped produce a Joint Wilderness Policy in 1980 that outlined the legislative status of wilderness in New Zealand, discussed social and physical criteria for such areas, proposed common definitions with which each agency would work, and identified general guidelines for establishing and managing wilderness.

FIGURE 3.9 Wilderness users need the highest levels of fitness, outdoor skills, and self-sufficiency to access and explore New Zealand's wilderness. Photo courtesy Gordon Cessford.

Following an FMC-sponsored conference on wilderness, in late 1981, the minister of Lands and Forests appointed a Wilderness Advisory Group (WAG) to advise on appropriate policies and action. The WAG reached a consensus based on the earlier Joint Wilderness Policy and provided a succinct statement of objectives for wilderness protection and criteria for wilderness areas, though allowing for flexibility in designation and management. Wilderness would be large enough to take at least two days' foot travel to traverse, have clearly defined topographic boundaries, be adequately buffered from human influences, and have no developments such as huts, tracks (trails), bridges, signs, or mechanized access. The WAG developed specific recommendations for designating ten areas, totaling more than 1.2 million acres (500,000 ha) (Molloy 1983).

Progress has been gradual since the initial provisions for zoning wilderness in the 1952 National Parks Act, followed by the 1977 Reserves Act, a new National Parks Act in 1980, and a new Conservation Act in 1987. In each of these statutes, the purpose of wilderness areas can be summarized as the preservation of a natural state, primarily for wilderness experiences (Fig. 3.9). More specifically, the Conservation Act of 1987 lists provisions that must apply to any wilderness area: its indigenous natural resources shall be preserved; no building, machinery, or apparatus shall be erected or maintained in it; no livestock, vehicles, or aircraft shall be allowed to be taken into or used in it; and no roads, tracks, or trails shall be constructed in it (Molloy 1997; Cessford and Reedy 2000).

Today there are few recreation-based threats to the integrity of New Zealand's currently designated wilderness. In 2023, relevant issues facing wilderness area expansion and management can be summarized as:

- The depletion of Indigenous species and habitat by introduced animals and plants remains the greatest overriding threat to conservation values. Most effort from the Department of Conservation remains focused on this issue through pest and weed control and species/habitat restoration.
- Well established legislative provisions continue to minimize any potential for compromises to designated wilderness areas from mechanized access or recreational infrastructure developments. Given this limited demand pressure and priorities to address other conservation pressures there is only

low-level advocacy for extended or new wilderness area designations.
- Maori associations with their traditional lands, values for natural areas, and customary uses of resources are now being more formally recognized.

The most significant recognition of Maori status specifically relating to a designated wilderness area has been the Te Urewera Act 2014 (https://www.environmentguide.org.nz/regional/te-urewera-act/). Under this legislation the "Te Urewera" area (incorporating the "Raukituri Wilderness Area") ceased to be a government national park and was vested in itself as its own unique legal private entity. It is co-governed by a board led by the Tūhoe people and comprised of a mix of government and majority-Tūhoe appointed representatives. Building a new direction from its National Parks Act (1980) and Conservation Act (1987) precursors, the Te Urewera Act's purpose is to:

- strengthen and maintain the connection between Tūhoe and Te Urewera;
- preserve as far as possible the natural features and beauty of Te Urewera, the integrity of its Indigenous ecological systems and biodiversity, and its historical and cultural heritage; and
- provide for Te Urewera as a place for public use and enjoyment, for recreation, learning, and spiritual reflection, and as an inspiration for all.

The formal "wilderness area" classification of Ruakituri was maintained by the 2014 Act and to date it appears to have remained as protected as it ever was. As stated by an FMC representative (Stewart 2015): "The significance of having our home-grown recreational concept of 'wilderness areas' accepted as part of a new, indigenous way of recognizing and managing our interactions with a significant natural area should not be underestimated. The Te Urewera Act may prove to be a one-off, but equally it may prove to be the harbinger of a new way of exercising guardianship of natural lands, of which wilderness is now an accepted part." Ongoing communication and consultation with local communities, recreational groups, and other stakeholders will be critical for Te Urewera's evolution, as apparent from current controversies over its recreational access provisions and facility infrastructure plans, although to date these do not appear to affect the designated wilderness area.

Zimbabwe

Zimbabwe is the first developing country to have proclaimed a wilderness area (Martin 1990). The Mavuradonna Wilderness Area was designated in Zimbabwe early in 1989 by tribal authority of the Mzaribani District Council on their own communal lands. This designation was made in conjunction with national policy direction under Zimbabwe's Communal Areas Management Programme for Indigenous Resources (CAMPFIRE), which develops economic, educational, and cultural linkages between natural-resource management and tribal authorities. The minister of State for Tourism officially opened the area, furthering national recognition of the wilderness concept and this area's designation.

The Mavuradonna Wilderness Area is approximately 192 square miles (500 sq km) in the escarpment of the Zambezi Valley; one road bisects the area. The area is used for a combination of wilderness "trails" (hiking), sport hunting (primarily bow hunting), and wildlife cropping, with a primary purpose of generating economic benefit for local villagers and farmers. The area produces income each year under the proposed uses, derived mostly from wildlife harvesting. As a result of the initial wilderness proclamation, other areas in Zimbabwe were under consideration for wilderness designation by other tribal authorities. Subsequent political turmoil in Zimbabwe has complicated this.

There has been continuing resistance from some sectors against designating wilderness in developing nations, based primarily on the argument that the wilderness concept is too elitist, exclusionary, and recreation-oriented to be of value in developing countries where higher-priority issues are about food, education, rising population, and poverty. Therefore, the Mavuradonna Wilderness Area signals a shift in attitude within a developing nation toward accepting the wilderness concept; the wilderness designation is linked to rural development with the management emphasis being on economic sustainability of the local people, and it is an example of a designation that is relative to the culture in which it exists (Martin 1990).

The Mavuradonna Wilderness Area was established with little or no public agitation. Anglo-oriented conservationists introduced the concept to the tribal authority, and it was welcomed with little resistance or controversy. Zimbabwe's adoption of the wilderness concept could mean that the concept is further evolving to include and incorporate the needs of more traditional and diverse people, thereby becoming more relevant to developing nations. However, no additional areas have been protected as wilderness in Zimbabwe and the WDPA only lists one area, preceding the Mavuradonna Wilderness Area in its existence, on the Chirinda State Forest (UCN and UNEP-WCMC 2022).

Tanzania

More than one-third of Tanzania is protected as national parks, game reserves, forest reserves, and other types of protected areas. Wilderness zoning in national parks is an important part of a tourism strategy in the country to promote low-impact, sustainable tourism development. Most of Tanzania's national parks have wilderness zones. Tanzania's largest national park, Serengeti National Park, has more than 1.28 million acres (517,000 ha) zoned as wilderness. This park's management plan suggests that promoting tourism for the wilderness zone is crucial to achieving the range of park conservation objectives (Kormos and Martin 2008). The WDPA indicates that Tanzania has eight total Category 1b areas designated for protection, two of them since 2008 (both in 2019 and all eight are on National Forest Reserves) (UCN and UNEP-WCMC 2022).

Italy

In Italy, a small but determined effort by the Wilderness Associazione Italiana per la Wilderness (AIW), founded by Franco Zunino—who was inspired by the Wilderness Society (U.S.) and by participating in the 3rd World Wilderness Congress (1983)—ultimately pioneered a new type of wilderness designation by private landowners and public authorities in the 1980s (Zunino 2007). The very first sector of Italian wilderness area, the Fosso del Capanno Wilderness Area, was established in 1988 through a management agreement with a private foundation and was 283 acres (118 ha) in size (later expanded to 1,877 acres or 750 ha). The largest wilderness is the Ausoni Wilderness Area, at 10,338 acres (4,230 ha). Most areas are now protected by municipalities, regional forestry authorities, or private landowners. Only one designation is by a national park authority. By the end of 2022, AIW reported that there were eighty wilderness areas covering 144,536 acres (58,491 ha) in twelve regions of Italy and twenty-five provinces. More wilderness designations are expected. The term *wilderness area* (first used in Italy, and the first official use in Europe) is now in common use in Italy, as much a new linguistic innovation as it is a new land-use and planning category in that country.

Although not as secure as the type of protection provided by national legislation, the creative approach in Italy demonstrates the important growth of the wilderness concept internationally, and that it can be a useful and politically viable concept even in highly developed European countries. An anomaly worth noting is that there are no officially designated IUCN Protected Area Category 1b areas in Italy (UCN and UNEP-WCMC 2022).

Japan

The Nature Conservation Law of 1972 provides the foundation for establishing wilderness areas and nature conservation areas in Japan (Caouette 2008). The Ministry of Environment in consultation with government land management agencies can declare both types of areas. Currently, there are five designated wilderness areas that protect 2,280 acres (5,631 ha) of national forestland. They are intended to be free from human activities and to protect ecosystems that remain in their original condition. These wilderness areas typically allow only research and science activities that require specific permits. While some of these areas are adjacent to national parks, they are not components of larger protected areas or systems, as seen in many countries. All eight Category 1b areas in Japan are federally designated, all since 2008. Four are offshore (UCN and UNEP-WCMC 2022).

Class III–Wilderness Protection Through Recognition in Conservation Programs

The Class III type protection for wilderness occurs where wilderness is "recognized" as a distinct component of resource-management programs and a larger conservation effort.

Namibia

Namibia is an arid country on the southwest coast of southern Africa. A high proportion of the country is desert or semi-desert, and it includes the great Namib Desert. The country is 40 percent formally protected under various categories, is the only country in the world whose entire coastline is under protected status, and has numerous, large, protected areas including the world-famous Etosha Pan and the Skeleton Coast National Parks. Though Namibia does not have any legally designated wilderness, most of these extensive protected areas contain within their borders some of the best-preserved de facto wilderness zones on the subcontinent (Cooper 2001).

Namibia has important areas of privately owned land that are either partly or completely managed as natural areas for wildlife production, or as private reserves and game farms. The natural communities in these areas are largely intact, and a full range of large and small wildlife species are present. Many valuable conservation projects, such as the Cheetah Conservation Fund, which owns and manages almost 250,000 acres (98,000 ha) for wildland value, take place on these privately owned areas and make important contributions to the economy through various forms of tourism, including photography, local education, and hunting (Barnes 1998).

One of the most important developments in Africa in recent years is the establishment of community-owned and managed nature conservancies in Namibia, all located on and operated by local tribes and communities. Pioneered largely by the NGO Integrated Rural Development and Nature Conservation (IRDNC), these lands are not necessarily categorized as wilderness. In IUCN terms, they may fall under a different category that directly addresses landscapes stewarded by Indigenous Peoples and Local Communities, but for all intents and purposes, they contain and are managed to maintain significant wilderness values. Some 20 percent of Namiba's lands are gazetted as community conservancies (half of the nation's total), affecting 8 percent of the nation's population. Namibia is the only country in Africa where all populations of wildlife are regarded as increasing, some dramatically, yielding considerable protein and tourism revenue directly to communities (IRDNC 2019). Important to note is that despite the progress in wilderness protection mentioned here, no Category 1b sites have been formally registered in Namibia (UCN and UNEP-WCMC 2022).

Philippines

In the Philippines, wilderness is recognized under a national legislative Act as a type of land classification, but it does not seem to be singled out for special management. The Philippine National Integrated Protected Areas System Act of 1992 established a national system of 200 potential areas (Pollisco and Meniado 1997). The intent is to maintain essential ecological processes and life-support systems, to protect genetic diversity, to ensure sustainable use of resources, and to maintain natural conditions. The act establishes protection of these places as national parks, game refuges, bird sanctuaries, wilderness areas, watershed areas, mangrove reserves, and virgin forests. Eight proposed areas in six regions were targeted for immediate inventory and intense public participation in planning, funded by a grant from the European Union (Rambaldi 1997).

In northeastern Luzon, the Palanan Wilderness Area is a relatively new example of Class III wilderness, in which use of the term is descriptive rather than part of a formal land-use management plan. It deserves note for several reasons. Originally designated a wilderness area, the Palanan rainforest was proclaimed a national park in 1997 and later renamed the Northern Sierra Madre Natural Park. It encompasses the towns of Divilacan, Dinapigue, Palanan, Maconacon, San Mariano, Ilagan, San Pablo, Cabagan, Santa Maria, and Tumauini. It is one of the most important and largest tracts of tropical rain forest in the Philippines (494,000 acres or 200,000 ha). It has also been home to the Agta forest people for more than 2,500 years. In 2003, this area was combined with the neighboring Peñablanca Protected Landscape and Seascape, increasing the size to closer to 300,000 ha and nearly 72,000 ha of marine areas. Like many of the protected wilderness areas discussed in this chapter, it is also the result of citizen action by a local NGO, the Northern Sierra Madre Wilderness Foundation. Conservation International, the U.S.-based NGO, is very active working with and supporting the local NGO and other private groups in a community-based natural-resource management plan including traditional people's rights, conservation enterprise, sustainable use, and scientific research. In the Philippines there are also five Category 1b areas, with three designated before 2008 and two since 2008, and all are under federal protection. One of the areas designated since 2008 is within a marine resource reserve and one is terrestrial. The Palanan Wilderness Area is not listed in the WDPA (UCN and UNEP-WCMC 2022).

PROTECTION OF WILDERNESS VALUES IN OTHER REGIONS

The nations discussed above, along with the United States, are primarily the only ones in which wilderness protection has achieved some type of formal status within the country. However, wilderness values are recognized more broadly, sometimes with formal protection in other regions as well. The distinction here is subtle; formal, explicit wilderness protection is still relatively limited around the world, but the use of Category 1b designation is growing rapidly. Further, if we broaden our perspective to embrace programs designed to protect wilderness values, albeit without the explicit labeling of areas as wilderness, we find additional important activity. The reasons for not labeling such areas as wilderness vary widely; in some places, wilderness is simply not a part of the language. Elsewhere, there is the belief that other designations, such as national parks or nature reserves, adequately protect wilderness values.

The distinguishing qualities of wilderness, as used in this book, includes areas where natural ecological processes operate generally free of human influence and can provide experiences with opportunities for solitude and primitive recreation. In most of the countries just reviewed, this general concept of wilderness prevails. But programs in many other countries protect natural conditions and processes and also the solitude and primitive recreation opportunities associated with wilderness (Fig. 3.10).

FIGURE 3.10 Trekking in the Annapurna Sanctuary near Pokhara, Nepal, provides the opportunity to see the interaction of mountain communities with wilderness environments. Photo courtesy Chad P. Dawson.

Latin America

The wilderness concept has generally not been adopted by Latin American land management agencies and institutions. As of 2008, no protected areas meeting the IUCN protected areas Category 1b definition or the protection criteria of the three classes of wilderness defined in this chapter exist in the region. Nevertheless, even though wilderness policy frameworks are scarce south of the U.S.-Mexican border, wilderness as a place persists, often in wild abundance, due to over a century of initiatives led by Latin American conservationists. The first national park in Latin America, Nahuel Huapi National Park (made possible by a land trust by Argentine scientist and explorer Francisco Moreno) was proposed in 1903 and achieved formal designation in 1922 (Scarzanella 2002), initiating a decades long movement to preserve the ecological qualities of this ecologically mega-diverse continent.

Two of the largest marine protected areas in the world are in Chile: the Rapa Nui Marine and Coastal Protected Area (~579,000 square km) and the Nazca-Desventuradas National Park (~300,000 square km). To the northeast of Chile, Brazil's vast and diverse ecology is home to the world's largest terrestrial protected areas, and not just in the Amazon. The Cerrado Biosphere Reserve, which lies almost entirely in Brazil but extends across the frontier to Paraguay and Bolivia, encompasses approximately 297,000 square kilometers of subtropical woodland and grassland. Chile, which has successfully protected 42 percent of its marine waters and 21 percent of its land, has expanded its National System of Protected Areas by more than 36,000 square kilometers since 2017, in no small part due to the support and generosity of Kris and Doug Tompkins of Tompkins Conservation. In 2021, two of those national parks joined the IUCN's Green List for Protected and Conserved Areas, which recognizes places that are managed to a high global standard (Germani and Sepúlveda 2021). Indeed, some of the most globally significant and intact wilderness remains in Latin America, including the Amazon rainforest and the Patagonian steppes and austral forests, despite the intensification threats posed by the expansion of agricultural lands and myriad political, economic, and geographic challenges to the enforcement of existing laws.

While formal government protection of wilderness is rare in Latin America, wilderness values are evident in other conservation strategies and actions carried out by government officials, private actors, and Indigenous leaders (Wakild 2014).

Mexico

Although Latin American countries currently do not officially register wilderness areas, since 2008, a few areas (for example in Argentina in 2018 and 2019, Brazil in 2017, and Mexico in 2009) have been registered as protected among IUCN Category 1b areas and listed in the WDPA (UCN and UNEP-WCMC 2022).

Agreement on the first designated wilderness area in Latin America, and the only Category 1b area listed for Mexico, was announced at the 8th World Wilderness Congress (in Alaska in 2005). The Tierra Silvestre Cañón del Diablo (El Diablo Wilderness Area) is listed in the WDPA at 224 square km (86.5 sq mi) (UCN and UNEP-WCMC 2022). El Diablo Wilderness Area is within the El Carmen area in the State of Chihuahua, 75,000 privately owned acres (30,500 ha), and the core of a large transfrontier area between Mexico and Texas. As a result of subsequent activities at the 9th World Wilderness Congress in Mexico (2009), by 2011 this larger, transfrontier area was formally recognized by the U.S. and Mexico governments under the Cooperative Action for Conservation in the Big Bend/Rio Bravo Region (USDI 2011). This 10-million-acre (4-million-ha) mosaic of protected areas (private and public), ranches, and *ejidos* (communal areas) reflects the vision of the Mexican conservation photographer Patricio Robles Gil and his organization, Agrupación Sierra Madre, and others working on the U.S. side of the border. The management, rehabilitation, and wildlife reintroduction work in the core area of the complex, including the El Diablo Wilderness Area, is funded and on land owned by CEMEX, a large international corporation. El Carmen is one of the fabled sky islands of the Chihuahuan and Sonoran Deserts of northern Mexico and is an international biodiversity hotspot. Numerous NGOs and experts on both sides of the border continue to work on a vision of a U.S./Mexico International Park in this region.

Brazil

Unrelenting threats, from national politics to the climate emergency, have not deterred Brazilian advocates and their international partners from efforts to protect the Amazon, which remains one of Earth's largest, most intact wildernesses, the size of Mexico, Mongolia, Peru, and Egypt combined. Mapping and administering such a vast and often impenetrable area (over 7 million square kilometers) are not without their challenges.

Though not recognized as "wilderness" officially, since 1965 Brazil has sought to keep the Amazon intact through the Forest Code, a law requiring landowners in the Amazon to maintain 80 percent of their properties under native vegetation. Stringent in spirit, this policy has proven difficult to enforce in the context of a social-ecological system characterized by remote, nearly impenetrable terrain and a powerfully entrenched political resistance.

Increasingly, Indigenous communities are leading regional efforts, working in concert with international partners, to maintain wilderness values at a landscape level. One little known, but potentially consequential, initiative in the Western Amazon could result in the use of corridors to interconnect three pristine Indigenous reserves in the heart of the rainforest. Known as the "umbilical cord," this area would encompass over 10,000 square kilometers, providing ecological and cultural protections for surrounding traditional communities, including one of the world's highest concentrations of uncontacted people (Lewis 2023).

China

China is Asia's largest single country (considering that Russia spans both Europe and Asia) and the fourth largest in the world, with a terrestrial area of 9.6 million sq km (3.7 million sq mi), and is also the world's most populous country, with 1.4 billion people. Despite its high population and long history of occupation, China is also one of the "mega-wild" countries in the world (Watson et al. 2018). China, however, is also facing potential loss of wild areas due to rapid changes in land use and land cover (Cao et al. 2022a; Cao et al. 2022b). In addition, China's national policy of "eco-civilization" (or promoting ecological progress—生态文明) is a national strategy to change land development patterns, promote resource conservation and utilization, protect natural ecosystems and the environment, and improve quality of life, creating new opportunities to consider the role of wilderness protection in this national strategy (Li 2017).

Since 2008 and the emergence of wilderness as a topic in China, Tin and Yang (2016) explored past Chinese reference to what we call wilderness. Tin and Yang (2016) described recent past use of the term 荒野 (huang ye) as the most common term adopted for wilderness in Chinese (Mandarin), though they questioned the accuracy of the meaning of the translation. They explained that 荒 (huang) commonly refers to uncultivated, unattended, uncivilized, or wild places (it can actually be used as an adjective or a noun), and 野 (ye) can be translated either as an adjective as "wild" or as a noun meaning "open country." Explaining 荒野 (huang ye) as a combination of adjectives and nouns helps to clarify some of the confusion it has caused and underscores the need for better translations of these terms. Since this translation into Chinese had mostly referred to an inhospitable place, a place without desirable features, it was not fully capturing the positive reasons a wilderness area, defined by IUCN Protected Area Category 1b, would be protected. In subsequent presentations and discussions in the international community (for example at the Global Land Project Third Open Science Meeting on October 24–26, 2016, in Beijing), they suggested using the term 荒(huang)野(ye)区(qu)——保持着原荒和野性的地区 (areas maintaining their original wildness and wild nature), which offers a more precise meaning of wilderness area, not just a vague reference to wilderness. They found this term already in use in academia in China. It offered a more complex but clarifying connotation by transforming both 荒 (huang) and 野 (ye) into adjectives. The overall reference now focuses on an area (a noun), providing an opportunity to introduce a new term for a previously undefined area of interest within environmental protection circles in China.

Wilderness-related policy, technical, and science papers related to China have been appearing more in national and international journals, such as the Chinese Journal of Landscape Architecture (Cao et al. 2017; Martin 2017; Watson and Carver 2017), the International Journal of Wilderness (Cao et al. 2018), and others. The rapidly rising popularity of the wilderness concept is also evident in academic exploration of the term "urban wilderness" in China (Martin and Hill 2021). There is also an effort to keep this popularity in perspective because of the inherent confusion and misappropriation in referring to wilderness within urban areas, although some wilderness areas around the world do exist in close proximity to urban areas.

One wilderness area mapping exercise in China (Cao et al. 2019) identified over 86,000 wilderness areas (or patches) with a total area of just over 4 million sq km, (1.5 million sq mi) covering 42 percent of China's terrestrial area. These wilderness areas were identified using the following three criteria: areas that contain only natural land cover, areas that are roadless, and areas that are free from permanent human settlements. In addition, extra-large wilderness areas are found within the provinces of Xinjiang, Tibet, Qinghai, and Inner Mongolia. Large wilderness areas are mainly distributed in western China, while medium, small, and extra-small wilderness areas are distributed throughout most of the provinces. Crucially, these areas can still be found in the densely populated and developed east of China. Landscapes in east and west China are significantly different, including the fact that the degree of human influence and landscape fragmentation in the west of China is far lower than that in the east (Cao et al. 2020).

The WDPA (UCN and UNEP-WCMC 2022) indicates the first designation and protection of IUCN Category 1b (wilderness areas) in China in 2018. All three areas listed are contained within Ramsar sites: Gansu Yanchiwan Wetlands in the Yanchiwan National Nature Reserve on the northern edge of the Tibetan plateau, Hubei Wang Lake in the Wang Lake Wetland Natural Reserve in the catchment of the Fu river, and the Heilongjiang Youhao Wetlands in the Youhao National Nature Reserve on the slopes of the Lesser Khingan mountains in north-east China (UCN and UNEP-WCMC 2022).

Antarctica

The earth's largest remaining wild area, Antarctica, is a world of superlatives—the coldest, iciest, windiest, and most remote of the world's continents, girded by the stormiest ocean. Equivalent in size to the combined area of the United States and Mexico, it expands in size each winter when the surrounding seas turn to ice.

Antarctica is vital for understanding our planet. Its ancient rocks, once at the core of the Gondwana supercontinent, hold the key to unlocking the earth's geological history. The vast Antarctic ice sheet, on average 6,560 feet (2,000 m) thick, is a window for observing changes in global climates over tens of thousands of years. The ice is a storehouse for 90 percent of the earth's freshwater resources—enough to raise the global sea level by many meters, if all of it were to melt. In this era of global warming, dramatic changes to the ice cap are already being observed. The continent and surrounding seas are a vital influence on the well-being of human life on the earth through the ways they regulate the global atmosphere and the oceans, thus affecting patterns of global weather and ocean circulation.

Antarctica remains mostly untrammeled wilderness—but it is not pristine. Soon after its discovery, human exploitation targeted the marine resources and in just a few decades of the nineteenth century, Antarctica's fur seals and whales were brought to the brink of extinction (though have now mostly recovered). Since the late 1960s, new fisheries commenced in the Southern Ocean for the massive numbers of surface-living krill (*Euphausia superba*), and for rock cod and ice fish, and the fisheries experienced classic boom-and-bust episodes when harvest levels were neither ecologically nor economically sustainable. A significant accomplishment in the Antarctic oceans region in recent years—requiring years of consensus-building among twenty-six nations—is the establishment of the Ross Sea Marine Protected Area. At 600,000 sq miles (1.6 million sq km) immediately adjacent to Antarctica and protected from commercial fishing for an initial thirty-five years, this MPA with its commercial "no-take" policy may well cause this area to be regarded as the world's largest so-called "marine wilderness" (Brooks et al. 2021).

On land, steadily expanding national research programs have established dozens of scientific stations, composed of buildings, stores, airstrips, and increased ship and air traffic to sustain thousands of scientists and support staff in Antarctica each year. A growing tourist industry brings many more thousands of people each year into Antarctica. Although Antarctic scientific and tourist activities generally exhibit a high level of environmental awareness, instances of pollution of land, sea, and air from fires, fuel spills, waste dumping, shipwrecks, and aircraft crashes occasionally damage the environment. Pollutants such as DDT, polychlorinated biphenyls (PCBs), and other compounds originating from distant industrialized regions are also found. Of even greater concern is the impact of "greenhouse" gases on global warming that are very rapidly increasing melting of Antarctic glaciers and ice.

Considered together, these impacts and growing risks of significant environmental damage highlight the value of stewardship of the Antarctic region. The initial Antarctic Treaty (1961) set the stage for the 1991 Environmental Protocol to the Antarctic Treaty (Madrid Protocol) that laid out a farsighted and comprehensive program for protecting and managing the Antarctic environment, and in which the word "wilderness" appears many times, but the treaty doesn't specify how to protect these areas. The Treaty declares Antarctica a natural reserve devoted to peace and science, creates the world's largest de-militarized zone, bans mineral extraction, and presents ambitious mandates requiring all treaty parties and private individuals to minimize their impact on Antarctica.

The Madrid Protocol provides for designating special protection for areas acknowledged to have significant natural, historic, scientific, or landscape values, and areas where multiple uses might cause undesirable environmental impact or disruption among conflicting activities. Management plans are required for most protected areas established over natural sites, such as breeding colonies of seabirds and seals, and historic sites, including stations from the heroic era of Antarctic exploration (Dingwall 1997).

The Antarctic Treaty has proven effective, practical, and responsive to changing global problems and circumstances. In agreeing to the Environmental Protocol, the treaty governments have signaled a strong commitment to protection and sustainable management of Antarctica's resources. But challenges lie ahead with the increasing human presence, growing demands for the use of its resources, and global climate change.

OTHER INITIATIVES TO PROTECT WILDERNESS VALUES

In addition to international definitions, the archival of designation data, and efforts by individual countries to inventory and initiate protection of wilderness character, there are several initiatives that have led to a better understanding of the importance of wilderness. These initiatives have introduced a wider range of definitions and protection methods, and other influences to increase both wilderness recognition and stewardship.

Protecting Biodiversity

Determining How Much Wild Nature Needs to be Protected

Efforts to protect biodiversity directly support wildland conservation and wilderness because it is most often in wild areas that diversity flourishes and species evolve as naturally as possible. Recognizing this, the Convention on Biological Diversity (CBD) was signed at the Earth Summit in Rio de Janeiro, Brazil, in 1992, and it is the world's premier treaty addressing biodiversity loss, with its three mutually reinforcing objectives: conservation, sustainable use, and equitable sharing of benefits from biodiversity (Faries and Cervigni 1998). The Global Environmental Facility (GEF) of the World Bank is the financial mechanism of the Convention on Biological Diversity and provides assistance to developing countries that are parties to the Convention. The protection of entire biomes preserves the biodiversity of species needed to maintain the productive capacities of ecosystems. Protected areas provide nesting, calving, spawning sites, and habitat for plant and animal species that local people rely on for food, medicine, or for sale to acquire income.

In 2023, the CBD is expected to have concluded its (pandemic-delayed) 10-year program that was agreed upon in Aichi, Japan, in 2010. This global framework has twenty-one targets and ten milestones for 2030, on the way to "living in harmony with nature by 2050" (CBD 2022). An element of these targets deals specifically with the amount of nature needed to protect in order to sustain life on Earth. The "Aichi Targets" called for the protection of 17 percent of land and 10 percent of oceans as milestones towards to-be-determined, increased targets. These 2010 targets were not quite realized, yet the call increases for more ambitious targets. As conservation science and remote sensing rapidly increased in accuracy in the past fifteen years, it is evident that minimum-effective sizes of intact ecosystems—functioning wildlands and seas—are necessary to support all life on Earth. Various NGO-driven processes emerged to drive this debate. The first process, called Nature Needs Half, was launched on the global scene at the 9th World Wilderness Congress (Mexico, 2009) and was developed by the WILD Foundation (Locke 2018). Later, the important Half Earth Project (Wilson 2016) was launched by the E.O. Wilson Biodiversity Foundation. Furthermore, the IUCN's Resolution 125, which set area-based conservation targets based on evidence of what nature and people need to thrive, was adopted at the (pandemic-delayed) World Conservation Congress in 2022. This resolution approves the science recognizing that protecting approximately half of intact nature is required to secure essential ecological services and also, of course, with full, prior, and informed consent of IPLC and recognition of Indigenous rights (WCC 2020).

The Role of Indigenous Peoples in Protecting Biodiversity

The critical and inherent role of Indigenous Peoples (and other Local Communities) in protecting what is essentially their own lands and seas was long ignored. As this continues to change and a new conservation paradigm emerges, it is important for the non-Indigenous protected area manager or policymaker to understand that IPLC are not a homogenous bloc. Though there are matters somewhat similar to them all—especially the nature-based culture of Indigenous societies—they are comprised of many societies and attitudes. Differences in policies and actions abound as one would expect in any grouping of humans and societies. However, most important is the certainty that they are key partners in the fate of the planet's biodiversity and remaining wildlands and seas.

Of the world's remaining high biodiversity areas most are in high-functioning wildlands, some scientists claim that 80 percent is under the direct stewardship of Indigenous Peoples (Raygorodetsky 2018). One example is the Xingu Indigenous Reserve in the central Amazon that is 25 million acres (10 million ha) of virtually intact tropical rainforest under the stewardship of the Kayapo people, the largest Indigenous Reserve in the world. How protection of these areas must be approached has begun to dramatically change. In contrast to the former, neocolonial approach of expropriating native lands to become new protected areas, much positive progress has been made since 2009 under the United Nations Declaration of the Rights of Indigenous Peoples (UNDRIP), which calls for "free, prior, and informed consent" when considering actions that affect the cultures and homelands of Indigenous Peoples (Cambou 2019). Examples are beginning to emerge in Central and South America, South and Southeast Asia, Australia, and parts of Africa where Indigenous Peoples' rights are recognized, and their well-being integrated into plans for sustainable wildland management. As is emphasized in the Guidelines for Category 1b Wilderness, recognizing their role as partners (not just another stakeholder) in protecting wildlands and biodiversity, and helping them protect their cultures and lifeways, also protects the wilderness. Progress is being made in this regard but much more is needed.

Wildland Corridors

Conservation biology is concerned with joining protected wild areas and low-impact areas, such as national forests or private forests and ranchlands, to create wildland habitat corridors to ensure migration of wildlife, enhanced biological intactness, and sustainable ecosystem functioning. One of the earliest and ongoing models, started in 1993, is Yellowstone to Yukon (Y2Y), a joint Canadian-U.S. network (www.y2y.net) of more than 800 organizations, institutions, foundations, and conservation-minded individuals who have recognized the value of working together to restore and maintain an unprotected area of the corridor between Yellowstone (in northwestern Wyoming) and the Yukon (northwestern Canada and northeastern Alaska) region. Y2Y opens new possibilities for ensuring the continued presence of North American wildlife and wildlands. In the wake of Y2Y came similar initiatives, many of them inspired or managed by The Wildlands Project (www.twp.org), such as Heart of the West, Southern Rockies, and Northern Sierra Madre in North America, and other initiatives on every continent. Recognizing the critical importance of corridors, the World Commission on Protected Areas, IUCN, established the Connectivity Conservation Specialist group.

Transfrontier Conservation

Some mega-wildland corridors are *transfrontier,* that is, crossing national political boundaries to embrace entire ecosystems. This is not a new concept, and the first such area was established in 1932 between Glacier National Park (Montana, U.S.) and Waterton National Park (Alberta, Canada). A 2005 survey reported in *Transboundary Conservation* (Mittermeier et al. 2005) indicates that there are 188 Internationally Adjoining Protected Areas (IAPA) or complexes involving 818 protected areas in 112 countries and representing approximately 17 percent of the global extent of high-biodiversity wilderness areas. Nearly half of the total land area of IAPAs is located in North America, followed by Africa. The Russian Federation has the greatest number of IAPAs (twenty-one), followed by China (fourteen) and Canada (twelve). Countries contributing the most area to the overall IAPAs are Greenland, the U.S., and Venezuela.

IAPAs, or transfrontier protected areas, involve important issues in political and land management cooperation, which is a challenging task even across a friendly border such as that between the U.S. and Canada with the Boundary Waters Canoe Area and Canada's Quetico Provincial Wilderness Park, or (as mentioned earlier in this chapter) between Mexico and the U.S. in the Big Bend/Rio Bravo region—North America's largest and most diverse desert ecosystem.

Elsewhere in the world, such tough issues may take years to resolve, if they are resolved at all. The Peace Parks Foundation (www.peaceparks.org), based in South Africa, has popularized this concept and is working to establish a network of "peace parks" throughout Southern Africa. Their work is extensive, large-landscape focused, and involves all levels of authority and responsibility from local communities to heads-of-state. Peace Parks areas exemplify what the world could be like with concern for wild nature at the center of human dialogue.

Wild Rivers and Marine Wilderness

At the 6th World Wilderness Congress in Bangalore, India, in 1998, Michael McCloskey presented the results of his most recent international survey, "Wild Rivers of the World" (McCloskey, Michael 2001). In his second reconnaissance-level global survey, McCloskey studied all rivers at least 50 kilometers long (31 miles). They earned the "wild" distinction if they were free of dams and their effects and had a pollution-free watershed, with no roads along the banks. Rivers that ran through lightly developed territory (small, scattered settlements), though not moderately or heavily developed, were also included. McCloskey's study identified 6,000 river segments in the world that appeared wild under these criteria. Added together, these segments amounted to 484,250 miles (781,051 km), averaging 81 miles (130 km) each segment. This amounts to 19.6 percent of all rivers studied, mostly found in developing nations.

North America had the highest percentage of wild rivers—37.9 percent—most of which were in the sub-Arctic and Arctic north above 53 degrees north latitude. South America still had 26.1 percent of its rivers wild, mostly in the Amazon Basin. Three other regions were below average: Eurasia with 17 percent, mostly in eastern Siberia; Australia and Oceania with 10.6 percent; and, surprisingly, Africa with the lowest percentage (7.6). Nearly 60 percent of remaining wild rivers are in just two countries, Russia and Canada. Another 20 percent are in the United States (mostly in Alaska), Brazil, and China.

There are evolving efforts to apply the wilderness concept to water and marine areas. Initially, this was done by incorporating portions of aquatic areas in wilderness designations with surrounding terrestrial areas, the Boundary Waters Canoe Area Wilderness in the United States being the best example. Internationally, an important example of this is the wilderness zone in Lake St. Lucia located in KwaZulu-Natal in South Africa. A low fence in the lake marks areas for nonmotorized use. There is also a terrestrial wilderness zone on the eastern shore between the lake and the Indian Ocean, which was the focus of one of the most dramatic conservation struggles in Africa during most of the 1990s. Conservation NGOs were pitted against the world's largest mining company, Rio Tinto Zinc, and eventually were successful when President Mandela's government proclaimed the Greater St. Lucia Wetland Park in 1998, which was declared South Africa's first World Heritage Site in 2000.

At the 8th World Wilderness Congress in Alaska in 2005, Mike Smith presented a new inventory of freshwater wilderness that extended the earlier McCloskey (2001) analysis to include lakes and wetlands, as well as streams, and identified large areas of the world where surface waters were most likely to remain in natural condition. This study found important areas of freshwater wilderness only at the northern extremities of the continents and in a few small watersheds of South America. The loss of wild water in much of the tropics is associated with the fact that rivers have often served as avenues for human penetration of tropical forests (Sanderson et al. 2002). Although Amazonia and the Congo Basin are considered terrestrial wildernesses (Mittermeier et al. 2003), they are impacted by river channel fragmentation and human management of flow regimes, which affect large areas within catchments. Southeast Asia, which has some of the world's highest levels of freshwater biodiversity, has long been impacted by agriculture and high human population density. It is now subject to high regulation of water flows at all scales. In the Northern Hemisphere, population density and agriculture eliminate wild waters at midlatitudes. Regulation of water flow extends human impacts to even higher latitudes, leaving only a few northward-flowing watersheds with large-scale wilderness conditions. These studies show that wild freshwater is the rarest form of wilderness and is the form that is disappearing most rapidly.

Carver (2018) suggested that protecting the wilderness qualities of river catchments is the key to protecting the wildness of rivers. Carver (2018) attempted a global assessment of wild rivers based upon connectivity between land and rivers and how much human activity within catchments (including settlement and transportation infrastructure) and modification via human land use (agriculture, forestry, etc.) affects that connectivity. Using purely global datasets, Carver (2018) mapped the hydrological functioning of watersheds, those that are providing natural water flows. There are many limitations associated with global datasets according to Carver (2018), including coarse resolution, generalization, variable accuracy, and consistency across regions and national boundaries, plus they were contending with fixed and arbitrarily defined catchment boundaries. Despite these limitations and suggestions to undertake more national, regional, and local assessments with more precise and varied sources, they concluded that many of the wildest rivers of the world fall mainly into permafrost or desert environments. As such, Carver (2018) clarifies that these rivers are likely to be ephemeral and only flow when fed either from seasonal ice/snowmelt or high magnitude/low frequency flash flood events in desert areas. Mid-latitude rivers in temperate or moist tropical regions are not well represented in the top 10 percent of wildest rivers at a global scale, the exception being those within the intact tropical rain forest areas of northern Amazonia and the larger rivers draining northern Australia.

There has also been movement toward recognition of oceanic or marine wilderness. The 2nd World Wilderness Congress, in Australia, featured discussion of a proposed wilderness zone in the Great Barrier Reef Marine Park. The Australian Park Authority designated zones in the Capricornia section for wilderness-type use and later degazetted them. Although small in terms of the overall size of the park, it should be emphasized that, in terrestrial terms, the zones are extensive (Kelleher 1982). The U.S. National Oceanic and Atmospheric Administration announced inquiry into the feasibility of oceanic wilderness (Foster and Lemay 1988) at the 4th World Wilderness Congress.

At the 6th World Wilderness Congress, Maxine McCloskey added considerably to the concept of marine wilderness with an innovative, global survey of potential wilderness areas on the high seas (McCloskey, Maxine 2001). One of former president Clinton's executive orders (May 2000) established a new Marine Protected Area in the northwestern Hawaiian Islands. One of its original intentions was to institute "no take" management, a widely accepted criterion for marine wilderness. In 2006, former president Bush signed a proclamation that created the Northwestern Hawaiian Islands Marine National Monument (renamed the Papahānaumokuākea Marine National Monument in March 2007). This national monument placed nearly 140,000 square miles (362,600 sq km) of the northwestern Hawaiian Islands under the highest form of marine environmental protection. The national monument will preserve access for Native Hawaiian cultural activities, provide for carefully regulated educational and scientific activities, enhance visitation in a special area around Midway Island, prohibit unauthorized access to the monument, phase out commercial fishing over a five-year period, and ban other types of resource extraction and dumping of waste. This marine national monument is the largest single area dedicated to conservation in the history of the United States, and the largest protected marine area in the world. It is more than one hundred times larger than Yosemite National Park. In the U.S., there are forty Category 1b areas that are classified as marine, including eleven predominantly or totally marine and twenty-nine mixed terrestrial and coastal marine (UCN and UNEP-WCMC 2022). Of the forty, one is BLM, two are NPS, and the rest are USFWS. Eight Category1b areas have been designated since 2008 (all are FWS, with none designated since 2010).

Rewilding

While it can be a crucial consideration in the feasibility of wilderness designation and protection strategies, rewilding is also a concept being applied well beyond the boundaries of wilderness today (Lorimer et al. 2015) to promote restoration of free-willed nature across the landscape. Rewilding has been receiving increasing worldwide attention, with particular focus in Europe and North America. In North America the original reason to use rewilding methods was to link lands in the National Wilderness Preservation System to larger landscapes (Hintz 2007). In Europe, it has been applied in a broader way, often in countries without established wilderness systems but where land abandonment is increasingly common (Corlett 2016). Lorimer et al. (2015) summarized that these efforts serve to maintain or increase biodiversity, while reducing the impact of present and past human interventions through the restoration of both species and ecological processes.

The concept of trophic levels has emerged as a central organizing theme for a great deal of rewilding research and projects because protecting the benefits of intact trophic cascades— relating to food and feeding (trophic) and enhancing the number of opportunities for animals, plants and other creatures to feed on each other (cascading)—is important in understanding natural systems (Monbiot 2014; Svenning et al. 2015). Wilderness areas, with free-willed ecosystems, are an important way to recapture these benefits. The emotional human aspects of freedom, self-will, and wildness are important goals of rewilding. Greater awareness of the importance of social and cultural factors in conservation have contributed to the popularity of rewilding (Corlett 2016). Monbiot (2014), in describing the emotional attraction and motivation of feral environments, suggests that his primary excitement about rewilding is not to protect natural processes, but to witness these processes in a self-willed environment (or simply know they exist in some places). Excitement about reconnecting landscapes, seascapes, rivers, and humans to free-willed nature is central to wilderness conservation through rewilding.

Soule and Noss (1998) emphasized the 3 Cs of rewilding in North America: core areas, corridors, and carnivores. The "3 Cs" was subsequently elaborated on by adding more "Cs": climate, compassion, and coexistence, according to Johns (2019). In Europe, there was realization that protection of special sites alone would not secure conservation goals, so more focus on connectivity through ecological corridors, core wild areas, restoration, and buffer zones emerged (Hobbs 2002; Jones-Walters 2007). A major aspect of European efforts in rewilding has been an emphasis on naturalistic grazing and effects on biodiversity (Regos et al. 2016). While predators have yet to be restored in large numbers there, focus has been more on the role of large herbivores in opening up forest canopy into more park-like settings. Rewilding is a popular concept in Europe, particularly in areas of rural depopulation and land abandonment (Lorimer et al. 2015; Regos et al. 2016; Corlett 2016) and a recent subject of the IUCNs Commission on Ecosystem Management (CEM) to propose guiding principles for rewilding, which clarified that rewilding exists on a continuum of scale, connectivity, and level of human influence and aims to restore ecosystem structure and functions to achieve self-sustaining autonomous nature (Carver et al. 2021).

Rewilding, however, does not have a single, simple definition. Similar to wilderness and wildness, it is defined differently in different cultural and geographical contexts. Jørgensen (2015) details multiple rewilding definitions as: (1) cores, corridors, carnivores; (2) Pleistocene megafauna replacement; (3) island taxon replacement; (4) landscape through species reintroduction; (5) productive land abandonment; and (6) releasing captive-bred animals into the wild. Jørgensen (2015) amusingly (but realistically) adds another use of the term by suggesting that rewilding has also been adopted by activists as a rallying "plastic" word. This plasticity might be celebrated by some as giving license to activate the emotional aspect of returning control to nature. Like the term wilderness, the plastic uses of rewilding illustrate the cultural attraction to wild nature, to nostalgic feelings, and the sense of loss created by the arrival and domination by humans in the Anthropocene era. Lorimer et al. (2015) emphasized that rewilding can be an approach that seeks to maintain or increase biodiversity and reduce or reverse past and present human impacts by restoring more functional ecosystems. Incorporating cultural landscape considerations into rewilding decisions, however, is also an important criterion. The level of human intervention (trammeling) during ecological restoration, required to achieve some prescribed set of conditions or biological components is often controversial, but the endpoint goal of freedom or self-will of a fully cascading trophic system is the ultimate goal. The costs in terms of level of intervention during ecological restoration to get to this ultimate goal remain a topic of great debate.

Restoration, by itself, can have the primary goal of better ecosystem functioning to produce human benefits of water, food production, etc. Just one major, positive demonstration of this is the 4-million-acre (1.6 million ha) Loess Plateau Watershed Rehabilitation Project in central China, guided by John D. Liu and summarized for the World Bank by Chen et al. (2004). But, beyond restoration, there is rewilding, with a primary goal of restoring self-sustaining and autonomous nature, often involving ecological restoration as a prerequisite step. The global importance of restoration and rewilding is evidenced by the United Nations' Decade on Ecosystem Restoration (DECADE; https://www.decadeonrestoration.org/), generally spanning 2021 to 2030. This initiative draws together global principles, on-the-ground projects, "community toolkits," and more.

The World Wilderness Congress

The World Wilderness Congress (WWC), a project of The WILD Foundation and associates (www.wild.org), has met on ten occasions (1977, South Africa; 1980, Australia; 1983, Scotland; 1987, the United States; 1993, Norway; 1998, India; 2001, South Africa; 2005, Alaska; 2009, Mexico; and 2013, Spain) to explore management, cultural, and scientific elements of wilderness protection (Martin 2001). The 11th World Wilderness Congress was planned for Jaipur, India, in March 2020, and was canceled just two weeks before convening due to the Covid-19 pandemic. Discussions are underway for the 12th World Wilderness Congress. The WWC has provided a continuing international forum for expanding the wilderness concept, and its plenary and technical symposia proceedings document important international perspectives and information on wilderness values, opportunities, management, science, and cultural issues. The WWC has also been adamant, since its inception, that Indigenous and traditional communities must be involved as full partners, not stakeholders, in international wildlands conservation.

The World Wilderness Congress has been a focal point for the worldwide evolution of wilderness definition and legislation. Each of the ten World Wilderness Congresses has worked toward a definition of wilderness and wilderness management guidelines acceptable to the world community, with each proceedings recording new perspectives. Ultimately a resolution and proposed definition from the 4th World Wilderness Congress (Martin 1988) was submitted and adopted by the IUCN Commission on National Parks and Protected Areas—now called the World Commission on Protected Areas, discussed earlier. The IUCN protected area classification (Category 1b) definition for wilderness was subsequently updated to *"A large area of unmodified or slightly modified land, and/or sea, retaining its natural character and influence, without permanent or significant habitation, which is protected and managed so as to preserve its natural condition"* (Kormos et al. 2008, 22).

There is a distinct difference between this evolving international definition and that contained in the 1964 U.S. Wilderness Act, "Where the earth and its community of life are untrammeled by man, where man himself is a visitor who does not remain." In developing countries, many wildland areas are either occupied by Indigenous people or are under pressure from subsistence farmers, people who engage in shifting cultivation, grazing, or other extractive and potentially damaging practices.

The World Wilderness Congress is very different from most conferences. Each new congress is conceived and implemented in and with a different country, and on a few occasions, it returns years later to a country to reconvene and update. Practical conservation and wilderness objectives are established, and a structure and process implemented to achieve them. The WWC has an impressive list of these accomplishments, some of which have been mentioned elsewhere in this chapter, such as helping initiate the Global Environmental Facility, establishing the International League of Conservation Photographers, initiating the Native Lands and Wilderness Council, pioneering the marine wilderness concept, establishing wilderness on private lands, creating the first World Wilderness Inventory, creating the first inventory of freshwater wilderness; designating the first wilderness area in Latin America (Mexico); helping launch a significant new focus on Rewilding Europe, and others.

Even though the 11th World Wilderness Congress was canceled, planning had proceeded to the point that practical outcomes were still accomplished, such as the Global Charter for Rewilding the Earth (Rewilding Charter Working Group 2020), out of which emerged, within a few months, the Global Rewilding Alliance (www.globalrewilding.earth).

The Aldo Leopold Wilderness Research Institute collaborated with several U.S. federal agencies and international cooperators to produce technical proceedings from the 6th World Wilderness Congress in India (Watson et al. 1998) and (Watson et al. 2000), the 7th WWC in South Africa (Watson and Sproull 2003), the 8th WWC in Alaska (Watson et al. 2007), the 9th WWC in Mexico (Watson et al. 2011b) and the 10th WWC in Spain (Watson et al. 2015) and are available online (https://leopold.wilderness.net/).

International Journal of Wilderness

Another initiative of The WILD Foundation and several academic, NGO, and government collaborators, in close partnership with the WWCs, is the International Journal of Wilderness (IJW). Launched in 1995, the IJW is sponsored by a combination of federal agencies, universities, and nongovernmental organizations (four of them in countries other than the United States). The IJW features articles on wilderness protection, allocation and designation, science, management, and uses worldwide. It services a professional network of wilderness professionals, managers, scientists, and advocates, and is another reflection of the international growth and acceptance of wilderness. As a forum for scientific and professional articles, opinions, and discussion about wilderness around the world, IJW is published in three issues per year (www.ijw.org) and includes peer-reviewed professional and science research articles as well as opinion and feature articles about wilderness designation, stewardship, and management concepts and issues. Some examples of wilderness-related special issues in recent years include a focus on wilderness in specific countries; achieving wilderness fire restoration; wild river inventories, science, and protection; long trail designation and protection across boundaries; the 50th anniversary of the Wilderness Act; and wilderness ecosystem services. In 2017, the journal initiated online-only publication. The website (www.ijw.org) provides access to past volumes that you can read online and are available to download PDF (from 1995 through the past year's volume).

SUMMARY

The idea of protecting wilderness character on a large scale began in the United States, but it has gained recognition and support around the world, both through official efforts to designate wilderness areas in some countries and through the designation of individual areas that meet IUCN Protected Area Category 1b in many more. In 1989, there were forty-four areas registered as wilderness within the IUCN system, (Eidsvik 1989) and by 2016 there were 2,992 marine and terrestrial wilderness areas registered with the IUCN as solely Category 1b sites (UCN & UNEP-WCMC, 2016). The American notions about wilderness are an influence, but not a determinant, on how other nations approach the idea. Differences of geography, culture, and economics alter the specific ways in which other countries approach designating and managing wilderness.

This chapter describes countries where wilderness is protected by law (Class I), as an administrative zone in a park or other protected area (Class II), or as part of another conservation program designation (Class III). The concept of wilderness varies around the globe in its approach to designation and protection criteria, ecological protection, and Indigenous populations, which are major issues regarding wilderness in the U.S., Australia, South America, the Philippines, and Finland. These are all issues that individual countries must confront and resolve considering their own social, economic, and cultural circumstances. In developing countries, the need for resolving wilderness and wildland protection with economic and rural development needs is a primary consideration.

International interest in wilderness has grown and continues to grow. Global progress has been made institutionally through the recognition of wilderness by the IUCN as a formal, distinct category of protected areas (Category 1b) and guidelines have been distributed that can guide designation and protection. Numerous countries around the world have begun to consider how wilderness might fit into their conservation management systems and have even initiated large-scale assessments of wilderness potential. Even in some countries where long-term occupancy and development have greatly modified the landscape, the possibility of restoring wilderness and natural conditions is being discussed. The wilderness idea reflects an increasing understanding that wilderness settings are a repository of a country's natural heritage and a symbol of that country's international responsibility to protect the environment and care about quality of life for future populations.

STUDY QUESTIONS

1. What role do you believe wilderness protection efforts in the United States played internationally in protecting wilderness?
2. What are some differences between wilderness protection in the United States and in other countries?
3. Identify and discuss some of the factors that account for differences in progress in wilderness preservation between the United States and Europe.
4. Identify some of the important relationships between wilderness and other international conservation designations.
5. As population increases throughout the world, do pressures for development in lesser-developed countries lead to undesirable changes in ecological conditions? What strategies and actions will be necessary to improve prospects for wilderness and wildland protection in developing nations?
6. Why was there interest in inventorying the freshwater and marine waters of the world for natural and wild conditions? What is the value of protecting some of these areas?
7. Explain the distinctions between Class I, Class II, and Class III wilderness designations presented in this chapter. What nations have Class I wilderness? How are they different in their mechanisms for protection?
8. How are Russian zapovedniks similar to wilderness in the United States? How are they different?
9. What are transfrontier conservation areas?
10. Where would you look to find information on IUCN Protected Area Category 1b Wilderness designations in the country where you live?

ACKNOWLEDGMENTS

This chapter was updated for the fifth edition by Vance Martin, founder/past president of The WILD Foundation; Alan Watson, formerly of the Aldo Leopold Wilderness Research Institute, USDA Forest Service, Rocky Mountain Research Station; Christopher Armatas of the Aldo Leopold Wilderness Research Institute; and Amy Lewis, CEO of the WILD Foundation. Their work updates and expands on Martin and Watson's coauthorship of the fourth and third editions, building on the second edition by George Stankey, Vance Martin, and Roderick Nash and the first edition by Roderick Nash.

The coauthors wish to thank Cyril Kormos, CEO of Wild-Heritage (and former vice president for policy at The WILD Foundation) for his previous contributions to the section on international wilderness law and policy. We also wish to thank our many colleagues for their reviews and contributions to this and earlier editions, but the interpretation, conclusions, errors, and omissions are the responsibility of the authors. We thank Bill Bainbridge, consultant (retired), Natal Parks Board, South Africa; Hugh Barr, past president, Federated Mountain Clubs of New Zealand; Ralf Buckley,

Griffith University, Australia; Julie Cajune of the Confederated Salish and Kootenai Tribes; CaoYue, National Tsinghua University in China; Gordon Cessford, formerly Department of Conservation, New Zealand; Paul Cryer, Applied Ecology Unit, African Conservation Trust; Paul Dingwall, formerly Department of Conservation, New Zealand; Grant Dixon, independent researcher, Australia; Brian Dych, BC Parks, Victoria, British Columbia, Canada; Tom Elliot, Parks Canada, White Horse, Yukon, Canada; Harold Eidsvik (retired), Parks Canada; Martin Hawes, independent researcher, Australia; Peter Helman, private consultant, Sydney, Australia; Steve Hollenhorst, University of Idaho; Evelyn Hurwich, The Antarctica Society, U.S.; Ian Lockwood, Overseas School of Colombo, Pelawatte, Sri Lanka; Alec Marr and Julie McGiness, The Wilderness Society, Australia; Metsähallitus, formerly Finnish Forest and Park Service, Finland; Brad Mills, Land Conservation Council of Victoria, British Columbia, Canada; Pers Nilsens, Parks Canada, Ottawa; Dave Ostergren, Northern Arizona University; Vicki Sahanatien, PhD, Bella Coola, British Columbia; George Wallace, International School for Forestry and Natural Resources, Colorado State University, Fort Collins; and Mikhail Yablokov, formerly Russian Ministry of the Environment and Natural Resources.

The findings and conclusions in this publication are those of the authors and should not be construed to represent any official U.S. Department of Agriculture or U.S. Government determination or policy.

CASE DISCUSSION: CANADIAN WILDERNESS—THE QUETICO PROVINCIAL PARK

The Quetico Provincial Park (QPP) in northwestern Ontario, Canada has a Wilderness Zone occupying 90 percent of QPP's approximately 1.18 million acres (471,878 hectares). The southern boundary is shared with the Boundary Waters Canoe Area Wilderness in the Superior National Forest in the state of Minnesota (U.S.). The QPP was created in 1913 and classified as a wilderness park in 1977, and it has remained remote with difficult access except for two access roads. Canoeists who enter the area to enjoy the more than six hundred lakes and waterways must access the QPP by obtaining a permit via one of six ranger stations that serve more than twenty access points.

Historically, the area was used by Native peoples and later as a travel route for European explorers, fur trappers, and traders. In 1996, the Canadian federal and provincial governments designated the QPP as part of Boundary Waters—Voyageur Waterway and Canadian Heritage River. This designation was a historic and cultural tribute to the voyageurs who paddled these waters as fur traders.

The QPP is in an ecological transition zone between the plains to the west, the boreal forest to the north, and the mixed hardwood and softwood forests to the south. Fire has been a historic natural influence throughout the area. Logging and some mining activities have impacted the forests but were prohibited by 1971. In 1979, all motorized vehicles including motorboats were banned, except for an agreement with part of the Lac La Croix First Nation, which were allowed to operate motorized guideboats with less than a 10-horsepower motor until the year 2015.

The QPP is representative of the wilderness values most often reported by Canadians in justifying their support of wilderness protection: a bequest to future generations and protection from the advancement of civilization. The QPP is one of the first protected areas in the province of Ontario and the second largest such area. International recognition of the importance of protecting this area was noted by the IUCN, which classified the QPP as a Category 2 protected area—a national park designated for ecosystem protection and recreation use.

Recreational opportunities for remote and primitive canoe camping abound, with more than 2,000 unimproved campsites spread over the lakes of the QPP. There is also one campground with 107 sites that is accessible by vehicle at the Dawson Trailhead. The interior recreational use is restricted to daily quotas and regulated by personnel at the ranger stations through a permit process, with group size limits and the prohibition of motorized and mechanized equipment and of nonburnable, disposable food and beverage containers. The interior has both day use and overnight visitors, with an annual average of 125,000 visitors and 122,000 camper nights.

The initial QPP master plan was approved in 1977 and then revised in 1982, 1988, and 1995. In 2007–2008, when the fourth edition of this textbook was published, the QPP was engaged in a review of its management plan. Revision of that plan concluded in 2018 (Quetico Provincial Park 2018). Some of the major issues they were addressing in 2008 included:

1. Zoning (e.g., creation of nature reserve zones, new access zones).
2. Policies for resource stewardship, operations, and development for:
   a. Natural resources, including fisheries management, wildlife management, fire management, and vegetation management.
   b. Cultural resources—First Nation values (e.g., spiritual sites, pictographs) and historical sites (e.g., Cabin 16, King's Point).
   c. Operations—natural heritage education (e.g., interpretation, information services); research (e.g., wildlife, natural fire, fisheries); recreation management (e.g., carrying capacity and quotas, new recreation activities, and winter uses); Atikokan-Quetico Tourism recommendations; tourism services (e.g., outfitting); and marketing (e.g., partnerships, relationship with Atikokan, Lac La Croix, Thunder Bay).
   d. Development—economic development (e.g., Lac La Croix and Atikokan); new access points (e.g., locations, facility requirements); roads (e.g., Batchewaung Lake); and other (e.g., backcountry hiking/skiing trails).
3. Implementation priorities for stewardship, operations, and development policies.
4. Social and economic impact analysis (e.g., job creation, infrastructure).
5. Management planning discussions will recognize existing Aboriginal uses and the aspirations of the Lac La Croix First Nation to achieve economic and employment opportunities associated with the park.

CASE DISCUSSION QUESTIONS

1. The complexities of the issues that were addressed were a challenge to the planning team. What types of representation should there be on the planning team in view of the various constituents and disciplines affected and the considerations being reviewed? Does that change when you think about the mission of the QPP?
2. When you read over the list of considerations the planning team had to address, does it seem as if the definition of wilderness should factor into what was considered further and what was dropped? Do you think the definition of wilderness at that time was based on what was currently allowed or prohibited in the QPP? Would each of the five issue categories above, being considered under the QPP plan revision, have negatively impacted or positively supported that definition of wilderness?
3. Some people argue that Canada has such vast wildlands and waterways in the northern area of the nation that parks in the south like QPP should be declassified as wilderness and further developed. How would you argue for maintaining strong protection of wilderness values in QPP? What could international recognition, such as the IUCN Category 1b, do to support your argument?
4. The role of Indigenous people in the history and current management plan for this area is extensive. Is that historic role evident in the QPP plan?

Please see QPP (2018) for the way these discussion questions were addressed by the planning team.

REFERENCES

Bainbridge, W. R.; Lax, I. 2008. South Africa. In: Kormos, Cyril, ed. *A Handbook on International Wilderness Law and Policy*. Golden, CO: Fulcrum, pp. 241–267.

Barnes, J. I. 1998. Economic value of wilderness in Namibia. *International Journal of Wilderness*. 4(1): 33–38.

Barnett, S.; Stewart, J.; Wilson, P. 2015. Wilderness Areas, Then and Now. FMC Bulletin. 200: 18-22. Official quarterly of the Federated Mountain Clubs of New Zealand.

Beltrán, Javier, ed. 2000. Indigenous and Traditional Peoples and Protected Areas: Principles, Guidelines and Case Studies. IUCN, Gland, Switzerland and Cambridge, UK and WWF International, Gland, Switzerland. xi + 133p. [Pre-publication]

Brown, B. 1982. The use and misuse of wilderness in SW Tasmania. In: Martin, Vance G., ed. *Proceedings of the 2nd WWC*. Fulcrum Publishing, pp. 81–86.

Brooks C.; Bloom, E.; Kavanagh, A.; Nocito, E.; Watters, G.; Weller, J. 2021. The Ross Sea, Antarctica: A highly protected marine protected area in international waters. *Marine Policy*. 134: 104795.

Brüggemann, Jens. 1997. National parks and protected areas management in Costa Rica and Germany: A comparative analysis. In: Ghimire, Krishna B.; Pimbert, Michael D., eds. *Social Change and Conservation: Environmental Politics and Impacts of National Parks and Protected Areas*. London: Earthscan, pp. 71–96.

Cambou, Dorothée. 2019. The UNDRIP and the legal significance of the right of indigenous peoples to self-determination: A human rights approach with a multidimensional perspective. *The International Journal of Human Rights*, 23(1–2): 34–50.

Cao, Yue; Long, Ying; Yang, Rui. 2017. Research on the identification and spatial distribution of wilderness areas at the national scale in Mainland China [J]. *Chinese Landscape Architecture*. 6: 26–33. [In Chinese; English translation available at https://ijw.org/2018-mapping-wilderness-in-mainland-china/]

Cao, Yue; Yang, Rui; Ying, Long; Carver, Steve. 2018. A preliminary study on mapping wilderness in Mainland China. *International Journal of Wilderness*. 24(2): 104–116.

Cao, Yue; Carver, Steve; Yang, Rui. 2019. Mapping wilderness in China: Comparing and integrating Boolean and WLC approaches. *Landscape and Urban Planning*, 192: 103636.

Cao, Yue; Yang, Rui; Carver, Steve. 2020. Linking wilderness mapping and connectivity modelling: A methodological framework for wildland network planning. *Biological Conservation*. 251: 108679.

Cao, Yue; Tseng, Tz-Hsuan; Wang, Fangyi; Jacobson, Andrew; Yu, Le; Zhao, Jianqiao; Carver, Steve; Locke, Harvey; Zhao, Zhicong; Yang, Rui. 2022a. Potential wilderness loss could undermine the post-2020 global biodiversity framework. *Biological Conservation*. 275: 109753.

———. 2022b. Identifying ecosystem service value and potential loss of wilderness areas in China to support post-2020 global biodiversity conservation. *Science of the Total Environment*. 846: 157348.

Caouette, B. 2008. Japan. In: Kormos, Cyril, ed. *A Handbook on International Wilderness Law and Policy*. Golden, CO: Fulcrum, pp. 171–181.

Carver, Steve. 2018. The river wild: Towards a global assessment of wild rivers. *International Journal of Wilderness*. 24(1).

Carver, Steve; Convery, Ian; Hawkins, Sally; Beyers, Rene; Eagle, Adam; Kun, Zoltan; Van Maanen, Erwin, et al. 2021 Guiding principles for rewilding. *Conservation Biology*.

Casson, Sarah A.; Martin, Vance G.; Watson, Alan; Stringer, Angie; Kormos, Cyril F.; Locke, Harvey; Ghosh, Sonali, et al. 2016. Wilderness protected areas: Management guidelines for IUCN Category 1b protected areas. Gland, Switzerland: IUCN. x + 92p. https://portals.iucn.org/library/sites/library/files/documents/PAG-025.pdf; accessed December 6, 2022.

Convention on Biological Diversity. 2022. Open-ended working group on the post-2020 global biodiversity framework. *Third Meeting (Part II)*. Geneva, Switzerland; March 14–29, 2022. https://www.cbd.int/article/draft-1-global-biodiversity-framework; accessed December 6, 2022.

Cessford, Gordan R.; Dingwall, Paul R. 1997. Wilderness and recreation in New Zealand. *International Journal of Wilderness*. 3(4): 39–43.

Cessford, Gordan R.; Reedy, Murray C. 2000. Wilderness status and associated management issues in New Zealand. In: Watson, Alan E.; Aplet, Greg H.; Hendee, John C., comps. *Personal, Societal, and Ecological Values of Wilderness: 6th World Wilderness Congress Proceedings on Research, Management and Allocation*, Vol. II; October 24–29, 1998; Bangalore, India. RMRS-P-14. Ogden, UT: U.S. Department of Agriculture, Forest Service, Rocky Mountain Research Station, pp. 185–192.

Cessford, Gordan R. 2008. Personal communication with Vance Martin.

Chen, S.; Wang Y., et al. 2004. Loess Plateau Watershed Rehabilitation Project. Scaling Up Poverty Reduction: *A Global Learning Process and Conference Shanghai*; May 25–27, 2004. https://web.worldbank.org/archive/website00819C/WEB/PDF/CHINA_LO.PDF.

Commonwealth of Australia. 2021. Collaborative Australian Protected Areas Database (CAPAD). 2020.

Commonwealth of Australia. 1998. The Malimup Communiqué. Native title and protected areas project. https://indig-enviro.asn.au/Support21.htm; accessed January 18, 2023.

Cooper, T. 2001. Wilderness in Namibia. In: Martin, Vance, ed. *Wilderness and Humanity: The Global Issue. Proceedings of the 6th World Wilderness Congress*; October 24–29, 1998; Bangalore, India. Golden, CO: Fulcrum, pp. 128–133.

Corlett, R. T. 2016. Restoration, reintroduction, and rewilding in a changing world. *Trends in Ecology & Evolution*. 31(6): 452–462.

Cryer, P. B. 2009. *The implementation of an environmental monitoring and management system in the wilderness area of the Hluhluwe-Imfolozi Park.* Thesis. Centre for Environment, Agriculture and Development, School of Applied Environmental Sciences, University of KwaZulu-Natal.

Confederated Salish and Kootenai Tribes (CSKT). 2005. Mission mountains tribal wilderness. A case study. *Native Lands and Wilderness Council.* https://wild.org/wp-content/uploads/2010/02/Mission-Mountains-Tribal-Wilderness.pdf; accessed May 8, 2023.

Cumberland, Kenneth B. 1981. *Landmarks.* Surry Hills, New South Wales, Australia: Reader's Digest Proprietary.

Dasgupta, Partha. 2021. The economics of biodiversity: The Dasgupta review. London: HM Treasury. http://www.gov.uk/official-documents.

Dingwall, Paul. 1997. Environmental management for Antarctic wilderness. *International Journal of Wilderness.* 3(3): 22–26.

Draper, M. 1998. Zen and the art of Garden Province maintenance: The soft intimacy of hard men in the wilderness of KwaZulu-Natal, South Africa, 1952–1997. *Journal of Southern African Studies.* 24(4): n.p.

Dudley, N., ed. 2013. Guidelines for Applying Protected Area Management Categories. IUCN, Gland, Switzerland.

Eidsvik, H. K. 1989. The Status of Wilderness: An International Overview. Natural Resources Journal 29: 57–82.

Environment and Climate Change Canada (ECCC). 2021. Canadian Protected and Conserved Areas Database. https://www.canada.ca/en/environment-climate-change/services/national-wildlife-areas/protected-conserved-areas-database.html.

———. 2022. Canadian Environmental Sustainability Indicators: Canada's conserved areas. Consulted on January 4, 2023. www.canada.ca/en/environment-climate-change/services/environmental-indicators/conserved-areas.html.

Environment Guide. 2014. "Te Urewera Act." https://www.environment-guide.org.nz/regional/te-urewera-act/.

European Commission. 2018. Innovative financial instruments. https://ec.europa.eu/info/business-economy-euro/growth-and-investment/financing-investment/innovative-financial-instruments_en; accessed December 12, 2022.

European Parliament. 2009. European Parliament resolution of 3 February 2009 on wilderness in Europe. https://www.europarl.europa.eu/doceo/document/TA-6-2009-0034_EN.html; accessed December 6, 2022.

European Wilderness Society. 2022. European wilderness quality standard and audit system: The need for a European wilderness quality standard. https://wilderness-society.org/european-wilderness-definition/european-wilderness-quality-standard-audit-system/; accessed December 6, 2022.

Faries, W.; Cervigni, R. 1998. Achieving sustainable conservation: The global environment facility. *International Journal of Wilderness.* 4(2): 32–33.

Fisher, M.; Carver, S; Kun, Z.; McMorran, R.; Arrell, K.; Mitchell, G. 2010. Review of status and conservation of wild land in Europe. Project commissioned by the Scottish Government. Final Report, Tender Ref: CR/2009/31, Project No: WRI/001/09. On file at The Wildland Research Institute, The University of Leeds, Leeds, UK.

Fletcher, Michael-Shawn; Hamilton, Rebecca; Dressler, Wolfram; Palmer, Lisa. 2021. Indigenous knowledge and the shackles of wilderness. *PNAS* 118(40): e2022218118.

Forest and Finance Policy Assessment. 2022. Is your money destroying rainforests or violating rights? https://forestsandfinance.org/wp-content/uploads/2022/10/FF2022-policy-briefer.pdf.

Foster, N.; Lemay, M. H. 1988. Ocean wilderness—Myth, challenge or opportunity? In: Martin, Vance G., ed. *For the Conservation of the Earth. Proceedings of the 4th World Wilderness Congress*; September 11–18, 1987; Denver and Estes Park, CO. Golden, CO: Fulcrum, pp. 171–174.

Germani, F. S.; Sepúlveda, M. 2021. Chile adopts international certification standards for management of protected areas: Inclusion of national parks on IUCN Green List would recognize country's work on conservation. Pew Charitable Trust. Retrieved May 2, 2023.

Glenday, J. 2014. Federal Government will respect UN committee's Tasmanian forest ruling. ABC. Australian Broadcasting Corporation. June 24, 2014, archived from the original on September 1, 2014. Retrieved December 11, 2014. https://www.abc.net.au/; accessed December 12, 2022.

Goodland, R. J. A. 1987. The World Bank's wildlands policy: A major new means of financing conservation? *Conservation Biology.* 1(3): 210–213.

Green, M. J. B.; Paine, J. 1997. State of the world's protected areas at the end of the twentieth century. Paper presented at the *IUCN World Commission on Protected Areas Symposium on Protected Areas in the 21st Century: From Islands to Networks*; November 24–29, 1997; Albany, Australia. Cambridge, England: IUCN Publications Services Unit. www.iucn.org.

Hannah, L.; Lohse, D.; Hutchinson, C.; Carr, J. L.; Lankerani, A. 1994. A preliminary inventory of human disturbance of world ecosystems. *Ambio.* 23: 246–250.

Hawes, Martin; Dixon, Grant; Ling, Roger. 2015. Assessing wilderness values—The Tasmanian Wilderness World Heritage Area, Australia. *International Journal of Wilderness* (21): 3.

Hawes, Martin; Dixon, Grant; Bell, Chris. 2018. Refining the definition of wilderness: Safeguarding the experiential and ecological values of remote natural land. Bob Brown Foundation, Hobart.

Hawes Martin; Dixon, Grant. 2021. What do we mean when we talk about wilderness? *Wild.* 18 (Spring 2021).

Hebblewhite, Mark; Hilty, Jodi A.; Williams, Sara; Locke, Harvey; Chester, Charles; Johns, David; Kehm, Gregory; Francis, Wendy L. 2022. Can a large-landscape conservation vision contribute to achieving biodiversity targets? *Conservation Science and Practice.* 4: e588.

Helman, P. M., Jones, A. D., Pigram, J. J., Smith, J. M. B. 1976. *Wilderness in Australia: Eastern New South Wales and Southeastern Queensland.* Armidale, Australia: University of New England, Department of Geography.

Hintz, John. 2007. Some political problems for rewilding nature. *Ethics, Place and Environment.* 10(2): 177–216.

Hobbs, Richard. J. 2002. Habitat networks and biological conservation. In: Gutzwiller, K. J., ed. *Applying Landscape Ecology in Biological Conservation.* New York: Springer, pp. 150-170.

Hummel, Monte. 1989. *Endangered Spaces: The Future for Canada's Wilderness.* Toronto, Ontario, Canada: Key Porter Books.

Indigenous Circle of Experts. 2018. We Rise Together. Achieving pathway to Canada Target 1 through the creation of Indigenous Protected and Conserved Areas in the spirit and practice of reconciliation.

Integrated Rural Development and Nature Conservation (IRDNC). 2019. IRDNC Strategy 2019–2024. https://www.irdnc.org.na/pdf/IRDNC-Strategy-2019%E2%80%932024.pdf; accessed December 12, 2022.

International Union for Conservation of Nature (IUCN). 1998. *1997 United Nations List of Protected Areas.* Cambridge, England. www.iucn.org.

———. 2016. IUCN Members' Assembly. Resolution 6.069: Defining Nature-Based Solutions, WCC.2016-Res-069. Gland, Switzerland.

Johns, David. 2019. History of rewilding: Ideas and practice. In: Pettorelli, N.; Durant, S.; du Toit, J., eds. *Rewilding.* Cambridge University Press, pp. 12–33.

Johnson, J. A.; Ruta, G.; Baldos, U.; Cervigni, R.; Chonabayashi, S.; Corong, E.; Gavryliuk, O.; Gerber, J.; Hertel, T.; Nootenboom, C.; Polasky, S. 2021. The economic case for nature: A global earth-economy model to assess development policy pathways. Washington, DC: World Bank. https://openknowledge.worldbank.org/handle/10986/35882.

Jones-Walters, Lawrence. 2007. Pan-European ecological networks. *Journal of Nature Conservation* 15: 262–264.

Jørgensen, Dolly. 2015. Rethinking rewilding. *Geoforum.* 65: 482–488.

Kajala, L. 2008. Finland. In: Kormos, Cyril, ed. *A Handbook on International Wilderness Law and Policy.* Golden, CO: Fulcrum, pp. 143–155.

Kelleher, Graeme. 1982. Management of the Great Barrier Reef. In: Martin, Vance, ed. *Wilderness.* Moray, Scotland: The Findhorn Press, pp. 136–141.

Kormos, C. F.; Bertzky, B.; Jaeger, T.; Shi, Y.; Badman, T.; Hilty, J.A.; Mackey, B.G.; Mittermeier, R.A.; Locke, H.; Osipova, E.; Watson, J. E. M. 2016. A wilderness approach under the World Heritage Convention. *Conservation Letters: Policy Perspectives.* 9(3): 228–235.

Kormos, C. F., ed. 2008. *A Handbook on International Wilderness Law and Policy.* Golden, CO: Fulcrum.

Kormos, C. F.; Martin, V.G. 2008. Southern and Eastern Africa. In: Kormos, Cyril, ed. *A Handbook on International Wilderness Law and Policy.* Golden, CO: Fulcrum, pp. 313–317.

Křenová, Z.; Kiener, H. 2012. Europe's wild heart—Still beating? Experiences from a new transboundary wilderness area in the middle of the old continent. *European Journal of Environmental Sciences*, Vol. 2, No. 2, pp. 115–124

Ledger, Sophie. E. H.; Rutherford, Claire A.; Benham, Charlotte; Burfield, Ian J.; Deinet, Stefanie; Eaton, Mark; Freeman, Robin; Gray, Claudia; Herrando, S.; Puleston, H.; Scott-Gatty, K.; Staneva, A.; McRae, L. 2022. Wildlife comeback in Europe: Opportunities and challenges for species recovery. Final report to Rewilding Europe by the Zoological Society of London, BirdLife International and the European Bird Census Council. London, UK: ZSL.

Lesslie, R. G.; Taylor, S. G. 1985. The wilderness continuum concept and its implications for Australian wilderness preservation policy. *Biological Conservation.* 32(4): 309–333.

Lesslie, R. G.; Maslen, M. 1995. *National Wilderness Inventory: Handbook of Principles, Procedures and Usage*, 2nd ed. Canberra: Australian Heritage Commission.

Lewis, A. 2023. Personal conversation with Chief Tashka Yawanawá.

Li, P. 2017. Proposing a National Protected River System in China: Ecological civilization. *International Journal of Wilderness.* 23(2): 64–70.

Locke, Harvey. 2018. The international movement to protect half the world: Origins, scientific foundations, and policy implications. *Elsevier: Reference Module in Earth Systems and Environmental Sciences.*

Lorimer, Jamie; Sandom, Chris; Jepson, Paul; Doughty, Chris; Barua, Maan; Kirby, Keith J. 2015. Rewilding: Science, practice, and politics. *Annual Review of Environment and Resources.* 40: 39–62.

Martin, Vance G. 1983. Personal conversation with Ian Player.

———, ed. 1988. Resolution of the Congress. In: *For the Conservation of the Earth: Proceedings of the 4th World Wilderness Congress*; September 11–18, 1987; Denver and Estes Park, CO. Golden, CO: Fulcrum.

———. 1990. International wilderness: Adapting to developing nations. In: Lime, David, ed. *Managing America's Enduring Wilderness Resource.* Conference proceedings; September 11–17, 1989; St. Paul: University of Minnesota Extension Service, pp. 252–266.

———. 1994. Wilderness designation as a global trend. In: Martin, Vance; Tyler, N., ed. *Arctic Wilderness: Proceedings of the 5th World Wilderness Congress*; September 1993. Tromso, Norway. Golden, CO: Fulcrum, pp. 8–19.

———. 2001. The World Wilderness Congress. *International Journal of Wilderness.* 7(1): 4–9.

———. 2017. Wilderness—International Perspectives and the China Opportunity. *Chinese Landscape Architecture.* 33(258): pp 5–9. [Abstract in English.]

Martin, Vance G.; Robles Gil, P. 2009. Wilderness. Collection Voces de la Tierra Sierra Madre. Sierra Madre, 123 p.

Martin, Vance G.; Hill, Melanie. 2021. Urban wildness—A more correct term than "urban wilderness" *Landscape Architecture Frontiers.* 9(1): 80–90.

Maxwell, S. L.; Cazalis, V.; Dudley, N.; Hoffmann, M.; Rodrigues, A. S. L.; Stolton, S.; Visconti, P.; Woodley, S.; Kingston, N.; Lewis, E.; Maron, M.; Strassburg, B. B. N.; Wenger, A.; Jonas, H. D.; Venter, O.; Watson, J. E. M. 2020. Area-based conservation in the twenty-first century. *Nature.* 586(7828): 217–227.

McCloskey, J. M.; Spalding, H. 1988. A reconnaissance-level inventory of the wilderness remaining in the world. In: Martin, Vance G., ed. *For the Conservation of the Earth: Proceedings of the 4th.* Golden, CO: Fulcrum.

McCloskey, Michael. 2001. Wild rivers of the world. In: Martin, Vance, ed. *Wilderness and Humanity: The Global Issue. Proceedings of the 6th World Wilderness Congress*; October 24–29, 1998; Bangalore, India. Golden, CO: Fulcrum, pp. 20–25.

McCloskey, Maxine. 2001. The high seas: Is there room for wilderness? In: Martin, Vance; Sarathy, M. A. Partha, eds. *Wilderness and Humanity: The Global Issue. Proceedings of the 6th World Wilderness Congress*; October 24–29, 1998; Bangalore, India. Golden, CO: Fulcrum, pp. 193–197.

McDonald, T. 1995. Mission Mountains Tribal Wilderness Area of the Flathead Indian Reservation. *International Journal of Wilderness.* 1(1): 20–21.

McDonnell, A. 2021. Affric Highlands, Scotland: Scotland's wild east-west corridor. https://rewildingeurope.com/landscapes/affric-highlands/; accessed December 6, 2022.

McLuhan, T. C., ed. 1971. *Touch the Earth: A Self-Portrait of Indian Existence.* New York: Outerbridge and Dienstfrey.

McNamee, K. 2008. Canada. In: Kormos, Cyril, ed. *A Handbook on International Wilderness Law and Policy.* Golden, CO: Fulcrum, pp. 91–117.

Mittermeier, R. A; Kormos, C. F.; Mittermeier, C. G.; Robles Gil, P.; Sandwith, T.; Besançon. C. 2005. *Transboundary Conservation: A New Vision for Protected Areas.* CEMEX-Agrupacion Sierra Madre-Conservation International, Mexico. 372p.

Mittermeier, R. A.; Mittermeier, C. G.; Gil, P. R.; Pilgrim, J. 2003. *Wilderness: Earth's Last Wild Places.* Conservation International and University of Chicago Press.

Molloy, L. F. 1983. Wilderness recreation—the New Zealand experience. In: Molloy, L. F., ed. *Wilderness Recreation in New Zealand: Proceedings of the FMC 50th Jubilee Conference on Wilderness*; August 22–24, 1981; Lake Rotoiti, Nelson Lakes National Park, New Zealand. Wellington, New Zealand: Federated Mountain Clubs of New Zealand, pp. 4–19.

———. 1997. Wilderness in New Zealand: A policy looking for someone to implement it. *International Journal of Wilderness.* 3(2): 11–45.

Monbiot, G. 2014. Feral: Rewilding the land, the sea and human life. Chicago: The University of Chicago Press.

Muir, K. 2007. Action toward wilderness protection in Australia. In: Watson, Alan; Sproull, Janet; Dean, Liese, comps. *Science and Stewardship to Protect and Sustain Wilderness Values: Eighth World Wilderness Congress Symposium*: September 30–October 6, 2005; Anchorage, AK. Proceedings RMRS-P-49. Fort Collins, CO: U.S. Department of Agriculture, Forest Service, Rocky Mountain Research Station.

Mutwa, C. Vusamazulu. 1999. *Indaba, My Children.* First American ed. New York: Grove Press, now Grove Atlantic.

Myers, Norman. 1998. Lifting the veil on perverse subsidies. *Nature.* 392: 327–328.

Nash, R. 1982. *Wilderness and the American Mind.* New Haven, Connecticut: Yale University Press.

Nelson, Michael P. 1998. An amalgamation of wilderness preservation arguments. In: Callicott, J. B.; Nelson, M. P., eds. *The Great New Wilderness Debate.* Athens: University of Georgia Press, pp. 154–200.

No'kmaq, M.; Marshall A; Beazley K. F.; Hum, J.; Joudry, S.; Papadopoulos, A.; Pictou, S.; Rabesca, J.; Young, L.; Zurba, M. 2021. "Awakening the sleeping giant": Re-Indigenization principles for transforming biodiversity conservation in Canada and beyond. *FACETS.* 6: 839–869.

Ostergren, D. M.; Shvarts, E. 2000. Russian zapovedniki in 1998: Recent progress and new challenges for Russia's strict nature preserves. In: Watson, A. E.; Aplet, G. H.; Hendee, J. C., comps. *Personal, Societal, and Ecological Values of Wilderness: 6th World Wilderness Congress proceedings on research, management, and allocation*; Vol. 2; October 24–29, 1998; Bangalore, India. RMRS-P-14. Ogden, UT: United States Department of Agriculture, Forest Service, Rocky Mountain Research Station, pp. 209–213.

Ostergren, D. M. 2001. An organic act after a century of protection: The context, content and implications of the 1995 Russian Federation law on specially protected natural areas. *Natural Resources Journal.* 41(1): 125–152.

Parks Canada Agency. 2000. Unimpaired for future generations? Protecting ecological integrity with Canada's National Parks. Vol. I. "A call to action." Report of the Panel on the Ecological Integrity of Canada's National Parks. Ottawa, Ontario: National Parks Canada.

Pietikäinen, S. 1995. Finland's wilderness act—a Scandinavian model. In: Martin, Vance; Tyler, N., eds. *Arctic Wilderness: Proceedings of the 5th World Wilderness Congress*; September 1993; Tromso, Norway. Golden, CO: Fulcrum, pp. 181–186.

Plumptre, A.; Baisero, D.; Belote, R. T.; Vázquez-Domíguez, E.; Faurby, S.; Jędrzejewske, W.; Kiara, H.; Kühl, H.; Benítez-López, A.; Luna-Aranguré, C.; Voigt, M.; Wich, S.; Wint, W.; Gallego-Zamorano, J.; Boyd, C. 2021. Where might we find ecologically intact communities? *Frontiers in Forests and Global Change.* (4): 626635.

Pollisco, W, S.; Meniado, A. P. 1997. Implementation of the National Integrated Protected Areas System Act. *Sylvatrop.* 7(1–2): 2–6.

Pryde, P. R. 1987. The distribution of endangered fauna in the USSR. *Biological Conservation.* 42(1): 19–37.

Quetico Provincial Park. 2018. Quetico Gwetaming Management Plan. https://files.ontario.ca/quetico-gwetaming-park-management-plan-2018.pdf; accessed December 12, 2022.

Rambaldi, G. 1997. RRA as a tool in integrating people's participation in protected areas management in the Philippines. *Sylvatrop.* 7(1–2): 28–39.

Raygorodetsky, G. 2018. Indigenous Peoples defend Earth's biodiversity: But they're in danger. *National Geographic.* November 16, 2018. https://www.nationalgeographic.com/environment/article/can-indigenous-land-stewardship-protect-biodiversity-.

Regos, A.; Domínguez, J.; Gil-Tena, A.; Brotons, L.; Ninyerola, M.; Pons, X. 2016. Rural abandoned landscapes and bird assemblages: winners and losers in the rewilding of a marginal mountain area (NW Spain). *Regional Environmental Change.* 16: 199–211.

Rewilding Charter Working Group. 2020. Global charter for rewilding the Earth. *The Ecological Citizen.* 4(A): 6–21.

Rosu, Alexandru-Cristian. 2014. Property restitution in Romania. *Global Economic Observer; Bucharest.* 2(2): 37–41.

Rutledge, R.; Vold, T. 1995. Canada's wilderness. *International Journal of Wilderness.* 1(2): 8–13.

Saarinen, J. 1998. Wilderness, tourism development, and sustainability: Wilderness attitudes and place ethics. In: Watson, A. E.; Aplet, G. H.; Hendee, J. C., comps. *Personal, Societal, and Ecological Values of Wilderness: Proceedings of the 6th World Wilderness Congress proceedings on research, management, and allocation*, Vol. I; October 24–29, 1998; Bangalore, India. Proc. RMRS-P-4. Ogden, UT: USDA, Forest Service, Rocky Mt. Research Station, pp. 29–36.

Sanderson, Eric W.; Jaiteh, Malanding; Levy, Marc A.; Redford, Kent H.; Wannebo, Antoinette V.; Woolmer, Gillian. 2002. The human footprint and the last of the wild. *Bioscience.* 52 (10): 891–904.

Sawyer, N. 2015. Wilderness quality mapping—The Australian experience. In: Watson, A.; Carver, S.; Křenová, Z.; McBride, B., comps. *Science and Stewardship to Protect and Sustain Wilderness Values: 10th World Wilderness Congress Symposium*; October 4–10, 2013; Salamanca, Spain. Proc. RMRS-P-74. Fort Collins, CO: USDA, Forest Service, Rocky Mt. Research Station, pp. 100–108.

Scarzanella, Eugenia. 2002. Las bellezas naturales y la nación: Los parques nacionales en Argentina en la primera mitad del siglo XX. [Natural beauties and the nation: National parks in Argentina in the first half of the 20th century.] *Revista Europea de Estudios Latinoamericanos y del Caribe.* 73: 5–21. https://www.jstor.org/stable/25675985.

Shah, Anup. 2014. Nature and animal conservation. *Global Issues: Social, Political, Economic and Environmental Issues That Affect Us All.* January 9, 2014. https://www.globalissues.org/article/177/nature-and-animal-conservation; accessed December 6, 2022.

Sippola, A-L. 2000. Biodiversity in Finnish wilderness areas: Aspects on preserving species and habitats. In: Watson, A. E.; Aplet, G. H.; Hendee, J. C., comps. *Personal, Societal, and Ecological Values of Wilderness: Proceedings of the 6th World Wilderness Congress proceedings on research, management, and allocation*, Vol. II; October 24–29, 1998; Bangalore, India. RMRS-P-14. Ogden, UT: U.S. Department of Agriculture, Forest Service, Rocky Mountain Research Station, pp. 48–56.

Somers, J. 2008. New Zealand. In: Kormos, Cyril, ed. *A Handbook on International Wilderness Law and Policy.* Golden, CO: Fulcrum, pp. 193–213.

Soulé, M. E.; Noss, R. 1998. Rewilding and biodiversity: Complementary goals for continental conservation. *Wild Earth.* 8(3): 18–28.

Stankey, G. H. 1989. Beyond the campfire's light: Historical roots of the wilderness concept. *Natural Resources Journal.* 29: 9.

Stewart, P. 2015. The possibilities of North Island Wilderness. *FMC Bulletin.* 200: 26–29. Official quarterly of the Federated Mountain Clubs of New Zealand.

Svenning, Jens-Christian; Pedersen, Pil B.; Donlan, C. Josh; Ejrnaes, R.; Faurby, S.; Galetti, M.; Hansen, D. M.; Sandel, B.; Sandom, C. J.; Terborgh, J. W.; Vera, F. W. M. 2015. Science for a wilder Anthropocene: Synthesis and future directions for trophic rewilding research. *Proceedings of the National Academy of Sciences of the United States of America.* 12: 1–7.

Sylvén, M., Wagner, K., & Martin, V.G. 2020. *The Story of Our Future: The Survival Revolution: The WILD11 Jaipur Declaration.* The WILD Foundation. www.wild.org/wilderness-specialist-group; www.wild11.org/outcomes; accessed December 12, 2022.

Tin, T.; Yang, R. 2016. Tracing the contours of wilderness in the Chinese mind. *International Journal of Wilderness.* 22(2): 35–40.

Trippensee, R. E. 1948. *Wildlife Management: Upland Game and General Principles.* New York. McGraw-Hill.

UCN and UNEP-WCMC. 2016. The World Database on Protected Areas (WDPA). http://www.protectedplanet.net; accessed January 10, 2023.

UCN and UNEP-WCMC. 2022. The World Database on Protected Areas (WDPA). http://protectedplanet.net; accessed September 2022.

UNEP-WCMC. 2017. World Database on Protected Areas User Manual 1.5. http://wcmc.io/WDPA_Manual.

U.S. Public Law 88–577. The Wilderness Act of September 3, 1964. 78 Stat. 890.

U.S. Department of the Interior. 2011. U.S., Mexico Announce Binational Cooperative Conservation Action Plan: Salazar, Elvira Underscore Commitment to Protecting and Preserving the Big Bend/Rio Bravo Region.

Ustaoglu, E.; Collier, M. J. 2018. Farmland abandonment in Europe: An overview of drivers consequences and assessment of the sustainability implications. *Environmental Reviews.* 26(4).

Wakild, E. 2014. Acts of courage, acts of culture: The Wilderness Act in Latin America. *Environmental History.* 19(4): 728–734.

Watson, A. E.; Aplet, G. H.; Hendee, J. C., comps. 1998. *Personal, Societal, and Ecological Values of Wilderness: 6th World Wilderness Congress Proceedings on Research, Management, and Allocation*, Vol. I; 1997; October; Bangalore, India. Proc. RMRS-P-4. Ogden, UT: U.S. Department of Agriculture, Forest Service, Rocky Mountain Research Station. 158 p. https://winapps.umt.edu/winapps/media2/leopold/pubs/342.pdf; accessed December 29, 2022.

Watson, A. E.; Aplet, G. H.; Hendee, J. C., comps. 2000. *Personal, Societal, and Ecological Values of Wilderness: 6th World Wilderness Congress Proceedings on Research, Management, and Allocation*, Vol. II; October 24–29, 1998; Bangalore, India. Proc. RMRS-P-14. Ogden, UT: U.S. Department of Agriculture, Forest Service, Rocky Mountain Research Station. 248 p. https://winapps.umt.edu/winapps/media2/leopold/pubs/375.pdf; accessed December 29, 2022.

Watson, A. E.; Carver, S.; Křenová, Z.; McBride, B., comps. 2015. *Science and Stewardship to Protect and Sustain Wilderness Values: 10th World Wilderness Congress Symposium*; October 4–10, 2013; Salamanca, Spain. Proceedings RMRS-P-74. Fort Collins, CO: U.S. Department of Agriculture, Forest Service, Rocky Mountain Research Station. 208 p. https://winapps.umt.edu/winapps/media2/leopold/pubs/846.pdf; accessed December 29, 2022.

Watson, A.; Matt, R.; Knotek, K.; Williams, D.; Yung, L. 2011a. Traditional wisdom: Protecting relationships with wilderness as a cultural landscape. *Ecology and Society.* 16(1): 36. http://www.ecologyandsociety.org/vol16/iss1/art36/.

Watson, A.; Murrieta-Saldivar, J.; McBride, B., comps. 2011b. *Science and Stewardship to Protect and Sustain Wilderness Values: 9th World Wilderness Congress Symposium*; November 6–13, 2009; Meridá, Yucatán, Mexico. Proceedings RMRS-P-64. Fort Collins, CO: U.S. Department of Agriculture, Forest Service, Rocky Mountain Research Station. 225 p. https://winapps.umt.edu/winapps/media2/leopold/pubs/740.pdf; accessed December 29, 2022.

Watson, A.; Carver, S.; Matt, R.; Waters, T.; Gunderson, K.; Davis, B. 2013. Place mapping to protect cultural landscapes on tribal lands. In: Stewart, W.; Williams, D.; Kruger, L., eds. *Place-Based Conservation.* Springer, Dordrecht.

Watson, A.; Matt, R.; Waters, T.; Gunderson, K.; Carver, S.; Davis, B. 2009. Mapping tradeoffs in values at risk at the interface between wilderness and non-wilderness lands. In: González-Cabán, Armando, ed. *Proceedings of the 3rd International Symposium on Fire Economics, Planning, and Policy: Common Problems and Approaches.* PSW-GTR-227. Albany, CA: USDA Forest Service, Pacific Southwest Research Station, pp. 375–387.

Watson, A.E.; Carver. S. 2017. Wilderness, rewilding and free-willed ecosystems: Evolving concepts in stewardship of IUCN protected category 1b areas. *Chinese Landscape Architecture.* 33(258): 34–38. [In Chinese. English translation available at https://winapps.umt.edu/winapps/media2/leopold/pubs/923.pdf.]

Watson, A.; Sproull, J., comps. 2003. *Science and Stewardship to Protect and Sustain Wilderness Values: 7th World Wilderness Congress Symposium*; November 2–8, 2001; Port Elizabeth, South Africa. Proc. RMRS-P-27. Ogden, UT: U.S. Department of Agriculture, Forest Service, Rocky Mountain Research Station. 272 p. https://winapps.umt.edu/winapps/media2/leopold/pubs/487.pdf; accessed 29 December 2022.

Watson, A.; Sproull, J.; Dean, L., comps. 2007. *Science and Stewardship to Protect and Sustain Wilderness Values: 8th World Wilderness Congress Symposium*; September 30–October 6, 2005; Anchorage, AK. Proceedings RMRS-P-49. Fort Collins, CO: U.S. Department of Agriculture, Forest Service, Rocky Mountain Research Station. 581 p. https://winapps.umt.edu/winapps/media2/leopold/pubs/623.pdf; accessed 29 December 2022.

Watson, J. E.; Venter, O.; Lee, J.; Jones, K. R.; Robinson, J. G.; Possingham, H. P.; & Allan, J. R. 2018. Protect the last of the wild. *Nature.* 563(7729): 27.

WCC. 2020. WCC-2020-Res-125-EN. Setting area-based conservation targets based on evidence of what nature and people need to thrive. https://portals.iucn.org/library/sites/library/files/resrecfiles/WCC_2020_RES_125_EN.pdf. A; accessed December 12, 2022.

Weiner, D. R. 1988a. *Models of Nature: Ecology, Conservation, and Cultural Revolution in Soviet Russia.* Bloomington: Indiana University Press.

———. 1988b. The changing face of Soviet conservation. In: Worster, Donald, ed. *The Ends of the Earth: Perspectives on Modern Environmental History.* Cambridge, England: Cambridge University Press, pp. 252–273.

Wild, R.; McLeod, C., eds. 2008. *Sacred Natural Sites: Guidelines for Protected Area Managers.* Gland, Switzerland: IUCN.

Wilson, E. O. 2016. *Half-Earth: Our Planet's Fight for Life.* Liveright/W.W. Norton.

World Bank. 1986. The World Bank's Operational Policy on Wildlands: Their Protection and Management in Economic Development.

World Commission on Environment and Development (WCED). 1987. *Our Common Future.* New York: Oxford University Press.

Wright, P.A.; Moghimehfar, F.; Woodley, A. 2019. Canadians' perspectives on how much space nature needs. *FACETS.* 4: 91–104.

Zealand, Clark T. W. 2007. Decolonizing Experience: An ecophenomenological investigation of the lived-experience of the Appalachian Trail thru-hikers. Thesis. University of Waterloo, Ontario, Canada. http://hdl.handle.net/10012/3036.

Zunino, F. 2007. A perspective on wilderness in Europe. *International Journal of Wilderness.* 13(3): 40–43.

II. U.S. Legal Authority and Process for Wilderness

CHAPTER

4

The Wilderness Act: Legal Basis for Wilderness Protection and Management

INTRODUCTION

In the United States, wilderness has evolved from a general, ill-defined, and historical concept to a specific and legal land designation. This evolution reflects the history of citizen concern for wilderness and its protection, and various actions taken by land-management agencies in response to these concerns. To fully understand the basis for wilderness protection in the United States, one must understand the history of the protection measures taken over the past seventy years and the reasons for those actions. We believe the same understanding is necessary to understand wilderness protection in other countries.

This chapter examines the early actions taken by the resource-management agencies to preserve wilderness, the reasons for these actions, and their outcomes. We also examine the growth of concern for statutory protection of wilderness that began following World War II and trace the way in which wilderness became a major item on the nation's political agenda. Finally, we discuss the 1964 Wilderness Act, the basic legislation underlying wilderness protection and management in the United States. All ensuing wilderness designation laws have their roots in the Wilderness Act. Some designation acts have contained special provisions that recognize or accommodate local situations, but such provisions are typically accompanied by specific recognition that they do not constitute precedents for other areas. In other cases, Congress has attempted to clarify the intent underlying certain Wilderness Act clauses, either through legislative language or through House or Senate committee reports. In the final analysis, the Wilderness Act has remained a stable base from which decisions to protect and manage wilderness in the United States can be made (Fig. 4.1).

THE L-20 REGULATION

As explained in Chapter 2, *the idea of wilderness protection may be a century old, depending on what is considered.* Although Yellowstone and Yosemite National Parks were not explicitly established to preserve wilderness, they nonetheless were major steps in protecting wilderness values. In the early 1900s, the individual efforts of Arthur Carhart and Aldo Leopold, both working within the United States Forest Service, brought specific recognition of wilderness as a land use. Yet there was much concern about the extent of the nation's remaining wilderness resource; that is, just how much roadless land remained, how much of it, where, and should it be protected from development?

In 1926, the USFS appraised the extent of wilderness remaining in the national forests. A national inventory of all areas greater than 230,000 acres reported that 74 tracts totaling 55 million acres still remained, with the largest tract about 7 million acres. Three years later, this inventory became a basis for *the first systematic program of wilderness protection; that is, Administrative Regulation L-20 promulgated by the USFS in 1929* (Roth 1984).

From a wilderness viewpoint, the L-20 Regulation was not very protective of the resource. It was primarily a list of permitted and prohibited uses. Timber harvesting was permitted

FIGURE 4.1 President Lyndon Johnson, surrounded by cabinet and congressional officials at the White House, signs the Wilderness Act, September 3, 1964. The Wilderness Act (U.S. Public Law 88-577) provides the basic framework governing the establishment and management of wilderness in the United States. Photo courtesy Lyndon Johnson Memorial Library, Austin, Texas.

in the so-called primitive areas in the belief that, if properly regulated, it might not be incompatible with the primitive-area designation. For example, logging occurred in about 80,000 acres of the South Fork Primitive Area, one of the three primitive areas that originally comprised what is now the Bob Marshall Wilderness in Montana.

Many uses generally considered incompatible with wilderness today were allowed. Although there were 72 primitive areas with a gross area of 13,482,421 acres in ten western states, the management plans for these areas allowed road construction in 15 areas, grazing in 62 areas, and logging in 59 areas. In only 4 primitive areas, totaling 297,221 acres, were logging, grazing, and roads absolutely excluded (Gilligan 1954).

Nor was the L-20 Regulation strictly enforced. The broad latitude in L-20 management stemmed from a belief that primitive-area status did not represent a long-term commitment of resources, but was intended to prevent road building and commercial development in areas of scenic and recreational attraction until such time as detailed management plans might be prepared (Outdoor Recreation Resources Review Commission 1962). Some have also suggested that the L-20 Regulation was a USFS strategy to combat the transfer of national forestlands to the National Park Service (Allin 1982; Roth 1984) by giving the USFS a land-use designation to compete with national parks and prevent transfer of such lands to NPS administration. Former USFS chief Richard McArdle (1952–1962) discounts this argument, pointing out that the creation of national forest wilderness has never prevented transfers of national forestland to the NPS. For example, the 226,000-acre Olympic Primitive Area, established by the USFS in 1938, was subsequently transferred to Olympic National Park in Washington (McArdle 1975).

The NPS was definitely concerned about the increasing awareness of recreation among some USFS officials. National Park Service director Stephen Mather frequently challenged the authority of the USFS to develop recreation programs, arguing that recreation was the sole responsibility of his agency (Baldwin 1972, 62–71).

The interagency rivalry was also reflected in the terminology used to identify wilderness. In the 1920s, the USFS called undeveloped areas "wilderness," while the NPS called them "primitive areas." Following USFS promulgation of the L-20 Regulation establishing primitive areas on the national forests, the NPS switched to calling its undeveloped areas wilderness.

THE U REGULATIONS

Dissatisfaction with the looseness of the L-20 Regulation led to its replacement with Regulations U-1, U-2, and U-3(a) in 1939 (Gilligan 1954) *as a means of preserving roadless lands on the national forests.* These new regulations were formulated largely under the influence of Robert Marshall, then chief of the Division of Recreation and Lands in the USFS. Marshall was a dynamic proponent of wilderness, and much of his career centered on efforts to strengthen wilderness preservation programs both inside and outside the federal government. Earlier, as director of the Forestry Division of the U.S. Office of Indian Affairs, Marshall was responsible for the designation of sixteen wilderness areas on Native American reservations (Nash 1982). Along with Aldo Leopold, Marshall was instrumental in establishing The Wilderness Society in 1935. Joining the USFS in 1937, Marshall pushed forcefully for expansion of wilderness reserves on the national forests and the U regulations culminated his efforts. Although Marshall's leadership for wilderness cannot be minimized, the role of his assistant in the Division of Recreation and Lands, John H. Sieker, has not been fully recognized or appreciated. Sieker, a forester educated at Yale and Princeton, enjoyed a close professional relationship with Marshall, endorsing his wilderness ideas, a key factor since he was well respected in the USFS. Marshall, while dynamic and energetic, was viewed by many as eccentric (Mitchell 1985). Politically, he was a liberal, advocating civil rights, civil liberties, and the redistribution of wealth, views that gained the enmity of conservative colleagues (Glover and Glover 1986). As an insider, Sieker became an important force in efforts to implement and gain acceptance for the U regulations (McArdle 1975).

The U regulations broadened the purpose of wilderness beyond what was earlier defined by the L-20 Regulation, and three land-use designations were recognized. Regulation U-1 established wilderness areas—tracts of land of not less than 100,000 acres. Acting on the recommendation of the chief of the USFS, the secretary of agriculture could designate such an area as wilderness, and only the secretary could modify or eliminate a wilderness area.

Regulation U-2 defined wild areas—tracts of land between 5,000 and 100,000 acres that could be established, modified, or eliminated by the chief of the USFS. Thus, in addition to their size, U-1 and U-2 areas were different in terms of who could establish, modify, or eliminate them, but they were managed identically.

Finally, Regulation U-3(a) established roadless areas to be managed principally for recreational use "substantially in their natural condition." Roadless areas larger than 100,000 acres could be established or modified only by the secretary of agriculture; areas less than 100,000 acres could be established or modified by the chief of the USFS. Timber cutting, roads, and other modifications were permissible if provided for in area-management plans.

The U regulations provided much more protection for wilderness than the L-20 Regulation they replaced. The U regulations were intended to be permanent, not just an interim measure to halt haphazard development (Allin 1982). They prohibited timber cutting, road construction, and special-use permits for such things as summer homes and hunting camps. Mechanized access, except where well established or in emergencies, was banned. Grazing and water-resource development was allowed. Mining was also allowed to continue, subject to existing mining and leasing laws. The USFS could insist that all such developments take pains to minimize impact on the wilderness (Roth 1984).

Because many of the seventy-six L-20 primitive areas had been established without adequate surveys, each was to be reviewed and reclassified under the new U regulation guidelines. A ninety-day public review period was called for. While under review, all primitive areas were to be managed according to U regulation requirements.

However, progress in reviewing the primitive areas was very slow; between 1939 and the outbreak of World War II, three areas were reclassified as wilderness, six as wild, and three were consolidated into the Bob Marshall Wilderness in

Montana. No reclassifications occurred during the war. After the war, reviews resumed, again slowly. By the late 1940s, only 2 million acres had become wilderness (Roth 1984). The slow progress and the loss of acreage in the reclassification process troubled conservationists. In some places, they argued, timbered lower-elevation areas within the old L-20 primitive-area boundaries were lost when the area was reclassified. Although the USFS argued that total primitive-area acreage had remained stable, critics countered that this was because high-elevation areas, devoid of any timber, had been substituted, thus lowering the quality and variety of the wilderness (Roth 1984).

STATUTORY PROTECTION FOR WILDERNESS

The L-20 Regulation and the U regulations were administrative designations; they were implemented at the discretion of the secretary of agriculture or the chief of the USFS. Because of the agency discretion under an administrative system of wilderness protection, wilderness proponents were concerned that roadless lands were not adequately protected. The answers seemed to lie in a legal, rather than an administrative, approach to wilderness protection in which the requirements and procedures for safeguarding wilderness values were prescribed by statute rather than left to the discretion of agency administrators.

Legal protection for wilderness was not a new idea. In the 1930s, Marshall supported the idea of establishing wilderness areas by congressional action (Nash 1982), as did H. H. Chapman, professor of forestry at Yale University (Chapman 1938), and Secretary of the Interior Harold L. Ickes (Mackintosh 1985).

Subsequently, the need for a legislatively protected wilderness system received additional support from a report issued by the Legislative Reference Service of the Library of Congress in 1949 (Keyser 1949). The report highlighted the widely disjointed programs of wilderness preservation; included opinions from a survey of numerous federal, state, and nonpublic organizations; and reported substantial concern for the future of wilderness and widespread support for wilderness protection as secure as that of the national parks.

In 1953, a dissertation by James P. Gilligan at the University of Michigan provided an appraisal of the condition and quality of the existing system of wilderness and primitive areas. He noted that although 13 million acres had been administratively set aside as wilderness, wild, or primitive areas, the conditions within them often left much to be desired. For example, he reported the existence of 200 miles of public roads, 145,000 acres of private inholdings, up to 500 mining claims, 60 mines, 24 airstrips, pasturage for 140,000 sheep and 25,000 cattle, and nearly 90 dams. Allin (1982) noted that Gilligan's information confirmed the concerns of many conservationists, that wilderness boundaries frequently had been modified, shifting land suitable to economic development outside the preserved area.

In 1962, the Wild Land Research Center at the University of California, with Gilligan as project director, in its report to the Outdoor Recreation Resources Review Commission (1962), recommended the enactment of legislation creating a wilderness system protected by law. Otherwise, the center predicted that the wilderness resource would be gradually lost, based on the following observations (Outdoor Recreation Resources Review Commission 1962):

1. Land-administering agencies could put wilderness to other uses.
2. Agencies lacked full jurisdiction over some land uses in wilderness, such as mining.
3. There was a lack of coordinated control over wilderness uses.
4. There was a lack of distinctive management policy.

One of the strongest advocates of a national wilderness system was Howard Zahniser (Fig. 4.2), executive director of The Wilderness Society (E. Zahniser 1984; H. Zahniser 1964; Scott 1976). His major theme was the need for a persisting program of wilderness preservation—a cohesive program that could eliminate the need for continual, fragmented holding actions against various threats. As early as 1949, he had outlined a wilderness system similar in structure to that eventually proposed in the first wilderness bill in 1956. At the Sierra Club's 1951 Biennial Wilderness Conference, Howard Zahniser (1964) remarked the following:

> Let's try to be done with a wilderness preservation program made up of a sequence of overlapping emergencies, threats, and defense campaigns. Let's make a concerted effort for a positive program that will establish an enduring system of areas where we can be at peace and not forever feel that the wilderness is a battleground.

Zahniser and the others who supported legal protection for wilderness *had three main concerns for the law that was*

FIGURE 4.2 Howard Zahniser, executive director of The Wilderness Society, made monumental contributions to establishing a wilderness system and passing the Wilderness Act. Zahniser died in June 1964, before the act was passed. Photo courtesy The Wilderness Society, Washington, DC.

to be written. First, they wanted a clear, unambiguous document. One of the oft-criticized features of the previous administrative approaches to wilderness protection was that they contained many loopholes through which protection could be controverted. *Second, they wanted to maintain the political coalition that had formed during the earlier efforts to protect Echo Park in Dinosaur National Monument* in Colorado and Utah from flooding. This coalition of groups had effectively resisted efforts in 1955 by the Bureau of Reclamation and Army Corps of Engineers to construct a dam within the monument as part of an ambitious Upper Colorado River Storage Project. Zahniser recognized the importance of maintaining this tested coalition of allies and was careful to circulate early drafts of the wilderness bill to the various organizations to maintain a uniform front of support. And *third, they wanted to minimize opposition to the legislation.* Zahniser and his colleagues sought to assure federal land-managing agencies that existing jurisdictions would be respected and that the proposed bill would not supersede the purposes for which the land was being administered, except for preserving its wilderness character. To commodity interests, Zahniser indicated that the bill would respect existing uses, and the termination of any uses considered nonconforming (e.g., stock use) would be done in a manner equitable to them (Roth 1984).

Difficulties were encountered immediately as both the USFS and the NPS opposed the legislation. The USFS argued that it was not urgently needed. Moreover, efforts to pass it might jeopardize agency efforts to expand and consolidate its existing program of wilderness protection by generating strong opposition to wilderness from within as well as outside government (McArdle 1975). Director Conrad L. Wirth of the NPS also opposed it because it was not necessary, and he claimed that it might endanger national park wilderness areas by lumping them with those of other agencies (Allin 1982; Roth 1984).

Nevertheless, Zahniser persisted in his pursuit for improved wilderness protection (Fig. 4.2). His goal was a congressionally established national wilderness system that would encompass areas of adequate size and numbers to meet future needs and to provide legal protection to ensure the perpetuation of their primeval character. To meet this objective, he and leaders from the Sierra Club, National Parks Association, National Wildlife Federation, and Wildlife Management Institute prepared a draft bill in 1955 at the urging of Senator Hubert Humphrey (MN). *In 1956, Senator Humphrey and nine other senators introduced the first wilderness bill. Representative John Saylor (PA) introduced similar legislation in the House.* The long congressional struggle for legislative establishment of a national wilderness preservation system was under way.

A BRIEF LEGISLATIVE HISTORY OF THE WILDERNESS ACT

It took eight years for the final Wilderness Act to be passed by Congress. During that time, sixty-five different wilderness bills were introduced. Eighteen hearings were held across the nation, many thousands of pages of testimony were printed (McCloskey 1966), *and the final bill was substantially changed from the initial version drafted by Zahniser* to the act signed into law by President Lyndon B. Johnson (U.S. Public Law 88-577) September 3, 1964. We discuss some of the issues that delayed passage as well as the major changes between the first draft and the final act; for a more detailed analysis read Allin 1982 and Scott 2004.

THE FIRST WILDERNESS BILL, INTRODUCED IN 1956

The first wilderness bill, introduced in 1956, proposed a wilderness system including lands in the National Forest System, National Park System, National Wildlife Refuges and Game Range System, and lands overseen by the Bureau of Indian Affairs. Altogether, about 65 million acres would have been subject to study; as many as 35 to 45 million acres might actually have been designated wilderness.

All thirty-seven national forest areas classified as wilderness, wild, or roadless under the U regulations were to be automatically included in the system under the first bill. In addition, the forty-four remaining primitive areas were to be temporarily included within the system, and the secretary of agriculture was given nine years to review the status of each and recommend an appropriate classification. Congress would then decide whether to extend permanent protection to each primitive area or exclude it from the system.

The secretary of the interior was directed to review the National Park System and the National Wildlife Refuges and Game Range System, also within nine years, and to recommend areas that should be designated as wilderness. Unlike the national forestlands, no Department of the Interior holdings were automatically included in the proposed wilderness system. Qualified areas under Bureau of Indian Affairs jurisdiction could be included, but only with the consent of the tribal councils. Moreover, protection would exist only so long as the appropriate Native American representatives concurred (Allin 1982).

The first proposed wilderness bill would have provided comprehensive protection from development. It would have prohibited logging, prospecting, dams, commercial enterprises, roads, motor vehicles, the landing of aircraft, the extension of motor boating to new areas, new mining, and new grazing. Some nonconforming but existing uses, such as grazing, and motorboat and aircraft use would have been respected (Allin 1982).

The first bill would also have established a National Wilderness Preservation Council composed of the heads of the USFS, NPS, U.S. Fish and Wildlife Service (FWS), Bureau of Indian Affairs, Smithsonian Institution, and six citizen representatives. Its functions would have been to receive and review all wilderness reports and recommendations from the secretaries of agriculture and the interior, to transmit these reports to Congress, and to advise Congress and the president during ensuing deliberations on the agency recommendations. The wilderness council was viewed as one means of checking the broad executive discretion of the administrative agencies.

The function of Congress in this earliest version of the wilderness bill was as a safeguard against any unwise and arbitrary action on the part of any secretary undertaking a measure that "disregarded conservation." This safeguard took the form of a legislative veto that could be used by either the House or Senate. When any secretarial recommendation came before it, Congress had 120 days within which to register its objection. Otherwise, the secretary's recommendation to support or oppose wilderness classification would become effective pursuant to the law. In other words, statutory protection of wilderness would come about in the absence of any *affirmative action* on the part of Congress.

It is important to note that, in this first bill, congressional authority to formulate and enact legislation was delegated to the secretaries of agriculture and the interior; that is, departmental recommendations became law if not vetoed. The idea of a legislative veto was subsequently dropped from the final version, but it had raised the more basic question: who should take the affirmative action in wilderness designations? In 1983, the U.S. Supreme Court ruled legislative vetoes unconstitutional.

This question—one of the crucial issues during the evolution of the wilderness bill—assumed particular importance during consideration of the bill's provisions for granting *temporary wilderness status* to Interior Department holdings identified as suitable for wilderness designation and USFS primitive areas, pending their review for permanent wilderness designation. Argument centered on two questions: (1) should these lands be included in the wilderness system initially (instead of remaining undesignated during their reviews), and (2) should affirmative congressional action—that is, a bill sponsored and introduced, debated, passed, and signed into law like any other legislation—be required to add any area to the wilderness system?

Representative Wayne Aspinall (CO), chair of the House Committee on Interior and Insular Affairs, argued strongly that each new area should be the subject of a separate congressional evaluation and an individual bill. Aspinall and other conservative legislators argued that this particular use of *affirmative congressional action* would help halt the erosion of congressional authority to the executive branch (Mercure and Ross 1970). Wilderness proponents, on the other hand, were concerned that requiring congressional approval of each individual area could prove to be a cumbersome barrier to rapid and equitable wilderness designation.

To force his position requiring affirmative congressional action on all wilderness proposals, Aspinall refused to allow hearings on the wilderness bill until legislation calling for a general review of all federal land-management policies was agreed on. Aspinall believed that the debate over affirmative action centered on the rules governing the withdrawal of wilderness-quality lands from multiple-use status. With more than 5,000 land-use statutes in place, he argued, an overhaul of the nation's land laws was necessary before making any extensive withdrawals for wilderness purposes (Baker 1985). As chair of the Interior and Insular Affairs Committee, Aspinall was able to make good on this threat. Consequently, wilderness proponents agreed to support his proposed legislation to create a Public Land Law Review Commission on the condition that Aspinall report out a wilderness bill that could be debated and amended on the floor of the House. In addition, Senator Clinton Anderson (NM), a major figure in the drive to pass wilderness legislation, agreed to the affirmative congressional action provision, if Aspinall would release the wilderness bill from committee (Baker 1985).

As the wilderness bill neared the end of its long journey through Congress, final refinements were worked out. USFS authority to declassify existing primitive areas by administrative action was eliminated, thereby giving the legislative branch full authority to control wilderness declassification. While the Senate acceded to the House provision requiring affirmative congressional action, the House agreed to reduce the time period in which new mineral exploration would be allowed in wilderness from twenty-five years to nineteen years. *The House passed the final version of the wilderness bill by a vote of 373 to one; the Senate, by a margin of 73 to 12.* President Lyndon Johnson, who said passage of the bill was "in the highest tradition of our heritage as conservators as well as users of America's bountiful natural endowments," signed the act on September 3, 1964.

CHANGES AND COMPROMISES TO PASS A WILDERNESS ACT

Contrasting the Wilderness Act with the original 1956 bill reveals several changes—compromises made to secure congressional support. Major changes included the following:

1. *Bureau of Indian Affairs lands were excluded from the bill.* In 1937, nearly 5 million acres of Native American land had been administratively reserved for wilderness purposes, largely through the efforts of Robert Marshall. However, because the tribal councils, rather than the federal government, held title to these lands, wilderness designation could occur only so long as the appropriate Native American representatives concurred.
2. *The National Wilderness Preservation Council was eliminated.* The USFS opposed the council from the beginning, arguing that it created an unnecessary step in the review process (McArdle 1975). Agency opposition, based partly on the six-to-five layperson–to–agency head representation on the council, remained even after the suggested number of laypersons was reduced to three in 1958. Opposition to the advisory council was also founded on the belief that it might very well end up making most of the final wilderness designation decisions, given its influential advisory role to Congress and the president. Zahniser, who had initially insisted on the council, agreed to its removal and counted on the president to check his cabinet officers' recommendations.
3. *The USFS primitive areas were not included within the initial wilderness system created in 1964, and their designation would require affirmative action by Congress.* Under the original bill, they would have been temporarily included in the system and a secretarial recommendation regarding their designation would have gone into effect in four months unless Congress took action to stop it (Roth 1984; Baker 1985).
4. *Prohibitions on uses of wilderness were less restrictive.* As initially conceived, for instance, all new mining in wilderness would have been prohibited on passage of the bill. Yet it became obvious that such a blanket restriction would mean no wilderness bill would pass (Roth 1984). So, as passed, the Act permitted prospecting to continue until December 31, 1983, and mining on claims established before this date would be allowed to continue indefinitely.

Many supporters of wilderness legislation were disappointed with the discrepancy between the act as passed and the original bill proposed by Zahniser and The Wilderness Society (Nash 1982). *The Wilderness Act clearly contained compromises, yet without them, it is unlikely the bill would have ever passed.* The bill enjoyed the support of several senators and congressional representatives, without whose help it would probably have failed (Baker 1985; McArdle 1975). The role of Zahniser as a committed citizen advocate and executive director of The Wilderness Society was also crucial (Scott 2004). Together, their efforts were rewarded with the passage of a piece of legislation unique in American, as well as international, conservation history—U.S. Public Law 88-577, the Wilderness Act.

THE WILDERNESS ACT OF 1964

The text of the Wilderness Act of 1964 is Appendix A. This act should be required reading for anyone involved in wilderness designation or management. Many of the arguments and much of the confusion surrounding wilderness stewardship stem from the lack of a careful reading and clear understanding of this important document. As the major piece of legislation guiding both wilderness designation and management, the Wilderness Act is the basic reference document for many questions regarding what can or cannot be done in wilderness.

Parts of the Wilderness Act are subject to differing interpretations, depending on one's particular wilderness philosophy and values (e.g., contrasting perspectives on how pure wilderness should be preserved are found in Foote 1973, Worf 1980, and discussed in Allin 1985).

We review the seven sections of the Wilderness Act to highlight important provisions and note ambiguities and contrasting interpretations. In some cases, subsequent wilderness designation laws, or committee reports on those laws, have clarified or elaborated specific wording of the Wilderness Act, such as livestock grazing and insect and disease control.

The Wilderness Act provides only broad guidelines and directions; detailed guidelines are contained in departmental regulations for implementing the law that are issued by the secretary of agriculture for wilderness on the national forests and by the secretary of the interior for wilderness administered by the National Park Service, U.S. Fish and Wildlife Service, and Bureau of Land Management.

The remainder of the chapter is devoted to explaining the intent and meaning of the law. Additional explanation of the Wilderness Act is found in an excellent handbook published by The Wilderness Society (2004).

The Wilderness Act is organized into seven sections.

SECTION 1—TITLE

The act shall be known as the "Wilderness Act."

SECTION 2—ESTABLISHES THE WILDERNESS SYSTEM

Section 2 provides a broad statement of policy, defines the term *wilderness*, and sets forth some of the conditions and implications of wilderness designation. Section 2(a) clearly states that the establishment and protection of wilderness is a policy of the U.S. Congress, reflecting a belief that because of population pressures, all areas of the nation will be occupied or modified—except those set aside in their natural condition.

Management is specifically referred to in this section, where the act states:

> Wilderness areas shall be administered for the use and enjoyment of the American people in such manner as will leave them unimpaired for future use and enjoyment as wilderness, and so as to provide for the protection of these areas, the preservation of their wilderness character, and for the gathering and dissemination of information regarding their use and enjoyment as wilderness.

Note that the phrase "as wilderness" appears several times. Although Congress fully intended wilderness to be for people's use and enjoyment, it is also apparent that such *"use and enjoyment" is to be contingent upon the maintenance of these areas "as wilderness," a condition the act later defines.*

Section 2(a) also specifies that only federal lands will be included in the National Wilderness Preservation System (NWPS) and that no federal lands except those protected by the act or by a subsequent act shall be designated "wilderness." This provision was included to prevent the executive branch of government from designating wilderness, thus reserving that responsibility for Congress. Section 2(a) merely prohibits *official designation* as wilderness by agencies other than Congress. It does *not* prohibit administrative agencies from *managing* lands for wilderness purposes. The USFS, for example, had such authority based on the Multiple-Use, Sustained-Yield Act of 1960. The Wilderness Act specifically indicates that none of its provisions shall interfere with the purposes of the Multiple-Use, Sustained-Yield Act.

The Multiple-Use, Sustained-Yield Act (U.S. Public Law 86-517) was enacted largely at the urging of the USFS as a reaffirmation of the agency's traditional multiple-use philosophy regarding resource management. The act was intended to counter pressures for dominant-use legislation, but it did provide general statutory sanction of wilderness preservation in the national forests, incorporated in the following statement: "The establishment and maintenance of areas of wilderness are consistent with the purposes and provisions of this Act." Robinson (1975) argues that inclusion of this provision helped lead to the withdrawal of USFS opposition to the proposed wilderness legislation (Allin 1982).

The Wilderness Act's provisions originally affected three federal agencies: the USFS in the Department of Agriculture, and the NPS and FWS, both in the Department of the Interior. Before 1964, the wilderness idea had been formally incorporated into USFS planning through the L-20 Regulation and U regulations. The NPS had used a zoning system to protect wilderness values in undeveloped areas more than one-half mile from roads (Outdoor Recreation Resources Review Commission 1962). Before the Wilderness Act, the FWS had not managed any areas specifically for wilderness purposes because habitat enhancement for wildlife often results in substantial modification of areas, thereby conflicting with wilderness values. Nevertheless, the FWS is charged with wilderness responsibilities under the 1964 Wilderness Act.

In 1976, Congress passed the Federal Land Policy and Management Act (U.S. Public Law 94-579), giving the Bureau of Land Management (BLM) in the Department of the Interior clear authority and direction for management of the public lands under its jurisdiction. FLPMA, as the act is known, instructs the secretary of the interior to review those roadless lands of 5,000 acres or more, as well as all roadless islands administered by BLM, and to make recommendations regarding the suitability of these areas for wilderness designation. This inventory used the wilderness characteristics specified in the Wilderness Act.

Before 1976, the BLM managed a system of areas for wilderness preservation purposes referred to as *primitive areas* (not to be confused with national forest areas of the same name designated under the L-20 Regulation) and *natural areas*. However, under the terms of the FLPMA, any area classified by BLM as a *primitive* or *natural* area before November 1, 1975, was to be reviewed by the secretary of the interior for its suitability for wilderness designation, and a recommendation made to the president by July 1, 1980.

In addition to specifying which federal lands shall constitute official wilderness, section 2(a) assigns management responsibilities. It specifies that each federal agency charged with jurisdiction of wilderness will continue to manage those areas originally under its jurisdiction after they have been made part of the NWPS. This clause was included so that no new agency would be created. In addition, budget appropriations to the wilderness system as a separate entity or appropriations based solely on the system's existence are prohibited.

The final subsection of section 2 defines wilderness, and it is this section that has probably led to more confusion and debate than any other. It first defines wilderness in an ideal, almost poetic, sense: "A wilderness, in contrast with those areas where man and his own works dominate the landscape, is hereby recognized as an area where the earth and its community of life are untrammeled by man, where man himself is a visitor who does not remain."

Zahniser specifically chose the word *untrammeled*, even though he was warned that it might confuse the definition. Not to be confused with *untrampled*, *untrammeled* means "not subject to human controls and manipulations that hamper the free play of natural forces."

Section 2(c) then goes on to define wilderness in a legal and workable sense as

> an area of undeveloped Federal land retaining its primeval character and influence, without permanent improvements or human habitation . . . and which (l) generally appears to have been affected primarily by the forces of nature, with the imprint of man's work substantially unnoticeable; (2) has outstanding opportunities for solitude or a primitive and unconfined type of recreation; (3) has at least five thousand acres of land or is of sufficient size as to make practicable its preservation and use in an unimpaired condition; and (4) may also contain ecological, geological, or other features of scientific, educational, scenic, or historical value.

The definition of wilderness in section 2(c) gives important clues to the congressional view of wilderness. Recognizing that the ideal did not exist, they added a working definition based on reality. Wilderness was clearly intended to be an area where humans' impact was minimal, and which was predominantly natural and unmodified. At the same time, the act accommodates reality by stating these areas "*generally* appear" to be "*primarily* affected" by nature with man's imprint "*substantially* unnoticeable" (emphasis added).

In the effort to accommodate reality by expressing the necessary conditions for wilderness in these more general terms, some difficult questions were created. For example, if a large tract of land were being considered for wilderness, with one portion in a pristine condition and another portion modified by previous developmental activity, would the criterion "substantially unnoticeable" be based on the aggregate area or on the modified portion? The answer obviously hinges on whether one looks at the whole area in its entirety, or at its individual parts.

Some have argued for keeping the system pure—that including or adding modified areas, even to large contiguous tracts in pristine condition, would compromise the quality of the entire system (Costley 1972). Yet, others have argued that the entire unit of land must be considered and that the inclusion of small areas of modified land is "substantially unnoticeable" when viewed within the larger context.

The intent of Congress was to establish a system of areas—an ideal—that embodied values espoused by early wilderness proponents, areas with naturalness and solitude. Yet, unreasonably rigid admission standards clearly were not the intent of Congress, either. The act first describes wilderness as an *ideal concept*, but then goes on to define wilderness *as it is to be considered* for the purposes of the act.

The definition of wilderness in section 2(c) cites the importance of "outstanding opportunities for solitude or a primitive and unconfined type of recreation." In recent years, in response to wilderness crowding, some have proposed that Congress meant that opportunities for either solitude *or* a primitive kind of recreation are required; that is, either one could qualify an area for wilderness designation and, thus, only one such quality need be maintained by wilderness management. As discussed in Chapter 1, it is our belief that both are required, broadly speaking—*naturalness and opportunities for solitude and primitive recreation are the distinguishing qualities of wilderness.*

The 5,000-acre minimum is often cited as absolute, but a careful reading of the act clearly shows that it is a suggested guideline. The intent is that the area designated be large enough to permit preservation objectives. In the conference committee that formulated the final wording of the Wilderness Act, the House recommended 5,000 acres as a minimum limit on the size of individual areas, but conferees decided that a statement of intent that a tract of "sufficient size as to make practicable its preservation and use in an unimpaired condition" would be satisfactory (Baker 1985).

Finally, the definition says wilderness *may* include ecological, geological, scenic, and other features. The important point to note is that these values are neither required for an area to be a wilderness nor by themselves sufficient criteria.

In conclusion, *our interpretation of the act's multifaceted definition of wilderness is that the criteria of naturalness and opportunities for solitude and primitive recreation are the distinguishing qualities of designated wilderness and are the principal criteria to guide wilderness management.*

SECTION 3—NATIONAL WILDERNESS PRESERVATION SYSTEM—EXTENT OF SYSTEM

Section 3, a five-part section of the Wilderness Act, describes the NWPS and the procedures for admitting areas to the system.

Section 3(a) defines the areas that formed the initial core of the NWPS. These included all USFS areas previously classified administratively as wilderness, wild, or canoe. Fifty-four areas were so designated, covering 9.1 million acres. It also instructed the secretary of agriculture to file accurate boundary descriptions of all these areas and make them available to the public. No Department of the Interior land was included in the initial wilderness system.

Section 3(b) instructs the secretary of agriculture, within ten years of the passage of the act, to review all USFS primitive areas for their suitability as wilderness and to make a report on the findings to the president. In turn, the president is to send Congress recommendations to support, oppose, or modify the secretary's proposal. It also established a timetable for the review of the thirty-four primitive areas (a total of 5.4 million acres at the time the act passed), with one-third to be reviewed within the first three years after the act's passage, two-thirds after seven years, and the remaining areas within ten years. Until Congress would take further action, primitive areas were to be administered as they were at the time of the act's passage.

This subsection also describes the president's latitude for enlarging an existing primitive area. At the time of the recommendation to Congress, the president may make an addition to any existing primitive area of not more than 5,000 acres, as long as no single unit of added land exceeds 1,280 acres. Additions beyond the 5,000-acre limit require congressional approval (McCloskey 1966).

Section 3(c) is similar to 3(b) in that it instructs the secretary of the interior to review all roadless areas of at least 5,000 acres in the NPS and similar size holdings and *every* roadless island within the National Wildlife Refuges and Game Range System, and to submit a report regarding the suitability of these areas for wilderness designation. The islands were specifically cited because, in spite of the small size of many of them, their isolation makes preservation a practicable alternative. A ten-year review period, along with the timetable described in section 3(b), was established; that is, one-third in three years, two-thirds after seven years, and the remainder at the end of ten years.

Section 3(d) provides guidelines for notifying public and local officials of recommendations that the secretaries of agriculture and the interior intend to submit to the president and provides for public hearings on these recommendations.

Finally, section 3(e) describes the procedures for modifying or adjusting any wilderness boundary. No modification of a wilderness boundary by an agency is permitted, even in cases where nonconforming or illegal uses are found. When exact boundary locations are not clear, the enabling legislation or conference or committee reports are used to establish legislative intent.

SECTION 4—USE OF WILDERNESS AREAS

Section 4 is subtitled "Limitation of Use and Activities." Subsection 4(a) indicates that the purposes of the Wilderness Act are "within and supplemental" to the purposes for which national forests and units of the national park and wildlife refuge systems are established. It specifically states that the Wilderness Act in no way interferes with several specific acts, such as the Multiple-Use, Sustained-Yield Act and the National Park Organic Act. By providing for wilderness preservation under the multiple-use umbrella, section 4(a) removes a major reason for conflict between these two potentially incompatible laws.

The responsibility of each managing agency to maintain the wilderness character of lands under its jurisdiction is reaffirmed in section 4(b). In addition, it states that, "except as otherwise provided in this Act, wilderness areas shall be devoted to the public purposes of recreational, scenic, scientific, educational, conservation, and historical use."

Section 4(c) describes facilities and activities that are not allowed in wildernesses designated by the act. This section must be carefully read, for it opens with the statement, "*Except as specifically provided for in this Act, and subject to existing private rights*" (emphasis added), before going on to catalog prohibited uses. These exceptions are quite substantial (Fig. 4.3).

Subject to the subsection's opening qualification, commercial enterprises and permanent roads are prohibited. Certain other uses (motorized vehicles and equipment, temporary roads, aircraft landings) are prohibited "*except as necessary*

FIGURE 4.3 The Wilderness Act allows agencies to use certain facilities and do activities essential to managing a specific wilderness. A remote automated weather station is essential to fire management in the Mount Logan Wilderness, on the Bureau of Land Management's Arizona Strip District. Photo courtesy Tom Folk, BLM.

to meet minimum requirements for the administration of the area for the purpose of this Act (including measures required in emergencies involving the health and safety of persons within the area)" (emphasis added).

Subsection 4(c) outlines illegal activities and administrative latitude. The section reads, in part, "*Except as specifically provided for in this Act . . . and, except as necessary to meet minimum requirements for the administration of the area for the purpose of this Act,*" and is followed by a list of prohibited activities such as temporary roads and motor vehicles. The phrase "minimum requirements" seems to mean "essential or necessary." The phrase "except as necessary to meet minimum requirements for the administration of the area for the purpose of this Act" permits administrators to carry out actions otherwise considered inappropriate in wilderness, *if it becomes apparent these actions are the minimum necessary to manage the area as wilderness.* This wording is the basis for the so-called minimum tool rule for wilderness management widely referenced in this book (see Chapter 7).

The basic interpretation of this section is that *administrators must provide evidence that any of their actions are the minimum necessary for managing the area as wilderness.* Simple convenience or cost advantage is not sufficient reason to justify, for example, the use of mechanized equipment. Worf (1980) contends that nonmechanized wilderness maintenance is often cheaper than mechanized maintenance. If, however, the use of mechanized equipment clearly reduces the amount of impact on the wilderness and its use (e.g., by reducing the level and extent of resource impact, and thus being the minimum tool), then administrators would have the flexibility to use such equipment.

McCloskey (1966) also points out a second ambiguity. Does the phrase "for the purpose of this Act" refer solely to the act's objective in section 2(a)—"to secure for the American people of present and future generations the benefits of an enduring resource of wilderness"—an objective limited to preservation, or does it include a broader purpose—preservation *and* compatible human enjoyment? Differing interpretations of the phrase could lead to different administrative actions. The wording of the act indicates that both use of wilderness and its preservation were intended, but it also suggests in section 4(c)(1) that any construction to facilitate use, such as bridges or shelters, must satisfy the criterion of *necessity.* In general, facilities for the convenience and/or comfort of users do not meet this criterion and have not been provided.

Important exceptions to the prohibited uses are described in section 4(c). These exceptions, which are subject to existing private rights, are outlined in section 4(d)(1) and 4(d)(2) and constitute what we call *allowable, but nonconforming uses*; they are legal, but they are generally considered incompatible with the wilderness protection goals of the act. These nonconforming uses reflect some of the compromises that were necessary to pass a Wilderness Act. Both subsections apply only to those areas of national forestland designated as wilderness by the 1964 act. However, in the interests of consistency, subsequent wilderness designation acts have called for the management of newly designated areas to follow the guidelines contained in the 1964 act, subject to any specific exceptions written into that subsequent legislation.

The following uses (nonconforming, but allowed uses) are expressly permitted in section 4(d):

1. Established uses of aircraft and motorboats.
2. Actions taken to control fire, insects, and disease outbreaks.
3. "Any activity, including prospecting, for the purpose of gathering information about mineral or other resources, if carried on in a manner compatible with preservation of the wilderness environment."
4. Continued application of the U.S. mining and mineral leasing laws to wilderness on national forestlands until December 31, 1983.
5. Water resource development (authorized by the president if determined that such use will better serve the national interest than would its denial).
6. Livestock grazing, where established prior to the act.
7. Management of the Boundary Waters Canoe Area under regulations previously laid down by the secretary of agriculture, which were generally less restrictive than those imposed by the Wilderness Act (resolved by subsequent legislation, as we shall discuss shortly).
8. Commercial enterprises necessary for activities that are appropriate in wilderness (e.g., outfitting and guiding) (Fig. 4.4).

In summary, section 4(d) of the act recognized that certain "nonconforming" existing uses—specifically aircraft and motorboats—could be permitted to continue in places where they were well established. However, it also noted that the secretary of agriculture might impose such restrictions as deemed necessary to protect the wilderness resource. So, although these "nonconforming" uses are "allowed where established" by the Wilderness Act, they can nonetheless be controlled by the administering agencies.

Overflights

Administrative action might become necessary to limit aircraft within the backcountry of Grand Canyon National Park in Arizona (the area has been proposed for wilderness designation and is managed as wilderness), where more than 117,000 commercial overflights occurred in 2000. Current regulations permit aircraft to fly as low as 500 feet above the surface, but only in "sparsely populated" areas.

Fire, Insects, and Disease

Although the secretary of agriculture can undertake measures to control fire, insects, and disease, this authority is *discretionary*; for example, it is not mandatory that fires be controlled.

The provision of section 4(d)(1) regarding actions to control fire, insects, and disease outbreaks has been elaborated on in subsequent legislative history. House Report 95-540, the committee report filed with the Endangered American Wilderness Act (U.S. Public Law 95-232) in 1978, emphasizes that any actions to control fire, insect, or disease outbreaks, such as the use of mechanized equipment, the building of fire roads, or the construction of firebreaks, are permissible if judged necessary for the protection of public health or safety (The Wilderness Society 2004).

FIGURE 4.4 A wilderness outfitter leads pack stock and a party on horseback in the Marble Mountain Wilderness, California. The Wilderness Act generally allows established commercial outfitting activities to continue. Photo courtesy U.S. Forest Service.

The discretionary latitude to control pest threats to wilderness values is exemplified in steps taken by the USFS to control outbreaks of the southern pine beetle in several wildernesses in Texas, Arkansas, Mississippi, and Louisiana. A variety of management actions were proposed, including the use of synthetic attractants (pheromones), cutting buffer strips around infected areas, and reducing stand density (Billings 1986; Billings and Varner 1986). The problem was confounded by the presence of several nesting sites of the red-cockaded woodpecker, an endangered species that nests only in mature, live pine trees. Harvesting buffer strips proved successful in halting expansion of the southern pine beetle infestations, but controversy remains as to whether this is the most appropriate technique or the *minimum tool necessary* to control insect outbreaks in wilderness. Ultimately the southern pine beetle epidemic subsided naturally, but not before several treatment strategies were applied, some of them in wilderness.

Mining

The discussion of mining in wilderness occupies a substantial proportion of section 4(d)(3). This part of the Wilderness Act refers to wilderness on the national forestlands because most Department of the Interior lands are withdrawn from mineral entry. Until 1976, however, six units of the NPS were open to mineral entry, including Death Valley National Monument (now Park), California and Nevada, and Crater Lake. In 1976, Congress passed legislation withdrawing these units from further mineral entry and placed existing claims under stringent regulations issued by the secretary of the interior (Fig. 4.5).

FIGURE 4.5 Mining is permitted in wilderness, under certain restrictions. Shown is a mining operation on the western edge of the Cabinet Mountains Wilderness in western Montana. Photo courtesy U.S. Forest Service.

Basically, the Wilderness Act specified that prospecting and mining could continue in wilderness on national forestlands until December 31, 1983. After that date, mining could continue only on valid claims existing before that date. In other words, one could not file a claim for mining after December 31, 1983, but could continue mining a claim that existed before that date. A patent conveying both surface and mineral rights may be taken on a valid claim located prior to the Wilderness Act; for a valid claim located after the date of the Wilderness Act, the patent conveys title to mineral rights only.

A claim permits an individual to occupy a site and use it for the purposes of developing the mineral values it contains; however, uses beyond those necessary for development of the mineral values are not allowed and title to the land remains with the U.S. government. A patented claim conveys full ownership of the site, including surface and subsurface values, to the claimant. Both a claim and a patent are considered property rights.

The secretary of agriculture is instructed to issue *reasonable regulations* governing access to claims and related facilities such as transmission lines, roads, and buildings. Generally, these regulations impose more stringent standards for mineral operations in wilderness than on other national forestlands. Under federal regulations, mining operators in national forests are required to prepare an operating plan that describes who is doing the work, where and when it will be done, the nature of the disturbance the work will create, and measures to be taken to protect other resources. Restoration is also called for when the operation ceases. Surface resources, such as timber, may be used if needed for the mining operation and if founded on sound principles of forest management. After January 1, 1984, or the date of any wilderness designated after this date, but subject to valid existing rights, minerals in wilderness were withdrawn from entry.

This portion of the act also instructs the secretary of the interior to develop, in consultation with the secretary of agriculture, a plan for recurrent surveys of the mineral values in any wilderness and to submit these findings to the public, Congress, and the president. A survey of 74 million acres of wilderness and wilderness study areas in the early 1980s revealed that only 2.7 million acres had a high probability of containing significant oil and gas reserves; half of this total was in western Montana (U.S. Department of the Interior 1984).

Mining in wilderness is a paradox, and its presence can make sense only when viewed as a necessary political compromise (Matthews et al. 1985). Nevertheless, many view its presence as an internal contradiction within the Wilderness Act.

Mining in wildernesses located in the eastern United States represents a further complication. The titles to most national forestlands in the East were acquired from private ownership, but mineral rights typically remained in private hands. Thus, the agency cannot prevent private mineral development. The scale of this problem is substantial; in 1984, 103 of the 192 designated and potential wilderness areas in the East contained private mineral rights covering nearly 1 million acres (General Accounting Office 1984). Short of acquiring these mineral rights (estimated in the BWCAW alone as nearly $100 million), the USFS can protect the wilderness qualities of these areas only by the conditions it imposes on the mining operation. At the time the Wilderness Act passed, oil and gas development in wilderness was simply not an issue, but since then, concerns with developing domestic sources of supply and rising prices have changed this situation. A large potential mineral supply lies within designated wildernesses in the Rocky Mountains.

Requests for oil and gas development on national forestlands are approved or rejected by the BLM, acting for the Department of the Interior. It is Department of the Interior policy to request a USFS recommendation on any application, and these recommendations typically have governed the Interior Department's decision (Short 1980). In the event a permit is issued by BLM, the actual site development would be administered by the USFS.

Lands under the administration of the BLM were not addressed in the 1964 Wilderness Act. The FLPMA (U.S. Public Law 94-579) provided authority for designating wilderness on the public lands administered by BLM. However, during the review of potentially suitable wilderness lands under BLM jurisdiction, the FLPMA instructs the secretary of the interior to manage such lands to not impair their suitability for preservation as wilderness. This "nonimpairment" standard is subject, however, to the continuation of existing mining and mineral leasing "in the manner and degree in which the same was being conducted" at the time the FLPMA passed. A legal opinion issued in 1978 interprets the nonimpairment language of the FLPMA as being a paramount consideration applying to all areas under BLM administration identified as having wilderness characteristics, exempting from the nonimpairment standard only those uses *actually taking place* at the effective date of the FLPMA (Short 1980).

Water Resource Development

Section 4(d) grants authority to the president to authorize programs for developing water resources within wilderness on the national forestlands, if it is determined that permission for such developments would better serve the interests of the United States than their denial. Actions involved here could include reservoir construction, power projects, and transmission lines. Current controversy surrounds the renovation, repair, and reconstruction of existing reservoirs in wilderness, some of them seventy-five to one hundred years old.

Grazing

Grazing is allowed in wilderness on the national forestlands where it was established before the signing of the Wilderness Act (Fig. 4.6). As with the several other nonconforming but allowed uses, the secretary of agriculture is permitted to impose "such reasonable regulations as are deemed necessary." Although the secretary of agriculture cannot prohibit grazing merely because of wilderness, he can restrict or eliminate such use based on principles of range management (McCloskey 1966).

Grazing in wilderness has led to some confusion. In 1980, Congress expressed the concern that certain national forest administrative policies and regulations were acting to discourage or unduly restrict on-the-ground activities necessary for proper grazing management. There was some pressure to amend the Wilderness Act to clarify the intent of Congress with regard to grazing in wilderness. In a House report on the Colorado Wilderness Act (U.S. Public Law 96-560), the committee observed that attempts to draft specific statutory language covering grazing in the entire NWPS would likely not be successful. Instead, the committee reaffirmed the existing language of the Wilderness Act and included a series of guidelines and specific statements

FIGURE 4.6 Livestock grazing, when established prior to an area's designation as wilderness, may continue. A drift fence in a desert wash in a BLM-administered wilderness in Arizona is a reminder that even in arid regions there are legal grazing leases where cattle feed on ephemeral vegetation. Photo courtesy Marilyn Hendee.

of legislative policy (The Wilderness Society 2004, 44–46) that included the following:

1. There shall be no curtailments of grazing in wilderness areas simply because an area is, or has been, designated as wilderness, nor should wilderness designations be used as an excuse by administrators to slowly "phase out" grazing.
2. The maintenance of supporting facilities, existing in an area prior to its designation as wilderness (including fences, line cabins, water wells and lines, and stock tanks), is permissible in wilderness.
3. The replacement or reconstruction of deteriorated facilities or improvements should not be required to be accomplished using "natural materials," unless the material and labor costs of using natural materials are such that their use would not impose unreasonable additional costs on grazing permittees.
4. The construction of new improvements or the replacement of deteriorated facilities in wilderness is permissible if it is in accordance with these guidelines and the management plans governing the area involved.
5. The use of motorized equipment for emergency purposes, such as rescuing sick animals or the placement of feed in emergency situations, is also permissible.

Subsequent legislation establishing wildernesses in Arizona (U.S. Public Law 98-406), Utah (U.S. Public Law 98-428), and Wyoming (U.S. Public Law 98-550) referred to the congressional guidelines on grazing contained in the Colorado Wilderness Act (U.S. Public Law 96-560), incorporating them as statutory directives for these states as well. It is clear that Congress sees grazing as a continuing, legitimate use in wilderness, and the guidelines in the Colorado Wilderness Act are intended to ensure appropriate treatment of grazing throughout the NWPS. Nevertheless, grazing continues to be a controversial issue in many proposals for new wilderness areas and in management of existing wilderness.

Boundary Waters Canoe Area

Another controversial portion of section 4(d) involved the management of the Boundary Waters Canoe Area Wilderness (BWCAW) on the Superior National Forest in Minnesota. What is now the BWCAW was originally three roadless areas designated under the U-3(a) Regulation: the Superior, Little Indian Sioux, and Caribou Roadless Areas; they were renamed the BWCAW in 1958. Section 4(d) specified that the BWCAW was to continue to be managed in accordance with regulations established by the secretary of agriculture. In general, the primitive character of the area was to be maintained, but certain other uses, including timber harvesting, vegetation manipulation, and road building, in certain locations were to be permitted, as the management of the area as prescribed under the earlier U-3(a) designation was to continue.

Obviously, the battle for preservation of the BWCAW had not been won. More than 500,000 acres in the area were still virgin forest (Heinselman 1973), the largest contiguous tract of virgin forest left in the eastern United States, and that uniqueness led to great concern about the appropriateness of logging in the BWCAW (Fig. 4.7). The history of resolution to conflicts over the apparent contradictions with wilderness in the act's provisions for the BWCAW is too much to cover here (Haight 1974; Proescholdt 1984, 1995, 2000).

FIGURE 4.7 Under the original terms of the Wilderness Act, many nonconforming uses, such as logging, mining, motorboats, and snowmobiles, were permitted in the Boundary Waters Canoe Area Wilderness. Legislative amendments have curtailed or eliminated most such exceptions in the area, thus enhancing the wilderness environment. Photo courtesy U.S. Forest Service.

Because management of the BWCAW under terms of the Wilderness Act was to be significantly different from other areas, and there were continuing pressures for mining, logging, and mechanized access, environmentalists pushed for separate legislation for the BWCAW that would put it on a more equal footing with other wildernesses. A citizens' advocacy group, Friends of the Boundary Waters Wilderness, was formed to push for the area's protection. Finally, in 1978, *Congress passed U.S. Public Law 95-495 designating the BWCAW. This law constitutes the only amendment to the Wilderness Act as it repealed paragraph 5 of section 4(d) of the act, which dealt with the special exceptions regarding management of the BWCAW.* However, its general effect was to strengthen the wilderness provisions regarding the area and to eliminate some of the complex exceptions that applied to it. The U.S. Court of Appeals rejected a 1981 court challenge of the act's constitutionality.

The new law prohibited motorboats within the area except for certain routes and lakes, specified the size of the motors permitted, prohibited snowmobiles except for a few selected routes, terminated all timber sale contracts existing within the wilderness with compensation to the affected companies, and prohibited all future logging in the virgin forest portions of the area. *The BWCAW is the United States' most visited wilderness, the largest wilderness east of the Rocky Mountains and north of the Everglades, and the largest lake and waterway canoe system in the NWPS.* With the adjoining Voyageurs National Park to the west and Canada's Quetico Provincial Park to the north, and La Verendrye Provincial Park to the east, the *BWCAW is part of a 2.5-million-acre international wilderness complex* (Proescholdt 2000).

The BWCAW was the only area in the NWPS that had an airspace reservation. President Harry Truman signed an executive order in 1949 prohibiting flights below an altitude of 4,000 feet, except for emergencies and for official business of the federal, state, or county governments (Andrews 1953). The 1978 law incorporates the terms of that executive order within it.

U.S. Public Law 95-495 added more than 45,000 acres to the BWCAW, bringing its total size to nearly 1.1 million acres, and established within the wilderness a 222,000-acre Boundary Waters Canoe Area Mining Protection Area. Within this area, exploration for and mining of minerals owned by the United States are prohibited, as is exploration or mining that would in any way affect navigable waters. The act also authorized the secretary of agriculture to acquire any minerals or mineral rights owned by persons other than the federal or state government. If necessary to acquire mineral rights, the secretary is authorized to use the secretary's power of eminent domain; however, congressional appropriation of money to secure these private mineral rights would be necessary (Proescholdt 1984).

Outfitting and Guiding, Wildlife, and Other Exceptions

Recognizing that certain commercial services might be necessary to realize fully the recreational values of wilderness, section 4(d) permits activities such as guiding and outfitting, despite the general prohibition of commercial enterprises cited in section 4(c).

After noting that the act does not touch on the question of federal exemption from state water laws, section 4 concludes with a standard disclaimer that the federal government recognizes state jurisdiction over wildlife and fish on national forests.

Language in the House report on the Endangered American Wilderness Act (U.S. Public Law 95-232) offered further guidelines regarding fishery-management activities within wilderness. The report notes that fisheries' enhancement activities and facilities are permissible, including fish traps, stream barriers, aerial stocking, and protection and propagation of rare species. These guidelines apply throughout the NWPS (The Wilderness Society 2004).

Congress has attempted, in passing subsequent wilderness legislation, to minimize the number of special exceptions to the Wilderness Act, as the need for consistency in wilderness management was a major rationale for the passage of the act. In some areas, special circumstances have led to exceptions to the general prohibitions in the Wilderness Act. For example, in the Central Idaho Wilderness Act (U.S. Public Law 96-312), establishing what is now the Frank Church–River of No Return Wilderness in 1980, aircraft use of landing strips "in regular use" shall be left open. This denies the USFS the discretion to close airstrips that are contained in the Wilderness Act, which reads "use of aircraft . . . where these uses have already become established, may be permitted to continue." This exception applies only to the Frank Church–River of No Return Wilderness, where nineteen such landing strips are located.

Major exceptions to the general management provisions of the Wilderness Act are also contained in the Alaska National Interest Lands Conservation Act, known as ANILCA (U.S. Public Law 96-487). Before passage of ANILCA in 1980, no designated wilderness existed in Alaska. At least one reason for this was the feeling that Alaska was different and that wilderness management there would have to be different than in the lower forty-eight states. When the legislation was drafted, Congress included several special provisions for the management of wilderness in Alaska but explicitly stated that such provisions applied only to wilderness in Alaska. These exceptions included the use of snowmobiles, motorboats, and aircraft; temporary fishing and hunting camps; and subsistence uses by natives and nonnatives (The Wilderness Society 2004). The ANILCA even allows log salvage along coastlines within wilderness, but the act also provided authority to the USFS to regulate or limit the above uses to protect an area's wilderness character.

In summary, then, *the direction contained in section 4 of the Wilderness Act primarily concerns the uses considered appropriate in wilderness, thus providing the guiding criterion for wilderness management.* As the preceding discussion suggests, these guidelines are fairly strict. Controversy continues over the application of these guidelines in the allocation of wilderness, as opposed to its management. This debate—the so-called purity argument—has centered on the extent to which previous human impacts can be accepted in wilderness. The USFS, in particular, took the position that evidence of previous human impact normally precluded an area from consideration as wilderness. This position was defended because to justify the substantial opportunity costs associated with wilderness, it was important that any area selected be of highest quality; that is, the area must exhibit the highest degree of "environmental purity" (Costley 1972; Worf 1980).

Some authors (Behan 1971; Foote 1973) have rebutted the purity doctrine. Former senator Frank Church (ID), floor manager for the wilderness bill at the time of its passage, charged that neither the letter nor the intent of the Wilderness Act called for exclusion of areas once impacted by human use, and that agency claims to the contrary were simply evidence of resistance to congressional will (Church 1977). Allin (1982, 1985) has further argued that USFS reliance on the "purity argument" derives from the organization's concern with preserving a greater degree of administrative control over forestlands. The increasing loss of administrative discretion over the management of the national forests led the agency to consider ways in which it could limit the impact of wilderness designation on commodity production.

The admission standards for designating wilderness are dealt with in section 2(c) of the Wilderness Act, not section 4. Preservationists as well as Congress have viewed reliance on section 4 to define wilderness as inappropriate at best and illegal at worst (Allin 1982). Congressional actions in designating wilderness also indicate that a rigid and stringently pure conception of wilderness is not necessary.

For example, in the Mission Mountains Wilderness in Montana and the Agua Tibia Wilderness in California, Congress overruled impacts judged by the USFS to preclude wilderness designation. In the Mission Mountains, about 2,000 acres of the old primitive area were salvage logged in the 1950s after an insect outbreak. The USFS proposed excluding this area, managing it to restore wilderness values, prior to its inclusion in the wilderness system. Congress, however, included the impacted area. In the Agua Tibia Wilderness proposal, the USFS excluded an area containing a road closed to public use but used for fire protection. However, Congress approved a bill including the road, ruling that such a development was permissible under the administrative exceptions clause of the Wilderness Act, section 4(c).

Not all impacts are acceptable to Congress, of course. Congress excluded a 6,000-acre area containing a 22-mile-long access road to a mining claim in the Emigrant Basin Primitive Area in California from the wilderness bill it approved for this area. The challenges and problems of wilderness management are closely tied to decisions regarding such questions as what constitutes a "substantially unnoticeable" impact; that is, the greater the latitude in admission standards, the more urgent will be the need for rigorous management to enhance restoration of the wilderness resource and prevent further damage of the impacted area.

Passage of the so-called Eastern Wilderness Act in 1975 (U.S. Public Law 93-622) provides perhaps the strongest statement of congressional intent regarding the conditions needed to constitute wilderness. This act (discussed in more detail in Chapter 5) explicitly accepted the notion that areas where previous human activity, such as logging and agriculture, had occurred could still be congressionally designated as wilderness. Nelson (1985) notes that House Report 99-262 on a bill to designate new wilderness in Michigan indicated that within areas proposed, two-thirds of the mineral rights were privately owned, motorboat and off-road vehicle use was present, developed camping facilities were located, and one area contained an experimental forest with structures located in it. Such congressional actions clearly repudiate the purity policy as applied to designation (Allin 1982). However, such steps also reflect an evolving perception on the part of the public as to what constitutes wilderness. The criteria of wilderness naturalness and purity standards over time, and in different parts of the country, appear to have been relaxed (Hendee 1986).

SECTION 5—STATE AND PRIVATE LANDS WITHIN WILDERNESS AREAS

Section 5 describes the rules and procedures for access to private or state-owned inholdings within national forest wilderness. It provides "such rights as may be necessary to assure adequate access" to such lands. The right of reasonable access is further endorsed in section 1110(b) of ANILCA: "The State or private owner or occupier shall be given by the Secretary such rights as may be necessary to assure adequate and feasible access." These rights are also taken to apply to wilderness lands outside Alaska (The Wilderness Society 2004). Section 5 also indicates that such lands may be exchanged for federally owned property in the same state and of approximately equal value. But, unless the state or private owner relinquishes the mineral interests in the surrounded land, the U.S. government will not transfer the mineral rights of any exchanged land.

The secretary of agriculture is permitted, subject to the appropriation of funds by Congress, to acquire private inholdings within wilderness if the owner agrees to such acquisition and if the acquisition is approved by Congress.

Condemnation is not permitted by the Wilderness Act, an authority that was important to the development of legislation establishing wilderness in the eastern U.S. (see Chapter 5). However, even in the case of the so-called Eastern Wilderness Act, the authority to condemn private land was limited to the sixteen areas designated by that legislation.

SECTION 6—GIFTS, BEQUESTS, AND CONTRIBUTIONS

Both the secretary of agriculture and the secretary of the interior are authorized to accept gifts in furtherance of the purposes of the act. Such gifts might be land, money, or both. If the gift involves land adjacent to, rather than within, an existing wilderness, Congress must be notified sixty days before acceptance. On acceptance, the land becomes part of the wilderness.

SECTION 7—ANNUAL REPORTS

The last section of the Wilderness Act instructs the two secretaries to report jointly to the president at the opening of each session of Congress on the status of the NWPS. These reports provide information on the number of areas and acres in the NWPS, changes in the NWPS, regulations in effect, and the status of areas under consideration. They are available to the public from the Government Printing Office and from senators and representatives. To date, however, they have fallen short of what was proposed for these reports in the Wilderness Act.

SOME EXCEPTIONS AND AMBIGUITIES IN WILDERNESS LEGISLATION

The act's directives and guidelines for area management pertain only to those areas designated by that legislation. So technically, the Wilderness Act imposes use restrictions only on the fifty-four national forest areas designated as wilderness on its passage. It also provides that, pending designation as wilderness, primitive areas were to be managed under regulations in effect at the time the Wilderness Act was passed. Left in some confusion, then, was the status of the Department of the Interior holdings, the wilderness study areas specified in other wilderness designation acts, and all other lands that might be added in the future to the NWPS.

Why Congress did not clearly specify the management direction for these other areas in the 1964 legislation is not entirely clear. The lack of definitiveness does allow Congress latitude for adding special management provisions to the legislation designating each individual area, perhaps in response to an area's unique features, problems, or uses, and this has been done more frequently during recent decades.

However, wide use of special management provisions could also undermine one of the major reasons a wilderness bill was initially proposed—*consistency* (Aspinall 1964; H. Zahniser 1964). *Consequently, in all subsequent legislation designating wilderness, Congress has affirmed that these new areas shall be managed in accordance with the 1964 act's provisions.* In other words, although the Wilderness Act did not specifically prescribe the management for areas that would be subsequently designated, Congress has extended the 1964 act's provisions to them. *Where special provisions and exceptions have been included in subsequent legislation, they have been restricted to specific areas and not used as a precedent for changes elsewhere.*

The Wilderness Act has many ambiguities, weaknesses, and omissions—more, or less, depending on one's views (e.g., permitting mining or grazing; failure to describe procedures for the review of national forest roadless lands). The general vagueness of many procedures has placed substantial burden on administrative agencies, citizen advocates, and, especially, the courts. Nevertheless, the act takes away much discretion from the management agencies by using "a statutory structure with detailed requirements—such as no vehicles, no structures, and no commerce—along with a strict, overarching mandate to preserve wilderness character" (Dettmann 2008, 4). And, when an agency fails to heed the act's direction, plaintiffs can appeal to the courts for relief. Numerous times over recent decades, courts have provided judicial opinions clarifying the act.

Notwithstanding all the unresolved issues, the Wilderness Act represents the principal statutory foundation for wilderness preservation and management in the United States. It defines the broad goals, objectives, policies, and procedures through which an enduring resource of wilderness is to be designated and managed. With one exception—addressing the management of the BWCAW—Congress has not amended the Wilderness Act since its passage in 1964.

THE EVOLUTION OF WILDERNESS PROTECTION

What did the evolution of wilderness protection culminating in the Wilderness Act accomplish beyond the administrative protection already being provided at the time of its passage?

First, the permanency of wilderness was substantially enhanced. The USFS L-20 Regulation afforded little if any permanency—its primitive-area designation was viewed as only an interim measure to halt haphazard development. Longevity of protection was improved by the U regulations, but administrative discretion to choose the level of protection was a major shortcoming. Similarly, although national park or national wildlife refuge designation ensured protection from many kinds of development, it did not necessarily guarantee permanent protection of wilderness values. The Wilderness Act brought increased assurance that such values would be protected.

Second, permitted uses of wilderness were increasingly restricted. The L-20 Regulation contained little in the way of prohibited uses. Logging and other forms of resource development were permitted. The U regulations were developed primarily to exclude some permitted uses—which they did. However, because the U regulations were instituted at the administrative rather than the legislative level, there remained the possibility that certain uses, inconsistent with wilderness, might be permitted. Only through legislation would certain use prohibitions be secure.

Finally, the evolution from the L-20 Regulation and U regulations to the Wilderness Act reflects a change in purpose. The L-20 Regulation was intended to establish a series of areas for the purposes of public education, inspiration, and recreation (Outdoor Recreation Resources Review Commission 1962). The U regulations emphasized the importance of protecting the natural environment (Marshall 1933) and of retaining the primitive quality of these lands, particularly regarding the style of travel permitted, permitted uses, and procedures for designating or modifying wilderness areas.

Similarly, the management guidelines for national parks and wildlife refuges before passage of the Wilderness Act did not explicitly define the purposes of wilderness preservation. Although a generally low level of development prevailed in many of these areas, the purposes for these areas lacked clear direction.

In the Wilderness Act, we find a new emphasis on the purpose of wilderness. Although public use and enjoyment are clearly provided for, wilderness, as defined by the act, is a landscape where the earth and its community of life are untrammeled by humans. Thus, *the Wilderness Act established a national policy and purpose of maintaining a system of areas where natural processes could operate as freely as possible.* Recreational use was an appropriate use of these areas, *only so long as it was consistent with this purpose.* The evolution of wilderness protection, from the L-20 Regulation to the U regulations to passage of the 1964 Wilderness Act, was a shift toward a more biocentric concept of wilderness, one focused on protection of the natural processes that created these settings.

This evolution of purpose in wilderness preservation is key, in our opinion. Wilderness preservation has evolved from a holding strategy for minimizing unplanned development until more carefully thought-out plans could be formulated to a carefully defined and legally established national system for protecting the ecological integrity of designated areas.

Our endorsement of a biocentric philosophy of wilderness management rests on the belief that framers of the Wilderness Act clearly intended to create a system of areas where people allowed nature's way, as far as possible, to continue unhampered. We share the lawmakers' recurring insistence that human use of these lands must not interfere with the preservation of the area as wilderness. *It is from these assumptions regarding purpose that our interpretation of the Wilderness Act flows and that our management recommendations in this book are founded.*

STUDY QUESTIONS

1. Identify and discuss some of the factors leading to the pressure to create a congressionally established and protected wilderness system.
2. Discuss the relationship between the L-20, the U regulations, and the Wilderness Act. How did these earlier administrative regulations affect the Wilderness Act? Consider both positive and negative influences.
3. What were some of the critical differences between the wilderness bill proposed in 1956 and the legislation passed in 1964? What were some of the reasons for these differences?
4. In your judgment, what are some of the major problems with the Wilderness Act in terms of how well it protects wilderness values? How can these problems be resolved?
5. What are some of the major changes that have occurred in applying the Wilderness Act since its passage in 1964? What forces have led to such changes? What future changes would you suspect might occur?

ACKNOWLEDGMENTS

George H. Stankey was lead author for this chapter in the first and second editions, and his major contributions remain. John C. Hendee, with help from Jay Watson of The Wilderness Society, provided revisions for the third edition and fourth edition.

CASE DISCUSSION: COMPARE THE 1956 BILL TO THE 1964 WILDERNESS ACT

Four major differences were noted between the original 1956 bill and the Wilderness Act of 1964 in the Chapter 4 section title: "Changes and Compromises to Pass a Wilderness Act." Consider how the NWPS might be different today if the 1956 bill had passed as proposed by comparing the four changes to the 1956 bill and the 1964 Wilderness Act as passed by Congress.

CASE DISCUSSION QUESTIONS

1. Bob Marshall had long advocated including wilderness designations on Native American lands. For example, the Mission Mountain Wilderness (managed by the USFS) adjoins the Mission Mountain Tribal Wilderness Area (managed by the Confederated Salish and Kootenai Tribes) on the Flathead Indian Reservation. How might the NWPS be different today if tribal lands had been part of the NWPS?
2. The proposed National Wilderness Preservation Council would have included five agency representatives and six laypersons to review wilderness designation proposals prior to congressional action and provide an advisory role. The Interagency Wilderness Policy Council today coordinates wilderness policy among the four federal agencies but does not serve the role envisioned by the National Wilderness Preservation Council. What impacts might a National Wilderness Preservation Council have had on wilderness designations in Congress?
3. All mining would have been prohibited on passage of the 1956 bill; however, the 1964 Wilderness Act allowed prospecting to continue until December 31, 1983, and mining claims before this date would be allowed to continue. See the discussion on mining under the Chapter 4 subsection "Section 4—Use of Wilderness Areas." How might the NWPS be different today if all mining had been prohibited from the NWPS?

REFERENCES

Allin, Craig W. 1982. *The Politics of Wilderness Preservation.* Westport, CT: Greenwood Press.

———. 1985. Hidden agendas in wilderness management. *Parks and Recreation.* 20(5): 62–65.

———. 1987. Wilderness preservation as a bureaucratic tool. In: Foss, Philip O., ed. *Federal Land Policy.* Westport, CT: Greenwood Press, pp. 127–138.

Andrews, Russell P. 1953. *Wilderness Sanctuary.* New York: Bobbs-Merrill.

Aspinall, Wayne N. 1964. Underlying assumptions of wilderness legislation as I see them. *Living Wilderness.* 86: 6–9.

Baker, Richard A. 1985. The conservation congress of Anderson and Aspinall, 1963–64. *Journal of Forest History.* 29(3): 104–119.

Baldwin, Donald L. 1972. *The Quiet Revolution: The Grass Roots of Today's Wilderness Preservation Movement.* Boulder, CO: Pruett Publishing.

Behan, R. W. 1971. Wilderness purism—here we go again. *American Forests.* 78(12): 8–11.

Billings, Ronald F. 1986. Coping with forest insect pests in southern wilderness areas, with emphasis on the southern pine beetle. In: Kulhavy, David L.; Conner, Richard N., eds. *Wilderness and Natural Areas in the Eastern United States: A Management Challenge;* May 13–15, 1985; Nacogdoches, TX. Nacogdoches, TX: Stephen F. Austin University, School of Forestry, pp. 120–125.

Billings, Ronald F.; Varner, Forest E. 1986. Why control southern pine beetle infestations in wilderness areas? The Four Notch and Huntsville State Park experiences. In: Kulhavy, David L.; Conner, Richard N., eds. *Wilderness and Natural Areas in the Eastern United States: A Management Challenge;* May 13–15, 1985; Nacogdoches, TX. Nacogdoches, TX: Stephen F. Austin University, School of Forestry, pp. 130–135.

Chapman, H. H. 1938. National parks, national forests, and wilderness areas. *Journal of Forestry.* 36(5): 469–474.

Church, Frank. 1977. Wilderness in a balanced land use framework. Wilderness Resource Distinguished Lectureship. Moscow: University of Idaho, Wilderness Research Center.

Costley, Richard J. 1972. An enduring resource. *American Forests.* 78(6): 8–11.

Dettmann, J. 2008. The need for wilderness litigation. *International Journal of Wilderness.* 14(1): 4–6.

Foote, Jeffrey P. 1973. Wilderness—a question of purity. *Environmental Law.* 3(4): 255–260.

General Accounting Office. 1984. *Private Mineral Rights Complicate the Management of Eastern Wilderness Areas.* RCED-84-101. Washington, DC: U.S. Government Printing Office.

Gilligan, James P. 1953. The development of policy and administration of Forest Service primitive and wilderness areas in the Western United States. Ann Arbor: University of Michigan. Dissertation.

———. 1954. The contradiction of wilderness preservation in a democracy. In: *Proceedings, Society of American Foresters*; October 24–27, 1954; Milwaukee, WI. New York: Society of American Foresters, pp. 119–122.

Glover, James M.; Glover, Regina B. 1986. Robert Marshall—portrait of a liberal forester. *Journal of Forest History.* 30(3): 112–119.

Haight, Kevin. 1974. The Wilderness Act: Ten years after. *Environmental Affairs.* 3(2): 275–326.

Heinselman, Miron L. 1973. Restoring fire to the canoe country. *Naturalist.* 24(4): 21–31.

Hendee, John C. 1986. Wilderness: Important legal, social, philosophical and management perspectives. In: Kulhavy, David L.; Conner, Richard N., eds. *Wilderness and Natural Areas in the Eastern United States: A Management Challenge;* May 13–15, 1985. Nacogdoches, TX: Stephen F. Austin University, School of Forestry, pp. 5–11.

Keyser, C. Frank. 1949. The preservation of wilderness areas—an analysis of opinion on the problem. Washington, DC: Library of Congress, Legislative Reference Service (renamed Congressional Research Service in 1970).

Mackintosh, Barry. 1985. Harold L. Ickes and the National Park Service. *Journal of Forest History.* 29(2): 78–84.

Marshall, Robert. 1933. *The Forest for Recreation.* Senate Document 12. Washington, DC: U.S. Government Printing Office, pp. 463–487.

Matthews, Olen Paul; Haak, Amy; Toffenetti, Kathryn. 1985. Mining and wilderness: Incompatible uses or justifiable compromise? *Environment.* 27(3): 12–17, 30–36.

McArdle, Richard E. 1975. Wilderness politics: Legislation and Forest Service policy. *Forest History.* 19(4): 166–179.

McCloskey, Michael. 1966. The Wilderness Act of 1964: Its background and meaning. *Oregon Law Review.* 45(4): 288–321.

Mercure, Delbert V., Jr; Ross, William M. 1970. The Wilderness Act: A product of Congressional compromise. In: Cooley, Richard A.; Wandesforde-Smith, Geoffrey, eds. *Congress and the Environment.* Seattle: University of Washington Press, pp. 47–64.

Mitchell, John G. 1985. In wildness was the preservation of a smile: An evocation of Robert Marshall. *Wilderness.* 48(169): 10–21.

Nash, Roderick. 1982. *Wilderness and the American Mind.* 2d ed. New Haven, CT: Yale University Press.

Nelson, Thomas C. 1985. Wilderness re-re-defined. *Journal of Forestry.* 83(12): 717–718.

Outdoor Recreation Resources Review Commission. 1962. *Wilderness and Recreation: A Report on Resources, Values, and Problems.* ORRRC Study Report 3. Washington, DC: U.S. Government Printing Office.

Proescholdt, Kevin. 1984. Boundary Waters: More obstacles for a troubled law. *Sierra.* 69(4): 18–21.

———. 1995. *Troubled Waters: The Fight for the Boundary Waters Canoe Area Wilderness.* St. Cloud, MN: North Star Press.

———. 2000. Boundary Waters Wilderness: An international story. *International Journal of Wilderness.* 6(1): 35–37.

Robinson, Glen O. 1975. *The Forest Service.* Baltimore: The Johns Hopkins University Press.

Roth, Dennis. 1984. The national forests and the campaign for wilderness legislation. *Journal of Forest History.* 28(3): 112–125.

Scott, Douglas. 1976. Howard Zahniser: Architect of wilderness. *Sierra Club Bulletin.* 61(9): 16–17.

———. 2004. *The Enduring Wilderness: Protecting Our Natural Heritage Through the Wilderness Act.* Golden, CO: Fulcrum Publishing.

Short, L. Rex. 1980. Wilderness policies and mineral potential on the public lands. *Proceedings of the Annual Institute of the Rocky Mountain Mineral Law Institute.* 26: 39–67.

U.S. Department of the Interior. 1984. Wilderness mineral potential: Assessment of mineral-resource potential in U.S. Forest Service lands studied, 1964–1984. Vol. 2. Geological Survey Professional Paper 1300. Washington, DC: U.S. Department of the Interior, Geological Survey.

U.S. Public Law 86-517. The Multiple-Use Sustained-Yield Act of June 12, 1960. 74 Stat. 215.

U.S. Public Law 88-577. The Wilderness Act of September 3, 1964. 78 Stat. 890.

U.S. Public Law 93-622. (Eastern Wilderness) Act of January 3, 1975. 88 Stat. 2096.

U.S. Public Law 94-579. (Wilderness Study) Federal Land Policy and Management Act (FLPMA) of October 21, 1976. 90 Stat. 2743.

U.S. Public Law 95-232. Endangered American Wilderness Act of February 24, 1978. 92 Stat. 40.

U.S. Public Law 95-495. (Boundary Waters Canoe Area Wilderness) Act of October 21, 1978. 92 Stat. 1649.

U.S. Public Law 96-312. Central Idaho Wilderness Act of July 23, 1980. 94 Stat. 948.

U.S. Public Law 96-487. Alaska National Interest Lands Conservation Act (ANILCA) of December 20, 1980. 94 Stat. 2371.

U.S. Public Law 96-560. (Colorado Wilderness) Act of December 22, 1980. 94 Stat. 3265.

U.S. Public Law 98-406. Arizona Wilderness Act of August 28, 1984. 98 Stat. 1485.

U.S. Public Law 98-428. Utah Wilderness Act of September 18, 1984. 98 Stat. 1657.

U.S. Public Law 98-550. Wyoming Wilderness Act of October 30, 1984. 98 Stat. 2807.

The Wilderness Society. 2004. *The Wilderness Act Handbook*, 40th anniversary ed. Washington, DC: The Wilderness Society.

Worf, William A. 1980. Two faces of wilderness—a time for choice. *Idaho Law Review.* 16: 423–437.

Zahniser, Edward. 1984. Howard Zahniser: Father of the Wilderness Act. *National Parks.* 58(1–2): 12–14.

Zahniser, Howard. 1964. How much wilderness can we afford to lose? In: Brower, David R., ed. *Wildlands in Our Civilization;* March 30–31, 1951; Berkeley, CA. San Francisco: Sierra Club, pp. 46–51.

CHAPTER

5

Management Implications in the Wilderness Designation Process, Wilderness Designation Acts, and Wilderness Litigation

INTRODUCTION

The wilderness designation process, implementing legislation, and litigating over agency decisions can have important management implications. *In this chapter, we consider how the wilderness designation process, legislation to designate additional wilderness, and litigation over wilderness decisions also influence management. We have organized this chapter into three parts.*

Part I. We review the evolution of the wilderness designation process, focusing on some key events in the evaluation of roadless lands for their wilderness suitability. For example, how roadless lands were reviewed, evaluated, studied, and designated in the national forests, national parks, and fish and wildlife refuges, and lands administered by the Bureau of Land Management influences management and the kinds of areas designated in the process.

Part II. We summarize the management implications of wilderness designation laws that set aside wilderness for protection. We focus on seven benchmark laws providing key policy direction and then summarize management direction and special provisions for wilderness uses and activities in other specific wilderness laws.

Part III. We review the management implications of litigation related to wilderness laws.

While all this information may seem tediously complex to review, it forms the legal foundation of the NWPS and the designation and management of the areas in the system.

EVOLUTION OF THE DESIGNATION PROCESS

EVALUATION OF ROADLESS LANDS FOR WILDERNESS SUITABILITY

Section 3 of the 1964 Wilderness Act provided procedures for designating wilderness. First, the act provided for an "instant" wilderness system by proclaiming that all the lands administered by the U.S. Forest Service as wilderness, wild,

FIGURE 5.1 A visitor in the Absaroka-Beartooth Wilderness in Montana kneels on headwaters of the Boulder River's East Fork. The Wilderness Act of 1964 instructed the secretary of agriculture to review within ten years all administratively designated primitive areas on the national forests to ascertain suitability of wilderness. Photo courtesy Richard Behan.

or canoe areas prior to 1964 would be known henceforth as (legally designated) wilderness areas.

Second, the Wilderness Act instructed the secretaries of agriculture and the interior to review roadless lands within their respective jurisdictions and to recommend to the president the suitability of these lands for designation as wilderness. The USFS, through the secretary of agriculture, was instructed to review those national forestlands previously classified administratively as *primitive areas* (a total of 34 areas and 5.4 million acres) within ten years after passage of the act (Fig. 5.1).

Similar instructions were addressed to the secretary of the interior. Within ten years, all roadless areas in the various units of the National Park System and the national wildlife refuges and game ranges in excess of 5,000 acres, as well as all roadless islands, were to be reviewed by the respective agencies, and recommended their suitability for wilderness to the president. The president would make his recommendations to Congress. Finally, Congress would propose and vote on legislation to designate each area as wilderness (or not).

However, three problems immediately surfaced for the agencies: (1) *what criteria to use* when making recommendations about the wilderness character of lands and the need for wilderness; (2) *what lands to consider*; and (3) *who should be the primary force in the review process*, allegedly to complete and round out the NWPS: the agencies or interest groups? These issues have permeated the wilderness review process from 1964 until the present, and no agency experienced this debate more dramatically than the USFS.

REVIEW OF NATIONAL FOREST ROADLESS LANDS

The USFS's roadless lands review began in late 1964 with a special task force of experienced wilderness managers whose assignment was to write wilderness policy and guidelines in accordance with the act. The task force believed their work would conclude quickly, but they found that the Wilderness Act set only broad philosophical definitions, that little consensus had developed in their own agency on what wilderness was, and that ultimately the input of citizens and congressional action, as required by the act, would strongly influence the nature of the wilderness system (Worf 1980).

The task force took a "pure" or "strict constructionist" approach to wilderness policy, believing Congress had the vast and pristine wildernesses of the West in mind for the NWPS. They knew of the many requests for commodity use of wilderness and decided that strict guidelines were needed to prevent the "cheapening" of the NWPS.

From a more practical standpoint, they believed that wilderness and wilderness management would be expensive, in terms of uses forgone and in terms of management to maintain primitive conditions and experiences. Expense would increase sharply if the agency had to remove evidence of previous human use (Costley 1972). The task force also worried that adding areas that had significant human-caused impacts to the wilderness system would be a dangerous precedent that would permit existing, pristine areas to decline to the lower standard of the newly admitted areas. Finally, the task force believed that consistent, clear, and precise guidelines were needed for a system of comparable, high-quality areas to be designated and managed (Worf 1980).

The task force might have expected the environmental groups that had lobbied for the passage of the act to support its strict constructionist approach toward implementing the act, but the environmentalists did so mainly with respect to wilderness management (Roth 1984), and not with respect to adding new areas to the NWPS. *As the USFS developed its guidelines and conducted its review of the thirty-four primitive areas, environmental groups and the public began to demand a greater role in the wilderness designation process, and their definition of wilderness was not always as pure as that of the USFS.* For each primitive area, the agency strongly advocated its position on whether an area should receive wilderness designation within the existing primitive-area boundaries, have larger or smaller wilderness boundaries, or be allocated to other multiple-use purposes because the area was deemed unsuitable for wilderness.

Environmental groups often disagreed with the USFS recommendations because the Wilderness Act stated that only a bill passed by both bodies of Congress and signed by the president could create a wilderness. This made wilderness designation of areas more difficult, and much more political. For the first time, Congress had given itself the power to determine how a piece of national forestland was to be used. In essence, the USFS became an advisor to Congress and citizen environmental groups about wilderness decisions on lands it managed. Thus, the political process would determine the ultimate size of the NWPS.

The review, debate, and final congressional decisions regarding the San Rafael and the Gore Range–Eagles Nest Primitive Areas in California and Colorado illustrate the increasingly political nature of the wilderness designation process and growing influence of environmental groups and the courts. *Thus, although the wilderness designation process was defined by the Wilderness Act, it was a little different, more complicated, and more political than anyone anticipated* (Fig. 5.2).

FIGURE 5.2 Some of the finest wilderness solitude is in southwestern deserts, where many areas and millions of acres were designated as wilderness in the 1990s. Wilderness designation under the Wilderness Act is a lengthy process requiring resource inventory, public involvement, formulating alternatives, congressional and public debate, passage through Congress, and finally, the president's signature. Photo courtesy Marilyn Hendee.

The San Rafael Primitive Area

The USFS began review of its primitive areas with the 74,990-acre San Rafael Primitive Area in California—an area with little timber or mineral value and known for its grassy balds, scenic rock outcrops, mountain lions, bears, eagles, condors, and Chumash Indian rock paintings. As directed by the Wilderness Act, the USFS examined contiguous roadless lands and proposed a wilderness area of 110,403 acres. Under the USFS proposal, 36,244 acres were to be added to the primitive area and 831 acres containing a fuel break and a lookout tower were to be withdrawn. Little controversy was expected.

At public meetings, environmentalists urged that five other contiguous areas be included in the wilderness proposal. They also opposed deletion of the 831 acres because this area contained favorite camping spots for backpackers and resting sites for California condors. This testimony advocated increasing the size of the wilderness proposal by almost one-third, to about 158,000 acres.

In response to public testimony and management considerations, the USFS then submitted a revised recommendation to designate 142,722 acres as wilderness, an increase of 32,000 acres over its initial proposal and almost double the size of the original primitive area. President Lyndon Johnson transmitted the recommendation, unmodified, to Congress on February 1, 1967, and strongly urged Congress to give it "early and favorable approval."

The temporary peace and goodwill between the USFS and environmentalists began to unravel later in 1967. Disagreements centered on 4,700 acres—an area that environmentalists maintained should be included in the wilderness proposal because it had wilderness character, was a condor flyway, and contained Chumash Indian pictographs. The USFS reported that 2,500 acres of the area contained an administrative site and road and included areas that had been converted from brush to grass for wildlife habitat, livestock forage, and fire control. The remaining 2,200 acres contained the only suitable site for construction of a larger firebreak. In fact, in June of the preceding year, a fire that ignited from the crash of a small airplane burned more than 90,000 acres, 70,000 of which were in the proposed wilderness. USFS officials believed that even under the most liberal interpretation of the Wilderness Act, the existing fuel break with its bulldozer and plow marks could not be considered wilderness.

The local Santa Barbara Sierra Club had already convinced California senator Thomas Kuchel to introduce a bill that was essentially the environmentalists' proposal. It called for a 158,000-acre wilderness and included the 2,200 acres that the USFS wanted to exclude for potential use as a firebreak. This action was early evidence that in questions of wilderness allocation, grassroots environmental organizations would help shape congressional decisions (Roth 1984). Ultimately, environmentalists indicated a willingness to accept a reduction of some 13,000 acres from their original proposal, but they would not give in on the 2,200-acre addition—the area where the USFS wanted to retain the option of building a firebreak.

The 2,200 acres quickly assumed symbolic political importance far beyond their intrinsic worth as wilderness. Debate was vigorous—for precedent was at stake. Environmentalists feared that if Congress accepted the USFS position, it would ever after simply rubber-stamp USFS wilderness proposals. The USFS believed its professionalism was at stake and was further distressed that its authoritative recommendations about fire management were being questioned. Congressional representatives from the district where the San Rafael Area was located supported the USFS proposal because fire danger was such a serious matter. Ultimately, the USFS position was upheld and the 2,200 acres were excluded when, in a special White House ceremony in March 1968, President Johnson signed *the San Rafael Wilderness bill into law* (U.S. Public Law 90-271). It was the first addition to the NWPS. It had passed through the various steps and procedures prescribed by the Wilderness Act; it received considerable debate in Congress; and *it further defined and demonstrated the wilderness designation process as follows:*

1. A preliminary agency recommendation, resting on substantial review (including field surveys and a mineral assessment), was revised following citizen review and input.
2. The revised recommendation was submitted through the secretary of agriculture to the president.
3. The president submitted the proposal to Congress, recommending passage.
4. Congressional committees considered several draft bills and held public hearings.
5. When the House and Senate approved different versions of the bill, a joint conference committee studied the differences and recommended the Senate version.
6. After a debate in the full House of Representatives, the House also accepted the conference committee–recommended bill, and Congress then passed it.
7. The president signed the legislation. More than three years had passed between initial field investigation and final passage.

The San Raphael Primitive Area review provided three important insights into the wilderness classification designation: (1) it began to define more precisely the kinds of lands Congress would accept for the NWPS; (2) it identified key players in the wilderness designation process, with grassroots environmental groups wielding much more influence than anticipated; and (3) it demonstrated an intensity of debate over final provisions of wilderness legislation that had not been anticipated.

The Gore Range–Eagles Nest Primitive Area

The study of the wilderness suitability of the Gore Range–Eagles Nest Primitive Area introduced another key player to the wilderness designation process—the courts. As the USFS reviewed the Gore Range–Eagles Nest Primitive Area on the White River National Forest in north-central Colorado, the contiguous undeveloped East Meadow Creek drainage quickly became the focus of attention. The area was rolling, timbered high country with small meadows and parklike stands of old-growth Engelmann spruce, lodgepole pine, and fir. It had high wildlife values, including an elk migration and nursing ground. Moreover, it had become a popular backpacking, horse-packing, and big game hunting area for those who lived in or began their trips from nearby Vail, a wealthy ski resort and town built in 1964.

A primitive truck road was constructed through the area in the 1950s for access to combat a bark beetle infestation. In 1962, the USFS drew up a plan for timber sales in the area and in 1964 built an access road to the edge of the area. USFS officials at the ranger district and forest levels recommended against wilderness designation of East Meadow Creek because of the "bug road," the existence of some private inholdings and unpatented mining claims, its location outside the original primitive area, and because established sawmill operators in the area depended on USFS timber (Roth 1984).

The proposed timber sale provided a dilemma. The primitive-area review process had not yet been completed. If the timber harvesting went forward, it would destroy the wilderness character of lands contiguous to a primitive area, effectively making a decision about wilderness designation—a judgment Congress reserved for itself. On the other hand, a decision against the timber sale would be seen as favoring the expansion of the wilderness system—something only Congress can do. Thus, a middle-ground decision was offered that included timber cuts in only six of the proposed fourteen timber-harvesting blocks, with a large buffer zone between the primitive area and the timber sale.

The compromise did not work. Neither environmental groups nor the forest-products industry liked the decision. Both felt it was a decision that only Congress could make. A third group, the citizens of Vail, was especially upset. They argued that their town depended on recreation dollars, that the timber sale was planned in 1962 before their town even existed, and that timber harvesting in the East Meadow Creek would significantly reduce a valuable, recreation-based economic resource.

The citizens of Vail, the Sierra Club, and several Colorado conservation organizations sought an injunction to stop the timber sale (Roth 1984). Their attorney was aware of the landmark Scenic Hudson decision of 1965 in which the judge gave local citizens' groups standing in court and eventually ruled in favor of their suit to halt the construction of a proposed hydroelectric plant. He also noted the language of the Wilderness Act that said, "Nothing herein contained shall limit the President in . . . recommending the addition of any contiguous area of national forest lands predominantly of wilderness value."

The district court heard the case, and the judge accepted the plaintiffs' arguments. He granted them the right to sue the government and listened to the testimony of the environmental witnesses who vouched for the wilderness character of the area and the USFS's arguments that the bug road disqualified it. On February 17, 1970, the judge ruled in favor of the plaintiffs and permanently enjoined the timber sale. On October 1, 1971, the Tenth Circuit Court upheld the ruling. The USFS continued to believe that a bad interpretation of the Wilderness Act had been made. It believed that the law did not *require* the review of every single contiguous acre of land with wilderness character by the president and Congress, although the act did not *prohibit* such consideration either. The Department of Justice, on behalf of the USFS, appealed to the Supreme Court, which refused to hear the case (Roth 1984).

Grassroots citizen groups had gained a major victory in the "precedent-setting" Gore Range–Eagles Nest Primitive Area review. The judicial decision in effect protected all undeveloped (roadless) lands contiguous to USFS primitive areas until Congress made the final decision; it demonstrated again that wilderness designation decisions would be shaped by actions beyond the USFS and the president; thus it freed the environmentalists to turn their attention to USFS lands that were not contiguous to wilderness or primitive areas, but which were still roadless.

DESIGNATING NONRESERVED NATIONAL FORESTLANDS

One major block of lands potentially available for wilderness designation was not specifically mentioned at all in the Wilderness Act. These were the so-called *de facto wilderness* lands in the National Forest System, areas that are in fact wilderness in the general sense of the term—roadless and undeveloped—but that lack explicit, legal designation and thus protection as wilderness.

Although the act itself did not specifically direct the USFS to review lands other than those designated as primitive areas, the Wilderness Act did not preclude such review. *Thus, the agency decision to review all its roadless lands, in addition to lands specifically noted in the Wilderness Act (primitive areas and their contiguous roadless lands), represented a judgment of agency personnel that the act did not prevent such a review. The agency also recognized that environmentalists would soon be demanding designation of additional wilderness lands not specified in the act. The agency wanted to seize the initiative—to act more than react and to approach wilderness review and evaluation from a multiple-use planning perspective. They would inventory and evaluate the wilderness potential of all roadless areas on the national forests.*

U.S. Forest Service Inventory of Potential Wilderness Additions

Three criteria were formulated to guide USFS managers in identifying potential additions to the wilderness system: suitability, availability, and need:

1. *Suitability was defined by minimum conditions set forth in the Wilderness Act and by conditions believed to enhance the opportunity for wilderness-dependent experiences.* These conditions are: (a) an absence of roads

or other development; (b) be at least 5,000 acres if not adjoining existing wilderness or primitive areas (smaller, undeveloped contiguous areas were also to be inventoried); (c) be free of present or foreseeable nonconforming uses or activities; (d) possess a wide range of subjective values for helping people discover freedom and spiritual renewal; (e) offer outstanding opportunities for challenge and primitive recreation like camping, ski touring, and hiking; (f) include abundant and varied wildlife; and (g) possess outstanding opportunities for formal and informal education and scientific study.

2. *Availability specified that wilderness designation must represent the highest and best use of the land over time.* The tangible and intangible values of wilderness designation had to offset the potential value of all resources that would be forgone, such as the value of timber, minerals, and non-wilderness recreation, and the costs of managing the area as wilderness.
3. *Need meant that the need for wilderness must be compared to that for other resources.* Need was to be determined by consideration of the location, size, type, and capacity of other wildernesses in the general vicinity, by local and national trends in wilderness use, and by the extent to which non-wilderness lands were available to provide dispersed recreation opportunities.

The USFS argued that no formula for reaching decisions on potential additions existed, but that the objective in each case was to determine the predominant public value within the meaning of the Multiple-Use, Sustained-Yield Act of 1960 (U.S. Public Law 86-517): "With consideration being given to the relative values of the various resources, and not necessarily the combination of uses that will give the greatest dollar return or the greatest unit output."

The initial inventory, in this "Roadless Area Review and Evaluation" (RARE; subsequently called RARE I because it would later be repeated as RARE II), identified 1,449 roadless areas on the national forests with wilderness potential, containing 56 million acres. The chief of the USFS planned to use this inventory, in combination with other data, to complete a list of "study areas" to receive more formal review and evaluation. During the review process, the USFS held some 300 public meetings and received more than 50,000 written and oral comments—at the time the largest public-involvement effort ever undertaken by the federal government (Hendee 1977). However, of the 1,449 areas, only two were in the East and one was in Puerto Rico. The lack of areas in the East was a disappointment to environmentalists and quickly became a source of controversy.

Roadless Area Review and Evaluation (Rare I)

The USFS wanted to avoid fights over individual areas, or studying all individual area management options, including wilderness. Fresh in mind was controversy in 1972 between the USFS and a citizen group over a roadless area near Lincoln, Montana, which resulted in the designation of the 240,000-acre Scapegoat Wilderness—the first USFS area designated under the 1964 act that had not previously been a primitive area (U.S. Public Law 92-395). The long and costly debate over this area demonstrated the need for a comprehensive review of all roadless lands.

Such a review would serve as a first approximation of which roadless lands deserved careful, detailed study for their potential for wilderness designation, and which lands appeared to be more suited for non-wilderness management. The agency also believed that a comprehensive national review and evaluation of all areas would constitute a national Environmental Impact Statement (EIS) under the recently signed National Environmental Policy Act (NEPA) of 1970 (U.S. Public Law 91-190), and thus avoid the need to write an EIS for each roadless area.

The USFS assembled an interdisciplinary team to complete the national Roadless Area Review and Evaluation of the 1,449 potential wildernesses identified in the inventory. The procedure developed and used by the team, called the Roadless Area Review and Evaluation (RARE), became the principal analytical tool used to recommend which of the inventoried areas should receive further intensive study for possible wilderness designation.

The principal objectives of the RARE selection process were:

1. To obtain as much wilderness value as possible relative to the cost and value of the forgone opportunities to produce other goods and services for society.
2. To disperse the future wilderness system as widely as possible over the United States.
3. To represent as many ecosystems as possible so that the scientific and educational purposes of wilderness preservation are best served.
4. To obtain the most wilderness value with the least relative impact on the nation's wood product output.
5. To locate some new wilderness areas closer to densely populated areas so that more people can directly enjoy their benefits.

To evaluate the undeveloped areas, the RARE process used several quantitative and qualitative judgments for each area, including:

1. The total gross acres of roadless area.
2. A quality index (QI). Field personnel rated each area on three factors: scenic quality (S), isolation and likely dispersion of visitors within the area (I), and variety of wilderness experiences and activities available (V). Each of these factors was weighted and used to calculate the quality index by the formula: $QI = 4(S) + 3(I) + 3(V)$.
3. An effectiveness index (EI). To derive this measurement, total gross acres were multiplied by the quality index.
4. Total opportunity costs index. This index was composed of the sum of the following:
   a. Estimated costs for studies, establishment, operation, and maintenance.
   b. Cost, if any, of acquiring private land.
   c. Cost of replacing special-use improvements.
   d. Mineral values.
   e. Potential water development values.
   f. Timber values.

The RARE draft EIS proposed 235 areas covering 11 million acres (about one-fifth of the acreage inventoried) as new wilderness study areas (WSAs) to receive intensive study and early consideration for addition to the NWPS (U.S. Department of Agriculture, Forest Service 1973). Until the studies were completed, no management programs could be undertaken that would alter their undeveloped state. More than 7,000 letters and documents were received in response to the draft EIS (Hendee 1977).

Following this analysis, 61 new wilderness study areas were added and 22 were deleted for a net gain of 39 new areas and a revised total (in the final EIS released in October 1973) of 274 areas and 12.3 million acres, an acreage gain of about 10 percent more than in the draft EIS.

The four principal variables—the ones on which marginal cases were decided and which the chief and staff repeatedly used in making final decisions: (1) *public input*, including sentiment of involved citizens and organizations and the views of legislators and government agencies; (2) *potential wilderness quality of the roadless areas* as measured by the quality index; (3) *cost effectiveness,* reflecting the value of other resource uses forgone compared to relative wilderness values; and (4) an *overall judgment factor* resting heavily, but not entirely, on the recommendations of local, regional, and national decision makers (U.S. Department of Agriculture, Forest Service 1973). *Many criticisms were leveled at the roadless area inventory and particularly at the RARE process* (Milton 1975), including timing and the RARE methodology—such as the subjective nature of the quality index (QI), the overwhelming influence of size of area on the effectiveness index (EI), and the validity of placing a monetary value in the opportunity cost estimate for preservation values many claimed to be nonquantifiable.

Environmentalists believed worthy roadless areas—most notably in the East—were not included among the 1,449 inventoried areas, and therefore were not adequately considered in the RARE process. Disappointment in the wilderness recommendations led the Sierra Club and others to file a suit in federal court, which was ultimately dismissed, but without prejudice, meaning it could be heard by the court again.

Both the Sierra Club and the USFS hailed the outcome of the lawsuit as a victory. The judge's decision did not prevent the conversion of roadless areas to multiple-use management; it simply stated that a proper EIS must be filed for *each* area and that in the EIS, wilderness must be considered one of the viable management options (Allin 1982).

The RARE I process had two important benefits for wilderness. First, the inventory provided a reasonably comprehensive list of remaining roadless lands on the national forests, and this highlighted the need for management guidelines and programs for these lands to protect their wilderness potential.

Second, the RARE process—for all its flaws—systematically reviewed and weighed the relative values of a variety of uses of roadless areas, taking a regional and national perspective.

An Eastern Wilderness Act

Even as the RARE I process was reaching its culmination in the early 1970s, the debate over what to do about national forestlands in the East (generally defined as east of the 100th meridian) was heating up. Environmentalists were disappointed with the amount and kind of land that had been protected in the East under the Wilderness Act. They sought protection for areas that, while not as pristine or as vast as areas in the West, still possessed a primitive character. Most of these acres had previously been logged or even farmed and roaded, but time was rapidly healing these wounds.

The USFS steadfastly held to its purity position that only areas that *retained* primeval conditions qualified for wilderness designation and thus pressed for the protection of lands in the East under an alternative wild areas system. Wild areas would be distinct from wilderness areas because they would be primarily for recreation enjoyment.

During 1972 and 1973, Congress, the USFS, and environmentalists debated the wild areas–versus–wilderness ideas (Allin 1982; Hendee et al. 1991; Roth 1984) and their implications for how much land would be set aside for protection, and to what standard it would be protected. By 1973, environmental interests were together in support of adding areas in the East to the NWPS. The USFS dropped the idea of a separate wild-areas system in the East and proposed designating sixteen eastern areas as instant wilderness areas and that another thirty-seven be studied for possible wilderness designation.

Senate bill 3433 was introduced into Congress—a bill that called for nineteen instant wildernesses to be established, and forty study areas to be reviewed by the secretary of agriculture within five years according to the procedures outlined in the Wilderness Act. It contained two particularly important clauses: (1) the secretary of agriculture was given the power of condemnation when private (inholding) landowners failed to use their land in a manner compatible with wilderness and were unwilling to sell voluntarily (this was an important authority, absent from the 1964 Wilderness Act, and was included because of many private inholdings in eastern national forests); and (2) all lands designated as wildernesses or as wilderness study areas (WSAs) were withdrawn from mineral entry.

To the dismay of the environmental community, an amendment reduced the number of wildernesses in the bill from 19 to 15, and study areas from 40 to 17, and was signed into law (U.S. Public Law 93-622) by President Gerald Ford on January 3, 1975 (Roth 1984).

This law (popularly called the "Eastern Wilderness Act") represented a major challenge for management. It stated the desire of Congress to locate wildernesses nearer population centers, and it included areas that are generally smaller in size and show more past evidence of human use than national forestlands previously placed in the NWPS.

The Endangered American Wilderness Act

Environmentalists had criticized the RARE I process for failing to select several undeveloped areas located near population centers for wilderness study. Sierra Club leaders decided that the time was right for an omnibus wilderness bill that would protect several areas that the RARE process had either excluded or had not recommended for future wilderness study. The bill would thus be symbolic of alleged defects in the USFS's RARE I process, would help build grassroots support for wilderness throughout the country, and would demonstrate this support to every congressional representative and senator. The environmental groups convinced Senator Frank Church (ID) and Representative Morris Udall (AZ) to introduce the Endangered American Wilderness bill in Congress in 1976. At first, congressional mail ran against the bill, but as the Sierra Club's campaign geared up, support began to build.

President Jimmy Carter signed the Endangered American Wilderness Act February 24, 1978 (U.S. Public Law 95-237), creating 1.3 million acres of wilderness in 17 new wilderness areas or additions to wilderness areas. It was the largest addition to the NWPS accomplished by a single wilderness designation act to that point. In one sense it repudiated the RARE, because it designated some areas that the RARE process had either ignored in their inventory or had not even recommended for wilderness study.

Another Roadless Area Review and Evaluation

RARE II—the second USFS Roadless Area Review and Evaluation—was a critical influence on the wilderness designation process and, thus, on wilderness management. *By the mid-1970s, there was considerable dissatisfaction with the first RARE process because it favored too much or too little wilderness, depending on the view of the various groups.* When Jimmy Carter was elected president in 1976, controversies over the validity and reliability of the first RARE process had slowed the wilderness designation process to a crawl.

The basic objective of RARE II was to accelerate the planning process mandated by the recently passed Forest and Rangeland Renewable Resources Planning Act (RPA) of 1974 (U.S. Public Law 93-378) and the National Forest Management Act (NFMA) of 1976 (U.S. Public Law 94-588). *RARE II called for a comprehensive study to identify roadless and undeveloped lands in the National Forest System and to determine their general uses both for wilderness and for resource management and development. The study would produce recommendations on which roadless lands should be designated wilderness, which lands should be released to uses other than wilderness, and which warranted further study for all uses, including wilderness.*

The study was to be a methodical, rational approach to planning in a national context. The goal of identifying roadless areas with important non-wilderness values was equal in importance to identifying areas with wilderness potential. *The desired outcome was a balance of land uses that would best meet the nation's needs* within the requirements of NEPA.

The *USFS began RARE II striving to improve on RARE I in two ways: first, to separate the identification or inventory, and evaluation of each roadless area; and second, to establish greater reliability and validity in the inventory process.* In doing this, the USFS wanted to strengthen public confidence in the RARE II process by separating the assessment from the evaluation components. The assessment component—which included area identification and inventory—had wilderness suitability as its primary criterion. *The assessment question was not whether a roadless area should or should not be wilderness, but rather whether it had attributes to meet minimum criteria for wilderness consideration. Judgments about which areas should be recommended for wilderness were left to the evaluation phase.*

RARE I had reviewed 1,449 areas containing 56 million acres. The roadless status of most of these areas had remained unchanged, and they became part of the RARE II inventory. The public was asked to comment and suggest additions and deletions for the RARE II study, and more than 50,000 comments were received.

Many more areas in the East were included in RARE II than had been previously studied in RARE I because somewhat less restrictive standards were used to select areas in the inventory there. This responded to the limited wilderness opportunities in the East, the need for wilderness close to people, and the faster regenerative capacity of ecosystems in the East. Areas were selected that did not exceed *more than one* of the following: one-half mile of improved road for each 1,000 acres, if road was under USFS jurisdiction; 15 percent of area in nonnatural planted vegetation; 20 percent of area with timber harvested in the past ten years; and an area could contain a few dwellings if the dwellings and access were obscured by natural features. *The final RARE II inventory list included 2,919 roadless areas and encompassed 62 million acres on national forests and national grasslands in 38 states and Puerto Rico* (U.S. Department of Agriculture, Forest Service 1979).

A variety of data were collected on the RARE II areas. Each area's renewable-resource potential was measured through a system called the Development-Opportunity Rating System (DORS). This system compared the benefits and costs of developing a roadless area for renewable multiple-use outputs, including saw timber, grazing, developed recreation, hunting, and fishing. In addition, site-specific information on potential timber, programmed harvest, and grazing was reviewed, and each area's mineral potential was assessed.

The RARE II process also sought to increase ecological diversity within the NWPS, with the goal of obtaining two or more distinct examples of each of the nation's ecosystems (U.S. Department of Agriculture, Forest Service 1979). Using the Bailey-Küchler mapping system, an approach using Bailey's (1976) ecoregion concept and Küchler's (1966) system of potential natural vegetation, the USFS identified 242 distinct ecosystems in the United States and Puerto Rico (later corrected to 261). The ecosystems were distinct in such physical environment factors as climate and soil and in biological environment factors such as vegetation (Davis 1989). The RARE II process gave preference for wilderness recommendation to those areas that contained an ecosystem not currently represented in the NWPS.

In late 1977, a USFS task force developed the RARE II Wilderness Attribute Rating System (WARS) to inventory the wilderness characteristics of each RARE II area. WARS established a procedure for identifying the attributes of wilderness as defined by the Wilderness Act and for rating their condition. The purpose of WARS was to provide a measure of the area's wilderness quality, just as DORS provided a development-opportunity rating system to measure its potential commodity and developmental values.

The *critical attributes used in WARS* (those mentioned in the Wilderness Act) were *natural integrity* of the area, *apparent naturalness* (amount and nature of perceptible impacts), *outstanding opportunities for solitude,* and *primitive-recreation opportunities.* WARS also permitted adjustment in natural integrity and apparent naturalness ratings if area boundaries were adjusted to remove serious intrusions. Each of the four critical attributes was rated on a scale of one to seven, ranging from the outstanding presence of the attribute to its virtual absence. Composite wilderness attribute (WARS) scores were determined by adding the ratings of each of the four critical attributes. Thus, each of the four attributes was weighted equally in the overall score.

There were two supplemental area ratings: a supplementary wilderness-attribute rating and a scenic value rating.

These ratings reflect section 2(c)(4) of the Wilderness Act, which indicates that an area "may also contain ecological, geological, or other features of scientific, educational, scenic, or historical value." This suggests that these features are not necessary for wilderness suitability, but that their presence in an extraordinary degree enhances wilderness values. These two variables were not part of an area's overall composite wilderness-attribute score but instead were viewed as supplemental information to help make marginal decisions or to identify areas that might be placed in other types of USFS protected-area management.

The inventory process of RARE II resulted in a numerical "wilderness-attribute rating" for each area ranging from four to twenty-eight, which indicated the extent to which the area contained wilderness attributes listed in the Wilderness Act. The question that remained was: Which of these areas should be recommended for wilderness designation, released for non-wilderness uses, or held for further study?

In the RARE II draft EIS published on June 15, 1978 (U.S. Department of Agriculture, Forest Service 1978), the public was asked to comment on three things: (1) what individual areas should be allocated to wilderness, non-wilderness, or further planning, and why; (2) what approaches should be used in reaching a decision on allocating the total roadless-area inventory; and (3) what decision criteria should be used in developing a proposed course of action. Response was massive: 264,093 inputs carrying 359,414 signatures, seven times as many public comments as received in RARE I (Hendee 1986; Hendee et al. 1980).

Ten general alternative allocation strategies, reflecting a range of mixes of wilderness and non-wilderness options, were offered to the public. The potential physical, biological, social, and economic effects of each alternative were quantified and evaluated by the USFS to the extent feasible. Of primary concern were the benefits realized and benefits forgone (opportunity costs), if more or fewer wilderness areas were recommended. This analysis, along with public input, existing laws and regulations, identified public needs, and USFS professional judgment of decision makers, resulted in the USFS's preferred alternative. In essence, the proposal called for the allocation to non-wilderness of areas with high potential for commodity resource output, and allocation to wilderness of those with high wilderness attributes. This responded to public input that expressed favor for economic values and jobs, timber production, accessibility, high scenery and diversity in the wilderness system, and high-quality additions to the NWPS.

Recommendations from RARE II

The final RARE II EIS was published on January 4, 1979, approximately 18 months after initiation of the study (USDA FS 1979). It called for the wilderness designation of 624 areas totaling 15,008,838 acres, the allocation to non-wilderness of 1,981 areas totaling 36,151,558 acres, and further planning for 314 areas totaling 10,796,508 acres. The document recommended the proposed wilderness areas to the Ninety-sixth Congress for legislative action and indicated that no activities would be permitted that might alter their wilderness qualities. Areas listed for further planning would remain undeveloped until their status could be reviewed more thoroughly under routine USFS land- and resource-management planning processes. Until a final recommendation was made on these further planning areas, no commercial timber harvesting would be allowed, and exploration and leasing for oil, gas, and energy minerals would be permitted only under rigid stipulations.

When the RARE II results came forward, environmentalists were disappointed. Only about one-fourth of the inventory of national forest roadless areas were being recommended for wilderness, with only one-sixth of the lands allocated to further planning—meaning wilderness was still considered a possibility. And environmentalists were even more critical about the distribution of proposed wildernesses—of the approximately 15 million acres recommended for wilderness, 5 million were on the Tongass National Forest in Alaska. Furthermore, the permanent loss of roadless areas recommended for release to non-wilderness uses was imminent. The final RARE II EIS called for the release to non-wilderness uses of the roadless areas recommended for non-wilderness during the first cycle of forest plans mandated by the NFMA. These "released" areas became the center of controversy in 1979 and in the 1980s.

In California, the general sense was that the wilderness issue should not be settled quickly, and RARE II recommendations for the state reflected this feeling. Of the approximately 6 million acres in California considered by RARE II, about 1 million were recommended for wilderness, 2.4 million for non-wilderness, and 2.6 million were placed in the further-study category.

Huey Johnson, director of the California Resources Agency, sued the USFS and sought a court injunction against the release of 46 California roadless areas to non-wilderness uses. Johnson, in *California v. Bergland, contended that the programmatic RARE II analysis did not meet the requirements of NEPA, and on January 8, 1980, the district court agreed. The court enjoined the 46 areas from development and indicated that before there could be any change in the status of these areas, an EIS for each area would have to be prepared.* In October 1982, the Ninth Circuit Court of Appeals upheld the lower court ruling (Roth 1984).

As a result, the assistant secretary of agriculture for Natural Resources and Environment in the new Reagan administration directed the USFS to reevaluate all RARE II recommendations within the ongoing land-use planning process. His inclusion of *all* areas caused environmental groups to fidget, knowing that the further analyses might cause areas recommended *for* wilderness by RARE II to be dropped.

U.S. FOREST SERVICE WILDERNESS DESIGNATION IN THE 1980S

By the early 1980s, RARE II had provided a lot of information, recommendations, and some agreement toward resolving the status of roadless areas on the national forests. A comprehensive and careful inventory of USFS roadless areas had been completed, including maps, lists of wilderness attributes, ecosystem representation, extent and content of public involvement, and descriptions of commodity trade-offs (Fig. 5.3). The USFS, the Carter administration, and the Reagan administration supported wilderness designation for about 15 million acres of national forestlands and further

FIGURE 5.3 Blodgett Canyon in the Bitterroot National Forest was identified during RARE II as having important wilderness values. Although the process had many criticisms and concerns, it did identify areas where important wilderness values were found. Photo courtesy Don Dodge.

study for 11 million acres. Just how legislation to protect these areas would come forward was unclear, and there was uncertainty over the final resolution of the areas released for non-wilderness uses. Suggestions for a third RARE study had little support.

How would the RARE II data reflecting its "objective approach" affect wilderness designation? One analysis of wilderness designation decisions in Arizona, Idaho, and Utah concluded that the RARE II information about the renewable and nonrenewable resource potential of areas was relatively unimportant in explaining the eventual designation of areas to wilderness or allocation to non-wilderness or further planning (Mohai and Verbyla 1987). Although public signatures favoring wilderness and WARS ratings were consistently among the best predictors of wilderness designation, many areas that were allocated to non-wilderness did not have high resource potential, whereas many areas with high resource development opportunity were recommended for wilderness (Mohai and Verbyla 1987). *This analysis confirmed that rational, empirical analyses based on hard data were playing only a limited role in the highly politicized environment. Politics were proving even more important than resource analysis in wilderness decisions.*

Three issues dominated wilderness allocations on the national forests in the 1980s. *First,* how to add areas to the NWPS? Should legislation be drawn up to add single RARE II areas or a few areas at a time to the system? Should one omnibus bill be passed that included all or most areas recommended by RARE II? Or, should each state with RARE II areas have its own wilderness bill? *Second,* how permanent would the "release to other uses" be for areas recommended for non-wilderness by RARE II? The timber industry and the USFS preferred permanent or *long-term release, or "hard release" as it came to be called;* this would permit long-term planning for management of commodity or non-wilderness recreational use of these lands. The environmental community favored *short-term release, or soft release*—the released areas could continue to be considered for wilderness during Forest Service land-management planning and would thus provide opportunities in the future to gain wilderness protection for some of the released land. A *third* question concerned the adequacy of the RARE II analysis as a programmatic EIS. Would a separate EIS have to be prepared for each area? It seemed so as the *California v. Bergland* court decision indicated that the final RARE II recommendations did not constitute an acceptable EIS for certain areas in California.

Statewide Wilderness Bills—The Prototype of the 1980s

As a result of RARE II, politicians were much more aware of the wilderness potential of national forest roadless areas throughout the country. With the data from RARE II, they had a much firmer grasp of where existing and potential areas were, the quality and characteristics of those areas, locations where need was the greatest, and trade-offs that could be made to meet commodity and wilderness needs. Thus, they were better prepared to deflect the emotional appeals of single-interest commodity or environmental groups. They were also better prepared to look at the big picture, how trade-offs in one area might be offset in another. In addition, because there were so many areas to consider, the sheer workload of an area-by-area approach for legislation to designate wilderness was daunting. The idea of omnibus legislation to designate a lot of areas at one time was appealing, but not without problems.

The environmental community opposed the omnibus bill approach from the beginning. They had a fifteen-year history of successful influence on the wilderness designation process with a strategy largely based on grassroots organizations that influenced local forest supervisors and congressional representatives. Most discussion of an omnibus bill included some sort of permanent release to commodity use of areas not recommended by RARE II for wilderness or further study. Environmental groups feared this most of all, for it would forever

FIGURE 5.4 The Wilderness Act directed the secretary of the interior to review all roadless areas of 5,000 acres or larger in the National Park System and to recommend which should become wilderness. The Hidden Lake area of Glacier National Park, Montana, was recommended to be designated wilderness, and it has since been under consideration for decades. Photo courtesy Richard Behan.

prevent them from seeking wilderness protection for many areas they felt still deserved wilderness consideration.

Soon, individual state congressional delegations, tired of inactivity on an omnibus RARE II bill, feeling pressure from their own constituencies, and wanting to seize the initiative, began to put forward statewide bills—bills that responded to the RARE II recommendations for their state. This approach was a logical extension of the way Congress had done business in the past, when local congressional representatives typically sponsored bills to designate wilderness areas in their district. With the more comprehensive data developed in RARE II, state delegation members could negotiate among themselves to devise a bill that was acceptable to them. Such *statewide bills rapidly became the norm in the 1980s; virtually all national forest wilderness areas added to the NWPS in the 1980s were included in state bills.*

Sufficiency-Release Language

The *California v. Bergland* decision, that an EIS must be prepared for any area whose status would change with release to non-wilderness uses, was, on the face of it, a clear victory for environmentalists. The court decision raised questions about the sufficiency of the entire RARE II process, and this could have set the stage for another national roadless area review (Allin 1982). Yet, the mood of Congress in the late 1970s and early 1980s was not in favor of further delay, study, or expense for more wilderness evaluation. Environmental groups feared a legislative backlash and perhaps a permanent release of non-wilderness RARE II lands. Representative Tom Foley (WA) had introduced legislation in late 1979 that would have mandated immediate non-wilderness management for the 36 million acres recommended for non-wilderness in RARE II and non-wilderness management by January 1, 1985, for any area in the national forests not given statutory wilderness protection by that time. This would have effectively fixed the final amount of wilderness on the national forests at the level set aside by that date (Allin 1982).

A California wilderness bill provided an opportunity for compromise. A meeting of representatives of industry, environmental groups, and the USFS was convened to settle the issues of "sufficiency" and "release." The eventual agreement was to place within each statewide wilderness bill a *sufficiency clause—a statement that said that the RARE II study was "sufficient" for NEPA purposes.* This was to protect against RARE II lawsuits. Implementing this recommendation regarding release language, however, became controversial. The compromise worked out in the meetings called for the (soft) release of lands not recommended for wilderness or further study for one planning cycle (ten to fifteen years) of the Forest Service planning process required by the NFMA of 1976. When the next generation of plans was prepared, the USFS could then study any of these lands still remaining roadless for their wilderness potential. This type of release (for one planning cycle only) became known as "soft release," or later as "Colorado release" when the terminology was included in the 1980 Colorado wilderness bill. Permanent release to non-wilderness uses became known as "hard release."

With the approach of December 31, 1983—which would mark the end of new mineral exploration and leasing in wilderness under terms of the Wilderness Act—mining companies mounted a campaign to get the Department of the Interior to lift its administrative moratorium on the processing of claims in existing wilderness. Such mineral exploration and leasing were still legal within wilderness, but because of costs, negative public opinion, and strict USFS regulations, no leases had been issued. Secretary of the Interior James Watt broke from this tradition and considered mineral leasing requests in the Bob Marshall Wilderness of Montana and Washakie Wilderness in Wyoming and actually approved a mineral lease in the Capitan Wilderness of New Mexico. The public outcry to this action was loud and reverberated through Congress. Congressional committees passed resolutions condemning the proposed mining activity and ordering it to be stopped, and bills were introduced in Congress to permanently ban wilderness mineral leasing.

In the early 1980s, the House had passed several wilderness bills that included the soft release language. USFS officials expressed concern that an amendment to a forest management plan might, with soft release terminology, be considered a revised plan. Thus, a second-generation plan might occur much before the anticipated ten to fifteen years.

The final negotiations of the release issue were feverish between USFS and Senate and House leaders. *The agreement reached was that future state bills would have no fixed time period for release but would simply refer to the forest plans typically lasting ten to fifteen years. Further, during this period, the USFS was not categorically prohibited from managing released land to preserve its wilderness character; it just was not required to do so.* Commodity interests had gained in the final solution by tightening up the definition of a plan revision; this was to ensure that a plan would really run the expected ten to fifteen years, during which released lands might be considered for commodity use, typically timber harvest. In addition, the agreement stipulated that roadless areas of less than 5,000 acres, as well as those areas examined in old USFS unit plans before the RARE II study and recommended for non-wilderness use, were released during the first generation of forest plans.

Agreement on the release issue opened the door to rapid wilderness designation. In the summer of 1984, 18 national forest wilderness bills were passed, adding 6.6 million acres to the NWPS. This was the largest addition to wilderness on the national forests in a single session of Congress since passage of the Wilderness Act. *About 13.6 million acres were released to USFS multiple-use management based on their respective land-management plans under the NFMA.*

DESIGNATING DEPARTMENT OF THE INTERIOR ROADLESS LANDS: NPS, FWS

The Wilderness Act of 1964 clearly described the National Park Service (NPS)–and U.S. Fish and Wildlife Service (FWS)–administered lands subject to wilderness consideration. All roadless lands more than 5,000 acres and all roadless islands were to be studied (Fig. 5.4). For those areas considered suitable for wilderness designation following agency study, wilderness proposals were to be developed for submission to Congress through the Department of the Interior (USDI), the Office of Management and Budget, and finally the president.

Operating with these guidelines, the Department of the Interior did not have to decide what lands to consider or how to propose them for wilderness designation. However, it did have to struggle with the problem of establishing wilderness criteria: among the available candidate areas, which ones should be recommended for wilderness designation? On what basis? It was eight years after the act passed before official selection criteria and guidelines were established in the Interior Department. In the meantime, a beginning had to be made.

Soon after passage of the act, the Interior Department estimated that 22.5 million acres of NPS land and 24.1 million acres of FWS holdings were subject to wilderness study—and this area grew as new national parks and refuges were created. Eventually, *during the ten-year period following the passage of the Wilderness Act, the NPS identified 63 areas covering more than 28 million acres for wilderness review, while FWS identified 113 areas totaling 29 million acres* (U.S. Congress 1973).

The review was very slow to reach Congress. Just why the NPS wilderness study process moved so slowly is not completely clear. Allin (1982) contends that the NPS had little enthusiasm for wilderness because such designation would limit its management discretion and prohibit future development for its primary clientele, touring motorists. In addition, early on, the NPS seemed to take a very strict stance on wilderness resource requirements. For example, Director George Hartzog made it clear that national parklands administratively designated within the NPS as Class III, "natural environment areas," would not meet the criteria for wilderness but instead would be viewed as transition or buffer zones between developed and wilderness lands.

Environmentalists pointed out that the sights and sounds of civilization that might be found in these threshold zones were now acceptable in wilderness, and that the NPS really was simply trying to retain the option of development for these zones. Subsequent statements by Director Hartzog confirmed these suspicions (Allin 1982). Moreover, the NPS developed a policy to exclude any private inholdings within a roadless area from a wilderness recommendation unless federal acquisition of the area was ensured. Areas with other nonconforming but legally permitted uses like water-monitoring devices and motorboat patrol zones were also eliminated from wilderness consideration.

The absence of an explicit wilderness designation procedure for USDI lands likely explains some of the delay and difficulty encountered by the NPS and FWS. Traditionally, the NPS had zoned roadless and undeveloped tracts in the individual parks' master plans. For example, much of Yellowstone's 2.2 million acres is zoned "roadless" in the park's master plan. The problem the USDI faced in meeting the obligations of the Wilderness Act was much the same as that faced by the USFS in its roadless area reviews—determining specific boundaries of areas for wilderness study.

Roadless Area Review Criteria: NPS and FWS

In June 1972, more than eight years after passage of the Wilderness Act, assistant secretary of the interior for Fish, Wildlife and Parks Nathaniel Reed issued a memo to the directors of the NPS and the FWS defining *criteria to be followed in determining an area's suitability for wilderness designation.* These criteria suggest a response to public opinion in being less pure in their requirements for wilderness recommendations. In particular, the memo specified that *roadless areas should not be excluded from wilderness consideration by the existence of the following conditions:*

1. Existing or proposed use of tools, equipment, structures, or facilities if these are necessary for the health and safety of wilderness users or the protection of the resource.
2. Prior rights or privileges such as grazing or limited commercial services that are proper for realizing recreational or other wilderness purposes.
3. The existence of unimproved roads, structures, installations, or utility lines that can and should be removed upon designation as wilderness.
4. Use of the area for research unless it requires permanent structures or facilities not needed for management.

In addition, the following nonconforming uses or facilities should not, by themselves, exclude an area from a wilderness recommendation so long as subsequent legislation designating the area specifically allows them: small structures like primitive boat docks or shelters, controlled burning, natural-appearing lakes created by water-development projects, hydrologic devices for monitoring water resources, underground utilities such as gas pipelines and transmission lines, and minimum tools and equipment necessary to maintain the water developments and utilities.

Special provisions should also be included in the wilderness legislative proposals for those areas that are currently surrounded by nonqualifying federal lands, but which in the foreseeable future will qualify, and thus permit the secretary of the interior to designate such lands as wilderness. Finally, no portion of a national park or wildlife refuge should be designated as wilderness unless the wilderness designation is compatible with the purposes for which the park or refuge was established.

Wilderness Study Process: NPS and FWS

The NPS and FWS began their wilderness review with the roadless lands that were under their jurisdiction in 1964. To the extent possible, wilderness study proceeded concurrently with the preparation of park or refuge General Management Plans (GMPs). The process typically had four phases: *preliminary study phase, public meetings, final recommendations, and legislative phase.* The *preliminary phase* included the collection and analysis of basic and field data, preliminary recommendations of wilderness boundaries (if wilderness was to be recommended), preparation of first draft of the preliminary wilderness-suitability study and draft EIS, and briefings on the potential draft to be proposed for higher agency and departmental officials and the appropriate congressional delegation(s).

Changes in the proposal sometimes resulted from agency, departmental, and Council on Environmental Quality (CEQ) review. The *public meetings* provided citizens and organizations an opportunity to comment on the draft proposal. Appropriate notices were listed in advance of public meetings and an informational packet on the issues and concerns regarding the draft wilderness proposal was prepared for distribution. Comments were then recorded and analyzed to aid in subsequent decision making.

The *final recommendations phase* involved preparation of the final wilderness recommendation with the benefit of public input, review, and approval of a final report by the agency at agency regional and Washington offices; submittal of wilderness recommendation and final EIS to the assistant secretary for Fish, Wildlife and Parks for approval; and then to the secretary of the interior. The approved proposal with its final EIS then went to the Office of Management and Budget (OMB) and to the CEQ for review and approval. The approved proposal is submitted by the secretary to the president.

During the final or *legislative phase,* the president submits his recommendation regarding the proposal to Congress; typically, the Washington office of the agency prepares the draft legislation and the transmittal letters from the secretary to the president and from the president to each branch of Congress. A bill or bills including the actual or modified wilderness proposal are then introduced in Congress by a House or Senate member. Then debate, negotiation, compromise, and eventual passage by committees and the full Congress would result in wilderness designation. This closely resembles the process for wilderness bills including USFS lands.

On the face of it, this appears reasonably straightforward, and NPS and FWS wilderness proposals typically have not been as controversial as those on national forests. The prospect of commodity development does not exist on these lands so environmental groups first focused their attention on USFS and BLM lands. Still, there have been controversies regarding Assistant Secretary Reed's guidelines, agency implementation of those guidelines, and frequent reluctance by state congressional delegations to carry a legislative proposal forward.

The relationship between the original legislation establishing a park or refuge and subsequent wilderness designation within some of these areas has also created problems, particularly on national wildlife refuges and game ranges. Many of the proposals submitted by the FWS following review of their roadless areas recommended against wilderness designation. For example, about 29,000 acres of the 40,000-acre Red Rock Lakes National Wildlife Refuge in Montana was recommended for wilderness (U.S. Department of the Interior, U.S. Fish and Wildlife Service n.d.). The remaining 11,000 acres were judged not suitable because of existing and planned developments to manage waterfowl, especially the trumpeter swan. Field studies of the 45,000-acre Laguna Atascosa National Wildlife Refuge in Texas revealed that, although a portion of the area did qualify as wilderness, such designation would conflict with the primary objective of the refuge, which is to provide habitat for waterfowl, so wilderness designation was not recommended (U.S. Department of the Interior, U.S. Fish and Wildlife Service 1970). Congress concurred in substance with both recommendations, designating 32,350 acres of the Red Rocks Refuge as wilderness and concurring with the FWS recommendation against wilderness designation at Laguna Atascosa.

Conflicts between legislative objectives exist despite the declaration in section 4(a) of the Wilderness Act that wilderness designation is "within and supplemental to" the purposes for which national forests, national parks, wildlife refuges, and game ranges were established. Where a legitimate conflict exists between the goals of wilderness and those of the park or refuge enabling legislation, the organic legislation generally takes precedence. For example, where wilderness designation might restrict necessary wildlife management practices on a game range, wilderness designation is limited, rejected, or made with special recognition of the potential intrusion.

The Department of the Interior also recognizes what are called "potential wilderness additions." This was originally conceived as a designation for areas *where clearly nonconforming uses were present (e.g., structures), but which would qualify for wilderness designation once the nonconforming use was removed.* In omnibus legislation passed in late 1976 (U.S. Public Law 94-567), eight national park units, containing 53,506 acres, were identified as potential wilderness additions. Most of these areas were so labeled because of grazing, and it is not altogether clear why they could not have been included in the wilderness because grazing is allowed in wilderness when it is a preexisting right. However, under this designation, the secretary of the interior has authority to establish them as designated wilderness when the nonconforming use ceases. For example, in California, 1,800 acres near Point Reyes were added to the Philip Burton Wilderness after the removal of power lines subsequent to the adjoining area's earlier designation (U.S. Public Law 94-544).

By 1974, the end of the ten-year study period, the NPS had completed its review and submitted wilderness proposals on 56 areas. Yet, controversy, an apparent lack of broad-based public support, lack of a concerted push by environmental groups, or reluctant state congressional delegations has caused millions of acres of NPS-recommended wilderness to languish before Congress. For example, Yellowstone National Park has a 2-million-acre wilderness proposal submitted to Congress in 1972, which is still awaiting congressional action. Although wilderness designation has been implemented in Washington state's Olympic National Park (U.S. Public Law 100-668) and Death Valley National Park in California (U.S. Public Law 103-433), proposals including many millions of acres still await congressional action in more than a dozen parks, including large tracts in Great Smoky Mountains, Grand Canyon, and Yellowstone National Parks. Although some may argue that wilderness protection is ensured under national park status, knowledgeable persons argue that important wilderness values are being lost in these backcountry areas without legal designation as wilderness (Crumbo 1996; Watson 1996).

Olympic National Park Wilderness Review

A summary of the wilderness study in Olympic National Park will help illustrate the review procedure prescribed by the Wilderness Act for roadless areas 5,000 acres or larger in the national parks, wildlife refuges, and game ranges.

It has been the policy of the NPS to prepare a general management plan (GMP, formerly called a master plan) for each area of the NPS to provide the framework for its overall management, public use, and physical development. To help determine future use, a land-classification plan based on an area's resources is included in the GMP. In classifying land, the NPS uses six land classes developed by the Outdoor Recreation Resources Review Commission (ORRRC 1962). These six classes, modified for applicability to the NPS, include the following:

- Class I—High-density recreation areas
- Class II—General recreation areas
- Class III—Natural environment areas
- Class IV—Outstanding natural areas
- Class V—Primitive areas
- Class VI—Historic and cultural areas

Roadless areas within a park, typically Class V, but sometimes Class III, IV, and VI lands, have usually been managed and preserved in a roadless, natural condition before review and proposal for designation as wilderness. The review of parklands to develop a recommendation for wilderness designation has usually been carried out in conjunction with a major public review and updating of the park GMP, as was the case in Olympic National Park. Along with the review and updating of the park GMP, roadless portions of the 870,200-acre park were formally reviewed in a wilderness study.

During 1972 and early 1973, with the help of an NPS planning team composed of members from the local park staff, the regional office, and the planning division in the NPS Denver Service Center, a new GMP and a preliminary wilderness proposal were prepared for Olympic National Park. The initial draft called for 93 percent of the park to be designated as wilderness. Following public hearings and agency review, the final wilderness recommendation was 862,139 acres due to minor boundary changes and included about 96 percent of the park.

This recommendation was submitted to the Washington office in June 1974 and was transmitted from the NPS to the Department of the Interior and the Office of Management and Budget. A recommendation for wilderness designation of the four units in Olympic National Park, along with fifteen other parks and wildlife refuges, was included in a White House communication to Congress.

By 1987, the wilderness proposal for the Olympic National Park backcountry still had not been acted on by Congress. This decade of inactivity demonstrates a point that we and others (e.g., Mohai and Verbyla 1987) *have made throughout this chapter: the wilderness designation process is very political, and much more than a rational planning process carried out by agency bureaucrats.*

On March 15, 1988, the Washington state delegation introduced the Washington Park Wilderness Act of 1988 (S. 2165), proposing 871,730 acres in Olympic National Park and large acreages in Mount Rainier and the North Cascades National Park complex for wilderness, and a companion House bill (H.R. 4146). The Olympic wilderness proposals were similar to the NPS recommendation of 862,139 acres back in 1974. The increase in size of about 9,000 acres was made possible by park boundary adjustments and land acquisition during the intervening fourteen years, most notably the acquisition of land at Lake Ozette and the extension of park ownership of the Pacific beach to the lowest tide line.

Hearings were held on the Senate and House bills during the summer of 1988, and there was general support for wilderness designation from affected agencies, the city of Seattle, the environmental community, and many public interest groups. The environmental lobby did, however, recommend a slightly larger wilderness proposal by calling for inclusion of the roadless portions of Olympic National Park's strip of land along the Queets River, the surface area of Lake Ozette, the north shore of Lake Crescent, and a small, isolated stretch of wild beach along the Pacific Ocean. Compromises were quickly reached. Congress accepted the Queets River and wild beach proposal but rejected the Lake Crescent addition because of inholdings and the Lake Ozette surface due to motorized access across the lake to private dwellings. On November 16, 1988, Congress passed U.S. Public Law 100-668, designating 876,669 acres of Olympic National Park as wilderness. The long process of review, study, debate, and wilderness designation of a national park backcountry was complete (Fig. 5.5).

FIGURE 5.5 A large part of Olympic National Park was designated wilderness in 1988, fourteen years after the initial National Park Service proposal. Photo courtesy National Park Service.

THE BUREAU OF LAND MANAGEMENT

The Wilderness Act of 1964 gave the USFS, the NPS, and the FWS the authority to study, protect, and manage "legal" wilderness on lands under their jurisdiction. The act failed to give comparable authority to the BLM, even though this agency managed 473 million acres of federal land, far more than the other agencies. These were the "forgotten lands," vast acreages of the public domain that were never disposed of through the various federal programs to place land in private ownership or never designated for special purposes like national parks or national forests. About 174 million acres were in the contiguous United States, virtually all in the West. Although grazing, mining, and primitive low-quality roads had impacted many of these lands, their wilderness values seemed immense. Estimates in the early 1970s of BLM lands with wilderness potential in the lower forty-eight states ranged from about 50 million acres to as high as 90 million acres. These lands, along with the BLM lands in Alaska, represented by far the largest block of potential additions to the NWPS.

The Federal Land Policy and Management Act (FLPMA) of 1976 (U.S Public Law 94-579) gave the BLM an organic act, a long-needed legislative statement of direction and purpose for the agency. *Two sections of FLPMA called on the BLM to study, make recommendations to the president, and ultimately manage legally designated wilderness. Section 603 instructed the agency to review its roadless lands* of 5,000 acres or more and its roadless islands and make recommendations regarding their suitability or nonsuitability for wilderness designation within fifteen years after passage of FLPMA.

The same was to be done for the BLM administratively classified primitive or natural areas that existed before November 1, 1975, with recommendations regarding these areas to go to the president by June 1, 1980. In ten western states, 53 BLM primitive or natural areas existed that, with their contiguous roadless lands, amounted to 1.2 million acres. They had been established by administrative designation, possessed qualities comparable to those of legally defined wilderness, and were being managed for wilderness values.

Section 202 of the FLPMA mandated the development, maintenance, and revision of land-use plans, which among other things *gave priority to the designation and protection of areas of critical environmental concern.* Under this authority, the BLM studied—and continues to study—and develop recommendations for wilderness designation. These lands also include roadless parcels less than 5,000 acres in size (and thus not qualifying for study under FLPMA section 603), but which lie adjacent to existing or proposed wilderness or national park areas.

The final legal directive assigning wilderness study initiatives to the BLM was the Alaska National Interest Lands Conservation Act (ANILCA) of 1980 (U.S. Public Law 96-487). This law withdrew BLM roadless lands in Alaska from wilderness review but stated that the secretary of the interior at personal discretion could periodically study and make wilderness recommendations to Congress.

James Watt, President Ronald Reagan's secretary of the interior, issued a memorandum on March 12, 1981, directing that no further wilderness inventory and review be done in Alaska. The one exception to this is a wilderness review of the Central Arctic Management Area specifically mandated by the ANILCA legislation.

BLM Roadless Area Review Process

The BLM's wilderness review process has three major phases: inventory, wilderness study, and reporting wilderness or non-wilderness recommendations.

Inventory

The inventory phase, begun in 1978, sought to identify lands with wilderness attributes—lands that met the minimum standards for wilderness as defined in the Wilderness Act. The BLM attempted to accomplish this task quickly and anticipated spending more time in the study and evaluation of areas that made the "first cut." The task was enormous, and the BLM selected a two-step approach. During the first step, called the *initial inventory* and conducted in 1978 and 1979, areas identified by the BLM staff and the public as *not* having wilderness attributes were eliminated from further review. This was done largely by using resource data and maps available in the BLM district offices, and this process reduced the total acreage under consideration to about 50 million acres.

The remaining lands then became the focus of the *intensive inventory.* BLM resource professionals conducted on-the-ground inspections of each area to assess the presence or absence of wilderness characteristics. The public was invited to participate in the field inspections and to review the agency's assessment procedures and recommendations for each area. Public response was considerable: more than 10,000 comments were received from across the country. *The BLM intensive roadless area inventory was essentially completed by the end of 1980, and areas found to possess the basic characteristics of wilderness were classified as wilderness study areas (WSAs). A total of 861 areas containing about 25 million acres were classified as WSAs.*

At the time the BLM was doing its inventory, the focus of environmental groups was on the USFS's RARE II study. As a result, there were fewer challenges to the BLM process than might otherwise have been the case. Still, the 25 million acres of wilderness study areas were far less than the environmental community expected, and there were accusations that the BLM was applying wilderness-attribute standards too strictly. For example, debate raged over what constituted a road and whether tire tracks across the desert should disqualify an area from wilderness consideration. The displeasure intensified in late 1982 and early 1983 when Secretary of the Interior James Watt removed approximately 85 WSAs, or about 1.5 million acres, from further wilderness study. These WSAs either contained *split-estate lands* (where the federal government owned the surface and someone else owned the subsurface estate), were areas of less than 5,000 acres, or were areas that contained more than 5,000 acres but received a high wilderness rating because of adjacent existing or proposed park or wilderness lands. The Sierra Club, having learned valuable lessons from its legal action against the USFS, sued. The U.S. District Court for the Eastern District of California, in the case of *Sierra Club v. Watt,* ruled in favor of the Sierra Club. The court, among other statements, noted that Watt had failed to use good judgment, and it restored the deleted areas to their wilderness-study-area status. *Once again, the courts supported citizen wilderness views opposing an agency.*

Wilderness Study

In February 1982, the BLM published its "Wilderness Study Policy: Policies, Criterion and Guidelines for Conducting Wilderness Studies on Public Lands" (USDI, Bureau of Land Management 1982). This document specified the BLM's national wilderness program policy, listed and described planning criteria to use in the evaluation of the WSAs, prescribed quality standards for analysis and documentation, and recommended a framework for integrating the wilderness study process into the agency's multiple-resource planning system. Through this process, the BLM hoped to complete a high-quality study of its WSAs, including full public involvement, by the end of 1987. This would allow four years for mineral review of those areas recommended for wilderness and for making reports to the president. However, when additional areas had to be reviewed because of the *Sierra Club v. Watt* decision, this schedule proved to be overly optimistic. In addition, the secretary of the interior, following the lead of the secretary of agriculture and the USFS, decided to submit all recommendations on a *statewide* basis, rather than on an individual-area basis. So a revised schedule called for completion of wilderness studies for submission starting in 1989 and continuing on into 1991, with all studies being submitted before the 1991 due date.

The overall goal of the WSA studies was to determine whether each area, or a portion thereof, was more suitable for wilderness or for other uses. The BLM's wilderness program policy provided the philosophical underpinnings for the review. Basically, this philosophy is that wilderness is one of the several multiple uses of public lands, and therefore the study should be more than just a review of the wilderness potential of the WSAs. The inventory phase had already established that these areas possessed wilderness attributes. The study then should assess the values, resources, and existing and potential uses of the WSAs for wilderness *and* for other commodity or recreational outputs. In addition, any recommendation for a WSA should reflect the mix of resource uses and benefits currently provided in the area. The BLM recognizes that wilderness itself provides multiple uses, offering benefits from primitive recreational use, wildlife habitat protection, watershed protection, and protection of cultural and archeological resources. The agency also views wilderness as a long-term allocation and management commitment. *Thus, the capability of the agency to protect wilderness values in the long run was to be a primary consideration in recommendations for a given WSA.*

From the foregoing BLM philosophy of wilderness, *two criteria* were used to direct and justify all BLM wilderness recommendations—both for those judged suitable and those unsuitable. *The first criterion was the amount and quality of an area's wilderness values, the second was manageability, and the third was quality standards.*

1. Quality of Wilderness Values

The first criterion acknowledged that although all WSAs had at least minimum wilderness attributes, some were of higher quality than others. To evaluate quality, four components or kinds of wilderness values were measured and recorded: (1) quality of the area's mandatory wilderness characteristics; (2) special features, or quality of the area's optional wilderness characteristics; (3) multiple-resource benefits—the benefits to other multiple resource values and uses that wilderness designation could ensure; and (4) diversity in the NWPS.

The Wilderness Act of 1964 provides a list of mandatory attributes to be evaluated—size, naturalness, and opportunities for solitude or primitive recreation. Objective information was gathered to fully document these and other attributes and criteria for each WSA. This information was summarized in matrix classifications and other descriptive methods, but arbitrary and subjective weighting or ranking was not done. The BLM had watched and learned from the Forest Service problems with weighting and ranking in the RARE I study.

The BLM recognized that a WSA needed only to "generally appear" natural and have human imprints "substantially unnoticeable" for it to receive a wilderness "suitable" recommendation. Yet, areas with fewer such impacts seemed to have higher wilderness potential than those with more. Thus, during the field study, a general description, the location, the size of the area, and the overall influence of human imprints in the WSA were documented. In addition, whether the imprint was the result of activities occurring inside or outside of the area was noted, and the potential for separating the imprinted portions from the rest of the area and recommending the remainder for wilderness designation was recorded. The BLM had also learned that sights and sounds of humans outside a WSA do not necessarily preempt its wilderness designation, thereby acknowledging the will of Congress expressed in House Report 95-546 accompanying the previously discussed Endangered American Wilderness Act of 1978 (U.S. Public Law 95-237).

The BLM used a slightly different interpretation of the Wilderness Act's call for the provision of outstanding opportunities for solitude or primitive and unconfined recreation than did the USFS. It defined these two attributes—solitude and primitive recreation—differently and directed that *if an area had one or the other (solitude or primitive recreation), it qualified for wilderness. If an area had both attributes, it would possess greater wilderness potential. Solitude* was defined as the state of being alone or remote from habitation; isolation; or in a lonely, unfrequented, or secluded place. Environmental features can enhance opportunities for solitude. Size and configuration of area, topographic screening, vegetative screening, presence of outside sights and sounds that enhance or diminish wilderness values, and opportunities for users to find a secluded spot were measured and recorded for each study area. *Primitive* and unconfined types of recreation were defined as activities characteristic of dispersed, undeveloped recreation that do not require facilities or motorized equipment.

The Wilderness Act indicates that a wilderness area "may also contain ecological, geological, or other features of scientific, educational, scenic or historical value." The BLM recognized that these attributes are supplemental and not mandatory for an area to be recommended for wilderness. Still, the agency recognized that such attributes enhance wilderness values and, in marginal areas, may affect the final recommendation. Thus, the study guidelines provide for the recording of the abundance of these attributes and the importance of each to the overall wilderness value of the area.

The BLM responded to previous direction and actions affecting wilderness designation by giving special consideration to multiple-resource values and uses, such as wildlife habitat and archeological sites, that would be threatened if the area were not given the protective status of wilderness. Moreover, those WSAs were given greater wilderness value if, protected in a natural state, they had a high likelihood of fostering the return of animals and fishes that were formerly found in the area, or of improving water quality or visual resources within and beyond the boundaries of the WSA.

The final wilderness value component considered was *diversity*—diversity in ecosystems and landforms represented in the NWPS, balancing the distribution of wilderness areas throughout the nation, and, to the extent possible, giving greater consideration to areas located closer to population centers. These initiatives reflect the previous actions of Congress in passing laws to locate wilderness areas closer to the people (Eastern Wilderness Act; Endangered Wilderness Act) and a long-standing goal of the federal wilderness resource agencies to preserve examples of all the nation's ecosystems. The BLM used the Bailey-Küchler system to map ecosystems, as the USFS did in its RARE II process, and if an area under study contributed to the diversity of natural systems and features in the NWPS, it received additional wilderness value. On the other hand, if its ecosystem or landform duplicated that of existing or proposed areas, it was given a lower ranking. The same was true for geographical distribution; areas located in regions with fewer wilderness areas received higher consideration. Finally, areas located within a five-hour drive of population centers of 100,000 or more were given special consideration.

2. Manageability

In its evaluation of WSAs, the BLM carefully considered whether it could manage and protect wilderness attributes over the long term. *If current or projected resource conditions or uses would make wilderness protection of an area difficult or impossible, its wilderness potential was believed to be substantially lessened.* Five factors or assumptions were used in making judgments about manageability. *First, management problems* anticipated from nonconforming but allowable uses were considered. Such uses included *mining and grazing*. If it were reasonably certain that the nature or intensity of such uses would substantially destroy wilderness values, the study recommendation would be for non-wilderness allocation. *Second, the status of the land* in the WSA was considered. If someone other than the BLM owned the surface or subsurface rights to the land, the BLM believed that wilderness management would be a problem. Thus, if the WSA contained *private inholdings, state lands,* or valid existing *mineral leases or claims,* the wilderness rating of the area was lowered. The same was true if a private individual or corporation owned *subsurface mineral rights. Third,* the BLM evaluated the impact of the guaranteed *right of access to any private inholding.* Such access generally included the existence or further development of roads through the WSA, over which the BLM had little control. Fourth, the agency adopted a policy of *no buffer zones around a WSA,* a policy that reflected the view of Congress when it passed the Endangered American Wilderness Act. Thus, an area would stand alone, and appropriate use and development of commodity resources could occur up to the boundaries of a wilderness area. If such existing or proposed uses would make the management of the wilderness infeasible, the wilderness potential of the area was lowered. *Fifth,* the BLM acknowledged that there would be *no change in air quality recommendations* because of a wilderness recommendation. If existing or proposed uses of land adjacent to a WSA caused or might cause air quality of the area to drop below Class II status, the area would less likely be recommended as wilderness. *These BLM manageability judgment criteria indicate that the BLM, like the USFS before it, subscribed to purity in wilderness management and recognized an integral link between wilderness designation decisions and ease and quality of management.* The USFS experience, as we have seen, suggests that the BLM's purity position on wilderness designation might be challenged by public opinion and environmental groups.

3. Quality Standards for Wilderness Study

In addition to the wilderness values and manageability criteria, *the BLM published six standards to be used in analysis and documentation of WSAs.* The purposes of the standards were fourfold: to ensure consistency in analysis and reporting across the WSAs, to ensure consideration of wilderness value within the agency's multiple-use mandate, to place the wilderness review within the agency's land-management planning process, and to ensure that resource analyses and recommendations in study reports meet the requirements of necessary EIS or environmental assessments (EA).

The *first standard* required *careful study of the WSA's energy and mineral values,* with a view of reducing the nation's dependence on foreign resources vital to our economy and security. Thus, the presence or potential presence of vital metals or fuels in a WSA figured prominently in its wilderness recommendation. Even after a study recommendation was made favoring wilderness designation, a formal evaluation of the mineral values contained in the WSA would be made by the U.S. Geological Survey/Bureau of Mines before the study report was forwarded to the president.

Standards two and three called for consideration and reporting on the range of alternative uses for all or part of the WSA, followed by a *review of the impact on other resource values if the area was recommended for wilderness, or upon wilderness values if the area was not recommended for wilderness.* In making these analyses, BLM planners reviewed the existence of wilderness and commodity resources within the WSA, and outside the WSA on public or private lands but within the same region. Consideration was given to the impact of land-use recommendations on local and regional economies and whether use and development of non-wilderness resources in the area are compatible with management of the area as wilderness.

Standard four described the *public involvement process* to be used in the analyses, evaluations, and recommendations regarding the WSAs. This process, which includes the general public and state and local governments, seeks to identify issues that the public believes should be considered with respect to a particular WSA, and any values and resources in the WSA that would augment BLM's current information base. It is also intended to tap public opinions on whether an area is suitable for wilderness designation or more suitable for other resource uses. Wilderness recommendations would not be based exclusively on a vote-counting majority-rule system, but BLM district managers and state directors would consider public input along with the WSA's multiple resources and social and economic values and uses.

The fifth standard required that consideration be given to *any adverse or favorable social and economic effects that designation of wilderness areas would have on local areas.* This is a standard planning requirement of all BLM unit analyses. Finally, FLPMA and BLM planning regulations required that wilderness study teams document the extent to which any recommendation is *consistent with fully approved resource-related plans of state and local governments.* If a state or local government informs the BLM that all or part of a WSA report is inconsistent with that government's policies, the BLM will respond to this comment, explaining how the inconsistency was resolved and why. If the current state or local plan is generally consistent with the BLM policies and programs, every effort is to be made to mitigate the impacts of inconsistencies caused by a wilderness study report.

Reporting Wilderness or Non-Wilderness Recommendations

The wilderness study process ends with the BLM state director's preliminary wilderness recommendation for WSAs in that director's state. The wilderness reporting process then involves the roles of the BLM national director, the secretary of the interior, and the president in acting on the state director's preliminary recommendation. A somewhat different process is used for those areas found *suitable* by the BLM state director for wilderness versus those the director decides

FIGURE 5.6 The Cottonwood Point Wilderness in Arizona, managed by the Bureau of Land Management, is a desert, an ecosystem largely missing in units designated wilderness in the 1964 Wilderness Act. BLM wilderness areas substantially increased representation of desert ecosystems in the 1990s. Photo courtesy Tom Folks.

are unsuitable. For those areas with a positive wilderness recommendation, a formal mineral survey is requested by the U.S. Geological Survey/Bureau of Mines. On completion of the survey, the state director then reviews his or her initially positive wilderness recommendation considering the mineral survey results. If no change is called for, the state director submits the preliminary wilderness recommendation along with the mineral report to the BLM director for final acceptance. With the BLM director's concurrence, the recommendation is then submitted to the assistant secretary of interior for Land and Water Resources, then to the secretary of the interior, and ultimately to the president (Fig. 5.6).

However, if the findings of the mineral survey suggest that the preliminary wilderness recommendation is inappropriate (because it would mean a loss of important mineral potential), then the state director returns the proposal to the appropriate district manager for a revised recommendation. The new recommendation with appropriate supporting information then proceeds up through the agency and departmental decision hierarchy. If the BLM national director, on the preliminary recommendation of the state director, also recommends against wilderness, the recommendation proceeds directly to the assistant secretary. Then with concurrence of the assistant secretary, the "unsuitable" proposal moves on to the secretary of the interior and then to the president.

The BLM had been very active during the 1990s in studying and reporting recommendations on millions of acres of roadless lands under their management. Millions of acres appear to have high wilderness value, but many of these lands also contain valuable non-wilderness resources. As of 2023, 9.9 million acres of BLM land have been designated as wilderness.

ALASKA—A SPECIAL CASE?

About 90 percent of Alaska is de facto wilderness and the epitome of the last frontier. This vast area (375 million acres) is worth special attention because it contains such rich natural resources, including wilderness. It is also a study in special problems related to the disposition of public lands to Alaskan natives and state and federal land-management agencies.

Disposition of the Alaskan Public Domain

Since its acquisition from Russia in 1867, most of Alaska has remained in federal ownership. However, with the impending passage of the Alaskan Statehood Act in 1958, pressures for the transfer of much of this land to the state to facilitate economic development began to grow. Congress responded by granting the state a generous package of land rights. The state was given full title to the submerged lands of the continental shelf, estimated at 35 to 45 million acres. It was allowed to select 104 million acres from the federal domain along with full title to all mineral rights on these state-selected lands. Moreover, Alaska was to receive 90 percent of the revenues from mineral leasing on all those lands remaining under federal jurisdiction, twice the percentage other states received (Allin 1982).

A major omission in the disposal of lands to the state was any mention of claims by Alaskan natives. A 1961 Supreme Court decision ruled that native claims against selections made by the state could continue, but it also ruled that because the federal government had clearly given the state the right to select, such selections could continue. The potential legal nightmare such rulings created was headed off by the decision of then Secretary of the Interior Stewart Udall to freeze all further land transactions until such time that Congress could settle the native claims issue. Udall's temporary freeze was replaced by Public Land Order Number 4582 halting the selection process until the end of 1970. Alaska governor Walter J. Hickel, arguing that the state's economic future was stymied because 95 percent of Alaska remained in federal ownership, attempted to have the Land Order set aside, but was unsuccessful (Allin 1982).

The issue was further confounded in 1968 with the discovery of vast oil reserves on Alaska's North Slope. To capitalize on these vast reserves, it became immediately apparent that a pipeline from the oil fields to an ice-free port on Alaska's southern coast would need to be constructed. The pipeline, estimated to be 800 miles long, would cross federal land subject to the Public Land Order. The pipeline also represented an intrusion on wilderness comparable to the construction of the railroads across the western United States a century earlier (Allin 1982). A legal resolution of the native claims was badly needed.

After extensive debate in Congress, the Alaska Native Claims Settlement Act (ANCSA) was passed in 1971 (U.S. Public Law 92-203). ANCSA terminated the land freeze imposed by the Land Order. The secretary of the interior was directed to withdraw up to 80 million acres suitable for addition to the four existing conservation systems; that is, national forests, national parks, fish and wildlife refuges, and wild and scenic rivers (these were the so-called [d][2] lands, referring to the section of ANCSA where the instructions for withdrawal were contained). This was to be completed within nine months and lands withdrawn by the secretary would generally not be available for selection by the state or by the natives. The legislation also contained language that allowed the secretary to withdraw an unspecified amount of land "to insure that the public interest in these lands is properly protected" (Allin 1982, 217) (these were referred to as the [d][1] lands). This led some conservationists to recommend that as much as 50 million acres be set aside in addition to the 80 million acres of (d)(2) lands specifically identified.

ANCSA also authorized the transfer of 40 million acres to the ownership of native villages or newly formed native corporations. Each Alaskan village was permitted to withdraw 23,040 acres (36 square miles). In addition, each of the twelve native corporations was allowed to withdraw an amount of land prescribed by formula in the law.

In late 1972, Interior Secretary Rogers C. B. Morton announced his plans for Alaskan land withdrawal. More than 240 million acres were involved; a little more than 80 million acres were withdrawn for addition to existing conservation systems, including nearly 19 million acres in three new national forests, 32 million acres in new national parks, and 32 million acres in new national wildlife refuges. In addition, twenty new units in the National Wild and Scenic River System were recommended. Moreover, 47 million acres were withdrawn under the (d)(1) authority to protect the public interest in these lands. Also, 4.5 million acres were tied up in pipeline corridors mandated by ANCSA and another 112 million acres were withdrawn as a pool from which the natives could select their 40 million acres.

Secretary Morton's recommendations drew fire from almost every quarter. The state was alarmed by the large withdrawal (47 million acres) under the (d)(1) provisions. Wilderness groups were opposed to the recommendation to establish three new national forests in Alaska; they contended that commodity exploitation under multiple-use management would continue on BLM lands remaining after selections were completed and therefore no new national forestlands were needed. They were also concerned that although the USFS had committed itself to the designation of WSAs on these new forests, there was no guarantee of large-scale wilderness protection for them (Allin 1982).

Therefore, despite ANSCA's legislation regarding disposition of the public domain in Alaska, there still remained sharp disagreements over the future management direction of the land and resources in the state. The continuing dispute triggered yet another effort to arrive at a legislative solution to the dilemma. Beginning in 1973 and extending until 1980, probably the most significant congressional and public debate over U.S. natural-resource allocations raged. On the one side, the State of Alaska and its allies argued for the assertion of state rights and more opportunity for economic development. On the other side, conservation organizations argued that the last great opportunity to preserve a major portion of the nation's wild heritage was at hand and that we could not afford to lose it.

Given the large stakes, it is not surprising that Congress found it difficult to reach a satisfactory resolution over the future of the Alaskan lands. Conservationists were highly organized. The heart of this group was the Alaska Coalition, a diverse collection of state and national organizations ranging from the Sierra Club and The Wilderness Society to the United Auto Workers and National Council of Senior Citizens. The State of Alaska led the opposition, along with local developmental and commodity interests.

The inability of Congress to reach a satisfactory compromise on the issue eventually led to an unusual series of events. In 1978, Secretary of the Interior Cecil Andrus announced that the Carter administration was losing patience over the lack of progress and that it was exploring ways in which it might take action on its own behalf to solve the problem. When Congress failed to respond with any positive proposal, the Carter administration made good on its promise. In late 1978, Secretary Andrus withdrew 110 million acres for three years under the authority of the FLPMA. Two weeks later, President Jimmy Carter invoked the Antiquities Act of 1906 to create 17 new national monuments covering 56 million acres of the 110 million withdrawn by Andrus. In one stroke, President Carter had more than doubled the size of the NPS. In addition, under presidential direction, Secretary Andrus began the process of creating 40 million acres of new national wildlife refuges in Alaska, and Secretary of Agriculture Bob Bergland withdrew 11 million acres of existing national forests to prevent mineral entry and state selection (Allin 1982). The executive branch had seized the initiative from the legislative branch in a dramatic fashion.

However, a legislative solution to the future of resource management in Alaska was still required. After two more years of congressional debate, it became apparent that some form of compromise had to be reached. On December 2, 1980, in the closing days of his administration, President Carter signed the Alaska National Interest Lands Conservation Act into law. The new law had profound implications for the future of wilderness.

The Alaska National Interest Lands Conservation Act (ANILCA)

The ANILCA (U.S. Public Law 96-487) set aside 104.3 million acres of national parks, wildlife refuges, wilderness areas, and other conservation units in Alaska. Of these acres, approximately 56 million were placed in the NWPS. Wilderness acreage by agency was 32.4 million for the NPS, 18.6 million for the FWS, and 5.4 million for the USFS. *ANILCA stated that designated wilderness shall generally be*

FIGURE 5.7 The Alaska National Interest Lands Conservation Act (ANILCA) passed in 1980 and ensured wilderness in Alaska. Certain uses, such as cabins and floatplanes shown here in Admiralty Island National Monument Wilderness, were permitted so visitors could cope with adverse weather and bears. Photo courtesy U.S. Forest Service.

managed according to the provisions of the Wilderness Act of 1964, but with several exceptions to recognize unique conditions in Alaska, minimize impacts on established uses, and permit some economic expansion (Fig. 5.7). Exceptions such as providing public cabins recognized the unusual severity of Alaska weather and the danger of being attacked by bears; provisions made for *subsistence hunting and gathering* recognized the special relationship between rural Alaska residents and the wild resources on which they depend. *ANILCA did state that its special provisions applied only to Alaskan wilderness;* they are not to be viewed as precedents to alter wilderness management and future wilderness legislation in the lower forty-eight states (The Wilderness Society 2000).

ANILCA upheld the sufficiency of the RARE II EIS for Alaska. However, it did specifically require an investigation of a 2.1-million-acre tract called the Nellie Juan–College Fiord WSA in the Chugach National Forest. In addition, the Chugach National Forest was to proceed with recommendations regarding the wilderness suitability of fourteen areas identified during RARE II for further study. The final EIS prepared by the forest recommended 1.7 million acres (29 percent of the total forest area) for wilderness. No further study areas are located on the Tongass National Forest and 30 percent of the forest is presently designated wilderness. ANILCA also specifies that unless Congress expressly authorizes it, the USFS shall not conduct any further statewide roadless review and evaluation, thereby precluding the possibility of a RARE III in Alaska.

MANAGEMENT IMPLICATIONS IN WILDERNESS DESIGNATION ACTS

This section focuses on the management implications of several benchmark laws and other wilderness designation laws that contain management direction. Wilderness designation laws are acts of legislation passed by Congress to add areas to the NWPS, and some of them contain management direction.

During the years since passage of the Wilderness Act, from 1964 through 2023, more than 170 additional wilderness laws were passed by Congress and approved by the president, primarily to add areas to the NWPS (wilderness designation laws). These laws can be searched at the website www.wilderness.net. This wilderness designation legislation has generally reaffirmed the management direction in the Wilderness Act of 1964, but some laws have provided additional interpretation, clarification, or special direction for wilderness management, as well as providing for wilderness study, adjusting wilderness boundaries, or changing a wilderness name. The same law may also provide for other public lands purposes not directly related to wilderness.

For our purposes, we define a wilderness law as any federal statute that designates new wilderness, adds land to and/or deletes land from existing wilderness (including boundary revisions), or provides for wilderness study. The term omnibus *is specifically used to signify a law designating more than one wilderness.* In several cases, wilderness designation, study, or management provisions are secondary purposes for a law.

Because of legislation, *the National Wilderness Preservation System has grown from the 54 areas and 9.1 million acres established by the Wilderness Act in 1964 to more than 806 wilderness units totaling 111.8 million acres by 2023.* Wilderness has been a topic consistently dealt with by Congress since the first wilderness bill was introduced in 1956.

BENCHMARK WILDERNESS LAWS

Seven benchmark wilderness laws have set standards for wilderness designation and management, and/or have interpreted or clarified the wilderness concept or provided management direction. These laws are important milestones in wilderness legislation; we call them *benchmark* wilderness laws, beginning with the Wilderness Act (Browning et al. 1988).

Wilderness Act of 1964

This act defined wilderness for the purposes of establishing and managing the National Wilderness Preservation System, designated 9.1 million acres of national forestlands as wilderness, and directed wilderness study of national forest primitive areas and roadless areas within the national park and national wildlife refuge systems. A key element of the act was that the agency's future wilderness study and review include public involvement in the form of public notices and hearings. In addition, the law specified that only Congress had the authority to designate additional wilderness areas in the NWPS.

The Wilderness Act provided management direction by prohibiting certain uses and allowing others. Prohibited were permanent and temporary roads, most commercial enterprises, motorized equipment and mechanical transport, landing of aircraft (where not an established use prior to designation), and structures and installations—with exceptions for necessary administrative purposes and to protect preexisting private rights.

Special provisions of the Wilderness Act allowed for mining on valid claims and mineral development on leases established before December 31, 1983, with mineral prospecting and surveys permitted to provide information on mineral resources; water development projects with presidential approval; livestock grazing; fire, disease, and insect control; and aircraft landings and motorboat use where already established. Also allowed were certain commercial uses deemed compatible with the wilderness concept, such as outfitting and guiding. See The Wilderness Society handbook (2000) and Chapter 4 for additional explanation of the Wilderness Act.

Eastern Wilderness Act of 1975

U.S. Public Law 93-622, the so-called Eastern Wilderness Act of 1975, added 16 wildernesses in the East, where only 4 wildernesses on national forests previously existed. The act also designated 17 "wilderness study areas" to be managed to preserve their wilderness qualities until their suitability for wilderness designation could be evaluated and until Congress had sufficient time and opportunity to designate the selected areas as wilderness.

While debating proposed wilderness bills, Congress addressed the issue of whether lands in the East should be managed in a "wild-areas system" under different standards from the National Wilderness Preservation System. The argument for a separate system was based mostly on the premise that roadless areas in the East had been severely modified by previous human use and consequently did not qualify for wilderness designation under criteria of the Wilderness Act. This argument failed, and Congress eventually determined that designated roadless areas in the East should be included and managed as part of the National Wilderness Preservation System (U.S. Senate Report 1975, 93-803). Thus, it is correct to refer to such designated areas as "wilderness in the East," and not as "Eastern wilderness." *The legislation is also referred to as the so-called Eastern Wilderness Act because it does not have a formal title and does not establish a separate system of eastern wilderness areas.*

This act authorized the secretary of agriculture to use condemnation to acquire private lands in the 17 wildernesses designated by the Eastern Wilderness Act. Condemnation authority had not been provided in the Wilderness Act, but wilderness areas designated in the eastern U.S. contained many private inholdings. Congress believed that condemnation authority would help ensure that private owners used these lands in a manner compatible with the wilderness concept and would provide a means of acquisition if they did not.

Federal Land Policy and Management Act of 1976 (FLPMA)

The Federal Land Policy and Management Act of 1976, or FLPMA (U.S. Public Law 94-579), provided for wilderness study and designation of lands managed by the Bureau of Land Management, a provision not included in the Wilderness Act. This law addressed many aspects of the management of these public lands and directed that roadless areas having wilderness characteristics be inventoried within fifteen years to determine their suitability for wilderness designation. Suitable areas could be recommended to Congress for wilderness designation under the provisions of the Wilderness Act of 1964. The BLM thus joined the Forest Service, National Park Service, and U.S. Fish and Wildlife Service as a partner in wilderness stewardship.

The BLM's wilderness study areas were to be managed to preserve their wilderness characteristics, but existing uses such as mining, mineral leasing, and grazing were permitted to continue subject to regulations set by the secretary of the interior. Once designated as wilderness, the management provisions of the Wilderness Act pertaining to wildernesses on the national forest would generally apply.

Endangered American Wilderness Act of 1978

U.S. Public Law 95-237 was passed in 1978, partly in response to the perceived shortcomings of RARE I, the Forest Service's first "post–Wilderness Act roadless area review and evaluation" of remaining national forest roadless areas to identify those suitable for wilderness study or designation. *The Endangered Wilderness Act added 16 areas to the NWPS, primarily including areas either excluded from the RARE I inventory of roadless lands or not recommended for wilderness study or designation after their review. Thus the name, "endangered wilderness."* The Forest Service's "purity" requirements for wilderness designation, including its "sights and sounds doctrine" that wilderness should be out of sight and sound of civilization, came under intense congressional scrutiny during committee hearings. The "sights and sounds doctrine" was deemed contrary to Congress's desire to establish wildernesses near large cities (U.S. House of Representatives 1977). Much of the congressional debate focused on RARE I's criteria for recommending only 274 wilderness study areas totaling about 12 million acres from the 56 million acres inventoried in 1,449 roadless areas. Environmentalists claimed that much more of the national forest roadless land qualified for wilderness designation. *By passing the Endangered Wilderness Act, Congress further established that areas previously influenced by people should not be precluded from consideration for wilderness designation, nor should roadless areas near major cities,* as they could provide much-needed, primitive recreation for the nearby population. Congressional committees addressed the interpretation of the Wilderness Act concerning certain uses, activities, and management, and endorsed the Forest Service's plan to conduct RARE II, a second comprehensive Roadless Area Review and Evaluation (Weaver and Cutler 1977). Generally, these Congressional committees supported a less stringent view of designation and management criteria than were being applied by the Forest Service at the time.

Alaska National Interest Lands Conservation Act of 1980 (ANILCA)

The Alaska National Interest Lands Conservation Act, or ANILCA (U.S. Public Law 96-487), was passed in 1980 after many years of debate *and provided management direction for the large tracts of federal land in Alaska.* ANILCA was the first act to pass both bodies of Congress that contained RARE II sufficiency-release language (Gorte 1987). The law declared that Congress had performed its own evaluation of RARE II roadless areas in Alaska and had decided to designate some lands as wilderness, to release some lands to non-wilderness multiple uses, and to hold other areas for wilderness study or further planning. *The law added more than 56 million acres to the NWPS, most of it in units of the National Park System and National Wildlife Refuge System. Millions of additional acres were authorized for further wilderness study* within the national parks and wildlife refuges and the Chugach National Forest, while special guidelines were included for wilderness recommendations of public lands in Alaska managed by the BLM.

Congressional committee hearings closely examined the unique natural environments in Alaska, exemplified by the state's vast size and predominately undeveloped condition (U.S. House of Representatives 1979). Subsequently, *Congress provided special provisions in ANILCA for wildernesses in Alaska* to allow certain motorized use and access, along with maintaining existing wilderness cabins and establishing new cabins where administratively authorized. Provisions for subsistence uses of natural resources applicable to other wildlands in Alaska were also included for wilderness, for establishment and maintenance of structures for aquaculture purposes and temporary construction and use of facilities for hunting and fishing, and for modification of existing timber sales contracts that applied to certain newly designated wildernesses on national forestland. The special provisions for wilderness under ANILCA were an attempt to compromise between the wilderness definition and the nonconforming uses (e.g., motorized access) and cultural heritage of the landscape in Alaska (e.g., subsistence activities); however, the legislation left wilderness managers with a complex stewardship situation to navigate (Tanner 2004; U.S. Department of Agriculture, Forest Service 2005; Hummel 2007).

Colorado Wilderness Act of 1980

U.S. Public Law 96-560 was important because Congress referred to the 1977 House of Representatives Committee Report 96-617 for explicit management direction for livestock grazing. This management direction had far-reaching effects, because the committee report required that livestock grazing in all national forest wildernesses should be managed according to the report's management provisions that were offered as interpretation of the Wilderness Act grazing provisions (Wilkinson and Anderson 1987). *The law also prohibited establishing buffer zones around wildernesses; directed a review of fire, disease, and insect control measures and policies in Colorado's wildernesses; and included RARE II sufficiency-release language for roadless areas in Colorado.* The Colorado Wilderness Act, by including reference to the accompanying committee report on management direction, thus brought a degree of closure to congressional debate over certain wilderness management policy and its application that had begun with consideration of the Endangered Wilderness Act several years earlier.

California Desert Protection Act of 1994 (CDPA)

The California Desert Protection Act (U.S. Public Law 103-433), passed in 1994, designated 69 new BLM wilderness areas encompassing 7.6 million acres (Fig. 5.8). The CDPA more than doubled the number of wilderness areas managed by the BLM and increased BLM's wilderness acreage by 70 percent, thus greatly expanding that agency's wilderness responsibilities (Watson and Brink 1996). This act transferred more than 3 million acres of land from BLM jurisdiction to NPS jurisdiction for the expansion of Death Valley National Park, Joshua Tree National Park, and the creation of the Mojave National Preserve. The three NPS units contain 3.9 million acres of designated wilderness.

The California Desert Protection Act continued a trend of recent, prior wilderness legislation, including the Nevada Wilderness Act (U.S. Public Law 101-195) and the Arizona Desert Wilderness Act (U.S. Public Law 101-628), to not restrict or preclude low-level overflights of military aircraft or designate new, special airspace. The airspace above these areas was seen as critical for the training, research, and development of the U.S. armed forces, and there was a lack of alternative sites. The act stipulates military overflights are not incompatible with the protection and management of the natural, environmental, and cultural resource values of designated wilderness in the California desert. Another key provision was that the act provided for administrative access by mechanized means to all 69 wilderness areas designated, including vehicle access for law enforcement agencies and for wildlife management and restoration by appropriate state agencies.

MANAGEMENT DIRECTION AND SPECIAL PROVISIONS IN WILDERNESS LEGISLATION

In the Wilderness Act, Congress provided a legal foundation for wilderness management, recognizing that wilderness requires management to protect its wilderness character and values that are defined in the act. As the NWPS evolved, Congress has refused to amend the Wilderness Act of 1964 to clarify management direction or for any other reason, except for a minor amendment made in the Boundary Waters Canoe Area Wilderness Act of 1978 to provide new management direction in that area. Mostly, Congress has clarified management direction or made special provisions for particular areas through legislation and committee reports. For example, the Colorado Wilderness Act (U.S. Public Law 96-560) further defined grazing management through a House committee report (U.S. House of Representatives 1979), but the act and the accompanying committee report expressly stated that the provisions did not amend the Wilderness Act.

Management direction as defined here includes those provisions in wilderness laws that affirm, interpret, or modify the management direction of the Wilderness Act, for application to a specific wilderness or group of wildernesses in the case

FIGURE 5.8 The California Desert Protection Act (1994) added more than seventy areas and millions of acres to the National Wilderness Preservation System. The Bureau of Land Management administers Kingston Peak (above) in the Kingston Range. Photo courtesy Jay Swetech.

FIGURE 5.9 The first BLM-designated wilderness to become a unit of the National Wilderness Preservation System in 1983 was Bear Trap Canyon Wilderness, in Montana, and it is a unit of the Lee Metcalf Wilderness. Photo courtesy Bureau of Land Management.

of omnibus laws (Fig. 5.9). Following is a review of management direction included in wilderness legislation subsequent to the Wilderness Act of 1964 (Browning et al. 1988).

Mining

Mining in wilderness is an extremely complex legal issue and is a topic in numerous legal books and articles (Coggins and Wilkinson 1987; Loop 1986; Matthews et al. 1985; Wilkinson and Anderson 1987). Following is a general overview of mining provisions in wilderness designation laws; however, other laws and agency authority and regulations complicate the issue.

Mining and mineral leasing are most often addressed in national forest and BLM wildernesses because national parks and wildlife refuges were normally withdrawn from such activities when established. Mineral exploration on national forestlands is even more complex in the East, where subsurface rights are mostly privately owned and the Forest Service must estimate the impacts of exploration and potential mining in designated wildernesses as individual cases arise (U.S. Senate 1984f, Report 98-614). The U.S. Government Accounting Office (1984) reported that in future wilderness legislation, Congress may need to consider the acquisition of private mineral rights in existing wildernesses in the East or permit mining in these areas, and the Forest Service may need to assess and inform Congress of the potential acquisition costs for private mineral rights of potential wildernesses in the East.

The Wilderness Act permitted mineral prospecting and surveys to continue in national forest wilderness if "compatible" and "consistent" with protection of the wilderness resource, to provide information on the mineral resources in such areas. The law provided for mining and mineral leasing on national forest wilderness until December 31, 1983, but mining on valid claims and mineral development on leases established before that date could also continue or begin at a future date (Gorte 1988). *Mining on valid claims, and mineral development on leases existing at the time of wilderness designation, may also occur on those wildernesses established after 1983.*

Any mining activities in national forest wilderness today are subject to certain requirements and regulations set by the secretary of agriculture to define a valid claim and to protect wilderness characteristics (Loop 1986; Matthews et al. 1985; The Wilderness Society 2000; Wilkinson and Anderson 1987). FLPMA (U.S. Public Law 94-579) directed that mining be administered in BLM wilderness in generally the same manner as in national forest wilderness. Congress typically excludes from wilderness designation those national forest and BLM areas with high mineral potential, but exceptions have been made in some wilderness laws, as the following examples indicate. These exceptions and special provisions were usually compromises that allowed wilderness designation of an area to proceed despite conflict over its mineral potential or mining activity.

The Boundary Waters Canoe Area Wilderness Act established a BWCA mining protection area that prohibited mining of federally owned minerals in the wilderness and on adjacent non-wilderness lands. This act also set certain restrictions on mining of non-federally owned minerals within the mining protection area. The Central Idaho Wilderness Act of 1980 (U.S. Public Law 96-312) established a special mining-management zone in the Frank Church–River of No Return Wilderness to allow for cobalt exploration and mining. In establishing that wilderness, Congress elected to try to minimize the adverse environmental effects of mining while also protecting critical bighorn sheep habitat rather than excluding the land in question from the wilderness.

Coal deposits were a major concern in the Cranberry Wilderness of West Virginia, and provisions were made in the West Virginia Wilderness Act of 1983 (U.S. Public Law 97-466) to acquire all non-federally owned mineral interests and to permit exploration activities and drilling in the wilderness to determine the value of non-federally owned minerals, subject to guidelines set by the secretary of agriculture. Congress also provided special provisions for phosphate leasing and mining in the Florida Wilderness Act of 1984 (U.S. Public Law 98-430), with procedural steps for presidential and congressional approval of the need to mine phosphate. A Senate committee report for the Florida Wilderness Act expressed hope that future restoration technology would be better able to reduce environmental deterioration caused by phosphate mining (U.S. Senate 1984d, Report 98-580). In the Texas Wilderness Act of 1984 (U.S. Public Law 98-574), Congress designated the Indian Mounds Wilderness, despite active oil and gas drilling, thus leaving the Forest Service to mitigate the damage to wilderness qualities caused by such activities (Evans 1986).

Motorized Use

The Wilderness Act provided for private and state government inholders to maintain their existing access rights and to permit minimal motorized use for necessary administrative purposes. The California Desert Protection Act (U.S. Public Law 103-433) makes clear provision for motorized access for administrative purposes, including law enforcement and wildlife management by appropriate state agencies (Watson and Brink 1996). Wilderness designation laws and accompanying committee reports have also provided specific qualifications and instances where the use of planes and helicopters, motorboats, snowmobiles, and other types of motor vehicles will be allowed. For example, motorboat use is specifically allowed for recreational purposes by the Okefenokee Swamp Act (U.S. Public Law 93-429; U.S. House of Representatives 1974), for private access in one wilderness by the Florida Wilderness Act (U.S. Public Law 98-430), and for both purposes by the BWCA Wilderness Act (U.S. Public Law 95-495); the BWCA Wilderness Act also permits snowmobile use for grooming ski trails near resorts, and the landing of aircraft where previously established is affirmed by specific wording in the Central Idaho Wilderness Act (U.S. Public Law 96-312; Meyer 1999). Helicopter use is permitted to service vault toilets in certain wildernesses by the Endangered Wilderness Act (U.S. Public Law 95-237) and the Utah Wilderness Act (U.S. Public Law 98-428), and other forms of motorized use are mentioned in the Arizona Wilderness Act (U.S. Public Law 98-406), California Wilderness Act (U.S. Public Law 98-425), Utah Wilderness Act (U.S. Public Law 98-428), and Wyoming Wilderness Act (U.S. Public Law 98-550). ANILCA (U.S. Public Law 96-487) provided for a wide array of motorized use, including snowmobiles, motorboats, and aircraft.

Committee reports have also addressed congressional intent and interpretation of motorized use associated with wilderness recreational access (U.S. House of Representatives 1974), fish and wildlife management (U.S. House of Representatives 1979b; U.S. House of Representatives 1976), watershed protection and flood control projects (U.S. House of Representatives 1984; U.S. Senate 1984e), and grazing activities (U.S. House of Representatives 1979a).

Grazing

Livestock grazing and other wilderness management policies in national forest wildernesses were reviewed by congressional committees in the 95th (1978–1979) and 96th (1980–1981) Congresses. The result was, in 1979, U.S. House Committee Report 96-617, which accompanied the Colorado Wilderness Act (U.S. Public Law 96-560) and provided interpretation and clarification of grazing provisions in the Wilderness Act. The Colorado Wilderness Act directed that grazing management in Colorado wildernesses be guided by House Report 96-617, which stressed its interpretation that the Wilderness Act provided for continuation of existing grazing use; the maintenance and construction of supporting facilities, including fences, line cabins, water wells and lines, and stock tanks; and temporary use of motorized equipment to repair facilities and for emergency purposes. As mentioned earlier, these provisions were mandated by the committee report to apply to grazing activities in all national forest wildernesses.

Several subsequent laws, including the Arizona Wilderness Act, Utah Wilderness Act, Wyoming Wilderness Act, Nebraska Wilderness Act, and El Malpais Wilderness Act in New Mexico, also indirectly refer to the guidelines in House Report 96-617. The El Malpais Wilderness Act grazing provisions are important because they apply to a BLM wilderness, and not a national forest wilderness. In addition, several of these laws provide for administrative review of existing grazing policies to ensure that they are consistent with that report. The New Mexico Wilderness Act addressed grazing directly in authorizing additional fencing as provided in the grazing-allotment management plan for livestock in the Cruces Basin Wilderness.

Buffer Zones

In 1980, a congressional committee examined the issue of buffer zones around national forest wildernesses (U.S. House of Representatives 1980, House Report 96-1126) and pointed out that Congress takes great care in determining and establishing wilderness boundaries with the intent that only lands within the boundaries be managed as wilderness. Similarly, a Senate committee report (U.S. Senate 1984c, Senate Report 98-465) stressed that non-wilderness activities should not be restricted or prevented in areas adjacent to wildernesses simply because such activities can be seen or heard from within the wildernesses. Such restrictions applied by the Forest Service were formerly referred to unofficially as the "sights and sounds doctrine," part of the agency's "pure criteria" for wilderness to ensure high standards for designation and management. The New Mexico Wilderness Act (U.S. Public Law 96-550) in 1980 was the first law to prohibit buffer zones, and subsequent laws contain similar provisions, including the Colorado Wilderness Act, Oregon Wilderness Act, Washington Wilderness Act, Arizona Wilderness Act, Utah Wilderness Act, Arkansas Wilderness Act, Wyoming Wilderness Act, Pennsylvania Wilderness Act, Virginia Wilderness Act, and Michigan Wilderness Act.

Fish and Wildlife

Fish and wildlife protection has long been associated with wilderness designation, and the presence of fish and wildlife is considered an integral part of wilderness for hunting, fishing, viewing, and other recreational purposes. Fish and wildlife professionals often argue against wilderness designation, because it is perceived to preclude the use of many modern wildlife-management techniques, and wildlife-management controversies in wilderness are common (Czech and Krausman 1999). Many committee reports stress the importance of preserving and managing fish and wildlife in individual wildernesses.

One House of Representatives committee report (U.S. House of Representatives 1983) addressed the need to balance management activities with the protection of wilderness characteristics. This report stresses that management agencies have authority to maintain water supply facilities, restore natural vegetation, utilize prescribed burning, enhance and restore fish populations, and use motorized equipment to fulfill fish and wildlife management objectives. Congress reaffirmed such intent in ANILCA (U.S. Public Law 96-487) and provided for fisheries research and enhancement, and motorized use and access for subsistence hunting and fishing purposes. The Central Idaho Wilderness Act (U.S. Public Law 96-312) permitted access to manage bighorn sheep along mining roads authorized in the special mining management zone in the Frank Church–River of No Return Wilderness. The Wyoming Wilderness Act (U.S. Public Law 98-550) permitted occasional motorized use to manage bighorn sheep in the Fitzpatrick Wilderness, and the Endangered Wilderness Act (U.S. Public Law 95-237) authorized a fish and game research program to help protect these resources in the Gospel-Hump Wilderness in Idaho and surrounding non-wilderness lands. The California Desert Protection Act (U.S. Public Law 103-433) specifically provides for mechanized access for wildlife management in the 69 wilderness areas it designated (Watson and Brink 1996).

Fire, Insect, and Disease Control

The Wilderness Act specifically permits fire, insect, and disease control measures, but concern has been expressed about limitations that might be placed on control measures in wilderness. One House committee report (U.S. House of Representatives 1977) stressed that control measures include "the use of mechanized equipment, the building of fire roads, fire towers, fire breaks or other pre-suppression techniques where necessary, and other techniques for fire control." Wilderness legislation also affirms that such protective measures can be applied in wilderness. The Endangered Wilderness Act (U.S. Public Law 95-237) stipulated that fire control measures should be properly utilized to protect watersheds in two California wildernesses. The Colorado Wilderness Act (U.S. Public Law 96-560) directed administrative review of current policies for fire, disease, and insect control in Colorado wildernesses to ensure that these policies were consistent with congressional intent and were adequate for protection of adjacent non-wilderness lands.

Facilities and Structures

In several wilderness designation laws, Congress permitted facilities and structures for different purposes. In 1969, two existing dams were included in the Desolation Wilderness established in California (U.S. Public Law 91-82). A congressional committee previously debated whether the use and management of these dams would severely degrade the wilderness or detract from its surroundings (U.S. Senate 1969). Two reservoirs were also included in the Indian Peaks Wilderness in Colorado in 1978 (U.S. House of Representatives 1978). The use of sanitary facilities, such as vault toilets, was specifically provided for in the Endangered Wilderness Act (U.S. Public Law 95-237) and the Utah Wilderness Act (U.S. Public Law 98-428) to protect watersheds in several wildernesses. The Central Idaho Wilderness Act (U.S. Public Law 96-312) permitted construction of water supply facilities in certain areas. The installation and use of "weather modification special equipment," such as snow gauges, and water quality and quantity measuring instruments were endorsed in a House committee report as beneficial to furthering the "scientific, educational, and conservation purposes" of wilderness and to protecting watersheds and "preserving the wilderness character" in some cases (U.S. House of Representatives 1977). The Vermont Wilderness Act (U.S. Public Law 98-322) permitted the maintenance of trails and shelters along the Appalachian Trail and associated trails. "Hydrologic, meteorological, and telecommunications facilities" needed for flood warning and control purposes were permitted in specified wildernesses in the Arizona Wilderness Act (U.S. Public Law 98-406) and Utah Wilderness Act (U.S. Public Law 98-428).

ANILCA (U.S. Public Law 96-487) permitted the construction and maintenance of new and existing cabins in Alaska wildernesses, temporary construction and use of facilities for hunting and fishing, and also provided for fish hatcheries and weirs to improve fisheries. As mentioned earlier, the New Mexico Wilderness Act (U.S. Public Law 96-550) permitted construction of additional fencing for grazing purposes in the Cruces Basin Wilderness.

SPECIAL PROVISIONS AND CONGRESSIONAL DIRECTION

Special provisions have been included in wilderness laws and congressional committee reports to provide specific guidelines for designation and management based on unique circumstances of local or regional concern. Understanding the relationship among the designating legislation, the Wilderness Act of 1964, and legislative intent is a complex matter that often requires additional management information about the situations that lead to the need for the special provision (Watson et al. 2004; Fig. 5.10). Following is a discussion of some of these additional special provisions.

Sufficiency-Release Language

Through sufficiency-release language in wilderness designation laws, Congress has assumed a partial role in resolving the question of which remaining national forest roadless areas are eligible for further wilderness review and which lands are released for other uses. Since its first legislative use in 1980, Congress has consistently applied sufficiency-release language on a state-by-state basis. ANILCA was the first law to contain sufficiency-release language, but as such language evolved it was applied in twenty-seven subsequent laws.

Sufficiency language was constructed by Congress to respond to the *California v. Block* lawsuit, which invalidated the RARE II Final Environmental Impact Statement information for California and prevented development on

FIGURE 5.10 Bureau of Land Management signs mark a wilderness boundary and closure of an old road to vehicles. The California Desert Protection Act (1994) designated 69 BLM-managed wilderness areas. Wilderness designation closed many old mining and prospecting access roads. Photo courtesy John C. Hendee.

those roadless lands recommended for non-wilderness uses (Baldwin and Gorte 1984; Gorte 1987). *Sufficiency language states Congress's conclusion that the information in the RARE II Environmental Impact Statement for a particular state or section of a state is adequate for Congress's review, and that no further statewide roadless area reviews will be conducted by the Forest Service in that state, nor will there be judicial review of the decision releasing the national forestlands for non-wilderness uses. In other words, roadless area reviews have been "legally and factually sufficient," and land not designated for wilderness, wilderness study, or further planning is "released" for possible non-wilderness use.*

Release language has generally been debated in two forms: "hard" and "soft." In its strictest form, *hard release language would permanently release roadless lands not designated as wilderness or for wilderness study from further wilderness consideration unless authorized by Congress* (Baldwin and Gorte 1984). Hard release language might also require non-wilderness multiple use of released lands, although such interpretation might be argued. Although hard release language has not yet been included in wilderness legislation, various industry groups have advocated permanent or hard release in testimony on wilderness bills.

Soft release language provides that the wilderness option for RARE II non-wilderness roadless areas will not be considered again by the Forest Service during the development of the initial forest plans in accordance with the National Forest Management Act (U.S. Public Law 94-588)*, but may be considered during the revision of these plans* (Baldwin and Gorte 1984; U.S. Senate 1980). Before the initial plans are revised, the Forest Service is permitted to manage lands not designated as wilderness, wilderness study, or for further planning, for non-wilderness multiple use. Moreover, released lands do not necessarily have to be managed to protect their wilderness characteristics. Under provisions in the National Forest Management Act and Forest Service regulations, roadless areas that remain when the forest plans are revised at the end of each subsequent ten-year planning period must then be reconsidered for wilderness designation in the planning process for the next forest plan (The Wilderness Society 2000; U.S. Senate 1980).

From 1980 through 1983, soft release language of this type appeared in six wilderness laws, including ANILCA, the Alaska National Interest Lands Conservation Act, New Mexico Wilderness Act, Colorado Wilderness Act, Charles C. Deam Wilderness Act in Indiana, Paddy Creek Wilderness Act in Missouri, and West Virginia Wilderness Act.

In 1984, continued pressure from development interests led to some alteration and clarification of soft release language that put more responsibility on the forest planning process to protect or develop a released roadless area (Gorte 1987). *The result was compromise release language stating that released lands not designated as wilderness would be managed for non-wilderness multiple use in accordance with the National Forest Management Act, but protection of a roadless area's wilderness characteristics would be allowed if this decision were made in the planning process* (Gorte 1987; U.S. Senate 1984a) Thus, protection of these released lands for potential wilderness consideration was placed on the planning process. Congressional committees also clarified the circumstances under which forest plans could be revised, which further clarified the timing of wilderness reviews (Gorte 1987; U.S. Senate 1984a; U.S. Senate 1984b). Since 1984, most wilderness laws have contained this *compromise* version of soft release language (Gorte and Baldwin 1987).

Military Overflights

Congress took additional steps in the 1990s to ensure wilderness designation does not preclude low-level military overflights or designation of new special airspace within wilderness, with wording in the Nevada Wilderness Protection Act of 1989, Arizona Desert Wilderness Act of 1990, and California Desert Protection Act. Section 4(d)(1) of the Wilderness Act makes special provisions for the use of aircraft where these uses have become established, and they may be permitted to continue based on the opinion and restrictions of the secretary of agriculture. The original intent of this provision was aimed at wilderness airstrips already established in wilderness. More recently Congress's special provisions for military aircraft to continue making low-level training flights through designated wilderness, and tourist overflights as well, have become controversial based on the probable impacts on nature and disrupting opportunities for visitor solitude. General (nonmilitary) airspace reservation below 5,000 feet was provided for by presidential executive order in the Boundary Waters Canoe Area Wilderness, and the BWCA Wilderness Act directed that the earlier presidential directive control airplane flyovers.

Miscellaneous Special Provisions

These include a variety of provisions not failing under other categories. For example, the Sawtooth Wilderness Act and the Hells Canyon Wilderness Act both designated wilderness within national recreation areas and provided condemnation authority to acquire nonfederal lands, while also withdrawing the areas from further mining, subject to valid existing claims. The Alpine Lakes Area Management Act of 1976 authorized a special study of the Enchantment Area within the wilderness to determine the area's best management. A special study to consider including the Indian Peaks Wilderness in Rocky Mountain National Park was authorized by the Indian Peaks Act. The Virginia Wilderness Act called for a combined state and federal air quality study of designated wilderness study areas. The El Malpais and Cebolla Wilderness Act reserved water rights for the wildernesses designated by the act with certain special provisions.

WILDERNESS LEGISLATION: A VIEW OF THE FUTURE

Wilderness legislation is likely to remain a major topic for congressional action in the decades to come. Additional designations are likely in NPS and FWS units, national forest roadless lands, and BLM wilderness study areas (Fig. 5.11). As national forest planning cycles are completed, there will be renewed opportunity to consider the wilderness potential of the released national forest roadless areas. *We believe that future wilderness legislation will continue the trend of containing language that affirms and/or clarifies management direction in ways that address local concerns.* There will also be occasional legislation to adjust boundaries, change

FIGURE 5.11 Hoodoos in New Mexico's Bisti/De-Na-Zin Wilderness, managed by the Bureau of Land Management, show the variety of landforms and ecosystems within the National Wilderness Preservation System. Photo courtesy Bureau of Land Management.

wilderness names, correct inconsistencies, and otherwise adjust to inevitable change (Landres and Meyer 2000).

Between 1964 and 2000, only 5 of the 132 laws passed had removed acres from wilderness designation (Landres and Meyer 2000). However, there continue to be proposals to remove wilderness designations or make major special provisions and exceptions for some units, as some recent designations demonstrate (e.g., Steens Mountain Wilderness, in Washington, designated under U.S. Public Law 106-399). We and others are concerned that future wilderness laws may include more such special provisions, nonconforming uses, and exceptions that will dilute the quality of the NWPS (Nickas and Proescholdt 2005).

The 111.8 million acres of wilderness already designated by 2023 would surprise the original authors of the Wilderness Act, who anticipated an NWPS of only 40 to 50 million acres (Church 1977). Yet, 50 million acres or more of wilderness may likely be added to the NWPS from the roadless areas in Alaska, national parks in the forty-eight contiguous states, remaining national forest roadless areas, and the BLM wilderness study areas. The Wilderness Society and other environmental groups have called for a 200-million-acre wilderness system.

For at least another decade, and perhaps for two or three decades, it is likely that wilderness legislation will continue to be a major natural-resource issue. *We expect a continuation of the trend in wilderness laws toward omnibus legislation covering more than one area, often in individual states, and the inclusion of more language to affirm and clarify management direction and address local concerns.* Congress will increasingly have to address smaller areas, those previously modified to some degree, and areas with strongly competing alternative uses. To resolve some of the very difficult wilderness designation and management issues in the future, a variety of special provisions, exceptions, and compromise provisions are likely to be proposed and included in wilderness designations. *However, the evidence suggests that Congress will generally hold the line on proposals for major exceptions and unique provisions in wilderness laws and/or amending the Wilderness Act.*

MANAGEMENT IMPLICATIONS FROM LITIGATION INVOLVING WILDERNESS

Congress has the authority to regulate federal lands based on the U.S. Constitution, and the Wilderness Act of 1964 (U.S. Public Law 88-577) awarded the authority for designating wilderness areas to Congress as the means to create a National Wilderness Preserve System. This approach to an NWPS means that the legislative branch of government designates the areas, the executive branch carries out the management as directed by the act, and the judicial branch resolves conflicts in application of the act.

Like litigation on any contentious issue, wilderness litigation has been varied in content and reason for the action as well as in outcomes and decisions. It is difficult to obtain information on some cases that do not have written decisions, resulted in an out-of-court settlement, or were withdrawn. Similarly, some cases have been through multiple levels of the court system and have been overturned for various reasons. Litigation has tested the Wilderness Act and wilderness designation laws, and generally the courts have upheld the intent of the Wilderness Act (Ruckel 1999).

Citizens can sue federal land-management agencies using the Administrative Procedures Act (APA) of June 11, 1946, to challenge decisions that are unlawful or unsupported by the legislation and to ensure that the laws are executed in a responsible and legally appropriate manner. Dettmann (2008, 4) has expressed an opinion on why the Wilderness Act lends itself to litigation more than some other laws:

> The Wilderness Act . . . is a remarkable federal lands statute because it provides specific and clear directives on what wilderness is and how a system of wilderness is to be created and maintained. For established areas, the Act imposes a statutory structure with detailed requirements—such as no vehicles, no structures, and no commerce—along with a strict, overarching mandate to preserve wilderness character. These directives effectively remove the discretion that land management agencies might normally enjoy.

The idea that litigation is an affirmation of the Wilderness Act and not a punishment of the federal land-managing agencies is firmly outlined by Dettmann (2008, 5). His perspective is that the judicial oversight is a necessary and equally important third branch of government that has its place in executing the intent of the Wilderness Act.

> Our system of government seeks to resolve conflict by creating a forum for intelligent, civilized debate moderated by a judge vested with the power of the sovereign. The whole idea of litigation is to allow two disagreeing parties to come to a neutral authority that reviews the evidence and decides the case. The whole point of the judicial branch is to resolve conflict, not to create it. What better opportunity is there to advocate for the values of wilderness than through the judicial process? What more reasoned approach exists by which wilderness advocates can air their disagreements with agencies to an authority that is obligated to follow and enforce the law as it is written? Such a forum lends itself particularly well to a statute such as the Wilderness Act. It is hard to find a clearer definitive statutory command than, for example, there shall be no use of motor vehicles in wilderness. . . . Litigation enables a broader public accounting of the issues as well as a broader understanding.

Whether litigation can be brought regarding wilderness under the Administrative Procedures Act (APA) has been narrowly construed by the U.S. Supreme Court in 2004 to relate to specific agency actions that are legally required (Konrad 2006). The issue was that an environmental group contended (*Norton v. Southern Utah Wilderness Alliance*) that the BLM had failed to adequately protect wilderness study areas from degradation by off-road vehicles as required under the Federal Land Policy and Management Act, NEPA, and the BLM Land Use Plan (Konrad 2006); the court found that the APA could not be used to comply with general statutory mandates. The Wilderness Act, as Dettmann points out, is much more directive and mandates agency action through statutory requirements.

BOUNDARY WATERS CANOE AREA

The Boundary Waters Canoe Area (BWCA) is an example of a wilderness area that was founded in controversy and litigation that helped shape two wilderness laws related to wilderness designation of the BWCA. The area has one of the longest histories of legal battles between wilderness protection advocates and those who want to access and use the resources of the area for motorized transportation, timber harvest, resort and tourism development, mining development, and other uses. Protection of the wild character of the BWCA through legislation, administration, and litigation, starting in 1909 with the establishment of the Superior National Forest, did not reduce the conflicts and legal issues (Duncan and Proescholdt 1999). Multiple controversies, wilderness advocacy, and litigation eventually resulted in designating the BWCA as an original unit of the NWPS in the 1964 Wilderness Act. However, continued controversy and legal actions following the political compromises to permit nonconforming uses in the BWCA under the act resulted in the need for additional legislative clarification and the passage of the Boundary Waters Canoe Area Wilderness Act of 1978 (U.S. Public Law 95-495).

The BWCAW Act expanded the original boundaries of the BWCA by 57,000 acres to more than 1 million acres and renamed the area the Boundary Waters Canoe Area Wilderness. The BWCAW Act further restricted certain permitted and nonconforming activities, such as it "ended logging entirely, restricted mining and the use of snowmobiles, and cut motorboat use from sixty-four percent to twenty-four percent of the water surface" (Duncan and Proescholdt 1999, 626), but numerous permitted and nonconforming activities remained. The way the BWCAW Act achieved these changes was through repeal of "section 4(d)(5) of the 1964 Wilderness Act, the only instance where the 1964 law was amended" (Duncan and Proescholdt 1999, 626).

The political compromises in the 1978 BWCAW Act brought the area into better compliance with other wildernesses in the NWPS, but the controversies remain and the lawsuits continue. A long history of litigation has shaped the BWCAW and will continue to affect the management of the area as the judicial process continues.

THE EXTENT OF WILDERNESS LITIGATION

Considering the more than 170 wilderness designation laws following the Wilderness Act and the more than 800 units and 111.8 million acres in the NWPS by 2023, it is surprising that the number of cases of litigation is small after more than fifty years since the act was passed into law. There are very few compilations of litigation regarding federal land-management cases; however, a published study reported that litigation brought against the USFS during the sixteen-year period from 1989 through 2004 found that of the 888 documented cases with published decisions, only twenty-six cases (2.9 percent) were based on the Wilderness Act and another twenty-nine were based on the Wild and Scenic Rivers Act (U.S. Public Law 90-542), and three of those cases were based on both the Wilderness and Wild and Scenic Rivers Acts (Malmsheimer et al. 2008).

No studies have yet been published that compile the entire time period since 1964, or compile litigation for the other three federal wilderness management agencies, the National Park Service, Bureau of Land Management, and U.S. Fish and Wildlife Service.

Between 1989 and 2004, the USFS won ten (38.5 percent) of the twenty-six lawsuits brought against it regarding the Wilderness Act, lost eleven cases (42.3 percent), and settled five cases (19.3 percent; Malmsheimer et al. 2008). The statutory basis for these cases is not solely the Wilderness Act and often involved NEPA and other statutes. The purposes of these cases were complex and Malmsheimer et al. (2008) break them down into only two categories: 77 percent of the cases were for less resource use (e.g., to restrict use or uses) and 23 percent were for greater resource use (e.g., to overturn restrictions on amount or type of use). Malmsheimer et al. (2008) note that twelve of the twenty-six cases involved some aspect of motorized vehicle use or access.

LITIGATION THEMES AND MANAGEMENT IMPLICATIONS

Three types of themes have emerged from litigation around the Wilderness Act: the definition and characteristics of wilderness, the scope of protection under the Wilderness Act, and issues involving wilderness management (Ruckel 1999).

Cases focusing on the definition of wilderness and the suitability of certain areas are exemplified by the *Parker v. United States* suit in 1971, which questioned whether a road leading into the proposed East Meadow Creek area disqualified the area to be designated as wilderness. This was in the Gore Range–Eagles Nest Primitive Area on the White River National Forest in north-central Colorado. Courts found that the area met the wilderness definition, such as the definitional phrase that an area "generally appears to have been affected primarily by the forces of nature, with the imprint of man's work substantially unnoticeable." Ruckel (1999, 613) concludes from this and other cases:

> The courts have adopted a practical definition, recognizing the primitive, untrammeled nature of wilderness, but acknowledging the often unavoidable presence of conflicting circumstances or uses. Purist notions of completely untrammeled, unblemished land have not been sympathetically received. An important effect of this approach has been to maximize the availability of lands, some of which may be minimally impacted by a conflicting circumstance or use, for eventual wilderness classification.

Two cases illustrate that the scope of protection under the Wilderness Act is broad. In the 2003 case of *Wilderness Society v. U.S. Fish and Wildlife Service*, the courts decided that a commercial sockeye salmon fish-stocking project in the Kenai Wilderness of Alaska was expressly prohibited under the Wilderness Act because the project's purpose and effect were to support the commercial fishing industry (Rasband et al. 2004; Ryan 2005; Miller 2005). In the second case in 2000, *Potlatch Corp. v. United States* a decision by the courts ruled that Congress had intentionally and directly reserved the unappropriated water rights upstream of three Idaho wilderness areas in their designation under the 1964 Wilderness Act, which by extension affirms that federal water rights reservation applies to lands designated under the Wilderness Act of 1964 (Latimer 2000; Miller 2000).

Not all cases have been decided in favor of wilderness protection especially related to valid preexisting mining claims, road access across wilderness study areas to private inholdings, road construction and timber harvesting on lands adjoining a wilderness study area, and other situations on adjoining lands (Ruckel 1999).

Controversies and issues about the management of wilderness areas have been the source of some of the most well-known court cases regarding the Wilderness Act. *Wilderness Watch v. Mainella, Sierra Club v. Lyng, High Sierra Hikers Assn., et al. v. Bernie Weingardt, et al.*, and a USFS case banning fixed rock-climbing anchors in the Sawtooth Wilderness of Idaho illustrate the implications for managers who have some discretion to make management decisions; however, their decisions and management actions can be challenged by special interest and environmental groups.

The 2004 case of *Wilderness Watch v. Mainella* was about the NPS practice of transporting visitors across a portion of Cumberland Island Wilderness Area to provide access to historic sites, which was decided by the courts as violating the Wilderness Act and NEPA (Powers 2005).

The *Sierra Club v. Lyng* case in 1987 was over the USFS practice of extensive tree cutting and chemical spraying to control beetle infestations in southern USFS wilderness areas; the courts ruled against the program because it was justified based on protecting the economic interests and resources of adjacent private landowners (Glicksman and Coggins 1999). The courts remanded the USFS to file an EIS and redesign the program to limit the program's impacts on wilderness conditions—which the USFS carried out and received approval for implementation.

A seven-year legal battle in *High Sierra Hikers Assn., et al. v. Bernie Weingardt, et al.* was about commercial livestock use in the John Muir and Ansel Adams Wildernesses in California. The court decided on October 30, 2007, that the USFS violated the Wilderness Act by creating a plan that did not meet the requirement to preserve wilderness character and to properly regulate and limit commercial services for pack-stock use.

The fourth case involved the USFS decision to ban permanent fixed-anchor installations for rock climbing in Idaho's Sawtooth Wilderness. An administrative appeals hearing resulted in the USFS decision to initiate a negotiated rule-making process that was to develop a national policy on permanent fixed-anchor installations for rock climbing in the wilderness areas (Cheever 1999; Dolan 2000).

The judiciary has had a substantial and formative impact on the NWPS by hearing cases brought regarding suitability for wilderness designation, the breadth and scope of wilderness protection, and the complex and conflicting uses that Congress permits in wilderness as part of political compromises.

Ruckel (1999, 619) concludes after reviewing the published litigation on the Wilderness Act and designation legislation:

> It has taken strong action by the judiciary to strike balances and make determinations that have ensured that qualifying lands were all reviewed, and then, when formally dedicated to wilderness, protected. The courts have succeeded by carefully scrutinizing the facts of the cases brought before them and vigorously safeguarding wilderness principles and lands. They have maintained balance and objectivity by recognizing individual conflicting activities sanctioned by the Wilderness Act itself, or by specific legislation establishing individual wilderness areas.

In the final analysis, the attempt to litigate a disagreement over wilderness policies or management actions indicates much about the convictions of the parties involved regarding the issue in question. Litigation is very expensive and invariably follows a great deal of discussion and effort to resolve the matter without going to court. Thus, the threat of litigation brings a greater sense of urgency and importance to attempts to resolve wilderness issues than if the courts were never involved.

WILDERNESS DESIGNATION LESSONS

Important lessons have been learned. Although the wilderness designation process requires a sound technical basis to identify areas and inventory their wilderness attributes and other values, it is largely a political process. Thus, the American public has been the ultimate judge of the disposition of roadless lands. Environmental groups learned this early on, and they have had a far greater influence on wilderness designation than initially expected.

Agency planners and policy makers have struggled to understand and institutionalize the wilderness ideal, to integrate wilderness within their traditional resource programs, and to retain a position of leadership regarding wilderness. There have been and continue to be occasional problems, controversies, and even impasses. When this occurs, the courts increasingly have been the final arbitrator. Gradually, the agencies have found a more even and defensible course. They have done this by taking a comprehensive planning approach, integrating wilderness into legal planning processes, recognizing that wilderness study involves two very different issues (assessment and evaluation), anchoring assessment in wilderness attributes defined in law, keeping in touch with public values when making evaluative judgments, and working closely with individual members of Congress and congressional committees who make the final designation decisions.

The most difficult task facing the wilderness resource agencies remains keeping in touch with the slowly but inevitably evolving cultural definition of wilderness in the United States and applying it to the management of the remaining roadless areas so these values are protected. Such sensitivity seems necessary when recommending which lands should be wilderness, when integrating wilderness and the other multiple uses of public lands, and ultimately when making individual wilderness management decisions. *We have seen the drift of public opinion away from rigidly applied, pure wilderness standards in wilderness designation—and ultimately in wilderness management.* In the final analysis, this evolving process has led to, and is leading to, diversity in the NWPS. This, in turn, can lead to broader political support and a resource capable of providing for a variety of human needs.

STUDY QUESTIONS

Evolution of the Designation Process

1. Discuss the factors that led to the RARE I process. What were some of the deficiencies of RARE I and how might they have been corrected? What were some of the benefits of RARE I?
2. In your judgment, how well did RARE II correct the deficiencies of RARE I? What problems still remain?
3. What were some of the factors that led to the increasing use of "omnibus" state wilderness legislation?
4. Describe the concepts of sufficiency, hard release, and soft release as they apply to the wilderness designation process. How have they been employed in the state where you live?
5. Discuss the problems in obtaining wilderness designation in national parks like Yellowstone and Yosemite. What are some of the reasons underlying the lack of progress?
6. Describe the various opportunities and problems facing the wilderness designation of lands under the administration of the BLM.
7. Describe the relative roles and importance of environmental groups versus the agencies' wilderness study processes in securing wilderness designation.
8. How did the RARE II lawsuit (*California v. Bergland*) influence the wilderness designation process, and thus the size and nature of the NWPS?
9. Why was designating wilderness in Alaska such a challenge? How did Congress resolve the unique conditions in making wilderness possible?

Management Implications in Wilderness Designation Acts

1. How do wilderness designation laws affect wilderness management?
2. How can legislative history influence the interpretation of a wilderness designation law? Where is the legislative history found?
3. Why are seven laws identified as "benchmark laws" in this chapter?
4. How were grazing provisions of the Wilderness Act clarified?
5. What are some management provisions of the Wilderness Act that have been clarified by subsequent wilderness designation, and what general direction have the clarifications taken?
6. When and where can you mine in wilderness?

Management Implications of Litigation

1. What place does the judicial branch of government have in executing the intent of the Wilderness Act?
2. Should the amount of litigation involving the Wilderness Act be described as small or large? Has that litigation challenged the intent of the Wilderness Act? Has that litigation helped clarify the issues surrounding compromises for permitted uses in wilderness?
3. What are the three types of themes that have emerged from litigation around the Wilderness Act? Give an example of a case for each theme.

ACKNOWLEDGMENTS

Chad Dawson revised this chapter for the fifth edition. John C. Hendee and Chad P. Dawson revised this chapter for the fourth edition, integrating and updating the material from previous editions (Part I: wilderness designation process, including agency roadless area reviews; and Part II: wilderness designation laws) and adding a new third part (management implications of litigation involving wilderness). The chapter builds on contributions to earlier editions from our colleagues Jay Watson, Joseph W. Roggenbuck, George H. Stankey, Dennis M. Roth, Jim Browning, and Ed Krumpe.

CASE DISCUSSION QUESTIONS: ALASKA WILDERNESS

Read the section in this chapter entitled "Alaska—A Special Case" and the two subsections on "The Alaska National Interest Lands Conservation Act (ANILCA)." The case outlines the entire process by which wilderness studies and protection occurred in Alaska. Then answer the following discussion questions:

1. What were the major impediments to the eventual passage of the Alaska National Interest Lands Conservation Act (ANILCA)? Briefly list them.
2. What history and influences caused the passage of the Alaska Native Claims Settlement Act (ANCSA)? Why did it need to proceed prior to passage of ANILCA?
3. In the case of Alaska, how much did professional expertise contribute to wilderness designation, and how much did political involvement influence passage of legislation to designate wilderness?

REFERENCES

Administrative Procedure Act of June 11, 1946 (60 Stat. 237), and amended in 1966.

Allin, Craig W. 1982. *The Politics of Wilderness Preservation.* Westport, CT: Greenwood Press.

Bailey, Robert G. 1976. Ecoregions of the United States. Ogden, UT: U.S. Department of Agriculture, Forest Service, Intermountain Region. Map.

Baldwin, Pamela; Gorte, Ross W. 1984. National forest wilderness bill release alternatives: A legal analysis and policy issues. Congressional Research Service Report No. 84-542 ENR, February 21, 1984. Washington, DC: U.S. Library of Congress.

Brown, Perry. 2000. Issues in the quality of U.S. wilderness management. *International Journal of Wilderness.* 6(2): 3–4.

———. 2002. A summary of the report: Ensuring the Stewardship of the National Wilderness Preservation System. *International Journal of Wilderness.* 8(1): 10–12, 32.

Browning, James A.; Hendee, John C.; Roggenbuck, Joe W. 1988. *103 Wilderness Laws: Milestones and Management Direction in Wilderness Legislation,* 1964–1987. Bulletin 51. Moscow: University of Idaho, Idaho Forest, Wildlife and Range Experiment Station.

Cheever, Federico. 1999. Talking about wilderness. *Denver University Law Review.* 76(2): 335–345.

Church, Frank. 1977. Wilderness in a balanced land use framework. *First Annual Wilderness Resource Distinguished Lectureship.* Moscow: University of Idaho, Wilderness Research Center. 18p.

Coggins, George C.; Wilkinson, Charles F. 1987. *Federal Public Land and Resources Law,* 2nd ed. Mineola, NY: The Foundation Press.

Costley, Richard J. 1972. An enduring resource. *American Forests.* 78(6): 8–11.

Crumbo, Kim. 1996. Wilderness management at Grand Canyon—"Waiting for Godot?" *International Journal of Wilderness.* 2(1): 19–24.

Czech, Brian; Krausman, Paul. 1999. Controversial wildlife management issues in southwestern U.S. wilderness. *International Journal of Wilderness.* 5(3): 22–28.

Davis, George D. 1989. Preservation of natural diversity: The role of ecosystem representation within wilderness. In: Freilich, Helen R., comp. *Wilderness Benchmark 1988: Proceedings of the National Wilderness Colloquium*; January 13–14, 1988; Tampa, FL. General Technical Report SE-51. Athens, GA: U.S. Department of Agriculture, Forest Service, Southeastern Forest Experiment Station, pp. 76–82.

Dettmann, Joe. 2008. The need for wilderness litigation. *International Journal of Wilderness.* 14(1): 4–6.

Dolan, Timothy. 2000. Fixed anchors and the Wilderness Act: Is the adventure over? *University of San Francisco Law Review.* 34: 355–378.

Dombeck, Mike. 1999. A wilderness agenda and legacy for the U.S. Forest Service. *International Journal of Wilderness.* 5(3): 4–6.

Duncan, Richard A.; Proescholdt, Kevin. 1999. Protecting the Boundary Waters Canoe Area Wilderness: Litigation and legislation. *Denver University Law Review.* 76(2): 621–658.

Evans, Kent E. 1986. Indian Mounds Wilderness Area: Perceived wilderness qualities and impacts of oil and gas development. In: Kulhavy, David L.; Conner, Richard, N., eds. *Wilderness and Natural Areas in the Eastern United States: A Management Challenge.* Nacogdoches, TX: Stephen F. Austin University, pp. 156–165.

Glicksman, Robert L.; Coggins, George C. 1999. Wilderness in context. *Denver University Law Review.* 76(2): 383–411.

Gorte, Ross W. 1987. History of release language in wilderness legislation, 1979–1984. CRS Report No. 87-559 ENR, June 12, 1987. Washington, DC: U.S. Library of Congress, Congressional Research Service.

———. 1988. Wilderness: Overview and statistics. Congressional Research Report No. 8846, ENR, January 5, 1988. Washington, DC: U.S. Library of Congress.

Gorte, Ross W.; Baldwin, Pamela. 1987. Wilderness release language history and effectiveness. Unpublished Congressional Research Service Memorandum, June 25, 1987. Washington, DC: U.S. Library of Congress.

Hendee, John C. 1977. Public involvement in the United States Forest Service roadless area review: Lessons from a case study. In: Coppock, Terrance; Sewell, Derrick, eds. *Public Participation in Planning.* New York: Wiley, pp. 89–103.

———. 1986. Wilderness: Important legal, social, philosophical and management perspectives. In: Kulhavy, David L.; Conner, Richard N., eds. *Wilderness and Natural Areas in the Eastern United States: A Management Challenge;* May 13–15, 1985; Nacogdoches, TX. Nacogdoches, TX: Stephen F. Austin University, School of Forestry, pp. 5–11.

Hendee, John C.; Smith, Zane G.; Lake, Robert. 1980. Public involvement in resource decisions: RARE I and II and implications for the future. *Proceedings: Multiple Use Symposium,* September 1979. Clemson, SC: Clemson University.

Hendee, John C.; Stankey, George H.; Lucas, Robert C. 1991. *Wilderness Management.* 2nd ed., rev. Golden, CO: North American Press/Fulcrum.

Hummel, M. 2007. The Stikine-LeConte Wilderness in Alaska: Twenty-six years of management under ANILCA. *International Journal of Wilderness.* 13(2): 8–13, 17.

Konrad, Justin C. 2006. The shrinking scope of judicial review in *Norton v. Southern Utah Wilderness Alliance. University of Colorado Law Review.* 77: 515–547.

Küchler, A. W. 1966. Potential Natural Vegetation, Map. Natural Atlas of the United States. Washington, DC: U.S. Department of the Interior, Geological Survey, pp. 89–92.

Landres, Peter; Meyer, Shannon. 2000. National Wilderness Preservation System database. *Key attributes and trends 1964–1999,* 2nd ed., rev. RMRS-GTR-18. Fort Collins, CO: U.S. Department of Agriculture, Forest Service, Rocky Mountain Research Station. 97p.

Latimer, K. J. 2000. Federal reserved water rights doctrine under the Wilderness Act: Is it finally here to stay? *Journal of Land Resources and Environmental Law.* 20: 335–356.

Leshy, John D. 2005. Contemporary politics of wilderness preservation. *Journal of Land Resources and Environmental Law.* 25(1): 1–13.

Loop, Donna J. 1986. Claiming the Cabinets: The right to mine in wilderness areas. *The Public Land Law Review.* 7: 45–77.

MacCracken, John G.; O'Laughlin, Jay; Merrill, Troy. 1993. Idaho roadless areas and wilderness proposals. Report 10. Moscow: University of Idaho, Idaho Forest, Wildlife, and Range Policy Analysis Group.

Malmsheimer, R. W.; Falco, C.; Anderson, A. M.; Keele, D. M.; Floyd, D. W. 2008. U.S. Forest Service litigation—The Wilderness Act and Wild and Scenic Rivers Act: 1989–2004. *International Journal of Wilderness.* 14(1): 7–14.

Matthews, Olen P.; Haak, Amy; Toffenetti, Kathryn. 1985. Mining and wilderness: Incompatible uses or justifiable compromises? *Environment.* 7(3): 12–36.

McClure, James. 1985. Congressional directives and expectations. In: Frome, Michael, ed. *Issues in Wilderness Management: Proceedings National Wilderness Management Conference*; October 1983; Moscow, ID. Boulder, CO: Westview Press.

Meyer, Shannon. 1999. The role of legislative history in agency decision making: A case study of wilderness airstrip management in the United States. *International Journal of Wilderness.* 5(2): 9–12.

Miller, J. 2005. Six million sockeye salmon and the Kenai Refuge: Analysis of the Ninth Circuit Court of Appeals' two decisions in *The Wilderness Society v. United States Fish and Wildlife Service. Journal of Land Resources and Environmental Law.* 25(1): 85–97.

Miller, R. C. 2000. Water Law—Wilderness areas and federal reserved water rights: Unlimited appropriations of unallocated water endorsed by Idaho Court, *Potlatch Corp. v. United States (in re* SRBA) No. 39576, 1999 WL 778325 (Idaho 1999). *Land and Water Law Review.* 35(2): 375–394.

Milton, William John, Jr. 1975. National forest roadless and undeveloped areas: Develop or preserve? *Land Economics.* 51(2): 139–143.

Mohai, Paul; Verbyla, David L. 1987. The RARE II wilderness decisions. *Journal of Forestry.* 85(1): 17–23.

Nickas, G.; Proescholdt, K. 2005. Keeping the wild in wilderness: Minimizing nonconforming uses in the National Wilderness Preservation System. *International Journal of Wilderness.* 11(3): 13–18.

O'Laughlin, Jay; Freemuth, J. 2000. Roadless area policy, politics and wilderness potential. *International Journal of Wilderness.* 6(1): 9–12.

Outdoor Recreation Resources Review Commission. 1962. *Wilderness and Recreation: A Report on Resources, Values and Problems.* ORRRC Study Report 3. Washington, DC: U.S. Government Printing Office.

Powers, S. 2005. *Wilderness Watch v. Mainella:* The Wilderness Act and National Environmental Policy Act are given meaning by the Eleventh Circuit. *Journal of Land Resources and Environmental Law.* 25(1): 99–108.

Rasband, James; Salzman, James; Squillace, Mark. 2004. *Natural Resources Law and Policy.* New York: Foundation Press.

Roth, Dennis M. 1984. *The Wilderness Movement and the National Forest: 1964–1980.* FS-391. Washington, DC: U.S. Department of Agriculture, Forest Service.

Ruckel, H. Anthony. 1999. The Wilderness Act and the courts. *Denver University Law Review.* 76(2): 611–619.

Ryan, K. D. 2005. Preservation prevails over commercial interests in the Wilderness Act: *Wilderness Society v. United States Fish and Wildlife Service. Ecology Law Quarterly.* 32: 539–573.

Tanner, R. J. 2004. Subsistence, inholdings, and ANILCA: The complexity of wilderness stewardship in Alaska. *International Journal of Wilderness.* 10(2): 18– 22.

U.S. Department of Agriculture, Forest Service. 1973. Final environmental statement: Roadless and undeveloped areas (RARE). Washington, DC: United States Department of Agriculture, Forest Service.

———. 1978. Roadless area review and evaluation (RARE II). Draft environmental impact statement 78-04. Washington, DC: U.S. Department of Agriculture, Forest Service.

———. 1979. Final RARE II environmental impact statement. Washington, DC: U.S. Department of Agriculture, Forest Service.

———. 2005. What can I do in wilderness? Alaska National Interest Lands Conservation Act (ANILCA) and wilderness in national forests in Alaska. Ketchican, AK: United States Department of Agriculture, Forest Service.

U.S. Department of the Interior, Bureau of Land Management. 1982. Wilderness study policy: Policies, criteria and guidelines for conducting wilderness studies on public lands. *Federal Register*. 47(23): 5098–5122.

U.S. Department of the Interior, U.S. Fish and Wildlife Service. n.d. Red Rock Lakes Wilderness proposal.

———. 1970. Laguna Atascosa Wilderness study area.

U.S. Government Accounting Office. 1984. Private mineral rights complicate the management of eastern wilderness areas. GAO/RCED-84-101. July 26, 1984. Washington, DC: U.S. Government Accounting Office.

U.S. House of Representatives. 1973. *Ninth Annual Wilderness Report*. House Document 93-194. Washington, DC: U.S. Government Printing Office.

U.S. House of Representatives. 1974. House Report 93-872. 93rd Congress, 2nd Session. Washington, DC: U.S. Government Printing Office.

———. 1976. House Report 94-1562. 94th Congress, 2nd Session. Washington, DC: U.S. Government Printing Office.

———. 1977. House Report 95-540. 95th Congress, 1st Session. Washington, DC: U.S. Government Printing Office.

———. 1978. House Report 95-1460. 95th Congress, 2nd Session. Washington, DC: U.S. Government Printing Office.

———. 1979a. House Report 96-97, Part 1. 96th Congress, 1st Session. Washington, DC: U.S. Government Printing Office.

———. 1979b. House Report 96-617. 96th Congress, 1st Session. Washington, DC: U.S. Government Printing Office.

———. 1980. House Report 96-1126. 96th Congress, 2nd Session. Washington, DC: U.S. Government Printing Office.

———. 1983. House Report 98-40. 98th Congress, 1st Session. Washington, DC: U.S. Government Printing Office.

———. 1984. House Report 98-1019, Part 1. 98th Congress, 2nd Session. Washington, DC: U.S. Government Printing Office.

U.S. Public Law 86-517. Multiple-Use Sustained-Yield Act of June 12, 1960. 74 Stat. 215.

U.S. Public Law 88-577. The Wilderness Act of September 3, 1964. 78 Stat. 890.

U.S. Public Law 90-271. (San Rafael Wilderness) Act of March 21, 1968. 82 Stat. 51.

U.S. Public Law 90-542. Wild and Scenic Rivers Act of October 2, 1968. 82 Stat. 906.

U.S. Public Law 91-82. (Desolation Wilderness) Act of October 10, 1969. 83 Stat. 131.

U.S. Public Law 91-190. National Environmental Policy Act (NEPA) of January 1, 1970. 83 Stat. 852.

U.S. Public Law 91-504. (Omnibus Wildlife Refuge, National Park, and National Forest Wilderness) Act of October 23, 1970. 84 Stat. 1104.

U.S. Public Law 92-203. Alaska Native Claims Settlement Act of 1971 (ANCSA) of December 18, 1971. 92 Stat. 203.

U.S. Public Law 92-395. (Scapegoat Wilderness) Act of August 20, 1972. 86 Stat. 578.

U.S. Public Law 92-400. (Sawtooth NRA Wilderness Designation and Wilderness Study) Act of August 22, 1972. 86 Stat. 612.

U.S. Public Law 93-378. The Forest and Rangeland Renewable Resources Planning Act (RPA) of 1974. 88 Stat. 476.

U.S. Public Law 93-429. (Okefenokee Wilderness) Act of October 1, 1974. 88 Stat. 1179.

U.S. Public Law 93-622. (Eastern Wilderness) Act of January 3, 1975. 88 Stat. 2096.

U.S. Public Law 94-199. (Hells Canyon NRA and Wilderness Designation and Wilderness Study) Act of December 31, 1975. 89 Stat. 1117.

U.S. Public Law 94-357. Alpine Lakes Area Management (and Wilderness Designation) Act of July 12, 1976. 90 Stat. 905.

U.S. Public Law 94-544. (Point Reyes Wilderness) Act of October 18, 1976. 90 Stat. 2515.

U.S. Public Law 94-567. (National Park and Monument Omnibus Wilderness Designation and Coronado NF Wilderness Study) Act of October 20, 1976. 90 Stat. 2693.

U.S. Public Law 94-579. (Wilderness Study) Federal Land Policy and Management Act (FLPMA) of 1976. 90 Stat. 2743.

U.S. Public Law 94-579. Federal Land Policy and Management Act (FLPMA) of October 21, 1976. 90 Stat. 2743.

U.S. Public Law 94-588. National Forest Management Act (NFMA) of October 22, 1976. 90 Stat. 2949.

U.S. Public Law 95-237. Endangered American Wilderness Act of February 24, 1978. 92 Stat. 40.

U.S. Public Law 95-450. The Indian Peaks Wilderness Area, the Arapaho National Recreation Area and the Oregon Islands Wilderness Area Act of October 11, 1978. 92 Stat. 1095.

U.S. Public Law 95-495. (Boundary Waters Canoe Area Wilderness) Act of October 21, 1978. 92 Stat. 1649.

U.S. Public Law 96-312. Central Idaho Wilderness Act of July 23, 1980. 94 Stat. 948.

U.S. Public Law 96-487. Alaska National Interest Lands Conservation Act (ANILCA) of December 20, 1980. 94 Stat. 2371.

U.S. Public Law 96-550. (New Mexico Wilderness) Act of December 19, 1980. 94 Stat. 3221.

U.S. Public Law 96-560. (Colorado Wilderness) Act of December 22, 1980. 94 Stat. 3265.

U.S. Public Law 97-384. (Charles C. Dean Wilderness) Act of December 22, 1982. 96 Stat. 1942.

U.S. Public Law 97-407. Paddy Creek Wilderness Act of January 3, 1983. 96 Stat. 2033.

U.S. Public Law 97-466. (West Virginia Wilderness) Act of January 13, 1983. 96 Stat. 2538.

U.S. Public Law 98-322. Vermont Wilderness Act of June 19, 1984. 98 Stat. 253.

U.S. Public Law 98-328. Oregon Wilderness Act of June 26, 1984. 98 Stat. 272.

U.S. Public Law 98-339. Washington State Wilderness Act of July 3, 1984. 98 Stat. 299.

U.S. Public Law 98-406. Arizona Wilderness Act of August 28, 1984. 98 Stat. 1485.

U.S. Public Law 98-425. California Wilderness Act of September 28, 1984. 98 Stat. 1619.

U.S. Public Law 98-428. Utah Wilderness Act of September 28, 1984. 98 Stat. 1657.

U.S. Public Law 98-430. Florida Wilderness Act of September 28, 1984. 98 Stat. 1665.

U.S. Public Law 98-508. Arkansas Wilderness Act of October 19, 1984. 98 Stat. 2349.

U.S. Public Law 98-550. Wyoming Wilderness Act of October 30, 1984. 98 Stat. 2807.

U.S. Public Law 98-574. Texas Wilderness Act of October 30, 1984. 98 Stat. 3051.

U.S. Public Law 98-585. Pennsylvania Wilderness Act of October 30, 1984. 98 Stat. 3100.

U.S. Public Law 98-586. Virginia Wilderness Act of October 30, 1984. 98 Stat. 3105.

U.S. Public Law 99-504. Nebraska Wilderness Act of October 20, 1986. 100 Stat. 1802.

U.S. Public Law 100-184. Michigan Wilderness Act of December 8, 1987. 101 Stat. 1275.

U.S. Public Law 100-225. (El Malpais and Cebolla Wilderness) Act of December 31, 1987. 101 Stat. 1539.

U.S. Public Law 100-668. Washington Park Wilderness Act of November 16, 1988. 102 Stat. 3961.
U.S. Public Law 101-195. Nevada Wilderness Protection Act of December 5, 1989. 103 Stat. 1784.
U.S. Public Law 101-628. Arizona Desert Wilderness Act of November 28, 1990. 104 Stat. 4471.
U.S. Public Law 103-77. Colorado Wilderness Act of August 13, 1993. 107 Stat. 756.
U.S. Public Law 103-433. California Desert Protection Act of October 31, 1994. 108 Stat. 4473.
U.S. Public Law 106-399. Steens Mountain Cooperative Management and Protection Act of 2000 of October 30, 2000. 114 Stat. 1655.
U.S. Senate. 1969. Senate Report 91-97. 91st Congress, 1st Session. Washington, DC: U.S. Government Printing Office.
———. 1975. Senate Report 93-803. 93rd Congress, 2nd Session. Washington, DC: U.S. Government Printing Office. 47p.
———. 1980. Senate Report 96-914. 96th Congress, 2nd Session. Washington, DC: U.S. Government Printing Office. 64p.
———. 1984a. Senate Report 98-416. 98th Congress, 2nd Session. Washington, DC: U.S. Government Printing Office. 28p.
———. 1984b. Senate Report 98-463. 98th Congress, 2nd Session. Washington, DC: U.S. Government Printing Office. 37p.
———. 1984c. Senate Report 98-465. 98th Congress, 2nd Session. Washington, DC: U.S. Government Printing Office. 37p.
———. 1984d. Senate Report 98-580. 98th Congress, 2nd Session. Washington, DC: U.S. Government Printing Office. 16p.
———. 1984e. Senate Report 98-581. 98th Congress, 2nd Session. Washington, DC: U.S. Government Printing Office. 26p.
———. 1984f. Senate Report 98-614. 98th Congress, 2nd Session. Washington, DC: U.S. Government Printing Office. 27p.
Watson, A. E.; Patterson, M.; Christensen, N.; Puttkammer, A.; Meyer, S. 2004. Legislative intent, science and special provisions in wilderness: A process for navigating statutory compromises. *International Journal of Wilderness*. 10(1): 22–26.
Watson, Jay. 1996. Wilderness in the National Parks—now more than ever. *International Journal of Wilderness*. 2(1): 24.
Watson, Jay; Brink, Paul. 1996. The California Desert Protection Act: A time for desert parks and wilderness. *International Journal of Wilderness*. 2(2): 14–17.
Weaver, James; Cutler, Rupert. 1977. A colloquy between Congressman Weaver and Assistant Secretary Cutler. *Journal of Forestry*. 75(7): 392–394.
The Wilderness Society. 1974–75. The wilderness system: A report covering every existing or proposed wilderness. *The Living Wilderness*. 38(128): 38–47.
———. 2000. *The Wilderness Act Handbook*, 4th rev. ed. Washington, DC: The Wilderness Society.
Wilkinson, Charles F.; Anderson, H. Michael. 1987. *Land and Resource Planning in the National Forests*. Washington, DC: Island Press.
Worf, William A. 1980. Two faces of wilderness—a time for choice. *Idaho Law Review*. 16: 423–437.

CHAPTER

6

The National Wilderness Preservation System and Complementary Conservation Areas

INTRODUCTION

The Wilderness Act in 1964 (U.S. Public Law 88-577) created a National Wilderness Preservation System (NWPS) of areas protected in their natural condition at the primeval end of the environmental modification spectrum. The EMS concept, described in Chapter 1, is a continuum of environmental conditions from "the paved to the primeval." At one end of the continuum, cities and urbanized landscapes dominate; at the other end, more primeval conditions such as naturalness, solitude, and wildness characterize the setting. However, protection of areas with outstanding natural quality is not limited to the NWPS. Through a series of other federal and state laws and administrative actions, additional natural and wild areas also have been protected and managed to retain their unique natural qualities. For example, rivers designated as "wild" within the federal Wild and Scenic Rivers System (U.S. Public Law 90-542) are assured protection against dams and other developments along their free-flowing river segments.

Several states have also established state wilderness systems to protect state-owned wilderness areas that complement the federal program. Numerous but smaller areas of outstanding scientific value have been preserved as research natural areas (RNAs) under various administrative authorities. Protection of landscapes and biodiversity has increased with the support of private, nonprofit organizations such as The Nature Conservancy that own and manage reserves and often purchase sensitive flora and fauna land areas to hold for subsequent government purchase. Because these other areas may be near the primeval end of the environmental modification spectrum, it is important to understand how they complement and relate to the NWPS.

THE NATIONAL WILDERNESS PRESERVATION SYSTEM

Upon passage of the Wilderness Act, a core of 54 areas, totaling 9.1 million acres, was brought into the NWPS as "instant wilderness" (Fig. 6.1). The wilderness system grew slowly during the first few years following passage of the act. As discussed in Chapter 5, this was largely because it took the agencies time to develop procedures to carry out reviews of potentially suitable areas. During the first decade after passage of the Wilderness Act, the number of areas grew from 54 to 89 and the acreage from 9.1 million to slightly more than 11 million acres. By the end of the 1970s, the number of areas had doubled, to 118, as had the acreage, to 18.5 million.

In 1980, the NWPS experienced tremendous growth. Several omnibus bills designated a number of wilderness areas at once in individual states, but the system was expanded primarily by passage of the Alaska National Interest Lands Conservation Act (ANILCA; U.S. Public Law 96-487); that single act designated 35 units in Alaska and encompassed more than 56 million acres. In 1980 alone, 83 units were designated, covering more than 61 million acres, and the number of areas in the NWPS grew more than fourfold while acreage grew to nearly nine times that of the system created instantly in 1964.

The second major growth was 1984, when 223 units totaling 8.3 million acres were added. Many of these additions were in the eastern and southern United States, and the average size was small, about 37,000 acres. Only one area was added in 1985—the 13,300-acre Clifty Wilderness in Kentucky. With these additions, the NWPS almost reached a total of 89 million acres.

The third growth period was in 1990–1994 when numerous additions were designated with an emphasis on the southwestern U.S. The Arizona Desert Wilderness Act of 1990 (U.S. Public Law 101-628) added 39 units and over 1 million acres to the NWPS. The California Desert Protection Act of 1994 (U.S. Public Law 103-433) added 69 units and nearly 8 million acres to the system. The addition of these Arizona and California areas greatly increased the amount of low- and high-elevation desert environments in the NWPS, a component previously underrepresented.

The most recent growth period was in 2000 when 15 areas, totaling more than 1 million acres, were added in five new wilderness designation laws. The single largest addition was 10 areas in northwestern Nevada totaling 757,500 acres and managed by the BLM (U.S. Public Law 106-554). By 2023, the NWPS included more than 806 units (Table 6.1 shows 840 units because some areas are managed by one or more of the four federal agencies), covering more than 111.8 million acres.

The distribution of the NWPS among the four federal land-managing agencies is uneven. More than half of the wilderness areas are managed by the USFS, but the NPS manages more wilderness acreage. The BLM manages only a very small percentage of the NWPS acreage (8.9 percent), but its role expanded during the 1990s and more growth is possible.

The average wilderness-area size has increased greatly since 1964; this reflects the addition of some very large areas in Alaska, such as the Wrangell–Saint Elias Wilderness,

FIGURE 6.1 Oregon's Three Sisters Wilderness was one of the 54 U.S. Forest Service–managed "instant wildernesses" designated upon passage of the Wilderness Act of 1964. Photo courtesy U.S. Forest Service.

TABLE 6.1 Wilderness Units and Acres in the U.S. National Wilderness Preservation System (NWPS) by Federal Managing Agency in 2023 (www.wilderness.net, accessed June 2023)

Agency	Units[1]	Federal Acres *(millions)*	Percentage of NWPS acreage
Bureau of Land Management	260	10.0	8.9
U.S. Forest Service	448	36.7	32.8
U.S. Fish and Wildlife Service	71	20.7	18.5
National Park Service	61	44.3	39.7
Total	840	**111.8**	**100.0**

[1] The number of units managed by each agency does not equal the total because some areas are managed by more than one agency.

totaling 9.4 million acres, and the Noatak and Gates of the Arctic Wilderness Complex, which together total 12.9 million acres. In all, 52 percent of the total NWPS acreage was in Alaska by 2023.

The largest wilderness areas in the lower forty-eight states are Death Valley Wilderness in California and Nevada, managed by the NPS (3.2 million acres); the Frank Church–River of No Return Wilderness in Idaho, managed by the USFS and BLM (2.4 million acres); and the Selway-Bitterroot Wilderness in Montana and Idaho, managed by the USFS (1.3 million acres).

The total area of the NWPS (111.8 million acres) comprises approximately 4.5 percent of the entire U.S. land area. The bulk of the acreage is located in Alaska and the western United States; however, by 2023 some wilderness areas were located in all but six states: Connecticut, Delaware, Iowa, Kansas, Maryland, and Rhode Island.

The smallest wilderness is the five-acre Rocks and Islands Wilderness in California, administered by the BLM. As discussed in Chapter 4, the Wilderness Act defines a wilderness as containing at least 5,000 acres, but it qualifies this by stating "or is of sufficient size as to make practicable its preservation and use in an unimpaired condition." In all, more than 70 areas with fewer than 5,000 acres have been added to the NWPS and many, like 5.5-acre Pelican Island in Florida, are islands where protection of their wilderness character is possible, despite their size (Landres and Meyer 2000). Most wilderness areas (42 percent) are from 10,000 to 50,000 acres in size, even though the areas range from less than 100 acres to millions of acres (Landres and Meyer 2000).

The distribution of the NWPS is not relative to the percentage of the total U.S. land area and federal land area found in each of the four census regions of the U.S. (Table 6.2). The NWPS is located predominantly in the western United States and Alaska, and away from most of the nation's population. The western United States has only about 22 percent of the population, but more than 95 percent of the NWPS. The disparity between population and wilderness is most dramatic in Alaska; less than 1 percent of the U.S. population lives there, but the state contains 53 percent of the entire area of the NWPS. Wilderness makes up 16 percent of all land in Alaska and, of the federal lands in Alaska, 26 percent are in the NWPS (Table 6.2), reflecting the relatively limited development and settlement in Alaska. Conversely, less than 0.5 percent of the total land area in the south, midwest, and northeast regions is in the NWPS and less than 10 percent of the federal lands in those three regions is in the NWPS.

TABLE 6.2 Percentage of Total U.S. Land Base and Federal Land Area Designated as Wilderness by U.S. Census Region and Alaska, 2004 (Cordell et al. 2005a, 64)

Census Regions	Percentage of Total U.S. Land Area	Percentage of Federal Land Area
West (less Alaska)	5.8	12.1
Alaska	15.9	26.3
South	0.5	9.6
Midwest	0.3	5.7
Northeast	0.2	8.3

What does the future hold for further growth in the NWPS? Although the future growth is not clear, some general observations can be made. First, there is potential for growth in the number of areas and acreage in the NWPS. Large areas and numerous wilderness proposals and study areas remain to be acted on, especially in national parks and on BLM-administered land. Furthermore, many millions of acres more will likely undergo careful study for wilderness designation in Alaska. Second, although the growth of the NWPS has slowed, the newly added and potential additions may be especially valuable from various perspectives, such as ecological diversity. Finally, growth of the NWPS will be affected by decisions regarding the management of other lands. It appears likely that many additions will be made to the NWPS, perhaps 25 million acres (Hendee and Dawson 2004).

FEDERAL AGENCY MANAGEMENT IN THE NWPS

Since passage of the Wilderness Act, the role of the four federal agencies managing wilderness in the United States have changed with the growth of the NWPS.

National Forests

The number of wilderness areas and acreages designated in the national forests and, thus, under USFS administration has grown considerably since 1964. The USFS managed more than 36.7 million acres of wilderness by 2023 (Table 6.1), with wilderness representing an important management category on the national forests. Among their 448 wilderness units, the USFS shares jurisdiction on only a few units with the BLM, FWS, or NPS.

National Parks

Although wilderness was not an administrative designation used in the national parks before the Wilderness Act, it was an important concept underlying the establishment of many parks. It was informally defined as land 0.5 miles beyond any roads; many parks were predominantly de facto wilderness despite the lack of any formal administrative designation. Because no formal system of wilderness reservation like the USFS's primitive-area program existed in the national parks, no NPS lands were included in the "instant" wilderness system created in 1964. Over recent decades, national park representation in the NWPS has expanded. By 2023, there were 61 national park units in the NWPS covering more than 44.3 million acres (Table 6.1). The large NPS wilderness units of 1 million acres or more are in Alaska, California, or Florida. Several parks have large roadless areas recommended for wilderness designation and managed as backcountry, but they still await congressional action to add them to the NWPS (Crumbo 1996).

National Wildlife Refuge System

Passage of ANILCA (U.S. Public Law 96-487) in 1980 resulted in a vast expansion of the U.S. Fish and Wildlife

Service component of the NWPS. Under that legislation, more than 18 million acres of FWS land were added to the NWPS. By 2023, the FWS agency managed 71 units in the NWPS totaling more than 20.7 million acres, the majority of it in Alaska (Table 6.1).

Bureau of Land Management—Public Domain Lands

The BLM was not included in the Wilderness Act when it was passed in 1964. At that time, the BLM was primarily a custodial agency, charged with disposal of the public domain, and had no basic authority to manage public lands. However, a national review of public-land laws led to developing organic legislation giving BLM a mandate to manage the public lands under its jurisdiction, and it required the secretary of the interior to review areas of the public lands to determine if they had wilderness characteristics and to recommend to the president their suitability or unsuitability for preservation as wilderness by 1991.

Until final congressional decisions were made, such lands were to be managed so as not to impair their suitability for wilderness designation. By 2023, the wilderness areas legislatively designated and administered by BLM had grown to 260 units with a total of 10 million acres (Table 6.1).

ECOLOGICAL AND LANDSCAPE DIVERSITY WITHIN THE NWPS

The NWPS should be examined in terms of the kinds of lands that it contains, because a major goal, albeit implied, of the Wilderness Act was to help ensure ecological and landscape diversity within the nation's wildlands. Many supporters of wilderness have cited preserving a diverse, representative sample of ecological variety in the United States as a major reason for having the NWPS. However, ecological diversity has never been an explicit criterion driving the wilderness designation process, and as a result, the diversity within the system is largely accidental; there has been little systematic decision-making to broaden the NWPS's ecological coverage. One exception to this general neglect was in the USFS's second Roadless Area Review and Evaluation (RARE II), which used ecosystem representation as a criterion for evaluating potential wilderness additions.

Nevertheless, some efforts have been made to assess the extent to which the NWPS includes representatives of the nation's ecological diversity; the most complete analysis revealed major shortcomings in the 1980s. By merging Bailey's ecoregion concept and Küchler's mapping of potential natural vegetation, the United States was divided into 261 distinct ecosystems (Davis 1989). In 1989, 157 of the nation's 261 ecosystems (60 percent) were adequately represented within the NWPS, but representation was poor within the grasslands, deserts, eastern hardwoods, and coastal lowland ecosystems. The addition in the 1990s of extensive desert areas in Arizona, California, and Nevada improved the representation to some extent.

Cordell et al. (2005b) reported that by 2003, forty-three of the fifty-two different provinces in *Bailey's province-level ecosystem classification system* were represented to some degree in the NWPS. The extent or area protected in some provinces is small and the least well represented are in the eastern U.S., the prairies of the Midwest, the tropical forests of Puerto Rico, and the Hawaiian Islands (Cordell et al. 2005b). Overall, the fifty-two provinces in *Bailey's province-level ecosystem classification system* are organized into four domains; the percentage of total land area of the U.S. within each domain that is protected within the NWPS includes humid tropical domain (19.67 percent), polar domain (13.07 percent), dry domain (3.36 percent), and humid temperate domain (2.82 percent; Cordell et al. 2005b). Even with further designation, recommendations by the four federal wilderness management agencies to include more examples of each basic ecosystem under their jurisdiction, many ecosystems and provinces are still not adequately represented in the NWPS or in any equivalent state wilderness system.

COMPLEMENTARY CONSERVATION AREAS

In addition to formally designated wilderness, other kinds of conservation areas complement or supplement the purposes of the NWPS. Although some areas serve purposes similar to wilderness, others are quite different. Nevertheless, at least a portion of these other systems is located at the primitive end of the environmental modification spectrum, and their relationship to designated wilderness merits attention.

THE NATIONAL TRAILS SYSTEM

In 1968, Congress established a National Trails System (U.S. Public Law 90-543) to provide for the ever-increasing outdoor recreation needs of an expanding population and to promote public access to travel within, and enjoyment and appreciation of, the open-air, outdoor areas of the nation. The system includes four different types of trails: (1) national recreation trails, (2) national scenic trails, (3) national historic trails, and (4) connecting or side trails. Some of these trails are within wilderness or areas proposed for wilderness designation and pose a potential conflict to the extent that they may attract use inappropriate to wilderness.

National recreation trails are intended to provide a variety of outdoor recreation uses near urban areas. The secretary of the interior or the secretary of agriculture may establish a national recreation trail with the consent of any other jurisdiction whose lands would also be involved, such as a state or local government. Thus, trails on state lands may also be designated as national recreation trails by the secretary of the interior, with consent of the state. Criteria for designating national recreation trails have been adopted by the secretaries of the interior and agriculture. Such trails may be relatively short (perhaps one-half mile) but must be continuous. National recreation trails are to be available to large numbers of people; consequently, locations such as stream valleys, utility rights-of-way, abandoned railroad rights-of-way, and levees or dikes are likely candidates. They may be designed solely for one use, such as hiking or recreational vehicle use, but opportunities for multiple use are to be explored. Before designation of a trail, the administering agency must provide the appropriate secretary (interior or agriculture) assurance that the trail will be available to the public for at least ten consecutive years. By 2023, there were more than 1,300 national recreation trails spanning all 50 states and Washington, DC.

National scenic trails differ from recreation trails in several respects. They can be designated only by an act of Congress. U.S. Public Law 90-543 established two such trails: the 2,170-mile Appalachian Trail and the 2,650-mile Pacific Crest Trail. National scenic trails must possess superior scenic, historic, natural, or cultural qualities in combination with maximum outdoor recreation potential. They should avoid contact with developments such as transmission lines, highways, and industrial facilities; have adequate public access; and follow principal historic routes. Generally, they will be several hundred miles in length. Use of motorized vehicles on these trails is prohibited. By 2023, there were eleven national scenic trails authorized by Congress.

Portions of the National Trail System pass through some units of the NWPS. For instance, the Pacific Crest Trail runs through several wildernesses in the Cascade Range and Sierra Nevada, and the Continental Divide National Scenic Trail follows the crest of the Rocky Mountains and runs through wilderness areas in several states.

Managers of these areas are faced with conflicting objectives when national scenic trails pass through wilderness because wilderness is managed for naturalness and solitude; national scenic trails can attract many users whose primary interest is not in wilderness-related values, and whose impacts may dilute wilderness values. Generally, the more restrictive management standards for wilderness will prevail, but the conflicting objectives and the heavier visitor use attracted to national trails does create management problems. For example, where the Pacific Crest Trail passes high mountain lakes, these scenic alpine locations can become overused by hikers. The Appalachian National Scenic Trail traverses twenty-one designated wilderness areas in the eastern U.S. and the presence of huts along the Appalachian Trail poses management problems when the trail crosses wilderness, such as in the Great Gulf Wilderness on the White Mountain National Forest (Fig. 6.2).

National historic trails differ from scenic trails in that many partially follow roads or are "highway" trails for motorized traffic, with only some segments for hiking use. By 2023, there were nineteen trails authorized by Congress. Most of these trails do not cross into NWPS areas although some small historic segments do enter wilderness or are near wilderness, such as the Mormon Pioneer National Historic Trail, which passes near the North Maricopa Mountains Wilderness (BLM) in Arizona.

THE WILD AND SCENIC RIVERS SYSTEM

The Wild and Scenic Rivers Act of 1968 (U.S. Public Law 90-542) protected certain rivers throughout the nation to ensure their free and unimpaired flow and to preserve their outstanding scenic, recreational, geological, fish and wildlife, historic, and cultural values. Originally promoted as a means of countering federal dam-construction programs, the act evolved into an effort to limit development in general along rivers and their banks in the name of recreation (Tarlock and Tippy 1970).

FIGURE 6.2 Long-distance hikers take a break while backpacking on the Appalachian Trail (upper). The AT runs, in part, along the crest of the Presidential Range and along the upper boundary of the Great Gulf Wilderness (lower) on the White Mountain National Forest in Maine and New Hampshire. Photos courtesy Chad P. Dawson.

The act designated portions of eight rivers and adjacent lands as immediate members of the National Wild and Scenic Rivers System. Another twenty-seven rivers were identified for study within ten years for possible inclusion in the system, eighteen to be studied by the Department of the Interior and nine by the Department of Agriculture.

Designation of rivers under the Wild and Scenic Rivers Act can occur in two ways. First, specific congressional legislation can be enacted to protect a river or river segment. Second, a river can be added to the system by the secretary of the interior, provided certain nomination procedures are followed by the state in which the river is located, and provided there is no administrative cost to be borne by the federal government (Peters 1975).

The Klamath River in California and Oregon is an example of a river added under secretarial designations in 1981 and 1994. Six agencies, including the BLM, NPS, USFS, States of California and Oregon, and Hoopa Valley Indian Reservation, are responsible for management along 297 miles of the Klamath River designated as mainly recreational river, but with four river segments of wild and scenic river designation. The Allagash River in Maine was the first river designated in the national river system at the request of a state and with the state government then assuming primary management responsibility.

The Wild and Scenic Rivers Act recognizes three designations of rivers and provides guidelines for management (NPS and USFS 1982):

1. Wild Rivers—rivers or sections of rivers that are free of impoundments and generally inaccessible except by trail, with watersheds or shorelines essentially primitive and waters unpolluted. Many of these may border or be contained within wilderness areas.
2. Scenic Rivers—rivers or sections of rivers that are free of impoundments, with shorelines or watersheds still largely primitive and shorelines largely undeveloped, but accessible in places by roads.
3. Recreational Rivers—rivers or sections of rivers that are readily accessible by road or railroad that may have some development and agriculture along their shorelines, and that may have undergone some impoundment or diversion in the past.

Designating a river under the Wild and Scenic Rivers Act, combined with wilderness designation of the river and adjacent lands, could have very important complementary benefits (Fig. 6.3). Wilderness designation generally affects broad reaches of land and would protect the broader watershed of which the river is a part. Wild and scenic river designation, on the other hand, provides some important protection not afforded by the Wilderness Act. First, designation of a river segment as wild provides complete protection against dam construction and other water-development projects. Such facilities can be developed in designated wilderness if judged by the president to be in the public interest. Second, the Wild and Scenic Rivers Act prohibits constructing power transmission lines. Third, although the Wilderness Act does not allow condemnation of private land inholdings, the Wild and Scenic Rivers Act permits administering agencies to condemn private land, if less than 50 percent of the entire river is owned by federal, state, or local government. Land within a city, village, or borough cannot be condemned if valid zoning ordinances protecting the river areas are in effect. Finally, although lands in the NWPS were open to mineral entry until 1984, there was complete withdrawal from mineral entry of lands within 0.25 miles of the bank of any river designated for management under the wild category of the Wild and Scenic Rivers System (Tarlock and Tippy 1970). In general, the Wild and Scenic Rivers Act is more restrictive than the Wilderness Act, and in cases where the two laws overlap the more restrictive provisions prevail.

By July of 2022, 226 river segments had been added to the National Wild and Scenic Rivers System (Table 6.3), covering

FIGURE 6.3 Rafters on Idaho's Selway River, a wild and scenic river, ride heavy white water. The Wild and Scenic Rivers Act was originally intended to prevent dams on some of the nation's free-flowing streams; it became an instrument to limit development along river corridors and to protect the quality of water-oriented recreation. Photo courtesy U.S. Forest Service.

TABLE 6.3 River Mileage Designations for the U.S. National Wild and Scenic Rivers System in 2022 (NPS, July 2022)

River Designation	Classified River Mileage	Percentage by Miles
Wild	6,452.1	48.2
Scenic	2,971.4	22.2
Recreational	3,972.2	29.6
Total	**13,395.7**	**100.0**

more than 13,395 miles of river. The two states with the greatest mileage in the system are Alaska and Oregon.

Designating rivers as wild, scenic, recreational, or a varying combination of the three designation types plays an important role in preserving diverse wild river communities. Continued support and interest in protecting free-flowing wild rivers will ensure their preservation. Congress continually adds new river sections to the National Wild and Scenic Rivers System. For example, in 2000, six congressional acts designated 363 miles of river, with some designations in all three categories.

In addition to the National Wild and Scenic Rivers System, under authority of the federal government, 23 state river-protection systems were established by 1984. Nearly 200 rivers were protected under state legislation involving more than 5,800 miles and with additional river segments under study for possible future state designation. In 1984, nearly one river in five had some form of protective management under state legislation (Anderson and Morck 1984). Most state programs are fairly restrictive; they typically prohibit instream modification, establish land-use controls, and provide for the management of river use and users (Leatherberry et al. 1980).

STATE WILDERNESS SYSTEMS

The 1964 Wilderness Act pertains only to lands in federal ownership. Many states have also undertaken actions to preserve lands possessing wilderness qualities. This is an important step, for it represents another way to increase the geographic and ecological diversity of areas given wilderness protection. Many types of ecosystems simply are not found on federal lands; thus, even the most lenient approach to wilderness designation would fail to protect them. Where the states are important landowners, especially in the Midwest and in the East, state wilderness protection helps to expand ecosystem representation in protected areas and extend wilderness protection and stewardship to more lands. Moreover, by preserving areas as wilderness, state programs also help ensure provision of a kind of primitive-recreation opportunity that might otherwise not be available.

A national survey undertaken in 1983 examined state-level protection of wilderness (Stankey 1984). Eighty-nine agencies in the fifty states plus Puerto Rico, Guam, and the Virgin Islands were queried. To qualify for recognition as state wilderness, programs had to satisfy five criteria: (1) statutory or administrative recognition of the program; (2) provision for preserving natural qualities and for providing primitive-recreational opportunities; (3) prohibition of resource-development activities; (4) establishment of area size, as specific acreage qualitative description (how large an area must be to meet objectives); and (5) recognition of other values, such as features of historic or scientific interest, considered consistent with management as wilderness.

Nine states were found to have wilderness preservation programs meeting these criteria (Stankey 1984). In addition, three states had designated areas for wilderness protection and purposes but did not meet all the criteria. Most states have modeled their wilderness programs on the federal legislation. However, some important differences exist. For example, in some states, such as Alaska, wilderness was a zoning designation applied in 1984 to units of the state park system. In California, the state legislature can designate wilderness but, in lieu of action by that body, proposals can be brought before the California Park and Recreation Commission (Trumbly and Gray 1984).

By 1994, there were eight state wilderness programs still operating from the original nine studied by Stankey in 1984. Florida had its wilderness legislation repealed when it came up for reauthorization in 1989 and the ten wilderness areas were transferred to other state land-management programs (Peterson 1996).

In 2007, Propst and Dawson (2008) used the five criteria developed by Stankey (1984) and added a sixth criterion: a state had a wilderness program provided it had developed management plans to formally define wilderness area objectives that guided managers in activities and decision making. The 2007 study reported that the number of states had declined to seven state wilderness programs from 1984 (Florida had withdrawn its program and Minnesota did not qualify as having a wilderness program due to the sixth criterion); however, the number of designated state wilderness areas had increased from forty-eight to eighty-four areas and the total acreage had increased from 1.7 million to 2.9 million acres (Table 6.4), with important progress made in all seven states. About 96 percent of the acreage was in New York, California, and Alaska. Additionally, Propst and Dawson (2008) reported that they located five states (including Minnesota) that had one or two wilderness areas, but no formally defined wilderness program, and they included seven areas with 275,677 acres.

State activity in wilderness preservation represents an important complementary activity to federal efforts. Although federal legislation has set the general direction in defining wilderness and setting management guidelines, the states have adopted and modified these notions to apply to their situation. In addition, many states are actively engaged in other programs designed to protect the quality of their natural heritage through a variety of protected area designations other than wilderness designation. To see how state wilderness programs function, the California and New York systems are described in more detail (Dawson and Thorndike 2002).

California

In 1974, the California legislature established the California Wilderness Preservation System. Three basic criteria govern admission to the system: (1) the land must be state-owned;

TABLE 6.4 States with Wilderness Programs or States with One or Two Wilderness Areas, But No Program, in 2007 (Propst and Dawson 2008)

State	Year Established	Number of Areas	Total Acreage
States with Wilderness Programs			
Alaska	1972	5	1,133,400
California	1974	11	475,725
Maryland	1971	30	43,733
Michigan	1972	4	57,733
Missouri	1977	11	22,993
New York	1972	22	1,214,217
Wisconsin	1973	1	5,939
Subtotal		84	2,953,740
States with Wilderness Areas, but No Program			
Minnesota	1975	1	18,000
Hawaii	1981	2	30,857
Maine	1966	2	204,733
Ohio	1988	1	8,000
Oklahoma	1918	1	14,087
Subtotal		7	275,677
TOTAL		**91**	**3,229,417**

(2) the area must remain in, or have been returned to, or have substantially reestablished its principal, natural character and influence; and (3) the area must be of sufficient size to make its preservation practicable.

Although the system is legislatively founded, new areas can be added either administratively or legislatively. Legislatively established areas are called wilderness areas and are fully protected by law. State wildernesses also can be administratively designated on lands in the state park system by the California Park and Recreation Commission, a body appointed by the governor and overseeing management of the state park system. Although both types of areas are subject to protective management requirements of the act, those established by administrative designation can be reclassified and removed from the state wilderness system without approval of the legislature (Trumbly and Gray 1984).

The 1974 legislation created two wilderness areas: the Santa Rosa Mountains Wilderness Area, of about 87,000 acres, and the 10,000-acre Mount San Jacinto Wilderness Area, abutting the federally designated San Jacinto Wilderness. By 2007, the California State Wilderness System contained more than 475,725 acres (Propst and Dawson 2008).

Although the California Wilderness Act closely parallels the federal legislation, there are some important differences. The California legislation explicitly states that wilderness areas need not be pristine but only "substantially restored to a near natural appearance" (California Public Resources Code 5093.30). It also contains language prohibiting aircraft flights below an altitude of 2,000 feet; no such prohibition is contained in the federal law. Special provisions permit the unobtrusive use of equipment for collecting hydrometeorological data and conducting weather modification activities; this latter activity in particular is one that has raised concerns with regard to its potential impact on wilderness. Finally, the California law prohibits any vehicle use within wilderness except for emergencies involving the health and safety of individuals. The Wilderness Act of 1964 is more permissive in this regard, permitting permanent roads and access if they are determined to be necessary for the minimum requirements for administration of the area as wilderness.

The California Wilderness System is small compared to the federal wilderness acreage in the state (475,725 acres in state ownership by 2007 as compared to 14.3 million acres administered by federal agencies in 2007), but the types of areas complement those in the NWPS. Much of the land in the state system is in the Anza-Borrego Desert State Park (Fig. 6.4); other areas are along the coastline and nearby coast ranges. The state wilderness system adds to the broad diversity of ecosystems under wilderness protection.

New York

The wilderness preservation movement in New York state began in 1885 with legislation to create forest preserve lands. Numerous areas within the New York system had been logged, and there was concern about creating watershed protection and maintaining a forested landscape. The citizens of the state passed a referendum in 1894 to add constitutional protection to the forest preserve lands set aside within the Adirondack and Catskill Mountains. The most often-quoted portion of the legislation is Article XIV, which, in part, states: "The lands of the state, now owned or hereafter acquired, constituting the forest preserve as now fixed by law, shall be forever kept as wild forest lands. They shall not be leased, sold or exchanged, or be taken by any corporation, public or private, nor shall the timber thereon be sold, removed or

FIGURE 6.4 A palm oasis in the Borrego Palm Canyon wilderness unit benefits from protection as part of the California State Wilderness Act of 1974. The unit is one of twelve in the 386,000 acres of designated wilderness in Anza-Borrego Desert State Park. Photo courtesy Marilyn Hendee.

FIGURE 6.5 Lake Colden and Avalanche Pass, in the High Peaks Wilderness of New York's Adirondack Mountains, are heavily used. The High Peaks Wilderness is the largest wilderness in the New York state system, at more than 226,400 acres, and it has the greatest recreation use per year, with 140,000 visits. Photo courtesy Chad P. Dawson.

destroyed." The state-owned lands within the Adirondack and Catskill forests were subsequently termed the "Forest Preserve." These lands, in combination with extensive private landholdings, were established as regional planning and management areas labeled the Adirondack and Catskill Parks.

The specific designation of some of the Adirondack Forest Preserve lands as "wilderness" was first proposed by the state legislature in 1960 and finally adopted in 1972. By 2019, there were 21 wilderness management units (including the St. Regis Canoe Area) in the Adirondack Forest Preserve, totaling more than 1.2 million acres (Fig. 6.5). In 1985, four wilderness units in the Catskill Forest Preserve were created by state agency action and, by 2014, there were five units totaling more than 142,000 acres.

The New York state definition of wilderness is nearly identical to the federal wilderness definition, except New York state requires a minimum size of 10,000 acres. Some of the guiding management phrases, in both the New York state and national wilderness definitions, relate to the resources and environment being of "primeval character" and manifesting only "natural conditions" within the wilderness. The guiding management phrases for recreation management are providing "outstanding opportunities for solitude" and providing for a "primitive and unconfined type of recreation." Management approaches for recreation visitors and resources are similar in New York state to those of federal agencies and include many parallel planning and management activities for wilderness areas. New York state allows few nonconforming uses or facilities compared to the federal system.

RESEARCH NATURAL AREAS

In 1966, the Department of the Interior, as part of the United States' participation in the International Biological Program (IBP), established the Federal Committee on Research Natural Areas. The committee was composed of representatives of the major federal land-management agencies along with liaison representatives from the Department of Defense, Atomic Energy Commission, and Tennessee Valley Authority. Its purpose was to inventory and prepare a directory of natural areas on federal lands.

Research natural areas (RNAs) are related to wilderness because one of their key objectives is to maintain natural processes and because they serve an important research and education role. Their specific objectives are:

1. To assist in preserving examples of all significant natural ecosystems for comparison with those influenced by humans.
2. To provide educational and research areas for scientists to study the ecology, successional trends, and other aspects of the natural environment.
3. To serve as gene pools and preserves for rare and endangered species of plants and animals (Federal Committee on Research Natural Areas 1968).

Moir (1972) has identified six basic characteristics of research natural areas:

1. They are examples of the natural environment.
2. Their natural features have been disturbed as little as possible by humans.
3. They are defined by ecological criteria.
4. They are assured the greatest possible degree of preservation and permanency.
5. Their withdrawal is for scientific and educational purposes.
6. They harbor genetic stock of possible value to society.

RNAs in the United States are administered by federal land-managing agencies and range in size from less than 1 acre to more than 100,000 acres. However, given that one purpose of the RNA system is to preserve a representative array of the nation's ecosystems, much remains to be done (Burns 1984). For example, 570 areas were managed by the USFS, 192 areas managed by BLM, and 210 areas managed by USFWS.

Most RNAs are surrounded or buffered by federal land. Research conducted on these areas must be essentially nondestructive and consistent with the purpose and character of the surrounding land. Studies involving manipulating the environment are generally not permitted.

Recreational use of RNAs is, by definition, limited and subordinate to the scientific and educational objectives for these areas. They nevertheless do serve as important recreational settings for activities that focus on learning and environmental awareness. More important, they supplement the scientific and educational objectives of the NWPS.

BIOSPHERE RESERVES AND WORLD HERITAGE SITES

In Chapter 3, international progress in protecting wilderness values was discussed and the authors noted that biosphere reserves and world heritage sites were two important sources of such protection. In the United States, both designations have been used in conjunction with other conservation classifications, such as national parks, to further recognize and protect important natural values. Like the other types of designations we have discussed, *biosphere reserves and world heritage sites represent another mechanism through which wilderness values can be protected. Because biosphere reserves have no legal standing, the protection they confer is through the international recognition that such designation conveys.*

The Biosphere Reserve Project was established in 1973 under the auspices of the United Nations Educational, Scientific, and Cultural Organization (UNESCO). *The basic objective of the program is to foster developing an international network of representative ecosystems* for use in research and education and to promote improved land-management practices. *Biosphere reserves* typically include portions where little or no human impact has occurred, known as a core zone, as well as areas of human use and impact, referred to as multiple-functions zones and cultural zones.

In the United States, the Biosphere Reserve Project is jointly coordinated by the NPS and the USFS and the first areas were established in 1976. *By 2023, 47 areas in the United States had received recognition as biosphere reserves.* Portions of several reserves are also designated as wilderness, as, for example, in the Noatak National Arctic Range in Alaska where about one-half of the nearly 6-million-acre wilderness is recognized as a biosphere reserve. Nearly 81,000 acres of the 285,000-acre Three Sisters Wilderness in Oregon is similarly classified. The Admiralty Island Wilderness and Glacier Bay Wilderness have been classified as part of the Glacier Bay–Admiralty Island Biosphere Reserve.

World heritage sites, on the other hand, are products of the International Convention for the Protection of the World Cultural and Natural Heritage (World Heritage Convention) adopted by the UNESCO Geneva Convention in 1972. Rather than focusing on "representative ecosystems" as the Biosphere Reserve Project does, *areas qualifying for world heritage site listing must contain outstanding or superlative qualities of the natural environment.* Management of world heritage sites focuses on the protection of these outstanding qualities. In some cases, areas may be recognized by both designations; however, both of these international-level conservation classifications are typically applied in conjunction with an existing designation in the country where the area is located (e.g., national forest, national park).

By 2023, a total of 24 areas in the United States had been recognized as world heritage sites. Several world heritage sites overlap areas where wilderness designations are also in effect, such as Yosemite National Park in California. Some areas in the United States are both biosphere reserves and world heritage sites, such as Everglades National Park, Great Smoky Mountains National Park, and Yellowstone National Park.

The selection of these areas as biosphere reserves, world heritage sites, or both reflects international recognition of their value as repositories of natural values, as sources of knowledge and understanding about natural processes, and as contributors to society by providing a source of pleasure and inspiration. Such designations represent another complementary form of recognition and, thus, protection of wilderness values.

PRIVATE LAND PROTECTION EFFORTS

The Nature Conservancy (TNC) is one of the largest and best-known private organizations that protect ecosystems and areas that would otherwise be lost to development. Its origins trace from 1917 when a group of scientists, under the auspices of the newly established Ecological Society of America, formed a committee for the preservation of natural environments. The early work of this group led to publication of *The Naturalists Guide to the Americas*, detailing the known outstanding and unprotected natural areas of North and South America. Later, in 1950, the Ecologists Union, one of the organizations arising from the Ecological Society, took the name "The Nature Conservancy" and launched a program to achieve better protection of wildlife and plants, including an effort to attain nonprofessional participation (Blair 1986).

Using a program of revolving funds—using grants and other monies to acquire land followed by fundraising efforts to replenish the fund—TNC has become an important force in protecting natural environments; *it manages the largest privately owned group of nature sanctuaries in the world with professional managers who are assisted by volunteers.*

Because a major goal of TNC is to preserve natural diversity, the organization has developed four objectives: (l) *identifying* which endangered natural systems and species have not been preserved; (2) *protecting* the best areas where they are found; (3) *stewardship*, managing the protected sites; and (4) *funding*, repaying the amount spent from the revolving fund.

Since the early 1950s, TNC has been involved in protecting more than 12 million acres of habitat, either individually or with partnerships of government, private landowners, and corporations in the United States (Blair 1986). By 2023, TNC owned and managed more than 1,500 preserves, throughout the U.S., which comprises more than 9 million acres. In most cases, TNC has acquired critical habitats through purchases, gifts, and public-private partnerships. The organization also relies on volunteers who assist professional managers.

Areas under TNC management are managed primarily to protect natural values. Nevertheless, certain uses are allowed, including recreation. However, great care is taken to ensure that such use and supporting facilities like trails do not conflict with critical species or natural communities.

SUMMARY

After a period of relatively slow growth following passage of the Wilderness Act, the National Wilderness Preservation System has grown rapidly. As the discussion in this chapter indicates, many key ecosystems are still not represented in the NWPS—but working toward representation of all ecosystems is nevertheless a worthwhile goal.

It remains clear, however, that wilderness alone cannot supply all the environmental values and recreational opportunities the public seeks at the primitive end of the environmental modification spectrum. Other roadless, non-wilderness settings must also be provided to more adequately meet public desires. The expansion of the National Trails System, the Wild and Scenic Rivers System, and state wilderness systems, as well as the provision of roadless recreation areas managed intensively for recreation, are needed to adequately cope with growing public interest and demand. The quality of tomorrow's wilderness will depend as much on our success in fully developing these alternative opportunities as on our achievements in developing and implementing innovative wilderness management programs.

STUDY QUESTIONS

1. In what important ways do state wilderness systems complement the federal wilderness system?
2. In what ways do programs such as the National Trails System and the Wild and Scenic Rivers System complement the NWPS? In what ways do they conflict with the NWPS? How might some of these conflicts be resolved?
3. What additional protection can wild and scenic river designation provide for a river in wilderness that is not provided by wilderness designation? Which law is more restrictive and which provisions prevail?
4. In what ways are the goals of the research natural area (RNA) program similar to those of the NWPS? In what ways do they differ?
5. Compare and contrast the biosphere reserve and world heritage site programs. What similarities do they share; how do they differ? Discuss the relationship between both programs and the goals of the NWPS.
6. Explain how biosphere reserve and/or world heritage recognition can help protect a site or area.

CASE DISCUSSION: RECREATIONAL USE ON THE COLORADO RIVER IN GRAND CANYON NATIONAL PARK

The U.S. Congress established the Grand Canyon National Park (GCNP) in 1919 in northwestern Arizona and then enlarged it to 1.2 million acres in 1975. The GCNP was created for resource protection and enjoyment of the public. The GCNP includes 277 miles along the Colorado River. The most notable features of the GCNP are the spectacular landscape vistas that include the extensive and spectacular geological formations of the inner and outer Colorado River canyon and the numerous side canyons and waterfalls. While the rim areas are vegetated by conifer forests, the animal and plant communities of the canyon form a desert ecosystem intersected by a narrow riparian habitat along the river in the narrow inner canyon. The canyon system includes numerous historical and archeological features and sites.

While the GCNP is most often visited by those who come to look at the geological formations from the canyon rim or to hike down in the canyon, the river has had documented recreational boating use since the 1930s. Prior to 1960, recreational use included regularly scheduled motorized river trips and professionally outfitted trips as well as private, noncommercial trips on the river. The primary motorized and nonmotorized boating use occurs along a 226-mile segment from Lees Ferry to Diamond Creek.

The Wilderness Act of 1964 required a survey of all federal lands to evaluate and identify lands that could be recommended for possible designation as wilderness by Congress. The GCNP was included in the NPS review of these lands and was subsequently recommended for wilderness designation. Thus, the GCNP was required under the Wilderness Act to be managed as wilderness to preserve its unimpaired wilderness character and values in anticipation of its potential designation. The act in section 4(c) prohibits certain uses, including motorized use and mechanical transport. However, section 4(d) allows some exceptions that include nonconforming uses, such as previous existing motorboat use and commercial outfitting and guiding activities related to wilderness travel.

The issue of whether the inner canyon and the Colorado River should be designated by Congress as wilderness has been a pivotal debate that has kept any of the land area in the GCNP from yet being designated wilderness. A variety of wilderness designation proposals and plans have been developed by the National Park Service since 1970; some have recommended phasing out motorized use and others have not. Some have recommended that the river corridor be designated wilderness and others have recommended that it only be included as potential wilderness (due to motorized use). A plan to manage the Colorado River for recreational use has been developed, revised, and challenged in court by river user groups and preservationists. The various proposals to designate the majority (more than 90 percent) of the GCNP, about 1.1 million acres, as wilderness have stalled due to controversy over the recreational motorized use on the Colorado River through the GCNP. The GCNP remains in the category of potential wilderness.

The current version of the "Colorado River Management Plan" (CRMP) was signed in 2006 and allows motorized boat use on the Colorado River with regulations in four zones for the amount and type of use permitted (U.S. Department of the Interior 2006). The largest zone is along the popular 226-mile segment from Lees Ferry to Diamond Creek and it permits motorized and nonmotorized use by both commercial and private (noncommercial) user groups (Fig. 6.6). Each of the four zones has different regulations for these user groups.

The main season in Zone 1 (the longest and most heavily used zone) is 5.5 months, with another 6.5 months for nonmotorized use. The maximum group size allowed varies between eight and thirty-two participants per group and the maximum number of groups allowed to launch per day varies from one to six every other day, with no more than sixty groups (985 passengers) on the river at one time. Maximum trip length varies from ten to twenty-five days depending on type of group and season of the year. An estimated 24,657 users spend 228,986 user days on the river each year. The numbers of allowable total annual recreational users, launches, and user days have increased substantially since the previous CRMP plan in 1989 (for example, 169,950 user days vs. 228,986 user days) (Fig. 6.7).

FIGURE 6.6 Launching at Lee's Ferry to start the trip down Zone 1 on the Colorado River in Grand Canyon National Park. Photo courtesy of the National Park Service.

FIGURE 6.7 Rafting on the Colorado River in Grand Canyon National Park. Photo courtesy Brian Dawson.

CASE DISCUSSION QUESTIONS

1. What are the pros and cons of designating the GCNP as wilderness and *prohibiting motorized boat use* on the Colorado River? Which wilderness values are protected and which are not protected?
2. What are the pros and cons of designating the GCNP as wilderness and *allowing motorized boat use* on the Colorado River? Which wilderness values are protected and which are not protected?
3. Discuss which types of wilderness visitor or user are affected by the two different approaches to wilderness designation—such as visitors looking into the canyon from the drive-up sites on the rim, backpackers hiking into the inner canyon to camp, river runners, and hikers along the canyon rim. Compare how they would each react to what you reported for the previous two questions.
4. Should the majority of the GCNP be designated wilderness and the river corridor excluded from wilderness designation so that motorized use can continue? How would that proposal affect the wilderness experience of wilderness users in the rest of the GCNP? Are there adverse conditions created for wilderness experiences or not?

REFERENCES

Anderson, Dorothy H.; Morck, Victoria L. 1984. The state of federal river recreation management. In: Popadic, Joseph S.; Butterfield, Dorothy I.; Anderson, Dorothy H.; Popadic, Mary R., eds. *Proceedings of the 1984 National River Recreation Symposium;* October 31–November 3, 1984; Baton Rouge, LA. Baton Rouge: Louisiana State University, School of Landscape Architecture, pp. 466–473.

Blair, William D., Jr. 1986. The Nature Conservancy: Conservation through cooperation. *Journal of Forest History.* 30: 37–41.

Bureau of Land Management, Federal Highway Administration, National Endowments for the Arts, National Park Service, USDA Forest Service. 2005. *National Historic and Scenic Trails: Accomplishments 2001–2005.* Washington, DC: Bureau of Land Management.

Burns, Russell M. 1984. Importance of baseline information to the research natural area program. In: Johnson, Janet L.; Franklin, Jerry F.; Krebill, Richard G., coords. *Research Natural Areas: Baseline Monitoring and Management; Proceedings of a symposium*; March 21, 1984; Missoula, MT. General Technical Report INT-173. Ogden, UT: U.S. Department of Agriculture, Forest Service, Intermountain Forest and Range Experiment Station, pp. 50–52.

Cordell, H. Ken; Murphy, D.; Riitters, K.; Harvard, J. E., III. 2005a. The human context and natural character of wilderness lands. In: Cordell, H. Ken; Bergstrom, J. C.; Bowker, J. Michael. *The multiple values of wilderness.* State College, PA: Venture Publishing, pp. 57–89.

———. 2005b. The natural ecological value of wilderness. In: Cordell, H. Ken; Bergstrom, J. C.; and Bowker, J. Michael. *The multiple values of wilderness.* State College, PA: Venture Publishing, pp. 205–249.

Crumbo, Kim. 1996. Wilderness management of Grand Canyon—"Waiting for Godot." *International Journal of Wilderness.* 2(1): 19–23.

Davis, George D. 1989. Preservation of natural diversity: The role of ecosystem representation within wilderness. In: Freilich, Helen R., comp. *Wilderness Benchmark 1988: Proceedings of the National Wilderness Colloquium*; January 13–14, 1988; Tampa, FL. General Technical Report SE-51. Asheville, NC: U.S. Department of Agriculture, Forest Service, Southeastern Forest Experiment Station, pp. 76–82.

Dawson, C. P.; and Thorndike, P. 2002. State-designated wilderness programs in the United States. *International Journal of Wilderness.* 8(3): 21–26.

Federal Committee on Research Natural Areas. 1968. *A Directory of Research Natural Areas.* Washington, DC: U.S. Government Printing Office.

Hendee, J. C.; Dawson, C. P. 2004. Wilderness progress after forty years under the U.S. Wilderness Act. *International Journal of Wilderness.* 10(1): 4–7.

Landres, Peter B.; Meyer, S. 2000. *National wilderness preservation system database: Key attributes and trends, 1964 through 1999.* RMRS-GTR-18. Rev. ed. Ogden, UT: U.S. Department of Agriculture, Forest Service, Rocky Mountain Research Station.

Leatherberry, Earl C.; Lime, David W.; Thompson, Jerrilyn LaVarre. 1980. Trends in river recreation. In: *The 1980 National Outdoor Recreation Trends Symposium;* April 20–23, 1980; Durham, NH. General Technical Report NE-57. Broomall, PA: U.S. Department of Agriculture, Forest Service, Northeastern Forest Experiment Station, pp. 147–164.

Moir, William H. 1972. Natural areas. *Science.* 177(4047): 396–400.

National Park Service; U.S. Forest Service. 1982. Wild and Scenic Rivers Guidelines; National Wild and Scenic Rivers System; Final Revised Guidelines for Eligibility, Classification and Management of River Areas. *Federal Register.* 47(173): 39,454–39,461.

Peters, Clay E. 1975. A national system of wild and scenic rivers. *Naturalist.* 26(1): 28–31.

Peterson, M. R. 1996. Wilderness by state mandate: A survey of state-designated wilderness areas. *Natural Areas Journal.* 16(3): 192–197.

Propst, B. M.; Dawson, C. P. 2008. State-designated wilderness in the United States: A national review. *International Journal of Wilderness.* 14(1): 19–24.

Stankey, George H. 1984. Wilderness preservation activity at the state level: A national review. *Natural Areas Journal.* 4(4): 20–28.

Tarlock, A. Dan; Tippy, Roger. 1970. The Wild and Scenic Rivers Act of 1968. *Cornell Law Review.* 55(5): 707–739.

Trumbly, James M.; Gray, Kenneth L. 1984. The California Wilderness Preservation System. *Natural Areas Journal.* 4(4): 29–35.

U.S. Department of the Interior, National Park Service. 2006. Colorado River Management Plan. Grand Canyon, AZ: Grand Canyon National Park.

U.S. Public Law 88-577. The Wilderness Act of September 3, 1964. 78 Stat. 890.

U.S. Public Law 90-542. Wild and Scenic Rivers Act of October 2, 1968. 82 Stat. 906.

U.S. Public Law 90-543. National Trails Act. Act of October 2, 1968. 82 Stat. 919.

U.S. Public Law 96-487. Alaska National Interest Lands Conservation Act (ANILCA) of December 20, 1980. 94 Stat. 2371.

U.S. Public Law 101-628. Arizona Desert Wilderness Act of November 28, 1990. 104 Stat. 4471.

U.S. Public Law 103-433. California Desert Protection Act of October 31, 1994. 108 Stat. 4473.

U.S. Public Law 106-554. Black Rock Desert–High Rock Canyon Emigrant Trails National Conservation Area Act of December 21, 2000. 114 Stat. 2763.

III. Wilderness Management and Planning Concepts

CHAPTER

7

Principles of Wilderness Management

INTRODUCTION

Wilderness management is complex. When a problem arises, many solutions are possible; but seldom is there a single, unequivocal answer at hand. Instead, managers must devise and choose from alternative solutions, but it is important that the decision-making rationale produce solutions that are compatible with the wilderness idea. *This chapter offers a set of principles—fundamental assumptions and directions—that will help managers make consistent wilderness stewardship decisions.*

These thirteen principles can guide managers in solving day-to-day problems and in creating long-term policies. The principles offer perspectives on the nature of the wilderness resource, its uses, and its place in the spectrum of land uses. The list does not cover every situation, and managers may want to add a principle or two of their own. Nevertheless, applying these principles will usually result in appropriate action. The principles will often be cited and/or applied to specific management situations covered in subsequent chapters.

These thirteen principles have been developed with the U.S. wilderness system in mind, but many of them may also apply to wilderness in other countries.

PRINCIPLE 1: MANAGE WILDERNESS AS THE MOST PRISTINE EXTREME ON THE ENVIRONMENTAL MODIFICATION SPECTRUM

The concept of the environmental modification spectrum describes a continuum of settings that range from the "paved to the primeval" (Nash 1982); that is, from the totally modified landscape of modern cities to the most remote, pristine, and wild rural locations. U.S. society encouraged Congress, through the Wilderness Act of 1964 (U.S. Public Law 88-577), to preserve selected areas at the undeveloped and most pristine end of the environmental spectrum as part of the National Wilderness Preservation System (NWPS): Wording in the act describes its intent as "to assure that an increasing population, accompanied by expanding settlement and growing mechanization, does not occupy and modify all areas within the United States . . . leaving no lands designated for preservation and protection in their natural condition."

Several other countries (see Chapter 3) and U.S. state legislatures (see Chapter 6) have also decided by law or administrative policy that their most remote and pristine remaining areas should be protected as wilderness.

Principle 1 recognizes that lands range from the most pristine wilderness to those that are successively more modified than wilderness—from roadless but non-wilderness lands to wildlands with roads set aside for park, recreation, or scenic purposes; multiple-use lands serving resource production goals as well as public recreation, cultivated rural lands; and so on. A variation of this continuum is known as the recreation opportunity spectrum (ROS) and provides a guiding framework for land-use planning for recreation.

In the environmental modification spectrum, the wildest lands are in a relatively undisturbed condition, and the attributes of naturalness and solitude are indicative of wilderness (Fig. 7.1). Uses that alter naturalness and solitude reduce wildness and thereby reduce the total spectrum of environmental conditions, opportunities, and ecosystem services available to meet public interests, needs, and desires. Such uses also threaten to erode the threshold along this spectrum that separates wilderness from other lands. Thus, this principle mandates that a fundamental objective of wilderness management is to maintain the distinctive qualities that define and separate wilderness from other land uses—the relative naturalness and solitude (the wildness) of designated wilderness compared to adjacent lands.

To achieve this fundamental objective, we need to resist pressures to modify environmental conditions in wilderness. Wilderness simply cannot, and should not, meet all of the demands that might be placed on it. To do so would directly violate provisions of the Wilderness Act and lead to a loss of those environmental qualities that prompted passage of the act in the first place; that is, the naturalness and solitude—the wilderness conditions—that such areas offer.

Yet it is impossible to protect these distinguishing attributes of wilderness, in the long run, unless other opportunities along the environmental modification spectrum are also provided as alternatives so that commodity and recreation demands not requiring wilderness qualities can be met. For example, some non-wilderness but very wild areas may have some wilderness-like qualities that can meet some recreation demands and ecosystem services otherwise destined for wilderness, but they also allow for limited environmental changes and more intensive use and management. Such areas could thus accommodate many primitive-recreation uses outside of wilderness—thereby relieving pressures on wilderness.

FIGURE 7.1 Two mountaineers enjoy solitude and the primeval character of Wolf Peak, in Montana's Absaroka-Beartooth Wilderness. Photo courtesy Richard Behan.

A very controversial alternative proposed in 2000 by the U.S. Forest Service, under the title of "New Wilderness Recreation Strategy," for the national forests, would relax limiting use in some heavily used wilderness zones but maintain higher standards of naturalness and solitude in more remote and less visited zones (Oye 2001). Opponents of this view claim that it violates the Wilderness Act and the nondegradation principle of maintaining each area's wilderness standards existing at the time they were designated and improving naturalness and solitude in areas below a minimum standard (Worf 2001). Others supported the new strategy, claiming that it recognizes use levels already existing in popular locations and politically impossible to roll back, and that physical impacts, not solitude, should govern use levels. "If we lock people out who will fight to save the wilderness" is an expression of this view (Spring 2001, 17).

We support a careful, targeted implementation of such a wilderness recreation strategy, while recognizing the merits of both pro and con arguments. However, some special response

FIGURE 7.2 Installing, maintaining, and monitoring wilderness boundary protects wilderness thresholds. A "No Vehicles" sign marks road's end in the Kofa Wildlife Refuge, Arizona. Photo courtesy Marilyn Hendee.

is needed in areas where there is unusually high demand for wilderness visitation and there are few alternatives (Cole 2001). The challenge will be in holding the line against physical impacts on naturalness, even while relaxing standards of solitude in popular, high-use areas. The strategy will only be worthwhile if it allows protection of high standards of naturalness and solitude in the surrounding wilderness—thereby maintaining overall the threshold that distinguishes wilderness as the most pristine extreme on the environmental modification spectrum (Fig. 7.2).

PRINCIPLE 2: MANAGE WILDERNESS COMPREHENSIVELY, NOT AS SEPARATE PARTS

To label something as a resource is to say that it normally is useful to society, that it meets a vital need, and usually that it has economic value. Most of our wildland resources—timber, forage, water, and minerals—can be appraised in this fashion, but a few noncommodity resources, like air and especially wilderness, cannot.

Wilderness has only recently acquired value in our culture, and increasingly, we are recognizing economic values derived from protected and natural areas such as clean air, water, sequestered carbon, and other benefits.

Wilderness has clearly achieved the status of a resource in our society by virtue of its cultural values and ecosystem services, and the Wilderness Act refers to "an enduring resource of wilderness."

From a management standpoint, one important attribute of the wilderness resource is the natural relationship among all its ecosystem parts: vegetation, water, forage, wildlife, and geology. Because wilderness is a composite resource with interrelated parts, its management must be focused on the whole and not on its parts. For wilderness, therefore, one should not develop separate management plans for vegetation, wildlife, or recreation. Rather, wilderness planning must deal comprehensively with the interrelationships between and among these and all other components of the wilderness resource.

Criteria for managing and controlling the use of wilderness should be based on maintaining natural relationships. For instance, fishing regulations might be guided by effects of anglers on shoreline vegetation and soil and the solitude afforded visitors, as well as by the impact on the fish populations. Likewise, wilderness recreation might be managed or controlled not only for site capacity, but also for the capacity of vegetation, wildlife distribution and behavior, and water quality to maintain their natural relationships. Recognizing the need to manage wilderness comprehensively is the foundation of ecosystem management. The idea is that if one part of the ecosystem is allowed to go out of balance, the rest will quickly follow. By strictly maintaining naturalness and solitude in wilderness through truly comprehensive, ecosystem-based management, it becomes possible to identify ecosystem benchmarks.

If wilderness is viewed as a resource, what can be said about its renewability? Clearly, commercial exploitation—logging, mining (both mineral and hydrocarbon development)—would temporarily destroy what we define as the wilderness resource. However, just as clearly, the passage of time, coupled with restrictions on further disturbance, can eventually lead to restoring primitive, naturally appearing conditions—although perhaps without some plants or animals that were a part of the original ecosystem. In this sense, then, wilderness is renewable only to a degree.

Depending on the type of environment, renewing or restoring wilderness conditions could require many generations or even centuries. In the East, numerous areas have been designated as wilderness that were once farmed, had roads, and had been logged but are now reverting to wildness. Thus, for some of these lands under today's conditions, wilderness is being judged as their highest and best use (Hendee 1986). One might ask, which are more unnatural or permanently changed—wildernesses in the West that have had fire protection for eighty years or wildernesses in the East that were once extensively farmed and logged, but have now been returned to the influences of natural processes?

Renewability of the wilderness resource was a major theme and debate in the Wilderness Act. Many proponents have pointed to the importance of the act as a mechanism that might allow the wilderness characteristics of disturbed sites to recover. One of the great promises of the Wilderness Act is that, within reasonable limits, we can designate as wilderness formerly disturbed areas because natural forces and processes can restore the primeval scene. Thus, it is not easy to describe a precise status of wilderness on the renewable-to-nonrenewable continuum. *The idea that lands can revert to a wilder state and once again become wilderness is widely held. However, even when management seeks to restore wilderness conditions, the focus must be on wilderness, allowing natural forces and processes to freely operate, thus protecting naturalness as a defining quality of wilderness.*

PRINCIPLE 3: MANAGE WILDERNESS, AND SITES WITHIN, FOLLOWING A CONCEPT OF NONDEGRADATION

Not all areas designated as wilderness are identical in their primeval qualities; wilderness areas vary in the degree to which naturalness has remained unspoiled or to which opportunities for solitude remain undiminished by current, established uses. Such variations may occur within each wilderness, thus providing a "wilderness opportunity spectrum (WOS)." Expectations and definitions of wilderness may change with the condition of the areas surrounding it. The relativity of wilderness qualities and conditions was reflected in debates over the so-called Eastern Wilderness Act of 1975 (U.S. Public Law 93-622) and the subsequent designation of wilderness areas in the East that perhaps did not meet the more rigid, early criteria of the Wilderness Act of 1964 applied in the West (U.S. Public Law 88-577).

How are managers to work for a reasonably uniform standard for the NWPS when qualities such as naturalness and solitude vary widely, not only by region but also among and within individual wilderness areas? Wilderness managers have found the answer in the concept of nondegradation, which has been generally applied to management of air and water quality for decades. Basically, the nondegradation concept calls for maintaining existing environmental conditions, if they equal or exceed minimum standards, and for restoring conditions that are below minimum levels. For example, where air quality is currently higher than required by legislation, the higher level of quality should be preserved and not be allowed to deteriorate to minimum legal standards. Thus, the minimum legal standards of air quality do not constitute an acceptable level to which air quality everywhere will be allowed to deteriorate; rather, the nondegradation objective is to maintain currently high standards, to prevent further degradation, and to restore below-minimum conditions to acceptable levels (Meyers and Tarlock 1971; Mihaley 1972).

As applied to wilderness, the nondegradation principle recognizes that naturalness and solitude vary among individual wildernesses. The objective is to prevent degradation of current naturalness and solitude in each wilderness and to restore substandard naturalness and solitude to minimum levels, rather than letting all areas in the NWPS deteriorate to a common minimum standard. For example, the more pristine areas in Alaska or the northern Rocky Mountains should not be allowed to decline to the less wild conditions and greater levels of impact found in some urban-proximate wilderness in the southwestern or eastern U.S. Likewise, wilderness designation of heavily impacted areas in the eastern United States does not mean that the level of naturalness and solitude found in those areas should constitute an acceptable level for other areas in the East or in the West.

Generally, under the nondegradation principle, the conditions prevailing in areas when they were designated wilderness or identified as wilderness study areas for potential future designation, establish the benchmark of naturalness to be sought by management—unless conditions are deemed below standard, and then the objective is to restore naturalness and solitude to some minimum, acceptable standard.

The nondegradation concept also provides for the opportunity to upgrade or to restore wilderness quality. Where existing conditions are judged to be below minimum acceptable levels of naturalness and solitude, an appropriate priority of management is to promote restoration of wilderness to minimum quality levels. This does not automatically imply that wilderness restoration will involve activities such as planting grass and shrubs, fertilizing, and watering (unless that is the minimum necessary approach). Numerous management activities and policies can promote natural restoration by methods such as controlling visitor numbers, regulating the timing of use, excluding use in some areas, and other measures. How purely to manage each wilderness has been a long-standing issue; the principle of managing wilderness and sites within, under a nondegradation concept, helps resolve this issue.

PRINCIPLE 4: MANAGE HUMAN INFLUENCES, A KEY TO WILDERNESS PROTECTION

The need for managing wilderness visitors hardly has to be argued. Wilderness visitation has grown steadily, and many new areas that already sustain substantial primitive recreation have been added to the NWPS. Specific examples of site impacts and deterioration can be found in virtually any area. Human influence extends even to the most remote wilderness environments (Fig. 7.3). Indirect influence of humans is evident, for example, in the unnatural vegetation patterns resulting from fire prevention and suppression, which in turn contribute to unnatural distributions of wildlife. The fragile, sometimes irreplaceable, qualities of these areas are easily lost unless thoughtful, deliberate stewardship protects against the direct impacts of use and the indirect but pervasive influences of civilization and global change.

Increasing the size of the NWPS offers only short-term protection. Ultimately preserving wilderness resources and the values and benefits they offer people will depend on the stewardship and management of wilderness areas after they have been legally designated as wilderness or identified as wilderness study or candidate areas.

The principal goals of wilderness preservation are to maintain long-term ecological processes and to provide opportunities for solitude and primitive forms of recreation. Where these processes and opportunities have been lost, people are generally the cause. Thus, wilderness management is basically concerned with managing human use and influences to preserve naturalness and solitude (Fig. 7.4).

Recreation and permitted uses, like grazing, cause impacts that are currently among the most critical, unnatural influences in wilderness. Therefore, managers are challenged to guide, modify, and, if necessary, as a last resort, directly control wilderness uses and influences to minimize their impact.

FIGURE 7.3 Threats to wilderness resource conditions such as solitude and quality visitor experiences can be marred by heavy recreation use and congestion on mountain summits and overlooks. High Peaks Wilderness in New York State Adirondack Park. Photo courtesy Chad P. Dawson.

FIGURE 7.4 An important principle of management is to favor wilderness-dependent recreational activities—those that require naturalness and solitude. A rider (top) in the Sycamore Canyon Wilderness, Arizona, and a solitary angler (bottom) in the Absaroka-Beartooth Wilderness, Montana, experience naturalness and solitude. Top photo courtesy Starr Jenkins; bottom photo courtesy Richard Behan.

This has become an accepted tenet of wilderness management, although specific controls on visitation and uses (who and where) are usually controversial.

To date, wilderness management has emphasized the management of people. However, ecological problems are also becoming more important; wilderness managers are increasingly challenged to monitor the naturalness of wilderness ecosystems and to provide counterinfluences to human impacts. Fire, for example, needs to be restored to a closer approximation of its historical role and status in many areas (see Chapter 11). Noxious weeds like spotted knapweed, yellow star thistle, and leafy spurge have gained a foothold in many areas. Endangered wildlife may depend on the natural conditions and solitude of wilderness (see Chapter 12). The growing number of impacted sites in wilderness, our increasing knowledge about how they occur, and our skill in controlling or correcting them are also extending wilderness management activities to deal directly with these ecological and physical impacts (see Chapters 15 and 16). Furthermore, the NWPS has grown and now includes some previously impacted areas that need protection so that their natural wilderness qualities can return.

PRINCIPLE 5: MANAGE WILDERNESS BIOCENTRICALLY TO PRODUCE HUMAN VALUES AND BENEFITS

Wilderness is designated not just to protect its flora and fauna, but also for its enjoyment, values, and benefits (as wilderness) for people. The preservation goals for wilderness areas are clear in the Wilderness Act (U.S. Public Law 88-577): "it is . . . the policy of the Congress to secure for the American people of present and future generations the benefits of an enduring resource of wilderness." Thus, wilderness managers are legally mandated to produce the important and—in the judgment of some—necessary benefits that wilderness provides to people.

While benefits to wilderness users such as psychological restoration, physical challenges, or spiritual renewal are not easily measured, there also is no evidence that such benefits are produced only in wilderness. However, realizing that benefits are derived from wilderness preservation and use is implicit in the wilderness ideology and philosophy that led to passage of the wilderness acts and the subsequent designation of federal land as U.S. wilderness (Scott 2004). These anticipated benefits reflect one of the central beliefs of the U.S. wilderness system: that the character-building values of wilderness are vital to society (D. W. Scott 1984). The continued protection of wilderness naturalness and solitude as the source of these human values and benefits, and access to wilderness to experience them, are fundamental goals of wilderness stewardship.

There are two widely differing wilderness management philosophies; both are aimed at enhancing human benefits from wilderness. The biocentric perspective emphasizes environmental integrity as the basis for human benefits, a hands-off approach. The anthropocentric perspective promotes active management of natural processes and provides facilities to increase aesthetic pleasure and facilitate wilderness use (see discussion in Chapter 1).

The federal wilderness management agencies—the National Park Service (NPS), U.S. Forest Service (USFS), U.S. Fish and Wildlife Service (FWS), and Bureau of Land Management (BLM)—have established wilderness management policy manuals and/or guidelines reflecting a commitment to maintain high standards of naturalness and opportunities for solitude as mandated by the Wilderness Act. These agency policy manuals and handbooks provide direction for handling specific wilderness issues to preserve its contrasts to environments already modified by human activity. The agencies, in response to the Wilderness Act, might be considered generally biocentric, but there are differences in emphasis among them, with plenty of discretion remaining for agency wilderness managers in making day-to-day decisions.

The anthropocentric approach attempts to facilitate wilderness enjoyment by improving access to make visitation easier, more convenient, or simultaneously accessible to a larger number of people and can ultimately diminish wilderness's unique values and benefits. To us, this reality calls for a biocentric emphasis in wilderness management but it should be applied with common sense.

The biocentric-anthropocentric concepts are abstractions that facilitate the discussion of management alternatives. The biocentric approach emphasizes keeping the wild in wilderness but allowing as much use as is consistent with that goal, so people can directly benefit from the wilderness experience.

PRINCIPLE 6: FAVOR WILDERNESS-DEPENDENT ACTIVITIES

Wilderness is the setting for activities ranging from scientific study to recreational pursuits such as fishing, backpacking, day hiking, hunting, mountain climbing, rafting, and picnicking, some of which depend on a primeval setting. Some types of scientific study depend on access to a substantially unaltered ecosystem, perhaps covering a large area, and such conditions may only be found in wilderness. Thus, such use is wilderness dependent. Conversely, other activities, such as certain kinds of fishing, are not dependent on a wilderness situation at all, although such a setting can enhance them, for example catching a native trout at a high mountain lake and preparing it for dinner over an open fire. *Whenever one or more uses conflict or compete for access, the principle of wilderness dependency, which calls for favoring activities most dependent on wilderness conditions, is used to guide visitor management toward preventing overuse (Fig. 7.5).*

This principle is intended to ensure optimum use of wilderness resources, because activities that are not wilderness-dependent can, by definition, be enjoyed in other settings. Thus, any conflicts that arise should be resolved in favor of wilderness-dependent uses. This policy avoids displacing those seeking true wilderness experiences, whose opportunities are in short supply compared to those satisfied with non-wilderness experiences. This point of view is not new and was championed long ago by Robert Marshall (1930, 1937). Thus, providing an adequate range of non-wilderness recreation opportunities is important, as discussed under Principle 1, so persons not dependent on wilderness can use alternative areas for their activities.

FIGURE 7.5 More than a dozen trail riders with pack stock enjoy Washington's Glacier Peak Wilderness. Wilderness managers must assess visitor use and consider their area's carrying capacity. Too many visitors can depreciate the quality of the wilderness environment and solitude. Photo courtesy Robert DeWitz.

Defining an activity as wilderness-dependent can be difficult. Often, it is not the activity itself that is dependent, but the particular style in which it is pursued. For example, hunting is not necessarily wilderness dependent. However, certain styles of hunting, such as pursuing game under the most natural conditions away from roads, or stalking a bighorn sheep among high peaks, are highly dependent on wilderness settings. The importance of naturalness and solitude to the experience, not the mere quest for game, defines certain kinds of hunting as wilderness dependent.

Likewise, although people can fish in many settings, certain styles of fishing may be wilderness dependent. Those who desire remote, difficult-to-reach lakes, where one can fish under natural conditions without meeting other people, may rely more and more on wilderness for such opportunities.

Some higher-risk and adventure-recreation activities are not wilderness-dependent because their focus may be on the physical challenge and excitement and not the wilderness conditions and setting. In contrast, some commercial wilderness-experience programs use the wilderness environment as part of their method for education, therapy, or spiritual renewal.

Thus, favoring wilderness-dependent activities might call for reducing or discouraging—rather than eliminating—certain forms of some activities and encouraging other forms that are more wilderness-dependent. One is reminded again of the interdependency of wilderness with the rest of the recreation opportunity spectrum (ROS). One key to being able to emphasize wilderness-dependent activities in designated wilderness is to provide alternative non-wilderness lands to which non-wilderness-dependent uses can be diverted.

PRINCIPLE 7: GUIDE WILDERNESS MANAGEMENT USING WRITTEN PLANS WITH SPECIFIC AREA OBJECTIVES

Wilderness management and visitor management actions must be guided by formal plans that state specific area objectives and explain in detail how they will be achieved. Without such clear statements, management can become incremental, uncoordinated, and even counterproductive to the goals of the Wilderness Act. Local managers and the public need wilderness management plans to consider whether activities and strategies are appropriate for specific areas and are consistent with legislative goals and national policy. Public involvement is an important part of wilderness management planning and can be an educational process for both managers and citizens.

Wilderness management plans focus increasingly refined policy and action guidelines on the stewardship of individual areas (see Chapter 8). There is a hierarchy of guidance, from the general outline of goals from the Wilderness Act to narrower departmental regulations in the Code of Federal Regulations, to more explicit agency guidelines expressed in policy manuals, and to plans that translate into specific objectives for given areas and the policies and management actions to be used to achieve them. The planning process helps local agency officials and citizens review the varying situations of individual areas and develop local management strategies consistent with broad legislative goals. The legislative history and law designating a wilderness provide additional direction (Meyer 1999).

Wilderness management plans must include specific objectives—clear statements of desired wilderness conditions—so proposed management actions can be evaluated for their potential contribution toward a specific objective. Clear objectives and the commitment to be guided by them are important because management actions can have enduring—even irreversible—results. Philosophies, perceptions, and definitions of wilderness can vary widely among managers. Plans with clear, formally stated objectives are needed to guide judgments about what management actions are necessary; to provide continuity when managers change; and to prevent potential damage from ill-conceived plans, no matter how well intended.

Excessive or poorly conceived management actions can be as damaging to wilderness values as the absence of management. For example, a series of relatively minor decisions outside an overall plan might result in too many trails being built to unnecessarily high standards, an excessive number of signs, or unnecessary restrictions on user activities. The combined impact of small decisions can depreciate wilderness values. Only the minimum actions necessary to achieve objectives should be planned.

Articulating objectives gives benchmarks for measuring progress toward goals. If an objective is not met, an evaluator needs to ask why: Lack of feasibility? The need for different policies or implementing actions? The need for different kinds of administration or enforcement? Because the goals of the Wilderness Act are so broad, it is difficult to write clear objectives for all the various aspects of wilderness management. However, it is crucial to develop, through an orderly planning process, the clearest and most specific objectives possible and to use them as constant guides to management actions.

PRINCIPLE 8: SET CARRYING CAPACITIES AS NECESSARY TO PREVENT UNNATURAL CHANGE

Wilderness has limited capacity to absorb the impacts of use and still retain its wilderness qualities. As use increases, or as damaging patterns of use develop at specific places or during particular times, wilderness qualities may disappear, either over time or rapidly. Carrying capacity is the amount of use an area can tolerate without unacceptable change in conditions. The concept of carrying capacity offers a framework for managing use to preserve wilderness qualities (see Chapter 9 on carrying capacity and the limits of acceptable change [LAC] and visitor use management [VUM]).

Change due to natural ecological processes inevitably occurs in wilderness. The purpose of wilderness management is not to freeze ecosystems at a point in their natural succession—such as a vignette of earlier times—but to allow natural processes and change to occur with an absolute minimum of human manipulation and influence. Ecosystems are in constant change and the goal of wilderness management is to see that such change is natural—staying within its historical range of variation. Both ecological conditions and available wilderness experiences may thus also change. Standards of naturalness (the ecological integrity of processes and conditions) and human solitude that are established for an area—and the specific area-management objectives that express these standards—help define the carrying capacity of an individual wilderness.

Applied to wilderness, the concept of carrying capacity has two important parameters: (1) physical-biological and (2) social-psychological. Physical and biological dimensions describe the amount and kind of use an ecosystem can sustain without undue evidence of unnatural impacts. Unnatural physical and biological impacts include such things as campsite deterioration and soil compaction, trampling vegetation, proliferation of paths to and from locations of concentrated human use, exposed and protruding tree roots, and the unnatural behavior and distribution of wildlife (see Chapter 15).

Social or psychological dimensions refer to the levels and concentrations of human use an area can accommodate before the solitude of wilderness experiences is unacceptably diminished. "Outstanding opportunities for solitude" is a wilderness condition specified in the Wilderness Act. Several studies document that solitude—privacy from persons in other parties, particularly from large parties and other users camping near one's campsite—is an important attribute of the wilderness experience. Concentrations of visitors at popular campsites and other attractions such as hot springs, alpine lakes, or shelters might indicate that the social carrying capacity is being approached or exceeded (see Chapter 9 for more detail).

PRINCIPLE 9: FOCUS MANAGEMENT ON THREATENED SITES AND DAMAGING ACTIVITIES

Wilderness management is mainly concerned with maintaining naturalness and solitude, and impacts on these qualities are usually concentrated in time and space. These include physical impacts of use at popular locations, crowding impacts on group or individual visitor solitude, and the social-psychological experience of visitors. This calls for a site-specific orientation of management, rather than an across-the-board approach that would impose restrictions everywhere in a wilderness to solve problems that might be concentrated at popular locations, along high-use corridors, and/or at certain times or seasons.

Restrictions on use have been implemented for decades in some more heavily used wilderness areas and at certain popular locations in many others. Whenever such restrictions are contemplated, several difficult questions must be addressed: Are use restrictions the minimum tool to solve the problem? Where and when should use be restricted? Who (what kind of users) should be restricted? Under what conditions and criteria? How should restrictions be implemented?

Obviously, not all uses produce equal impacts. All types of wilderness use and activities can generally be ranked according to their relative physical, biological, and social-psychological impacts. When restrictions are necessary, those activities having the greatest long-term impact can be the first ones controlled. For example, various types of wilderness use might be ranked, even in the absence of an actual study, in the following order of decreasing environmental and social-psychological impact: large parties of horse users, small parties of overnight campers building fires, small parties of overnight campers not building fires, and small parties of day hikers. Additional criteria—such as visitors' knowledge or skill levels—might also be used to establish or modify priorities among users as well as the degree of wilderness dependency of their planned activities (see Chapters 14 and 16).

The unacceptable impact from various users might require regulation in only a few locations in a wilderness. Furthermore, the vulnerability of the resource is greater at different times, such as in early spring when vegetation is emerging and is easily damaged or on peak weekends when use is heavier. Thus, to minimize excessive environmental and social-psychological impacts, restrictions should be selective—to times, places, users, and activities having the greatest potential for damage.

Selective restriction strives to promote equity by specifying those conditions under which uses will be regulated, thus minimizing overall restrictions. Discrimination against certain types of use, such as pack stock or large, organized groups of hikers, is based on their respective impacts on the wilderness environment. It relieves wilderness managers of deciding arbitrarily on a certain mix of horse users, organized groups, hikers, and so forth; or alternatively, it keeps managers from deciding that because they have no criteria for making such choices, everyone or no one must be restricted. Many judgments must be made in wilderness management and, even with principles and criteria to guide decision logic, it is not an exact science. Sensitive managers willing to make decisions are required.

PRINCIPLE 10: APPLY ONLY THE MINIMUM TOOLS, REGULATIONS, OR FORCE TO ACHIEVE WILDERNESS-AREA OBJECTIVES

Freedom, spontaneity, and escape are recognized as important qualities of the wilderness experience. However, restrictions must sometimes be imposed to prevent these fragile attributes from being destroyed by overuse or misuse. Ironically, regulation itself can diminish the quality of the wilderness experience unless it is carefully implemented, and managers must always balance visitor freedom and resource protection (Cole 2000, 2001). The guiding principle is that only the minimum tools, regulation, or force necessary to achieve established wilderness-area objectives are justified. This principle is sometimes called the minimum tool rule—apply only the minimum tools, equipment, device, force, regulation, action, or practice that will bring the desired result. A "Minimum Requirements Decision Guide" and "Minimum Requirements Framework Analysis" are available online at www.wilderness.net.

Wilderness management actions fall on a continuum, ranging from subtle, light-handed, and indirect options (education and information) to direct and authoritarian options—such as telling visitors where they can travel and camp each day and how long they can stay. However, a key goal of wilderness management is to use indirect methods wherever and whenever possible to delay and minimize the need for direct controls (see Chapter 16). Which approach is appropriate depends on a manager's judgment about the degree of regulation necessary to achieve objectives and the likely effectiveness of various regulatory and nonregulatory actions in a certain situation.

FIGURE 7.6 A U.S. Forest Service ranger contacts wilderness visitors in the Three Sisters Wilderness in Oregon. Educating the public is one of a wilderness manager's most important tools. Direct visitor contact and guidance can help prevent undesirable impacts and postpone initiating stringent regulations. Photo courtesy U.S. Forest Service by Tom Iraci.

Subtle, indirect controls include visitor education, appeals for self-regulated and low-impact camping practices, and locating trailheads and trails to minimize impacts on fragile areas (Fig. 7.6). Direct methods might begin with managing specific, overused sites, where efforts would be made to educate, or sometimes to disperse users (keeping in mind that dispersing use may spread the impacts). Only as a last resort, when an array of specific and successively more restrictive measures has been exhausted, would direct control of visitation be considered.

If, for example, managers wish to distribute visitors, they might first use an indirect, educational approach—provide information about current visitor distribution, alternative trailheads and routes, times when use is low, and so on. However, if this approach appears to be inadequate or fails to redistribute use as desired, then a more direct, restrictive approach might be in order. A manager might then limit camping at heavily impacted sites, set entry quotas at each trailhead, or even assign campsites.

Ultimately, for every wilderness area, there is a point beyond which the wilderness experience becomes so diluted by restrictions that the feelings of freedom, spontaneity, and escape are in danger of being lost. It is at this point—which must be subjectively determined by managers, though perhaps guided by research findings—that rationing of entry, routes, and campsites must begin.

PRINCIPLE 11: INVOLVE THE PUBLIC AS A KEY TO THE SUCCESS OF WILDERNESS MANAGEMENT

The Wilderness Act was one of the first major laws mandating public involvement in natural-resource management—in selecting and designating areas to be set aside as wilderness. Nevertheless, during the first decade or so following passage of the Wilderness Act, many wilderness managers thought that the public did not have enough expertise to participate in wilderness management decisions. They thought that only professional resource managers could understand the legal, ecological, and social complexities involved. However, subsequently, with the passage of other laws requiring public involvement, such as the National Environmental Policy Act (NEPA) in 1969, the public rapidly learned how to participate in wilderness and public-land management (U.S. Public

FIG 7.7. Involving the public invests their interest in aspects of managing wilderness. Volunteers clean up trash at an old mining and homestead site in California's Old Woman Mountains Wilderness managed by the Bureau of Land Management. Photo courtesy Marilyn Hendee.

Law 91-190). Wilderness managers also became more adept at working with people and providing for public involvement as an essential step in planning for wilderness designation and wilderness management. Today, public involvement is recognized as perhaps the most important tool for successfully developing and implementing wilderness management plans and actions—and all other management of public lands. Laws increasingly require public involvement for more and more decisions. *Clearly, any proposed wilderness management plan or visitor management plan needs public involvement as a source of practical information and essential public support, without which its implementation will fail.*

Public involvement is also extending beyond planning and decision making to wilderness management work (Fig. 7.7). Volunteerism is a key to federal land management everywhere, and on state land too. Wilderness duty has been a favorite with volunteers for decades. Managing volunteers and group service projects is a key wilderness management skill and task. For example, managing volunteer rangers, wilderness information specialists, wilderness "cleanup" and "adopt-a-trail" projects, private contracting for projects such as trail construction and maintenance, and wilderness field-management projects with volunteers and partnerships with organizations all help increase citizen involvement in wilderness work. These volunteer efforts supplement diminishing budgets, but their greatest value goes beyond that: they increase recognition and appreciation of wilderness values—and that is one of the ultimate purposes of the NWPS (Hendee 1986).

Public involvement in wilderness management is firmly established and growing in importance. Increasingly, career positions will deal with managing volunteer projects, coordinating cooperative work with conservation and service organizations, contracting with private firms, and educating the public.

PRINCIPLE 12: MONITOR WILDERNESS CONDITIONS AND EXPERIENCE OPPORTUNITIES TO GUIDE LONG-TERM WILDERNESS STEWARDSHIP

Any management plan or program needs a monitoring system to evaluate progress toward stated objectives and to guide its long-term revisions, adjustments, and refinements. Devising monitoring systems remains one of the major challenges for advancing wilderness management. A good plan describes what objectives are to be achieved. Only through monitoring—the systematic gathering, comparing, and evaluation of data—can one tell whether those objectives are being realized. For example, under the limits of acceptable change (LAC) or visitor use management (VUM) planning approaches, data are needed to establish baseline conditions and then a system to monitor indicators to determine the degree of change occurring in various management areas (see Chapter 9).

Just what data should be collected, by what methods, and to what precision are important concerns for managers and scientists studying how to measure success (Fig. 7.8). The challenge is to measure and evaluate specific indicators under LAC or VUM that can be compared to certain standards for biological, physical, and social conditions in wilderness (Merigliano and Krumpe 1986). Biological conditions include categorizing vegetation, mammals, and fish conditions to consider potential indicators, which might singly or in combination reveal change. Physical conditions include potential soil, water, and air, indicators. Social conditions included describing visitors (e.g., day hikers, anglers, backpackers) and qualities of wilderness experiences with indicators. More recent work has focused on monitoring wilderness character, a hard-to-measure but important approach to check on the quality of the wilderness in the NWPS (Landres et al. 2005). Monitoring is important to wilderness management, yet much work remains to integrate monitoring into it.

Wilderness monitoring is also important for realizing the ultimate values of the NWPS for science and environmental assessment. Critical questions facing humanity concern how human activity has modified natural climate and weather processes globally as well as regionally and locally. Acid rain and related air pollution impacts, the effects of human activity on water quality and volume, climatic variation, heating and cooling of the earth, and other subtle global changes are all topics of concern. Such impacts may be reflected first in changes in vegetation and fish and wildlife habitats, and, subsequently, in loss of species. For example, in Alaska where climate change is pronounced, species migration and changes in

FIGURE 7.8 The National Wilderness Preservation System gained 516,200 acres with the designation of the Kofa Wildlife Refuge Wilderness in western Arizona, administered by the U.S. Fish and Wildlife Service. Monitoring the numbers of visitors, how they use wilderness, and wildlife behavior and interaction is an important task in wilderness management responsibility. Photo courtesy Marilyn Hendee.

relative abundance attributable to global climate change are occurring with species such as the polar bear (Regehr et al. 2007).

Wilderness provides enclaves of the earth's most natural remaining areas and can be a benchmark source of information on the degree of distortion of natural processes elsewhere—but only if sufficiently detailed information is collected and made available for use as environmental baselines. Wilderness management and monitoring could partially provide such data, supplementing more detailed scientific studies of wilderness conditions elsewhere.

PRINCIPLE 13: MANAGE WILDERNESS IN RELATION TO ITS ADJACENT LANDS

The principle—that wilderness managers must consider how adjacent lands are managed—is related to the environmental modification spectrum. Here, however, the concern broadens from outdoor recreation opportunities to a variety of other resource uses and management practices. Simply put, wilderness does not exist in a vacuum—what goes on outside of, and adjacent to, a wilderness can have substantial impacts inside its boundary. Conversely, designating a tract of land as a wilderness can substantially affect the management of adjacent areas (Hendee and Dawson 2001).

These interrelationships are illustrated by the impacts resulting from timber harvesting to building roads to the edge of a wilderness boundary that can dramatically affect the amount and character of its recreational use by providing easier access. Use of trail systems and target destination areas may then change, and thus the location of wilderness management emphasis to maintain natural conditions may need to change or be expanded.

Impacts resulting from increased use can compromise the goal of maintaining natural conditions. Reducing the number of roads, lowering road standards, and closing roads after timber harvesting might attenuate such impacts. Efforts to reduce road building and close roads on the national forests may ease this problem, although road closures and roadless-area-protection initiatives by the Forest Service are very controversial (O'Laughlin and Freemuth 2000) and politically volatile.

Similarly, developing high-density recreational facilities (youth camps, picnic areas, paved roads, and parking lots) immediately adjacent to wilderness or undeveloped backcountry can bring serious management problems as users of these facilities access the wilderness for hiking, fishing, or other activities.

Impacts can also move from wilderness to nearby non-wilderness areas. For example, programs to reestablish natural fire regimes through "let burn" or "prescribed fire" policies in wilderness can result in smoke pollution and a risk of fire losses in adjacent areas managed for commodity production or even summer-home development. Wildlife that thrives on summer range in high wilderness meadows may compete with domestic stock for forage in lower valleys during winter. In the Southeast, where the southern pine beetle has been a constant threat to the region's pine forests, there is long-standing concern that wilderness areas may harbor infestations that will spread to adjacent lands (Warren 1985).

Relating the management of wilderness to that of adjacent lands is complex and controversial. One common suggestion early in the history of the U.S. wilderness system was to create buffer zones—a band of land around the periphery of wilderness that would absorb impacts and help avert conflicts. However, Congress spoke against buffer zones in the Endangered American Wilderness Act of 1978 (U.S. Public Law 95-237), so any buffer zone must now fall within the wilderness, which could dilute wilderness conditions (see Chapter 5 for more discussion). However, the best protection for wilderness from impacts originating on surrounding lands is through comprehensive land-use planning that anticipates potential conflicts and addresses the complementary and competitive relationships between wilderness and adjacent lands.

Managers planning to allocate various land uses should carefully define activities that are compatible with wilderness and take steps to reduce incompatibilities through their location. The recreation opportunity spectrum (ROS) is such an approach. A concentric circle application of the ROS could have high-density, facility-oriented sites located around the periphery of a planning region, and increasingly environment-oriented and primitive-style opportunities (wilderness) located at the core. This gradation of uses can help protect the environmental quality of the interior wilderness while keeping development-oriented opportunities accessible, but outside the wilderness (Fig. 7.9). To do otherwise invites problems, particularly when the most intensive uses impinge on—and then ultimately displace—the least developed areas.

Explicitly defined use zones containing specific kinds of uses in particular locations also help protect against the invasion, succession, and displacement of primitive-site recreation users by new and more numerous, crowding-tolerant, and developed-site recreationists. The displaced users then move on to seek more primitive sites and can start a new wave of invasion, succession, and displacement. If this recreation-use displacement progression continues, the most pristine areas will have increased use, less solitude, and decreased natural conditions—the wilderness opportunity spectrum will disappear.

Managers may inadvertently aggravate undesirable impacts from this recreational displacement by responding to every increase in use with development to accommodate it—for example, by expanding campgrounds or parking lots, or allowing resort developments at the edge of wilderness.

FIGURE 7.9 The majority of visitor use in the Stikine-LeConte Wilderness in southeast Alaska is along the Stikine River corridor or its saltwater area. The wilderness's interior is pristine and dominated by glaciers and rugged topography. Photo courtesy John Hendee.

Furthermore, naturalness and solitude in the most remote sections of a wilderness can be impaired if wilderness managers are not aware of use interrelationships. Actions such as building trails to areas that are currently visited only by cross-country travelers seeking the greatest possible naturalness and solitude may ultimately eliminate the most remote and pristine wilderness experiences.

SUMMARY

These thirteen principles provide a foundation for much of the detailed discussion about wilderness management in the following chapters. By themselves, the principles do not ensure quality wilderness stewardship, but they do provide basic concepts to guide that management. At the very least, for those in search of consistent policies and actions, the principles provide a framework for reviewing and evaluating approaches to wilderness management situations, issues, and problems.

STUDY QUESTIONS

1. How does the availability of recreational opportunities on non-wilderness land affect nearby wilderness?
2. Explain the environmental modification spectrum, the recreation opportunity spectrum (ROS), and the wilderness opportunity spectrum (WOS).
3. In what sense is wilderness a renewable resource? What are the implications for wilderness management?
4. What are the two major parameters or dimensions of carrying capacity that wilderness managers must consider? Explain each briefly.
5. What are wilderness-dependent uses? How can favoring them help protect wilderness from excessive impacts and at the same time protect users from unnecessary restrictions?
6. What is the minimum tool rule? Give an example of a light-handed or minimum tool approach to reducing damage to vegetation and crowding at a popular wilderness campsite.
7. What does the concept of nondegradation suggest for the management of wilderness in the eastern United States, compared to more wild and pristine wilderness areas in the West?
8. How has increased public involvement affected the job of professional wilderness managers? Overall, do you think the effects have been or will be positive or negative? Why?
9. Explain the recreation-use displacement idea, how it can dilute the most pristine wilderness, and how managers can guard against it.

CASE DISCUSSION: ASSISTED TROUT MIGRATION INTO WILDERNESS?

As environmental and climate changes at the global and regional levels affect wilderness natural processes and conditions, managers will increasingly be faced with deciding when and how to intervene and to what extent interventions should follow the historic range of variation and natural processes and conditions (see Chapter 10). The spectrum of management options ranges from nonintervention (i.e., allow ecosystems to adapt and change as they will) to active intervention (i.e., restore historic natural conditions, resist change by maintaining current system, conserve some aspect of biodiversity, or transform to more resilient systems). Cole et al. (2008, 42) provided an example of a difficult decision to conserve a species over maintaining natural conditions:

> Recently, in the Bob Marshall Wilderness, Montana, stewardship decisions have been made that some might consider inconsistent with wilderness. These decisions condone assisted migration—helping species relocate to places where they are more likely to persist—and they place more importance on species conservation than on naturalness. The decisions pertain to management of fish populations in about 20 lakes in the Bob Marshall Wilderness that historically were fishless but that have been stocked with nonnative trout for many decades.

The plan was to remove all nonnative trout from these lakes. Then, rather than leave the lakes fishless as they originally were, they were planned to be stocked with genetically pure westslope cutthroat trout. These lakes, which fish are unable to migrate to themselves, offer a refuge from other fish that hybridize with westslope cutthroat and change them genetically. Wilderness provides the most inviolate refuge and, therefore, is considered necessary to preserve the species, even though the requisite action compromises naturalness.

This situation is complicated by states' rights issues. This intervention, pushed by the Montana Fish, Wildlife and Parks Department, would probably not have been proposed if the species at risk were not a game fish. Nevertheless, it illustrates the potential to give precedence to a conservation goal other than preserving natural conditions, even in wilderness.

CASE DISCUSSION QUESTIONS

1. Which of the thirteen principles in this chapter *support* this intervention of conserving the westslope cutthroat trout population? Why?
2. Which of the thirteen principles in this chapter *do not support* this intervention of conserving the westslope cutthroat trout population? Why?
3. How do you decide between conserving (in this case) species biodiversity in a regional context and maintaining the natural conditions present in a wilderness prior to European settlement in the region?

REFERENCES

Cole, David. 2000. Natural, wild, uncrowded or free: Which of these should our wilderness be? *International Journal of Wilderness.* 6(2): 5–8.

———. 2001. Wilderness recreation management strategies: Balancing freedom and protection. *International Journal of Wilderness.* 7(1): 12.

Cole, David; Yung, L.; Zavaleta, E. S.; Aplet, G. H. et al. 2008. Naturalness and beyond: Protected area stewardship in an era of global environmental change. *The George Wright Forum.* 25(1): 36–56.

Hendee, John C. 1986. Wilderness: Important legal, social, philosophical, and management perspectives. In: Kulhavy, David L.; Connor, Richard N., eds. Wilderness and Natural Areas in the Eastern United States: A Management Challenge. Symposium; May 13–15, 1985; Nacogdoches, TX: Stephen F. Austin State University, School of Forestry, pp. 5–11.

Hendee, John C.; Dawson, Chad P. 2001. Stewardship to address the threats to wilderness resources and values. International Journal of Wilderness. 7(3): 5–10.

Landres, P., Boutcher, S.; Merigliano, L.; Barns, C.; Davis, D.; Hall, T.; Henry, S.; Hunter, B.; Janiga, P.; Laker, M.; McPherson, A.; Powell, D.; Rowan, M.; Sater, S. 2005. Monitoring selected conditions related to wilderness character: A national framework. General Technical Report RMRS-GTR-151. Fort Collins, CO: U.S. Department of Agriculture, Forest Service, Rocky Mountain Research Station. 38p.

Marshall, Robert. 1930. The problem of the wilderness. Scientific Monthly. 30(2): 141–148.

———. 1937. The universe of the wilderness is vanishing. Nature. 29(4): 235–240.

Merigliano, Linda; Krumpe, Ed. 1986. Scientists identify, evaluate indicators to monitor wilderness conditions. Park Science. 6(3): 19–20.

Meyer, Shannon. 1999. The role of legislative history in agency decision making: A case study of wilderness airstrip management in the United States. International Journal of Wilderness. 5(2): 9–12.

Meyers, Charles J.; Tarlock, A. Dan. 1971. Water pollution. In: Meyers, Charles J.; Tarlock, A. Dan, eds. Water Resource Management. Mineola, NY: Foundation Press, pp. 677–708.

Mihaley, Marc B. 1972. The Clean Air Act and the concept of nondegradation: Sierra Club v. Rukelhaus. Ecological Law Review. 2(4): 801–836.

Nash, Roderick. 1982. Wilderness and the American Mind, rev. ed. New Haven: Yale University Press. 425p.

O'Laughlin, Jay; Freemuth, John. 2000. Roadless area policy politics and wilderness potential. International Journal of Wilderness. 6(1): 9–12.

Oye, Garry. 2001. A new recreation strategy for national forest wilderness. International Journal of Wilderness. 7(1): 13–14.

Regehr, E. V.; Lunn, N. J.; Amstrup; Stirling, I. 2007. Effects of earlier sea ice breakup on survival and population size of polar bears in western Hudson Bay. *Journal of Wildlife Management.* 71(8), pp. 2673–2683.

Scott, D. W. 1984. The visionary role of Howard Zahniser. *Sierra.* 69(13): 40.

Scott, Douglas. 2004. *The enduring wilderness: Protecting Our Natural Heritage Through the Wilderness Act.* Golden, CO: Fulcrum.

Spring, Ira. 2001. If we lock people out who will fight to save the wilderness. *International Journal of Wilderness.* 7(1): 17–19.

Stankey, George H.; Cole, David N.; Lucas, Robert C. et al. 1985. The limits of acceptable change (LAC) system for wilderness planning. General Technical Report INT-176. Ogden, UT: U.S. Department of Agriculture, Forest Service, Intermountain Forest and Range Experiment Station.

U.S. Public Law 88-577. The Wilderness Act of September 3, 1964. 78 Stat. 890.

U.S. Public Law 91-190. The National Environmental Policy Act (NEPA) of January 1, 1970. 83 Stat. 852.

U.S. Public Law 93-622. (Eastern Wilderness) Act of January 3, 1975. 88 Stat. 2096.

U.S. Public Law 95-237. The Endangered American Wilderness Act of February 24, 1978. 92 Stat. 40.

Warren, B. Jack. 1985. Why we need to control pine beetles in wilderness areas. *Forest Farmer.* 44(4): 6–8.

White, David; Hendee, John C. 2000. Primal hypotheses: The relationships between naturalness, solitude and the wilderness experience benefits of development of self, development of community, and spiritual development. In: McCool, S.; Cole, D.; Borrie, William T.; O'Loughlin, Jennifer, comps. *2000 proceedings: Wilderness Science in a Time of Change,* Vol. 3; *Wilderness as a Place for Scientific Inquiry;* May 23–27, 1999; Missoula, MT. RMRS-P-15. Fort Collins, CO: U.S. Department of Agriculture, Forest Service Proceedings, pp. 223–228.

Worf, William. 2001. The new Forest Service recreation strategy spells doom to the National Wilderness Preservation System. *International Journal of Wilderness.* 7(1): 15–16.

CHAPTER

8

Wilderness Management Planning

INTRODUCTION

Wilderness management needs to be guided by formal plans that clearly state goals, objectives, and standards, and prescribe specific actions necessary to meet the objectives and standards. *Wilderness management translates the general directions of legislation, departmental regulations, and agency policy into guides for management in the field.*

In this chapter we discuss the background and need for planning; review the wilderness management planning processes of the National Park Service (NPS), U.S. Forest Service (USFS), U.S. Fish and Wildlife Service (FWS), and Bureau of Land Management (BLM); outline tools and techniques for developing plans, including a framework and format for preparing wilderness management plans; provide excerpts from actual plans to show how the suggested framework is applied; and discuss criteria for evaluating management plans.

Because wilderness management plans are developed under the larger land-use planning organization of the different management agencies, there are differences in agency format and process for wilderness management planning (Fig. 8.1). *Regardless of the particular agency process, wilderness management needs to be guided by clear statements of goals and objectives, by selecting management actions necessary to achieve the goals and objectives, and by monitoring to ensure that the goals and objectives are being met.*

THE NEED FOR PLANNING

Since the passage of the Wilderness Act of 1964 (U.S. Public Law 88-577) the United States has embraced the concept of identifying, protecting, and managing vast amounts of land in the National Wilderness Preservation System (NWPS). By 2023, there were more than 111.8 million acres protected in over 800 wilderness areas nationwide, or about 4.5 percent of the United States. Many of the political battles fought to protect and preserve these landscapes have been lengthy and intensely debated, and the resulting legislation often includes compromises that pose problems for future management (see Chapter 5). *Wilderness designation alone is not sufficient to protect and perpetuate the human and ecological values for which these areas have been set aside.* An ongoing stewardship program is necessary to deal with the human influences from both outside and within wilderness, and their accompanying undesirable impacts. Along with recognizing that management was necessary, even in wilderness, came realization of the need to develop plans that would direct wilderness management toward long-range goals. The need for wilderness planning has been an integral part of wilderness management since the inception of the NWPS (Krumpe 2000).

FIGURE 8.1 Wilderness, such as the Selway-Bitterroot on the Idaho-Montana border (above), is managed under constraints of the Wilderness Act and agency direction. Management direction is contained in specific wilderness area management plans, or in resource management plans that include both wilderness and non-wilderness land. Photo courtesy U.S. Forest Service.

Three realities underlie the need for wilderness management planning: (1) human-caused change is inevitable; (2) recreation use is not self-limiting; and (3) good planning anticipates problems.

1. *Human-caused change is inevitable: we cannot prevent conditions from changing in wilderness.* Wilderness resources are not static but are constantly evolving naturally. All human use within and nearby wilderness areas causes some impact to natural wilderness conditions. Given our growing global population, threats to wilderness conditions (see Chapter 13), global climate change, and increasing participation in a variety of wilderness activities, resource conditions and experiences will continue to change. Planning is the logical process to anticipate change and to consider how much human-caused change is acceptable in the areas where it occurs.
2. *Recreation use is not self-limiting.* As wilderness use increases, people who seek quieter, less crowded conditions will typically modify their behavior to visit during less busy times or may decide to visit other, less crowded areas (hence, be displaced). These visitors are replaced by people willing to accept more crowded conditions and more evidence of use. As use continues to increase, this cycle repeats itself again and again. Thus, many wilderness places gradually become more crowded with more impacts and conflicts, unless a strategy is developed to stop this cycle with actions to protect and provide the desired wilderness settings and experiences. Thus, the question is not whether limits and management actions are needed, but rather which wilderness settings an area should provide and which management actions are needed to achieve them.
3. *Good planning anticipates problems: it is far better to begin planning early.* The best plans are those developed before unacceptable wilderness conditions are widespread. It is better to define when "corrective" management actions will be needed than to try to restore wilderness conditions or change use patterns or amounts of use when the situation is perceived to be causing undesirable impacts.

Without management plans developed by an orderly process, wilderness management may be no more than a series of uncoordinated reactions to immediate problems. This can lead to practically irreversible impacts on wilderness conditions.

Managers need to look ahead to identify undesirable trends and implement actions to achieve desired conditions and to prevent, reverse, or restore undesirable conditions. For example, the complete suppression of fires, trail building in sensitive wildlife habitat, overuse of popular locations, or introduction of exotic fish, wildlife, and plant species can lead to impacts, which are cumulative and could take decades to reverse.

Management-by-reaction works against wilderness preservation goals because management direction can easily be shaped by a succession of minor decisions. The cumulative results of such incremental decisions may be undesirable and hard to reverse. Unplanned management can be recognized by a shifting of focus from problem to problem; inconsistent, conflicting actions; and a loss of overall direction toward wilderness preservation goals.

Through planning, managers can reconcile differences in management philosophy and ideas before taking actions that have long-range effects on the wilderness resource. Good plans have a stabilizing influence on management, despite changes in personnel or the influence of multiple administrative units in a wilderness where several managers may have different philosophies, perceptions, and definitions of wilderness. *Plans that establish clear, attainable, measurable, and acceptable objectives and standards for an area, and the actions by which such objectives will be pursued, are essential for guiding wilderness management toward desired outcomes.*

Successful wilderness management depends on the quality of plans that guide management actions for individual areas. In an era of heightened concern about cost effectiveness and competition for scarce federal dollars, the potential for obtaining the personnel and money essential for wilderness management, or optimizing effectiveness without them, will be increased by plans that identify clear objectives and the management actions necessary to achieve them.

Furthermore, involving the public in the planning process will attract and focus volunteer efforts to help accomplish essential tasks. The planning process gives the interested public an opportunity to learn about, evaluate, provide input, and become involved in a wilderness area's management. Effectiveness and consistency of wilderness management, as well as the public's acceptance of that management and its implementation, are highly dependent on involving them in plans and the planning process.

TRENDS AND EVOLUTION OF WILDERNESS PLANNING DIRECTION

Wilderness management planning among the four federal wilderness management agencies has changed and evolved to reflect concurrent changes in regulations, policies, legislation, and society that affect planning. In addition, wilderness managers and planners have gained experience and expertise and have embraced evolving directions in natural-resource planning.

There are four trends in resource planning:

1. *Use allocation versus land allocation.* The first wilderness plans typically dealt with allocating land areas to various uses, primarily focusing on designating land as wilderness or non-wilderness. Planning attention has now shifted to directing stewardship of lands allocated to various uses, including wilderness. The current trend focuses on allocating resource use among various compatible human activities and excluding noncompatible uses from wilderness.
2. *An ecosystem approach to planning.* Planning in the 1960s tended to focus on separate aspects of the wilderness, such as recreation, vegetation, water, forage, wildlife, or fire within wilderness boundaries. More recently planners recognize that the greatest value of wilderness is its naturalness, and plans must address wilderness as a composite resource (see Chapter 7). This, of course, is the key value of all ecosystem planning. Furthermore, the conditions of naturalness and solitude in wilderness can provide benefits to the larger landscape. *The ultimate goal of planning is to protect and perpetuate interrelated natural ecosystem processes that often function on a landscape scale and that extend beyond wilderness boundaries.*
3. *A focus on desired resource and social conditions rather than outputs.* Early plans focused on quantifying "outputs" to be produced, such as recreation visitor days, animal-unit months of grazing, miles of trail developed, etc. *Plans present desired future resource and social conditions as goals for which specific objectives are developed.* Outcomes are now portrayed as protecting or restoring resource conditions, which produces desired services or benefits such as recreation, personal growth, improved health, species diversity, naturally functioning aquatic systems, natural fire regimes, endangered species recovery, and so forth.
4. *A focus on collaborative decision-making processes.* Traditional planning models relied heavily on experts and data and used rational and comprehensive approaches. Such planning was often paralyzed by the lack of "good" data, and implementation was often hampered by lack of public understanding and acceptance. Today, planning is a collaborative effort with the public and users of the wilderness. In wilderness planning, the question has shifted from "How much use is too much?" to a focus on the desired resource and social conditions and "How much change in conditions is acceptable?" This raises the question "Acceptable to whom?" which is a value-based rather than technical question. Science and data play important roles in understanding how an ecosystem works and identifying the consequences of various options, but the decision about what is desired or acceptable is based on society's values and beliefs (Krumpe and McCool 1997). Recent wilderness plans have placed greater emphasis on utilizing processes that involve the public in discussion throughout the planning process.

PLAN FLEXIBILITY AND EFFECTIVENESS

Wilderness management planning processes are constantly evolving, and several approaches and styles of plans have been developed during more than five decades of designated wilderness management. One approach, called a "goal-achievement framework," is described later in the chapter. Increasingly, managers face conflicting goals, such as providing freedom from restrictions versus protecting resource conditions and solitude. In these situations, the limits of acceptable change (LAC) approach and the Interagency Visitor Use Management Framework (VUM) are useful and are discussed in Chapter 9. Other planning approaches and frameworks may also be useful (McCool et al. 2007); for example, Wilderness Character Monitoring (Landres et al. 2015) and the Minimum Requirements Analysis Framework (www.wilderness.net).

There is room for flexibility in format and level of detail in wilderness plans as long as they effectively translate overall management direction—laws, regulations, and policy—into specific objectives and necessary management actions to achieve them. Whatever the format, well-reasoned plans are necessary to explain management intentions, to guide public discussion, to achieve public understanding of necessary management actions, to justify budget requests and meet legal requirements, and to ensure protection and stewardship of an enduring wilderness resource.

WILDERNESS PLANNING UNDER THE NATIONAL ENVIRONMENTAL POLICY ACT OF 1970

Following passage of the National Environmental Policy Act (NEPA; U.S. Public Law 91-190) in 1970, the federal wilderness managing agencies incorporated environmental analysis procedures into their planning processes. *NEPA requires that environmental impacts be considered through an analysis of a proposed action and its alternatives, and that the public be allowed to comment on the actions under consideration. Under the guidelines for implementing NEPA, agencies must assess the environmental and social impacts of alternative management actions before implementing them.* Thus, whatever wilderness management planning approach is used, the plan must conform to NEPA's requirements that alternatives be considered and analyzed, the effects of alternatives be disclosed, and the public have an opportunity to provide input.

These alternatives must include a range of actions, including taking no action at all. All alternatives must meet the need for taking action and respond to the major issues. During this NEPA analysis process, managers have an opportunity to consider prospects for the type of wilderness environment described in section 2(c) of the Wilderness Act of 1964 (U.S Public Law 88-577)—one featuring naturalness and outstanding opportunities for solitude and primitive recreation—considering conditions in the area under consideration and the impacts of actions proposed in the plan or proposed project. A significant requirement under NEPA is that all federal plans are required to develop an environmental impact statement (EIS) or environmental assessment (EA).

WILDERNESS MANAGEMENT PLANNING IN U.S. FEDERAL AGENCIES

Wilderness management planning is an integral part of the overall management-planning approaches for the natural-resource units managed by the four federal wilderness managing agencies. Thus, plans for wilderness, wilderness study areas, or proposed wilderness are subordinate to—meaning they fall within the larger planning framework of the agency jurisdiction—a national park or national forest plan, BLM resource-management plan, or FWS refuge plan. For example, the White Mountain National Forest Land and Resource Management Plan (U.S. Department of Agriculture, Forest Service 2005) incorporated broad, strategic consideration of the five wilderness areas within the forest throughout the forest plan and addressed the specific management direction for those areas within a separate appendix (Wilderness Management Plan). The forest plan included recommendations for additional wilderness areas and acreage as some of its specific recommendations. Congress subsequently passed the New England Wilderness Act of 2006 (U.S. Public Law 109-382) and added a new area, Wild River Wilderness, to the NWPS and expanded the existing Sandwich Range Wilderness based on recommendations in the 2005 forest plan.

The following discussion of planning procedures used by the federal agencies assumes the completion of the environmental analysis requirements mandated by NEPA. Most agencies have now combined NEPA's environmental analysis of environmental impact statements (EIS) and environmental assessment (EA) requirements and their land-use planning procedures, so that all NEPA requirements are met when a plan is completed.

NATIONAL PARK SERVICE

Management direction for wilderness in the National Park System comes from the Wilderness Act, the National Park Service Act of 1916 (U.S. Public Law 235), legislation establishing individual national park units, and wilderness designation laws designating wilderness areas within particular parks. Interpretation of this legislation for wilderness preservation and management is found in the revised Management Policies of the NPS (U.S. Department of the Interior, National Park Service 2006) and is supplemented by the NPS Director's Order #12 related to NEPA Planning and Order #41 on Wilderness Stewardship, and the Management and Reference Manual 41, which was included with that order. The congressional intent reflected in these laws and policies is implemented predominantly through multiple levels of planning applicable to all national park units, including national recreation areas and national monuments administered by the NPS. The overall plan is the General Management Plan (GMP), with an EIS for each NPS unit.

1. *General management plans* (GMPs) are required for each national park under the National Parks and Recreation Act of 1978 (U.S. Public Law 95-625). The NPS program standards explain (U.S. Department of the Interior, National Park Service 2004, 6):

   > General management planning results in a shared understanding among NPS managers and the public about the kinds of resource conditions and visitor experiences that will best fulfill the purpose of the park. General management plans zone the park for some variety of resource conditions and experiences (consistent with the discretion allowed under the NPS Management Policies) based on the intrinsic qualities of particular locations and considering the range of stakeholder interests and concerns. The plan looks at the park as a whole and as a part of larger ecological, cultural, and socioeconomic systems. This comprehensive approach helps ensure that the decisions made through general management planning are widely supported and sustainable over time. General management plans direct park managers to focus on achieving the conditions and experiences prescribed by the plan, but they do not provide direction for specific actions, recognizing that managers may have to continuously adapt their approaches to current situations based on changing information or conditions, including changing staffing, budgeting, and scheduling opportunities and constraints.

2. The GMP is supported by three other levels of planning as needed: (a) a program plan to identify and recommend the best strategies for achieving the long-term desired conditions and visitor experiences in that park unit; (b) a strategic plan to make decisions about how to achieve the desired conditions in the next three to five years of those identified in the park program plan; and (c) an implementation plan to describe in detail how to achieve the elements of the strategic plan over the next several years (U.S. Department of the Interior, National Park Service 2004, 2005c).

National parks with a potential or actual designation as wilderness address these opportunities and requirements in the GMP. A park with a potential wilderness must go through a formal wilderness study process prior to proposing an area or park for wilderness designation by Congress. NPS managers consider the extent to which the requirements of wilderness stewardship planning could be met when developing a park GMP. *The Park Planning Source Book* (U.S. Department of the Interior, National Park Service 2005c, 39) suggests:

> A wilderness stewardship plan done along with the GMP should address zoning and desired conditions and establish indicators and standards for achieving the desired conditions. These are GMP requirements already and would not result in extra work for the GMP team. Decisions about trails and other public facilities, campfires, user capacity, etc., can be addressed through zoning and desired conditions without mentioning each trail or cabin. For parks with few issues or little wilderness use, this level of wilderness planning may meet most of their needs. This plan would then provide the broad framework for more detailed implementation plans, such as a fire management plan or trail plan. Parks with complex wilderness issues, such as heavy overnight or day use, commercial pack trips, or climbing, will likely need a separate wilderness stewardship plan to provide management guidance for these issues.

NPS policy requires each administrative unit of the National Park System to address user capacity and impact issues and identify and offer mitigating actions for problems, such as overcrowding, stock damage to trails and tundra, disposal of human waste, threats to visitor safety, or danger from or to grizzly bears. The Park Service historically developed a planning process called Visitor Experience and Resource Protection (VERP) in response to the mandate to develop a carrying-capacity framework for each park and, more recently, adopted the Interagency Visitor Use Management (VUM) framework (see Chapter 9 and Interagency Visitor Use Management Council 2016).

Wilderness areas have been designated in more than 40 units of the National Park System and more than 30 wilderness study areas are under NPS management. Although environmentalists take heart from the protection these areas receive from being in the NPS system, completion of the proposed wilderness designation, and subsequent completion of wilderness management plans for each designated area, are a goal for many citizens and NPS managers.

U.S. FOREST SERVICE

Management direction for wilderness in the National Forest System comes from several laws: the Multiple-Use, Sustained-Yield Act of 1960 (U.S. Public Law 86-517); the Wilderness Act of 1964 (U.S. Public Law 88-577); the Forest and Rangeland Renewable Resources Planning Act of 1974 (U.S. Public Law 93-378); the National Forest Management Act (NFMA) of 1976 (U.S. Public Law 94-588); the Clean Air Act and amendments of 1977 (U.S. Public Law 95-95); and sometimes specific wilderness designation laws establishing areas on national forests. The U.S. Department of Agriculture Regulations 36 *Code of Federal Regulations*, parts 219, 228, 261, and 293, and USFS policy guidelines for wilderness planning (Chapter 2322 of the USFS manual) contain further direction (U.S. Department of Agriculture, Forest Service 2007).

The USFS has two primary land and resource plans that relate to wilderness: (1) the Renewable Resources Assessment and Program, a nationwide plan that guides the overall management of the National Forest System; and (2) the Forest Land and Resource Management Plan usually called the Land Management Plan or Forest Plan, which includes management direction for wilderness areas of the national forest.

1. *The Renewable Resources Assessment and Program* is required by the 1974 Forest and Rangeland Renewable Resources Planning Act (RPA), as amended by the 1976 National Forest Management Act (NFMA). Together, these laws provide a comprehensive framework for planning management of the National Forest System. The RPA calls for (a) preparation of an assessment of the supply and demand for the nation's forest and rangeland resources and (b) development of a management program for national forests that considers alternative management directions and the role of the national forests. The RPA program proposes (subject to annual appropriations by Congress) national direction and output levels for the National Forest System. These proposals and associated management implications are based on an assessment of supply and demand and fiscal and political considerations. Wilderness is one of the resources analyzed.
2. *The Forest Land and Resource Management Plan,* commonly known as the Forest Plan, defines the management emphasis for different areas of the forest and sets forest-wide and area-specific goals and objectives. This is the primary document providing long-term management direction for a particular wilderness. Six types of decisions are made in Forest Plans: (a) forest-wide goals and objectives, (b) forest-wide management requirements, (c) management-area direction, (d) suitability of land for resource use and production, (e) monitoring and evaluation requirements, and (f) recommendations for wilderness and wild and scenic river designation.

Each wilderness is a separate management area for which the Forest Plan specifies desired conditions, standards, and guidelines for management activities and monitoring requirements. The types of decisions made in a Forest Plan are programmatic in nature, meaning that the plan establishes what is to be achieved and parameters to guide the management and human uses that may occur, but it does not prescribe specific actions or commit to how goals and objectives will be achieved.

To implement the Forest Plan, a number of field-level documents are prepared. The Forest Service Manual (U.S. Department of Agriculture, Forest Service 2007) sets forth the guidelines for planning and managing wilderness areas with the National Forest System. The main objectives for wilderness planning are to ensure that wilderness resource considerations are integrated in the Forest Plan and that all resources and activities within a designated wilderness are in line with USFS wilderness policies and direction. A fish and wildlife management policy provides guidance for USFS planning activities in wilderness (U.S. Department of Agriculture, Forest Service and U.S. Department of the Interior, Bureau of Land Management 2006).

U.S. FISH AND WILDLIFE SERVICE

Management direction for wilderness in the National Wildlife Refuge System comes from the Wilderness Act; the National Wildlife Refuge System Administration Act of 1966 (U.S. Public Law 89-669), as amended; the Alaska National Interest Lands Conservation Act (ANILCA) of 1980 (U.S. Public Law 96-487); specific authorities (including executive orders) establishing individual units of the refuge system; and specific legislation designating wilderness-areas within particular refuges. Further national direction comes from FWS regulations for wilderness preservation and management (50 *Code of Federal Regulations,* part 35), and the agency's wilderness area management policy (U.S. Department of the Interior, U.S. Fish and Wildlife Service 2000). The wilderness-area management policy provides guidelines for administrative and public use and preparation of wilderness management plans.

Wilderness management direction is developed and implemented through the refuge planning process. Two types of planning documents guide management of refuge wilderness areas: comprehensive conservation plans (CCPs) and step-down management plans.

All refuges are to be managed in accordance with an approved CCP. The CCP is a land-use plan that describes the desired future conditions of a refuge or planning unit and provides long-range guidance and management direction to achieve the purpose of the refuge; helps fulfill the mission of the refuge system; maintains and, where appropriate, restores the biological integrity, diversity, and environmental health of each refuge and the refuge system; helps achieve the goals of the NWPS; and meets other mandates. The CCP sets forth goals, objectives, and strategies to guide management decisions and accomplish these ends. Public involvement and NEPA compliance are integrated into the CCP process. CCPs are prepared by the refuge manager and refuge staff and approved by the regional director.

Step-down management plans are prepared for a specific management program or resource; they provide detailed strategies and implementation schedules for meeting the broader goals and objectives identified in the CCP. Wilderness stewardship plans (WSPs) are a form of step-down management plan that guide the preservation, management, and use of a particular designated wilderness in a refuge or unit. The WSP describes specific management goals, objectives, and strategies based on the refuge or unit's purposes; the principal wildlife conservation mission of the refuge system; and wilderness stewardship principles.

BUREAU OF LAND MANAGEMENT

The Federal Land Policy and Management Act (FLPMA) of 1976 (U.S. Public Law 94-579) called for the review of all roadless areas and roadless islands on BLM-administered land by 1991 to determine their suitability for wilderness designation (see Chapter 5). At the end of the review, the secretary of the interior was to make recommendations to the president as to which wilderness study areas (WSAs) were suitable or unsuitable for wilderness designation. The timetable for wilderness designation of USDI Bureau of Land Management (BLM)–administered land has not progressed as fast or as far as FLPMA anticipated.

Wilderness areas administered by the BLM are managed under direction contained in BLM's wilderness management regulations and policy (43 *Code of Federal Regulations,* part 6300) (U.S. Department of the Interior, Bureau of Land Management 2000), which replaced the BLM's 1983 and 1984 *Management of Designated Wilderness Areas* manual (sections 8560 and 8561). WSAs are managed to protect their wilderness values until Congress either drops them from further wilderness consideration or designates them as wilderness. The wilderness management policy addresses a wide range of potential uses and management actions within wilderness areas and provides guidance for preparing wilderness management plans. A fish and wildlife management policy provides guidance for BLM planning activities in wilderness (U.S. Department of Agriculture, Forest Service and U.S. Department of the Interior, Bureau of Land Management 2006). The BLM's approach to writing wilderness management plans is based on concepts of the goal-achievement planning framework described later in this chapter.

BLM wilderness management plans are developed within the context of the bureau's overall planning system. *The BLM has two types of planning documents for wilderness areas: Resource Management Plans (RMPs) and activity plans. Wilderness management plans are one kind of activity plan.* Activity plans are also prepared for other BLM-managed resources, such as livestock grazing allotments, wildlife habitats, and recreation.

1. The Resource Management Plan (RMP) is a type of land-use plan designed to guide the management of BLM lands administered by a field office and may include several million acres of public land. RMPs may address both wilderness management and wilderness designation, that is, making wilderness suitability recommendations. Each alternative considered in the RMP contains a description of how the various resources and uses of WSAs would be managed. Planning decisions inconsistent with WSA protection requirements may not be implemented unless Congress determines that an area should not be designated a unit of the NWPS. In the case of areas that are eventually designated as wilderness, RMP decisions would be included in a wilderness management plan where they are consistent with the Wilderness Act.
2. An activity plan—specifically a wilderness management plan—is prepared for a specific wilderness area or two or more closely related areas. The plan may also apply in some cases to areas adjacent to a wilderness, such as trailheads. Wilderness management plans help to implement the BLM's wilderness management policy. They show the actions that will be taken to preserve the wilderness resource, wilderness character, and the connection between these actions and wilderness management objectives. However, because of the interrelationships between

wilderness and other lands, a coordinated activity plan that addresses many issues may be prepared in place of a wilderness-specific activity plan.

3. BLM wilderness management plans have three functions: (a) to explain the goals, objectives, policies, and specific actions for management of the wilderness resource and all associated resources; (b) to establish the general sequence for implementing necessary management actions; and (c) to establish time frames and procedures for monitoring and revising plans.

PLANNING TOOLS AND TECHNIQUES

TERMINOLOGY AND LOGIC

Regardless of the organization or format used in a wilderness management or stewardship plan, it should include certain basic planning concepts. When labeled, these concepts provide a terminology for discussing management direction, ranging from general goals and objectives to specific actions. Goals and objectives are hierarchical, with one or more objectives under each goal to provide more specific direction to the plan. *The relationship among the components of the framework—the planning logic—is important, that is, goals and objectives are pursued through planned actions.*

***Goals** are broad statements of intent, direction, and purpose, found in (l) legislation, for example, the Wilderness Act; (2) departmental regulations; and (3) agency national policy and philosophy that interpret legislation and establish management direction.* Goals limit the range of potential objectives by providing direction and purpose. Goals are often lofty statements of intent. One example from the Wilderness Act is "to secure for the American people the benefits of an enduring resource of wilderness" (section 2). Sometimes there is confusion about whether a particular statement is a goal or an objective. Distinction should rest on specificity and attainability. *Objectives are attainable in the short term and are more specific than goals.* Broad goals for the NWPS are found in the Wilderness Act, although, as explained earlier, other legislation can also shape goals and influence direction for specific agencies and particular areas.

***Objectives** are statements that describe specific conditions sought in a particular wilderness, serve as criteria for deciding what management actions are needed, and are used as the basis for later evaluation of the effectiveness of management.* In the suggested planning framework that follows, objectives are used to describe wilderness conditions to be achieved and/or maintained through management. Objectives are shaped by the goals they serve. They are descriptions of the field conditions sought through management and serve as criteria for identifying necessary management policies and actions. *Clearly stated objectives are the key to effective* management *plans.* Managers may need to develop objectives for all-important aspects of the wilderness resource and its use, such as visitor access, wildlife, fire management, and recreation.

***Situation and assumption statements** define local conditions, situations, and predictions about changes.* Important information regarding the current situation and assumptions about how things will change in the future should be identified because wilderness areas have different physical characteristics, attractions, and types and levels of use. This information can be helpful in setting feasible objectives and in identifying measures necessary to achieve them.

***Management actions** specify policies, programs, and actions through which objectives for a given wilderness are achieved. Policies* are explicit expressions of intent describing what will be done to attain objectives and sometimes describing what will not be done or otherwise prescribing constraints on management activity. *Programs* are sets of related actions that are combined to help achieve particular objectives under established policy, such as the Leave No Trace program (LNT). *Actions* are specific management practices applied to achieve objectives within the constraints of agency policy and regulations.

***Standards** are measurable statements—based on the objectives—that define minimally acceptable conditions.* They serve as reference points *during monitoring* that can trigger corrective management actions if conditions are not met.

The internal logic reflected in the planning framework is important. Managers develop feasible *objectives* consistent with wilderness goals suited to local conditions in specific areas and acceptable under agency management guidelines that interpret the Wilderness Act. The *objectives* and standards of quality established in the plan, considering current *situations* and *assumptions* about the future, lead to *actions,* which are designed to achieve desired conditions described under the *standards* of quality.

GOAL-ACHIEVEMENT FRAMEWORK—WRITING WILDERNESS MANAGEMENT OR STEWARDSHIP PLANS

The planning framework described in this chapter adapts basic planning principles to wilderness management. It emphasizes planning as a decision-making process that seeks to attain clearly stated management goals and objectives. Goals and objectives stated in plans serve two purposes: (1) they are intentions and aims for deciding which management policies and actions are necessary and (2) they are the targets against which the effectiveness of wilderness management is judged.

Table 8.1 shows a *goal-achievement framework* for organizing and writing wilderness management plans using the previously described concepts and terminology. The approach is a type of management-by-objectives framework. A variety of plan formats can adapt the logic of this framework. Plan organization and format are normally dictated by agency guidelines. This outline is merely one format for focusing wilderness management direction through a logical planning framework.

The following examples are from existing management plans that use concepts described in the goal-achievement framework and the integration of wilderness management and stewardship into the overall planning document for the forest, park, refuge, or planning area. The goals from a plan for the BLM's Bear Trap Canyon unit of the Lee Metcalf Wilderness in Montana (U.S. Department of the Interior, Bureau of Land Management 1985) were prepared as a prototype for future BLM wilderness management plans using the goal-achievement planning framework:

1. To provide for the long-term protection and preservation of the area's wilderness character under a principle of nondegradation. The area's natural condition;

TABLE 8.1 A Goal Achievement Framework for Wilderness Management or Stewardship Plans

Introduction	Brief description of the area, purpose, and organization of the plan. An overview or summary of current conditions affecting management, such as use levels and patterns, special situations, personnel, and general management strategy.
Goals	Concise summary of legislative requirements, departmental guidelines, and national agency policy and philosophy. A statement of goals for the management of the particular wilderness
Objectives	Specific wilderness conditions sought for all important aspects of the wilderness, such as vegetation, recreation, wildlife, fire, trails, and travel. (Topics may vary by agency and area.)
Current Situation and Assumptions	Summary of trends, conditions, and assumptions pertinent to each objective. Judgments about future trends, pressures, and problems pertinent to each objective.
Management Actions	Guiding policies that—considering current situation and assumptions about the future—are necessary to guide actions toward established objectives. Specific practices, programs, and actions that are judged necessary to achieve established objectives.
Monitoring and Standards	Minimally acceptable and measurable standards that trigger corrective management actions, if not met following monitoring assessments.

opportunities for solitude; opportunities for primitive and unconfined types of recreation; and any ecological, geological, or other features of scientific, educational, scenic, or historical value will be managed so that they will remain unimpaired.

2. To manage the wilderness area for the use and enjoyment of visitors in a manner that will leave the area unimpaired for future use and enjoyment as wilderness. The wilderness resource will be dominant in all management decisions where a choice must be made between preservation of wilderness character and visitor use.
3. To manage the area using a minimum of tools, equipment, or structures to accomplish the objective successfully, safely, and economically. The chosen tool, equipment, or structure should be the one that least degrades wilderness values temporarily or permanently. Management will seek to preserve spontaneity of use and as much freedom from regulation as possible.
4. To manage nonconforming but accepted uses permitted by the Wilderness Act and subsequent laws to prevent unnecessary or undue degradation of the area's wilderness character. Nonconforming uses are the exception rather than the rule; therefore, emphasis is placed on maintaining wilderness character.

Another example is the integrated approach to the Land and Resource Management Plan (LRMP) in the White Mountain National Forest (WMNF) of New Hampshire and Maine (U.S. Department of Agriculture, Forest Service 2005). Wilderness is addressed throughout the LRMP for the WMNF and more specifically in a separate appendix of the LRMP for the wilderness areas in the WMNF.

The Wilderness Management Plan specifically addresses all five wilderness areas within the WMNF: Great Gulf (5,500 acres), Presidential Range–Dry River (29,000 acres), Pemigewasset (45,000 acres), Sandwich Range (25,000 acres), and Caribou-Speckled Mountain (14,000 acres). The Wilderness Management Plan (U.S. Department of Agriculture, Forest Service 2005, E-3) begins by setting the stage for implementing the LRMP within wilderness areas:

> These lands are managed to allow natural processes to continue with minimal impediment, to minimize the effects and impacts of human use, to provide primitive and unconfined recreation opportunities, to foster appreciation of the qualities of wilderness landscapes, to continue use for educational and scientific purposes, and to recognize their evolving roles in the history of the landscape.
>
> This management plan describes processes and actions aimed toward further realizing these goals. Our intent is to provide strong, clear management, in order to maintain Wilderness character. These values include a balance of use and preservation, an understanding of and support for protection of these lands, and a perpetuation of Wildernesses' roles as representatives of landscapes minimally affected by the impacts of human use.

The WMNF Wilderness Management Plan uses the components of the goal-achievement framework as a second level of planning detail within the LRMP to explicitly state the standards, monitoring methods, and management actions necessary to achieve the overall goals and objectives.

WILDERNESS PLANNING AND MANAGEMENT FRAMEWORKS

The goal-achievement framework for wilderness management and stewardship planning presented in this chapter are frequently combined with other frameworks related to wilderness planning (McCool et al. 2007). The frameworks that are compatible with the overall goal-achievement framework all include one common component: monitoring using standards and assessments that compare current conditions with goals and objectives. The purpose is to measure the effectiveness of the management actions and for accountability in the expenditure of agency resources and effort. Each framework has a different geographic scale of consideration from landscape level to localized area situations and conditions and each framework has a different time scale from long-term to short-term. While each framework presented here can be used compatibly with the goal-achievement framework, understanding the strengths and utility of each framework requires more in-depth knowledge than can be presented in this chapter. The brief summaries that follow are intended to suggest additional frameworks for further consideration and are presented only as they to relate it to the goal-achievement framework.

Limits of Acceptable Change (LAC)

Limits of acceptable change (LAC) has become established as a useful wilderness management and planning tool, especially where there are conflicts between uses, trade-offs between goals, and concerns about the control of unacceptable change (Cole and Stankey 1997; McCoy et al. 1995; Stankey et al. 1985). The LAC approach is discussed in detail in Chapter 9. The goal-achievement framework for wilderness management planning and the LAC method focuses on developing wilderness management actions to achieve desired conditions. However, the two planning frameworks differ mainly in that the LAC approach uses very specific standards rather than more general objectives. Some of the differences between goal-achievement framework objectives and LAC standards are:

1. Goal-achievement framework objectives define a target to work toward (a "directional statement") based on what is considered desirable (what you are striving to achieve), do not clearly identify when there is a serious problem, may be written to reflect minimally acceptable management performance, and do not identify an "end point"; thus, once an objective is reached, there is no decision on whether to continue current management actions or not.
2. LAC standards define an absolute limit and an end point (a "state") based on what are considered minimally acceptable conditions (the maximum amount of deviation from "ideal" conditions considered acceptable), explicitly define the compromise between conflicting goals, clearly identify when there is a serious problem (conditions are no longer acceptable), and provide assurance to the public that "restrictive" actions will not be implemented unless the standard is not being met.

Goal-achievement framework objectives and LAC standards are on different tiers or levels in planning time, with long-term versus short-term frames, and different scales of area being planned (large forest and park size units compared to specific management areas). Use the goal-achievement framework objectives for general management planning and LAC standards for more specific implementation plans. Especially consider using the LAC approach when there is an actual or potential conflict between two or more goals, and when there is a need to balance between those goals to some extent. For example, when there is an opportunity to obtain solitude versus providing freedom from access restrictions, it may be appropriate to use a standard of solitude with a 90 percent probability of encountering no more than ten groups per day and a maximum of twenty people observed at one time (at a particular attraction area).

The LAC process consists of four major components: (1) specifying acceptable and achievable resource and social conditions, defined by a series of measurable parameters; (2) analyzing the relationship between existing conditions and those judged acceptable; (3) identifying management actions judged to best achieve these desired conditions; and (4) monitoring and evaluating management effectiveness. These four components, in turn, are broken down into ten planning steps to facilitate application (see Chapter 9). For example, managers (with help from the public) identify the issues and concerns that need to be resolved through the planning process and use indicators of resource and social conditions to determine standards for comparison (Merigliano 1989; Merigliano and Krumpe 1986).

Both the goal-achievement and LAC frameworks incorporate monitoring and management-action revision procedures. Management actions are not fixed but can be revised if monitoring shows that desired wilderness conditions are not being achieved. The LAC framework adds more specificity in the monitoring and evaluation of whether the use impacts have exceeded the limits defined as the minimum acceptable conditions for a given wilderness area. The LAC process can be used independently for a specific management situation, such as conflicts in use between wilderness hikers and packstock users, or as a component in an overall goal-achievement planning process (that is, the LAC framework for specific use issues or conditions is subordinate to the overall wilderness goal-achievement plan).

Visitor Use Management (VUM)

VUM is a multi-agency approach to outdoor recreation management developed by the Interagency Visitor Use Management Council (2016) to address management of visitor experiences and protect resources on federally managed lands. VUM has application in wilderness areas since, like the goal-achievement framework, it is guided by the legislative and agency policy purposes for which the federal lands are being managed.

> Visitor use management is fundamental for maximizing benefits for visitors while achieving and maintaining desired resource conditions and visitor experiences on federally managed lands and waters. By using this Visitor Use Management framework, managers collaboratively develop long-term strategies for providing access, connecting visitors to key visitor experiences, protecting resources, and managing visitor use. The purpose of the framework is to provide cohesive guidance on four major elements for analyzing and managing visitor use on federally managed lands and waters. It is also intended to provide a legally defensible, transparent decision-making process that meets law and policy requirements, ensures agency accountability, and provides sound rationales upon which to base management decisions and actions. A common thread throughout the framework is the use of the sliding scale. A sliding scale is used to ensure the investment of time, money, and other resources for a project is commensurate with the complexity of the project and the consequences of the decision. Issues with clearly small impacts usually require less depth and breadth of analysis than those with impacts of greater significance. Overall, this framework is meant to be adaptable to different agencies' policies and regulations and yet allow for a professional, comprehensive, and consistent approach to visitor use management on federally managed lands and waters. (IVUMC 2016)

The VUM framework focuses on outdoor recreation management, similar to many of the LAC framework uses in wilderness and is compatible with the goal-achievement framework and subordinate to it. The VUM framework can be used at different geographic and time scales to address either specific issues or more generalized concerns. The VUM framework is described in more detail in Chapter 9.

Wilderness Opportunity Spectrum

Recreation managers and researchers have made a concerted effort to classify the recreation opportunity spectrum on public lands and the kinds of facilities, management practices, and visitor behavior appropriate to each type of opportunity (Buist and Hoots 1982; Clark and Stankey 1979). This *recreation opportunity spectrum* (ROS) classifies *opportunity* from primitive, semiprimitive nonmotorized, semiprimitive motorized, roaded and natural, rural, and urban zones. *Not surprisingly, as managers began to address increasingly complex wilderness management situations, the idea of a wilderness opportunity spectrum (WOS) emerged as part of the management strategy to address finer gradations of primitive and semiprimitive classes. The WOS includes, for example, pristine, primitive, and portal (entry) designations, indicating decreasing degrees of naturalness and solitude. The WOS, like the ROS, is a kind of zoning—delineating particular areas where different management prescriptions or restrictions on visitor behavior apply.* Other examples include no-camping zones; trailless zones, where only cross-country travel and special minimum-impact camping practices are allowed; and special management zones having particular problems, high-use sites, impacted locations, or perhaps sensitive wildlife areas. The WOS framework can help focus the goals and objectives of a goal-achievement plan and is a subordinate to such a plan.

The WOS approach is appropriate in certain situations. First, remember the cardinal rule of wilderness management—*do only what is necessary.* Implement area or zoning restrictions only when required to protect the wilderness resource and consider the impact that the zoning would have on visitor perceptions and use. Second, *make area or zoning restrictions clear to users*, along with its rules and expectations. Third, *zoning should not be used to justify non-wilderness conditions that presently exist*; when such conditions exist, they should be corrected, *rather than* assigned to a "special management zone." Congress established wilderness as a land designation, and agencies should not try to change its status through zoning that allows for standards lower than those intended under the Wilderness Act.

Wilderness Character Monitoring

"*Keeping It Wild*" (Landres et al. 2015) is an interagency framework for monitoring trends in selected attributes of wilderness character and is intended as a long-term and large area approach. The wilderness character monitoring framework can help focus the goals and objectives of a goal-achievement plan and can address if management and stewardship activities affect attributes of wilderness character over long time periods. Since it is a monitoring approach used across the National Wilderness Preservation System, it facilitates comparison with other wilderness areas at similar WOS classes.

Minimum Requirements Analysis Framework

The Minimum Requirements Analysis Framework (MRAF) is the result of an interagency project to make the predecessor decision guide (Minimum Requirements Decision Guide) easier to use and more broadly applied as a framework (www.wilderness.net). The MRAF was intended to assist managers and planners to: (1) evaluate management actions proposed in wilderness involving a use otherwise prohibited by Section 4(c) Wilderness Act; and (2) consider wilderness appropriate choices about management actions to address an issue or situation. While some agencies also require or encourage the use of the MRAF for non-prohibited uses in wilderness, others (Cole 2007) have suggested that the MRAF should be used more often to address planning at larger spatial and temporal scales. This approach involves consideration of desired future conditions, especially in relation to goals and objectives impacted by global climate change. The MRAF can help focus the goals and objectives of a goal-achievement plan and can address whether proposed management actions are appropriate under the Wilderness Act.

PREPARING WILDERNESS MANAGEMENT PLANS

Regardless of the planning approach and frameworks selected, the following advice may be relevant in improving plans.

The actual writing of a wilderness management or stewardship plan has some pitfalls, one of which is the *potential for investing too much effort in stating current situations and assumptions about the future.* Some of each are needed, but if they are too detailed, they can become the focus of the plan. In extreme situations, a plan can attain a problem-solution focus, with current situation statements and assumptions describing problems in detail and presenting dire forecasts of worsening situations. This orientation can lead to heavy-handed management prescriptions to fulfill assumptions, whether the assumptions are well founded. In many cases, accurate data are lacking, so that assumptions about the future are no better than guesses. *A good plan needs to balance clear descriptions of current situations against explicit assumptions, both supported with field experience and data.*

Situation and assumption statements are easy to dwell on because they are easier to write than management objectives. However, a plan that focuses too much on situation and assumption statements will lack clear objectives as criteria for prescribing management policies and actions, and standards for judging their attainment. For example, a current situation statement describing heavy use, combined with assumptions predicting substantial increases in use, could lead indirectly to very restrictive policies and actions. Although these might ultimately be necessary, a recreational-use objective that describes the specific experience opportunities to be provided and standards for acceptable conditions would provide a better basis for identifying necessary actions. This approach is also easier to explain and to justify to critics, because debate and disagreement over wilderness management may be focused on the stated objectives and actions necessary to achieve them, rather than on the accuracy of situation and assumption statements or on individual management actions that, considered in isolation, might not appear necessary.

Clearly stated objectives are the key to good wilderness management plans. Objectives are particularly difficult to write for wilderness because of the subjective nature of the resource and the experiences it offers. Nevertheless, it is important to have clear objectives so that they can guide policies and actions and set standards for measuring attainment. *It is helpful to think of objectives as positive statements of wilderness condition or as experience opportunities that management seeks to preserve or provide (Fig. 8.2).*

FIGURE 8.2 Wilderness management plans may identify locations where special management is needed, such as travel corridors, heavy-use locations, unusual attractions, or vegetation zones where special restrictions apply. A no-campfire regulation in high alpine zones may help preserve aesthetic features such as this ancient whitebark pine snag in the John Muir Wilderness, California; the tree might otherwise be used as firewood. Photo courtesy U.S. Forest Service.

Not all objectives can be described with equal specificity. The precision of individual objectives varies with the aspect of wilderness under consideration—but usually the more quantifiable the objective, the better. Objectives ranging from *general* to *specific* can still suggest management direction. However, *vague objectives give few clues to necessary management direction. Compare the following examples that focus on planning:*

- *A Specific Objective*—To maintain lakes and watercourses in their current natural condition, with a high degree of water clarity and water quality suitable for primary human contact, subject to natural forces and free of human-caused contaminants.
- *A General Objective*—To maintain lakes and watercourses free of human-caused contaminants.
- *A Vague Objective*—To protect lakes, watercourses, and water quality.

Understanding the distinction between general and vague objectives is important. Objectives can be stated in broad, general terms and still retain fairly definite implications about the kinds of actions needed to achieve them. Vague objectives, on the other hand, are abstract and lack clear implications for management actions. *Although an objective might be general, its management implications should be clear.*

It is also important to recognize that the specificity of goals and objectives is directly related to their location in the planning hierarchy or framework. *The closer an objective is to the field-action level, the more specific it should be.* Compared to the lofty goals of the Wilderness Act ("to preserve, unimpaired, a wilderness resource"), management objectives for one particular aspect of wilderness will seem quite specific—for example, "to provide for primitive recreation only to the extent that naturalness, outstanding opportunities for solitude, and physical and mental challenge are preserved." Farther down the hierarchy, a field manager's objectives may be to implement very specific actions derived from goals, objectives, and policies formulated higher up in the planning process. *If a plan has internal logic, it should be possible to trace field actions to higher levels of the planning process (goals, objectives, and policies) by asking why the activity is being carried out.*

Management plans are most useful if they are straightforward, well organized, and readable. They should clearly inform managers and users of the management direction for the wilderness, what this entails in policies that govern field actions, and the kind of major actions that will be carried out at specific locations. However, if plans are so complex that their internal logic is not apparent to the public, the advantages they gain through complexity will be lost.

The level of detail that wilderness management plans should contain is an important consideration. A plan containing 200 pages of single-spaced typing might be so overwhelming that managers and the public may never read it. On the other hand, except for the simplest circumstances, a plan of only ten pages will not contain enough detail, even though it might give general direction. Again, balance of generality and detail is needed, depending on the size, complexity, levels of visitation, degree of public interest, and problem situations in the specific wilderness. *The plan should contain sufficient detail to describe all objectives, policies, and the what and where of particular actions; but the when, how, and by whom level of detail is more appropriately included in action, operating, or annual work plans.*

Public Involvement and Plan-Review Process

Public participation is especially important for managing wilderness. The public must have an opportunity to help formulate wilderness management or stewardship plans if these documents are to have credibility and acceptance. Plans not supported by the public will be difficult to implement—in fact they may be legally appealed and their implementation blocked under NEPA regulations requiring public involvement in a significant environmental action such as a wilderness management or stewardship plan.

Involving the public is an accepted part of the decision-making process in the federal agencies with wilderness management responsibilities. The 1964 Wilderness Act was one of the first resource-management laws that required public involvement in decision-making (U.S. Public Law 88-577, Sec. 3[d][l][d]) and was followed in 1970 by NEPA (U.S. Public Law 91-190), which mandated public involvement in all natural-resource decisions having potential for significant environmental impact.

The public is a source of ideas, a sounding board for the acceptability of proposed management direction and policy, and a potential partner in its formation and implementation. Furthermore, wilderness planning takes place in a political arena, in which consensus and negotiation are every bit as important as scientific data and logic (Krumpe and Stokes 1993). Planners and managers recognize that dual conditions are required for effective planning. First, a technically sound planning process is required for explicitness and to facilitate the search for reasonable alternatives by systematically working through a logical sequence. This is a necessary but not sufficient condition for effective planning. Managers understand that they need informed consent among those affected by the plan about the proposed course of action. In the politicized settings in which wilderness planning take place, the values in conflict are often well articulated, expressed, and pursued by the various contending interest groups. The arena of conflict may shift over time, but it still encompasses the agency and its perceived mission. Indeed, one or several groups may in reality hold the power of implementation rather than the planning agency (Krumpe and McCool 1997).

Planners and wilderness managers can become frustrated when politics gets in the way of rational planning. They become frustrated when decisions seem to be motivated more by political considerations than by purely scientific or philosophical considerations of fairness, equity, or other idealized values they hope would guide the management of publicly held natural resources (Krumpe 2000).

The public, on the other hand, experiences equal frustration at the extensive time and effort going into planning that often result in no change, or in plans that do not address the needs of a particular interest or may be so legalistic and bureaucratic as to confuse and alienate readers. As a result, both managers and the public have become disillusioned that science does not, or often cannot, give them the facts they need to answer the thorny questions raised in wilderness planning. In fact, it is a common delaying tactic for one user group or another to simply question the legitimacy of (and thus dismiss) any science that runs counter to their values or expectations.

Most managers who have successfully used public involvement are impressed with its helpfulness in making better and more acceptable decisions. Many user groups are affected by wilderness management: hiking and climbing clubs, conservation groups, hunters and anglers, horse riders, photographers, youth organizations, outfitters, and others (Fig. 8.3). Public involvement in wilderness matters extends beyond wilderness planning and management to include involvement in wilderness work as well. Volunteers, conservation group work parties, adopt-a-trail efforts, and even private contracting of wilderness work projects all increase the involvement of the public in wilderness management.

As mandated by the NEPA, the four federal wilderness management agencies have adopted procedures that integrate public involvement with the planning process. Agency staff routinely prepare environmental assessments and environmental impact statements during the wilderness planning process and before making land-use decisions. Such documents, with their sections detailing potential alternatives and environmental consequences, provide multiple opportunities for public involvement, from the initial identification of issues

FIGURE 8.3 An early spring visitor in the Selway-Bitterroot Wilderness on the Idaho-Montana border plunge-steps down a snow slope. Wilderness area management plans need to consider all types of visitor use throughout its entire area and through the year, not just seasonal heavy use. Photo courtesy Chad P. Dawson.

(called scoping) to lengthy public comment periods on the draft and final publications of EIS and wilderness plans.

Many different strategies may be successful for securing public involvement. Some may have widespread applicability. Beginning in the late 1990s, the Internet has been widely used to facilitate public involvement, with plans posted for downloading from World Wide Web sites and comments received over electronic mail. Face-to-face methods are still the heart of good public involvement. Public-involvement procedures are flexible to fit different circumstances. Both managers and the public must be prepared for the time required to get public input, incorporate it into new drafts, and then get agency staff and the public response to the revisions. One to two years, or more, may be required to adequately involve the public in a wilderness management and EIS planning process.

Bear Trap Canyon (BLM)

The public-involvement process used during development of the Bear Trap Canyon unit's wilderness management plan in 1984 is an early example of a successful program that began by establishing a five-member advisory group. This Bear Trap Canyon Wilderness Committee, representing a variety of interests both private and commercial, met as a group in January. BLM staff provided the group with a list of management issues to be addressed in the management plan and requested comments.

In March, members of the committee, key individuals, and agency officials were sent a copy of an in-house draft of the plan to review. Committee members were asked to disseminate pertinent portions of the document for comment to their constituencies. Next, individuals on the BLM's district office wilderness mailing list were sent postcards and asked to return them if they wanted copies of the draft Bear Trap plan for review.

A printed draft, revised in response to concerns of reviewers, was sent to members of the public who had indicated an interest in reviewing the plan, to the advisory group, and to agencies. Public notices in local newspapers informed the public that copies of the draft were available. The document sent to the public consisted of a draft plan and draft EA.

Due to public interest, the BLM extended the closing date of the original forty-five-day comment period from August 31 to October 12, 1984. During the comment period, the agency hosted two open houses in local communities. Twenty-seven written comments were received; in general, they supported continuing existing management of the area. Most respondents supported long-term protection and preservation of wilderness values, and they believed that existing use levels did not warrant restrictive measures.

What has become a standard procedure is for managers to print the public comments on a draft environmental analysis document and their response to such comments as part of the final version of the environmental analysis. In the Bear Trap plan, the planning team organized its response to public comment issue by issue.

The following issues illustrate the variety of concerns that can emerge from public comment: management for long-term protection of wilderness values versus maximizing visitor enjoyment, prohibition of overnight camping for rafting parties, trail closures and maintenance, prohibition of horse use, application of a let-fires-burn policy, construction of backcountry toilets, the number of commercial outfitters, use of paddleboats, and Air Force overflights.

In response to public comment and additional in-agency review, the BLM revised portions of the draft document before printing the final draft of the plan. The final plan and EA were published and distributed to the public in January 1985, one year after the first public-involvement efforts began.

Denali National Park and Preserve (NPS)

The 6-million-acre Denali National Park and Preserve includes 2.1 million acres that were designated by Congress as wilderness in 1980. The designated wilderness and much of the park are protected and managed collectively as backcountry. Development of the Backcountry Management Plan for Denali National Park and Preserve (U.S. Department of the Interior, National Park Service 2005a) included an amendment to the General Management Plan for Denali National Park and an EIS and used an extensive public-involvement and consultation process with federal, state, and local governments and native tribal governments throughout the scoping and planning processes.

NPS planners began this planning process by publishing their Notice of Intent to prepare an EIS and then held four public scoping meetings that were attended by 150 people. The planners received sixty-five written comments during fall 1998 through the spring of 1999 regarding issues and impacts that should be addressed in the Backcountry Management Plan. As a follow-up to the scoping meetings, a summary document describing the primary range of activities and alternatives to be evaluated in the plan and EIS was sent to more than 2,000 individuals, agencies, and organizations in January 2001. The alternatives for management and impact topics were further refined as part of five open house meetings in February 2001.

The NPS consulted with federal agencies, state and local governments, native tribal governments, and numerous user groups and organizations. During the winter and spring of 2002, collaborative planning workshops were held in two communities to discuss alternatives and issues associated with climbing and mountaineering, airplane use, subsistence, and other uses, especially those uses allowed under ANILCA (Fig. 8.4).

FIGURE 8.4 Wilderness planning in Alaska and some other wilderness areas in the western U.S. needs to consider activities such as sightseeing by aircraft. Photo courtesy Chad P. Dawson.

When NPS planners and managers released the Draft Backcountry Management Plan and EIS in February 2003, they held informational workshops in six communities. The public comment period on the draft plan was initially open for seventy-five days, from February 25 to May 7, 2003, and then was extended to May 30, 2003. During that time, public hearings were held in seven communities and 9,370 written comments were received from two agencies, 27 organizations, and 9,341 individuals. As a result of the extensive public comments, the NPS staff revised the draft plan and solicited some additional public comment before releasing the final plan in April 2005.

These examples are only two of many strategies that could be followed to involve the public in preparing wilderness management or stewardship plans. Each of the federal land-management agencies has developed its own public-involvement procedures based on their experience. *To summarize: (1) public involvement in wilderness management is essential and valuable, (2) many approaches are possible,* and (3) *adequate time (one to two years or more) is required for public involvement.*

Although this discussion focuses on how the public may assist with the review of draft plans, considerable technical and policy review of plans takes place at different levels within the agency and within other public agencies. It is essential that any other governmental units, such as fish and wildlife, forestry, environmental, or other resource-protection and management agencies, have a chance to review and provide input to plans. The challenge is to secure the optimum review and input from the public and agencies to afford managers the benefit of diverse opinions and to better represent the public and wilderness users in plan development. All input is advisory to the wilderness planners who must ultimately develop a plan within the legal requirements set by the Wilderness Act, the legislation designating the wilderness area, and agency policies and regulations.

CRITERIA FOR EVALUATING WILDERNESS MANAGEMENT PLANS

Growing public involvement ensures that wilderness management or stewardship plans will come under increasing public scrutiny; moreover, the growth of wilderness management as a resource specialty means that other managers will review plans. What criteria are appropriate for evaluating plans regardless of their format, specific agency requirements, or type of public involvement? We suggest the following questions to guide the review and evaluation of wilderness management or stewardship plans:

1. Does the plan summarize and explain the relationship of management goals to the Wilderness Act, legislation designating the area, departmental guidelines, national agency policy that guides and directs management of the area, and any higher-order plan of which it is a part, such as a forest plan in the USFS or a general management plan in the NPS? Including such goals will help relate management of the individual wilderness to the larger agency land-management system and the NWPS, of which it is a part.
2. Are local conditions, user activities, and internal and external threats relevant to management of the wilderness concisely described and explained?

3. Is the general management strategy concisely explained? This strategy might include (a) differing administrative methods by various units of the agency, the numbers of managers, and their responsibilities; (b) a description of user requirements such as a permit system; (c) use of the WOS or other zoning scheme; and (d) other details essential to implementing the management strategy.
4. Does the plan have an internal logic that links objectives to prescribed management policies and actions? It is essential that some kind of framework guide prescription of management policies and actions based on their *intention* to achieve objectives or some desired condition in the area.
5. How well does the plan consider alternative actions for meeting management objectives? The NEPA requires managers to analyze alternatives and how each alternative would attain goals or desired conditions.
6. Does the plan address the need for coordination of its resource-management activities and nonconforming uses, such as a national scenic trail crossing a wilderness, grazing activities, access to private land inholdings, or endangered species protection? Coordinating with other affected parties, such as adjacent landowners or state wildlife departments responsible for managing wildlife populations, is essential.
7. Were the managers who administered the wilderness area directly involved in the preparation of the plan? Involvement builds commitment to implementing the plan, because the best plan is of little value if it is not fully understood by managers and used to guide everyday stewardship.
8. Does the wilderness management direction contained in the plan help to resolve issues and management concerns facing field managers? Does the plan respond to important issues, concerns, and opportunities raised during public involvement efforts? To be effective and useful, plans need to provide site-specific application of wilderness management policy to resolve the everyday challenges of field managers seeking to achieve the plan's objectives.
9. Does the plan provide for a monitoring system, using field measures of indicators of wilderness conditions—biological, physical, and social—to determine if objectives are being achieved to the desired level of the standards? How often is monitoring done, and how will the data be used? The plan is not complete unless it includes provisions for objective data collection to see if the plan's objectives and standards are being achieved (Fig. 8.5).
10. Key measures of a plan's success are found in answers to these four questions:
    - Does the plan address and resolve key problems/conflicts/potential threats?
    - Does the plan contain an explicit rationale for decisions?
    - Is there an agency commitment to move forward with plan implementation?
    - Will implementing the plan's direction improve the current wilderness situation and condition?
11. Does the plan specify when and by whom the plan will be reviewed and updated? Does it identify what conditions or situations might prompt an earlier review? This information will help keep a plan from becoming obsolete.

FIGURE 8.5 Monitoring wilderness conditions and experiences involves studying the resources and its visitors; here visitors are asked to fill out surveys about their overnight camping experiences on a canoe trip. Photo courtesy Chad P. Dawson.

SUMMARY

Good planning is essential to good management. Planning is a formal process of thinking ahead about what conditions are desired and how to achieve them, the management problems likely to be encountered, and alternative methods for resolving them. This chapter described how wilderness management plans fit into the planning processes of federal agencies, and suggested some elements of good planning, a framework for developing plans, and several examples of plans. A planning process should prescribe site-specific policies and actions implementing the framework of principles of wilderness management outlined in Chapter 7.

STUDY QUESTIONS

1. Why is it important to have a management plan for a wilderness area?
2. How does the NEPA affect wilderness management plans?
3. Each of the four federal wilderness managing agencies (BLM, FWS, USFS, and NPS) has two or more levels of planning applicable to its wilderness areas. Briefly describe them.
4. Define the following basic planning concepts: goals, objectives, situation and assumption statements, management actions, and standards.
5. How can too strong a focus on describing current situations and assumptions about the future cause problems in preparing and implementing a wilderness management plan?
6. What are three criteria for deciding whether zoning restrictions are useful and appropriate in a particular wilderness area?
7. Why is it important to involve the public in the development of a wilderness management plan?
8. What are some questions to consider in evaluating wilderness management plans?
9. How is the goal-achievement framework compatible with other planning frameworks? Give one example.

ACKNOWLEDGMENTS

This chapter was written for the first and second editions by John C. Hendee and Russell von Koch, recreation planner, Bureau of Land Management, Moab, Utah. John C. Hendee and Chad P. Dawson, with help from Ed Krumpe, professor of resource recreation and tourism at the University of Idaho, and Linda Merigliano, wilderness coordinator, Bridger-Teton National Forest, Jackson, Wyoming, revised and updated this chapter for the third edition.

Chad P. Dawson and John C. Hendee updated this chapter for the fourth edition and Chad Dawson updated the fifth edition.

CASE DISCUSSION: COMPREHENSIVE CONSERVATION PLANNING FOR THE KENAI NATIONAL WILDLIFE REFUGE, ALASKA

The Kenai National Wildlife Refuge (KNWR) extends over 1.98 million acres on the Kenai Peninsula of Alaska and includes 1.3 million acres of designated wilderness. The refuge was originally established as the Kenai National Moose Range to protect the Alaska-Yukon moose, and it has brown and black bears, caribou, Dall sheep, mountain goats, wolves, lynx, wolverines, eagles, and thousands of shorebirds and waterfowl. The waters of the refuge include habitat for chinook, sockeye, coho, and pink salmon; Dolly Varden char; rainbow trout; and arctic grayling. Landscapes vary from the Harding Ice Field to mature spruce forests and glacial lakes. The KNWR's scenic landscape and wildlife include what has been described as a miniature version of many of the habitats in Alaska and, because of its proximity to Anchorage, draws half a million visitors a year, more than any other wildlife refuge in Alaska.

All refuges are required to be managed in accordance with an approved comprehensive conservation plan. The KNWR was involved in the planning process of revising the refuge's comprehensive conservation plan (CCP), which had been in place since 1985 (U.S. Department of the Interior, U.S. Fish and Wildlife Service 1985). Developing a CCP required following guidance found in the National Wildlife Refuge System Improvement Act of 1997 (U.S. Public Law 105-57), ANILCA, the Wilderness Act, and NEPA. Revising the KNWR CCP provided direction for management of the refuge for fifteen years after being approved. During 2003 to 2007, the KNWR staff progressed in developing the revised CCP and EIS through the leadership of the refuge's *interdisciplinary planning team*, consisting of U.S. Fish and Wildlife Service, Alaska Department of Fish and Game, and Alaska Department of Natural Resources staff. Distribution of the public review draft was published in May 2008 and a final plan was completed in 2010. As part of the public input process in 2005, the *interdisciplinary planning team* requested public input on a variety of issues and concerns and requested public feedback (U.S. Department of the Interior, U.S. Fish and Wildlife Service 2005).

One of the preliminary management issues in 2005 was how the KNWR could enhance wildlife-oriented recreation such as wildlife viewing opportunities. The public was asked to comment on four general alternatives (not specific to wilderness or non-wilderness areas of the KNWR), which were summarized as:

1. Maintain the current management situation where opportunities and facilities are limited and habitat projects with associated wildlife viewing facilities are not provided.
2. Improve existing wildlife viewing facilities (e.g., pullouts on roads and waysides) only.
3. Improve existing wildlife viewing facilities and develop new additional wildlife viewing facilities (e.g., overlooks, towers, blinds, boardwalks).
4. Improve existing wildlife viewing facilities and develop additional wildlife habitat improvement projects with new additional wildlife viewing facilities (e.g., overlooks, towers, blinds, boardwalks).

Wildlife viewing is challenging on the KNWR for several reasons: the vegetation is dense and the line of sight is limited except on roads, water bodies, and glacial areas; wildlife generally do not concentrate for easy viewing; wildlife are not habituated to humans in their environment and generally move away from humans; and some large animals can be a threat to human safety if startled or approached too closely.

CASE DISCUSSION QUESTIONS

1. Is the range of the four preliminary management alternatives presented to enhance wildlife viewing opportunities acceptable or unacceptable to you as a local visitor, as a tourist from outside Alaska, as a wilderness advocate, and as a wilderness manager? Why might the responses of different people vary between being acceptable and unacceptable?
2. To what degree should the KNWR enhance wildlife-oriented recreation such as wildlife viewing opportunities in the wilderness areas of the refuge and the non-wilderness areas of the refuge?
3. Balancing between human enjoyment and protecting the resource is a complex process. What planning framework may be helpful to the interdisciplinary planning team? Why?

REFERENCES

Buist, Leon; Hoots, Thomas A. 1982. Recreation Opportunity Spectrum approach to resource planning. *Journal of Forestry*. 80: 84–86.

Clark, Roger N.; Stankey, George H. 1979. The Recreation Opportunity Spectrum: A framework for planning, management, and research. General Technical Report PNW-98. Portland, OR: U.S. Department of Agriculture, Forest Service, Pacific Northwest Forest and Range Experiment Station.

Cole, David N.; Stankey, George H. 1997. Historical development of limits of acceptable change: Conceptual clarifications and possible extensions. In: McCool, Stephen F.; Cole, David, comps. *1997 Proceedings: Limits of Acceptable Change and Related Planning Processes: Progress and Future Directions;* May 20–22, 1997; Missoula, MT. INT-GTR-371. Ogden, UT: U.S. Department of Agriculture, Forest Service, Rocky Mountain Research Station, pp. 5–9.

Cole, David N. 2007. Scaling-up the minimum requirements analysis for big wilderness issues. International Journal of Wilderness 13(1): 8-12.

Interagency Visitor Use Management Council (IVUMC). 2016. Visitor use management framework: A guide to providing sustainable outdoor recreation. 130p.

Krumpe, Edward E. 2000. The role of science in wilderness planning—A state-of-knowledge review. In: Cole, D. N.; McCool, S. F.; Borrie, William T.; O'Loughlin, Jennifer, comps. *Wilderness science in a time of change conference*, Vol. 4; *Wilderness Visitors, Experiences, and Visitor Management;* May 23–27, 1999; Missoula, MT. RMRS-P-15-Vol-4. Ogden, UT: U.S. Department of Agriculture, Forest Service, Rocky Mountain Research Station, pp. 5–12.

Krumpe, Edward E.; McCool, Stephen F. 1997. Role of public involvement in the limits of acceptable change wilderness planning system. In: McCool, Stephen F.; Cole, David, comps. *1997 Proceedings: Limits of Acceptable Change and Related Planning Processes: Progress and Future Directions.* May 20–22, 1997; Missoula, MT. INT-GTR-371. Ogden, UT: U.S. Department of Agriculture, Forest Service, Rocky Mountain Research Station, pp. 16–20.

Krumpe, Edward E.; Stokes, Jerry. 1993. Application of the limits of acceptable change planning process in United States Forest Service wilderness management. In: Hendee, J. C.; Martin, V. G., comps. *Proceedings of the 5th World Wilderness Congress Symposium on international wilderness allocation, management and research;* September 24–October 2, 1993; Tromso, Norway. Fort Collins, CO: International Wilderness Leadership Foundation, pp. 186–191.

Landres, Peter; Barns, Chris; Boutcher, Steve; Devine, Tim; Dratch, Peter; Lindholm, Adrienne; Merigliano, Linda; Roeper, Nancy; Simpson, Emily. 2015. Keeping it wild 2: An updated interagency strategy to monitor trends in wilderness character across the National Wilderness Preservation System. Gen. Tech. Rep. RMRS-GTR-340. Fort Collins, CO: U.S. Department of Agriculture, Forest Service, Rocky Mountain Research Station. 114 p.

McCool, Stephen F.; Clark, R. N.; and Stankey, G. H. 2007. An assessment of frameworks useful for public land recreation planning. General Technical Report PNW-GTR-705. Portland, OR: U.S. Department of Agriculture, Forest Service, Pacific Northwest Research Station.

McCoy, K. Lynn; Krumpe, Edwin E.; Allen, Stuart. 1995. Limits of acceptable change planning evaluating implementation by the U.S. Forest Service. *International Journal of Wilderness.* 1(2): 18–22.

Merigliano, Linda L. 1989. Indicators to monitor the wilderness recreation experience. In: Lime, D. W., ed. *Managing America's Enduring Wilderness Resource.* Conference Proceedings; September 11–17, 1989; Minneapolis, MN. St. Paul: University of MN. Extension Service, pp. 156–162.

Merigliano, Linda L.; Krumpe, Edwin E. 1986. Scientists identify, evaluate indicators to monitor wilderness conditions. *Park Science.* 6(3): 18–19.

Stankey, George H.; Cole, David N.; Lucas, Robert C.; et al. 1985. The limits of acceptable change (LAC) system for wilderness planning. General Technical Report INT-176. Ogden, UT: U.S. Department of Agriculture, Forest Service, Intermountain Forest and Range Experiment Station. 37p.

U.S. Department of Agriculture, Forest Service; U.S. Department of the Interior, Bureau of Land Management. 2006. Policies and Guidelines for Fish and Wildlife Management in National Forest and Bureau of Land Management Wilderness. Washington, DC: U.S. Department of Agriculture, Forest Service; U.S. Department of the Interior, Bureau of Land Management.

U.S. Department of Agriculture, Forest Service. 2005. White Mountain National Forest Land and Resource Management Plan. Laconia, NH: U.S. Department of Agriculture, Forest Service, Eastern Region.

———. 2007. Wildernesses, primitive areas, and wilderness study areas. In: *Forest Service Manual,* chap. 2320. Washington, DC: U.S. Department of Agriculture, Forest Service.

U.S. Department of the Interior, Bureau of Land Management. 1985. Wilderness management plan for the Bear Trap Canyon Unit of the Lee Metcalf Wilderness. BLM-MT-ES PS-003-4332. Billings, MT: U.S. Department of the Interior, Bureau of Land Management.

———. 1995. Maricopa Complex Wilderness Management Plan, Environmental Assessment and Decision Record. EA AZ-026-94-20. Phoenix, AZ: U.S. Department of the Interior, Bureau of Land Management. Phoenix District, Lower Gila Resource Area.

———. 2000. *Wilderness Management.* 43 *Code of Federal Regulations,* Parts 6300 and 8560. *Federal Register 65* (241): 78,358–78,376; December 14, 2000. Washington, DC. U.S. Department of Interior, National Park Service. 1993. Wilderness Task Force: Report on improving wilderness management in the National Park Service. Washington, DC: U.S. Department of the Interior, National Park Service.

———. 2004. Program Standards: Park Planning. Washington, DC: U.S. Department of the Interior, National Park Service.

———. 2005a. Denali National Park and Preserve Revised Draft Backcountry Management Plan: General Management Plan Amendment and Environmental Impact Statement. Denali Park, AK: U.S. Department of the Interior, National Park Service.

———. 2005b. The National Parks: Index 2005–2007 Revised to include the Actions of the 108th Congress ending December 31, 2004. Washington, DC: U.S. Department of the Interior, National Park Service.

———. 2005c. *Park Planning Program Sourcebook: General Management Planning.* Washington, DC: U.S. Department of the Interior, National Park Service.

———. 2006. Wilderness Preservation and Management. In: *NPS Management Policies Handbook.* Washington, DC: U.S. Department of the Interior, National Park Service.

U.S. Department of the Interior, U.S. Fish and Wildlife Service (USDI, FWS). 1985. Kenai National Wildlife Refuge Final Comprehensive Conservation Plan, Environmental Impact Statement and Wilderness Review. Anchorage, AK: U.S. Department of the Interior, U.S. Fish and Wildlife Service.

———. 2000. Refuge Planning Policy. In: *Refuge Manual.* Part 602, Chap. 1, 3, 4. Washington, DC: U.S. Department of the Interior, U.S. Fish and Wildlife Service.

———. 2005. Kenai National Wildlife Refuge: Planning Update Winter–Spring 2005. Anchorage, AK: U.S. Department of the Interior, U.S. Fish and Wildlife Service.

U.S. Public Law 235 (codified at 16 U.S.C. sec. 1-18f). National Park Service Act of August 25, 1916. 39 Stat. 535.

U.S. Public Law 86-517. Multiple Use Sustained-Yield Act of June 12, 1960. 74 Stat. 215.

U.S. Public Law 88-577. The Wilderness Act of September 3, 1964. 78 Stat. 890.

U.S. Public Law 89-669. National Wildlife Refuge System Administration Act of October 15, 1966. 80 Stat. 927.

U.S. Public Law 91-190. National Environmental Policy Act (NEPA) of January 1, 1970. 83 Stat. 852.

U.S. Public Law 93-378. Forest and Rangeland Renewable Resources Planning Act of August 17, 1974. 88 Stat. 476.

U.S. Public Law 94-579. (Wilderness Study) Federal Land Policy and Management Act (FLPMA) of 1976. 90 Stat. 2743.

U.S. Public Law 94-588. National Forest Management Act of October 22, 1976. 90 Stat. 2949.

U.S. Public Law 95-95. Clean Air Act Amended as of August 7, 1977. 42 Stat. 7401.

U.S. Public Law 95-625. National Parks and Recreation Act of November 10, 1978. 92 Stat. 3467.

U.S. Public Law 96-487. Alaska National Interest Lands Conservation Act (ANILCA) of December 12, 1980. 78 Stat. 890.

U.S. Public Law 103-433. California Desert Protection Act of October 31, 1994. 108 Stat. 4473.

U.S. Public Law 105-57. National Wildlife Refuge System Improvement Act of October 9, 1997. 111 Stat. 1252.

U.S. Public Law 109-382. New England Wilderness Act of December 1, 2006. 120 Stat. 2673.

CHAPTER

9

Managing for Appropriate Wilderness Conditions: Visitor Use Management Process

by Robert G. Dvorak

INTRODUCTION

Most wilderness qualities and conditions are sensitive to the type and amount of use an area receives, and excessive use can impact the quality of the natural setting as well as the sense of solitude that visitors experience. Management concern with overuse by visitors is not a recent phenomenon, and there is strong legislative intent to provide for public use while at the same time preventing degradation from overuse and inappropriate use. This intent is stated clearly in the Wilderness Act of 1964 (U.S. Public Law 88-577, sect. 2[a]): that wilderness areas "shall be administered for the use and enjoyment of the American people in such a manner as will leave them unimpaired for future use and enjoyment as wilderness, and so as to provide for the protection of these areas, the preservation of their wilderness character."

As related in Chapter 2, there was concern with uncontrolled overuse and its associated impacts in the high country of California's Sierra as early as the mid-1930s. Lowell Sumner's call in 1942 for the restriction of use within an area's carrying capacity or recreational saturation point reflects both concern and recognition that continued increases in use could destroy the area's wild qualities (Sumner 1942). Following the COVID-19 pandemic, visitor overuse concerns have accelerated. While some of the most remote wilderness areas saw decreased use, urban-proximate wilderness areas saw visitation numbers significantly higher than in previous years (Tin 2022).

Most wilderness visitors and managers can relate to personal experiences with undesirable conditions: heavy visitor use on some trails, campsites devoid of vegetation, or mud and horse manure on trails. Wilderness integrity has been lost when conditions like these prevail. Although such conditions do exist in some areas, pristine settings and outstanding opportunities for solitude are found in much of the NWPS. In the face of substantial variations in use intensity within and between areas, how does one approach the matter of deciding what is an acceptable level and type of use in wilderness?

Establishing appropriate use levels in wilderness typically has been addressed through the concept of carrying capacity (Fig. 9.1). In previous chapters, we argued that preserving qualities essential to wilderness required a carrying-capacity constraint as a fundamental principle of wilderness management. Our concern is to determine at what point environmental and social conditions within wilderness become inconsistent with the qualities required by the Wilderness Act. Each set of wilderness conditions and visitor experiences sought as management objectives implicitly carries with it some limit in the kinds and amounts of recreational use that can be considered acceptable; it also implies the need for various policies and actions to see that desired conditions and acceptable thresholds are not exceeded. This approach to managing visitor impacts is effectively operationalized in the Visitor Use Management Framework (VUMF).

In this chapter, we explore the origins of the carrying-capacity concept, consider limitations in how the concept has been applied in wilderness settings, and propose the VUMF as a highly flexible and adaptable model designed to provide managers with a comprehensive and defensible approach for managing acceptable wilderness conditions. Brief case studies examining how such guidelines have been developed for several wilderness areas are also described.

FIGURE 9.1 Carrying capacity of wilderness is the amount, kind, and distribution of use that can occur without leading to unacceptable impacts on either the physical-biological resource or the available wilderness experience. This concept was used by the Bureau of Land Management in Arizona's North Maricopa Mountains Wilderness to ensure that hikers would be able to experience solitude. Photo courtesy Chad P. Dawson.

CONSIDERATIONS IN MANAGING VISITOR USE AND IMPACTS

Several basic considerations are necessary in managing recreational use and its impact. First, three components must be considered to determine the kinds of impacts that will occur and their possible implications, (Manning 2007):

- *Natural resources*—The physical and biological characteristics of the natural resource base greatly influence the amount of environmental change resulting from recreation activities. Although recreation use inevitably causes change in the environment, some resources are inherently more fragile than others. These changes can include such effects as wildlife disturbance, habitat degradation, and habitat modification of sensitive ecosystems (Marion 2019).
- *Social*—The quality of the visitor experience in wilderness is at risk if people feel crowded or are negatively affected by the presence of other visitors (Allen 2019). Because crowding and conflict are subjectively perceived, managers consider the perceptions of visitors to determine when social conditions in a wilderness setting lead to unacceptable social impacts. Therefore, user perceptions and opinions of what types and levels of use are preferred are an essential element in developing prescriptions of appropriate visitor use.
- *Managerial*—Legal directives and agency missions play a major role in determining appropriate and feasible resource, social-political, and management conditions. Managerial factors help identify what conditions should be maintained, what infrastructure is permissible, what recreation-related impacts need to be managed to maintain desired conditions, and what management actions and strategies might be used to achieve those conditions (Manning 2010).

Second, any recreation use of an area leads to change in the condition of these three components. *The inevitability of change is a critical consideration* that managers must consider in prescribing carrying capacities related to visitor management. Predicting acceptable maximum use levels is challenging (Cole 2019) (Fig. 9.2). For example, many studies of

FIGURE 9.2 Use limits can protect quality wilderness environment such as the Three Sisters Wilderness, Oregon. Excessive use, however, can lead to resource damage and to a degraded wilderness experience. Photo courtesy U.S. Forest Service.

ecological impact related to recreation activities point to the fact that most of the total impact recorded occurs under low-use levels (see Chapter 15). It may be necessary to establish lower use limitations than what would be acceptable to some public groups. Rather than attempting to define how much visitor use is too much, the Visitor Use Management Framework, as developed by the Interagency Visitor Use Management Council (IVUMC), focuses attention on the desired conditions for the resource conditions and visitor opportunities to be provided. Management strategies and actions then influence the amount of change that will be allowed to occur.

Third, determining the amount of change that will be considered acceptable involves a *subjective value judgment*, and such judgments are derived from the management objectives prescribed for the area (see Chapter 8). These objectives provide the framework within which judgments regarding desired conditions for an area—resource, social, and managerial—can be defined. These desired conditions need to be expressed explicitly and quantitatively to reduce disputes over interpretation and to allow for determination if objectives have been met or not. Management objectives contain explicit thresholds that define the extent to which conditions will be allowed to change, if at all. The Wilderness Act provides general guidelines as to what management objectives are to be achieved. For instance, environmental conditions must be highly natural, whereas visitor use conditions must provide opportunities for primitive and unconfined recreation. The exact meaning of these conditions can be defined through policy, such as wilderness character definitions (see Landres et al. 2005). Meanings can also be constructed through collaborative decision-making and are open to interpretation as people will have different ideas about their implications for management.

Finally, although the historic focus in visitor use management was on the number of users, the impacts of use—on resources and other users—are often related to other factors, such as the type of use, timing and location of use, and visitor behavior. The management approach of regulating visitor-use numbers may not sufficiently control impacts. More attention needs to focus on management strategies that more directly prevent or mitigate visitor impacts defined as unacceptable in wilderness.

These basic considerations underlie much of the discussion in this chapter. The goal of management is to identify the desired resource, social, and managerial conditions to be maintained or restored in wilderness, with these desired conditions expressed as explicit, measurable thresholds. Thus, the focus of management attention shifts from defining maximum recreation use to identifying desired resource conditions and visitor opportunities to be provided (Shelby and Heberlein 1986; IVUMC 2016).

THE CARRYING-CAPACITY CONCEPT

Carrying capacity is a fundamental concept in natural resource and environmental management (Dasmann 1964; Godschalk and Parker 1975). It can be defined as the maximum level of use an area can sustain as set by natural factors of environmental resistance such as food, shelter, or water. Beyond this natural limit, no major increases in the dependent population can occur (Odum 1959). In the field of wildlife management, carrying capacity describes the number of animals of a particular species that can use the range on a sustained basis, given available food, shelter, and water. If the balance between animals and the range's capacity is upset, either by an increase in the number of animals or through loss of the range's productive capacity (e.g., drought), problems will occur. Food resources become depleted as animals search for nutrition. Even though animal numbers might increase in the short term, loss of range productivity means that, in the long term, fewer animals can be supported; in extreme cases, irreversible environmental impacts can occur.

This description may imply that carrying capacity is an uncomplicated, straightforward notion. In fact, it is not. The capacity of the range would be quite different if populated by elk or sheep rather than deer; were it a mix of species, it would be even more complicated. The dynamic nature of ecosystems makes a static determination of wildlife carrying capacity difficult to calculate. On any given area of land, calculations of available forage and cover may vary widely within a single decade based upon perturbations such as wildfire, the stage of plant succession, presence or absence of drought conditions, and abundance or absence of predators, as well as changes in hunting seasons and regulations. Furthermore, nature can be influenced by resource managers when food crops are planted, water storage reservoirs created, and so forth. Thus, wildlife carrying capacity can be increased or decreased by management actions; it is not an inherent, fixed value of the land.

Practitioners and researchers in recreation management and wilderness management were early to adopt carrying capacity as a guide to making management decisions to identify how much use a resource could sustain. They sought to determine how many people could use a given recreational setting before unacceptable impact occurred. In 1964, Wagar published one of the first substantive discussions of the carrying-capacity concept in recreational management.

A major contribution of this analysis was the inclusion of the impacts of use on social or experiential considerations in addition to the typical environmental concerns (Manning 2007, 2010). As Wagar (1964) noted, "The study . . . was initiated with the view that the carrying capacity of recreation lands could be determined primarily in terms of ecology and the deterioration of areas. . . . It soon became obvious that the resource-oriented point of view must be augmented by consideration of human values."

Under this broadened concept of carrying capacity, recreational areas had not only an *ecological or biological capacity but a social capacity* as well. Use could impact not only an area's biophysical resources, such as soils and vegetation, but also the character and quality of the recreational experience. The recognition of a *social dimension* of carrying capacity implies that determination of carrying capacity is a sociopolitical process as well as a biophysical one (Burch 1984).

Wagar's monograph marked the beginning of a major effort to define the carrying capacity of recreational areas. This effort was driven, in part, by the great surge in recreational use following the end of World War II. Wilderness in the United States, once remote and little used, began to receive increasing amounts of use as what Clawson (1985) called the *four fueling factors*—leisure time, income, access, and population—grew substantially. Between 1955 and 1964, for instance, the average annual growth of recreational use on national forest wilderness exceeded 10 percent (Stankey and Lucas 1986). Since the passage of the Wilderness Act, wilderness recreation use is still increasing and will continue (Cole 1996; Tin 2022). Increasing use has generated concerns about ecological and social impacts that might jeopardize the essential wilderness qualities of naturalness and solitude.

Such concern over visitor impacts is the impetus for continued interest in using carrying capacity as a framework for decision-making to control impacts that would otherwise erode the qualities and conditions of wilderness. Literature on carrying capacity and the associated management techniques for applying the concept of carrying capacity has grown considerably (Graefe et al. 1984; Manning 2010; Manning and Lime 2000; Manning 2007; Manning et al. 2017; Marion 2016; IVUMC 2019b).

One the most serious problems in using carrying-capacity analysis in wilderness management was that some people considered the carrying capacity of an area as an inherent value of the resource base, one that could be determined through careful measurement and research. The concept of carrying capacity implied to some people that complex, controversial decisions about recreation visitor use could be easily formulated.

Attempting to apply a carrying-capacity concept to wilderness management also produced two serious kinds of backlash in the public. First, the carrying-capacity concept focused the public's attention on the wrong question: "How many users are too many?" Even though the concept's originators never meant it to be, the public invariably linked the term *carrying capacity* with controlling visitor numbers rather than managing for desired resource conditions and visitor opportunities. Second, the public often did not accept the carrying-capacity decisions made by managers. People who wanted more access and use and those who wanted use and impacts to be lessened protested any attempt to establish recreation carrying capacity. This forced researchers and managers to reexamine the carrying-capacity concept and its application in wilderness management.

Perhaps one of the most important developments in understanding the carrying-capacity concept over the past fifty years is that *carrying capacities are the product of value judgments as well as science*. This means that determining carrying capacity becomes a process by which biophysical and social research is integrated with agency policies and the values of managers and users to reach a collective judgment. Additionally, carrying capacity has been more specifically described as "visitor capacities" in the VUMF. The framework emphasizes that the term "visitor capacity" is one component of visitor use management and, therefore, should not be used interchangeably with the term "visitor use management." Identifying visitor capacity is just one of many visitor use management strategies available to achieve and maintain desired conditions (IVUMC 2019b).

In addressing the criticisms of the carrying capacity concept, resource managers have used the results of research on crowding and other social impacts to contribute to decisions about visitor capacity. One lesson learned is that it is more appropriate to ask, "What are acceptable resource and social conditions?" than to simply ask, "How much use is too much?" This places the emphasis on management goals and desired social and environmental conditions (Allen 2019). Visitor capacities can be important to identify when the amount and types of visitor use directly relate to achieving and maintaining desired conditions (IVUMC 2016) (Fig. 9.3). Additionally, identifying visitor capacities can be legally required. The 1960 Wild and Scenic Rivers Act, 1978 National Parks and Recreation Act, and 1968 National Trails System Act require the identification of visitor capacities and the development of management strategies to manage use within visitor capacities (Cahill 2018; IVUMC 2016). Wilderness managers, in areas where these mandates apply, have a legal obligation to incorporate visitor capacities into their management strategies and actions.

FIGURE 9.3 Horses typically have more impact on wilderness ecosystems than an equal number of hikers, especially on fragile sites such as lakeshores. To prevent irreversible damage, horses were prohibited at Reflection Pond in the Glacier Peak Wilderness, Washington. Photo courtesy U.S. Forest Service.

The evolving recognition of carrying capacity, and more specifically visitor capacity, as a *conceptual approach, derived from social and ecological judgments about appropriate wilderness conditions*, has led to increasing attention to the factors that shape and influence these judgments. The concern for applying the carrying-capacity concept (Graefe et al. 1986; Shelby and Heberlein 1986) has led to a series of similar planning efforts to develop more effective frameworks for managing visitor recreation use and associated impacts: the process for Visitor Impact Management (VIM), the Visitor Experience and Resource Protection (VERP) approach, the Management Process for Visitor Activities (known as VAMP), and the concept of the limits of acceptable change (LAC) all represent efforts to correct limitations in applying the carrying-capacity concept to recreation management (Nilsen and Tayler 1997; McCool et al. 2007). These frameworks share a common focus on identifying measurable objectives regarding desired conditions and on the distinction between steps involving objective description and analysis and those involving judgmental evaluations.

In this chapter, we describe how the LAC concept was historically used as a framework for managing wilderness within a carrying-capacity concept (McCool and Cole 1997; Stankey et al. 1985). Several examples are provided to demonstrate how the LAC concept was historically applied. We will then describe the VUMF that has been developed to provide a flexible and adaptable approach for analyzing and managing visitor use on federally managed lands.

THE LIMITS OF ACCEPTABLE CHANGE (LAC) CONCEPT

The concept of the *limits of acceptable change* (LAC) was developed for resolving the carrying-capacity issue. The LAC process recognizes that change in response to visitor use is inevitable and that decisions must be made regarding how much change will be permitted to occur. Lime (1970), for instance, called for establishing standards in the Boundary Waters Canoe Area Wilderness (MN) that defined acceptable limits to the impacts that would be permitted to occur on both the various resource elements and visitors. Frissell and Stankey (1972) outlined the basic framework for the LAC concept. They focused attention on the control of human-induced change on conditions. The goal of management, they argued, is to halt the character and rate of change that would lead to conditions judged as unacceptable in legally defined wilderness.

The LAC concept recognizes and seeks to enhance and protect a variety of wilderness conditions (Clark and Stankey 1979; Driver and Brown 1978; U.S. Department of Agriculture, Forest Service 1982). Variability in wilderness conditions is inevitable and desirable. In some areas, pristine physical-biological conditions and outstanding opportunities for solitude are found. Elsewhere, the evidence of visitor use is more apparent and contact with others is more common. No single condition constitutes wilderness—a range exists from the pristine to slightly modified. Establishing a clear measure of what constitutes acceptable wilderness conditions in the form of explicit, measurable standards is the intent of the LAC process.

THE HISTORICAL LAC PROCESS

The LAC process consists of four major components: (1) specifying acceptable and achievable resource and social conditions, defined by a series of measurable parameters; (2) analyzing the relationship between existing conditions and those judged acceptable; (3) identifying management actions judged to best achieve these desired conditions; and (4) monitoring and evaluating management effectiveness. These four components, in turn, are broken down into planning steps to facilitate the application of LAC.

The original LAC process was a nine-step planning process (Stankey et al. 1985), with each step designed to achieve a particular task and provide the basis for later activities. The LAC process requires managers to identify where, and to what extent, varying degrees of change are appropriate and acceptable. Once the appropriate and acceptable degree of change has been identified, managers can select from an array of management techniques that target the causes of impacts to ensure that the desired conditions are maintained or restored. These techniques range from indirect actions, such as providing information to visitors on appropriate camping methods, to more direct actions like restricting the number of visitors to the area. The conditions that characterize a particular area of wilderness and distinguish it from others are specified by measurable objectives. Figure 9.4 portrays the essential elements in the LAC process (a), including several important decision points when monitoring conditions and evaluating management actions (b).

By 1992, the LAC process had become widely applied, reportedly being used by 92 percent of 50 western national forests that contained 116 wilderness areas (McCoy et al. 1995). Many of these LAC plans did not strictly adhere to the original nine-step process because some of the steps were cumbersome for public work groups participating in the planning processes, and some elements seemed to be missing, such as defining goals and desired conditions.

The LAC process developed and adapted important visitor use management concepts that have been further refined in the VUMF to be described later. One of the concepts was the development of acceptable zones or opportunity classes for the wilderness (*Step 3: Define and Describe Acceptable Conditions*). An *opportunity class* defines the resource, social, and managerial conditions considered desirable and appropriate within the wilderness. The designation of opportunity classes follows the basic recreation opportunity spectrum (ROS) system and represents the range of wilderness recreation settings for which to manage within any given area. Cole and McCool (1997) suggest that the intent of this step was to develop "prescriptive management zones" with the emphasis on acceptable, nondegradation conditions rather than present conditions. The range or diversity of acceptable conditions could be divided into two or more classes, which implies that opportunity classes represent a continuum of social, resource, or managerial variables.

Managers decide not only how each opportunity class is defined, but also how many are appropriate for each wilderness. For example, smaller wildernesses may have only one or two classes, whereas larger areas may have up to six classes. The question of how many classes to designate can be answered only after analysis of the issues, the current range of

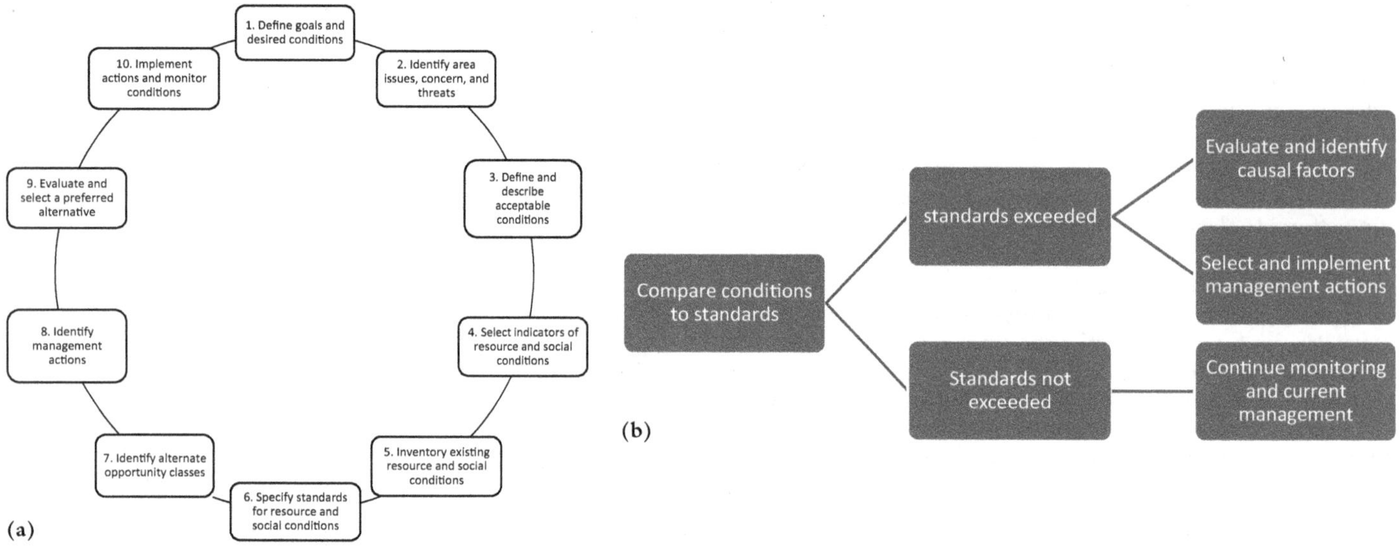

FIGURE 9.4 The essential elements of the ten-step limits of acceptable change process (a) lead to several important decision points (b) when monitoring conditions and evaluating management actions in the limits of acceptable change process.

conditions, the manageability and demands for wilderness recreation, and the regional supply of different wilderness settings. Historically, the ROS *defines six categories of classes; however, opportunity classes in LAC represent only those primitive and semi-primitive ROS categories permitted in wilderness.*

The opportunity class description represents the varied wilderness setting conditions that are acceptable and appropriate. These setting descriptions are broad and qualitative, and within each it is possible to describe several classes. For example, at major entry points, use levels may be relatively high, with frequent contact among visitor parties. Similarly, resource impacts can be moderately visible in these areas. Elsewhere in the same wilderness, there are areas where few visit and where ecological conditions are almost undisturbed. Between these extremes, a continuum of conditions exists, all within the same wilderness.

Collectively, the prescriptions of the resource, social, and managerial conditions for each opportunity class constitute the *management objectives* for the area. They prescribe the conditions sought in the wilderness and serve as criteria for identifying what and where specific management actions are needed. These objectives serve throughout the process to determine what types of information are needed, what standards need to be developed, the appropriateness of various activities, and what management actions are needed.

With the selection of management actions, the management program must be implemented, and its performance assessed. Monitoring provides systematic feedback on how well management actions are working and identifies trends in conditions that may require changes in management actions. The need for monitoring was a key component of the LAC process, and one that has been emphasized and further described in the VUMF. Monitoring reveals existing conditions at the time of monitoring and the trend in that condition. For instance, existing conditions might be better than what objectives call for but are nevertheless degrading over time, or existing conditions might be worse than those called for but their trend could be improving. *Monitoring alerts managers to the current conditions and trends in relation to desired conditions.*

LIMITATIONS OF LAC AND CARRYING CAPACITY PROCESSES

While these LAC and carrying capacity frameworks did much to improve visitor use management (IVUMF 2016; McCool et al. 2007), they had both real and perceived limitations. First, the frameworks were perceived as not fully integrated into current federal planning processes. In some situations, not all steps were needed, leading to confusion about how the legislatively mandated identification of visitor capacities fit in the frameworks. Second, there was the limitation of the perceived complexity and costliness of developing and implementing the frameworks. Managers perceived that the frameworks could not be implemented without extensive public involvement, new visitor survey research, or difficult to find monitoring resources. These concerns created barriers to flexibility and application. Third, there was a lack of emphasis that visitor use management frameworks were intended to be implemented in a proactive rather than reactive manner (IVUMF 2016). Identifying management actions were meant to predict what might be necessary to achieve management objectives. Monitoring was intended to identify trends and help understand what actions might be necessary to maintain desired conditions. However, previous frameworks tended to default to current conditions and management strategies reacted with resources currently available.

LAC PROCESS EXAMPLES

Important lessons in wilderness management and planning have been learned through the application of the LAC process. In the follow section, we examine how the LAC process was applied in the Bob Marshall Wilderness Complex and the White Mountain National Forest. Lessons learned from years of implementation in these contexts has provided guidance for wilderness planners and managers to further develop flexible and adaptable visitor use management frameworks.

The Bob Marshall Wilderness Complex

The LAC process was first applied and developed in the Bob Marshall Wilderness Complex (BMWC) planning and management process. The BMWC is composed of three wildernesses lying astride the Continental Divide in Montana: the Bob Marshall Wilderness, the Scapegoat Wilderness, and the Great Bear Wilderness. Together they comprise more than 1.5 million acres, one of the largest roadless areas in the lower forty-eight states. Four national forests (Flathead, Helena, Lewis and Clark, and Lolo) within the complex are administered through five ranger districts (Fig. 9.5).

Recreation use of the complex totaled approximately 200,000 visitor days annually in the early and mid-1980s. Most of the visits to the complex were hikers, comprising approximately 57 percent of the people, with horseback riders totaling about 36 percent of the visitors (Lucas 1985). This is a change from 1970 when horseback riders outnumbered hikers about two to one. Horseback riders tend to come in larger group sizes and stay longer than hikers, so most total use in visitor days came from horseback riders. Between 1970 and 1982, party sizes and lengths of stay tended to drop for both groups.

A wilderness management plan for the Bob Marshall Wilderness (designated in 1964) was developed in 1972, but much of it was never implemented because it lacked public support. The limited implementation of the 1972 Bob Marshall Wilderness Management Plan, the designation of the Scapegoat Wilderness in 1972 and Great Bear Wilderness in 1978, and the need for updated comprehensive management direction for the entire BMWC prompted the four national forests to develop a coordinated approach to all three wildernesses through their respective Forest Land and Resource Management Plans beginning in 1982.

FIGURE 9.5 Woodward Lake, in the Bob Marshall Wilderness, is one of three separate but contiguous wildernesses comprising the Bob Marshall WildeFrness Complex, which covers approximately 1.5 million acres in western Montana. Conditions range from pristine to heavily impacted, presenting a difficult challenge to managers. Photo courtesy U.S. Forest Service.

LAC was selected to provide the technical framework for developing common wilderness management direction for the entire wilderness complex. A comprehensive management document for the BMWC would then be appended to each forest's respective forest land and resource management plan. A major concern to managers was the lack of vested interest in and support for the 1972 Bob Marshall Wilderness Plan by wilderness users and interest groups. Thus, the public-involvement process was considered a critical element to ensure success of the LAC.

One major question facing managers of the BMWC was how to solicit public comment and review through the planning process. They selected a working group approach in the early 1980s, in which participants would share their wilderness knowledge and use experiences with the planners, who would share information about the wilderness planning model approach and systematic data analysis with the citizens. Through the dialogue and learning processes, the working group, using its accumulated knowledge, was intended to make an informed decision about a course of action (McCool et al. 1986). These working groups had the potential to form a viable political coalition that could ensure the plan has adequate political support to achieve implementation.

Issues confronting the BMWC were similar to those in other wildernesses: visitors were concerned about campsite impact and opportunities for solitude; trail conditions were frequently identified by both managers and visitors as unacceptable; conflict between hikers and horseback riders had increased; recreational stock and wildlife competed for forage; and the entire complex was a grizzly bear habitat. In response to these issues, the U.S. Forest Service, with the Flathead National Forest serving as the lead forest, initiated a planning effort in 1982 to develop more specific and additional direction than that established in the 1972 plan.

The approach to public involvement chosen was to develop a task force for the BMWC planning project. The task force was composed of managers, researchers, and citizens. This composition allowed the opportunity to share technical/scientific knowledge and personal knowledge among participants. The LAC Task Force for the BMWC consisted of up to fifty people. Over a period of nearly four years, with an independent facilitator helping in the last two years, the Recreation Management Direction for the BMWC was established. In total, more than eighty formal and informal meetings and information and working sessions occurred in the four-year period.

The LAC steps were used to focus task force input. For example, the task force developed the preliminary list of issues, identified desired conditions, mapped opportunity-class allocations, and identified potential management actions. In addition, because the LAC process was new and still being developed, the core team was continuously being educated. The LAC process resulted in four wilderness recreation-opportunity classes. Class I, the most pristine, permitted almost no evidence of human occupancy and use, whereas Class IV allowed somewhat more change to occur.

TABLE 9.1 Proposed Standards for Resource and Social Indicators for Each Opportunity Class in the BMWC (U.S. Department of Agriculture, Forest Service 1987)

Indicators	Opportunity Class I	Opportunity Class II	Opportunity Class III	Opportunity Class IV
		Social		
Number of trail encounters with other parties	80 percent probability of 0 encounters per day	80 percent probability of 1 or fewer encounters per day	80 percent probability of 3 or fewer encounters per day	80 percent probability of 5 or fewer encounters per day
Number of other parties camped within sight of parties or continuous sound	80 percent probability of 0 parties per day	80 percent probability of 0 parties per	80 percent probability of 1 or 0 parties per	80 percent probability of 3 or fewer parties per
		Human-Impacted Sites		
Number of human-impacted sites per 640-acre area[1]	1 permitted	2 permitted	3 permitted	6 permitted
		Range		
General range trend	Static or improving	Static or improving	Static or improving	Improving
Overall range condition	Excellent	Excellent	Generally good or better	Generally good

[1] Human-impacted sites defined as any site with evidence of human impact, normally centered on a fire ring, regardless of its prior use for camping.

Indicators of conditions and their thresholds were also developed through task force participation. Examples of these are shown in Table 9.1. Management actions were developed to address potential or existing unacceptable resource and social conditions. A general list of actions identified by type of problem and by opportunity class was developed.

The BMWC was the first place where LAC was attempted as a full process. Applying the LAC process to the planning of the BMWC focused on managing the area's recreational use because much of the impact on the area's quality stemmed from such use. However, managers must be concerned about many issues other than recreation. For example, the BMWC contains both diverse wildlife populations and key threatened and endangered species such as the grizzly bear and gray wolf. Although the LAC process did not directly address wildlife concerns, nonetheless it was possible to examine how the opportunity class allocations would affect the wildlife values contained within them.

The White Mountain National Forest Wilderness Management Plan

The Land and Resource Management Plan (LRMP) in the White Mountain National Forest (WMNF) of New Hampshire and Maine (U.S. Department of Agriculture, Forest Service 2005) addressed wilderness issues and management direction throughout the document. The LRMP specifically outlined the LAC process for all five wilderness areas within the WMNF: Great Gulf (5,500 acres), Presidential Range–Dry River (29,000 acres), Pemigewasset (45,000 acres), Sandwich Range (25,000 acres), and Caribou-Speckled Mountain (14,000 acres). This WMNF Wilderness Management Plan used the LAC and four opportunity class zones to monitor progress toward the plan goals.

The plan delineated four different wilderness zones and mapped the zones across the five wilderness areas (Fig. 9.6). The opportunity class zones were labeled A, B, C, and D and based on ecological characteristics, social conditions, and management needs. Although visitor use levels were not the determining factor in applying the zoning scheme, use estimates clarified the distinctions among zones; the zones generally run from least (Zone A) to most heavily used (Zone D).

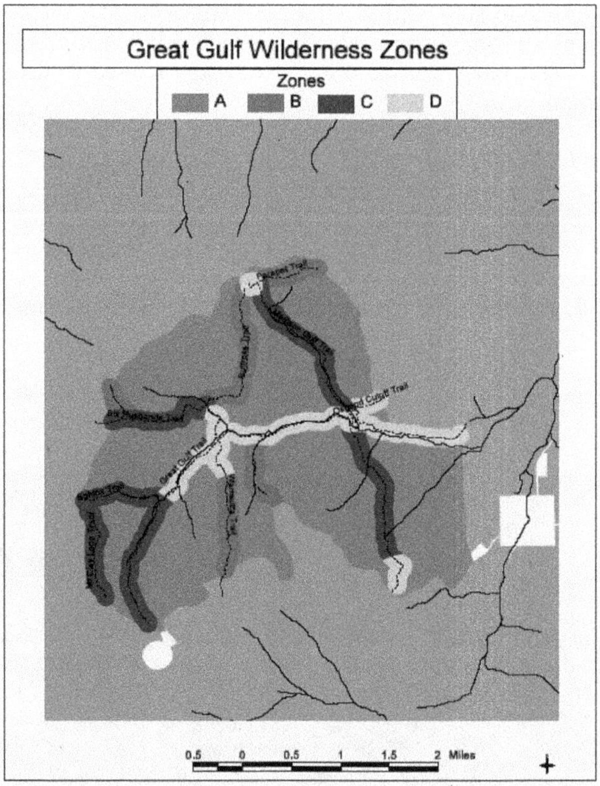

FIGURE 9.6 Four opportunity class zones in the Great Gulf Wilderness, part of the White Mountain National Forest, New Hampshire. Courtesy U.S. Department of Agriculture, Forest Service (2005, E-35).

The following is from the U.S. Department of Agriculture, Forest Service plan (2005, E-5 to E-9):

Zone A—Areas 500 feet or more from all trails
This zone includes trailless areas and represents the largest area of WMNF Wilderness. The landscape appears largely unmodified, supports no maintained trails or facilities, has few restrictions, has low managerial regulation, has little direct management activity, and has exceptional opportunities for visitors to experience both solitude and a very primitive and unconfined recreation.

Zone B—Areas within 500 feet of low-use trails
This zone includes the lowest-use, least developed trails within WMNF Wilderness. It offers the greatest opportunity for solitude and/or an unconfined recreation experience along a maintained trail system. The landscape appears largely unmodified, supports minimally maintained trails but no other facilities, and has regular opportunities for visitors to experience both solitude and a primitive recreation confined only by the presence of the trail system.

Zone C—Areas within 500 feet of moderate-use trails
This zone includes the moderate-use, moderately developed trails within WMNF Wilderness. It is generally more highly used and more highly developed than Zone B. Despite this, Zone C offers visitors an opportunity to experience escape from more highly developed landscapes while still being able to access a maintained trail system. In most places, the landscape appears largely unmodified. Exceptions include the trail system and associated structures and lasting campsites, including some designated sites. Facilities such as bridges may exist, but shelters and toilets do not. The area is likely to have site-specific regulations, with generally frequent managerial presence. Direct management activity including enforcement of regulations occurs.

Zone D—Areas within 0.25 mile of developed facilities or 500 feet of high-use trails
This zone includes the most heavily used and most highly developed trails and areas within WMNF Wilderness. The landscape within this zone is modified by the developed trail system and associated structures, and may include bridges, primitive shelters and/or toilets, designated campsites, and impacts resulting from recurring recreation use. To manage use and protect resource conditions the area likely has site-specific regulations with frequent managerial presence. Direct management activity including enforcement of regulations occurs. This area has occasional opportunities for visitors to experience solitude as well as primitive and unconfined recreation bounded by the presence of the trail system, existing regulations, shelters, toilets, and campsites.

The WMNF Wilderness Management Plan used the LAC process to develop four categories of indicators and thresholds: biophysical, social, aesthetic, and ecosystem process indicators. These indicators also include some monitoring efforts for wilderness conditions that are not part of the LAC process and address broader concerns about wilderness conditions such as suggested in a forest service framework for monitoring conditions related to wilderness character (Landres et al. 2005). The wilderness management plan concluded with an education plan to address concerns and issues present in each of the four opportunity class zones and outlines management actions (Table 9.2).

THE VISITOR USE MANAGEMENT FRAMEWORK

Visitor use management (VUM) in wilderness areas is fundamentally necessary for achieving and maintaining desired resource conditions and visitor experiences. As change occurs, managers need to be responsive to a diversity of needs, values,

TABLE 9.2 LAC Standards. Monitoring Methods, and Management Action for Off-Trail Campsites in the WMNF (U.S. Department of Agriculture, Forest Service 2005, E-17)

	Zone A	Zone B	Zone C	Zone D
Condition Threshold	0 lasting campsites with no visible impacts lasting more than 1 year.	0 sites within 500 feet of each other. 0 sites within 200 feet of trail.	0 sites within 200 feet of each other, maximum total of 2 sites within 500 feet of each other.	3 sites within 200 feet of each other, maximum total of 5 sites within 500 feet of each other.
Method of Measure, Frequency	Survey along 1 selected stream drainage within each wilderness each year. Survey 1 trailless peak above 2,999 feet within each wilderness each year, as appropriate.		Complete inventory once during the life of the plan.	
Management Action	1. Active site revegetation. Written reminder to all VIS centers reinforcing the established education message for this zone. Examine management that may contribute to a change in use patterns. 2. Increase focused patrols in the affected area. 3. If initial actions do not resolve issue, conduct focused management assessment to consider: ○ Enact closure order for affected area. ○ Implementation of limited overnight-use system.		4. Post revegetation signs. Written reminder to all VIS centers reinforcing the established education messages for this zone. Examine management that may contribute to a change in use patterns. Analyze group-use policies and act accordingly. 5. Increase focused patrols in the affected area. 6. If initial actions do not resolve issue, conduct focused management assessment to consider: ○ Enact or expand closure order for affected area. ○ Implement limited overnight-use system.	

and conditions. In 2011, the Interagency Visitor Use Management Council (IVUMC) was chartered to develop best practices for visitor use management on federally managed lands and waters (IVUMC 2016). The vision of the Council was to provide a unified voice for excellence in visitor use management to sustain resources and quality visitor experiences. It was committed to proactive, professional, and science-based visitor use management through a consistent approach to VUM that better served the public by creating seamless connections between lands and waters managed by different federal agencies. In this section, we describe the core elements of the VUM Framework and the steps included in each part of the process. It is important to recognize that a significant portion of the information that follows had been adapted and summarized from the *Visitor Use Management Framework* (IVUMC 2016), the Council website, (https://visitorusemanagement.nps.gov/), and various guidance documents provided by the Council.

The Interagency Visitor Use Management Council defines VUM as the proactive and adaptive process for managing the characteristics of visitor use and the natural and managerial setting using a variety of strategies and tools to achieve and maintain desired resource conditions and visitor experiences (IVUMC 2016). Visitor use management is more than minimizing or mitigating impacts that result from use. It is the consideration and deliberate provision of a range of opportunities and setting characteristics to facilitate appropriate and high-quality visitor experiences while protecting natural and cultural resources (Cahill et al. 2018). VUM is further defined by Cahill et al. (2018, p. 33) as:

- identifying desired conditions for resources, visitor experiences and opportunities, and facilities and services;
- gaining an understanding of how visitor use influences achievement of desired conditions; and
- committing to active/adaptive management and monitoring of visitor use to meet overall goals.

The Council designed the Framework in a flexible manner to be applied across a wide spectrum of situations that vary in extent and complexity, ranging from site-specific decisions to large-scale, comprehensive management (Cahill et al. 2018). The *Visitor Use Management Framework* (IVUMC 2016) offers a broadly applicable process and toolkit for making decisions at a variety of scales. Its purpose is to provide cohesive guidance for managing public use on federal lands and waters using a process that is a legally defensible and transparent planning and decision-making process. It is also intended to be proactive to determine management actions before unacceptable impacts occur.

SLIDING SCALE APPROACH

An important step in visitor use management is to determine the level of analysis required to adequately address visitor use management issues and opportunities. In the VUMF, this is addressed using a sliding scale of analysis. This approach ensures that the investment of time, money, and resources in the planning effort is commensurate with the complexity of the project and consequences of the decisions. It matches the investment of analysis with the level of uncertainty and risks associated with the issues being addressed.

Various factors influence where given visitor use management issues land on the sliding scale. The VUMF (IVUMC 2016) includes four criteria to consider when establishing the necessary level of analysis: issue uncertainty, impact risk, stakeholder involvement, and level of controversy/potential for litigation (Fig. 9.7).

- *Issue uncertainty*—How much uncertainty exists for the issue being addressed?
- *Impact risk*—Are there considerable threats to resource conditions and visitor experiences? Are threats imminent to irreplaceable resources? What is the scale of potential impacts?
- *Stakeholder involvement*—What is the level of stakeholder interest? Are special interest groups engaged? How well organized are groups?
- *Level of controversy / potential for litigation*—How likely are issues to be litigated?

If managers answer "yes" to these questions, a higher level of analysis may be appropriate to address concerns. Likewise, if there is low risk, little stakeholder involvement, or less uncertainty, then the investment of time and resources might be decreased for certain projects. Figures 9.8 and 9.9 demonstrate two examples of determining the level of analysis using the slide scale. One considers managing overnight use in a popular wilderness area ("High" on the sliding scale) and the other examining visitor capacity on a low-use remote river setting ("Low" on the sliding scale).

It is important to recognize that careful assessment and professional judgment is a key part of determining the level of analysis on the sliding scale. Whether high or low levels of analysis are needed, managers still must consider stakeholder participation, impacts risks, and varying uncertainty. History suggests that agency decisions are more likely to withstand

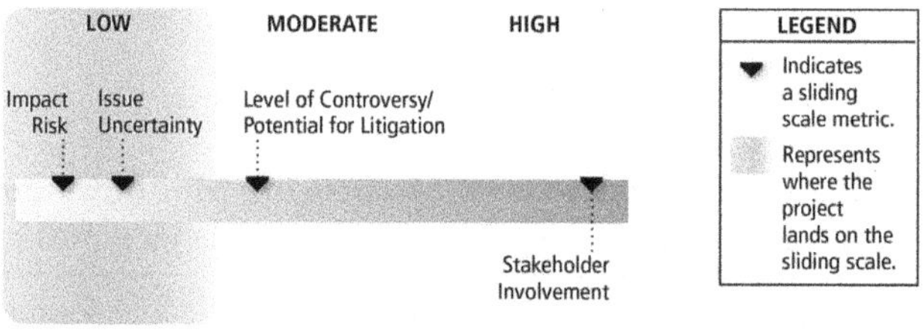

FIGURE 9.7 Representation of the four criteria included in the IVUMF sliding scale of analysis. Courtesy of the IVUMC (2019a).

	CRITERIA	RATIONALE	HIGH MODERATE LOW
A	Issue Uncertainty	Resource conditions and potential impacts are well understood; however, how the issue will affect visitor experiences and surrounding communities is less certain.	Moderate
B	Impact Risk	There are significant natural resources, cultural properties, and visitor experiences involved with this issue.	High
C	Stakeholder Involvement	The process is being closely followed by multiple stakeholder groups, each with a different opinion about what is appropriate.	High
D	Level of Controversy	All potential solutions will be considered controversial by one or several stakeholders, and there is a history of litigation for this park.	High
	Location on the Sliding Scale		High

FIGURE 9.8 Example of the four criteria that combine to yield a "High" qualitative rating on the sliding scale. Courtesy of IVUMC (2016).

	CRITERIA	RATIONALE	HIGH MODERATE LOW
A	Issue Uncertainty	While there is some uncertainty about current conditions associated with use patterns, the difficultly of access and low levels of use mitigate this criteria.	Low
B	Impact Risk	Low levels of use limit the risk of impacts.	Low
C	Stakeholder Involvement	Low number of stakeholders and no significant change for river users and outfitters.	Low
D	Level of Controversy	No controversy is expected.	Low
	Location on the Sliding Scale		Low

FIGURE 9.9 Example of the four criteria that combine to yield a "Low" qualitative rating on the sliding scale. Courtesy of IVUMC (2016).

challenges when such a process is followed, well documented, and well explained (IVUMC 2016).

FRAMEWORK ELEMENTS

The VUMF is divided into four major elements: (1) build the foundation; (2) define visitor use management direction; (3) identify management strategies; and (4) implement, monitor, evaluate, and adjust (Fig. 9.10). These represent the critical elements of the framework used to successfully manage desired conditions for natural resources and visitor experiences. Each element includes several steps that provide additional direction and address important management concepts to achieve the overall outcome of transparent and legal defensible decisions.

In the following section, we further describe each of the four elements and the important steps in the process. While these steps are presented in a linear order, it is important to recognize that visitor use management and the implementation of the framework is a highly iterative, adaptable, and flexible process. As managers and stakeholders navigate through the process, it may be necessary to revisit management directions or redefine goals. Emerging issues may create new uncertainty or require new stakeholders to be brought to the table. Therefore, a flexible and adaptable process allows the greatest opportunity for success.

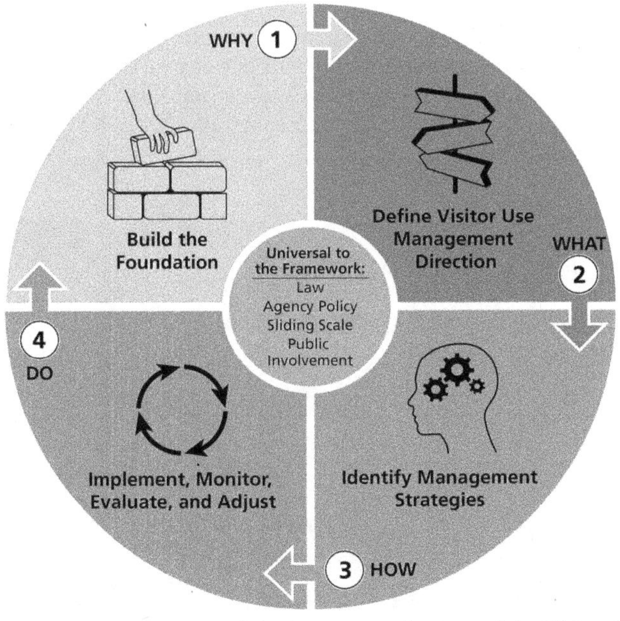

FIGURE 9.10 Overview of the four major elements of the Visitor Use Management Framework. Courtesy of IVUMC (2016).

Element 1—Build the Foundation

The purpose of the first element of the process is to understand why the project is relevant and develop how best to approach the project. Managers identify what needs to be done, how to organize the project, and how to define the resources available for completing the project. During this element, managers utilize the sliding scale of analysis approach to determine the effort needed across the project. Outcomes of this element in the framework include developing the project approach and establishing a solid process. Below are the four steps of the first element.

Step 1—Clarify the project purpose and need. A project's success can often be traced back to how well the purpose and need is described and understood (IVUMC 2016). In this step, managers may write a purpose and need statement based on the area's purpose and legislative mandates. This statement can clarify and determine which issues are within the scope of the project. Managers should consider what is creating the issue or opportunity? What values or desired conditions are be affected? And what resources or who could be affected by the issue?

Step 2—Review the area's purpose and applicable legislation, agency policies, and other management direction. This step identifies and reviews the area's stated purpose and the authority for its establishment. This may include legislation (e.g., the Wilderness Act of 1964, individual wilderness designation), planning documents (e.g., backcountry management plans, comprehensive conservation plans), cooperative agreements, or executive orders. This examination provides an interpretation to understand guidance and intent for the management of the area.

Step 3—Assess and summarize existing information and current conditions. This step is when current conditions of natural and recreation resources are examined. Significant threats to resources and experiences are described. Through this process, managers identify how to obtain existing data and classify where gaps in knowledge exist. This assessment information is then summarized and documented in a useful format to be utilized across the other elements of the VUMF.

Step 4—Develop a project action plan. This step is to organize and develop a project action plan that identifies who will accomplish the project and when. This plan identifies project team members (e.g. recreation planners, social scientists, communication specialists, senior leadership), deliverables, timelines, and budgets. This plan also develops a public involvement strategy. Developing a public involvement and outreach strategy ensures the proper level of participation and meaningful input (IVUMC 2016).

Element 2—Define Visitor Use Management Direction

This element describes the direction for visitor use management for the area and how natural resource and visitor experience conditions will be observed. Managers are allowed to communicate their positive vision for the future and establish a defensible link to future management actions. This element also guards against allowing incremental or haphazard change to occur when managers are only reacting to issues. Outcomes of this element include defined desired conditions to be achieved and maintained and how these conditions will be tracked over time. Defining desired conditions also provides long-term direction for resource conditions and visitor opportunities (IVUMC 2016). Below are several critical steps in the second element.

Step 5—Define desired conditions for the project area. *Desired conditions* are defined as "statements of aspiration that describe resource conditions, visitor experiences and opportunities, and facilities and services that an agency strives to achieve and maintain in a particular area" (IVUMC 2016, p. 30). These descriptions paint a picture of what an area "should look like," not necessarily what conditions exist currently. This step of defining desired conditions had evolved from the historical LAC process of identifying *opportunity classes* previously described in this chapter. Identifying classes or management areas within a process creates management zones. Each of these zones may have different desired conditions and provide a spectrum of opportunities that can meet the diverse interests and needs of visitors.

Desired conditions are developed from the foundation information collected and described in Element 1. They represent the area's purpose, values, and role in the regional and national context. Examples of desired condition statements are (adapted from Cahill 2003):

- Results-oriented to the "what" not the "how"
- Focused on fundamental opportunities, resources, and experiences
- Integrated into the physical, social, and managerial attributes of the area
- Future-oriented towards dynamic changes and trends
- Responsive to input from public engagement
- Useful in creating a shared understanding of short- and long-term outcomes

Step 6—Define appropriate visitor activities, facilities, and services. This step provides more specific direction for the types of visitor activities, facilities, and services that are consistent with proposed desired conditions. It helps managers understand what may be appropriate versus what is not appropriate for a specific area. For example, motorized use, developed facilities, and intensively managed trails and campsites are not considered appropriate for designated wilderness areas per policy in the Wilderness Act of 1964, wilderness designation language, and agency policy. Regardless of the setting, an exhaustive list of activities and services is not necessary or feasible. Thus, managers should consider where foreseeable activities may or may not be accommodated to maintain desired conditions.

Step 7—Select indicators and establish thresholds. A critical component of visitor use management is identifying acceptable levels of impact from visitor use. This step selects indicators for impacts to be monitored and establishes a threshold or standard for each indicator. Indicators translate the broad description of desired conditions into measurable attributes that can be tracked over time to evaluate changes in conditions. Thresholds ensure that conditions remain acceptable for those selected indicators (IVUMC 2016).

Indicators are defined as specific resource or experiential attributes that can be measured to track changes in conditions to assess progress towards achieving and maintaining desired conditions (IVUMC 2016). Examples include:

Resource examples:

- Trail conditions
- Campsite conditions
- Water quality

Social examples:

- Solitude while traveling
- Crowding
- Noise

Within these broad categories, managers need to identify one or more indicators that reflect the overall condition.

Once indicators have been selected, thresholds are determined for each indicator. *Thresholds* are defined as "the minimally acceptable conditions associated with each indicator" (IVUMC 2016, p. 38). Thresholds represent a "line in the sand," identifying when conditions would become unacceptable and showing where corrective action must be taken. Thresholds are informed by science but are established as a value-based management decision based on an area's purpose and desired conditions.

In more complex areas, triggers may be established along with thresholds. A *trigger* is defined as a condition of a concern for an indicator that prompts a management response to ensure that desired conditions are maintained before a threshold is crossed (IVUMC 2016). Sensitive resources or conditions, such as impacts to threatened and endangered species, might have multiple triggers as preventative management to provide responses before thresholds are reached (Fig. 9.11).

The following are examples of indicators, thresholds, and triggers for a wilderness zone:

Social Condition

- Indicator: Number of trail encounters with other groups per day.
- Threshold: Individual groups encounter no more than three other groups on the trail per day at least 90 percent of the summer use season (May 1–Sept 30).
- Trigger: An observed increase in encounters to two groups per day.

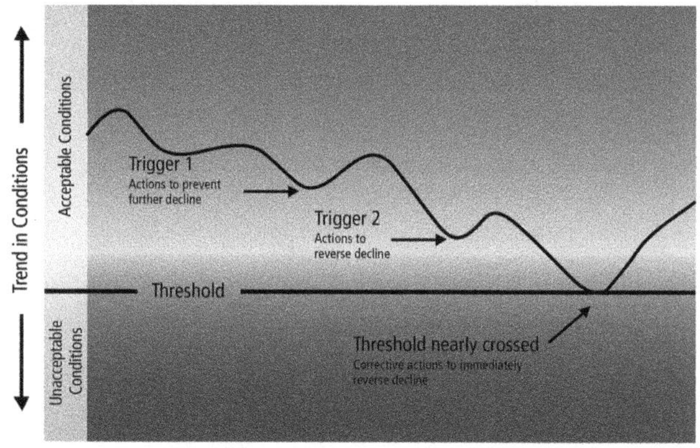

FIGURE 9.11 Management triggers and thresholds related to trends in monitored conditions. Courtesy of IVUMC (2016).

Resource Condition

- Indicator: Rating for soil erosion into Bear Creek at section along the trail.
- Threshold: Soil erosion rating will not exceed "Low" for 90 percent of trail zone.
- Trigger: An observed increase bank instability over one year's time.

Element 3—Identify Management Strategies

After indicators and thresholds are selected to monitor conditions, managers must identify which management strategies and actions will be used to achieve and maintain desired conditions. Management strategies are general approaches to address visitor use management issues while actions are ways of implementing strategies in an area. One possible management strategy in this element is the identification of visitor capacities. This chapter has described *visitor capacity* as the maximum amount and type of visitor use that an area can accommodate while maintaining desired resource conditions and visitor experience. Identifying visitor capacities and other management strategies in this element help managers understand the relationship between existing and desired conditions.

Step 8—Compare and document the differences between existing and desired conditions, and, for visitor use-related impacts, clarify the specific links to visitor use characteristics. This step compares the existing situation to desired conditions and indicators. Comparisons between existing and desired conditions demonstrate one of the following: (1) desired conditions are being achieved and existing conditions are within thresholds; or (2) desired conditions are not achieved or could fail to be achieved soon and existing conditions are close to thresholds.

Comparing existing conditions with desired conditions can include examining trend data (when available), existing information, public comments, and other information. If desired conditions are not being achieved, managers should explore and analyze the probable causes of changes (Cole et al. 1987). These causes may include:

- Types of visitor activity
- Visitor behaviors
- Inappropriate attitudes and expectations
- Location of use
- Amount of use

Step 9—Identify visitor use management strategies and actions to achieve desired conditions. In this step, managers develop managements strategies and actions to address or prevent the issue. Management strategies are selected before identifying the specific management actions for implementation. These strategies can specifically address the probable causes of changes described by Cole et al. (1987) in Step 8. Strategies include:

- Modify type of use
- Modify visitor behavior
- Modify attitudes and expectations
- Modify the location of use
- Reduce use or increase supply

Once a strategy has been selected, managers choose specific actions to implement the selected strategies. These actions fall into three fundament categories: site management/engineering, information and education, and regulations/enforcement. These can be referred to as the three "Es": engineering, education, and enforcement.

Engineering is the most common means to modify the location and spatial distribution of use. This can include site design and construction, hardening sites, and using vegetation and other barriers to direct visitor use. *Education and information* are the most common ways to modify visitor behavior and change attitudes and expectations. Visitors contacts, signage, and programs such as Leave No Trace are common examples of these actions. *Enforcement* through regulations can be used to implement all types of management strategies. Restricting access, prohibiting activities and equipment, and managing length of stay and group size can all be enforced through appropriate actions.

Step 10—Where necessary, identify visitor capacities and additional strategies to manage use levels within capacities. *Visitor capacity* is defined as "the maximum amounts and types of visitor use that an area can accommodate while achieving and maintaining the desired resource conditions and visitor experiences that are consistent with the purposes for which the area was established" (IVUMC 2016, p. 50). Managers may identify and implement a visitor capacity when the amounts and types of visitor use are directly related to achieving and maintaining desired conditions (IVUMC 2019b). For example, a wilderness area close to an urban population may be in high demand and create concerns about the loss of opportunities for solitude. Identifying a visitor capacity as a direct management strategy may be the preferred option to maintain desired conditions. Identifying visitor capacity may be a legal and statutory requirement of the wilderness unit (e.g., Wild and Scenic Rivers Act, National Parks and Recreation Act, National Trails System Act). Additional guidelines for developing visitor capacities can be found in the "Visitor Capacity Guidebook" (IVUMC 2019b).

Visitor capacity strategies should describe who or what is being managed. For example, the people at one time on a trail (PAOT), the number of canoes/boats at one time on a river (BAOT), or the number and percentage of campsites occupied. Managers can describe when these capacities are in place (during peaks season, year-round, weekends/weekdays) and where geographically they are applied in the area. Finally, visitor capacity decisions should include a rationale that is reasonable, well-articulated, and defensible. Managers should be able to provide stakeholders and the public with information that demonstrates that visitor capacity strategies and subsequent actions were implemented with the goal of maintaining desired conditions and visitor experiences.

Step 11—Develop a monitoring strategy. The final step of Element 3 is the development and implementation of an effective monitoring strategy. Monitoring is a key concept in visitor use management and integral to the application of the framework. Monitoring provides feedback of conditions to managers and helps determine whether actions to protect resources and experiences are effective. Well-planned monitoring strategies reduce the uncertainty of existing conditions and can make management actions more effective (Cole 2006; IVUMC 2016).

Broom and Hall (2008) describe a checklist of questions to answer to develop an effective monitoring strategy. Some of these questions include:

- What is going to be monitored (indicators, use levels, patterns)?
- Where will monitoring occur?
- What techniques will be used for each indicator?
- Who will collect the data?
- How will the data be managed and analyzed?
- How will findings be reported and to what audiences?

Element 4—Implement, Monitor, Evaluate, and Adjust

In the final element of the framework, management actions are implemented, conditions are monitored and evaluated, and strategies and actions are adjusted to achieve desired conditions. The efforts of this element build upon all the previous steps of the framework as decision-making and planning are put into action. It also prompts managers to take action to achieve desired conditions if the monitoring of indicators show that thresholds are being approached or current conditions trigger action.

Step 12—Implement management actions. This step implements management actions that were identified and designed in Steps 9 and 10 of Element 3. These actions are implemented to help ensure that thresholds are not being exceeded and that desired conditions are achieved. Managers can prepare for implementation by ensuring funding, staffing, and training are available to initiate actions.

Step 13—Conduct and document ongoing monitoring and evaluate the effectiveness of management actions in achieving desired conditions. This step follows the monitoring strategy developed in Step 11. The monitoring program assesses the selected indicators of desired conditions and compares results to determined thresholds. Monitoring alerts managers if current conditions are changing and identifies if triggers or thresholds are nearly reached. If the analysis of monitoring data documents downward trends in conditions, corrective and timely actions can be taken to protect resources and visitor experiences. For further information on monitoring programs, see the Council's Monitoring Guidebook (IVUMC 2019a).

Step 14—Adjust management actions if needed to achieve desired conditions, and document rationale. The purpose of the final step of the framework is to adjust management actions based on what is learned during the monitoring process. If this step determines that corrective actions are necessary, it is important to revisit if the probable cause of the issue are the same as previously determined. If causes are different, the level and type of action required may change to maintain desired conditions. Finally, managers should document the rationale for any adjustments, and these changes may differ from the original justification for management strategies and actions.

It is important to again emphasis after describing the four elements of the VUMF that this process is meant to be an iterative process where managers and planners may go back and adjust throughout the process. Advances in technology and new recreation activity trends may emerge that could influence the definition of appropriate activities, facilities, and services. Monitoring may show that indicators are not effectively representing current conditions. New ways of implementing management actions to address visitor capacities may emerge that are more effective and efficient. As these potential changes occur, managers can return to any step in the framework to reevaluate and assess what adjustments should be made to protect resources and visitor experiences. It is this flexibility and transparency that makes the VUMF a successful approach for visitor use management in wilderness settings.

IMPLEMENTING THE VUM PROCESS

The VUM process provides a general framework within which desired wilderness conditions can be prescribed and the management strategies and actions to maintain or restore these conditions are identified. However, like any planning framework, the value of the VUMF lies in its ability to help direct wilderness planning and management in the real world. In the following examples, we examine how the VUMF has been applied in the development of plans to guide wilderness management for entire wilderness areas and specific management issues.

Black Canyon of the Gunnison National Park

Over the past two million years, the Gunnison River, a major tributary of the Colorado River, has sculpted the vertical wilderness of the Black Canyon of the Gunnison (National Park Service 2022c). This area exposes some of the steepest cliffs, oldest rocks, and craggiest spires in North America. Born from multiple episodes of uplift and erosion, the Black Canyon reveals 2,000 feet of Precambrian basement rock. Black Canyon of the Gunnison National Park is located on the Western Slop of the Rocky Mountains in west-central Colorado between Gunnison and Montrose, Colorado. Black Canyon of the Gunnison National Monument was established by Presidential Proclamation 2033, signed by President Herbert Hoover on March 2, 1933, and was subsequently designated a national park through an act of Congress in 1999 (Public Law 106-76).

The purpose of Black Canyon of the Gunnison National Park is to preserve spectacular gorges, the Gunnison River, and breathtaking landscapes by protecting its natural, cultural, and wilderness integrity for public benefit, inspiration, and enjoyment (NPS 2022b). Situated in the transition between the Colorado Plateau and Rocky Mountains, the park includes a diversity of ecosystems from pinyon pine, juniper, and scrub oak forests to sagebrush hillsides, steep canyon walls, and riparian habitats along the Gunnison River. The Black Canyon is extremely deep and narrow and is a destination that offers vast opportunities to expert rock climbers in designated wilderness. The depth of the canyon at Warner Point (the deepest section of the canyon) is 2,722 feet. The Painted Wall is the tallest vertical wall in the State of Colorado with a height of 2,250 feet. In the area of the North and South Chasm Walls, where most climbing activity takes place, the depth of the canyon is 1,820 feet (NPS 2022b).

The National Park Service released a draft wilderness and backcountry management plan (WBMP) for public review in 2011. Following public responses and comments from stakeholders and members of the public, the National Park Service released alternatives to the plan in 2015. Following the release in 2015, the National Park Service updated the plan to account for current conditions and planning needs and released a new draft wilderness and backcountry management plan and environmental assessment in June 2022. This plan built upon previous evaluations and included updated management zoning, a climbing management plan, resource condition monitoring strategies, and identification of visitor capacities for wilderness areas (NPS 2022b). The project area encompasses the 15,599-acre Black Canyon of the Gunnison Wilderness, including 11,180 acres designated by Congress in 1976 (Public Law 94-567) and an additional 4,419 acres added in 1999 (Public Law 106-76) and the undeveloped backcountry of Black Canyon of the Gunnison NP (approximately 10,370 acres),

Following the IVUM framework, specifically *Element 1: Build the Foundation*, the park identified the purpose, need, and issues associated with the planning process. They identified that this wilderness and backcountry management plan was needed to provide an integrated framework for decision-making for the stewardship of wilderness and backcountry lands. This plan was needed because underlying trends, such as increasing visitation, could result in the establishment of incompatible uses and entrenched harmful practices, which could impact natural resources, high-quality visitor opportunities, wilderness character, and backcountry conditions (NPS 2022b). Goals for the plan included: (1) restoring, preserving, and enhancing wilderness character; (2) providing a range of visitor opportunities while protecting fundamental resources and values; and (3) enhancing public understanding of wilderness character and backcountry values and opportunities, with an objective to provide public information to promote backcountry skills and Leave No Trace outdoor ethics.

Additionally, the draft WBMP provided details on key issues to be addressed, including:

- Increasing visitation and its effect on wilderness resources, such as soundscape and viewshed impacts, inner canyon and upland hiking access, and fishing and climbing access.
- Lack of consistent management of overnight use.
- The lack of a formal climbing management plan.
- Changes in types of use and emerging uses on the Gunnison River in Black Canyon of the Gunnison National Park.
- Lack of indicators, thresholds, visitor capacities, and monitoring guidance and protocols for wilderness and backcountry lands.

Step 3 of the VUMF is to assess and summarize existing information and current conditions. In the draft WBMP, descriptions of existing conditions focused on the four wilderness character qualities of untrammeled, natural, undeveloped, and opportunities for solitude or primitive and unconfined recreation.

- The existing *untrammeled quality* was described as trending downward due to management actions to address climate change and actions required to control invasive species, clean up waste, and restore native plants.
- The existing *natural quality* was considered good and overall stable. Programs to remove invasive species were improving ecosystems and communities of proposed or listed threatened and endangered species were being protected. Water and air quality was overall in good condition.
- The existing *undeveloped quality* was rated as good and not likely to change, with the possibility for improvement. Unnecessary structures and installations may be removed over time, and the rugged terrain of the inner canyon naturally deters developments.
- The existing *opportunities for solitude or primitive and unconfined recreation* was rated as stable but may be showing downward trends in some areas. Rugged and primitive routes provide visitors access to the inner canyon and backcountry permit systems provide opportunities for solitude. Increased demand for backcountry use may reduce solitude and ongoing managed access may reduce unconfined qualities.

For the VUMF, *Element 2: Define Visitor Use Management Direction* begins with Step 5 and the defining of desired conditions and management zones for the study area. Three management zones were proposed for the wilderness and backcountry lands of Black Canyon of the Gunnison National Park (and an additional two zones for the adjacent backcountry lands in Curecanti National Recreation Area) (Table 9.3; Fig. 9.12). Within the proposed management zones, five *subzones* were identified to guide management of the canyon uplands and inner canyon areas more effectively, subdivided into separate "primitive" and "pristine" subzones. Generally, the distinction is that areas within pristine subzones feature a wild, rugged, untamed landscape and are isolated and especially challenging to access because of the geography and terrain. Although areas that fall within pristine subzones tend to be wilder and more isolated, the preservation and protection of wilderness character is described as vitally important within both primitive wilderness and pristine wilderness subzones.

The inner canyon wilderness zones include the canyon rims, walls, and canyon floor of the Gunnison River. Desired conditions for this zone are described as wild, rugged, and providing visitors with adventure. Managers desire experiences to be challenging where visitors are self-reliant and depend upon wilderness skills. The area should foster people's connection to wild places as the untrammeled wilderness character quality predominates.

The uplands zone's desired condition is described as wild, natural, and with experiences for solitude. Some trammeling

TABLE 9.3 Proposed Wilderness and Backcountry Zones for Black Canyon of the Gunnison National Park (National Park Service 2022b)

Management Zone	Subzones
Inner Canyon Wilderness Zone	• Inner Canyon Primitive Wilderness Subzone • Inner Canyon Pristine Wilderness Subzone
Uplands Zone	• Uplands Backcountry Subzone • Uplands Primitive Wilderness Subzone • Uplands Pristine Wilderness Subzone
Gunnison Gorge Interface Wilderness Zone	(No applicable subzones defined)

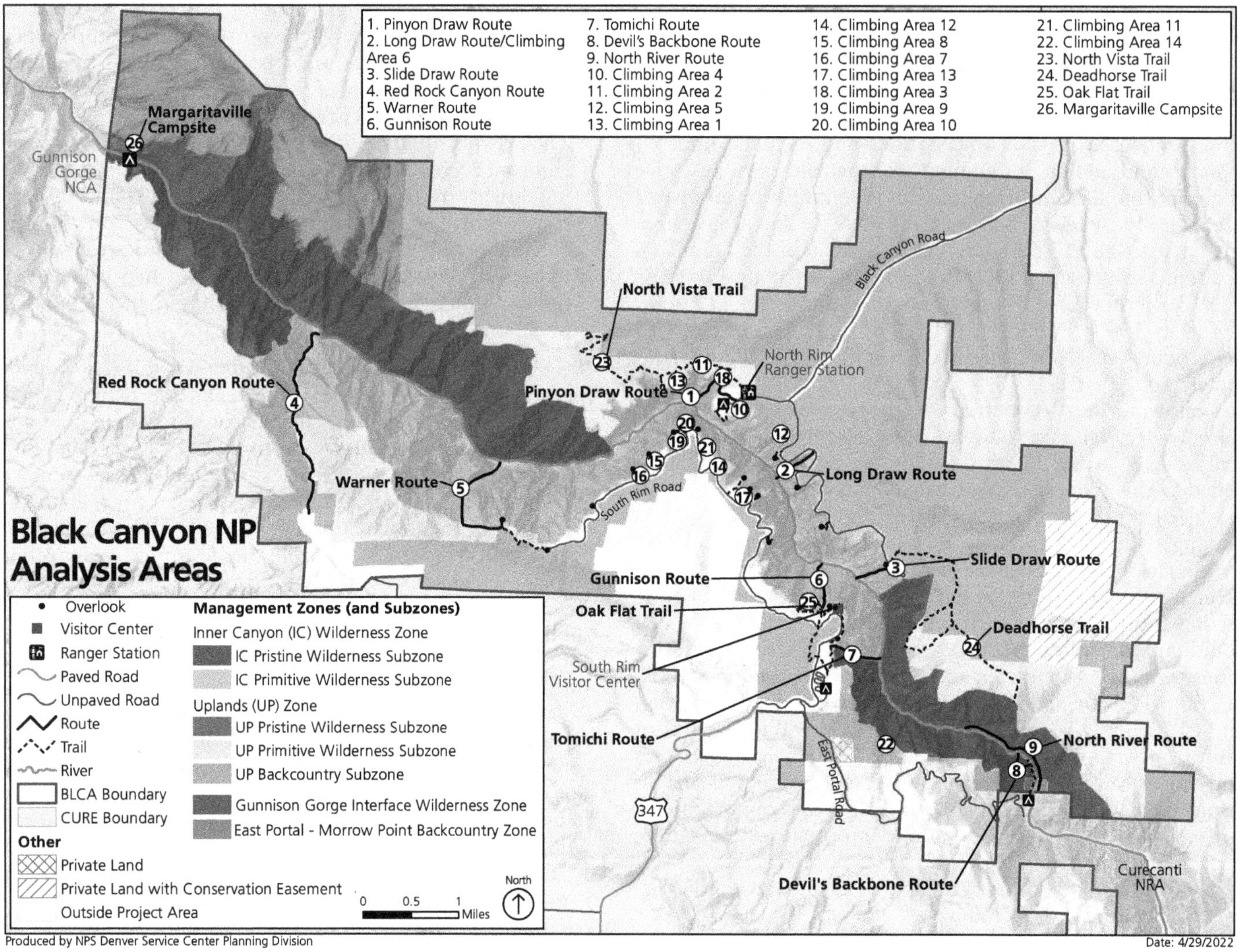

FIGURE 9.12 Proposed management zones for Black Canyon of the Gunnison WBMP. Courtesy National Park Service (2022b).

actions (e.g., controlling invasive species) may be accepted to provide long-term benefits to the natural qualities. These areas are remote yet approachable. Efforts are made to maintain wildness, improve naturalness, and provide outstanding opportunities for self-reliance and adventure.

The Gunnison Gorge Interface Wilderness Zone relates to lands adjacent to the BLM-managed Gunnison Gorge National Conservation Area and designated wilderness. Desired conditions would be wild, natural, rugged, remote, unspoiled, and uncrowded. Visitor experiences are to be challenging and provide opportunities for self-reliance and the use of wilderness skills (Fig. 9.13).

After describing desired conditions and allocating them across the management zones, the next step in the visitor use management process is to identify how conditions will be tracked over time. For Step 7 of the VUMF, the draft WBMP identified indicators, measures, and thresholds based on wilderness character qualities. Measures were described as the specific elements under each indicator where data would be collected during monitoring activities to assess trends in the indicator. Several examples of indicators, measures, and thresholds for the wilderness character qualities are shown in Table 9.4.

FIGURE 9.13 Former Ranger Mike O'Mahoney on Maiden Voyage route in the Black Canyon of the Gunnison. Photo courtesy Tom Schaefer.

The draft WBMP identified several management strategies to achieve and maintain desired conditions. This represents Step 9 of Element 3 from the VUMF. Strategies were described specifically to the three management zones and their

TABLE 9.4 Examples of Indicators, Measures, and Thresholds for Black Canyon of the Gunnison WBMP (National Park Service 2022b)

Wilderness Character Quality	Indicator	Measure	Threshold
Natural	Plants	Number of nonindigenous plant species/percent cover of invasive species	≥ 5% decrease in area affected by invasive plant species
Solitude or primitive and unconfine recreation	Remoteness from sights and sounds of human activity	Number of encounters with individuals from other groups	Not exceed site specific encounter rate standards by site
Undeveloped	Presence of nonrecreational structures, installations, and developments	Number of authorized nonrecreational structures, installations, and developments	0 new authorized without MRA approval
Untrammeled	Actions authorized that manipulate the biophysical environment	Number of fish stocked per body of water	0 stocking within the wilderness

included subzones. For the *Inner Canyon wilderness zone*, management strategies were to maintain and enhance existing opportunities for solitude for hikers, anglers, backcountry campers, climbers, and kayakers. The plan also suggested that the existing wilderness use permit system laid out in the 1993 Resource Management Plan for the Inner Canyon wilderness zone would be updated to achieve desired resource and experiential conditions. The primary difference in the management of the Inner Canyon pristine and primitive wilderness subzones would be the management of overnight use and climbing. Campsites would be designated in the inner primitive wilderness subzone and would not be designated in the pristine subzone. Visitors would have the freedom to choose a camping spot in the pristine subzone, would be encouraged to camp on durable surfaces and stay no more than one night, and would be provided with information about Leave No Trace principles (NPS 2022b).

Management strategies for the *Upland zone* were to emphasize restoration of natural qualities, including intensive programs for invasive species removal. Unnecessary structures and facilities would be removed, and the zone would remain relatively free of recreational developments with no designated campsites or toilet facilities. The park would implement guidelines for research and scientific installations in wilderness that would be consistent with the preservation and management of wilderness. The main difference between the uplands primitive and pristine wilderness subzones would be the presence of trails. The uplands primitive wilderness subzone has existing trails, a few signs, and more trails could be added in the future, while the pristine wilderness subzone has no trails and would remain trail-less and without signs.

Identifying visitor capacities was a legal requirement for Black Canyon of the Gunnison National Park. Visitor capacities were identified using best practices and other plans and projects across the National Park Service. As per the *Visitor Capacity Guidebook* (2019b), capacities were identifying following the guidelines:

- *Determine the analysis area*—These included the wilderness and backcountry lands represented by the three management zones and five management subzones previously described. The draft WBMP described the analysis area as hiking routes and climbing routes within the park. These included the Pinyon Draw Route, Slide Draw Route, BLCA Climbing Area 4 (Cruise Gully), and BLCA Climbing Area 7 (Dragon Point Gully).
- *Review existing direction and knowledge*—Managers reviewed the 1993 Resource Management Plan, 1997 General Management Plan, and other overarching guidance. Previous daily use limits for climbing routes and inner canyon routes were examined. Wilderness use permit and special use permit data was analyzed to compare amount, timing, and distribution of use to thresholds for resource conditions and visitor experiences. For example, the Pinyon Draw Route is the suggested route for first time North Rim inner canyon hikers and the most popular route in the wilderness zones. Visitation had increased from thirty-five overnight visitors in 2014 to 306 in 2018. Similarly, BLCA Climbing Area 4 (Cruise Gully) was the most heavily used climbing area with six to ten climbing parties per day during the peak season. In 2019, there were fifteen to thirty climbers per day on weekends in the peak season.
- *Identify the limiting attributes*—The feature that most constrains the ability of inner canyon hiking and climbing routes to accommodate use is the wilderness quality of *solitude or primitive and unconfined recreation.* Site-specific encounter rates observed during patrols represent the condition of this quality. Another limiting attribute for routes was the actual physical space along routes that are out and back recreation experiences. This creates multiple encounters with other groups. Likewise, the climbing areas have limited physical space on climbing routes and a fixed number of rappel areas to access routes.
- *Identify visitor capacities and implementations strategies*—Visitor capacities were most frequently expressed as people in one day. Appropriate amounts of use were defined at key areas based on current conditions and were related to the goals and objectives for that area. They were also identified based on the best available information and are continuously reevaluated with future planning efforts as informed by monitoring information. Examples of visitor capacities and associated management strategies are found in Table 9.5.

TABLE 9.5 Examples of Visitor Capacities in Black Canyon of the Gunnison National Park (National Park Service 2022b)

Analysis Area	Visitor Capacity	Management Strategies
Pinyon Draw Route	Current use levels at Pinyon Draw, 23 people in one day (PIOD), are achieving and maintaining the wild, natural, and rugged desired conditions for the inner canyon primitive wilderness subzone. The wilderness use permit system will not change with implementation of the plan. Therefore, use levels are to be maintained at 23 PIOD.	Continue implementing the wilderness use permit system to manage the timing and location of use and distribute visitors so that encounter rates are not exceeded. If capacities are exceeded, management strategies include increasing patrols to ensure compliance with the permit system and reevaluating and reducing capacities if encounter rate thresholds are exceeded.
BLCA Climbing Area 4 (Cruise Gully)	Current use levels, 15–30 climbers per day on peak weekend days, are maintaining the wild, natural, and rugged desired conditions for the inner canyon primitive wilderness subzone. Opportunities for solitude are key to the visitor experience, in addition to limited space on the wall and safety concerns associated with climbing. Therefore, use levels are to be maintained at 20 PIOD.	Continued use of the wilderness use permit system and managing the timing and location of use to distribute climbers with the white board at the North Rim Ranger Station. *Related potential management strategy:* Increasing patrols to ensure compliance with the wilderness use permit system.

The iterative process of the VUMF continues into Element 4 where managers for Black Canyon of the Gunnison National Park implement, monitor, evaluation, and adjust the wilderness and backcountry management plan. Managers implement proposed actions when indicators reveal that thresholds are being approached or triggers for conditions have been reached. Managers evaluated the effectiveness of zones, managements strategies, and actions towards achieved desired conditions, specifically wilderness character qualities in the Black Canyon of the Gunnison National Park wilderness area. And they made a commitment to adjust and revaluate management decisions through future planning efforts that engage stakeholders and the public.

Glacier Bay National Park and Preserve

Glacier Bay National Park (Haa Léelk'w Hás Aaní, *Our Grandparents' Land*) encompasses more than 3.2 million acres of rugged mountains, dynamic glaciers, temperate rainforests, coastlines, and fjords in remote southeast Alaska. Originally established as a national monument by presidential proclamation in 1925, the land became a national park and preserve with designated wilderness in 1980 under the Alaska National Interest Land Conservation Act (ANILCA; NPS 2022a). Very importantly, Glacier Bay National Park and Preserve recognizes and honors the rich cultural tapestry of indigenous use and occupation in the wilderness and supports the enduring connection between the Tlingit and their Homeland.

Glacier Bay Wilderness (Kawayéil' Aan, *Quiet, Peaceful Land*) is exceptional within the National Park system having the highest ratio of terrestrial area as wilderness (99 percent; 2.9 million acres) and a unique designation of wilderness, *marine wilderness*. The marine wilderness includes 53,000 acres of marine wilderness waters that are managed as a globally important sanctuary for ecologically important species, including threatened species, such as humpback whales and Steller sea lions (NPS 2022a).

The park lies within the ancestral Homelands of two Tlingit tribes, the Huna and Yakutat Tlingit, who sustained themselves for many generations from the area's abundant resources. According to the Tlingit, the park has been peopled from time immemorial and this human presence has shaped, and been shaped by, the landscape and ecological systems (NPS 2022a). The relationship between the park's land and waters and the ancestral and living Tlingit is integral to the park's intact ecosystem.

Annually, an average of 6,000 wilderness visitors (2009–2019 average) participate in some form of camping or are aboard small vessels that may offer shore landings. These visitor numbers have been stable for many decades, due in part to the remoteness, expense and skill required, and NPS-regulated marine vessel quotas to protect park wildlife and other resources while providing a range of recreational opportunities to park visitors (NPS 2022a) (Fig. 9.14).

In 2022, a draft backcountry and wilderness management plan was released for public comment. The purpose of the plan was to outline strategies to respond to increasing and changing visitor demands and to provide guidance for the terrestrial and marine wilderness areas. This plan was to provide protection for the natural and cultural resources and values of the park and preserve while providing high-quality visitor experiences within wilderness. The plan highlights Indigenous Homeland values embraced by Glacier Bay's original people and advance agencywide conversations about diverse cultural perspectives and values. Park staff

FIGURE 9.14 Annually, an average of 6,000 wilderness visitors to Glacier Bay National Park and Preserve participate in some form of camping or are aboard small vessels that may offer shore landings. Courtesy National Park Service Photos and Tania Lewis.

collaborated with Tlingit elders and speakers to incorporate, where appropriate, Tlingit language in the planning documents and process to encourage readers to consider concepts through the world view of those who consider Glacier Bay National Park Homeland.

The draft plan (NPS 2022a) provided guidance for several planning needs, including:

- *Preserving wilderness character*—Managers stated that wilderness management guidelines were outdated and were inadequate for the park to monitor and manage impacts to wilderness character. Routine backcountry and wilderness patrols or regular monitoring were not possible. Thus, guidance was needed to proactively preserve wilderness character considering increasing and changing visitor uses.
- *Providing visitor access to tidewater glaciers*—A core component of Glacier Bay's enabling proclamation was providing visitor access to tidewater glaciers. Tidewater glaciers are a defining feature of the landscape and a significant draw for park visitors. But increasing visitation to tidewater glacial areas and the dynamic nature of the glacial environment creates challenges for maintaining and improving visitor experience and resource conditions. Specific guidance on preserving wilderness character while providing high-quality visitor experience was needed for tidewater glacier areas.
- *Incorporating Tlingit Homeland values in wilderness management*—The wilderness character of Glacier Bay National Park is defined by the sustained connection between past, present, and future generations of Tlingit and the lands and waters they call *Homeland*. Tlingit interactions with Homeland have shaped the ecology of the area for countless generations and the park must continue to increase its understanding of Tlingit relationships to Homeland and incorporate that understanding in managing Homeland. Importantly, the park must also clearly articulate Homeland values for backcountry visitors.
- *Understanding intact, complex terrestrial and marine ecosystems*—Glacier Bay National Park and Preserve's enabling proclamation identifies "scientific inquiry" as a primary park purpose. As climate change and other stressors impact wilderness resources, Glacier Bay's legacy of research make it an ideal living laboratory to study natural processes and the human potential to affect them, including impacts from anthropogenic climate change. The park provides a place where dynamic glacial systems, successional landscapes, and the interplay between the terrestrial and marine systems can be studied.
- *Addressing desired conditions for resources and visitor experiences within marine wilderness areas*—Protecting marine wilderness presents unique challenges and opportunities given the considerable connectedness of marine ecosystems and the importance of the ocean's most productive and biologically diverse areas. With changes to visitation patterns and activities and the desire to access those designated marine wilderness areas and adjacent areas, there is a need to address desired conditions for resources and visitor experiences within marine wilderness.

To provide management direction for the wilderness lands and waters within the park, the draft plan proposes five management zones to best reflect current park needs. These include the Remote Wilderness Zone, the Shoreline Access Zone, the Wilderness Waters Zone, the Glacier Access Zone, and the Front Country Access Zone.

As previously described, *Element 2: Define Visitor Use Management Direction* of the VUMF focuses on Step 5 and the defining of desired conditions for the study area. The desired conditions for the Remote Wilderness Zone of Glacier Bay National Park were to provide for intact ecosystems processes where natural processes and patterns continue without intervention or alteration. Visitors were provided primitive experiences requiring self-reliance, high skill, and the ability to mitigate risk. The zone was void of development, trails, and visitors had low encounter rates with others in this zone.

The Wilderness Waters Zone represents most waters in the park. The draft plan describes the desired conditions as prioritizing the quality and natural processes within the marine ecosystem with minimal human influence. Visitors should experience solitude, low contact with other vessels, and experience remote wilderness waters that were less impacted by visitors than other main travel routes.

The Glacier Access Zone includes those areas of the park where glaciers are more accessible by various means of travel. This zone experiences some of the most concentrated use in the park and visitors expect encounters with other groups. The plan described the zone as managed to support a range of high-quality immersion experiences in a dynamic landscape, protecting wilderness character while providing access to the glacial environment as specified in the park's enabling proclamation (NPS 2022a). Desired conditions suggest that resource impacts from informal trails and campsites are likely and that visitors should expect to see an implementation of various management tools and strategies to ensure high-quality visitor experiences and resource protection.

To maintain these desired conditions, the draft plan presented several management actions for implementation across all zones. First, visitor information and education would have focused messaging on wilderness, Leave No Trace, and themes associated with the Tlingit Homeland. Next, to preserve wilderness character, the park would implement a registration system for mountaineering trips and backcountry camping. This would include year-round registration for all overnight wilderness use. Next, there would be an increased focus on ethnographic resources and traditional cultural properties. Ethnographic resources would be entered into the Cultural Resource Inventory System–Ethnographic Resources (CRIS-ER) database. Structures, archeological sites, and other features that are part of traditional cultural properties would be managed as ethnographic resources. Finally, the park would establish quotas for visitors' numbers in those areas where visitation was high enough that other mechanisms were unlikely to achieve desired conditions.

The VUMF iterative process was represented in the Glacier Bay National Park and Preserve Backcountry and Wilderness Management Plan as managers committed to implement proposed actions, evaluate and monitor conditions, and adjust actions and strategies when corrective action is needed to maintain and achieve desired conditions.

POTENTIAL BARRIERS TO THE VUM PROCESS

Several barriers can exist for the successful implementation of the visitor use management process. First, adequate staff and funding may not exist to conduct good planning and management. A lack of funding can impede the collection of baseline information on resource and social conditions in wilderness, can limit the implementation of management actions, and can prevent sufficient monitoring of current conditions and action effectiveness. Second, integrating stakeholders and involving the public can be challenging for agency-driven approaches. While new collaborative public engagement processes are developing, it can still be difficult to create capacity and consensus to manage wilderness resources and visitor experiences. Additionally, wilderness planning and management are a political process involving competing and sometimes polarized groups. These groups can influence federal agencies involved in wilderness planning and management to be unwilling to make needed but controversial wilderness management decisions. Finally, external factors and new emerging issues can compete for resources, staffing, and attention. Climate change, increased urbanization, overcrowding, and changing fire regimes can be issues competing with visitor use management as the priority for wilderness managers and planners.

POSITIVE BENEFITS TO WILDERNESS FROM THE VUM PROCESS

From the historical use of the LAC process to the development of the VUMF, the positive benefits of using the VUM process outweigh the potential barriers to conducting such a process. The VUM planning process is an adaptable, flexible, and effective approach to understanding the carrying-capacity concept and impacts to wilderness related to use. It is a proactive approach that describes the desired resource conditions and visitor experiences and monitors conditions to determine if management strategies and actions are effective at maintaining desired conditions. Data and information provide feedback on how to adjust management actions and more general information about the effectiveness of those management actions in similar situations in other wilderness areas.

The VUM process describes opportunity classes, or prescriptive management zones, allowing for more pristine areas within wilderness to be maintained and human-caused impacts to be limited. By describing zones with natural, social, and managerial setting attributes, managers can create opportunities to address a diversity of values and needs. Planning can find some compromise between the competing goals inherent in wilderness management under current wilderness legislation: wilderness resource preservation and wilderness recreation use. However, the VUM process makes explicit what the acceptable desired conditions are, how to manage for those conditions, how to monitor success toward those goals, and, if thresholds are exceeded, the management strategies and actions to improve wilderness conditions.

Finally, the VUM process involves public engagement throughout the planning process. This has several beneficial outcomes for wilderness and wilderness management. By making the planning and management processes more evident, wilderness managers make stakeholders and the public more aware of the costs and benefits of wilderness management. Such awareness extends from the diverse issues and concerns of various users to the consequences of different management actions, and to the effectiveness and equity of management actions to achieve the desired conditions. Furthermore, public engagement can lead to more informed discourse on how wilderness values can be protected and, potentially, to more public support for agency wilderness programs.

DIRECT VISITOR MANAGEMENT MAY BE NECESSARY WHEN THRESHOLDS ARE APPROACHED

The VUMF describes indicators of desired conditions and thresholds to represent the minimally acceptable conditions associated with each indicator. Monitoring these indicators serves as feedback to the wilderness managers, indicating when implemented management actions have or have not succeeded in maintaining acceptable desired conditions.

When wilderness conditions begin to degrade toward the threshold that was defined as acceptable and management actions have proved ineffective, managers must consider how an area's conditions can be protected. Direct management (e.g., visitor use limits directly addressing visitor capacity) may remain the only alternative available. In Chapters 15 and 16, specific techniques to directly manage visitors are discussed. In general, *the concept of minimum regulation should guide management.* As the level of use increases, it will become necessary at some point to use direct visitor management techniques, such as limiting the number of recreation users and types of activities, so that unacceptable impacts and conditions do not occur or can be improved.

SUMMARY

Three key concepts form the theme of this chapter. First, the challenge facing wilderness managers in dealing with the issue of *carrying capacity* is not a matter of developing maximum numbers that describe how much use is too much. Rather, it is a matter of prescribing what kinds of social and resource conditions are desired, comparing these desired states against existing conditions, and identifying management strategies and actions needed to maintain or restore the desired conditions. *Managerial judgment* is one key element in this process. Judgment is a product of experience, research data, basic inventory information, public input, careful analysis, and common sense.

The second concept is that establishing visitor use management direction depends on formulating explicit *desired conditions* that (1) are evaluated by specific, measurable indicators and (2) are defined by thresholds for the resource and social elements identified as critical indicators. One key factor in the success or failure of a wilderness management plan is the process by which management actions are proactively identified to address changes to desired conditions and implemented when monitoring efforts demonstrated that actions are needed to preserve or maintain conditions.

Third, either *resource or social indicators* can determine if an area has reached or exceeded its recreation visitor capacity. Managers may need to employ different management actions to control impacts. Some of these management actions will be more effective than others, and it may become necessary to implement direct visitor management. At some level of visitor use, it will become necessary to restrict further increases in the number of users to protect desired conditions, but it is the condition of the area, not the number of visitors, which is the focus of management attention.

The VUMF is a powerful planning tool that can be used as a process for managing wilderness within a carrying-capacity concept. *The goal of management is to identify the desired resource, social, and managerial conditions to be maintained or restored in wilderness, with these attributes expressed as explicit, desired conditions. Thus, the focus of management attention shifts from defining maximum visitor use to identifying desired conditions and managing recreation use levels and impacts so that resource, social, and managerial conditions are what was intended under the Wilderness Act.*

STUDY QUESTIONS

1. Identify four considerations in managing visitor recreation use and impacts in wilderness.
2. Discuss some of the factors that will influence determining the level of analysis on the sliding scale for visitor use management.
3. What are desired conditions and how are they applied in the VUMF? Discuss several characteristics that make for good, desired condition statements.
4. What are the differences between indicators, thresholds, and triggers? Provide several examples of indicators, thresholds, and triggers for resource and social conditions in a wilderness zone.
5. Visitor capacities define the maximum amounts and type of use that an area can accommodate while maintaining desired conditions. Describe basic guidelines for developing visitor capacities.

ACKNOWLEDGMENTS

This chapter is a major revision by Robert G. Dvorak, professor at Central Michigan University, of the fourth edition by Chad P. Dawson and John C. Hendee; we recognize and thank George H. Stankey, Stephen F. McCool, and Gerald L. Stokes, for their seminal efforts on these topics in the previous editions. For contributions and review comments on this and previous editions of this chapter, we thank the following colleagues: Ed Krumpe, professor of Resource Recreation and Tourism, University of Idaho, Moscow, Idaho; Liese Dean, wilderness program coordinator, Stanley Ranger Station, Sawtooth National Recreation Area and National Forest, USDA, Forest Service; Gregory Hansen, wilderness staff officer, Mesa Ranger District, Tonto National Forest, USDA, Forest Service, Mesa, Arizona; Richard Hanson, supervisory outdoor recreation planner, USDI, BLM Field Office, Phoenix, Arizona; James Mahoney, wilderness manager, USDI, BLM Field Office, Phoenix, Arizona; Linda Merigliano, wilderness, trails, and recreation planner, Bridger-Teton National Forest, USDA, Forest Service; R. Gregory Thompson, Rio Grande National Forest, USDA, Forest Service; Aleks Pitts, planning portfolio manager, USDI Regions 3, 4, and 5, National Park Service; Mary Ellen Emerick, natural resource specialist, USDA Washington Office Enterprise Program, Forest Service.

CASE DISCUSSION: CLIMBING MANAGEMENT IN BLACK CANYON OF THE GUNNISON NATIONAL PARK

Black Canyon of the Gunnison National Park is a destination that offers vast opportunities for expert rock climbers in designated wilderness. Located in southwest Colorado, the park's steep-walled canyon includes the Painted Wall, the tallest vertical wall in the State of Colorado with a height of 2,250 feet. Peak climbing season at the Black Canyon of the Gunnison National Park begins in late April and runs through the early part of June and then from mid-September through late October.

The National Park Service through Director's Order 41 (NPS 2013) recognizes that climbing is a legitimate and appropriate use of wilderness. If climbing activities occur in wilderness, climbing management strategies are included as part of the park's wilderness stewardship planning. As part of that process, climbing practices with the least negative impact on wilderness resources and character will always be the preferred choice. "Clean climbing" techniques should be the norm in wilderness and backcountry. This involves the use of temporary equipment and anchors that can be placed and removed without altering the environment (e.g., slings, cams, nuts, chocks, and stoppers). It is recognized that the use of removable anchors may reduce, but does not, in every case, completely eliminate the need for fixed anchors. Therefore, it is recognized in Director's Order 41 that the occasional placement of a fixed anchor for belay, rappel, or protection purposes does not necessarily impair the future enjoyment of wilderness or violate the Wilderness Act.

As part of the draft wilderness and backcountry management plan, the Black Canyon of the Gunnison National Park developed a climbing management plan (NPS 2022b). The goal of the climbing management plan includes preserving the fundamental resources and values in the Black Canyon of the Gunnison NP while providing for recreational climbing activities, and setting clearly defined guidelines that allow for the preservation of wilderness character.

In the development of a climbing management plan, several important topics need consideration as part of the larger visitor use management process. These include:

- *Wilderness use permits*—How should day use and overnight use related to climbing be managed?
- *Fixed anchors and equipment*—Within federally designated wilderness in the national park system, all new fixed equipment placements or replacements require authorization. Where could these be authorized?
- *Development of new climbing routes*—When and where should new routes be created or developed?
- *Rock alteration*—How should rock alterations be addressed, if allowed (e.g., safety)?
- *Vegetation alteration*—How should vegetation resources best be protected (e.g., vegetation, trees)?
- *Human waste*—How should human waste in the canyon be addressed?

Management strategies and actions address these important topics to maintain desired conditions for wilderness character while providing appropriate recreational opportunities for visitors to Black Canyon of the Gunnison National Park.

CASE DISCUSSION QUESTIONS

1. As part of the VUMF, what would be key issues, needs, and areas of concern to identify for a climbing management plan in wilderness?
2. What are examples of desired conditions for wilderness character and visitor experiences related to climbing management?
3. What would make good indicators of resource and experience conditions for climbing activities in Black Canyon of the Gunnison National Park?
4. What management strategies and actions would you recommend for climbing management in Black Canyon of the Gunnison National Park?

REFERENCES

Allen, Stewart. 2019. The relationship between amount of visitor use and social impacts. Interagency Visitor Use Management Council. March 2019. Edition One. Retrieved from https://visitorusemanagement.nps.gov/Content/documents/Contributing%20Paper_Social%20Impacts_Visitor%20Capacity_Edition%201.pdf

Broom, T. J.; Hall, Troy E. 2008. A guide to monitoring encounters in wilderness. University of Idaho, College of Natural Resources, Department of Conservation Social Sciences, Moscow, ID.

Burch, William R., Jr. 1984. Much ado about nothing—some reflections on the wider and wilder implications of social carrying capacity. *Leisure Sciences.* 6(4): 487–496.

Clark, Roger N.; Stankey, George H. 1979. The recreation opportunity spectrum: A framework for planning, management, and research. General Technical Report. PNW-98. Portland, OR: U.S. Department of Agriculture Forest Service, Pacific Northwest Forest and Range Experiment Station. 32p.

Clawson, Marion. 1985. Outdoor recreation: Twenty-five years of history, twenty-five years of projection. *Leisure Sciences.* 7(1): 73–100.

Cahill, Kerri L. 2003. Exploring the structure and development of management prescriptions for public lands. PhD diss. Virginia Polytechnic Institute and State University, Blacksburg, VA.

Cahill, Kerri; Collins, Rachel; McPartland, Susan; Pitt, Aleksandra; Verbos, Rose. 2018. Overview of the Interagency Visitor Use Management Framework and the uses of social science in its implementation in the National Park Service. *The George Wright Forum.* 35(1): 32–41.

Cole, David N. 1996. Wilderness recreation use trends, 1965 through 1994. Research Paper INT-RP-488. Ogden, UT: U.S. Department of Agriculture, Forest Service, Intermountain Research Station.

———. 2006. Visitor and recreation impact monitoring: Is it lost in the gulf between science and management? *The George Wright Forum.* 23(2): 11–16.

———. 2019. The relationship between amount of visitor use and environmental impacts. Interagency Visitor Use Management Council. March 2019. Edition One. Retrieved from https://visitorusemanagement.nps.gov/Content/documents/Contributing%20Paper_Environmental%20Impacts_Visitor%20Capacity_Edition%201.pdf

Cole, David N.; McCool, S. F. 1997. The limits of acceptable change process: Modifications and clarifications. In: McCool, Stephen F.; Cole, David N., comps. *Proceedings: Limits of Acceptable Change and Related Planning Processes: Progress and Future Directions;* May 20–22, 1997; Missoula, MT. INT-GTR-371. Ogden, UT: U.S. Department of Agriculture, Forest Service, Rocky Mountain Research Station, pp. 61–68.

Cole, David N.; Petersen, Margaret E.; Lucas, Roger C. 1987. Managing wilderness recreation use: Common problems and potential solutions. Gen Tech Rep INT-230. U.S. Department of Agriculture, U.S. Forest Service, Intermountain Research Station, Ogden, UT.

Dasmann, Raymond F. 1964. *Wildlife Biology.* New York: John Wiley and Sons.

Driver, Beverly L.; Brown, Perry J. 1978. The opportunity spectrum concept and behavioral information in outdoor recreation resource supply inventories: A rationale. In: Lund, G. H.; LaBau, V. J.; Folliott, P. F.; Robinson, D. W., tech. coords. Integrated inventories of renewable natural resources. General Technical Report RM-55. Fort Collins, CO: U.S. Department of Agriculture, Forest Service, Rocky Mountain Forest and Range Experiment Station, pp. 24–31.

Frissell, Sidney S., Jr.; Stankey, George H. 1972. Wilderness environmental quality: Search for social and ecological harmony. In: *Proceedings of the 1972 National Convention;* October 1–5, 1972; Hot Springs, AR. Washington, DC: Society of American Foresters, pp. 170–183.

Godschalk, David R.; Parker, Francis H. 1975. Carrying capacity: A key to environmental planning? *Journal of Soil and Water Conservation.* 30(4): 160–165.

Graefe, Alan R.; Kuss, Fred R.; Loomis, Laura. 1986. Visitor impact management in wildland settings. In: Lucas, Robert C., comp. *Proceedings: National Wilderness Research Conference: Current research;* July 23–26, 1985; Fort Collins, CO. INT-GTR-212. Ogden, UT: U.S. Department of Agriculture, Forest Service, Intermountain Research Station, pp. 432–439.

Graefe, Alan R.; Vaske, J. J.; Kuss, R. R. 1984. Social carrying capacity: An integration and synthesis of twenty years of research. *Leisure Sciences.* 8(3): 395–431.

Interagency Visitor Use Management Council (IVUMC). 2016. Visitor use management framework: A guide to providing sustainable outdoor recreation. Edition One. 130p.

———. 2019a. Monitoring guidebook: Evaluating effectiveness of visitor use management. Edition One. 118p.

———. 2019b. Visitor capacity guidebook: managing the amounts and types of visitor use to achieve desired conditions. Edition One. February 2019. 126p.

Landres, P.; Boutcher, S.; Merigliano, L.; Barns, C.; Davis, D.; Hall, T.; Henry, S.; Hunter, B.; Janiga, P.; Laker, M.; McPherson, A.; Powell, D.; Rowan, M.; Sater, S. 2005. Monitoring selected conditions related to wilderness character: A national framework. General Technical Report RMRS-GTR-151. Fort Collins, CO: U.S. Department of Agriculture, Forest Service, Rocky Mountain Research Station.

Lime, David W. 1970. Research for determining use capacities of the Boundary Waters Canoe Area. *Naturalist.* 21: 8–13.

Lucas, Robert C. 1985. Visitor characteristics, attitudes, and use patterns in the Bob Marshall Wilderness Complex, 1970–82. Research Paper INT-345. Ogden, UT: U.S. Department of Agriculture, Forest Service, Intermountain Research Station.

Manning, Robert E. 2007. *Parks and carrying capacity: Commons without tragedy.* Washington D.C.: Island Press. 313 p.

Manning, Robert E. 2010. *Studies in Outdoor Recreation: Search and Research for Satisfaction.* 3rd ed. Corvallis: Oregon State University. 368p.

Manning, R. E.; Lime, D. W. 2000. Defining and managing the quality of wilderness recreation experiences. In: Cole, D. N.; McCool, S. F; Borrie William T.; O'Loughlin, Jennifer, comps. *Proceedings: Wilderness Science in a Time of Change Conference,* Vol. 4: *Wilderness Visitors, Experiences, and Visitor Management;* May 23–27, 1999; Missoula, MT. RMRS-P-15-Vol-4. Ogden, UT: U.S. Department of Agriculture, Forest Service, Rocky Mountain Research Station, pp. 13–52.

Manning, Robert. E.; Anderson, Laura. E; Pettengill, Peter, R. 2017. *Managing outdoor recreation: Case studies in the National Parks.* 2nd ed. Boston: CABI. 236 p.

Marion, Jeffrey L. 2016. A review and synthesis of recreation ecology research supporting carrying capacity and visitor use management decision making. *Journal of Forestry.* 114 (3): 339–351.

———. 2019. Impacts to wildlife: Managing visitors and resources to protect wildlife. Interagency Visitor Use Management Council. March 2019. Edition One. Retrieved from https://visitorusemanagement.nps.gov/Content/documents/Impacts%20to%20Wildlife_Managing%20Visitors%20and%20Resources%20to%20Protect%20Wildlife_March%202019_Edition%20One.pdf

McCool, Stephen F.; Ashor, Joseph L.; Stokes, Gerald L. 1986. An alternative to rational-comprehensive planning: Transactive planning. In: Lucas, Robert C., comp. *Proceedings: National Wilderness Research Conference: Current Research*; July 23–26, 1985; Fort Collins, CO. INT-GTR-212. Ogden, UT: U.S. Department of Agriculture, Forest Service, Intermountain Research Station, pp. 544–545.

McCool, Stephen F.; Clark, R. N.; Stankey, G. H. 2007. An assessment of frameworks useful for public land recreation planning. General Technical Report PNW-GTR-705. Portland, OR: U.S. Department of Agriculture, Forest Service, Pacific Northwest Research Station.

McCool, Stephen F.; Cole, D. N. 1997. *Proceedings: Limits of Acceptable Change and Related Planning Processes: Progress and Future Directions*; May 20–22, 1997; Missoula, MT. INT-GTR-371. Ogden, UT: U.S. Department of Agriculture, Forest Service, Rocky Mountain Research Station.

McCoy, L.; Krumpe, E. E.; Allen S. 1995. Limits of acceptable change planning—Evaluating implementation by the U.S. Forest Service. *International Journal of Wilderness*. 1(2): 18–22.

National Park Service. 2013. *Director's Order 41: Wilderness Stewardship*. National Park Service supplement to Management Policies. Retrieved from https://www.nps.gov/policy/DOrders/DO_41.pdf

National Park Service. 2022a. Backcountry and wilderness management plan: A renewed vision for wilderness stewardship, a diversity of backcountry visitor experiences, and an enduring commitment to Tlingit Homeland values. Glacier Bay National Park and Preserve, Alaska.

———. 2022b. Draft wilderness and backcountry management plan and environmental assessment. Black Canyon of the Gunnison National Park-Curecanti National Recreation Area

———. 2022c. Foundation Document for Black Canyon of the Gunnison National Park, Colorado. Retrieved from http://npshistory.com/publications/foundation-documents/blca-fd-overview.pdf

Nilsen, P.; Tayler, G. 1997. A comparative analysis of protected area planning and management frameworks. In: McCool, S. F.; Cole, D. N., comps. *Proceedings: Limits of Acceptable Change and Related Planning Processes: Progress and Future Directions*; May 20–22, 1997; Missoula, MT. INT-GTR-371. Ogden, UT: U.S. Department of Agriculture, Forest Service, Rocky Mountain Research Station, pp. 49–57.

Odum, Eugene P. 1959. *Fundamentals of Biology*. Philadelphia: W. B. Saunders.

Shelby, Bo; Heberlein, Thomas A. 1986. *Carrying Capacity in Recreation Settings*. Corvallis: Oregon State University Press.

Stankey, George H.; Cole, David N.; Lucas, Robert C.; Petersen, Margaret E.; Frissell, Sidney S. 1985. The limits of acceptable change (LAC) system for wilderness planning. INT-GTR-176. Ogden, UT: U.S. Department of Agriculture, Forest Service, Intermountain Forest and Range Experiment Station. 37p.

Stankey, George H.; Lucas, Robert C. 1986. Shifting trends in backcountry and wilderness use. Unpublished paper on file at U.S. Department of Agriculture, Forest Service, Intermountain Research Station, Forestry Sciences Laboratory, Missoula, MT.

Sumner, E. Lowell. 1942. The biology of wilderness protection. *Sierra Club Bulletin*. 27(8): 14–22.

Tin, Tina. 2022. Urban proximity and visitor numbers of four wilderness areas in 2020 during the COVID pandemic. *International Journal of Wilderness*. 28(2): 38–55.

U.S. Department of Agriculture, Forest Service. 1982. ROS users' guide. Washington, DC: U.S. Department of Agriculture, Forest Service. 38p.

———. 1987. Opportunity class allocation for the Bob Marshall, Great Bear, and Scapegoat Wildernesses and map. Forest Service, Flathead, Lolo, Helena, Lewis and Clark National Forests.

———. 2005. White Mountain National Forest Land and Resource Management Plan. Laconia, NH: U.S. Department of Agriculture, Forest Service, Eastern Region.

Wagar, J. Alan. 1964. *The Carrying Capacity of Wild Lands for Recreation*. Forest Science Monograph 7. Washington, DC: Society of American.

IV. Wilderness Resources, Values, and Threats to Them

CHAPTER **10**

Wilderness Ecosystems

by Gregory H. Aplet, Jerry F. Franklin, and Andrew J. Larson

INTRODUCTION

Wilderness ecosystems have long been revered as places where nature is allowed to run free, unhindered by modern mechanized society. Their often-large size, relatively intact biota, and high water and air quality have made them important places for ecosystem study, and they are among the last places where large predators still roam. Wilderness ecosystems are found to be among the healthiest systems left in the United States and world, befitting their role, in Aldo Leopold's words, as a "base datum of normality" for a "science of land health" (Leopold 1941). However, despite this long-standing reputation, not all wilderness areas are ecologically healthy. Many suffer threats from both inside and outside their boundaries (see Chapter 13).

Some readers may think that ecosystem concepts have only a marginal relationship to wilderness management since so much of this book's content concerns people management. However, a large body of literature provides evidence that ignorance of ecosystem concepts—internal linkages among environment, plants, and animals, and successional dynamics—is at the root of many wilderness and park stewardship problems. Many concerns about fire-management policies are based on ecosystem considerations (see Chapter 11). Management of large ungulates, such as bighorn sheep, elk, and caribou, and carnivores, such as grizzly bears and wolves, inevitably must be based on ecosystem concepts (see Chapter 12). Even limits on human use under the naturalness constraint of the Wilderness Act—biophysical carrying capacity—rest on ecosystem concepts (see Chapters 7, 8, and 9). The idea of regulated human use and management of wilderness so as not to distort naturalness and the idea of humans as an integral but not dominant part of wilderness are important ecosystem concepts. All wilderness managers should be intimately familiar with ecosystem concepts and their implications for wilderness management.

Major features of wilderness ecosystems and the qualities that distinguish them are described in this chapter. The dynamic nature of ecosystems is emphasized, as are their strong internal linkages and ways that various human activities have affected and continue to affect the "naturalness" of wilderness ecosystems. The concepts of ecosystem integrity and dynamics and their relevance to wilderness management are reviewed, with special attention to the role of Indigenous stewardship and the emerging realities of global climate change. The role of wilderness in the larger landscape is discussed and an approach to land management is proposed that includes wilderness as part of a portfolio of strategies intended to spread the risks from climate change and other uncertainties. A discussion of the importance of monitoring conditions in wilderness and feeding the results into a comprehensive, landscape-scale adaptive management plan, developed and implemented by citizens representing the full range of stakeholders, closes the chapter.

THE NATURE OF ECOSYSTEMS

ECOSYSTEM FUNDAMENTALS

Wilderness ecosystems, like all ecosystems, include all the organisms of an area, the physical environment, and the network of linkages or interactions among them. As Odum (1971, n.p.) summarizes, "The ecosystem is the basic fundamental unit in ecology, because it includes both organisms and abiotic environments, each influencing the properties of the other and both necessary for the maintenance of life." The abiotic environment includes climatic conditions, such as temperature and moisture regimes, and inorganic substances supplied by mineral soil. The biotic community contains all organisms within an ecosystem, living or dead. Green plants provide the energy base of both terrestrial and freshwater ecosystems by capturing solar energy and using simple inorganic substances to build up complex organic substances. Animals, microbes, and fungi use the complex substances produced by plants as a food base and rearrange or decompose them.

All ecosystems can be characterized in terms of their composition, structure, and function. Composition refers to the biota present in various proportions and their interactions. Each biological component contains genetic information, as well as given amounts of biomass or energy, nutrients, and other materials. Exemplary interactions are food webs and symbiotic relationships. In any ecosystem, biotic elements have a spatial arrangement, or structure, that produces ecosystems we recognize as forest, meadow, or savanna, and structural changes over time cycle back to affect composition. For example, complex, multilayered older forests provide habitat that is different from the early, post-disturbance environment and therefore support different biotas in terms of both species and abundance.

Ecosystem function refers collectively to the various kinds of work that ecosystems carry out, including production, numerous and diverse processes associated with hydrologic regimes, nutrient retention and soil building and stabilization, and provision of habitat for the biota (see Franklin, Johnson and Johnson 2018). The production function includes photosynthesis, respiration, consumption, and decomposition going on within an ecosystem. These processes are characterized by linkages, or energy and material flows between parts of an ecosystem (Fig. 10.1). Photosynthesis is carried out exclusively by organisms. Hydrologic processes include interception, transpiration, and snow-related phenomena. Biota strongly affect accumulation and retention of nutrients and erosional processes but physical processes are also important and may be dominant. Environmental factors, such as light, temperature, and moisture, are important in controlling the rates, and periodic disturbances, such as fires and floods, occasionally disrupt these processes—increasingly as the climate warms (Fig. 10.2). All these processes are elements of ecosystem function.

In affecting function, the structure of the ecosystem can be equal in importance to its biological composition. For example, the diversity of structural or architectural features in a forest ecosystem, including live, dead, and down trees of different ages and their spatial arrangement, can affect all the processes described in the previous paragraph. Many natural forests display aggregated patterns of tree dispersal and only rarely uniformly distributed patterns; this has important consequences for how the ecosystem functions, such as for habitat or in response to disturbances, such as wildfire.

Composition, structure, and function are all tightly linked and reinforce each other. For example, the species composition of the plant community determines the kinds of fuels that may burn in a fire, and their structure contributes to the flammability of the fuel present on a site. Weather and flammability determine the behavior of fire as it moves through a forest stand and across a landscape. Low-intensity fire may have little effect on live vegetation, but high-intensity fire may convert much of the living biomass to dead organic matter as well as combust modest amounts (10 to 20 percent) of the material. Here, composition affects structure, structure affects function, and function affects composition and structure.

All ecosystems are results of complex interactions. So interdependent are these elements that some researchers consider them "inseparable" and have linked them together through the concept of "ecosystem integrity" (Roche and Campagne 2017; Rogers et al. 2022). Ecosystem integrity describes the ability of an ecosystem to "maintain its optimum operating state, given the prevailing environmental drivers and perturbations, and continue its processes of self-organization and regeneration" (Rogers et al. 2022). Franklin describes an ecosystem that has integrity as one that has all of its essential elements and capabilities (Franklin, Johnson and Johnson 2018). The U.S. Forest Service (2012) adopted the following definition of integrity: "the quality or condition of an ecosystem when its dominant ecological characteristics . . . occur within the natural range of variation and can withstand and

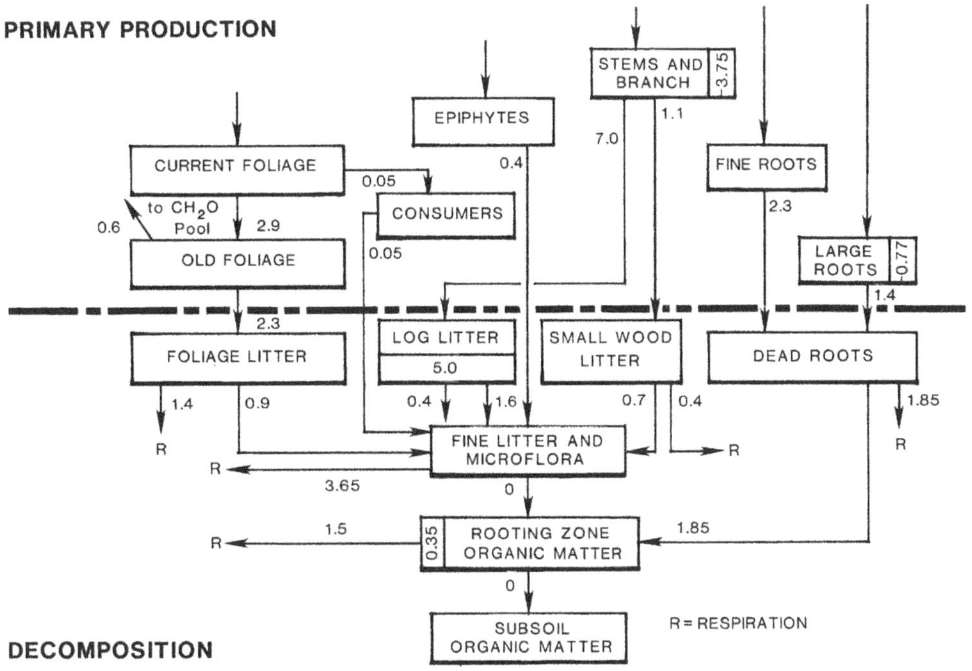

FIGURE 10.1 Any ecosystem, such as forest or meadow, contains many linked components, each possessing given amounts of biomass, energy, nutrients, or other material.

FIGURE 10.2 Clusters of subalpine fir invade a meadow near Big Agnes Mountain, Mount Zirkel Wilderness, Colorado. Wilderness ecosystems are constantly changing as a result of normal successional processes and patterns of periodic disruption. Wilderness management should ensure that natural processes proceed with minimal disruption. Photo courtesy Jay Higgins.

recover from most perturbations imposed by natural environmental dynamics or human influence."

In this sense, ecosystem integrity is similar to concepts of ecosystem development introduced in the twentieth century (Jenny 1941), whereby ecosystems are considered to be whole and bounded in their composition, structure, and function, as long as the major factors controlling their state (climate, the species present in a region, soil parent material, topography, and the disturbance regime), remain within an envelope of conditions described as the "historical range of variability" (HRV; Aplet and Keeton 1999; Landres et al. 1999). The problem today, of course, is that many of these factors have changed and will continue to change, most notably the climate, species pools, and disturbance regimes, thus eroding ecosystem integrity.

Ecosystems are fundamentally open systems with permeable boundaries. This means that material, biota, and information are constantly exchanged across ecosystem boundaries, whether those are natural boundaries (e.g., the edge between a lake and the adjoining upland forest) or a political boundary, like the legislatively determined border of a wilderness area. Consequently, ecosystem composition, structure, and function are not insulated from what goes on outside the wilderness border. Quite the opposite! The exchange of liquids and gases between the atmosphere and terrestrial ecosystems through the hydrologic cycle, and the exchange of carbon dioxide between the atmosphere and ecosystems as it is taken up by green plants during photosynthesis and expelled back to the atmosphere through decomposition of organic matter are a few examples. Mobile organisms are another example, such as seed transport and caching by birds (Lorenz et al. 2011), and wide-ranging ungulate herbivores (Middleton et al. 2020). Anadromous salmonid fish provide a compelling example: they accumulate biomass in the ocean, then return to their natal streams to spawn, sometimes hundreds of miles inland, where their decomposing bodies contribute marine-derived nutrients like nitrogen to freshwater food webs (Tonra et al. 2015) and fertilizing riparian trees (Quinn et al. 2018). In this way, conditions hundreds of miles away in the ocean that affect salmon survival can have profound effects on the structure, composition, and function of wilderness ecosystems.

Flow of biological information across ecosystem boundaries need not involve whole organisms. Genetic information is exchanged across long distances too, for example, through wind dispersed pollen in conifer forests. The same is true of disturbance processes like wildfire. Extinguishing a lightning-ignited fire outside of a wilderness area can directly impact forest composition, structure, and function inside the wilderness by precluding the spread of fire into the wilderness area and its effects on the wilderness ecosystem

(Hessburg et al. 2015). The practical implication of permeable ecosystems is that their conservation is not assured by simply designating a boundary—actions entirely outside of wilderness can alter wilderness ecosystem composition, structure, and function when they disrupt or augment exchanges of material, biota, and information across boundaries, including introduction and spread of non-native plants, animals, and pathogens. Conversely, wilderness ecosystems strongly influence ecosystems outside formal wilderness areas, for example by providing headwaters for large river ecosystems and source populations of dispersing species.

DRIVERS OF ECOSYSTEM CHANGE

For the past several centuries in North America, the major drivers of ecosystem condition (e.g., climate, the species present in a region, dominant disturbance regimes) remained relatively consistent, as did the ecosystems they shaped. Ecosystems fluctuated with the influence of both natural and human-caused disturbances but remained bounded within what has been called their "historical range of variability" or HRV (Morgan et al. 1994; Landres et al. 1999). Arising coincident with the concept of ecosystem management, HRV was embraced as a goal of management because it recognized the historical dynamics of ecosystems as essential to sustaining biodiversity into the future (Manley et al. 1995; Aplet and Keeton 1999; Safford et al. 2012). In the years since, though, mounting evidence of pervasive change in the drivers of ecosystem condition has rendered HRV untenable as a management goal, despite the continued value of ecological history to inform future management (see Wiens et al. 2012 for a comprehensive review). Here, we review some of the major pervasive drivers of ecosystem change and their consequences. See Chapter 13 for a more complete evaluation of threats to wilderness ecosystems.

Climate Change

Climate has a profound effect on ecosystems, shaping the appearance of regions and determining the species that can live there. Throughout the past millennium, the climate of North America has fluctuated between the relatively warm Medieval Warm Period and the relatively cool Little Ice Age of the seventeenth to nineteenth centuries AD, both of which had measurable effects on the range of some species, but the overall distribution of vegetation remained relatively stable. In recent decades, however, patterns of temperature, precipitation, and moisture stress are departing from known HRV. "Hot drought" is killing trees that were able to withstand previous cooler droughts (Williams et al. 2013). Wildfires are burning with larger high-severity patches that affect wildlife habitat and challenge recovery (Falk et al. 2019). Extended droughts and more frequent rain-on-snow events threaten to alter the ecology of whole river systems. The amount of future change already "baked into" the climate system ensures that future ecosystems will not be like the past.

Land Use and Habitat Fragmentation

According to a recent article by James Watson et al. (2018), a century ago, only 15 percent of Earth's surface was used to grow crops and raise livestock. Today, more than 77 percent of the terrestrial surface (outside Antarctica) and 87 percent of the ocean has been modified by direct human activities (Watson et al. 2018). The alteration of so much habitat leaves the remainder fragmented and degraded. Fragmentation reduces habitat area, increases the isolation of remaining patches, and increases the amount of edge habitat, which has been shown to expose birds and other wildlife to increased predation (Haddad et al. 2015). Increased isolation requires species to move through degraded habitat, which exposes them to risk, reducing population viability and ecosystem integrity of remaining patches (Barnett and Belote 2021; Belote et al. 2022).

Altered Disturbance Regimes

Fires at characteristic intervals are a natural feature of almost all forest and rangeland ecosystems. Elimination of fire, when periodic low-intensity fires were historically part of an ecosystem's environment, can have catastrophic consequences. Generally, with fire exclusion through suppression and elimination of Indigenous burning, forests become denser with less herbaceous undergrowth. In the eastern U.S., elimination of fire in drier forests and the resulting change in dominant tree species led to the phenomenon of "mesophication," in which the entire forest microclimate became cooler and wetter (Nowacki and Abrams 2008). In the absence of fire, not only does the plant composition of the forest change, but the habitat for many animal species is also altered with consequent changes in the animal composition of an area. Organic matter and nutrients accumulate in slowly decomposing woody plant material on the forest floor. Continuity develops between the crown and surface fuels as the density of saplings and small trees increases. When fire does inevitably return, it burns at higher severity, killing trees that historically would have survived frequent fire. Even in vegetation types that historically burned infrequently with high severity, fire exclusion has likely altered the vegetation mosaic of large landscapes, creating a more continuous fuel bed and larger patches of mortality in the event of fire, which is increasingly likely under a changing climate (Falk et al. 2019). In other places, novel ignition sources (e.g., sparks from automobiles and power lines) have increased the frequency of wildfire above historical levels, driving the replacement of native vegetation with communities of invasive species (Keeley et al. 2009).

Species Loss

Eliminating species is another way modern humans have altered, and continue to alter, the integrity of wilderness ecosystems. The best-known examples are the elimination of predatory animals—for example, wolves, wolverines, grizzly bears, cougars, and various raptors. Elimination or reduction of these species directly affects species composition of wilderness ecosystems. Furthermore, loss of large predators, such as wolves, can have significant ecological impacts when these natural controls on large herbivore populations and medium-sized predators are removed (Silliman and Angelini 2012). Irruptions of herbivores can result in overgrazing, and "meso-predator release" (e.g., raccoons, opossums) can have devastating impacts on forest-nesting birds.

Plant species have also been eliminated or drastically reduced in some areas. Losses of plant species are most frequently caused by human introduction of exotic pathogens. Examples include the nearly complete loss of American chestnut to the nonnative chestnut blight, and substantial losses of five-needle pines to exotic white pine blister rust. The loss of whitebark pine to blister rust is of great concern and contributed to the pine's listing as a threatened species under the Endangered Species Act in 2022. Another catastrophic example is the destruction of eastern hemlock forest by the introduced hemlock woolly adelgid, which is advancing farther north due to warmer winter weather patterns. Exotic insects and pathogens can have extremely profound and often poorly perceived effects on otherwise intact landscapes. Once introduced, exotic pathogens are virtually uncontrollable, because they lack natural enemies to keep their populations in check.

Species Introductions

The introduction of invasive, exotic plants and animals also degrades ecosystem integrity. Invasive plants cause a host of problems, including displacement of native species, promotion of nonnative animals, increased erosion, and alteration of nutrient cycles and fire regimes (Randall 2000). Particularly vexing are the speed and permanence of some of these invasions. For example, in northern California, yellow star thistle is estimated to have spread from 1 million acres in 1977 to 10 million acres in 1995, and in the Selway-Bitterroot Wilderness in Idaho, spotted knapweed continues to spread, although grazing, which is often implicated in the spread of exotic weeds, has not occurred for more than fifty years (Asher and Harmon 1995).

Fish and animals introduced into new landscapes, often to make the areas more attractive recreationally, also affect ecosystem integrity. The introduction of sport fish to originally "barren" lakes and streams is an outstanding example. There can be no question that when fish introductions of this type are successful, the affected aquatic ecosystem is markedly changed. An entirely new component may be added if a fish at the same trophic level was not previously present, and native fish may be displaced or eliminated entirely. An extensive study of stocked and unstocked lakes in the Sierra Nevada mountains of California documented important impacts on the rare mountain yellow-legged frog after the introduction of exotic sport fish (Matthews and Knapp 1999). In the Everglades, home to the largest wilderness area in the East, a host of introduced species as exotic as the Nile monitor lizard and the Burmese python voraciously consume native species and contribute to their endangerment.

Pollution

Air and water pollution can also have pervasive impacts on ecosystems and readily cross boundaries far from their source (Lovett et al. 2009; Greaver et al. 2012). Sulfur, nitrogen, ozone, and mercury affect ecosystems from estuaries to alpine tundra, acidifying lakes, causing eutrophication of estuaries and coastal waters, accumulating in aquatic food, acidifying soils, and altering biogeochemical cycles. Pesticides and other complex organic compounds can have detrimental effects up and down the food chain. At acute levels, ozone has damaged and killed trees far from its source in the Los Angeles basin (Miller 1973), and acid precipitation traced to power plants was blamed for accelerated mortality in spruce and fir stands in the northeastern and southeastern United States. No place on Earth is beyond the influence of these pollutants (Vitousek et al. 2000).

In addition to chemical pollutants, there is increasing concern about the effects of light and sound pollution on ecosystems. A recent study (Kyba et al. 2023) showed that the brightness of the night sky more than doubled in the previous decade, potentially affecting ecological processes from bird migration to plant pollination to the reproductive strategies of corals. Sound, too, from transportation, industry, and other sources, can have far-reaching impacts on animal physiology, reproductive success, communication, and movement, all of which can be expressed through changes to ecosystem species composition and distribution (Sordello et al. 2020).

THE ROLE OF PEOPLE IN ECOSYSTEMS

Up to this point, we have described ecosystems as though they do not include people, but the Americas have been home to people for thousands of years—and even longer in most of the rest of the world. Native people built settlements, raised crops, hunted, fished, tended wild food, and used fire extensively to shape their surroundings (Denevan 1992; Mann 2006; Stewart 2009). Nowacki et al. (2012) reviewed the influence of Native Americans on pre-Columbian ecosystems and found it to vary, not surprisingly, with population density. In northern hardwood-conifer forests, effects were largely local, limited to settlements and travel corridors. In the Southeast and Mississippi River floodplain, vast areas were cleared for maize production, and settlements were extensive. Where the eastern forest encroached upon the prairie, grasslands depended on extensive Indigenous burning. Farther west, where rainfall tapered off, Native fire use augmented natural ignitions to maintain grasslands and open forest, with influence generally diminishing with elevation. In all regions, human activity varied across the landscape, with greater influence near settlement, whether permanent or seasonal, and less influence farther from home and popular travel routes (Vale 1998; Anderson 2005; Thomas-Van Gundy et al. 2015; Kipfmueller et al. 2021; Roos et al. 2022).

With Euro-American colonization arrived plows, saws, guns, livestock, and, tragically, epidemic disease and genocide against the Native population, resulting in wholesale transformation of large parts of North America, even before the industrial revolution (Cronon 1983; Whitney 1994). Since then, modern machinery and chemicals have added to the scope and scale of change (Brinkley 2022). The slaughter and forced relocation of Native people virtually eliminated their influence from North American ecosystems, and the spread of settler colonization converted forest to farmland, drained wetlands, and very nearly exterminated large game and carnivores from most of the continent. With the dawn of the twentieth century, the vast central grasslands were nearly eliminated, most rivers were dammed, and centuries-old forests were logged off from the Southeast to the Great Lakes to the West Coast. By the middle of the century, it could be assumed that what was not protected in national parks would eventually be converted to tree farms, pastures, oil fields, and urban development (Brinkley 2022). As these forces transformed the land, a wilderness movement was born to ensure that the last of the original ecosystems were not lost.

THE NATURE OF WILDERNESS ECOSYSTEMS

The biophysical environment and role of humans are important to understanding ecosystems in general, but what is it about an ecosystem that makes it a *wilderness* ecosystem? Since at least Thoreau's time, scholars have differed over what gives wilderness its character, but recent reviews have settled on two essential qualities (Cole 1996; Aplet 1999; Aplet and Cole 2010; Kelly and Landres 2022). Wilderness character is composed of both the concept of ecological wholeness, or ecosystem integrity, as described above, and the freedom from intentional human control, or "trammeling." These two qualities are enshrined in the definition of wilderness included in the Wilderness Act. Section 2(c) describes wilderness as "an area of undeveloped Federal land retaining its primeval character and influence, without permanent improvements or human habitation, which is protected and managed so as to preserve its *natural conditions*," reflecting generations of authors who saw natural conditions as essential to wilderness character (Aplet 1999; Aplet and Cole 2010). Howard Zahniser (1956), principal author of the Wilderness Act, described wilderness poetically as "a piece of the long ago that we still have with us." This conception of "natural" as dynamic but bounded within historical limits incorporates both of Cole's (2000) notions of wilderness as historically whole and uninfluenced by modern technological humans (Aplet and Cole 2010).

The same section of the Wilderness Act also notes that "[a] wilderness, in contrast with those areas where man and his own works dominate the landscape, is hereby recognized as an area where the earth and its community of life are *untrammeled by man*, where man himself is a visitor who does not remain." Zahniser chose the word "untrammeled" carefully to reflect his concern that, in an era of massive dam construction, harnessing of the atom, rampant pesticide use, and other actions to dominate nature, we needed places that we simply left alone to their own devices. He wrote, "I believe . . . we have a profound, a fundamental need for areas . . . of the earth within which we stand without our mechanisms that make us immediate masters over our environment" (Zahniser 1956). In describing wilderness as a place "where man himself is a visitor who does not remain," Zahniser was not denying the place of Indigenous people in the landscape; rather, he was asserting that we, as a society, should identify some places that we do not try to conquer with roads, chainsaws, recreational cabins, and other implements of the control of nature.

Thus, it can be seen that wilderness character is at its highest where both ecosystem integrity and freedom from control are at their highest. Aplet (1999) represented this relationship as a two-dimensional space in which wilderness character increases along an axis of naturalness (or ecosystem integrity) from "artificial" (novel) to "pristine" (intact or whole) and along an axis of freedom that runs from "controlled" to "self-willed" (Fig. 10.3). The wildest places, those with wilderness

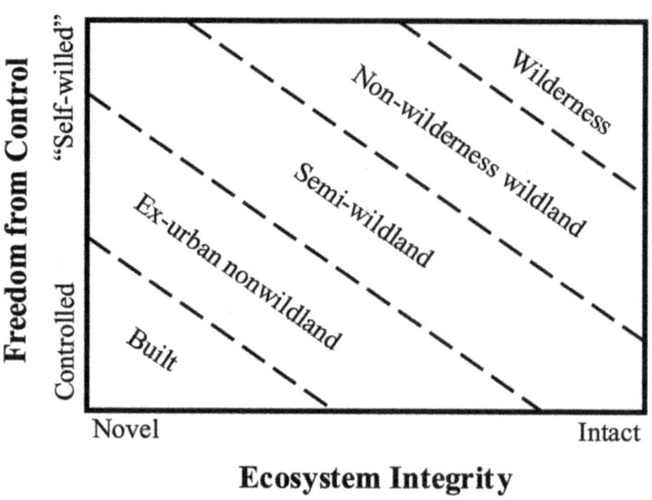

FIGURE 10.3 Wilderness character (wildness) can be said to increase in two dimensions: along a gradient of ecosystem integrity from intact to novel and along a gradient of human influence from controlled to untrammeled (adapted from Aplet 1999).

character, occupy those parts of the landscape where ecosystem integrity and freedom from human control are at their highest. It is important to note that this gradient is not absolute but, rather, relative. The poet Gary Snyder (1990) says, "Every region has its wilderness. There is the fire in the kitchen and the place less traveled. In most settled regions, there used to be some combination of prime agricultural land, orchard and vineland, rough pasturage, woodlot, forest, and desert or mountain 'waste.' The de facto wilderness was the extreme backcountry part of all that." These are the places we have traditionally recognized as *wilderness ecosystems*.

Wilderness ecosystems are not necessarily places that have been designated as such. They are places that have been relatively left alone in a natural condition. Alaska is well known to have vast areas of undesignated wilderness. The U.S. National Forest System has over 50 million acres of "Inventoried Roadless Areas" that are not designated wilderness but have abundant wilderness character. The history of wilderness designation in the eastern United States shows there is no threshold level of human alteration that defines a wilderness ecosystem. In the East, there was a great debate as to whether its apparently natural remaining lands qualified for wilderness designation or belonged in a separate "wild-areas" system (see Chapter 5). Subsequently, even formerly farmed and otherwise altered lands were included in the Wilderness System under provisions of the so-called Eastern Wilderness Act (U.S. Public Law 93-622). It is the wilderness character of a place *relative to its surroundings* that distinguishes it as a wilderness ecosystem (Aplet 1999).

IMPLICATIONS FOR MANAGEMENT

Representing ecosystems in two-dimensional space defined by ecosystem integrity and human control facilitates consideration of human agency in shaping landscape character (Fig. 10.4). Increased human control can drive systems away from historical conditions through transformation, as has typified the progress of civilization. Alternatively, human

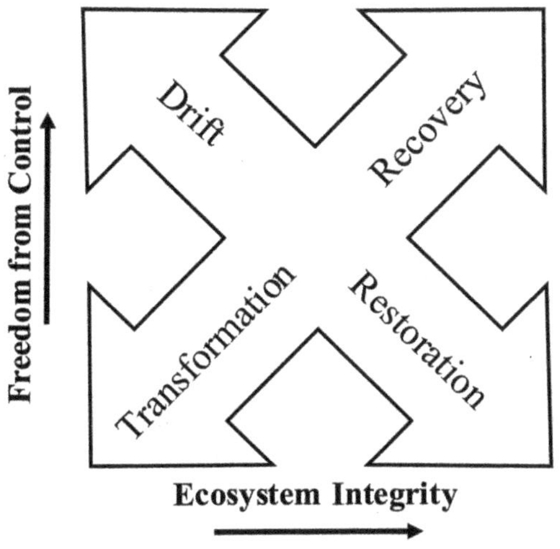

FIGURE 10.4 Management can move a landscape in any direction with regard to ecosystem integrity and human control. Restoration is the act of bringing an ecosystem under tighter control to increase its ecosystem integrity. Releasing an ecosystem from control can result in recovery of ecosystem integrity or drift into novel conditions (adapted from Aplet 1999).

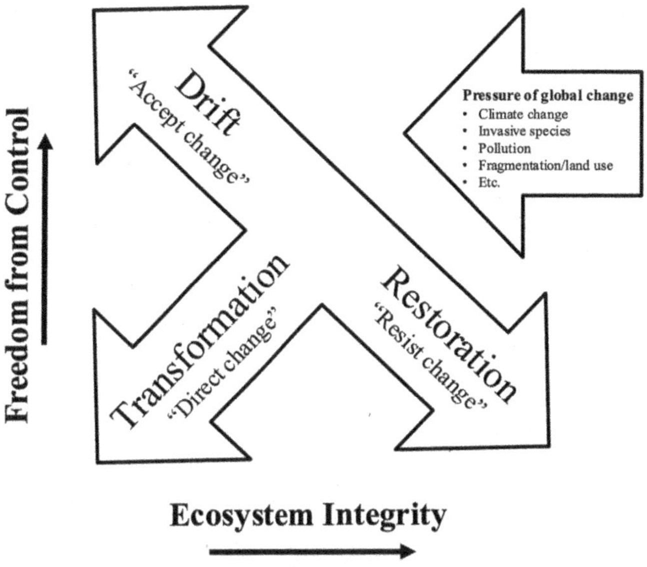

FIGURE 10.5 Global change can be thought of as a pressure on ecosystems driving them away from historical conditions. In the future, degraded sites cannot be expected to "recover" to their historical condition in the absence of human influence (though some short-term recovery is possible upon relief from existing stressors). The remaining options are to resist change, accept change and direct change, or "RAD" (adapted from Aplet and McKinley 2017).

effort can be exerted to increase historical fidelity and mitigate human impacts through the process of restoration. In the absence of active management, land freed from human control can either recover ecosystem integrity or drift toward a more novel condition. Historically, designation of a place as wilderness was assumed to result in recovery or maintenance of high ecosystem integrity. Alternatively, where restoration was necessary to improve ecosystem integrity or reverse its decline, the expectation in wilderness was that "acceptable" intervention would be "light touch" and temporary and that recovery would result if the intervention was successful (Cole 1996; Franklin and Aplet 2002). Such activities might include restoration of intensively used campsites or removal of stocked gamefish from a wilderness lake (Cole 2000).

With the realization of pervasive global change, it has become increasingly clear that it is unrealistic to expect a particular place to recover to historical conditions if freed from human control over the long term. This is particularly true where historical conditions were maintained through human action, such as cultural fire. The ideal of wilderness management, the perpetuation of whole historical ecosystems through passive untrammeling, is no longer tenable in the long run and drops out of the figure in the face of such pressure (Fig. 10.5). Of course, it is important to note that recovery (passive restoration or movement in the direction of historical conditions) will not disappear completely. There are many examples of even severely degraded ecosystems that have shown enormous capacity for resilience if simply left alone. Larson et al. (2013a) described what they called a "latent resilience" to wildfire in a ponderosa pine forest that had been degraded by fire exclusion but was allowed to burn again under a frequent-fire regime. Situations like this should be expected when missing processes are restored, but it is unrealistic to expect historical conditions to persist in the face of dramatically altered ecosystem drivers. In the future, we can expect to be left with only three management choices: accept change, resist change, and direct change (Fig. 10.5).

RESIST-ACCEPT-DIRECT

Recognition of this new reality has led several managers and scientists to propose a "new paradigm" for natural resource management that they call "RAD," shorthand for Resist-Accept-Direct (Magness et al. 2021; Schuurman et al. 2020, 2021; Lynch et al. 2021). These authors acknowledge the risk and uncertainty of managing in the face of novel conditions and propose that the future will require application of all three strategies.

To *resist* is to fight, or at least slow, the changes being wrought by the drivers of ecosystem change. It is aimed at achieving the fundamental, ethical goal of conservation: to pass on to future generations that which we have inherited from the past. The strategy is based on the recognition that the only ecosystems that we know sustained the biodiversity of the past are the ecosystems that functioned as they did in the past, especially with regard to their disturbance regimes (Aplet and Keeton 1999). This was the idea behind restoring historical range of variability as a target of management (Manley et al. 1995; Quigley and Arbelbide 1997), a target that we now understand to be increasingly difficult to achieve without heroic intervention (Millar et al. 2007; Safford et al. 2012; Schuurman et al. 2021). Still, in some cases, it may be worth investing in sustaining historical ecosystems to "buy time" for elements of composition, structure, or function for which we have not developed alternative conservation measures, especially endangered species or culturally valued plants. Reliance on HRV as a target evolved out of a recognition of our past hubris in trying to control ecosystems to serve commercial purposes and an admission that we don't have all the answers, a situation that is as true today as when it was first acknowledged (Swanson et al. 1994).

Acknowledging the need for humility, we also now know that the future will not be like the past, and we need to prepare for an altered future. Unless we are simply to give up, this requires that we begin to manage differently, *directing* ecosystems into conditions that will support the species, structures, and ecosystem services that we hope to keep with us. This is a "brave new world" in which society must choose winners and losers and manage ecosystems to sustain the winners in an altered future. In some ways, this is not a great deal different from the way people have always modified their environment to improve their condition. We have moved species around, some with disastrous effects and some benign. We have altered disturbance regimes and cultivated conditions to support species of our choosing. In the future, we will need to be much more thoughtful about what we want from our ecosystems and what the unintended consequences are of those choices, but, if the fundamental driving forces of ecosystems are changing, we have to recognize the consequences of those changes and manage accordingly.

Resistance and direction are logical, rational responses to change, but in adopting these strategies, we must not reject humility and return to a belief that we have all the answers. There is still a role for *acceptance* of change. Landres (2010) makes a powerful case for a "hands-off" approach to preserving wildness, including invoking the need for social and scientific humility, maintaining places for continuing evolution outside of human control, sustaining species that are not the target of management, as a hedge against mistakes made elsewhere, and simply to provide examples of what unmanipulated systems looks like—Aldo Leopold's (1941) "base datum of normality"—as scientific "controls" for management experiments elsewhere (Belote et al. 2015). In the process, Landres makes a convincing case for the conservation value of protected areas, a track record that has been demonstrated by others (Gray et al. 2016) and should not be trivialized. Wilderness, because of the statutory intent that it remain "untrammeled," makes an obvious candidate for allocation to "accept change."

Some elements of the RAD literature seem to suggest that any of these three approaches may be appropriate, with selection depending on financial, social, and ecological or management criteria (Magness et al. 2011; Clifford et al. 2021; Lynch et al. 2021). However, simply treating RAD as a menu of options fails to account for the benefits of a portfolio approach in which all options are applied, with risk spread among them (Aplet and McKinley 2017; Magness et al. 2021). As Schuurman et al. (2021, 24) observe, "Pursuing different options in different places can—if the efforts are well coordinated—promote landscape-scale diversity and redundancy and therefore hedge the risks associated with climate change." Magness et al. (2021, 36) explain, "[A] portfolio approach assumes that some assets will perform poorly, and seeks to minimize risk for a given level of performance . . . [which] is assessed at the scale of the portfolio, rather than the project scale." If the full benefit of the RAD approach is to be realized, all three strategies will need to be applied simultaneously on portions of landscapes allocated to each strategy (Aplet and McKinley 2017). Gallo et al. (2020) provide a means to model such an allocation.

WILDERNESS AS PART OF THE LARGER LANDSCAPE

It is important to keep in mind that there is no scale that defines a wilderness ecosystem. An ecosystem can be large or small, based on how we define its boundaries. It can be an aquarium, forest stand, watershed, entire park or wilderness area, biome, or the entire world. It is necessary only that the area have the characteristics of an ecosystem—constituent organisms interacting with their environment, energy flow, and material cycles, and all that is necessary for it to be a wilderness ecosystem is that it be largely untrammeled and whole relative to the past.

Historically, wilderness ecosystems were defined by the boundaries that surround wilderness, but research over the past several decades has shown that ecosystems vary in scale depending on the element or process of interest, and the choice of scale affects our understanding of composition, structure, and function. For example, scientists and managers often find a watershed to be a useful unit for ecosystem study because it is relatively easy to define the physical boundaries and to measure many of the flows into and out of the ecosystem; it also incorporates both terrestrial and aquatic elements and allows study of their interaction. Nevertheless, ecosystems can be of widely varying size or scale, depending on which level of watershed is selected as the focal scale: subwatersheds of tributary streams are nested inside the watershed boundary of larger streams, which are nested inside of the basin boundary of large rivers, and the scale of analysis affects the behavior of each of these ecosystems. A localized rainstorm may cause a small stream to flood, but the increase in stream discharge may have no noticeable effect on the big river downstream. Conversely, heavy snows over the region may cause no flooding in headwater streams, but the cumulative effect of spring snowmelt may cause large rivers to rise downstream (Poff et al. 1997). Similarly, historically small patches of high-severity fire may have had significant effects on the dynamics of old-growth forests at the sub-watershed scale, but those influences were dampened at the scale of the region (Fig. 10.6). Only when changes affect the scale of ecosystem drivers, such as the effect of climate change on regional weather or the influence of pervasive fire exclusion on fuel continuity, do we see consequent change in the scale of ecosystem composition, structure, and function (Hessburg et al. 2015).

In addition to increased appreciation of the effect of scale on the observation of ecosystem dynamics, the past several decades have demonstrated the importance of ecosystem scale to the sustainability of ecosystem elements and processes. For most of the history of wilderness and primitive-area designation, decision makers and managers relied on a belief that ecosystem protection required only that we designate a protected-area boundary and stop activities inside that threatened its integrity. With time, though, we have come to understand that boundaries do not necessarily protect ecosystems. Many threats to wilderness are oblivious to boundaries, and, in some cases, boundaries can create problems (see Chapter 13; Knight and Landres 1998; White et al. 2000).

Fire is an excellent example. Managers may deem that inside of wilderness, fire should be allowed to play its natural role in the ecosystem, but outside, it should be suppressed to protect non-wilderness resources and private property. This policy ignores the fact that many fires start at low elevations, outside of wilderness, and burn upslope. Fire suppression outside of wilderness may prevent some natural fires from ever entering wilderness, thus disrupting a historically natural and basic ecosystem function. The problem also works in the

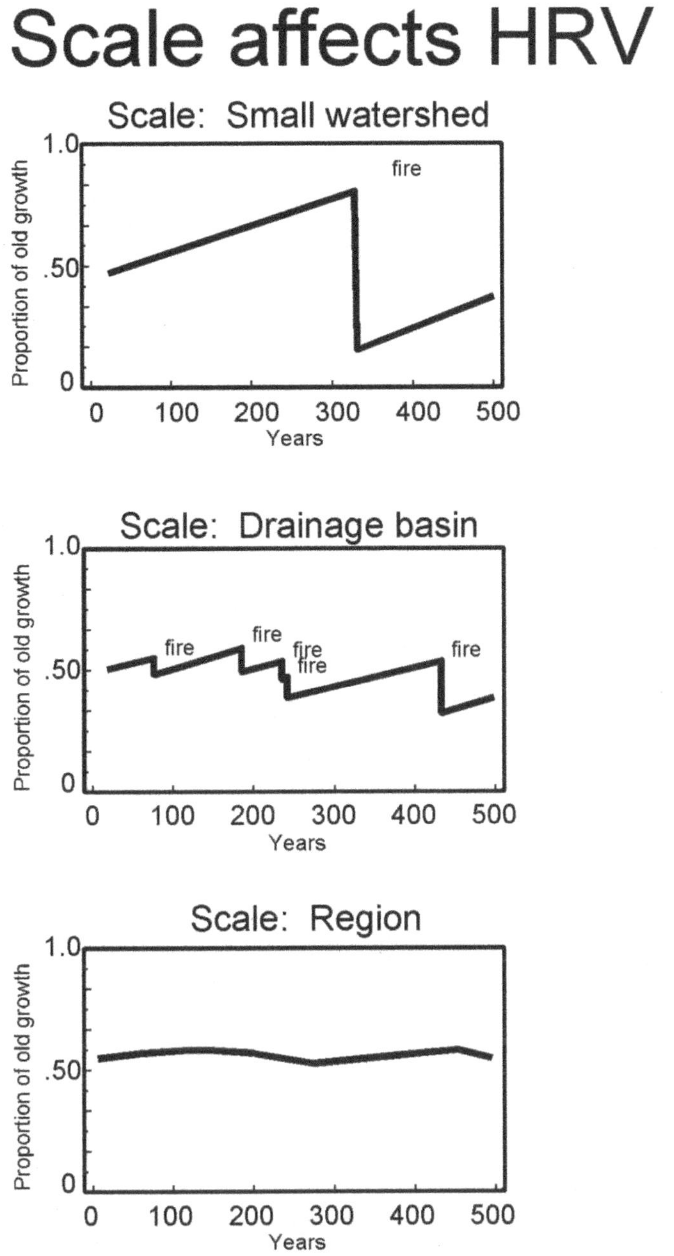

FIGURE 10.6 The range of variability in composition, structure, or function changes with the scale of observation. Here, variability in the proportion of an ecosystem in old-growth forest declines (top, small watershed) as the size of the ecosystem increases (middle, drainage basin; bottom, region) (Aplet and Keeton 1999).

opposite direction, where fires inside of wilderness are suppressed for fear they will spread outside (Barnett et al. 2016).

Another example is wildlife movement. Wilderness may be established to protect certain places, but if it does not also protect migratory corridors, wildlife will continue to leave the protected area and thereby be subject to unnatural influences (see Chapter 12). Such is the case near Yellowstone National Park, where hunters may shoot native wildlife as they migrate from the protected park area in search of winter range. In Arizona, desert bighorn sheep are regularly killed in automobile accidents when they leave wilderness to reach traditional watering holes.

Realizing that existing protected areas are rarely large enough to protect all of the components of the ecosystem has given rise to the "greater ecosystem" concept, which accounts for all of the resources outside a protected area that are necessary to sustain the protected area. One of the first applications of this concept was in the region of Yellowstone National Park. There, after studying the movement patterns of grizzlies, bison, and elk, ecologists realized that the park, even at 2 million acres, was insufficient to sustain viable populations of many species. They found that animals were using resources beyond even the wilderness areas surrounding the park, including considerable private property, where they were vulnerable to many threats, ranging from habitat destruction to direct mortality. In response, managers and citizen activists began to promote the "Greater Yellowstone Ecosystem" (GYE), which is a way of thinking about the interconnections across the region rather than a fixed line on a map. Today, the GYE has spawned many programs and management reforms aimed at protecting all resources in the surrounding region necessary to sustain the park's ecosystem.

In the years since, scientists have expanded on the greater ecosystem concept to include disturbance processes that inform "minimum dynamic reserves" (Pickett and Thompson 1978; Leroux et al. 2007) as well as hydrologic and atmospheric flows, species home ranges, species-area effects, and edge effects to identify "Protected Area Centered Ecosystems," or "the zone around each protected area wherein human activities may influence important ecological processes as well as the viability of populations of native organisms within the protected area" (Hansen et al. 2011). Belote and Wilson (2019) gauged the "permeability of the landscape to species movement and ecological flows" to identify lands relatively ecologically connected to protected areas. All this work shows that wilderness and other protected areas are part of much larger ecosystems that must be managed together to address threats to their integrity. The portfolio of RAD strategies must be allocated from these greater landscape ecosystems, with wilderness providing the backbone of the "accept change" allocation.

PROFOUND VERSUS COSMETIC CHANGE AND THE ROLE OF INTERVENTION IN WILDERNESS

While the default position in wilderness should be to accept change, there may be times when it is appropriate to contemplate intervention. Landres et al. (2020) present 28 questions reflecting scientific, legal, and ethical concerns when considering intervention, reflecting the seriousness with which managers must take the issue. With the exception of legal questions, which are specific to the language of the Wilderness Act, the authors are concerned with whether the changes to be addressed are profound or cosmetic, pervasive or local, and permanent or temporary. They also ask about the likely success/effects of intervention and whether observed changes derive from use or actions taken within wilderness or whether they represent forces from the outside. The answers have profound implications for whether a proposed intervention should be considered a trammeling, thereby undermining the objective to accept change. To illustrate some of these considerations, consider the following examples.

Introduction of the chestnut blight fungus from Asia resulted in the functional extirpation of the American chestnut from U.S. forests (Shugart and West 1977). The hardwood ecosystems of which it was a part underwent a permanent change in composition and population structure, not only as a result of the loss of a dominant tree, but also because of adjustments in animal species dependent on chestnuts for food and changes in rates of energy and nutrient cycling, forest structure, and successional sequences. These are among the most profound, widespread, and permanent ecological effects an ecosystem can experience, but they also derive from forces outside the wilderness, and solutions are not clear. In such cases, it may be best simply to accept the change. To add to the rationale, despite these profound changes, visitors to forests from which chestnuts have been eliminated do not typically perceive these as unnatural or human-altered ecosystems—the space chestnuts once filled has been taken over by other hardwood tree species. In contrast, campsite overuse or minor trail degradation may represent more cosmetic and local concerns deriving from use of the resource, but ones that, if not addressed, could get worse and for which remedies are well understood and effective (Fig. 10.7; see Chapter 15). In these cases, intervention represents more of a repair or rehabilitation of damage done by users than a trammeling of wilderness, though humility and restraint must always guide action.

The issue becomes more complex where a threat may be local, such as the establishment of an invasive weed or the introduction of exotic sportfish, but if left unattended may grow into a profound change. In these cases, intervention may be seen less as rehabilitation and more as a protective action aimed at preventing an external threat from degrading ecological integrity without pervasive trammeling. Again, restraint and humility, as well as careful consideration of minimum requirements analysis (Cole 2007; see Chapters 8 and 9) and the questions posed by Landres et al. (2020), should guide any decision.

Even more complex are situations where change has occurred due to past or ongoing trammeling, such as the exclusion of fire. Despite the intention to let natural fires burn in wilderness (see Chapter 11), most wilderness fires are suppressed (Parsons 2000; Miller and Aplet 2016). Other fires that would have burned into wilderness are suppressed before they even get there. The result has been a change in the characteristic pattern of vegetation and a buildup of fuels that now exposes ecosystems to unprecedented fire behavior when fire inevitably returns—usually under extreme weather conditions (Hessburg et al. 2015). Given these circumstances—the fact that current conditions are the result of trammeling—the intentional reintroduction of fire (prescribed fire) might be seen as an act of "untrammeling," releasing the system from an intentional control, especially if the fire is subject to minimal control following ignition (Fig. 10.8). Similarly, the reintroduction of a species that had been eliminated historically could be seen not simply as contributing to the restoration of ecological integrity, but to the restoration of untrammeled nature. There is no question that camping in a wilderness inhabited by grizzly bears and wolves feels wilder than one without, not only because it feels more whole, but because it feels more out of human control. It should be noted though, that under the current interagency monitoring strategy, both actions are considered "trammeling" (Landres et al. 2015). Proposals to restore threatened species using intentionally selected or modified genetic stock, as is occurring for the American chestnut (Popkin 2018) and whitebark pine (Lundquist 2019), raise even more vexing questions.

MONITORING AND ADAPTIVE MANAGEMENT

It is tempting, for areas allocated to accepting change, simply to designate such places and leave them alone, but as the rest of this book clearly shows, that would be a mistake. History has shown that wilderness ecosystems are vulnerable to overuse and careless, and sometimes intentional, abuse that can degrade their ecological integrity and freedom. And, in a future that requires a portfolio of innovative approaches to be employed in response to climate change and other uncertainties, it will be essential to understand the condition of wilderness and manage it as an integral component of a broader strategy. Fortunately, the past several decades have

FIGURE 10.7 A campsite on Grassy Lake, John Muir Wilderness, California, shows impacts of human use, including pronounced trampling. Restoration of such damage is localized with well understood effects and should be considered repair, rather than a pronounced trammeling. Photo courtesy U.S. Forest Service.

FIGURE 10.8 Dense stands of small-diameter ponderosa pines encroach on the open, old-growth forest, Mount Trumbull Wilderness, in northern Arizona. BLM managers have proposed restoring the old-growth forest structure and disturbance regime of frequent low-intensity fires. Photo courtesy Gregory H. Aplet.

seen great improvement in our ability to manage in the face of uncertainty. *Adaptive management is a framework for planning and management that embraces the uncertainty inherent in ecological systems and applies a system for intentionally learning and improving management as knowledge is gained* (Larson et al. 2013b; Chapter 9). While different models exist, all involve a cycle of goal setting, identifying assumptions and uncertainties, implementing a range of management actions, monitoring the effects of actions, and evaluating the need for changes to goals and management actions based on monitoring results. When done with participation of a full range of stakeholders, it can help resolve conflict by building trust and mutual learning (Cheng et al. 2019). The explicit inclusion of diverse groups and individuals, especially those who have been marginalized from natural resource processes in the past, is the key to ensuring that a broad range of perspectives are brought to bear on decision-making and that the benefits of those decisions are distributed equitably.

In the context of RAD, wilderness and other lands allocated to the "accept change" strategy assume an especially important role as benchmarks against which to judge actions taken in the name of resisting and directing change. Adaptive management is essentially an experimental approach to management, and all good experiments need "controls" (Larson et al. 2013b). *Under RAD, restorative and innovative, future-looking treatments will need to be compared against untreated areas to determine their effects.* Belote et al. (2015) describe a method not just for allocating existing wilderness to serve as controls but for adding to the system to increase its diversity and improve its ability to compare against active management (Fig. 10.9 and Fig. 10.10).

For wilderness to provide the full range of ecosystem types, it must be surveyed for that purpose. The first attempt to do this in a formal way was during the RARE II process (see Chapter 5), in which the Forest Service determined that 131 "ecosystems" were "adequately represented" in the NWPS in 1978, accounting for 50 percent of the 261 ecosystems eventually recognized by Bailey (1980) in the U.S. Davis (1989) repeated the analysis almost a decade later, and found that the NWPS, despite having more than tripled in size, included only 157 ecosystems, or 60 percent of the total. Using a simplified system involving only thirty-five ecosystem "provinces" (Bailey 1995), Loomis and Echohawk (1999) determined that, despite the growth of the NWPS to more than 100 million acres, barely one-third of the ecosystems of the lower forty-eight states had more than 1 percent of their area protected as wilderness. More recently, Dietz et al. (2015) found that, of the 553 "ecological systems" (the finest category of ecoregion mapped for the U.S.) that occur on federal lands, 436 (79 percent) had some proportion in the National Wilderness Preservation System, but only 20 percent were what they called "well represented" (>20 percent protected). They also found that, despite additions to the system, diversity had not increased significantly in thirty years. Subsequently, Aycrigg et al. (2016) found that adding other potentially qualifying lands (non-wilderness national park land and roadless areas on national forests and Bureau of Land Management lands) would increase the NWPS area from 12.8 to 48.3 percent of federal lands, increase diversity by adding 46 ecological systems, and nearly triple the number of ecological systems on federal lands with >20 percent representation. Clearly, *there is still a tremendous opportunity to improve the ability of the wilderness system to serve as a scientific control for management experiments elsewhere in the RAD portfolio.*

In addition to serving as a control, wilderness has an important role to play as a scientific resource. As the "end member" of the land-use gradient, wilderness serves not only as a control for experiments done outside, but as an important "treatment" unto itself for scientific inquiry. From the earliest days of the wilderness movement, wilderness has been recognized for its potential contributions to science (Leopold 1941), and science is recognized explicitly in the Wilderness Act as an appropriate purpose for wilderness. A wilderness research review panel (Turner et al. 2007) identified several important roles of wilderness for science in addition to its role in understanding land use:

- Help scientists and managers understand changes in public wildlands and resources, independent of management intrusions.
- Serve as natural laboratories for studying effects of various environmental stressors including global climate change,

FIGURE 10.9 Ecosystems represented in the NWPS: semiarid Gila Wilderness, New Mexico (left); the seral landscape of Craters of the Moon National Monument, Idaho (center); and deciduous forest of the Presidential Range–Dry River Wilderness, New Hampshire (right). Left photo courtesy U.S. Forest Service; center courtesy National Park Service; right courtesy U.S. Forest Service.

FIGURE 10.10 Wilderness ecosystems exist at a variety of scales: from small islands in Washington's San Juan Islands Wilderness (top left), to vast landscapes in the Absaroka-Beartooth Wilderness, Wyoming (top right); to an arid desert game range in Nevada (bottom left), and a remnant roadless area in Monongahela National Forest, West Virginia (bottom right). Top left photo courtesy George Devan, FWS; top right courtesy L. J. Prater, U.S. Forest Service; bottom left courtesy David B. Marshall, FWS; and bottom right courtesy U.S. Forest Service.

invasive species, air pollution impacts and fire, in areas not confounded by management or direct human impacts.
- Assess long-term ecosystem/landscape changes occurring independent of management effects.
- Understand the importance of wilderness contributions to the quality of life and economies of nearby human communities and populations ranging from rural to urban.
- Explore social capital issues associated with public lands.

Doing wilderness science—including basic and discovery science, not just applied science to answer visitor use or similar questions—is part of a virtuous cycle of wilderness science and improving wilderness management (Kreider et al. 2023). Indeed, many of the changes to this chapter from the first to current edition are because of knowledge accumulated from ecological science conducted in wilderness areas. Currently, it can be difficult to conduct science in wilderness, as some managers view it as a "trammeling" and subject it to extensive review, but the value of wilderness to science demands that protocols be worked out to facilitate this important function.

Of course, for wilderness to play its role as a "sensitive sentinel of change" (Turner et al. 2007) and as a resource for adaptive management, its condition must be understood. Almost fifty years before the publication of this book, author Jerry Franklin (1978) concluded the first edition of this chapter:

> A final word on the present state of monitoring, or baseline data collection, in wilderness is in order. At best, it is woefully inadequate; in most areas, it is essentially absent. What little monitoring is being done is generally not part of a systematic, comprehensive plan. Documentation is poor. Most work is focused on "sores" which are immediate management problem areas; such sites obviously have high priority but perpetuate a tendency to lose perspective on monitoring and baseline data needs over the wilderness as a whole. No other natural resource managers would ever tolerate such an inadequate inventory base in their programs. In timber management programs, for example, there are extensive systems of continuous inventory plots, comprehensive stand examinations, simulation models, and complex data storage and retrieval systems. Certainly, this type and intensity of inventory and monitoring is not advocated for wilderness. But it should make the wilderness manager aware of the total inadequacy of past efforts and the imperative for improving a data base in the future. Even provision for periodic high quality aerial photography of wilderness tracts, something sorely lacking for many areas, would be a major improvement.

In the years since, many improvements have been made. The timber inventory Franklin referenced has been expanded in scope and extended into wilderness. High quality aerial imagery is widely available. A detailed strategy for monitoring trends in wilderness character was developed by an interagency committee (Landres et al. 2015), and baseline data collection is now well underway. But, by and large, Franklin's critique still stands. The national inventory is maintained on an ad hoc basis by committed individual wilderness managers who understand the value of such a system, but it relies on self-reporting, not on systematic sampling. As Smith and Gray (2021) note, Forest Inventory and Analysis plots could be used for strategic monitoring of wilderness condition, but they are limited to forested areas, and non-forest vegetation is unevenly measured across the system. They encourage improving access to currently unmeasured areas, expanding measurements to include more non-forest locations, and adding complementary surveys (e.g., plant diversity, wildlife, invertebrates, soils, climate) to increase the utility of all data sets, but these improvements are left for the future. Monitoring programs based on statistically valid sampling of ecosystem attributes are an essential component of managing wilderness ecosystems (Smith and Gray 2021). Even with these improvements, there will still be a need for finer-scale and issue-specific monitoring. *Every wilderness management plan should include a collaboratively developed and implemented ecological monitoring strategy as part of its adaptive management program.*

SUMMARY

Ecosystems can be described on a scale of "integrity," which is often defined in terms of wholeness relative to a historical standard that includes the influence of Indigenous peoples who called these places home for thousands of years. Wilderness ecosystems occur in the part of a landscape that has relatively high integrity and is mostly left alone. The old assumption was that leaving an ecosystem alone would increase or maintain its integrity, but global change has proven that assumption false. We are left with only three management options: resisting change, accepting change, and directing change, which has been codified as "RAD," an approach to wildland management under which multiple strategies are applied as a portfolio to spread risk in the face of uncertainty. Experience has revealed the need to contemplate management actions at the scale of large landscapes to account for the needs of species across boundaries, and applying RAD at this "greater ecosystem" scale will require allocating large landscapes to each of the RAD options. *Wilderness ecosystems logically belong in the "Accept Change" allocation, but we will need to monitor and apply adaptive management, developed collaboratively and involving all those with a stake in the land because decisions about the future management of landscapes affect all people and the process must be done equitably.*

STUDY QUESTIONS

1. Define the term *ecosystem*. What are the three main components of an ecosystem?
2. What are some of the influences that people have had on ecosystems since time immemorial? How have conditions changed recently?
3. What are two qualities that distinguish wilderness ecosystems from other ecosystems?
4. Describe what is meant by "historical range of variability." What is its relevance to wilderness management?
5. What are the components of the RAD strategy? Why is it limited to just these three?
6. What is the relationship of wilderness to the surrounding landscape? How must that be accounted for in wilderness management?
7. What kinds of considerations must managers consider when contemplating intervention in wilderness?
8. What is the value of wilderness to science?
9. What is the value of wilderness to management of the larger landscape?
10. Why is it important to monitor the condition of wilderness, and how can monitoring be improved?

ACKNOWLEDGMENTS

Jerry F. Franklin, Professor Emeritus of Ecosystems Analysis, College of Forest Resources, University of Washington, Seattle, wrote this chapter for the first two editions of the book and coauthored a major revision for the third edition with Gregory H. Aplet, forest ecologist with The Wilderness Society, Denver, Colorado. The chapter was thoroughly revised for this fifth edition with the added coauthorship of Andrew J. Larson, Director of the Wilderness Institute at the University of Montana, Missoula.

We are indebted to the following colleagues who provided review of and/or input to earlier drafts of this chapter: Peter Landres, research ecologist, Aldo Leopold Wilderness Research Institute, Missoula, Montana; Pete Morton, resource economist, The Wilderness Society, Denver, Colorado; Travis Belote, senior science director, The Wilderness Society, Bozeman, Montana; Tim Fullman, wildlife ecologist, The Wilderness Society, Anchorage, Alaska; Jason Leppi, fisheries biologist, The Wilderness Society, Anchorage, Alaska; Wendy Loya, Assistant Regional Director, U.S. Fish and Wildlife Service, Anchorage, Alaska; Malcolm North, ecologist, U.S. Forest Service, Davis, California; Kathy Zeller, research biologist, Aldo Leopold Wilderness Research Institute, Missoula, Montana.

CASE DISCUSSION: IMPLEMENTING RAD AT THE KENAI NATIONAL WILDLIFE REFUGE

The Kenai National Wildlife Refuge in south-central Alaska is 805,000 hectares (1.92 million acres) of mostly intact but rapidly changing wildlands, about half of which is designated wilderness, in an urbanizing landscape of private, tribal, and other public lands. Since the late 1960s, the climate has been warming and water availability decreasing. Beginning in the late 1980s the area experienced a historically unprecedented bark beetle outbreak that killed 100 percent of the mature spruce canopy that occupied the western half of the Refuge when it was established, and a subsequent wildfire resulted in the establishment of a novel ecosystem dominated by the native rhizomatous grass, bluejoint, where boreal forest once stood. The Kenai Refuge occupies a location on the cusp of transition between boreal forest, temperate forest, and grassland biomes, and refuge managers, who are legally bound to conserve biodiversity, worry that this change is only the leading edge of further shifts that will permanently change refuge ecosystems.

Thus far, managers have mostly accepted what seem like inevitable changes, such as drying peatlands, rising treelines, and grassland invasion, while resisting other changes through familiar management activities, such as spraying invasive weeds, reintroducing extirpated species, and maintaining landscape connectivity with highway overpasses and land acquisitions. Outreach has largely focused on helping neighbors understand the fact of climate change and potential impacts, but the reality facing managers and adjacent landowners is opening new conversations. Among the interventions being considered are the introduction of the more drought- and fire-adapted lodgepole pine to replace spruce in a novel forest, the introduction of wood bison and Sitka black-tailed deer, neither of which is native to the Kenai peninsula, to graze the newly established grassland now considered unfavorable to the native moose, and using prescribed fire to create habitat diversity in the novel grassland.

While some managers at the Kenai Refuge are among the leaders in contemplating the Resist-Accept-Direct strategy, RAD has not yet been formally adopted due to internal agency resistance and the novelty of the approach. Certainly, there has been no designation of a spatial portfolio, though understanding of the place of the refuge in the larger landscape makes it nearly certain that such spatial planning will be more explicit in the future. And there are signs that things are changing. For example, managers are already considering manipulating beaver populations to restore drying peatlands, which, whether or not it is understood in the context of RAD, represents an explicit decision to move from accept to resist.

Complicating things further is the fact that much of the transition from forest to grassland is taking place in wilderness. Managers are keenly aware of the importance of untrammeled conditions to wilderness character, and they understand that some management activities, such as planting non-native species, would compromise those values. Other potential activities, such as the reintroduction of caribou or the introduction of non-native bison, even if it occurs outside wilderness, may have similar effects. As yet, no decision has been made as to how to handle wilderness in the context of RAD. [This case study is summarized from Magness et al. (2021).]

CASE DISCUSSION QUESTIONS

1. What ecological outcome do you think should be pursued at the Kenai National Wildlife Refuge?
2. Given the specific ecological conditions of the Kenai Refuge, how do you think wilderness at the refuge should be managed? What should be its place in the RAD strategy?
3. How would you decide when to attempt restoration to a historic condition and when to allow human-caused disturbances to an ecosystem (that originate outside that wilderness area) to run their course and arrive at a new condition? How much trammeling do you think is acceptable to restore historic naturalness?

REFERENCES

Anderson, M. Kat. 2005. *Tending the Wild: Native American Knowledge and the Management of California's Natural Resources.* Berkeley and Los Angeles: University of California Press.

Aplet, Gregory H. 1999. On the nature of wildness: Exploring what wilderness really protects. *Denver University Law Review.* 76: 347–367.

Aplet, Gregory H.; Cole, David N. 2010. The trouble with naturalness: Rethinking park and wilderness goals. In: Cole, David N.; Yung, Laurie, eds. *Beyond Naturalness: Rethinking Park and Wilderness Stewardship in an Era of Rapid Change.* Island Press, Covelo, CA, pp. 12–29.

Aplet, Gregory H.; Keeton, William S. 1999. Application of historical range of variability concepts to the conservation of biodiversity. In: Baydack, Richard K.; Campa, Henry, III; Haufler, Jonathan B., eds. *Practical Approaches to the Conservation of Biological Diversity.* Covelo, CA: Island Press, pp. 71–86.

Aplet, Gregory H.; McKinley, Peter S. 2017. A portfolio approach to managing ecological risks of global change. *Ecosystem Health and Sustainability.* 3(2): e01261.

Asher, Jerry E.; Harmon, David W. 1995. Invasive exotic plants are destroying the naturalness of U.S. wilderness areas. *International Journal of Wilderness.* 1(2): 35–37.

Aycrigg, Jocelyn L.; Tricker, James; Belote, R. Travis; Dietz, Matthew S.; Duarte, Lisa; Aplet, Gregory H. 2016. The next 50 years: Opportunities for diversifying the ecological representation of the National Wilderness Preservation System within the contiguous United States. *Journal of Forestry.* 114: 396–404.

Bailey, Robert G. 1980. *Description of the eco-regions of the United States.* Miscellaneous Publication 1391. Washington, DC: U.S. Department of Agriculture, Forest Service.

———. 1995. *Description of the eco-regions of the United States,* 2nd ed. Miscellaneous Publication 1391. Washington, DC: U.S. Department of Agriculture, Forest Service.

Barnett, Kevin; Belote, R. Travis. 2021. Modeling an aspirational connected network of protected areas across North America. *Ecological Applications.* 31(6): e02387.

Barnett, Kevin; Miller, Carol; Venn, Tyron J. 2016. Using risk analysis to reveal opportunities for the management of unplanned ignitions in wilderness. *Journal of Forestry.* 114(6): 610–618.

Belote, R. Travis; Barnett, Kevin; Zeller, Katherine; Brennan, Angela; Gage, Josh. 2022. Examining local and regional ecological connectivity throughout North America. *Landscape Ecology.* 37:2977–2990.

Belote, R. Travis; Dietz, Matthew S.; Aplet, Gregory H. 2015. Allocating untreated "controls" in the National Wilderness Preservation System as a climate adaptation strategy: A case study from the Flathead National Forest, Montana. *Northwest Science.* 89:239–254.

Belote, R. Travis; Wilson, Melissa B. 2019. Delineating greater ecosystems around protected areas to guide conservation. *Conservation Science and Practice.* 2(6): e196.

Brinkley, Douglas. 2022. *Silent Spring Revolution.* New York: HarperCollins.

Cheng, Antony S.; Aplet, Gregory H.; Walz, Amy E. M. 2019. Challenges and opportunities for collaborative adaptive management in forest landscape restoration. In: Butler, William H.; Schultz, Courtney A., eds. *A New Era for Collaborative Forest Management.* New York: Routledge, pp. 119–136.

Clifford, Katherine R.; Cravens, Amanda E.; Knapp, Corrine N. 2021. Responding to ecological transformation: Mental models, external constraints, and manager decision-making. *Bioscience.* 72:57–70.

Cole, David N. 1996. Ecological manipulation in wilderness: An emerging management dilemma. *International Journal of Wilderness.* 4(3): 28–31.

———. 2000. Paradox of the primeval: Ecological restoration in wilderness. *Ecological Restoration.* 18(2): 77–86.

———. 2007. Scaling-up the minimum requirements analysis for wilderness issues. *International Journal of Wilderness.* 13(1): 8–12.

Cronon, William. 1983. *Changes in the Land: Indians, Colonists, and the Ecology of New England.* New York: Hill and Wang.

Davis, George D. 1989. Preservation of natural diversity: The role of ecosystem representation within wilderness. In: Freilich, Helen R., comp. *Wilderness Benchmark 1988: Proceedings of the National Wilderness Colloquium*; January 13–14, 1988; Tampa, FL. General Technical Report SE-51. Asheville, NC: U.S. Department Agriculture, Forest Service, Southeast Forest Experiment Station, pp. 76–82.

Denevan, William M. 1992. The pristine myth: The landscape of the Americas in 1492. *Annals of the Association of American Geographers.* 82:369–385.

Dietz, Matthew S.; Belote, R. Travis; Aplet, Gregory H.; Aycrigg, Jocelyn L. 2015. The world's largest wilderness protection network after 50 years: An assessment of ecological system representation in the U.S. National Wilderness Preservation System. *Biological Conservation.* 184:431–438.

Falk, Donald A.; Watts, Adam C.; Thode, Andrea E. 2019. Scaling ecological resilience. *Frontiers in Ecology and Evolution.* 7.

Franklin, Jerry F. 1978. Wilderness ecosystems. In: Hendee, John C.; Stankey, George H.; Lucas, Robert C., eds. *Wilderness Management.* USDA Forest Service Miscellaneous Publication No. 1365. Washington, DC: USDA Forest Service. pp. 191–212.

Franklin, Jerry F.; Aplet, Gregory H. 2002. Wilderness ecosystems. In: Hendee, John C.; Dawson, Chad, eds. *Wilderness Management: Stewardship and Protection of Resources and Values*, 3rd ed. Golden, CO: Fulcrum Publishing. pp. 263–286.

Franklin, Jerry F.; Johnson, K. Norman; and Johnson, Debora L. 2018. *Ecological Forest Management*. Long Grove, IL: Waveland Press.

Gallo, John A.; Aplet, Gregory H.; Greene, Randall; Thomson, Janice L.; Lombard, Amanda T. 2020. A transparent and intuitive modeling framework and software for efficient land allocation. *Land*. 9:444.

Gray, Claudia L.; Hill, Samantha L. L.; Newbold, Tim; Hudson, Lawrence N.; Börger, Luca; Contu, Sara; Hoskins, Andrew J.; Ferrier, Simon; Purvis, Andy; Scharlemannk, Jörn P. W. 2016. Local biodiversity is higher inside than outside terrestrial protected areas worldwide. *Nature Communications*. 7: 12306.

Greaver, Tara L.; Sullivan, Timothy J.; Herrick, Jeffrey D.; Barber, Mary C.; Baron, Jill S.; Cosby, Bernard J.; Deerhake, E.; Dennis, Robin L.; Dubois, Jean-Jacques B.; et al. 2012. Ecological effects of nitrogen and sulfur air pollution in the US: What do we know? *Frontiers in Ecology and the Environment*. 10(7): 365–372.

Haddad, Nick M.; Brudvig, Lars A; Clobert, Jean; Davies, Kendie F.; Gonzalez, Andrew; Holt, Robert D.; Lovejoy, Thomas E.; Sexton, Joseph O.; Austin, Mike P.; et al. 2015. Habitat fragmentation and its lasting impact on Earth's ecosystems. *Science Advances*. 1: e1500052.

Hansen, Andrew J.; Davis, Cory R.; Piekielek, Nathan; Gross, John; Theobald, David M.; Goetz, Scott; Melton, Forrest; and Defries, Ruth. 2011. Delineating the ecosystems containing protected areas for monitoring and management. *Bioscience*. 61:363–373.

Hessburg, Paul F.; Larson, Andrew J.; Churchill, Derek J.; Haugo, Ryan D.; Miller, Carol; Spies, Thomas A.; North, Malcolm P.; Povak, Nicholas, A.; Belote, R. Travis.; et al. 2015. Restoring fire-prone Inland-Pacific landscapes: Seven core principles. *Landscape Ecology*. 30:1805– 1835.

Jenny, Hans. 1941. *Factors of Soil Formation: A System of Quantitative Pedology*. Dover Publications, New York, 281 p.

Keeley, John E.; Aplet, Gregory H.; Christensen, Norman L.; Conard, Susan C.; Johnson, Edward A.; Omi, Philip N.; Peterson, David L.; Swetnam, Thomas W. 2009. *Ecological foundations for fire management in North American forest and shrubland ecosystems*. Gen. Tech. Rep. PNW-GTR-779. Portland, OR: U.S. Department of Agriculture, Forest Service, Pacific Northwest Research Station. 92 p.

Kelly, Patrick; Landres, Peter. 2022. Does wilderness matter in the Anthropocene? Resolving a fundamental dilemma about the role of wilderness in 21st century conservation. *Ethics, Policy and Environment*.

Kipfmueller, Kurt F.; Larson, Evan R.; Johnson, Lane B.; Schneider, Elizabeth A. 2021. Human augmentation of historical red pine fire regimes in the Boundary Waters Canoe Area Wilderness. *Ecosphere*. 12(7): e03673.

Knight, Richard L.; Landres, Peter B., eds. 1998. *Stewardship Across Boundaries*. Washington, DC: Island Press.

Kreider, Mark R.; Jaffe, Melissa R.; Berkey, Julia K.; Parks, Sean A.; Larson, Andrew J. The scientific value of fire in wilderness. *Fire Ecology*. 19:36.

Kyba, Christopher, C. M.; Altintas, Yigit Oner; Walker, Constance E.; Newhouse, Mark. 2023. Citizen scientists report global rapid reductions in the visibility of stars from 2011 to 2022. *Science* 379(6629): 265–268.

Landres, Peter. 2010. Let it be: A hands-off approach to preserving wildness in protected areas. In: Cole, David N.; Yung, Laurie, eds. *Beyond Naturalness: Rethinking Park and Wilderness Stewardship in an Era of Rapid Change*. Washington, DC: Island Press, pp. 88–105.

Landres, Peter; Barns, Chris; Boutcher, Steve; Devine, Tim; Dratch, Peter; Lindholm, Adrienne; Merigliano, Linda; Roeper, Nancy; Simpson, Emily. 2015. *Keeping it wild 2: An updated interagency strategy to monitor trends in wilderness character across the National Wilderness Preservation System*. Gen. Tech. Rep. RMRS-GTR-340. Fort Collins, CO: U.S. Department of Agriculture, Forest Service, Rocky Mountain Research Station. 114 p.

Landres, Peter; Hahn, Beth A.; Biber, Eric; Spencer, Daniel T. 2020. Protected area stewardship in the Anthropocene: Integrating science, law, and ethics to evaluate proposals for ecological restoration in wilderness. *Restoration Ecology*. 28(2): 315–327.

Landres, Peter B.; Morgan, Penelope; Swanson, Frederick J. 1999. Overview of the use of natural variability concepts in managing ecological systems. *Ecological Applications*. 9: 1179–1188.

Larson, Andrew J.; Belote, R. Travis; Cansler C. Alina; Parks, Sean A.; Dietz, Matthew S. 2013a. Latent resilience in ponderosa pine forest: Effects of resumed frequent fire. *Ecological Applications*. 23: 1243–1249.

Larson, Andrew J.; Belote, R. Travis; Williamson, Matthew A.; Aplet, Gregory H. 2013b. Making monitoring count: Project design for active adaptive management. *Journal of Forestry* 111:348–356.

Leopold, Aldo. 1941. Wilderness as a land laboratory. *The Living Wilderness*. 6(July):3.

Leroux, Shawn J.; Schmiegelow, Fiona K. A.; Lessard, Robert B.; Cumming, Steven G. 2007. Minimum dynamic reserves: A framework for determining reserve size in ecosystems structured by large disturbances. *Biological Conservation* 138(3): 464–473.

Loomis, John; Echohawk, Chris. 1999. Using GIS to identify under-represented ecosystems in the National Wilderness Preservation System in the U.S. *Environmental Conservation*. 26:53–58.

Lorenz, Teresa J.; Suillivan, Kimberley A.; Bakian, Amanda V.; Aubry, Carol A. 2011. Cache-site selection in Clark's Nutcracker (*Nucifraga columbiana*). *The Auk*. 128(2): 237–247.

Lovett, Gary M.; Tear, Timothy H.; Evers, David C.; Findlay, Stuart E. G.; Cosby, B. Jack; Dunscomb, Judy K.; Driscoll, Charles T.; Weathers, Kathleen C. 2009. Effects of air pollution on ecosystems and biological diversity in the eastern United States. *Annals of the New York Academy of Sciences*. 1162:99–135.

Lundquist, Laura. 2019. Should whitebark pine seedlings be planted in wilderness? *Missoula Current*, April 9, 2019. https://missoulacurrent.com/whitebark-pine/.

Lynch, Abigail J.; Thompson, Laura M.; Morton, John M.; Beever, Erik A.; Clifford, Michael; Limpinsel, Douglas; Magill, Robert T.; Magness, Dawn R.; Melvin, Tracy A., et al. 2021. RAD adaptive management for transforming ecosystems. *Bioscience*. 72:45–56.

Magness, Dawn R.; Morton, John M.; Huettmann, Falk; Chapin III, F. Stuart; McGuire, A. David. 2011. A climate-change adaptation framework to reduce continental-scale vulnerability across conservation reserves. *Ecosphere*. 2(10): 1–23.

Magness, Dawn R.; Hoang, Linh; Belote, R. Travis; Brennan, Jean; Carr, Wylie; Chapin, III, F. Stuart; Clifford, Katherine; Morrison, Wendy; Morton, John M. 2021. Management foundations for navigating ecological transformation by resisting, accepting, or directing social–ecological change. *Bioscience*. 72: 30–44.

Manley, Patricia N.; Brogan, Gary E.; Cook, Carolyn; Flores, Mary E.; Fullmer, Donald G.; Husari, Susan; Jimerson, Thomas M.; Lux, Linda M.; McCain, Michael E., et al. 1995. *Sustaining ecosystems: A conceptual framework*, R-5-EM-TP-01. San Francisco: US Department of Agriculture Forest Service, Pacific Southwest Region and Station.

Mann, Charles C. 2006. *1491: New Revelations about the Americas before Columbus*, 2nd edition. New York: Random House.

Matthews, Kathleen R.; Knapp, Roland A. 1999. A study of high mountain lake fish stocking effects in the U.S. Sierra Nevada wilderness. *International Journal of Wilderness*. 5(1): 24–26.

Middleton, Arthur D.; Sawyer, Hall; Merkle, Jerod A.; Kauffman, Matthew J.; Cole, Eric K.; Dewey, Sarah R.; Gude, Justin A.; Gustine, David D.; McWhirter, Douglas E., et al. 2020. Conserving transboundary wildlife migrations: Recent insights from the Greater Yellowstone Ecosystem. *Frontiers in Ecology and the Environment*. 18(2): 83–91.

Millar, Constance I.; Stephenson, Nathan L.; Stephens, Scott L. 2007. Climate change and forests of the future: managing in the face of uncertainty. *Ecological Applications*. 17(8): 2145–51.

Miller, Carol; Aplet, Gregory H. 2016. Progress in wilderness fire science: Embracing complexity. *Journal of Forestry*. 114(3): 373–383.

Miller, Paul L. 1973. Oxidant-introduced community change in a mixed conifer forest. *American Chemistry Society Advances in Chemistry.* No. 122. Washington, DC: American Chemistry Society, pp. 101–117.

Morgan, Penelope; Aplet, Gregory H.; Haufler, Jonathan B.; Humphries, Hope C.; Moore, Margaret M.; Wilson, Wade D. 1994. Historical range of variability: A useful tool for evaluating ecosystem change. *Journal of Sustainable Forestry.* 2: 87–111.

Nowacki, Gregory J.; Abrams, Marc D. 2008. The demise of fire and "mesophication" of forests in the eastern United States. *BioScience.* 58: 123–138.

Nowacki, Gregory J.; MacCleery, Douglas W.; Lake, Frank L. 2012. Native Americans, ecosystem development, and historical range of variation. In: Wiens, John A.; Hayward, Gregory D.; Safford, Hugh D.; Giffen, Catherine M., eds. *Historical Environmental Variation in Conservation and Resource Management.* John Wiley and Sons, Ltd., West Sussex, UK, pp. 76–91.

Odum, Eugene P. 1971. *Fundamentals of Ecology*, 3rd ed. Philadelphia: W. B. Saunders.

Parsons, David S. 2000. Restoration of natural fire to United States wilderness areas. In: Watson, Alan E.; Aplet, Gregory H.; Hendee, John C., comps. *Personal, Societal, and Ecological Values of Wilderness: 6th World Wilderness Congress Proceedings on Research, Management and Allocation*, Vol. II; October 24–29, 1998; Bangalore, India. RMRS-P-14. Ogden, UT: U.S. Department of Agriculture, Forest Service, Rocky Mountain Research Station, pp. 42–47.

Pickett, S. T. A.; Thompson, John N. 1978. Patch dynamics and the design of nature reserves. *Biological Conservation.* 13:27–37.

Poff, N. Leroy; Allan, J. David; Bain, Mark B.; Karr, James R.; Prestegaard, Karen L.; Richter Brian D.; Sparks, Richard E.; Sromberg, Juliet C. 1997. The natural flow regime: A paradigm for river conservation and restoration. *Bioscience* 47:769–784.

Popkin, Gabriel. 2018. To save iconic American chestnut, researchers plan introduction of genetically engineered tree into the wild. *Science* doi:10.1126/science.aav2573.

Quigley, Thomas M; Arbelbide, Sylvia J., tech. eds. 1997. *An Assessment of Ecosystem Components in the Interior Columbia Basin and Portions of the Klamath and Great Basins*; Vol. 2. PNW-GTR-405. Portland, OR: U.S. Department of Agriculture, Forest Service, Pacific Northwest Station, pp. 337–1055.

Quinn, Thomas P.; Helfield, James M.; Austin, Catherine S.; Hovel, Rachel A.; Bunn, Andrew G. 2018. A multidecade experiment shows that fertilization by salmon carcasses enhanced tree growth in the riparian zone. *Ecology.* 99:2433–2441.

Randall, John M. 2000. Improving management of nonnative invasive plants in wilderness and other natural areas. In: Cole, David. N.; McCool, Steven. F.; Borrie, William T.; O'Loughlin, Jennifer, comps. *Wilderness science in a time of change conference, Vol. 5: Wilderness Ecosystems, Threats, and Management*; May 23–27, 1999; Missoula, MT. RMRS-P-15. Ogden, UT: U.S. Department of Agriculture, Forest Service, pp. 64–73.

Roche, Philip K.; Campagne, C. Silvie. 2017. From ecosystem integrity to ecosystem condition: A continuity of concepts supporting different aspects of ecosystem sustainability. *Current Opinion in Environmental Sustainability.* 29:63–68.

Rogers, Brendan M.; Mackey, Brendan; Shestakova, Tatiana A.; Keith, Heather; Young, Virginia; Kormos, Cyril F.; DellaSala, Dominick A.; Dean, Jacqueline; Birdsey, Richard, et al. 2022. Using ecosystem integrity to maximize climate mitigation and minimize risk in international forest policy. *Frontiers in Forests and Global Change.* 5:25.

Roos, Christopher I.; Guiterman; Christopher H., Margolis, Ellis Q.; Swetnam, Thomas W., Laluk, Nicholas C.; Thompson, Kerry F.; Toya, Chris; Farris, Calvin A.; Fule, Peter Z., et al. 2022. Indigenous fire management and cross-scale fire-climate relationships in the Southwest United States from 1500 to 1900 CE. *Science Advances.* 8(49).

Safford, Hugh D.; Hayward, Gregory D.; Heller, Nicole E.; Wiens, John A. 2012. Historical ecology, climate change, and resource management: Can the past still inform the future? In: Wiens, John A.; Hayward, Gregory D.; Safford, Hugh D.; Giffen, Catherine M., eds. *Historical Environmental Variation in Conservation and Resource Management.* West Sussex, UK: John Wiley and Sons, pp. 46–62.

Schuurman, Gregor W.; Hoffman, Cat Hawkins; Cole, David N.; Lawrence, David J.; Morton, John M.; Magness, Dawn R.; Cravens, Amanda E.; Covington, Scott; O'Malley, Robin; Fisicheli, Nicholas A. 2020. *Resist-Accept Direct (RAD)—A Framework for the 21st-century Natural Resource Manager.* Natural Resource Report NPS/NRSS/CCRP/NRR—2020/2213. Fort Collins, CO: USDI National Park Service. 20 pp.

Schuurman, Gregor W.; Cole, David N.; Cravens, Amanda E.; Covington, Scott; Crausbay, Shelley D.; Hoffman, Cat Hawkins; Lawrence, David J.; Magness, Dawn R.; Morton, John M., et al. 2021. Navigating ecological transformation: Resist–Accept–Direct as a path to a new resource management paradigm. *Bioscience.* 72:16–29.

Shugart, Herman H., Jr.; West, D. C. 1977. Development of an Appalachian deciduous forest succession model and its application to assessment of the impact of chestnut blight. *Journal of Environmental Management.* 5(2): 161–179.

Silliman, Brian R.; Angelini, Christine. 2012. Trophic cascades across diverse plant ecosystems. *Nature Education Knowledge.* 3(10): 44.

Smith, Robert J.; Gray, Andrew N. 2021. Strategic monitoring informs wilderness management and socioecological benefits. *Conservation Science and Practice.* 3:e482.

Snyder, Gary. 1990. *The Practice of the Wild.* New York: North Point Press.

Sordello, Romain; Ratel, Ophélie; De Lachapelle, Frédérique Flamerie; Leger, Clément; Dambry, Alexis; Vanpeene, Sylvie. 2020. Evidence of the impact of noise pollution on biodiversity: A systematic map. *Environmental Evidence.* 9(20).

Stewart, Omer C. 2009. *Forgotten Fires: Native Americans and the Transient Wilderness.* University of Oklahoma Press. Norman.

Swanson, Frederick J.; Jones, Julia A.; Wallin, David O.; Cissel, John H. 1994. Natural variability—Implications for ecosystem management. In: Everett, Richard L., team leader. *Ecosystem Management: Principles and Application. Vol. II of the Eastside Forest Ecosystem Health Assessment.* PNW-GTR-318. Portland, OR: U.S. Department of Agriculture, Forest Service, Pacific Northwest Experiment Station, pp. 80–94.

Thomas-Van Gundy, Melissa A.; Nowacki, Gregory J.; Cogbill, Charles V. 2015. *Mapping Pyrophyllic Percentages across the Northeastern United States Using Witness Trees, with Focus on Four National Forests.* NRS-GTR-145. Newtown Square, PA: U.S. Department of Agriculture, Forest Service, Northern Research Station. 26 p.

Tonra, Christopher M.; Sager-Fradkin, Kimberly; Morley, Sarah A.; Duda, Jeffrey J.; Marra, Peter P. 2015. The rapid return of marine-derived nutrients to a freshwater food web following dam removal. *Biological Conservation.* 192:130–134.

Turner, Monica G.; Anderson, Dorothy H.; Aplet, Gregory H.; Dawson, Chad P.; Dennis, John G.; Noon, Barry R.; Waide, Jack B. 2007. *Report to USDA-Forest Service Research and Development: External Peer Review Panel for the Wilderness Research Program Area.* Washington, DC: USDA Forest Service. 23p.

U.S. Forest Service. 2012. National Forest System land management planning (36 CFR 219): Final rule and record of decision. *Federal Register.* 77(68): 21162–21276.

Vale, Thomas R. 1998. The myth of the humanized landscape: An example from Yosemite National Park. *Natural Areas Journal.* 18(3): 231–236.

Vitousek, Peter M.; Aber, John D.; Goodale, Christine L.; Aplet, Gregory H. 2000. Global change and wilderness science. In: Cole, David N.; McCool, Steven F.; Freimund, Wayne; O'Loughlin, Jennifer, comps. *Wilderness Science in a Time of Change Conference, Vol. 1: Changing Perspectives and Future Directions*; May 23–27, 1999; Missoula, MT. RMRS-P-15-Vol-1. Ogden, UT: U.S. Department of Agriculture, Forest Service, pp. 5–9.

Watson, James E. M.; Venter, Oscar; Lee, Jasmine; Jones, Kendall R.; Robinson, John G.; Possingham, Hugh P.; Allan, James R. 2018. Protect the last of the wild. *Nature.* 563:27–30.

Wiens, John A.; Hayward, Gregory D.; Safford, Hugh D.; Giffen, Catherine M., eds. 2012. *Historical Environmental Variation in Conservation and Resource Management.* West Sussex, UK: John Wiley and Sons, 337p.

White, Peter S.; Harrod, Jonathan; Walker, Joan L.; Jentsch, Anke. 2000. Disturbance, scale, and boundary in wilderness management. In: McCool, Steven F.; Cole, David N.; Borrie, William T.; O'Loughlin, Jennifer, comps. *Wilderness Science in a Time of Change Conference, Vol. 2: Wilderness Within the Context of Larger Systems*; May 23–27, 1999; Missoula, Montana. RMRS-P-15. Ogden, UT: U.S. Department of Agriculture, Forest Service, pp. 27–42.

Whitney, Gordon G. 1994. *From Coastal Wilderness to Fruited Plain: A History of Environmental Change in Temperate North America, 1500 to the Present.* New York: Cambridge University Press.

Williams, Alton Park; Allen, Craig D.; Macalady, Alison; McDowell, Nate. 2013. Temperature as a potent driver of regional forest drought stress and tree mortality. *Nature Climate Change.* 3:292–297.

Zahniser, Howard. 1956. The need for wilderness areas. *The Living Wilderness.* 59(Winter-Spring):58–43.

CHAPTER

11

Fire in Wilderness Ecosystems

by Sean A. Parks

INTRODUCTION

Periodic fires serve as a necessary ecological process in most wilderness ecosystems. In designated wilderness areas where the Earth and its community of life are to remain untrammeled by humans, fire is as natural and vital a process as rain, snow, or wind. From jack pine and aspen in the Boundary Waters Canoe Area Wilderness (BWCAW) of Minnesota; through lodgepole pine in Yellowstone National Park, Wyoming; to ponderosa pine in Grand Canyon National Park, Arizona; and to the giant sequoia of California's Sequoia–Kings Canyon and Yosemite National Parks, fire plays an essential role in structuring ecosystems. Many plant communities and even entire ecosystems actually depend on fire for their persistence and well-being and are thus called fire-dependent plant communities and ecosystems. However, land management agencies have a longstanding policy of preventing and controlling (or suppressing) most fires.

As area burned continues to increase in North America's wildlands, the impetus for conducting scientific studies involving fire has increased in parallel. There have been numerous important studies on fire frequency, effects of fires on ecosystems, the influence of weather, drought, and climate on fire, and Indigenous fire management practices that serve as the foundation for wilderness fire management and for this chapter.

Evidence of the historical role of fire is found in charred wood and cones in glacial deposits, in charcoal stratigraphy of laminated lake-bottom sediments, and in fire-scarred cross-sections, increment borings, or wedges from both recent and ancient trees (Gavin et al. 2007; Swetnam et al. 2009). Fires were ignited by both lightning and Indigenous peoples. Given an ignition source, grasslands, shrublands, and forest would be susceptible to fire during the dry season or over extended dry periods (i.e., drought).

The kind of fire activity that characterizes a specific region or vegetation community is referred to as its "fire regime." Examples of such regimes include ponderosa pine forests that generally experience frequent, low-intensity surface fires and spruce-fir forests that generally experience infrequent (300 years or more), high-intensity crown or surface fires that kills all or most trees. Such fires are sometimes called "stand-replacement" or "high-severity" fires (Fig. 11.1).

The goal of wilderness fire management is to restore fire as nearly as possible to its natural role. Prior to 1980 (and to a large extent today), our society and land management agencies perceived fire as a destroyer and attempted to ban it from nearly all lands, including wilderness. The past four decades have brought a more enlightened, biocentric approach to wilderness management that acknowledges fire must play its natural role to the greatest possible extent—consistent with the safety of people and adjacent non-wilderness resources. However, this recognition is still more an ideal than a practical reality. This chapter explores our current knowledge of fire ecology and alternatives for restoring fire as an important and necessary ecological process to wilderness.

Wildland fire management has become an important and prominent natural resource management specialty, with wilderness fire management as a subset. Federal and state agencies managing fire in wilderness recognize that well-trained professionals are essential to direct and implement the complex fire programs involved in wilderness fire management. These agencies have in-house and intra-agency training programs

FIGURE 11.1 Elk on the East Fork of the Bitterroot River during a fire on the Sula Complex, August 6, 2000. Fire has been a historic force shaping the character of the wilderness. Restoring fire to its natural role after years of a policy of suppression is a major challenge to wilderness managers. Photo courtesy John McColgan, Bureau of Land Management.

for managers, professionals, and technicians in fire behavior, fire effects, fire monitoring, and fire suppression concepts and techniques. The organizational training programs supplement college and university fire ecology and fire management curricula and short courses, and periodic interagency conferences on park and wilderness fire management. Wilderness fire is a prominent topic in natural resource and fire management journals and magazines, building an evolving information base on fire in ecosystem management and for training fire managers.

FIRE REGIMES

A fire regime is described by attributes of fire that recur over space and time for a given vegetation community. A fire regime describes the frequency, intensity, ecological effects (or severity), size, seasonality, and type (e.g., surface vs. crown) of fire that occurs within an ecological community over an extended period of time (Agee 1996). Fire frequency and fire severity are arguably the most important fire regime characteristics. Fire frequency can be quantified by the fire return interval (FRI); low FRI equates to frequent fire. Fire severity is a measure of fire's effects on ecosystems; for example, a high-severity fire kills all or most trees and may also be referred to as a "stand-replacing" fire. Stand-replacing fire can also occur in grasslands and shrublands where all or most aboveground vegetation is killed by fire. For this chapter, we use the following five fire regime groups (FRGs; Table 11.1) (LANDFIRE 2016).

Multiple fire regimes commonly occur across regions. For example, low-elevation dry forests in the Rocky Mountains dominated by ponderosa pine are characterized by a frequent, low-severity fire (FRG1), whereas higher elevation lodgepole pine in the Rocky Mountains is generally characterized by an infrequent, stand-replacing fire regime (FRG4).

It is worth noting that deviations within a particular fire regime are not uncommon (Fig. 11.2). For example, although dry ponderosa pine forests are characterized by frequent, low-severity fire, it is possible, and even probable, that high-severity, stand replacing fires can occur, albeit this is not the norm. Conversely, lodgepole pines forests characterized by infrequent, stand-replacing fire can experience low-severity surface fires on occasion.

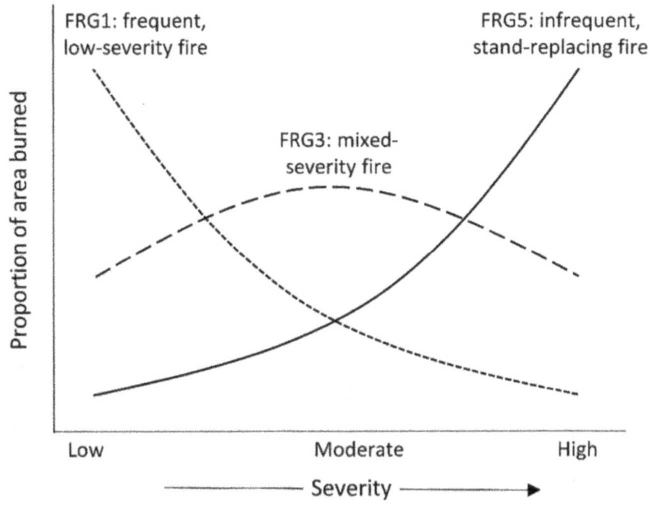

FIGURE 11.2 Within each fire regime group (FRG), fire severity exhibits its variation over space and time even though the largest proportion of area burned is concentrated in the appropriate severity class. The mixed-severity regime is the most complex, with a mix of low-, moderate-, and high-severity fires. Modified from Agee (1996).

HISTORICAL FIRE REGIMES

For the most part, descriptions of fire regimes often characterize the properties of fire that were common prior to European colonization of North America. As we will describe in the next section, historical fire regimes have been substantially altered across much of North America.

TABLE 11.1 Description and Examples of the Five Fire Regime Groups (FRGs) Discussed in This Chapter

Fire Regime Group	Description	Examples
FRG1	Frequent (≤35-year FRI), low-severity fire	1. Oak forest found in California, the Appalachian Mountains, and elsewhere 2. Ponderosa pine forest of western North America 3. Some shrublands
FRG2	Frequent (≤35-year FRI), stand-replacing fire	4. Mostly comprises prairie and grassland
FRG3	Infrequent (36–200-year FRI), mixed-severity fire	5. Mixed-conifer forest of the northern Rocky Mountains 6. Pinyon-juniper woodland common in western North America 7. Some subalpine forest
FRG4	Infrequent (36–200-year FRI), stand-replacing fire	8. Sagebrush shrubland common in western North America 9. Chaparral in California 10. Aspen forest in the Rocky Mountains
FRG5	Very infrequent (200+ year FRI), stand-replacing fire	11. Boreal forest common in northern North America 12. Hemlock forest in the Pacific Northwest 13. Hardwood forest in northeastern North America 14. Tundra

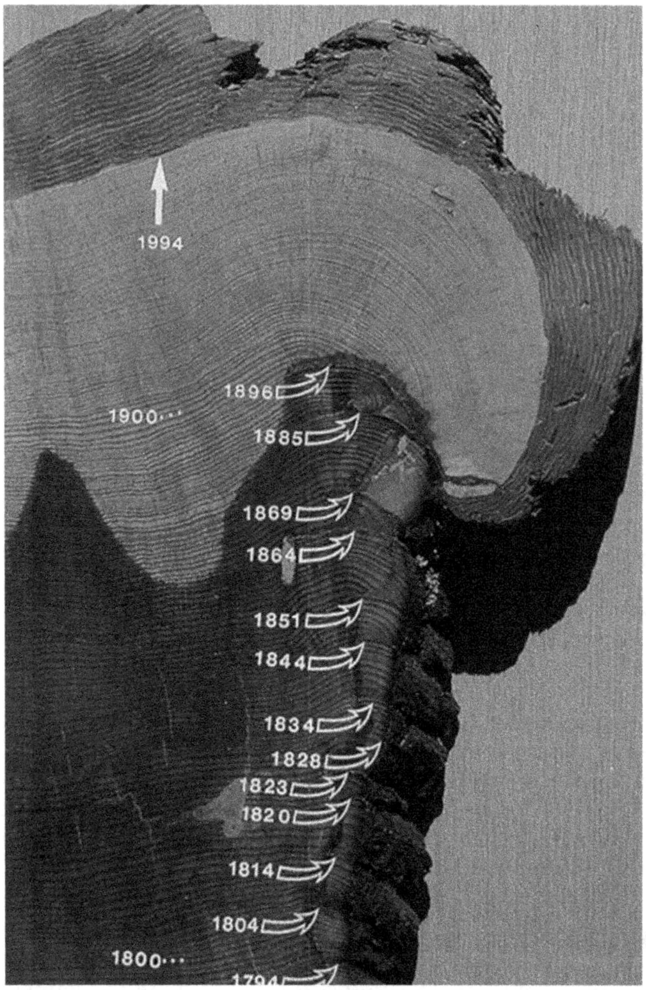

FIGURE 11.3 Fire-scarred cross section of a pine tree sampled in Yosemite National Park. This tree recorded fire about every eight years until 1896. The long fire-free period from 1897 to 1994 illustrates the fire deficit and is a result of policies and practices that excluded fire from the landscape. The lack of recurring fire allows for an increase in fuels, tree density, and fire-intolerant trees (see Fig. 11.4), thereby decreasing resilience to the inevitable fire. Photo courtesy Thomas Swetnam.

Prior to European colonization of North America, fire occurrence was fairly common across much of the continent (Heyerdahl et al. 2008, Marlon et al. 2012). Fires were ignited by both lightning and people, though the proportion of each ignition type likely varied according to Indigenous population density, the dominant vegetation communities, or along Indigenous travel routes (Kipfmueller et al. 2021). Although it is difficult to determine the relative contribution of lightning- vs. human-ignited fires prior to European colonization, several studies have concluded that the historical fire regimes could not have been achieved without humans augmenting lightning-ignitions (e.g., Knight et al. 2022). Regardless of the ignition source, fire was a dominant ecological process that shaped most North American Ecosystems.

CONTEMPORARY FIRE REGIMES AFTER 100+ YEARS OF FIRE EXCLUSION

The historical fire regimes in North America were substantially disrupted after the arrival of Europeans, though the timing of these fire regime changes vary across the continent (Margolis et al. 2022). Importantly, the arrival of European Americans resulted in a notable reduction in fire frequency and area burned through the cessation of Indigenous burning, livestock grazing that reduced fine fuels that are conducive to fire spread, and the direct suppression of fire (Lake et al. 2017). These activities effectively excluded fire from many North American ecosystems for over a century, resulting in a substantial "fire deficit" (Marlon et al. 2012) (Fig. 11.3).

Although many may applaud the overall reduction in area burned and associated wildfire smoke resulting from fire exclusion, several changes to fire-adapted ecosystems have been documented that, paradoxically, reduce the ability of ecosystems to withstand future wildfire events and even increase wildfire risk to communities (Parisien et al. 2020). Fire exclusion, as well as timber harvesting practices that removed large, fire-tolerant trees, has altered the composition and structure of many forests across North America (Hagmann et al. 2021). Specifically, increases in the prevalence of shade-tolerant, fire-sensitive trees, an increase in landscape homogeneity, and an overall increase in tree density and fuels are attributable to practices and policies that have excluded fire (Hessburg et al. 2019). Forests altered by 100+ years of fire exclusion exhibit reduced resilience to the inevitable fire (Fig. 11.4). This said, it is generally thought that forests characterized by infrequent fire (FRG5; Table 11.1) are less departed from historical conditions compared to forests that historically experienced more frequent fire (FRG1 and FRG3), and consequently, these forests may be less susceptible to enduring, fire-catalyzed change.

Climate change and other climate-mediated disturbances such as insect outbreaks that weaken or kill trees also contribute to contemporary fire regimes. For example, about half of the increased annual area burned experienced in recent decades in the western U.S. can be attributed to climate change (Abatzoglou and Williams 2016); the other half is likely attributable to naturally occurring drought and perhaps because fires are more difficult to effectively suppress given excessive live and dead fuels across forested landscapes. Warmer and drier conditions associated with climate change can directly weaken or kill trees or do so indirectly by facilitating conditions that are conducive to outbreaks of insects such as mountain pine beetle and spruce budworm. The cumulative and compounding effects of fire-, temperature-, and moisture-related tree mortality and insect-caused tree mortality threatens long-term forest persistence in some regions (Steel et al. 2022).

Are Recent Fires "Unprecedented"?

The last couple of decades were marked by several very active fire seasons in North America. In observing these extraordinary fire seasons, some might ask if they are "unprecedented" in terms of annual area burned. It turns out that some of these fire seasons are indeed unprecedented when considering the timeframe in which we have high quality fire perimeter data starting in the 1970s and 1980s. However, whether or not they are unprecedented when looking further back in time (i.e., prior to European colonization) is an active area of research and requires different datasets, such as fire-scarred trees and sediment cores from lakes.

FIGURE 11.4 Two photos, taken eighty years apart, in the Mariposa Grove, Yosemite National Park, California, illustrate plant succession in the absence of fire. The 1890 photo (left) shows few understory trees. By 1970 (right), a dense thicket of white fir has grown up. Such thickets provide a fuel ladder that could result in a fire intense enough to kill mature giant sequoias. Left photo courtesy Mrs. Dorothy Whitener and Mary and Bill Hood; right, courtesy Dan Taylor, National Park Service.

To date, evidence suggests that some of the recent fire seasons are unprecedented compared to historical time periods (i.e., prior to 1880) in certain regions and vegetation types that are characterized as FRG5 and shrub-dominated FRG4 (Table 11.1). For example, analyses of lake sediment cores suggest that recent burning in the boreal forests of interior Alaska exceeds that of the past 10,000 years (Kelly et al. 2013). Furthermore, Higuera et al. (2021) found that subalpine forests in the Rocky Mountains are burning more than at any point in the last 2000 years. It is also possible that some of the large areas burned and frequent fire that has been recently observed in parts of the sagebrush-dominated Great Basin and Columbia Plateau are unprecedented, largely driven by a widespread invasion by an exotic annual grass (cheatgrass) that is conducive to rapid fire spread. In other vegetation types representing FRGs 1–3 (Table 11.1), recent fire seasons are not likely unprecedented in terms of annual area burned (Mallek et al. 2013; Haugo et al. 2019).

THE INEVITABLE FIRE

Given that annual area burned has substantially increased in recent decades (Parks and Abatzoglou 2020), it is becoming painfully obvious that we are unable to suppress all fires in perpetuity. Simply put, fire will inevitably occur at some future time in nearly all natural ecosystems across North America. The previously described changes to forest structure and composition resulting from fire exclusion across much of the continent have triggered concerns that contemporary fires are burning in an uncharacteristic manner (i.e., more severely than expected compared to historical conditions). Evidence from some forested regions indicate that this is the case: greater proportions of contemporary fires are burning at stand-replacing severities (i.e., a fire that kills all or most trees) compared to historical reference conditions that were prevalent prior to widespread European American colonization (Haugo et al. 2019; Williams et al. 2023). As a result, we are starting to observe enduring fire-catalyzed shifts from forest to non-forest (Figure 11.5) (Coop et al. 2020). Increasing

FIGURE 11.5 Under the historical fire regime in which fire was not excluded, many forests were fairly open and tree density was low (top). As a result of fire exclusion, many forests changed with more fuels and higher tree density (middle). Forests affected by fire exclusion are less resistant to the inevitable fire (which occurred in 2002) and are at risk of enduring fire-catalyzed conversion to non-forest (bottom). Top and middle photo from Kaufman et al. (2006).

vulnerability to the inevitable fire has the potential to affect wildlife habitat, water quality and quantity, timber production, recreational opportunities, aesthetics, carbon storage, and spiritual renewal.

FIRE ECOLOGY

One of the most fundamental concepts on which all wilderness fire programs are based is that fire historically played an important ecological role in maintaining healthy ecosystems. Moreover, wilderness fire programs recognize that many species are adapted to fire and that some species and ecosystems are in fact fire-dependent.

Fire-adapted species are those that coevolved with fire and have adaptations to either survive fire, quicky reestablish after fire, or use resources created by fire. For example, ponderosa pine, western larch, and giant sequoia trees have thick bark that insulates live tissues from heat and allows them to survive fire. Lodgepole pines are often killed by fire, but their cones protect seeds from fire, which are subsequently released, thereby providing ample seeds for rapid lodgepole pine recovery, as demonstrated after the 1988 fires in Yellowstone National Park. Resprouting is yet another fire adaptation; trembling aspen and many shrub species across North America resprout after fire even though the aboveground vegetation may have been killed.

Fire-dependent species and ecosystems are those that depend on fire for their widespread persistence (Bond and Wilgen 2012). This is often because the previously described fire adaptations allow certain species to outcompete species that are not fire-adapted or are fire-intolerant. Conversely, the opposite is also true in the absence of fire, in that fire-dependent species and ecosystems may be replaced or highly altered by fire-intolerant species. For example, many ponderosa pine and oak forests experienced frequent fire and thus gained a competitive advantage over other fire-intolerant species that occur in mixed conifer communities. Grasslands, prairies, and savannas are also fire-dependent, and in the absence of fire, encroaching fire-intolerant trees and shrubs can substantially alter these ecosystems (Van Auken 2009).

Fire performs several roles in most ecosystems or plant communities across North America. Here, we describe the most important roles of fire.

Fire influences species composition of plant communities in many ways including (a) triggering release of seeds (jack pine, lodgepole pine); (b) stimulating flowering and fruiting; (c) altering seedbeds by reducing litter and humus and exposing bare soil, ash, or thin humus, thus favoring germination and survival (pines, Douglas-fir, giant sequoia, larch, and many other trees, shrubs, and herbs—including some nonnative species in certain systems); (d) stimulating vegetative reproduction of many species when the overstory is killed; (e) reducing competition for moisture, nutrients, heat, and light; (f) selectively eliminating parts of a plant community; and (g) controlling species and age composition for types reproduced by crown fires only (jack pine, lodgepole pine, and chaparral).

Fire interrupts, restarts, and alters succession. Early ecological concepts held that climax communities were inherently more diverse and stable than pioneer communities that follow fire. However, "undisturbed" natural systems apparently did not exist, with fire playing a prominent factor in shaping successional pathways in shrubland, chaparral, grassland, and forest ecosystems. Post-fire changes in vegetation may follow various pathways, depending on species-specific attributes related to reproduction and survival, different levels of fire severity, and the elapsed time between fires (Harvey and Holzman 2014). Fire thus creates an ever-changing, fire-created mix of successional stages, communities, and stand ages in the vegetation mosaic of most wilderness ecosystems. This pattern creates diversity and stabilizes the system as a whole (Hessburg et al. 2000).

For example, the giant sequoia–mixed conifer forests of the Sierra Nevada historically experienced low- or moderate-severity surface fires with mean fire intervals ranging from 7–15 years (FRG1) (Swetnam et al. 2009). These frequent fires prevented shrubs and fire-intolerant trees such as incense cedar from invading and establishing. They scarred, but seldom killed, the larger giant sequoias, and in fact, fires helped maintain this fire-dependent ecosystem through time. At long fire intervals, however, as we are now seeing as a result of the fire deficit, high severity fire is becoming more common, and a recent study showed that 84 percent of large, mature giant sequoia trees died as a result of recent fire events (Shive et al. 2022).

This pattern contrasts sharply with that of jack pine and lodgepole pine in the boreal and Rocky Mountain forests, where the historical regime was primarily crown fire or severe surface fire covering thousands of acres at intervals of 100–300 years (FRG3 and FRG5). Such fires often killed most or all of the trees and opened the closed cones of the pines to reseed the area (Schoennagel et al. 2003). When fire intervals are longer, lodgepole pine forests can be increasingly dominated by spruce or fir (Romme and Knight 1981). Conversely, if a lodgepole pine forest experiences two, short-interval fires, the trees that established after the first fire may not reach reproductive age (i.e., no cones) when the second fire occurs, thereby setting the site on a successional pathway towards a different forest type or non-forest (Turner et al. 2019). It is also worth noting that chance conditions soon after disturbance, as well as long-term climate changes, play an important role in successional change and contribute to heterogeneity in wilderness (Christensen et al. 1989).

Fire influences the mosaic of vegetation type and age class. Periodic fire occurrence results in a vegetation mosaic, a patchwork on the landscape in which the scale of the patches of contrasting age classes, species compositions, successional stages, or vegetation types reflecting past fire size and the kind of fire regime involved (McKenzie et al. 2011). The physiographic base on which fire operates also heavily influences scale. Steep and broken terrain often shows more complex patterns than level, gently rolling, or uniformly graded terrain, because fire and other important environmental factors are more varied in behavior and effects on complex slopes. In regions such as the boreal forest, where large crown fires or severe surface fires are common, individual patches of the mosaic may cover many thousands of acres in dynamic patterns. Although there are basic principles at work, the patterns of disturbance and recovery from fire are uncertain and—like the pieces in a kaleidoscope—are periodically rearranged by fire and succession, and a variety of configurations are possible.

Fire regulates fuel accumulations. In most coniferous forest ecosystems, the production of plant biomass exceeds decomposition for many years following initiation of a new plant-community succession. Although production of all plant biomass increases with temperature, precipitation, and fertility, so does the rate of decomposition. The overall result is that plant materials accumulate throughout an ecosystem in both living and dead trees, shrubs, and grasses, and in rotting organic debris comprising the ground litter and humus, fuel that is periodically reduced by fire.

All of these materials are potential fuel, but it is important to recognize that at any one time, some biomass, like living tree boles, are unlikely to burn. Even buildup of downed dead biomass occurs in an irregular manner as branches and tree boles accumulate on the ground in response to natural causes of mortality and downfall. Although the process is complex—neither circular nor linear—the general trend is toward an increase in fuels with elapsed time following fire (Dodge 1972; Calkin et al. 2015). Then when fire does occur again, all materials combustible under the prevailing weather and fuel moisture conditions may be reduced to ash or volatilized, and the fuel cycle starts again. Some trees and shrubs are killed but not consumed by a given fire; such snags (standing dead tree trunks) and dead shrub skeletons gradually fall and become part of a new ground fuel buildup for a subsequent fire.

Fire influences nutrient cycles and energy flows. In fire-dependent ecosystems, fire is a major factor influencing nutrient cycling and energy flow. The immediate effect of fire is to convert organic matter stored in litter, twigs, leaves, branches, and dead and down wood to ash and charred materials, with some loss to the atmosphere by volatilization and some material remaining incompletely burned. However, a universal model of nutrient cycling is unavailable, as recent research has concluded that reliable estimates of fire's effects on soil weathering, nitrogen fixation and nutrient export remain elusive because of complex interactions among the severity of the fire and post-fire biotic and abiotic factors (Smithwick et al. 2005; Johnson and Turner 2019).

Fire affects wildlife habitat. With the great differences in fire frequencies and severities found in various fire regimes, one would expect fire to have diverse impacts on the broad array of wildlife species and habitats. Direct, fire-induced mortality of wildlife species is rarely documented (Figure 11.6), though it likely occurs during large, high-severity events. Fire's largest effect on species is through changing habitat quantity and quality. Some species benefit from fire and some species are harmed; there are winners and there are losers (or there is no measurable effect) (Fontaine and Kennedy 2012). Similarly, it is important to recognize that habitat is species specific; one can't say whether "habitat" was improved or harmed by fire without specifying the species of interest. Variation in wildlife response to fire-induced habitat changes are generally related to the species of interest, differences in fire severity, fire season, and time since fire, the latter of which recognizes there may be differences in shorter- vs. longer-term effects.

The influence of fire on wildlife and biodiversity remains an active area of research (Jones et al. 2022), some of which is summarized here. In the hardwood forests in the central US and Appalachian region, low-severity fire at intervals of 2–7 years provides diverse understory structure, creates snags (dead trees), and increases browse, forage, and berries, thereby providing habitat and food for songbirds, cavity nesting birds, bats, and deer (Harper et al. 2016). Moving west, Canada lynx in the US Northern Rocky Mountains peaked about 25 years after wildfire (Olson et al. 2023), illustrating that, although fire may harm habitat in the short-run, habitat can be improved when considering longer timeframes. In California, there is strong evidence that pyrodiversity (spatial and temporal variation in fire severity) results in high bird diversity because the post-fire landscapes exhibit a variety of habitat types (Tingley et al. 2016).

FIGURE 11.6 This Columbia Spotted Frog was found in a severely burned patch less than one year after 2017 Lolo Peak Fire. Photo courtesy Sean Parks.

The influence of fires on wildlife habitat can be summarized with several broad principles: (a) quantity and quality of browse are often increased immediately after fires; (b) fires increase quantity or availability of berries and seeds; (c) in boreal forests, fires tend to eliminate some forage plants associated with older stands of timber, such as arboreal lichens; (d) fires may increase populations of surface and wood-boring insects that are important to quail, woodpeckers, and other insectivorous birds, but may decrease populations of other insects and animal parasites; (e) fires impact "cover" by changing the scale and pattern of vegetation mosaics, such as "edge" and diversity of related wildlife habitat through frequency, intensity, and size of burn; and (f) fires interrupt succession and alter plant-species composition and vegetation structure in ways that favor some wildlife species, harm other wildlife species, or have no effect.

As many species adapted to the historical fire regime, coevolved with the historical fire regime, or generally avoid fire (habitat includes moist lowlands), it is not likely that fire will negatively impact biodiversity over broad extents when operating within its historical range of variability. As previously described, however, fire regimes are changing. As a consequence, fire has the potential to negatively impact extant species, as shown in the Sierra Nevada in California where recent large, high-severity fires have reduced habitat for spotted owl and fisher (Steel et al. 2022).

Fire interacts with insects and diseases. At broad extents, fire (or its absence) regulates the total vegetative mosaic and the age structure of individual forest patches that, in turn, influence insect populations. For example, when extensive stands of Engelmann spruce, balsam fir, or lodgepole pine reach maturity, outbreaks of the spruce budworm or mountain pine beetle can kill trees and create fuel concentrations that make large-scale fires possible (Perrakis et al. 2014). Such fires then terminate the outbreak by eliminating the host trees until new stands attain susceptible ages. Stand-replacing fires may thus temporarily remove such pests as spruce budworm or mountain pine beetle, or plant parasites such as dwarf mistletoe. Thus, attempts to reduce disturbance by fire may lead to situations where large areas of forests are more susceptible to increasingly large and severe disturbances by insects, fire, or other factors. As a result of climate change, insect outbreaks are becoming more common and some insect species are spreading beyond historic ranges (i.e., expanding to higher elevations and latitudes) (Bentz et al. 2010).

In cases when fire does not kill trees, the damage caused by fire can weaken trees and make them more susceptible to pathogen- or fungus-induced decay, beetle attacks, and dwarf mistletoe, thereby resulting in more fuel that can contribute to high-severity fire (Parker et al. 2006). Conversely, however, ponderosa pine may actually increase its defenses to pine beetle attack after low severity fire (Hood et al. 2015).

Although interactions among fire, insects, and diseases are complicated and contextual, historical disturbance regimes from fire, native pathogens, and insects have functioned in a balanced manner across the landscape in many North American forests to create ecosystems that were resilient to perturbation (Hessburg et al. 2000). As is common theme in this chapter, this resilience is threatened by fire exclusion and climate change.

Fire affects soils, water yields, and water quality. Fire can affect the physical and chemical properties of soils, and because fire may change vegetation cover over large areas, it can also affect the rate and extent of surface water runoff and thereby the amount of minerals, contaminants, and sediment deposited in waterways. As with many factors discussed in this section, fire severity plays a prominent role. Low-severity fire's effect on soil characteristics, runoff quantity (i.e., streamflows), and water quality is minimal when compared to high-severity, stand-replacing fire events and even to other land use types. Overall, low-severity fire is a beneficial and low-impact disturbance that likely improves fire-adapted forest health (Klimas et al. 2020).

High-severity fire, on the other hand, can have substantial impacts on soil and water, the majority of which can be attributed to two processes, the first of which is soil destabilization. When high-severity fire kills all or most vegetation, soil on steep slopes can become highly unstable as rainfall is no longer intercepted by grasses, forbs, shrubs, and trees (increasing runoff) and as roots decay and can no longer serve to hold the soil in place (resulting in soil erosion) (Neary et al. 2012).

The second major process that affects water and soil is hydrophobicity (i.e., water repellency). Extreme heat associated with high severity fire consumes all or most organic matter and "bakes" the soil, making it hydrophobic. Hydrophobic soils repel water so it cannot infiltrate, thereby resulting in increased runoff and the potential for severe soil erosion on steep slopes and downstream flooding (Doerr et al. 2000). The timing, amount, and intensity of post-fire precipitation is a key factor affecting both processes (unstable soil and hydrophobicity).

FIRE BEHAVIOR

Over the past several decades, we have improved our capability to model and predict the behavior of wildland fire (Scott and Burgan 2005). However, fire models are not enough; personal knowledge of fire behavior and actual fire experience are still required to ensure valid input of data and interpretation of the output.

IGNITION AND FACTORS INFLUENCING FIRE SPREAD

Key fire behavior factors that determine whether a fire will ignite and spread are fuels, weather, and topography. Overall, most fires are fairly small, with a modest percent of fires being responsible for the majority of area burned. For example, the largest 3 to 5 percent of fires are responsible for 95 to 97 percent of total area burned (Stocks et al. 2002, Short 2014).

Ignition Source

A major natural ignition source is lightning. If thunderstorms are not accompanied by heavy rains, fires can result from lightning strikes in snags, trees with dry rot, flammable crowns, or ignition of surface fuels (Figure 11.7). Humans and infrastructure are the other ignition source in and in proximity to wilderness along trails, boundary roads, highways, railroads, power lines, and overall areas of high human use. In the U.S., the majority of fires (85 percent) are started by humans (Short 2014). However, lightning-ignited fires are responsible for 60 percent of the area burned. These sorts of statistics vary across regions and ecosystems. In Montana, for example, lightning ignitions are responsible for 43 percent of fires but over 80 percent of area burned, and in California, lightning ignitions are responsible for 12 percent of fires but 29 percent of area burned (Short 2014). Considering the number of fires without also considering area burned can provide a misleading and incomplete understanding of the cause and consequences of lightning- vs. human-ignited fire.

FIGURE 11.7 Lightning is the primary source of ignition in wilderness ecosystems. Photo courtesy John Cossett.

Historically, Native Americans commonly burned much of what is wilderness today (Kipfmueller et al. 2021). Indigenous peoples burned for many reasons, including signaling, hunting, managing forage and animal populations, managing vegetation, maintaining habitat diversity, protecting villages from fire, and reducing pest populations (Kimmerer and Lake 2001).

Fuels

The accumulation of some minimum level of smaller fuels, sufficiently dried, and properly arranged is critical to ignition and spread of fire. Even heavy fuel loads may not burn when exposed to a moving, self-perpetuating fire if they are discontinuous, poorly arranged in terms of heat transfer, or too moist. The chemistry of certain fuels such as chaparral, however, renders them more flammable when living than when dead and dry. In forests, unlike chaparral, crown fires usually require adequate surface fuels beneath the stand and some ladder fuels to carry fire up into the tree crowns.

Weather

Suitable burning weather is characterized by one or more factors that promote drying of available fuels: (1) a precipitation-free period long enough to reduce the moisture content of fine- to medium-sized fuels to the critical level that supports fire, (2) sufficiently high temperatures, and (3) low relative humidity. In level terrain, some wind is generally required to make wildfires spread rapidly. In mountainous terrain, fires create their own drafts uphill. In certain chaparral areas, like southern California, dry, warm Santa Ana or Foehn winds are important contributors to fire weather. Regardless of terrain, major fire spread events are usually associated with hot, dry, and windy conditions. "Fire weather" is a term that refers to weather conditions that are warmer, drier, and windier than average.

Topography/Landscape Features

Fire spread is also related to aspect, slope, and geology of local landforms. Fire tends to move upslope. Its movements are particularly influenced by aspect and steepness of slopes. The steeper south- and southwest-facing slopes in the northern hemisphere (winch have higher solar radiation and are drier) and infertile or dry landforms are particularly fire-prone. Dense forests on moist and fertile sites will burn only after long drying periods. Other landscape features that limit fire spread include nonflammable areas, such as lakes, rivers, swamps, barren rock areas, snowfields, timber lines, and recently burned areas where fuels are lacking.

SEASONAL AND CLIMATIC FACTORS

Wildfires most often spreads in the dry season when fuels are sufficiently dry to burn. The annual climatic regime has much to do with the expected occurrence of wildland fires and, therefore, the recommended timing of prescribed burning or wildfires that might be allowed to burn. Many North American wilderness areas at higher elevations have snow cover from early fall through late spring or early summer. Except for this common element, seasonal patterns in North America vary significantly from one region to another and can be broadly categorized into one of three major types: (1) wet winter and dry summers, (2) dry winters and wet summers, and (3) more complex patterns.

1. Wet winters and dry summers characterize the Mediterranean climates of the Pacific coast and the Cascade Range and Sierra Nevada from Washington to Baja California. Much of the total annual precipitation in the mountains is snow. Winters are not very cold; summers are clear, warm, and dry. Occasional summer thunderstorms occur, and they carry little rainfall; hence mid- to late-summer is the fire season.
2. Dry winters and wet summers are typical of the Northeast, the Great Lakes, the Midwest, and much of the Canadian boreal forest region. Summers have frequent frontal rainstorms, and thunderstorms are common from spring through fall, but particularly in midsummer. Enough rain to extinguish lightning ignitions accompanies most storms. Some dry storms, however, occur in most years, and prolonged summer droughts occur at intervals of five to thirty years. In the southwestern U.S. (primarily Arizona and New Mexico but extending into southern Colorado and Utah), the spring and early summer are quite dry, and fires are active during this time period; these fires are often extinguished by the monsoonal rains that arrive in early July.
3. More complex patterns are found in the Intermountain West, the Southeast, and Gulf states. Characteristics of both the Pacific maritime and continental climates are found from interior British Columbia south through eastern Washington and Oregon, all of Idaho, and extreme western Montana to Utah, and Nevada. Winter snowfall and snowpacks are minimal to heavy, and occasional periods of frontal summer rainfall occur. Extended summer dry periods are frequent, however, and dry thunderstorms are common. Lightning occurrence decreases northward. The southeastern states and the Gulf Coast have considerable maritime precipitation derived from the Gulf of Mexico and the Atlantic Ocean. Annual precipitation is heavy and well distributed over the year, but temperatures are high, and even short droughts create conditions conducive to burning. Winter fires are possible in most areas because vegetation is cured, and snow cover is usually lacking. Although thunderstorms are frequent, most (but not all) lightning ignitions are extinguished by rainfall. The vegetation of many areas is not very flammable except for the southern pine regions of the Coastal Plain and Piedmont.

MAJOR DROUGHT EPISODES

Stand-replacing fires evidently occurs during infrequent major regional or subcontinental droughts (Brunelle and Anderson 2003). Although the extensive Greater Yellowstone Area fires of 1988 seem to fall in this category (Christensen et al. 1989), as may other recent years in which we've experienced extensive area burned across North America, stand-replacing fire is normal in certain vegetation types, as previously described (FRG5; Table 11.1). Elsewhere, it is clear from studies of fire-scarred pines, Douglas-fir, and giant sequoia that surface fires burned at closely spaced intervals, but during major drought years, were regionally widespread (Falk et al. 2011). Certainly not all these fires occurred in only major drought years, but studies generally indicate that more large and intense fires in these areas may have coincided with such droughts. In 1748, for example, over 40 percent of the fire-scar record in Arizona and New Mexico recorded fire (remember that these trees lived to record the fire, indicating that the fires were not stand-replacing) (Swetnam and Betancourt 1990).

CLIMATE CHANGE AND FIRE

It is evident that warmer and drier fire seasons contribute to increases in annual area burned (Parks and Abatzoglou 2020). By logical extension, as climate becomes warmer and drier, we can expect increases in annual area burned under climate change (Abatzoglou and Williams 2016). An emerging body of scientific literature predicts an increase in the occurrence of fire-conducive weather and more annual area burned under a warming climate (Wang et al. 2015, Coop et al. 2022). As climate change interacts with excessive fuel accumulation resulting from the fire deficit, we may see an increased prevalence of fire-catalyzed conversions from forest to non-forest (Parks et al. 2019).

WILDERNESS FIRE MANAGEMENT

Considering the important roles that fire historically played in various wilderness ecosystems and that the impacts of fire exclusion have substantially impacted short-fire-interval systems over the past 100+ years, it is not surprising that restoring fire and letting fire play a "natural" role is an objective of most wilderness fire management programs. The basic objective can usually be stated as follows: "To restore fire to its natural role in the ecosystem to the maximum extent consistent with safety of persons, property, and other resources." Note that the goal is not to produce any specific mix of vegetation types, to create desirable wildlife habitat, to reduce fuels, to improve aesthetics, or to attain related specific benefits. Some of these benefits might accrue from a successful program, but the objective is to "restore the naturalness of the environment and let natural processes take over" to the maximum extent possible.

In large national parks and wilderness areas, the objectives also include perpetuating landscapes and landscape processes. Said another way, managers want to aim for a moving-picture vignette made up of a range of spatial patches that would have existed had we not interfered with the processes responsible for creating and maintaining such patches. The range of patches and relative importance of different disturbance processes should be regarded as variables likely to change in the future. Similarly, in national forest wilderness, the objective is to use fire to create desired future conditions in wilderness. These ideas are embraced in the concept of historical range of variation (HRV) that has emerged in importance in the last few decades: the notion that ecosystems may vary in structure and appearance over time, but they are still considered natural if they remain within their HRV or even their future range of variation (FRV). Fire is a key determinant of the HRV and FRV.

AGENCY FIRE POLICY AND OBJECTIVES IN WILDERNESS

Despite up to five decades of effort to restore fire in wilderness, the accomplishments lag far behind the acreage that would have been burned historically with no intervention by modern society. Although some good fire management plans have been implemented in larger wilderness areas, most wilderness areas are small and do not have fire plans. Every bad fire year or incident, such as the Yellowstone fires of 1988, temporarily tightens fire policies in the direction of suppression (Parsons and Landres 1998).

Aplet (2006) observes that while there are many excellent managers throughout the federal land-management agencies who support wilderness fire, some managers still struggle against three types of barriers: risk-averse agency administrators and a system that has historically suppressed fires; institutional complex planning and management requirements as well as the liability and responsibility when operating as the fire manager; and political and public concerns about the nuisance of smoke and fire, the fear for human health and welfare, and concern about the loss of structures and other economic losses. Similar barriers have been highlighted elsewhere (Doane et al. 2006, Black et al. 2008).

The naming convention of non-suppressed fires (i.e., those allowed to burn without suppression activities) has changed over the decades. The original term, "let burn" was later supplanted by "prescribed natural fire," "wildland fire use," and "resource objective wildfire" (Huffman et al. 2020). In 2009, however, the Guidance for Implementation of Federal Wildland Fire Policy was released to ensure a consistent approach to managing wildfire across agencies. The policy allows wildfire to be managed for resource benefit on public land in one area of the fire while actively suppressing it in another area where people, homes, infrastructure, or other values (e.g., ecologically important habitats, historically significant cultural sites) are potentially threatened. This policy offers a new term, "managed fire" and provides greater flexibility in managing lightning-ignited fires in wilderness and elsewhere. Although this policy increases flexibility in managing fire, some scientists have lamented the change because tracking fire management type and intent through time has become more difficult with this new policy and terminology. For the purpose of this chapter, we will revert to "prescribed natural fires" when referring to lightning-ignited fires that are primarily allowed to burn in the absence of aggressive fire suppression. These fires are generally allowed to burn under prescribed conditions. See this video about prescribed natural fires and wilderness fire management: https://www.nrfirescience.org/resource/21557.

The largest wilderness areas in the contiguous U.S. are managed by the USFS and the NPS. Wilderness fire management in these two agencies has changed tremendously in the past five decades toward allowing lightning-ignited fire to play a more natural, historic role in wilderness ecosystems—when conditions are right in terms of location within the wilderness, time of year, and expected fire weather (i.e., prescribed conditions). Sometimes the prescribed conditions are narrow.

EVOLUTION OF WILDERNESS FIRE MANAGEMENT

In 1968, the NPS modified its policy of immediate suppression of all fires, which had been standard practice since the NPS was created. This policy change was a response to the Leopold Report of 1963 (Leopold 1963) and to ample ecological research highlighting the ecological importance and historical relevance of fire. Despite the fact that the Leopold Report can be interpreted as seeing park ecosystems as static rather than dynamic entities, it was the first document to explicitly call for an ecologically based management philosophy: "Above all other policies, the maintenance of naturalness should prevail" (Leopold 1963).

In the early 1970s, the USFS also began allowing lightning-caused fires to play a more natural role in wilderness, an exception to its policy of suppressing all fires by 10 AM the following day. In fact, the first prescribed natural fire in the White Cap Fire Management Area (in the Selway-Bitterroot Wilderness) ignited on the same day that the Chief of the Forest Service signed off on the plan in 1972.

These policy changes were not immediately applied to all NPS and USFS wilderness units. For NPS wilderness units, the policy to allow prescribed natural fires was first implemented in Sequoia and Kings Canyon National Parks (California) in 1968, followed by Saguaro National Park (Arizona) in 1971 and Yosemite National Park (California) in 1972. For USFS wilderness units, the policy was first implemented in the Selway-Bitterroot Wilderness in 1972 followed by Gila Wilderness (New Mexico) in 1975. The specific policies have gradually changed over the decades and expanded to some other NPS and USFS wilderness areas as they developed their fire management plans, noting that any policy allowing for prescribed natural fires does not universally apply across all wilderness units in either agency.

As previously mentioned, exceptional individual or regional fire events gave agencies pause in terms of allowing prescribed natural fires and the risk they pose. Some of the fires that occurred in Yellowstone National Park and its vicinity in 1988 were initially allowed to burn as prescribed natural fires. During the same year, the Canyon Creek fire in the Scapegoat Wilderness (Montana) was initially managed as a prescribed natural fire. These fires defied expectations as the summer progressed and became very large, threatening infrastructure and communities. As a result, the Secretaries of the Interior and Agriculture convened a policy review team to evaluate agency fire policies. Although the team concluded that fire's ecological role is of fundamental importance, changes to the fire management plans were recommended. Ultimately, the use of prescribed natural fire was paused across NPS and USFS wilderness (van Wagtendonk 2007).

Starting in 1990, fire management plans allowing prescribed natural fires were beginning to come back online across the agencies, but the process was slow across many wilderness units. A devastating fire, the South Canyon Fire, then occurred in 1994 on BLM land in Colorado that resulted in the unfortunate deaths of fourteen firefighters. This incident prompted a comprehensive review and update of the federal wildland fire policy. The resulting report stated that, in regard to prescribed natural fires, "Wildland fire will be used to protect, maintain, and enhance resources and, as nearly as possible, be allowed to function in its natural ecological role." The report also reiterated that firefighter safety and public safety are the highest priorities when managing prescribed natural fires (van Wagtendonk 2007).

The Secretaries of the Interior and Agriculture reconvened the interagency federal wildland fire policy review working group after the 2000 Cerro Grande Fire. Initially a prescribed fire in Bandelier National Monument (New Mexico), the Cerro Grande escaped and destroyed over 200 homes in the town of Los Alamos and threatened the Los Alamos National Laboratory. The working group recommend several changes to the 1995 policy to clarify purpose and intent, resulting in an updated 2001 policy, which in subsequently led to an interagency implementation guide pertaining to prescribed natural fires (USDA and USDOI 2005). As previously mentioned, the 2009 Guidance for Implementation of [the 2001] Federal Wildland Fire Policy allows for flexibility in managing wildland fires, in which some portions of the fire can be managed for resource benefit (i.e., limited suppression) and other portions of the fire can be aggressively suppressed.

EXEMPLARY WILDERNESS FIRE MANAGEMENT PROGRAMS

Wilderness fire management programs and policies were extended to many other parks and wilderness areas with fire-dependent ecosystems in the 1980s and 1990s. However, the actual implementation of prescribed natural fires is fairly infrequent because either the prescribed conditions are too narrow, or the managing official is not comfortable with the risk (Seielstad 2015). Simply put, lightning-ignited fire is generally suppressed in national parks and wilderness areas, with some notable exceptions described below. Although not described below, the wilderness fire management plans of the Frank Church–River of No Return Wilderness, Saguaro National Park, and National Parks and USFS wilderness areas in the Greater Yellowstone Ecosystem are also considered exemplary.

Selway-Bitterroot Wilderness, Idaho and Montana

The Selway-Bitterroot Wilderness (SBW) is managed by the USFS and is one of the largest units (1,350,000 acres) of the NWPS in the contiguous U.S. The SBW is the first USFS unit having operational experience with a sizable wilderness fire management program. The White Cap fire management area in the Selway-Bitterroot in 1972 became the first approved exception to the USFS 10 AM (total suppression) policy.

The first major test of the White Cap fire management plan was the 1,200-acre 1973 Fitz Creek Fire (Berkey et al. 2021), although the fire had to be suppressed in one area where it had escaped the approved fire management area. The Fitz Creek Fire boosted managers' confidence in managing prescribed natural fires and, overall, the experiment was successful and presaged the incorporation of fire management considerations into wilderness planning throughout the USFS.

Between 1974 and 1979, additional fire management plans were developed for other portions of the SBW. The 1979

Independence Fire burned more than 16,000 acres during a three-month period, the largest pre-1988 fire allowed to burn under a wilderness fire management program by any agency. Fire managers who worked on the Independence Fire gained an understanding that fire is unpredictable, and that wilderness fire managers must both accept this risk and minimize it to the greatest extent possible with the tools available (Berkey et al. 2021).

Fire seasons were fairly mild until 1988 in the SBW when 53,000 acres burned (Figure 11.8). As 1988 was an exceptional fire year in SBW and elsewhere (i.e., the Yellowstone fires), public reaction to the 1988 fire season led to greater hesitancy around wilderness fire management on both USFS and NPS units. Until the mid-1990s, managers often labeled a fire as a suppression fire but took no direct management action, essentially treating it as a prescribed natural fire (Berkey et al. 2021). There have been three years in which area burned exceeded that of 1988 in SBW: 2007, 2012, and 2017. In particular, 2012 was exceptional in that over 150,000 acres burned in SBW.

As Berkey et al. (2021) note, the extensive experience with wilderness fire management in SBW has created a culture where the natural process of fire is valued, both as a land management tool and as an ecological process. Fire suppression is often viewed as a missed opportunity to capitalize on the ecological role of fire. The rich history of wilderness fire in the SBW has made it an important laboratory, both for studying the effects of fire and for identifying best practices for managing fire (Berkey et al. 2021).

Bob Marshall Wilderness Complex, Montana

The Bob Marshall Wilderness Complex (BMWC) is over 1.5 million acres in size and comprises three wilderness units managed by the USFS: the Great Bear Wilderness, Bob Marshall Wilderness, and Scapegoat Wilderness. Building on momentum and success in managing prescribed natural fires in the SBW, a wilderness fire management plan was adopted for a small portion of the BMWC in 1978; similar plans were adopted in another portion of BMWC in 1981 (Berkey et al. 2021). The 1981 Cigarette Creek Fire was the first prescribed natural fire in the BMWC; this fire received close ground and air monitoring and grew to 250 acres in size.

The wilderness fire management plan was expanded to the remainder of the BMWC in 1983. The 1984 Lodgepole Fire was the next test of the BMWC fire management plan; it grew to only 80 acres. However, the most formative fire for the budding wilderness fire management in the BMWC was the 1985 Charlotte Peak Fire (Berkey et al. 2021). Allowing this fire to burn was controversial, and it eventually grew to 5,500 acres in size. For the next couple of years, most fires were suppressed.

In 1988, however, two fires in the BMWC, the Gates Park Fire and the Canyon Creek Fire, were both managed, at least initially, as prescribed natural fires. The Gates Park Fire was generally considered a wilderness fire management success even though it burned over 42,000 acres. Although it took some suppression efforts on one of the fire's fronts, the fire stayed within the wilderness (Berkey et al. 2021).

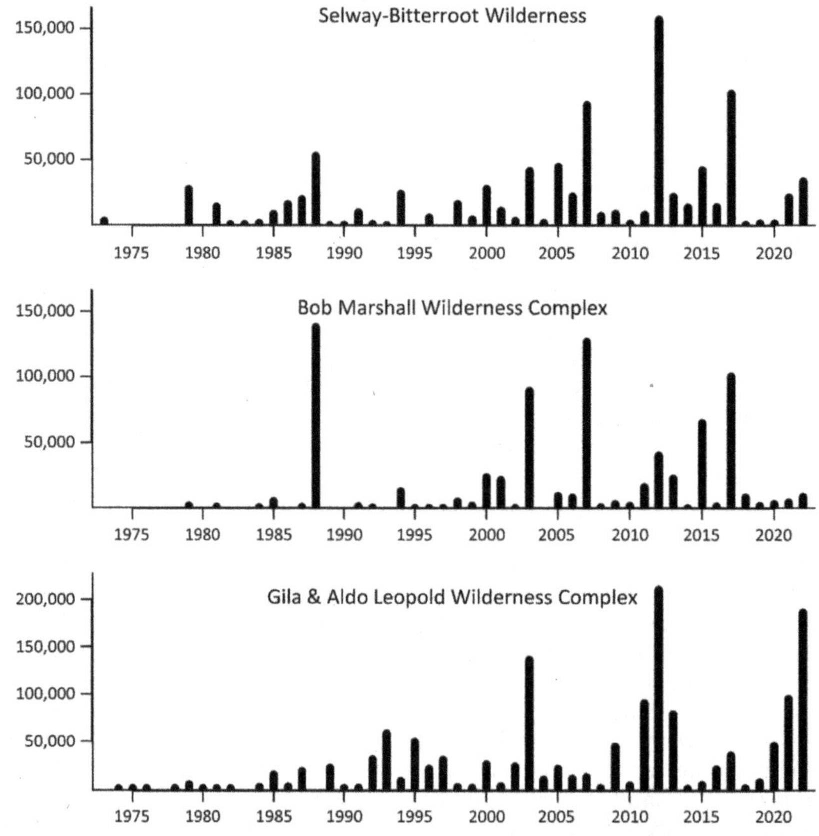

FIGURE 11.8 Annual area burned (acres) for three areas with exemplary wilderness fire management programs, noting that some fires are still suppressed in these areas.

In contrast, the Canyon Creek Fire was not considered a success because it crossed the wilderness boundary and ultimately threatened the town of Augusta. Incredibly strong winds resulting from a low-level jet stream caused explosive fire growth; the fire ultimately grew to about 200,000 acres. The Canyon Creek Fire was the first prescribed natural fire to escape wilderness boundaries, and many thought that it would mark the end of wilderness fire management (Berkey et al. 2021). As previously described, the 1988 fire season in this region (including the Yellowstone fires) motivated a review of wilderness fire management plans across the USFS and NPS.

As Berkey et al. (2021) note, the 1988 fire season was the first to demonstrate that not all wilderness fires would be small and easily managed. It became clear that large fire seasons, and large-fire events, are normal for the BMWC. Historical documents and newspaper accounts from the late-nineteenth century contained stories of equally large fires, including the well-known fire years of 1889 and 1910. At the time of this writing, 1988 remains the largest year in the BMWC from 1972–2022 (Figure 11.8).

Fire activity in the BMWC remained relatively mild until 2003, when over two dozen fires burned about 90,000 acres. Other large fire years include 2007, 2015, and 2017. With these increases in large fires in the BMWC, maintenance of trails and campsites emerged as a major challenge to wilderness fire management (Berkey et al. 2021).

Also of interest is that management-ignited prescribed fires were implemented in 2003, 2009, and 2011 in the BMWC. The main objective of these prescribed fires was to provide fuel breaks, which then allowed managers to treat lightning-ignitions as prescribed natural fires in the area and reduced the probability that fire would escape the wilderness (Berkey et al. 2021). Although the NPS regularly implements prescribed fire in wilderness, there are only a few examples of prescribed fire in USFS wilderness.

Gila and Aldo Leopold Wilderness Complex (GALWC), New Mexico

This wilderness complex comprises the Gila and Aldo Leopold Wilderness areas and is over 550,000 acres in size. The GALWC is arguably the best example of an unrestrained fire regime in dry conifer forests in the western U.S. Wilderness fire management officially started in 1975 in the Gila Wilderness under fairly narrow prescribed conditions (Hunter et al. 2014). Area burned was relatively low in the early years (1975–1989), burning about 63,000 acres total (Figure 11.8). The largest prescribed natural fire during this time period was the 1989 Shelly Fire that burned over 10,000 acres (Hunter et al. 2014).

Annual area burned started increasing in the early 1990s. By 1998, wilderness fire management plans were updated to allow for prescribed natural fires in the Aldo Leopold Wilderness and even areas outside of the GALWC (Hunter et al. 2014). Between 1990 and 1989, about 260,000 acres burned, with prescribed natural fires being responsible for over 60 percent of area burned (Hunter et al. 2014).

Three exceptional fire years have occurred since 2000 (Figure 11.8). About 135,000 acres burned in 2003, most of which burned in Dry Lakes Complex. About 210,000 acres burned in 2012, nearly all of which was burned by the Whitewater-Baldy Complex, which (at that point in time) was the largest recorded fire in New Mexico history. About 185,000 acres burned in 2022 as a result of the Black fire, the second largest recorded fire in New Mexico history (the largest recorded fire in New Mexico history also burned in 2022, the Calf Canyon/Hermits Peak Fire). These large fires in the GALWC, and also the 2013 Silver Fire, burned over extensive areas that had not burned for 50–100+ years, resulting in undesirable fire effects in some areas. Many of these fires were not prescribed natural fires (they were actively suppressed), but it is another reminder that when the inevitable fire burns through forests affected by fire suppression, undesirable effects are likely. The opposite may also hold, because when these large fires encountered areas that previously burned, the spread and severity of these fires were tempered (Hunter et al. 2014).

Sequoia–Kings Canyon and Yosemite National Parks, California

The first designation of a natural fire management zone in forested western wilderness, and the first breakthrough in prescribed natural fires in the West, occurred in Sequoia and Kings Canyon National Parks in 1968. At this time, lightning-ignited fires were allowed to burn under surveillance unless safety problems or resource damage was anticipated in a special management zone above 8,000 feet in the Middle Fork Kings River drainage. This zone was gradually expanded and now encompasses about 740,000 acres, or roughly 86 percent of these parks. The forests are generally open and subalpine, and much of the terrain is extremely rocky or above timberline. Some 305 fires were allowed to burn within this zone up to 1986, burning about 29,000 acres. Partial suppression was needed for only a few fires, and no serious problems developed (Bancroft et al. 1985).

In 1969, starting in the Redwood Mountain Area of Kings Canyon National Park, prescribed burning to reduce unnatural fuels and invading shade-tolerant understory trees began in the lower-elevation mixed-conifer forests, including some giant sequoia groves (Kilgore and Sando 1975). Initially, some cutting and piling of small trees and litter were used to avoid control problems and unnatural overstory damage. Similar work was done later on a twenty-acre unit of the Mariposa Grove of giant sequoias in nearby Yosemite National Park. Later, more extensive understory prescribed burning in mixed conifers was carried out without significant modification of fuels in both the Yosemite and Sequoia–Kings Canyon areas. This type of burning became part of a large-scale operational program (Bancroft et al. 1985).

In 1972, a natural-fire zone was also established in Yosemite National Park and, by 1982, it included 594,000 acres, or about 78 percent of the park. The judgment was that essentially any fire within this zone could safely be allowed to burn because of prevailing fuel, vegetation, and physiographic factors if specified weather and fuel prescriptions were met.

An additional 57,000 acres were in a conditional fire zone, an area of restricted burning because of the fuel accumulations generated by previous fire exclusion. As these fuels were reduced, parts of the conditional zone became areas in the natural-fire zone. In the remainder of the park, which included most of the park's developments, a full-scale fire-suppression program was in effect, but prescribed burning was used to reduce fuels in critical areas. Wilderness fire management, and its evaluation with research and monitoring, have continued at Sequoia–Kings Canyon and Yosemite National Parks, thereby providing valuable knowledge in support of wilderness fire management programs elsewhere (van Wagtendonk 2007).

WHAT IS "NATURAL" IN WILDERNESS FIRE MANAGEMENT?

The term natural does not yet have a common meaning in wilderness fire management or fire ecology. It can mean (1) fire occurring prior to colonization by European-Americans; (2) fire occurring without human past or present intervention or influence; (3) the attributes of the fire process known or presumed to be an intrinsic part of a given vegetative type; or (4) the role fire played in the evolution of an ecosystem. Wilderness managers are still seeking to clarify whether Native American ignitions should be considered as sources of "natural" fire. Consequently, we strive to avoid the term "natural" fire and instead use the term "lightning-ignited" fire when referring to fires that are, intuitively, ignited by lightning and "historical" fire regime to fires that were historically (prior to European-American colonization) ignited by both lightning and humans.

Nevertheless, philosophical and policy questions remain about the appropriateness of fire-restoration efforts. Should wilderness and park managers (1) simply allow lightning-ignited fires to burn, (2) reduce obvious fuel accumulations in certain zones with prescribed fires and then allow lightning-ignited fires to burn, or (3) carefully restore forest composition and structure to estimated historical conditions before allowing lightning-ignited fires to burn?

Allowing lightning-ignited to burn has a broad appeal, but it has many drawbacks, too. Prior to European-American colonization, many lightning fires originated outside and spread into areas now designated as wilderness. In addition, some lightning fires currently ignited in wilderness will be suppressed. Hence, an array of options need to be carefully evaluated. These options include allowing for lightning-ignited fire to burn where possible and using prescribed fire where lightning-ignited fires cannot be allowed to burn, where ignitions from outside wilderness are no longer possible, and where unnatural fuel accumulations must be reduced.

As we learn more about how fire operates, the definition of naturalness becomes more complex (Cole and Yung 2012). National park and wilderness ecosystems cannot be defined at a particular level that will unequivocally be perceived as "natural," because the word natural involves individual value judgments. Park and wilderness preservation goals will have to be stated in more precise system-component terms, depending on the values represented by the individual area.

Although it may be difficult to define such concepts as "wilderness," "wildness," and "natural," the management objectives for parks and wilderness require that we try. If not, we will have lost contact with the philosophy behind the original establishment of national parks and wilderness. In our approach to such objectives, we need to realize that whatever we learned from past research may be supplanted by newer knowledge. We need to be continually open to the best and latest thinking about the complex role that fire and fire suppression have had on ecosystems.

PRESCRIBED FIRE IN WILDERNESS

Prescribed fire (intentional management-ignited fire) is legally allowed in virtually all wilderness areas, regardless of managing agency, with varying goals of reducing fire hazard, restoring historical structure and habitat, or allowing lightning-ignited fire to return (Figure 11.9) . However, prescribed fire is rarely implemented in wilderness areas managed by the USFS and BLM, likely due to some of the same barriers faced when managing prescribed natural fires (as described earlier in this chapter). Another barrier is that prescribed fire could be considered trammeling and may seem at odds with the Wilderness Act of 1964. This said, our growing understanding of the importance and prevalence of Indigenous burning prior to European colonization (Kipfmueller et al. 2021, Knight et al. 2022) challenges whether we should consider prescribed burning as a trammeling action.

In an era of rapid climate change and after decades of fire exclusion, there is concern that wilderness ecosystems

FIGURE 11.9 Prescribed burns (left) are used in certain wildernesses and parks when lightning fires cannot be allowed to burn because of hazardous accumulations of fuel and changes in forest structure induced by fire suppression. Such prescribed burning may be appropriate in small wildernesses and near wilderness boundaries, where natural fire might threaten commercial developments and private lands (right). Photos courtesy Bruce Kilgore and William Jones, National Park Service.

are losing resilience and the capacity to recover from the inevitable fire, which is increasingly likely to occur under extreme and novel conditions. In this context, tension emerges between a perceived harm to wilderness values brought by the deliberate reintroduction of fire and the likely harm to wilderness ecosystems brought about by continued fire suppression. A thorough analysis of the costs and benefits of prescribed fire in wilderness is needed to understand the complex ways this management tool intersects with wilderness management values. This pressing need was recently expressed at a workshop focused on prescribed fire in wilderness held in December 2022 at Western Colorado University. The workshop synthesis report (https://www.centerforpubliclands.org/s/Wilderness_Rxfire_Synthesis.pdf) and associated StoryMap (https://tinyurl.com/wildernessandfire) are valuable resources.

WILDERNESS FIRE MANAGEMENT ALTERNATIVES AND THEIR CONSEQUENCES

Five broad alternatives are available when managing wilderness fire: (1) attempt to suppress all fires; (2) allow all fires to burn; (3) manage lightning-ignitions as prescribed natural fires; (4) management-ignited prescribed fires; and (5) manipulate vegetation and fuels without fire. Although these options can be combined, and often are, it is useful to describe each separately. Failure to pursue a specific policy may result in the unintended or haphazard implementation of one or more of these options.

ALTERNATIVE 1: ATTEMPT TO SUPPRESS ALL FIRES (FIRE EXCLUSION)

This alternative requires an immediate, aggressive attempt to suppress all fires, regardless of cause, location, or expected outcomes. It might fail in application, but if the policy requires prompt suppression, without exceptions, then the objective is to exclude fire from the ecosystem. Until the early 1970s, this was the standard policy in most wildernesses and national parks, and it remains the policy in many. At the very least, it is often defensible as a holding action until the expertise and equipment to implement a different alternative is available, or until a judgment concerning the best alternative can be made. Growing evidence suggests, however, that a strict policy of fire exclusion acts as a powerful form of vegetation manipulation in fire-dependent ecosystems.

If this alternative succeeds, there will be major changes in both vegetation and wildlife populations and perhaps also major changes in the productivity, diversity, and stability of the ecosystem through time. Often, excluding fire is only temporary (i.e., the inevitable fire), and thus, this alternative will eventually fail, and the deleterious effects will become obvious and long-lasting.

When fire is excluded, accumulation of fuels and changes in vegetation structure may occur. In many ecosystems, fire was one of nature's ways of reducing fuels and recycling biomass and nutrients. With both lightning and humans as sources of ignition, one wonders if fire exclusion is really a viable option in such ecosystems. By following this alternative, we may be setting the stage for major high-intensity fires that could not only endanger human life and property but also disrupt the very ecosystems we are purportedly protecting.

Fire exclusion often requires the use of bulldozers or other heavy line-building equipment, aircraft, smoke jumpers, and fire retardant. When major fires in wilderness threaten lives or property outside the wilderness, managers are tempted to use all available techniques, regardless of the consequences to natural resources. Fire exclusion, if practiced too long, can force managers to inflict major damage on the landscape in the name of saving the wilderness from fire. For example, tractor-built or hand-constructed fire lines can cause erosion, leading to lasting environmental damage. Old trails and temporary roads opened for access to the fire can also erode and can change subsequent access patterns to and in the area. Fire camps can leave impacts and fire retardant can pollute streams. Unless fire is not a significant factor in the ecosystem, excluding fire except as an interim measure could cause irreparable ecosystem harm. Aggressive fire suppression is undoubtedly the most expensive fire management policy. When feasible, other alternatives could reduce costs of fire management.

Instead of an aggressive suppression alternative, some wilderness fires are managed using the "confine and contain" tactic. Confine means to restrict a fire within predetermined boundaries; contain means to surround the fire with a fire line to limit its spread beyond these lines. Some fairly creative efforts at minimizing impacts on wilderness resources can be developed under this tactic, including careful observation and monitoring of the fire's progress to ensure that it stops at reasonable natural boundaries.

ALTERNATIVE 2: ALLOW ALL FIRES TO BURN

This alternative allows all fires to burn, regardless of ignition source, and is the opposite of the fire exclusion alternative. This alternative may endanger human life and property—including lives and property beyond wilderness boundaries. A high priority goal of natural resource management is to provide for the safety of persons and property—both outside the wilderness and for visitors and agency personnel within the area. Therefore, this alternative is usually rejected except in the most remote and large wilderness areas and those with exemplary wilderness fire management programs, as previously described.

ALTERNATIVE 3: MANAGE LIGHTNING-IGNITIONS AS PRESCRIBED NATURAL FIRES

This alternative involves managing lightning-ignited fires under prescribed conditions that depend on ignition location, time of year, and expected fire weather. This alternative rests on the belief that lightning-ignited fires were historically prevalent in wilderness ecosystems and maintain or create desirable forest composition and structure. This alternative aims to restore lightning-ignited fires in wilderness, to the greatest extent possible, while protecting lives and property and preventing unnatural fuel accumulations that, if ignited, might even damage wilderness resources. This alternative can use the previously described "confine and contain" tactic and involves selective fire control based on both safety and ecological considerations.

ALTERNATIVE 4: IGNITE PRESCRIBED FIRES

The goal of this alternative is to restore the historical fire regime by substituting deliberate ignitions for ignitions historically ignited by lightning or Native Americans. The commonly held assumption that the ecological effects of fire will be the same, whether human- or lightning-ignited, is not totally valid. Although it might be possible through skillful and well-planned ignitions to create significant burns that closely resemble lightning-ignited fires, it might also be possible to burn in seasons when lightning never occurs or at closer than natural intervals.

This said, it is well-recognized that the historical fire regime in many areas included fires ignited by humans, and such fires did not always coincide with the driest time of year or in seasons when lightning was common (Greenwood et al. 2022). As best as possible, knowledge of cultural burning practices should be considered when trying to replicate historical fire regimes when igniting prescribed fires.

The main advantage of this alternative is that fires can be managed best if the time and place of ignition are selected in advance, thus allowing time to ready personnel and equipment, monitor weather forecasts, assess fuels, and prepare control lines. Many situations exist—such as small management unit size and proximity to wilderness boundaries—where lightning-ignited fires are generally suppressed. Prescribed fire then becomes a reasonable substitute (Figure 11.9).

ALTERNATIVE 5: MANIPULATE VEGETATION AND FUELS WITHOUT FIRE

This alternative manipulates vegetation structure and type with mechanical or hand harvest and may even involve planting trees or other plants in certain situations. The alternative rejects fire, both lightning-ignited and prescribed, as an unacceptable or unsafe agent of change. Safety is usually given as the reason for favoring this option that may be best suited to areas immediately adjacent to park or wilderness infrastructure (e.g., national park visitor center or campground) or when a wilderness borders urban areas. Vegetation removal need not be commercially motivated, and in fact, it would usually be controversial in a wilderness or park to sell timber.

Although this alternative is probably best suited for areas adjacent to infrastructure or near the park/wilderness-urban boundary, it could be a viable option for restoring vegetation previously altered by commercial logging and by fire exclusion elsewhere. This is particularly the case when the intent of the vegetation manipulation is to then allow lightning-ignited or prescribed fire to occur in a manner consistent with the historical fire regime.

ALTERNATIVE 6: COMBINED ALTERNATIVES

Various combinations of the wilderness fire alternatives are possible, but the most practical combination—one now being used in several parks and wilderness areas—involves (a) allowing lightning-ignited fires to assume their ecological role as much as possible, (b) using prescribed fire where it is impractical to allow lightning-ignited fires to burn or where past suppression requires a reduction in unnatural fuel accumulations, and (c) suppressing fires near wilderness boundaries where valuable resources outside the area are at risk or if the ignition occurs under conditions not prescribed by the fire management plan.

UNANSWERED QUESTIONS AND CHALLENGES

Wilderness fire management requires that we continually learn from the past, properly follow policy, and adjust to changing climatic and vegetation conditions (Miller et al. 2011). The panel of scientists assessing the ecological consequences of the 1988 fires in the Greater Yellowstone Area pointed out that it was relatively easy to set broad objectives to protect a particular national park or wilderness landscape; however, it is far more difficult to develop objectives to preserve and protect particular natural processes. The next generation of managers will need to rewrite ideals into management objectives (Cole and Yung 2012) and answer the questions: What kind of intervention is acceptable? By what means? For what purposes? On what scale? According to what social and political processes?

Four questions related to managing wilderness fire must be faced: (1) Under what conditions are prescribed natural fire and management-ignited prescribed fire appropriate in parks and wildernesses? (2) Do we need to simulate the historic role of Native American burning in certain wilderness units? (3) How do we deal with the natural role of high-intensity, stand-replacing fires characteristic of many northern and high-elevation wilderness ecosystems? (4) If lightning-ignited fires are part of a given unit's fire plan, what scale of fire (patch size) is acceptable?

Lacking evidence on the importance of Native American burning and recognizing the difficulties of simulating it, the NPS has historically minimized such ignitions in management plans for parks in the Sierra Nevada (Parsons et al. 1986). However, as recent research has revealed the prevalence and importance of Indigenous burning prior to European-American colonization (Lake et al. 2017; Kipfmueller et al. 2021; Knight et al. 2022), perhaps this will change in future years. However, a major problem in trying to simulate Native American burning and in restoring fire in general is that fuel conditions have changed as a result of decades of fire suppression. Fuel has built up, with one consequence being that area burned in large, severe fires has increased in recent decades (Parks and Abatzoglou 2020). In vegetation types historically characterized by frequent low- or mixed-severity fire (FRG 1 and FRG 3; Table 11.1), contemporary high-severity fires now pose one of the most difficult park and wilderness policy and program dilemmas for the future.

WILDERNESS FIRE MANAGEMENT PLANNING: CONSIDERATIONS AND CONSTRAINTS

Wilderness fire management planning aims to provide a guide for all fire management actions within designated wilderness or areas with similar protections. Such planning involves determining the appropriate response to lightning-ignited fires and the use of management-ignited prescribed fires to accomplish management objectives. Important considerations include fire and ecosystem interactions, including fire history and fire effects, as well as special resource and use considerations, such as archeology, rare and endangered species, and administrative sites (Hunter et al. 2014).

Managers need a clear understanding of the historical fire regime in the area of concern for developing and evaluating a park and wilderness fire management plan. However, there is some question about whether a given historical fire regime would be appropriate in the future without allowance for the considerable yet difficult-to-predict variations that would occur under climate change.

In places like Sequoia and Kings Canyon National Parks, fire management is now focused primarily on restoring fire as an ecological process rather than restoring precise vegetation structure. Given our limited and imperfect knowledge about vegetation structure, fire regimes, the role of Native American burning, and the effects of climate change, the NPS has decided it can best recreate and maintain an ecosystem in which fire can function fairly naturally by using a strategy of minimum intervention.

At the same time, every effort is being made to define the term natural (Cole and Yung 2012), as it is one of the key characteristics of the Wilderness Act of 1964. As such, carefully monitoring the behavior and impact of lightning-ignited and prescribed fires is sometimes necessary. Continually re-evaluating both objectives and methods of wilderness fire management helps park and wilderness managers strive to achieve a natural wilderness character.

Modern management tools can help; computer mapping of fuels and vegetation with the use of GIS and remote-sensing tools help managers understand the landscape dimension of the potential fire regimes. Computer simulation of historical fire patterns, intensities, and frequencies can be used along with additional fire-history studies to obtain a better understanding of the historical range of predictable fire intensities (Keane et al. 2019).

ECONOMIC CONSIDERATIONS

Wilderness fire management programs have been planned and evaluated largely based on ecological factors rather than on cost effectiveness. Yet one of the arguments for allowing lightning-ignited fires to burn in parks and wilderness is that society will save costs of suppressing fires and gain positive ecological values by perpetuating wilderness character. Although identifying dollar value returned for dollar value spent in various wilderness fire programs is difficult, it is abundantly clear that prescribed natural fires cost substantially less than suppression fires (Dale et al. 2005). Although highly variable, costs of aggressive fire suppression average from $250 to $1000/acre (Lynch 2004, Gebert et al. 2007) and have continued to increase in recent fire years. In contrast, prescribed natural fires (an important wilderness fire management alternative) average $43/acre (Dale 2006).

INTERAGENCY PLANNING

One of the most important advances in wilderness fire planning has been the movement toward joint efforts across agency boundaries (Desmond 1994). This began in 1976 with the Interagency Wilderness Fire Management Program, involving one-half million acres of the Teton Wilderness in Wyoming (Bridger-Teton National Forest) adjacent to Yellowstone National Park. For the first time on a planned basis, fire on NPS lands would be allowed to cross the boundary onto USFS lands, and fires from the Teton Wilderness would be allowed to cross into Yellowstone National Park. Cooperative agreements were in effect in 1982 that involved more than 4 million acres on five national forests and two national parks in the Greater Yellowstone region. The 1988 Greater Yellowstone Ecosystem fires were the first test of this concept, and postfire reviews by the Interagency Fire Management Policy Review Team urged that such fire management planning be further strengthened. Subsequent interagency fire planning efforts involved Lassen National Forest and Lassen Volcanic National Park in California, Glacier National Park and Flathead National Forest in Montana, and extensive efforts in Alaska.

These experiences all indicated the importance of the informal as well as the formal aspects of the planning process, involving all levels of the organization to promote feelings of ownership, and getting the public involved early in the planning process. Integrating the planning team with interagency members was also important so that differences in procedures and terminology did not interfere with the overall objective of allowing fire to burn in the combined NPS and USFS wilderness ecosystems in a prescribed, rational pattern. Today, interagency park and wilderness fire planning (including fire suppression) is a widespread, accepted practice wherever park and wilderness jurisdictions adjoin.

In 1999 and 2000, the U.S. Government Accounting Office (USGAO) recommended that the USFS and NPS develop more cohesive strategies for agency response to wildland fire threats and events and then evaluated their program five years later (U.S. Government Accounting Office 2005). The 2005 USGAO report acknowledged that progress had been made by both agencies; however, the report highlighted the important shortcomings and challenges to complete a more cohesive strategy and plan of action to address wildland fires in a timely and cost-effective manner. The 2009 Guidance for Implementation of Federal Wildland Fire Management Policy attempts to address some of the shortcomings, as does the 2014 National Cohesive Wildland Fire Management Strategy (https://www.forestsandrangelands.gov/strategy/).

INSTITUTIONAL AND POLITICAL BARRIERS

As previously mentioned, most fire management plans for parks and wilderness allow for prescribed natural fire. However, they do not require land managers to choose this option. There are several institutional and political barriers that serve to limit the use of prescribed natural fires, which can at least partly explain why fires are frequently suppressed in parks and wilderness.

One of these barriers is the lack of incentives to allow for prescribed natural fire, and in fact, there are strong disincentives. As Aplet (2006) describes, the strongest of these disincentives is personal liability if a prescribed natural fire did not go as expected. In general, managers will not be reprimanded for suppressing a fire, but careers have ended as a result of decisions to allow fire to burn. Another disincentive is the lack of proper funding and that planning, including contingency planning, can be overwhelming to agency personnel who are already stretched thin (Dale 2006). These disincentives likely contribute to "suppression bias" within agencies (Aplet 2006).

Even in cases where a manager is supportive of allowing lightning-ignited fires to burn, there may be agency barriers that prevent the manager from choosing this option. In 2012, for example, the Deputy Chief of the USFS State and Private Forestry issued an agency-wide memo stating that all fires, even those in designated wilderness, were to be suppressed because of the potential for significant fire growth. This memo, later dubbed the "Hubbard Letter", applied to all fire starts regardless of local fuels and weather conditions. The public has also been slow to embrace wilderness fire management due to safety and smoke concerns. This can translate into political pressure to suppress all fires (Aplet 2006)

OTHER CONSTRAINTS ON PLANNING

Other constraints that need to be considered in the wilderness fire management planning process include (1) air-quality impacts, (2) unnatural buildup of fuels and changes in vegetation structure following extensive and extended suppression activities, (3) safety of wilderness users, (4) impacts of fire management on recreation and cultural resources, and particularly on archeological resources, and (5) fires "escaping" the wilderness.

1. Air Quality—Prescribed natural fires and prescribed fires must take into consideration public tolerance for smoke and the Clean Air Act. Both prescribed natural fires and prescribed fires produce varying quantities of smoke. Although a nuisance, smoke may actually contribute to healthy ecosystems, as seed germination of some species is actually triggered by smoke (Keeley and Fotheringham 1998). Wildfire smoke also has a cooling effect on stream and river temperatures with the potential to benefit cold-water species (David et al. 2018). Prescribed natural fires are exempt from the Clean Air Act. However, the public's reaction and tolerance of smoke may be a consideration and some managers may choose to suppress fire when the public expresses vocal opposition.

   Although prescribed fire may be required to meet wilderness management objectives, the short-term effects of fire on air quality may violate certain air-quality standards, either inside or outside the wilderness. The Environmental Protection Agency (EPA) has developed National Ambient Air Quality Standards (NAAQS) for six air pollutants, including particulate matter. The Clean Air Act requires all federal agencies to comply with all federal, state, and local air-quality regulations when implementing prescribed fire (Schultz et al. 2019). Smoke-management objectives and techniques must be carefully applied to keep local and regional air quality within acceptable limits. Land managers need to work closely with EPA and state air regulatory personnel to explain the importance of uses and control methods available for prescribed fire. Tools have been developed to predict and monitor air-quality impacts and to coordinate policy making with federal and state regulatory agencies (Liu et al. 2019).
2. Unnatural Fuels and Forest Structure—We have already discussed the impacts that fire exclusion had in creating an unnatural increase in fuels or a change in vegetation structure. The danger is in the possibility that such conditions could foster high-intensity crown fires, risking damage to resources beyond the wilderness, or to local wilderness resources from too intense heat. Thus, where unnaturally high fuel loads exist, managers may favor prescribed fire to reduce fuel before allowing lightning-ignited fires to burn.
3. Safety of Wilderness Users—As numbers of visitors increase in parks and wilderness areas where prescribed natural fires are allowed to burn for long periods of time (months in some cases), the likelihood of human contact with such fires also increases (Figure 11.10). Although public safety has not been directly threatened, visitors have been inconvenienced when routes are closed because of fire and anyone present would be irritated by smoke. The potential exists for life-threatening events involving wilderness visitors and wildfire. Managers aim to prepare visitors and agency personnel to avoid accidents and disasters. Yet, in wilderness, safety from risk is not guaranteed, and this is part of the appeal for some wilderness users. There is more risk in wilderness recreation than elsewhere because of the emphasis on natural processes such as fire and because wilderness tends to be remote and rugged.

   Safety procedures that can help prevent fire accidents include (a) interpretative contacts made with visitors near ongoing fires, (b) consistent, accurate monitoring and evaluation of fire behavior as a basis for effective briefings and contingency plans, (c) closing trails in the vicinity of ongoing fires, (d) informing nearby residents and visitors about fire occurrences, status, and actions,

FIGURE 11.10 Kayaker on the South Fork of the Flathead River watches smoke plume in the Bob Marshall Wilderness. Photo courtesy Andrew J. Larson.

(e) caution signs on roads and trails to warn travelers that a fire is in progress, (f) maps and brochures that instruct/inform about safety hazards and precautions, and (e) implement fire restrictions when fire risk is high.

Mutch and Davis (1985) describe the possibility of a disaster resulting from prescribed natural fire or wildfire in wilderness. Their scenario is that ten to twenty small lightning fires have gradually been allowed to burn under prescribed conditions during June and July in a large wilderness where numerous visitors, traveling on foot and on horseback, are scattered at unknown locations. Over a period of weeks, the weather turns hot and dry until one day in mid-August, "red flag" (dangerous conditions) weather is forecast: strong afternoon winds with gusts to fifty miles per hour, temperatures of 95 to 100 degrees Fahrenheit, and low humidity. The "prescribed conditions" under which the twenty small lightning fires had been allowed to burn have changed. The extreme, red flag conditions could not be forecast in June or July when the prescribed conditions for letting the fires burn were met.

Wilderness managers aim to recognize subtle warning signals during the early stages and constantly monitor conditions and adjust their wilderness fire program's information and policies to ensure the safety of both recreationists and agency personnel. Thus, the possibility for disaster is a constant "warning" influence on agency plans for lightning-ignited and prescribed fires in wilderness.

4. Impacts on Cultural Resources—Wilderness fires may damage cultural resources at archeological sites, and high-severity fires are more likely to damage cultural resources than low-severity fire and prescribed fire (Friggens et al. 2021). It is likely that fire suppression activities also threaten archeological resources. During the 1977 wildfire at Bandelier National Monument, New Mexico, archeologists guided fire-line construction to prevent damage to cultural resources. Managers of park and wilderness areas with important recreation and cultural resources consider their protection in fire management planning; they involve recreation and cultural resource specialists in prescribed fire and fire-suppression implementation when such resources may be damaged.
5. Fires "Escaping" the Wilderness—Allowing lightning-ignited fires to burn is usually not practical in smaller wilderness units, such as where distance from the ignition point to the boundary may be less than a mile or where topography would allow the fire to move quickly to lands beyond the boundary. This same limitation, of course, can apply to ignitions near boundaries of even larger wilderness units (Barnett et al. 2016).

This concern is heightened when there is a risk of fires escaping parks and wilderness into adjoining developed lands and residential areas. As human habitation and second home developments increase around wilderness areas, so does concern for the loss of structures and human life (Miller 2006). There are many examples of wildland fires burning into urban and suburban areas, resulting in fatalities and the loss of structures and homes. For example, the 2018 Camp Fire (California) burned nearly the entire town of Paradise, caused over 80 fatalities, and destroyed over 18,000 structures. Other examples include the 2021 Marshall Fire (Colorado) that destroyed over 1,000 structures in Boulder County and the wildfires that occurred in Oregon in 2020 that destroyed over 4,000 homes. As these disasters are becoming more common (Higuera et al. 2023), fire managers are becoming increasingly concerned about escapes from wilderness. Although most of the most damaging fires did not originate as prescribed natural fires in wilderness, there are some examples. For example, the 2021 Tamarack Fire (California) ignited in the Mokelumne Wilderness and was initially managed as a prescribed natural fire because of firefighter safety concerns and the natural fuel breaks in the area, but the fire ultimately escaped the wilderness due to the extreme fire weather and at least twenty structures were damaged or destroyed.

FIRE BEHAVIOR PREDICTION

As previously mentioned, fire-behavior and prediction models have advanced rapidly over the last several of decades, and in fact, some were initially developed for use in wilderness or used wilderness fire data to parameterize and test predictions (Finney et al. 1994, Parisien et al. 2005). The potential for intense crown fires, effects on air quality, and visitor safety are important outputs from many of these models. Fire-behavior predictions are also useful for understanding historical fire behavior in wilderness, and "gaming" techniques can be used to compare predictions of fire size, intensity, and rates of spread with the actual results or to evaluate how fire suppression has affected contemporary fires (Davis et al. 2010). All the research tools available to land managers, however, can only supplement—not replace—the planner/manager's basic fire behavior knowledge and experience. Judgments based on available data, but grounded in manager judgments, are fundamental components of wilderness fire management plans.

LIMITING FIRE-SUPPRESSION IMPACTS

The special characteristics of parks and wilderness require that routine fire suppression actions be adjusted to minimize signs of human activities. Tractor fire lines, felled snags, helispots, and areas clear-cut of standing trees and snags can cause longer-lasting adverse impacts on the wilderness resource than the wildfire itself (Figure 11.11). Decisions must be made about what equipment will be allowed in wilderness and under what circumstances. Special concern is needed for impacts on wilderness resources during suppression actions, such as human life, threatened and endangered species, recreation, historical or archeological sites, and Native American religious sites.

Fire-suppression strategies and tactics to minimize damage to the wilderness are now common. Both pre-fire training in special suppression techniques and post-fire monitoring and evaluation of the results can help improve efforts to be easy on the land with fire suppression efforts. However, it all starts in the wilderness fire management plan.

MONITORING AND EVALUATING WILDERNESS FIRES

As in most planning systems, wilderness fire planning calls for monitoring and evaluating as a final step. Operational monitoring of prescribed natural fire calls for collecting and recording data on fire behavior, weather, fuels, topography, air

FIGURE 11.11 Efficient fire control, using helicopters (top left) and fire-retardant bombing (top right), has contributed to less than natural incidence of fire in some wildernesses. Photos courtesy USFS. Light-handed tactics are being used in areas such as the Eagle Cap Wilderness, Wallowa-Whitman National Forest, Oregon (bottom), whereby fire is allowed to burn in low-fuel areas and to either go out or be extinguished with minimal effort and disturbance of the natural setting. Photo courtesy Francis Mohr, U.S. Forest Service.

quality, and fire effects to provide a basis for immediate evaluation of whether the fire is still within prescribed conditions, or later evaluation of whether objectives were achieved. The primary purposes for monitoring fire behavior and weather on wilderness fires are to ensure the fire remains in prescription and inside park or wilderness boundaries and that human life and property are not threatened. The main reasons for gathering data on fuels, fire effects, and air quality are to ensure that resource management objectives are met and to minimize impacts on resources, such as air and water. The objective in wilderness and parks usually involves the restoration of natural processes, including intensities and frequencies of fire, to maintain or restore a natural range of conditions. A partial list of resources for monitoring and evaluating active wilderness fires and post-fire effects include the NPS Fire Monitoring Handbook (USDI National Park Service 2003) and the Fire Effects Monitoring and Inventory System (FIREMON) (Lutes et al. 2006); satellite data can also be useful (Roy et al. 2013).

LEARNING FROM PAST MISTAKES

Although the use of prescribed natural fire and management-ignited prescribed fire will always incur some risk, we can and must learn from ongoing programs, particularly examples where fires have not gone as planned. Such situations may be the result of mistakes in judgment or lack of knowledge and training, but they are often because fire is inherently difficult to predict and because fire-conducive weather can unexpectedly arise during long-duration events. Reviewing case studies of fires that did not go as planned can be instructive and the fire literature provides many of them (e.g., Black et al. 2020).

Several agencies have had prescribed natural fire or prescribed fire that went awry, some in wilderness and some in outside of wilderness. Examples include the 1978 Ouzel Fire (NPS; Colorado), and the 2000 Cerro Grande Fire (NPS; New Mexico), the 2021 Tamarack Fire (USFS; California), and the 2022 Calf Canyon/Hermits Peak Fire (USFS; New Mexico).

The 1978 Ouzel Fire was a wilderness fire ignited by lightning in Rocky Mountain National Park, and the NPS decided to allow the fire to burn. Ouzel began on August 9, 1978, in a fire management unit above 10,000-feet elevation in spruce-fir forest. Although it initially smoldered and crept along, high winds caused crowning and spotting by early September. On September 15 and 16, it made substantial runs outside the fire management zone in the direction of Allenspark, a small community near the park's boundary. The fire had to be suppressed and no lives or homes were lost.

The 2000 Cerro Grande Fire was started as a prescribed burn of 900 acres by nineteen staff and crew in Bandelier National Monument (NPS) on May 4, 2000. The fire spread, due to weather changes, into the Santa Fe National Forest and threatened adjoining structures. By May 19, the fire had been substantially contained by more than a thousand firefighters, but not before the fire had burned 48,000 acres, caused the evacuation of 18,000 residents, destroyed or damaged 280 residences and forty structures at the Las Alamos National Laboratory, and caused over $1 billion in damages.

The 2021 Tamarack Fire was ignited by lightning and was discovered on July 4 on a rocky ridgetop in the Mokelumne Wilderness. The fire was allowed to burn because of firefighter safety concerns and the natural barriers on the ridgetop that were expected to limit fire growth. On July 16, however, the fire began to spread as winds picked up (fuel moisture was also low). The fire grew to 21,000 acres by July 17. The fire continued to grow, threatening numerous communities, including the town of Markleeville. In total, 2,439 people were evacuated and at least twenty structures were damaged or destroyed.

The 2022 Calf Canyon/Hermits Peak Fire resulted from two wildfires originally started as an escaped prescribed burn on April 6 (Hermits Peak) and from leftover burn piles from a prescribed burn earlier in the year (January; Calf Canyon Fire). The two fires merged during strong winds on April 22. Ultimately, this fire became the largest and most destructive fire in New Mexico History, burning about 340,000 acres. The fire destroyed over 900 structures, including several hundred homes. As a result of this fire, the Chief of the U.S. Forest Service announced a prescribed fire "pause" while the agency conducted a 90-day review of protocols, decision support tools, and practices before prescribed fire operations were to resume in the fall.

Although these and similar incidents are unfortunate, fire managers can use them as a learning opportunity to reduce the potential for escapes in the future. However, there will always be some risk of escapes. Nevertheless, many scientists and land managers firmly believe that prescribed fire and wilderness fire management programs are an important and necessary tool for maintaining and restoring forest resilience to the inevitable fire (Kolden 2019).

TRENDS AND FUTURE NEEDS

Wilderness fire management is complex and demanding work, requiring professional competence and intimate knowledge of the specific ecosystem and land unit to be managed. More people trained in ecology, botany, and forestry with an understanding of fire behavior and an interest in preservation management are needed to manage these programs.

The need for wilderness fire management programs is generally understood by informed supporters of national parks and wilderness. A challenge, however, is that much of the public has negative views of fire. This is probably because (a) the media tends to focus on mistakes and destructive fires and rarely reports on the successes of prescribed natural fires or prescribed fires and (b) wildfire-prevention campaigns, while well-intended, do not provide the counterpoint that fire has been an important ecological process for millennia and helped maintain healthy and resilient forests. Elected officials can be expected to follow public opinion and generally take a conservative view of wilderness fire management and prescribed fire. The legacy from fire-prevention campaigns can be overcome only by new and innovative public information and education efforts with the media and schools. For example, the FireWorks Program developed by the Missoula Fire Sciences Laboratory is an educational program about the science of wildland fire designed for students in grades 1–12 (https://www.firelab.org/project/fireworks-educational-program). Also, here are two excellent StoryMaps that serve as examples of innovative public outreach and education efforts:

describing how Indigenous fire use shaped the Boundary Waters Canoe Area Wilderness: http://z.umn.edu/peoplefirepine

about adapting western US forests to climate change and wildfires (Hessburg et al. 2021): https://tinyurl.com/Ten-Questions

The more people understand about the function of fire in maintaining the naturalness of wilderness ecosystems, the more likely that they will support wilderness fire management programs and prescribed fire. However, the task facing wilderness fire programs is daunting because of the legacy of fire prevention and control that have allowed fuels to accumulate, thus setting the stage for the inevitable fire that may have undesirable consequences. Fire in wilderness ecosystems is an important forest research topic, as the numerous references in this chapter attest. The research has documented fire histories, frequencies, ecosystem effects, and emerging concerns such as climate change, fire impacts on local populations, and wilderness and recreation use.

WILDERNESS FOR SCIENCE

Wilderness fire management is important not only because it can maintain the natural landscapes and biota of wilderness as a cultural and recreational resource, but also because it can maintain large-scale functioning ecosystems that will contribute to basic scientific knowledge. Wilderness ecosystems can serve as an important benchmark, classroom, and laboratory for ecosystem knowledge. Our understanding of natural ecosystems—as systems—is still growing. Future scientists will require whole, functioning, natural ecosystems as research laboratories to probe questions vital to the future of humans. Many of the most widespread natural ecosystems of the Earth—including most coniferous forests, savannas, glades, and grasslands—are fire-dependent. In addition, many of our most important domestic plants and animals and useful forest trees evolved in such ecosystems. It is chiefly in our larger wildernesses and national parks that natural ecosystems can be studied because people are rapidly converting the rest of our planet to farms, pastures, commercial forests, mines, highways, and urban areas. As this book has emphasized throughout, the value of natural wilderness processes and conditions will surely increase in the next few decades. And we know that it holds many secrets important to our future on this planet.

The best place to study the causes and consequences of fire in the absence of human-induced confounding factors is wilderness (Figure 11.12). This is particularly true in national

FIGURE 11.12 Scientists need whole, functioning, natural ecosystems with active fire regimes to better understand the causes and consequences of fire. Here, scientists collect post-fire data in the Saguaro Wilderness (Arizona; left) and Nahanni National Park (Northwest Territories, Canada; right). Photos courtesy Sean Parks.

parks and wilderness areas that have active wilderness fire management programs. To date, the scientific and management community has learned a great deal about fire resulting from studies conducted in parks and wilderness. A very broad summary of some of this research is provided below, but also check out this short film about conducing fire research in the Selway-Bitterroot Wilderness: https://tinyurl.com/WildernessFireResearch.

To better understand how the biophysical environment and fire exclusion influences fire regime characteristics in landscapes with low human influence, several scientific studies have been conducted within national parks and wilderness areas. Findings suggest that topography and vegetation plays a major role in influencing fire severity (Kane et al. 2015) and that some wilderness landscapes in the western United States are more likely to be resilient to reintroduced fire because of the absence of previous logging (Larson et al. 2013). These unlogged wilderness landscapes can also provide valuable information on fire-induced tree mortality (Belote et al. 2015) and on historical fire interval, forest structure, and forest composition (Barth et al. 2015).

In protected areas that have allowed fire to play a more natural role (for example, the Selway-Bitterroot Wilderness and Gila Wilderness), a natural experiment is unfolding that is allowing scientists to better understand the role of past fires in influencing subsequent fire behavior, size, and occurrence. Scientists are using observed fire patterns to evaluate the effectiveness and longevity of wildland fire to act as a fuel treatment and evaluate factors that maintain or increase resilience to future fire events. Collectively, these studies have concluded that previous fire indeed limits the occurrence, size, and severity of subsequent fire, but that the strength and longevity of this effect varies depending on the productivity and fire regime characteristics of each region (e.g., Collins et al. 2009; Parks et al. 2018). These studies also concluded that previous fire is less effective at regulating subsequent fire characteristics under extreme weather conditions (Parks et al. 2015a). These studies complement and expand on previous studies that used simulation models to look at similar questions (e.g., Davis et al. 2010).

On a more regional level, recent studies conducted within the United States and internationally have shown that fire activity and climate are tightly linked in national parks and wilderness, but that this relationship weakens, or decouples, as landscapes become more human influenced (Archibald et al. 2010; Parks et al. 2014). As such, protected areas serve as a valuable benchmark that provides insight as to how fire naturally responds to climate and has allowed scientists to quantify how much fire "should" be expected within various climate domains. In fact, this reasoning has helped scientists identify where there has been more or less fire than the climate would dictate, not just within national parks and wilderness, but covering the western United States irrespective of protected area status (Parks et al. 2015b). Evaluations of the relationship between fire and climate in protected areas may also provide insight into how climate change will influence fire regime characteristics. Also see the publication list by the Aldo Leopold Wilderness Research Institute on numerous wilderness fire topics, http://leopold.wilderness.net/publications/. The findings of these and other studies conducted within national parks and wilderness have relevance far beyond protected areas and are applicable to lands of all management designations.

STUDY QUESTIONS

1. What factors describe a fire regime? Name some typical regimes and contrast at least two fire regime groups (FRGs).
2. List some roles that fire plays in most ecosystems or plant communities. Explain the importance of each.
3. What impact has the exclusion of fire had on fuels and forest structure of wilderness ecosystems? And what is the potential outcome when an area inevitably burns?
4. What is the main objective of wilderness fire management in national parks and wilderness?
5. How do scientists determine the fire history of an area?
6. List some wilderness fire policy alternatives and the advantages and disadvantages of each. Which alternatives can or should be combined in a wilderness fire program?
7. What role did Indigenous peoples play in the historical fire regime that was present prior to European colonization of the North America?
8. Name some constraints on planning for wilderness fire management. Indicate how each impacts development of wilderness fire plans.
9. What are things to consider in addressing visitor safety with respect to wilderness programs?
10. What has been learned by scientists conducing fire research in wilderness? Can you think of any fire-related research questions where wilderness can serve as a natural laboratory?

ACKNOWLEDGEMENT

This chapter was written for this fifth edition by Sean A. Parks, research ecologist, Aldo Leopold Wilderness Research Institute, Missoula, MT and replaces previous versions of this chapter.

CASE DISCUSSION: PRESCRIBED FIRE IN WILDERNESS

Land management agencies often resort to fire suppression even though there is widespread evidence and recognition that this practice has had unintended consequences that reduce ecosystem resilience to the inevitable fire (Calkin et al. 2015). Even in national parks and wilderness, fire suppression is the default fire management strategy in many cases. As a result, fire-dependent ecosystems have accumulated excessive fuels and vegetation that is contributing to uncharacteristically severe fire. Annual area burned has increased six-fold in western U.S. forests since 1985, but perhaps more importantly, high severity area burned has increased eight-fold over this time period (Parks and Abatzoglou 2020).

When considering how to solve problems resulting from human-caused changes to fire regimes, tension exists between a perceived threat of management-ignited prescribed fire to wilderness values, the harm of continued fire suppression, and the risk of anomalous fire behavior from unplanned ignitions burning under novel conditions. The Wilderness Act does not prohibit prescribed fire. Rather, it requires that prescribed fire represent the minimum action required to sustain wilderness character. Policy for all four federal wilderness managing agencies also permits use of prescribed fire in wilderness. However, interpretations of the suitability of prescribed fire in wilderness vary between agencies, individuals, and among the public.

Morton et al. (2006) report that there are five situations in which prescribed fire might be an appropriate tool in wilderness:

1. To restore or enhance habitats of federally listed threatened and endangered species.
2. To control or eradicate invasive flora.
3. To increase the likelihood of naturally ignited fire to burn unimpeded (by reducing hazardous fuel loads around structures and the urban interface).
4. To restore a natural fire regime that has been temporarily altered (e.g., extreme fuel loads due to blowdown from a hurricane).
5. To mimic a natural fire regime that has been altered and is not expected to be restored due to constraints on wildfire management.

Even though most lightning-ignited fires are suppressed in wilderness and national parks, lightning-ignited fires are, for the most part, the only process by which managers have historically relied on to restore fire regimes and forest conditions. Human-ignited fires are universally suppressed. Generally speaking, fire activity in recent decades still lags behind what was experienced prior to European colonization of North America.

This raises the question whether management-ignited prescribed fire could be used to augment lightning-ignited fire in wilderness and parks as a means to improve resilience to subsequent fire and other disturbances such as insect outbreaks (Miller 2006). The idea of management-ignited prescribed fire that is carefully managed in wilderness by following well-designed fire-management plans is not philosophically acceptable to many managers and raises serious questions about to what extent we manage natural processes like fire.

CASE DISCUSSION QUESTIONS

1. Outline the role of fire in fire-dependent communities. What changes can be expected when fire is removed from an ecological community of your choosing?
2. Does the Wilderness Act permit the use of management-ignited prescribed fire in wilderness?
3. What are some philosophical objections to implementing prescribed fire in wilderness?
4. Implementing prescribed fire in wilderness can be contentious, but does recognizing the historical importance and prevalence of Indigenous fire use change the conversation?
5. What degrades wilderness character more in the long run: continued fire suppression or implementing prescribed fire? Why?
6. What are the pros and cons of management-ignited prescribed fire and prescribed natural fire in (a) small wilderness areas, (b) near the boundary of a large wilderness area, (c) when non-extreme fuel moisture conditions are prevalent, and (d) when fuels are exceptionally dry?

REFERENCES

Abatzoglou, J. T., and A. P. Williams. 2016. Impact of anthropogenic climate change on wildfire across western U.S. forests. Proceedings of the National Academy of Sciences 113:11770–11775.

Agee, J. K. 1996. *Fire Ecology of Pacific Northwest Forests*. Island Press.

Aplet, G. H. 2006. Evolution of wilderness fire policy. *International Journal of Wilderness.* 12:9–13.

Archibald, S.; Nickless, A.; Govender, N.; Scholes, R. J.; and Lehsten, V. 2010. Climate and the inter-annual variability of fire in southern Africa: A meta-analysis using long-term field data and satellite-derived burnt area data. *Global Ecology and Biogeography*. 19:794–809.

Bancroft, L.; Nichols, T.; Parsons, D.; Graber, D.; Evison, B.; and Van Wagtendonk, J. 1985. Evolution of the natural fire management program at Sequoia and Kings Canyon National Parks, pp. 174–180. Proceedings of the Symposium and Workshop on Wilderness Fire. U.S. Department of Agriculture, Forest Service, Intermountain Forest and Range Experimental Station. Ogden, Utah.

Barnett, K.; Miller, C.; and Venn, T. J. 2016. Using risk analysis to reveal opportunities for the management of unplanned ignitions in wilderness. *Journal of Forestry*. 114:610–618.

Barth, M. A. F.; Larson, A. J.; and Lutz, J. A. 2015. A forest reconstruction model to assess changes to Sierra Nevada mixed-conifer forest during the fire suppression era. *Forest Ecology and Management*. 354:104–118.

Belote, T. R.; Larson, A. J.; and Dietz, M. S. 2015. Tree survival scales to community-level effects following mixed-severity fire in a mixed-conifer forest. *Forest Ecology and Management*. 353:221–231.

Bentz, B. J.; Régnière, J.; Fettig, C. J.; Hansen, E. M.; Hayes, J. L.; Hicke, J. A.; Kelsey, R. G.; Negrón, J. F.; Seybold, S. J. 2010. Climate change and bark beetles of the western United States and Canada: Direct and indirect effects. *BioScience*. 60:602–613.

Berkey, J. K.; Miller, C.; Larson, A. J. 2021. A history of wilderness fire management in the Northern Rockies. Gen. Tech. Rep. RMRS-GTR-428. Fort Collins, CO: U.S. Department of Agriculture, Forest Service, Rocky Mountain Research Station. 88p. 428.

Black, A. E.; Hayes, P.; and Strickland, R. 2020. Organizational learning from prescribed fire escapes: A review of developments over the last 10 years in the USA and Australia. Current Forestry Reports. 6:41–59.

Black, A.; Williamson, M.; Doane, D. 2008. Wildland fire use barriers and facilitators. *Fire Management Today*. 68:4.

Bond, W. J.; van Wilgen, B. W. 2012. *Fire and Plants*. Springer Science & Business Media.

Brunelle, A.; Anderson, R. S. 2003. Sedimentary charcoal as an indicator of late-Holocene drought in the Sierra Nevada, California, and its relevance to the future. *Holocene*. 13:21–28.

Calkin, D. E.; Thompson, M. P.; Finney, M. A. 2015. Negative consequences of positive feedbacks in U.S. wildfire management. *Forest Ecosystems*. 2:9.

Christensen, N. L.; Agee, J. K.; Brussard, P. F.; Hughes, J.; Knight, D. H.; Minshall, G. W.; Peek, J. M.; Pyne, S. J.; Swanson, F. J.; Thomas, J. W.; Wells, S.; Williams, S. E.; Wright, H. A. 1989. Interpreting the Yellowstone Fires of 1988. *BioScience*. 39:678–685.

Cole, D. N.; Yung, L. 2012. *Beyond Naturalness: Rethinking Park and Wilderness Stewardship in an Era of Rapid Change*. Island Press.

Collins, B. M.; Miller, J. D.; Thode, A. E.; Kelly, M.; van Wagtendonk, J. W.; Stephens, S. L. 2009. Interactions among wildland fires in a long-established Sierra Nevada Natural Fire Area. *Ecosystems*. 12:114–128.

Coop, J. D.; Parks, S. A.; Stevens-Rumann, C. S.; Crausbay, S. D.; Higuera, P. E.; Hurteau, M. D.; Tepley, A.; et al. 2020. Wildfire-driven forest conversion in Western North American landscapes. *BioScience*. 70:659–673.

Coop, J. D.; Parks, S. A.; Stevens-Rumann, C. S.; Ritter, S. M.; Hoffman, C. M. 2022. Extreme fire spread events and area burned under recent and future climate in the western USA. *Global Ecology and Biogeography*. 31:1949–1959.

Dale, L. 2006. Wildfire policy and fire use on public lands in the United States. *Society & Natural Resources*. 19:275–284.

Dale, L. A.; Aplet, G.; Wilmer, B. 2005. Wildland fire use and cost containment: A Colorado case study. *Journal of Forestry*. 103:314–318.

David, A. T.; Asarian, J. E.; Lake, F. K. 2018. Wildfire smoke cools summer river and stream water temperatures. *Water Resources Research*. 54:7273–7290.

Davis, B. H.; Miller, C.; Parks, S. A. 2010. Retrospective fire modeling: Quantifying the impacts of fire suppression. Gen. Tech. Rep. RMRS-GTR-236. Fort Collins, CO: U.S. Department of Agriculture, Forest Service, Rocky Mountain Research Station. 40 p. 236.

Desmond, J. 1994. Interagency wilderness fire management. The Biswell Symposium: Fire Issues and Solutions in Urban Interface and Wildland Ecosystems, Walnut Creek, CA, USA, pp. 15–17.

Doane, D.; O'Laughlin, J.; Morgan, P.; Miller, C. 2006. Barriers to wildland fire use. *International Journal of Wilderness*. 12:37.

Dodge, M. 1972. Forest fuel accumulation—A growing problem. *Science*. 177:139–142.

Doerr, S. H.; Shakesby, R. A.; Walsh, R. P. D. 2000. Soil water repellency: Its causes, characteristics and hydro-geomorphological significance. *Earth-Science Reviews*. 51:33–65.

Falk, D. A.; Heyerdahl, E. K.; Brown, P. M.; Farris, C.; Fulé, P. Z.; McKenzie, D.; Swetnam, T. W.; Taylor, A. H.; Van Horne, M. L. 2011. Multi-scale controls of historical forest-fire regimes: New insights from fire-scar networks. *Frontiers in Ecology and the Environment*. 9:446–454.

Finney, M. A.. 1994. FARSITE: A fire area simulator for fire managers. The Proceedings of the Biswell Symposium, Walnut Creek, California. Citeseer, 7.

Fontaine, J. B.; Kennedy, P. L. 2012. Meta-analysis of avian and small-mammal response to fire severity and fire surrogate treatments in U.S. fire-prone forests. *Ecological Applications*. 22:1547–1561.

Friggens, M. M.; Loehman, R. A.; Constan, C. I.; Kneifel, R. R. 2021. Predicting wildfire impacts on the prehistoric archaeological record of the Jemez Mountains, New Mexico, USA. *Fire Ecology*. 17:18.

Gavin, D. G.; Hallett, D. J.; Hu, F. S.; Lertzman, K. P.; Prichard, S. J.; Brown, K. J.; Lynch, J. A.; Bartlein, P.; Peterson, D. L. 2007. Forest fire and climate change in western North America: Insights from sediment charcoal records. *Frontiers in Ecology and the Environment*. 5:499–506.

Gebert, K. M.; Calkin, D. E.; Yoder, J. 2007. Estimating suppression expenditures for individual large wildland fires. *Western Journal of Applied Forestry*. 22:188–196.

Greenwood, L.; Bliege Bird, R.; Nimmo, D. 2022. Indigenous burning shapes the structure of visible and invisible fire mosaics. Landscape Ecology. 37:811–827.

Hagmann, R. K., Hessburg, P. F.; Prichard, S. J.; Povak, N. A.; Brown, P. M.; Fulé, P. Z.; Keane, R. E.; et al. 2021. Evidence for widespread changes in the structure, composition, and fire regimes of western North American forests. *Ecological Applications*. 31:e02431.

Harper, C. A.; Ford, W. M.; Lashley, M. A.; Moorman, C. E.; Stambaugh, M. C. 2016. Fire effects on wildlife in the Central Hardwoods and Appalachian regions, USA. *Fire Ecology*. 12:127–159.

Harvey, B. J.; Holzman, B. A. 2014. Divergent successional pathways of stand development following fire in a California closed-cone pine forest. *Journal of Vegetation Science*. 25:88–99.

Haugo, R. D.; Kellogg, B. S.; Cansler, C. A.; Kolden, C. A.; Kemp, K. B.; Robertson, J. C.; Metlen, K. L.; Vaillant, N. M.; Restaino, C. M. 2019. The missing fire: Quantifying human exclusion of wildfire in Pacific Northwest forests, USA. *Ecosphere*. 10:e02702.

Hessburg, P. F.; Miller, C. L.; Parks, S. A.; Povak, N. A.; Taylor, A. H.; Higuera, P. E.; Prichard, S. J.; et al. 2019. Climate, environment, and disturbance history govern resilience of western north American forests. *Frontiers in Ecology and Evolution*. 7.

Hessburg, P. F.; Prichard, S. J.; Hagmann, R. K.; Povak, N. A.; Lake, F. K. 2021. Wildfire and climate change adaptation of western North American forests: A case for intentional management. *Ecological Applications*. 31:e02432.

Hessburg, P. F.; Smith, B. G.; Salter, R. B.; Ottmar, R. D.; Alvarado, E. 2000. Recent changes (1930s–1990s) in spatial patterns of interior northwest forests, USA. *Forest Ecology and Management*. 136:53–83.

Heyerdahl, E. K.; McKenzie, D.; Daniels, L. D.; Hessl, A. E.; Littell, J. S.; Mantua. 2008. Climate drivers of regionally synchronous fires in the inland Northwest (1651–1900). *International Journal of Wildland Fire*. 17:40–49.

Higuera, P. E.; Cook, M. C.; Balch, J. K.; Stavros, E. N.; Mahood, A. L.; St. Denis, L. A. 2023. Shifting social-ecological fire regimes explain increasing structure loss from Western wildfires. PNAS Nexus:pgad005.

Higuera, P. E.; Shuman, B. N.; Wolf, K. D. 2021. Rocky Mountain subalpine forests now burning more than any time in recent millennia. *Proceedings of the National Academy of Sciences.* 118:e2103135118.

Hood, S.; Sala, A.; Heyerdahl, E. K.; Boutin, M. 2015. Low-severity fire increases tree defense against bark beetle attacks. *Ecology.* 96:1846–1855.

Huffman, D. W.; Roccaforte, J. P.; Springer, J. D.; Crouse, J. E. 2020. Restoration applications of resource objective wildfires in western U.S. forests: A status of knowledge review. *Fire Ecology.* 16:18.

Hunter, M. E.; Iniguez, J. M.; Farris, C. A. 2014. Historical and current fire management practices in two wilderness areas in the southwestern United States: The Saguaro Wilderness Area and the Gila-Aldo Leopold Wilderness Complex. Gen. Tech. Rep. RMRS-GTR-325. Fort Collins, CO: U.S. Department of Agriculture, Forest Service, Rocky Mountain Research Station. 38p. 325.

Johnson, D. W.; Turner, J. 2019. Tamm Review: Nutrient cycling in forests: A historical look and newer developments. *Forest Ecology and Management.* 444:344–373.

Jones, G. M.; Ayars, J.; Parks, S. A.; Chmura, H. E.; Cushman, S. A.; Sanderlin, J. S. 2022. Pyrodiversity in a Warming World: Research Challenges and Opportunities. Current Landscape Ecology Reports.

Kane, V. R.; Cansler, C. A.; Povak, N. A.; Kane, J. T.; McGaughey, R. J.; Lutz, J. A.; Churchill, D. J.; North, M. P. 2015. Mixed severity fire effects within the Rim fire: Relative importance of local climate, fire weather, topography, and forest structure. *Forest Ecology and Management.* 358:62–79.

Keane, R. E.; Gray, K.; Davis, B.; Holsinger, L. M.; Loehman, R.; Keane, R. E.; Gray, K.; Davis, B.; Holsinger, L. M.; Loehman, R. 2019. Evaluating ecological resilience across wildfire suppression levels under climate and fuel treatment scenarios using landscape simulation modelling. *International Journal of Wildland Fire.* 28:533–549.

Keeley, J. E.; Fotheringham, C. J. 1998. Smoke-induced seed germination in California chaparral. *Ecology.* 79:2320–2336.

Kelly, R.; Chipman, M. L.; Higuera, P. E.; Stefanova, I.; Brubaker, L. B.; Hu, F. S. 2013. Recent burning of boreal forests exceeds fire regime limits of the past 10,000 years. *Proceedings of the National Academy of Sciences.* 110:13055–13060.

Kilgore, B. M.; Sando, R. W. 1975. Crown-fire potential in a Sequoia forest after prescribed burning. *Forest Science.* 21:83–87.

Kimmerer, R. W.; Lake, F. K. 2001. The role of Indigenous burning in land management. *Journal of Forestry.* 99:36–41.

Kipfmueller, K. F.; Larson, E. R.; Johnson, L. B.; Schneider, E. A. 2021. Human augmentation of historical red pine fire regimes in the Boundary Waters Canoe Area Wilderness. *Ecosphere.* 12:e03673.

Klimas, K.; Hiesl, P.; Hagan, D.; Park, D. 2020. Prescribed fire effects on sediment and nutrient exports in forested environments: A review. *Journal of Environmental Quality.* 49:793–811.

Knight, C. A.; Anderson, L.; Bunting, M. J.; Champagne, M.; Clayburn, R. M.; Crawford, J. N.; Klimaszewski-Patterson, A.; et al. 2022. Land management explains major trends in forest structure and composition over the last millennium in California's Klamath Mountains. *Proceedings of the National Academy of Sciences.* 119:e2116264119.

Kolden, C. A. 2019. We're not doing enough prescribed fire in the Western United States to mitigate wildfire risk. *Fire.* 2:30.

Lake, F. K.; Wright, V.; Morgan, P.; McFadzen, M.; McWethy, D.; Stevens-Rumann, C. 2017. Returning fire to the land: Celebrating traditional knowledge and fire. *Journal of Forestry.* 115:343–353.

LANDFIRE. 2016. Biophysical Setting Layer, LANDFIRE 2.0.0. U.S. Department of the Interior, Geological Survey, and U.S. Department of Agriculture.

Larson, A. J.; Belote, R. T.; Cansler, C. A.; Parks, S. A.; Dietz, M. S. 2013. Latent resilience in ponderosa pine forest: Effects of resumed frequent fire. *Ecological Applications.* 23:1243–1249.

Leopold, A. S. 1963. Wildlife management in the national parks. U.S. National Park Service.

Liu, Y.; A. Kochanski; Baker, K. R.; Mell, W.; Linn, R.; Paugam, R.; Mandel, J.; et al. 2019. Fire behaviour and smoke modelling: Model improvement and measurement needs for next-generation smoke research and forecasting systems. *International Journal of Wildland Fire.* 28:570–588.

Lutes, D. C.; Keane, R. E.; Caratti, J. F.; Key, C. H.; Benson, N. C.; Sutherland, S.; Gangi, L. J. 2006. FIREMON: Fire effects monitoring and inventory system. Gen. Tech. Rep. RMRS-GTR-164. Fort Collins, CO: U.S. Department of Agriculture, Forest Service, Rocky Mountain Research Station. 1 CD. 164p.

Lynch, D. L. 2004. What do forest fires really cost? *Journal of Forestry.* 102:42–49.

Mallek, C.; Safford, H.; Viers, J.; Miller, J. 2013. Modern departures in fire severity and area vary by forest type, Sierra Nevada and southern Cascades, California, USA. *Ecosphere.* 4:art153.

Margolis, E. Q.; Guiterman, C. H.; Chavardès, R. D.; Coop, J. D.; Copes-Gerbitz, K.; Dawe, D. A.; Falk, D. A., et al. 2022. The North American tree-ring fire-scar network. *Ecosphere.* 13:e4159.

Marlon, J. R.; Bartlein, P. J.; Gavin, D. G.; Long, C. J.; Anderson, R. S.; Briles, C. E.; Brown, K. J.; et al. 2012. Long-term perspective on wildfires in the western USA. *Proceedings of the National Academy of Sciences.* 109:E535–E543.

McKenzie, D.; Miller, C.; Falk, D. A. 2011. The Landscape Ecology of Fire. Springer Science & Business Media.

Miller, C. 2006. Wilderness fire management in a changing world. *International Journal of Wilderness.* 12:18–21.

Miller, C.; Abatzoglou, J.; Brown, T.; Syphard, A. D. 2011. Wilderness fire management in a changing environment. In: McKenzie, D.; Miller, C.; Falk, D. A., eds. *The Landscape Ecology of Fire.* Springer Netherlands: Dordrecht, pp. 269–294.

Morton, J. M.; Berg, E.; Newbould, D.; MacLean, D.; O'Brien, L. 2006. Wilderness fire stewardship on the Kenai National Wildlife Refuge, Alaska. *International Journal of Wilderness.* 12:15.

Mutch, R.; K. Davis. 1985. Visitor protection in parks and wildernesses: Preventing fire-related accidents and disasters. USDA Forest Service General Technical Report INT-Intermountain Forest and Range Experiment Station, USA.

Neary, D. G.; Koestner, K. A.; Youberg, A.; Koestner, P. E. 2012. Post-fire rill and gully formation, Schultz Fire 2010, Arizona, USA. *Geoderma.*191:97–104.

Olson, L. E.; Crotteau, J. S.; Fox, S.; Hanvey, G.; Holbrook, J. D.; Jackson, S.; Squires, J. R. 2023. Effects of compound disturbance on Canada lynx and snowshoe hare: Wildfire and forest management influence timing and intensity of use. *Forest Ecology and Management.* 530:120757.

Parisien, M.-A.; Barber, Q. E.; Hirsch, K. G.; Stockdale, C. A.; Erni, S.; Wang, X.; Arseneault, D.; Parks, S. A. 2020. Fire deficit increases wildfire risk for many communities in the Canadian boreal forest. *Nature Communications.* 11:2121.

Parisien, M.-A.; Kafka, V.; Hirsch, K.; Todd, J.; Lavoie, S.; Maczek, P. 2005. Mapping wildfire susceptibility with the BURN-P3 simulation model.

Parker, T. J.; Clancy, K. M.; Mathiasen, R. L. 2006. Interactions among fire, insects and pathogens in coniferous forests of the interior western United States and Canada. *Agricultural and Forest Entomology.* 8:167–189.

Parks, S. A.; Abatzoglou, J. T. 2020. Warmer and drier fire seasons contribute to increases in area burned at high severity in Western U.S. forests from 1985 to 2017. *Geophysical Research Letters.* 47:e2020GL089858.

Parks, S. A.; Dobrowski, S. Z.; Shaw, J. D.; Miller, C. 2019. Living on the edge: Trailing edge forests at risk of fire-facilitated conversion to non-forest. *Ecosphere.* 10:e02651.

Parks, S. A.; Holsinger, L. M.; Miller, C.; Nelson, C. R. 2015a. Wildland fire as a self-regulating mechanism: The role of previous burns and weather in limiting fire progression. *Ecological Applications*. 25:1478–1492.

Parks, S. A.; Miller, C.; Parisien, M.-A.; Holsinger, L. M.; Dobrowski, S. Z.; Abatzoglou, J. 2015b. Wildland fire deficit and surplus in the western United States, 1984–2012. *Ecosphere*. 6:1–13.

Parks, S. A.; Parisien, M.-A.; Miller, C.; Dobrowski, S. Z. 2014. Fire activity and severity in the Western U.S. vary along proxy gradients representing fuel amount and fuel moisture. *PLOS ONE*. 9:e99699.

Parks, S. A.; Parisien, M.-A.; Miller, C.; Holsinger, L. M.; Baggett, L. S. 2018. Fine-scale spatial climate variation and drought mediate the likelihood of reburning. *Ecological Applications*. 28:573–586.

Parsons, D. J.; Graber, D. M.; Agee, J. K.; Van Wagtendonk. J. W. 1986. Natural fire management in National Parks. *Environmental Management*. 10:21–24.

Parsons, D. J.; Landres, P. B. 1998. Restoring natural fire to wilderness: How are we doing. Tall Timbers Fire Ecology Conference Proceedings, pp. 366–373.

Perrakis, D. D. B.; Lanoville, R. A.; Taylor, S. W.; Hicks, D. 2014. Modeling wildfire spread in mountain pine beetle-affected forest stands, British Columbia, Canada. *Fire Ecology*. 10:10–35.

Romme, W. H.; Knight, D. H. 1981. Fire frequency and subalpine forest succession along a topographic gradient in Wyoming. *Ecology* 62:319–326.

Roy, D. P.; Boschetti, L.; Smith, A. M. S. 2013. Satellite Remote Sensing of Fires. Fire Phenomena and the Earth System. John Wiley & Sons, Ltd, pp. 77–93.

Schoennagel, T.; Turner, M. G.; Romme, W. H. 2003. The influence of fire interval and serotiny on postfire lodgepole pine density in Yellowstone National Park. *Ecology*. 84:2967–2978.

Schultz, C. A.; McCaffrey, S. M.; Huber-Stearns, H. R.; C. A. Schultz, C. A.; McCaffrey, S. M.; Huber-Stearns, H. R. 2019. Policy barriers and opportunities for prescribed fire application in the western United States. *International Journal of Wildland Fire*. 28:874–884.

Scott, J. H.; Burgan, R. E. 2005. Standard fire behavior fuel models: A comprehensive set for use with Rothermel's surface fire spread model. Gen. Tech. Rep. RMRS-GTR-153. Fort Collins, CO: U.S. Department of Agriculture, Forest Service, Rocky Mountain Research Station. 72. p. 153.

Seielstad, C. 2015. Reconsidering wildland fire use: Perspectives from the Northern Rockies. In: Keane, Robert E.; Jolly, Matt; Parsons, Russell; Riley, Karin. Proceedings of the large wildland fires conference; May 19-23, 2014; Missoula, MT. Proc. RMRS-P-73. Fort Collins, CO: U.S. Department of Agriculture, Forest Service, Rocky Mountain Research Station. p. 207–212. 73:207–212.

Shive, K. L.; Wuenschel, A.; Hardlund, L. J.; Morris, S.; Meyer, M. D.; Hood, S. M. 2022. Ancient trees and modern wildfires: Declining resilience to wildfire in the highly fire-adapted giant sequoia. *Forest Ecology and Management*. 511:120110.

Short, K. C. 2014. A spatial database of wildfires in the United States, 1992–2011. *Earth System Science Data*. 6:1–27.

Smithwick, E. A. H.; Turner, M. G.; Mack, M. C.; Chapin, F. S. 2005. Postfire soil N cycling in northern conifer forests affected by severe, stand-replacing wildfires. *Ecosystems*. 8:163–181.

Steel, Z. L.; Jones, G. M.; Collins, B. M.; Green, R.; Koltunov, A.; Purcell, K. L.; Sawyer, S. C.; et al. 2022. Mega-disturbances cause rapid decline of mature conifer forest habitat in California. *Ecological Applications*. e2763.

Stocks, B. J.; Mason, J. A.; Todd, J. B.; Bosch, E. M.; Wotton, B. M.; Amiro, B. D.; Flannigan, M. D.; Hirsch, K. G.; Logan, K. A.; Martell, D. L.; Skinner, W. R. 2002. Large forest fires in Canada, 1959–1997. *Journal of Geophysical Research: Atmospheres*. 107:FFR 5-1-FFR 5-12.

Swetnam, T. W.; Baisan, C. H.; Caprio, A. C.; Brown, P. M.; Touchan, R.; Anderson, R. S.; Hallett, D. J. 2009. Multi-millennial fire history of the Giant Forest, Sequoia National Park, California, USA. *Fire Ecology*. 5:120–150.

Swetnam, T. W.; Betancourt, J. L. 1990. Fire-southern oscillation relations in the Southwestern United States. *Science*. 249:1017–1020.

Tingley, M. W.; Ruiz-Gutiérrez, V.; Wilkerson, R. L.; Howell, C. A.; Siegel, R. B. 2016. Pyrodiversity promotes avian diversity over the decade following forest fire. *Proceedings of the Royal Society B: Biological Sciences*. 283:20161703.

Turner, M. G.; Braziunas, K. H.; Hansen, W. D.; Harvey, B. J. 2019. Short-interval severe fire erodes the resilience of subalpine lodgepole pine forests. *Proceedings of the National Academy of Sciences*. 116:11319–11328.

USDA and USDOI. 2005. Wildland fire use: Implementation procedures and reference guide. U.S. Department of Agriculture (USDA) and the U.S. Department of the Interior (USDOI).

USDI National Park Service. 2003. Fire monitoring handbook. Fire Management Program Center, National Interagency Fire Center Boise (ID).

Van Auken, O. W. 2009. Causes and consequences of woody plant encroachment into western North American grasslands. *Journal of Environmental Management*. 90:2931–2942.

van Wagtendonk, J. W. 2007. The history and evolution of wildland fire use. *Fire Ecology*. 3:3–17.

Wang, X.; Thompson, D. K.; Marshall, G. A.; Tymstra, C.; Carr, R.; Flannigan, M. D. 2015. Increasing frequency of extreme fire weather in Canada with climate change. *Climatic Change*. 130:573–586.

Williams, J. N.; Safford, H. D.; Enstice, N.; Steel, Z. L.; Paulson, A. K. 2023. High-severity burned area and proportion exceed historic conditions in Sierra Nevada, California, and adjacent ranges. *Ecosphere*. 14:e4397.

CHAPTER

12

Wildlife in Wilderness: A North American and International Perspective

INTRODUCTION

The presence of certain indigenous wildlife, and their natural distribution, abundance, and behavior, reflect wilderness conditions. Some wildlife are icons of certain wilderness areas.

Internationally, Tibetan antelope and Mongolian gazelle symbolize the vast Asian steppe; wildebeest and white-eared kob symbolize the trackless East African wilderness; the deep taiga of the Amur evokes images of the Siberian tiger; the Serengeti's sere landscape evokes images of leopards, cheetahs, and African lions; Nepal's Chitwan National Park is rhino and tiger country.

In North America, the Bob Marshall Wilderness in Montana is grizzly country; the canyon lands evoke images of the cougar; caribou symbolize the vastness of the Arctic wilderness; and the wolf is now successfully reestablishing its species as a symbol of wilderness in the Yellowstone ecosystem.

Without wilderness, many species of wildlife could not survive or behave as truly wild animals. Wilderness without wildlife, and wildlife without the freedom of wilderness, are virtually unthinkable (Fig. 12.1). *However, wilderness stewardship must be comprehensive because wilderness is more than wildlife habitat.* Thus, important as it is, *wildlife is only one component of the wilderness resource and must be managed as such—one consideration in an overall wilderness stewardship scheme* (see Chapter 7).

This chapter categorizes and describes wilderness wildlife, some important interrelationships between wildlife and wilderness, and the many wildlife-related problems in wilderness stewardship, including legal, administrative, and cultural constraints that influence managing wildlife in wilderness (Fig. 12.2). The perspective is international, because some of the world's most dramatic wilderness wildlife are in developing regions. *The chapter proposes some stewardship objectives for wildlife in wilderness and guidelines for attaining them in the United States and North America. Basic principles of wilderness management (see Chapter 7) should prevail.* That is, *when management actions are necessary, only the minimum tools, methods, and force should be used to meet planned area objectives. In designated wilderness, natural forces should normally be allowed to shape wildlife habitat, populations, and behavior to the fullest possible extent.*

FIGURE 12.1 Wildlife is a part of all wilderness ecosystems; its distribution, abundance, and behavior reflect the naturalness of a wilderness. Photo courtesy Richard Behan.

FIGURE 12.2 Wild horses roam adjacent to a BLM roadless area in Nevada; they compete for forage with native wildlife. Photo courtesy Bureau of Land Management.

THE WILDERNESS WILDLIFE RESOURCE

A literal definition of "wilderness" would be "place of the wild beasts" (Nash 1970). But what wild beasts?

From ecological and economic perspectives, several definitions of wilderness wildlife have been suggested. Aldo Leopold (1933) considered wilderness wildlife as species harmful to or harmed by economic land uses specifically—wapiti (elk), caribou, bison, grizzly bear, moose, mountain (bighorn) sheep, and mountain goat. A. Starker Leopold (1966) identified five North American ungulates associated primarily with wilderness climax-forage types: caribou, bighorn sheep, mountain goat, musk ox, and bison. Dasmann (1966) defined wilderness species as those that are obligate members of a climax community or wilderness area, for example, the passenger pigeon, caribou, musk ox, bighorn sheep, and grizzly bear. Durward Allen (1966) called the cougar, grizzly bear, and wolf true wilderness animals because they are capable, wide-ranging, and at odds with the livestock industry. Hochbaum (1970) includes many species of migratory waterfowl, like swans and geese, that seek undisturbed (wilderness) wetlands for nesting purposes.

Mattson (1997) considered wilderness species to be those vulnerable to contact with humans or, like Leopold (1933), fundamentally incompatible with traditional human economic activities. By these standards, megaherbivores, such as elephants and rhinos, and large carnivores, such as bears, wolves, and large cats, depend on wilderness naturalness and solitude for their survival and viability. Migratory ungulates such as caribou, wildebeest, and Mongolian gazelle also require wilderness conditions with unimpeded access to forage over vast areas.

Theoretically, salamanders and butterflies are as important in wilderness as grizzly bears, mountain lions, eagles, and even smaller and less familiar creatures—what Edward O. Wilson (1984, 12) calls "the miniature wilderness that can take almost forever to explore." *However, as a practical matter, the ungulates and carnivores at or near the top of their food chains have historically depended on or symbolized wilderness.* Thus, this chapter focuses on these large species.

CATEGORIES OF WILDERNESS WILDLIFE

Wilderness wildlife may be categorized as (1) wilderness-dependent wildlife, (2) wilderness-associated wildlife, (3) common native wildlife found in wilderness, and (4) nonnative species found in wilderness. We recognize the difficulty of placing a particular species into any single category because of overlap, regional and local variation, and exceptions that do not fit neatly into compartments. However, these categories are useful for discussion because they suggest something about the relationship of the species to conditions found in wilderness.

Wilderness-Dependent Wildlife

Wilderness-dependent wildlife includes species that require wild habitat found in wilderness and that are vulnerable to contact with humans, including those species whose relationship with humans is intractable because they threaten human safety or chronically damage livestock or crops. Such species require refuge from humans or the absence of human development such as roads, railways, and fences.

Although a host of organisms, large and small, are a part of wilderness ecosystems, many of these species are not themselves dependent on wilderness, though they may benefit by wilderness conditions. Species that depend in some way on wilderness are a potential focus of management attention in wilderness settings, and thereby are a symbol of wilderness itself.

Polar bears, Siberian tigers, African elephants, and Pantanal jaguars are examples of wilderness-dependent wildlife, although even these species may, in places, lead a tenuous existence in modified environments outside wilderness. In any particular area, locally or regionally rare species that are dependent on the continued naturalness of their habitats can also be placed in the wilderness-dependent category. In the U.S. northern Rocky Mountains, for example, such locally or regionally rare species as grizzly bear, mountain caribou, native "west slope" cutthroat trout, Canada lynx, wolverine, mountain goat, Richardson's blue grouse, fisher, marten, peregrine falcon, bald and golden eagles, osprey, mountain sheep, and northern white-tailed ptarmigan are associated with wilderness. In the Southwest, desert tortoise and desert bighorn sheep would be on the list.

Internationally, megaherbivores like rhinos, elephants, African Cape buffalo, and large carnivores such as tigers, leopards, African lions, brown bears, pumas, jaguars, Komodo dragons, and large crocodiles are wilderness-dependent—meaning they also depend on naturalness and solitude. In the case of rhinos, elephants, bears, and tigers, problematic relations with humanity are exacerbated by poaching motivated by the high value placed on their body parts.

Wilderness-Associated Wildlife

Wilderness-associated wildlife includes species commonly associated in human perception and ecological reality with wilderness conditions and species displaced to wilderness by human activities. For example, this category includes wildlife common in the high-elevation habitat of the western United States (e.g., whistling marmots) and species associated with conditions characteristic of wilderness in the East, including southern swamps (alligators) and hardwood forests (black bears).

The relationship of wilderness-associated species with humans is not as intractable as that of wilderness-dependent wildlife. Human impacts on them are potentially subject to remedy. For example, construction of the Ulaanbataar-Beijing railroad severed traditional migratory routes of the Mongolian gazelle, resulting in massive declines in gazelle populations due to lost access to forage and reduced opportunities to escape harsh winter conditions (Lhagvasuren et al. 1999). By contrast, with judicious forethought and design modifications providing for migration, construction of the oil pipeline from the North Slope of Alaska to ports to the south did not interfere with migrations of caribou, which have since flourished (Cronin et al. 1998).

Wilderness-associated wildlife also includes species displaced to wilderness for protection by local demand for meat and by dependence on habitat conditions lost with the expansion of agriculture and domestic livestock. A host of ungulates, primates, and their interdependent mid-size or large carnivores in Africa, Asia, and South America, as well as prairie dogs and black-footed ferrets in North America, have been relegated to wilderness for these reasons. A solution? Alternate sources of protein or modest changes in husbandry and farming practices could alleviate some displacement of these kinds of wilderness-associated species, and thus their human-created dependence on wilderness conditions. Such is less often the case for wilderness-dependent wildlife. This is not to say that wilderness is unimportant to the long-term survival of animals such as the orangutan, spectacled bear, red deer (wapiti), or Tibetan antelope. Rather, compared to wilderness-dependent wildlife, humans are graced with more options for coexisting with wilderness-associated wildlife.

Common Native Wildlife Found in Wilderness

Common native wildlife found in wilderness includes species that happen to be found in wilderness but that also often live in more modified environments. Their relationship to wilderness is incidental. They are not necessarily associated in ecological fact or in human perception with especially wild places. Some examples include deer, coyotes, bobcats, raccoons, rabbits, muskrats, and minks; a host of rodents like squirrels, field mice, and rats; and many kinds of birds, such as some raptors, grouse, woodpeckers, sparrows, juncos, and thrushes.

Unlike the other two categories, common native wildlife species do not have specific relationships to wilderness. They neither are dependent on such areas ecologically nor have any real or perceived association with wilderness in the human mind. However, when found in wilderness, they are no less important in their natural roles. In fact, these common species, because they are adapted to more modified environments (and may even be benefited by them), may reveal through their natural place in the wilderness scheme an important comparison with non-wilderness conditions. Thus, keeping wilderness wild provides opportunity for keeping "wildness" in species otherwise adapted to civilization.

Nonnative Species Found in Wilderness

Nonnative species are, unfortunately, a fourth category that is increasingly a threat to wilderness character (see Chapter 13). This category would include *those introduced species that have adapted to wilderness habitat and compete with resident native wildlife, leading to changes in native ecosystems.* This would include such species as wild (feral) hogs, horses and burros, feral goats, nutria, chukkar partridge, and eastern brook trout in wilderness in the western United States. Whether or not a species is native to a particular wilderness can be controversial, such as with mountain goats in the High Uintas of Utah and in Olympic National Park in Washington. The ecological impacts of many nonnative species have been well documented in some proposed or designated wilderness areas, so wilderness wildlife managers are often involved in conceiving, implementing, and evaluating alternatives for the management, impact mitigation, or control of nonnative species.

These four categories are useful guides to inventories of wilderness wildlife and the characteristics of natural habitat on which they depend. Such information is essential for management as well as useful in assessing some of the values dependent on an area's wilderness designation. In the following, *some of the most dramatic examples to illustrate the wilderness wildlife categories come from outside the United States.*

WILDERNESS WILDLIFE RELATIONSHIPS

Wildlife relates to wilderness, and vice versa, in many ways. The presence of certain wildlife can be a measure of wilderness character; wildlife also play a role in the wilderness ecosystems they inhabit and wilderness certainly plays a role in wildlife preservation; wilderness wildlife can serve as an environmental baseline and laboratory; finally, wilderness wildlife has recreational, educational, aesthetic, economic, and political values.

Wildlife as a Measure of Wilderness Character

The distribution, numbers, diversity, and behavior of wildlife species can be a measure of the naturalness and solitude of a wilderness. Wildlife reflects ecological conditions and their changes over time, so wildlife can serve as indicators of wilderness character and quality—in fact as well as in human perception.

Culturally, for many people the concept of wilderness is linked to some form of wildlife. For example, when Margaret and Olaus Murie (1966) wrote of their American elk studies in Wyoming, they called their book *Wapiti Wilderness.* Andy Russell (1971) titled his book on Canadian wilderness *Grizzly Country.* The presence of particular kinds of wildlife suggests the relative absence of human influence and the existence of primitive harmonies, for example, grizzlies and wolves. For many people, simply knowing that such wildlife are present is important to the meaning of wilderness. If key wildlife are removed, although everything else remains visibly the same, the intensity of the sense of wilderness may be diminished (Nash 1970).

Leopold described the diminishment of wilderness after the demise of an old grizzly, the last of its kind roaming the Arizona high country. When a government hunter shot the bear for bounty, Leopold wrote a eulogy: "(Mount) Escudilla still hangs on the horizon but when you see it you no longer think of bear. It's only a mountain now" (Leopold 1949, 137). The extent to which wildlife is perceived as a wilderness criterion is striking. For example, John Milton (1972) described wildlife species, especially caribou, as carrying the soul of the Alaskan wilderness because they require freedom and vast space in which to range. To Milton, Alaska's wildlife symbolizes its wilderness—the space, the openness, and the silence of the North. Sigurd Olson (1963) wrote of the wild-laughing choruses of the common loon as the sound that more than any other typifies the rocks and waters and forests of wilderness. These examples suggest the importance of wildlife as a cultural reference to wilderness character. However, just as important—some would say more important—is wildlife as a biological indicator of wilderness conditions.

Wildlife that potentially indicate wilderness conditions include megaherbivores, large carnivores, and migratory ungulates, that is, certain wilderness-dependent wildlife. Species that survive only where there are few humans, minimal human activity, and little agriculture logically denote wilderness in a sense that is meaningful to most industrialized societies. Thus, animals such as grizzly bears, wolves, rhinos, lions, tigers, and saltwater crocodiles are prime indicators of wilderness conditions in their native habitats.

Wide-ranging wilderness-dependent species potentially serve as an indicator of wilderness conditions, and can have *umbrella effects,* that is, the protection of natural processes and nontarget organisms as a consequence of managing for a few "featured" species (Caro and O'Doherty 1999). For example, in the case of wide-ranging species vulnerable to contact with humans, protection of wilderness areas (especially large ones) may also preserve natural migratory movements and disturbance regimes and, thus, a full complement of biodiversity.

Organisms that combine wide-ranging movements with wilderness dependence are readily identified. In the Far North, these animals include large carnivores such as wolves, polar bears, and interior brown or grizzly bears. An individual male grizzly bear in the Northwest Territories of Canada can annually range over 2,600 square miles whereas an individual polar bear can wander in excess of almost 18,000 miles across the Arctic ice in search of aquatic prey such as seals, walruses, and beluga whales. Everywhere they still survive, ungulates that undertake long-range migrations are associated with the most extensive primeval conditions still to be found on earth (Fig. 12.3). This holds for wildebeest and white-eared kob in East Africa, springbok and eland in southern Africa, Mongolian gazelle and Tibetan antelope on the steppes of east-central Asia, and caribou and wild reindeer in circumpolar Arctic and boreal regions. Vast herds of these herbivores can still move hundreds of miles between seasonal ranges only because humans have spared their traditional migratory routes.

FIGURE 12.3 Elk browse on summer range on the Gallatin National Forest near Yellowstone National Park, Montana. Animals, such as elk and deer, may migrate seasonally in and out of designated wilderness areas, which makes their management a multiagency stewardship challenge. Photo courtesy W. E. Steuerwald, U.S. Forest Service.

Wildlife's Role in Wilderness

Wildlife is an inseparable part of the wilderness resource. It plays a vital role in developing, maintaining, and modifying soil, vegetation, and ecosystems that cover wilderness topography; in dispersal, planting, and germination of seeds; in pollination; in fertilization; in distribution of nutrients; and in conversion of dead plants into organic matter that is more usable by living plants.

The alligator and its relentless search for water in the dry season are an example from the Everglades wilderness of Florida. Alligators seek low places where the water table lies just below the surface and work either to deepen the existing waterholes or to excavate new ones, breaking the caked earth with their powerful tails and shoveling away the debris with their broad snouts. During the worst droughts, alligators have been known to dig their way down through four feet of compacted mud and peat before coaxing water from the porous substrate. Such gator holes, found throughout the parched glades, attract many thirsty creatures ranging from otters to herons, and soon become biological microcosms of the whole region. The alligators, conserving energy and living on their own fat, largely ignore these boarders; the refugees sustain their lives on the gator holes' remaining fish, insect life, and vegetation, and live side by side in a relative state of truce. When the rains finally return, it is from these gator-made oases that the various species go forth to repopulate the Everglades (Carr 1973).

Alligators in Florida are a good example of a *"keystone" species—wildlife that have a disproportional effect on the structure and function of the ecosystems within which they live* (Simberloff 1998). Such species are consequently of great importance. Large carnivores have keystone effects in wilderness settings attributable to a cascade of influences down the trophic ladder, from top predator, to herbivores, to vegetation, to microorganisms. This has been shown for wolves and moose on Isle Royale, in Lake Superior in the United States (D. L. Allen 1979); for orcas and sea otters along Pacific coastal North America; and for jaguars, puma, and their prey in tropical Central and South America (Terborgh et al. 1999).

Elephants also have a well-documented ability to dramatically transform ecosystems, especially where their natural movements are curtailed by confinement to artificially small protected areas. Elephants have virtually deforested some of these areas and consequently changed fire regimes and ecosystem productivity (Dublin 1995). Prairie dogs and termites can have similarly transforming effects on ecosystems where they exist in sufficient densities. Anadromous salmon are important vectors for transporting nutrients from productive oceanic systems to impoverished terrestrial ones; bears preying on the fish finish the job by depositing nutrients contained in fish upslope from streams either as carcasses or bear feces. Finally, some wilderness animals, such as bears, can have inordinate effects at a smaller scale. Grizzlies are well-known diggers and can dramatically alter soil structure, nutrient flows, and vegetation composition at sites that are habitually dug for roots and rodents (Tardiff and Stanford 1998). The rootings of wild pigs impact deciduous forests in the Great Smoky Mountains in the southeastern United States in much the same way (Singer 1984). Although all organisms affect the ecosystems within which they live, keystone species—those that have a disproportionate impact on their environment—are of special interest in managing wilderness ecosystems.

A Wilderness Role in Wildlife Preservation

Wilderness is critical to the survival of wilderness-dependent wildlife and a major factor in conserving wilderness-associated species. Wilderness may directly enhance survival of species with certain specialized habitat needs and those that are vulnerable to contact with humans by providing them with security. Although many species with an affinity for wilderness conditions may in fact survive in less pristine habitats, they live and behave most naturally in wilderness. Certain animals injure or kill humans regularly enough to preclude peaceful coexistence near even sparse human settlement. These animals include grizzly bears, polar bears, lions, tigers, leopards, pumas, crocodilians, Komodo dragons, and elephants. Often, it is young, injured, or senescent animals that prey on or otherwise kill humans, but human antagonism is often generalized to the entire species. Among humans' cultures, there is widespread intolerance of animals that kill and potentially eat people. Perhaps more than any, it is these species that depend on wilderness sanctuaries—or areas where the intrusions of people are tightly controlled for their survival.

For many of these same animals, as well as for a host of others, wilderness conditions can be critical to the preservation of essential food and cover. For example, the 64,000-acre Badlands Wilderness in Badlands National Park, South Dakota, is the primary site for reintroducing endangered black-footed ferrets to the wild. This wilderness is one of the last locations with extensive intact prairie dog towns—home of the ferret's primary prey.

Over large parts of the world, humans often exacerbate conflicts with larger carnivores and herbivores by reducing or eliminating prey and forage. In many instances, this is accompanied by dramatic increases in numbers of domesticated livestock. Substituting domesticated animals for native prey leads to spiraling conflict with predators because they resort to preying on cattle or sheep as wild ungulates decline. The spread of intensive agriculture and pastoralism also typically escalates crop damage by large herbivores forced to seek out forage in agricultural areas as forage in adjacent or former rangelands is depleted by livestock or converted to crops. This scenario is pervasive in Africa, India, the Himalayas, Southeast Asia, and parts of South and Central America, involving predators such as jaguar, snow leopards, wild dogs, cheetahs, and lions, and herbivores as diverse as Cape buffalo, eland, red deer, kiang, Nilgiri tahr, guanaco, and elephants. For animals like these, wilderness conditions can be critical to their survival, especially where economic and cultural influences do not foster or allow for changes in human behavior or tolerance. Such is the case in many impoverished countries experiencing rapid population growth.

Wildlife that are prone to overharvest by humans also benefit from wilderness conditions. For some, wilderness is even critical to their survival. In most industrialized, developed countries, wildlife harvests are closely regulated with an eye to sustainability. Elsewhere, this is often not the case. Overharvest of vulnerable animals occurs in developing countries for two primary reasons—because body parts of the animal are highly valued or because the animal is an important source of protein. For some animals, such as the African elephant, these causes are combined. Most residents of Western countries do not think of wild game as a primary source of protein—yet in many tropical regions of Africa and South America, wildlife is an important source of food. In fact, protein from wild game is considered to be the factor most limiting to human population growth in African rain forests (Barnes and Lahm 1997). It is thus not surprising that tapirs and larger-bodied primates and antelopes are overharvested apace with the destruction of tropical forests and the local growth of human populations on several continents. Short of the unlikely event that alternate sources of protein will be found, the fates of these animals are inextricably tied to vast tracts of unexploited tropical forest, that is, wilderness conditions. The emphasis is on large areas, because, in the absence of enforced protected areas, harvests of larger-bodied primates or antelope are often unsustainable within about seven miles of tropical human settlements (Muchaal and Ngandju 1999; Wilkie et al. 1998).

The high monetary value placed on the body parts of some species can make the wilderness dependence of these animals even more extreme and can jeopardize their survival. Although the ranges of tigers, rhinos, elephants, and tropical bears have declined primarily because of competition with humans and related losses of habitat, poaching or unregulated harvest driven by demand for bones, horns, tusks, or gall bladders is a grave threat, even with current restrictions on trade under the Convention on International Trade in Endangered Species (CITES). To a lesser extent, markets, legal or otherwise, for their skins and pelts also threaten animals like crocodiles, caimans, and leopards. The potential wealth derived from an elephant tusk or rhino horn is so great that space often does not provide much protection from highly motivated and well-outfitted poachers. Antipoacher patrols and extreme regulations can give a false sense of security and are often disrupted by even a short period of civil unrest. The precariousness of civil order in many countries around the world argues for the importance of very large wilderness reserves for the long-term survival of elephants, rhinos, tigers, and tropical bears.

Extensive wilderness conditions can be important to maintaining natural evolutionary selection and related behaviors among wilderness-dependent or wilderness-associated species, because wilderness protects habitats that have been least modified from conditions under which their biotic communities evolved. For example, selective harvest of large-tusked African elephants has led to increasing numbers of tuskless or small-tusked animals, along with alteration of mating and foraging behaviors related to tusk size (Jachmann et al. 1995). Strategies to reduce poaching on African rhinos by the deliberate removal and sale of rhino horns also promise to change selective pressures affecting survival and breeding (Berger and Cunningham 1994). The evolution of pachyderms has come to increasingly reflect the influences of international human commerce rather than their natural habitat. Elephants and rhinos may persist, but without wilderness they will no longer be the products of a natural and wild environment. The same could be said of the remaining migratory ungulates. Although all of these species could survive in small populations in small protected areas, their numbers and behaviors that define both a presence on the landscape and effects on whole ecosystems would have been lost, as has already happened with American bison and the black wildebeest of South Africa.

Wilderness can protect conditions, such as wetlands, grasslands, or old-growth primary forest, that are potentially important to numerous species at some stage in their life cycle, but prone to be eliminated in non-wilderness settings. For example, large, hollow trees are highly preferred for natal dens by American and Asiatic bears and are more abundant in wilderness settings (Mattson 1990). Similarly, although the old-growth forests that spotted owls depend on in the U.S. Pacific Northwest can persist in settled areas, they are highly associated with wilderness conditions because of historical timber harvests. The list of species benefited in this way could go on, including snow geese, brant, eiders, scaups, tundra swans, and white-fronted geese that nest in Arctic wetlands. *The point is that, although humans can choose not to drain wetlands or cut down old trees anywhere, such protection of natural conditions is guaranteed in areas set aside as wilderness.* Along with such protection come many benefits to many species dependent on the natural patterns and cycles of nature, whether involving water, wind, or fire.

Finally, for all of those species that benefit from wilderness conditions, wilderness can serve as a wildlife bank, especially valuable following times of overharvest and depleted populations. For example, between 1892 and 1962 many of the elk ranges of the western United States were restocked, after several decades of massive elk declines, with animals live-trapped in the backcountry of Yellowstone National Park. Similarly, wildlands of northern Minnesota in the United States and of Spain, Italy, and Poland were the source of wolves that have established themselves widely in surrounding semiwild areas since protections were instituted in the 1950s and 1970s.

Having made the case for the benefits of wilderness to wildlife conservation, wilderness conditions are obviously not of equal benefit to all species or of any benefit to some. Many species, including crows, starlings, robins, Norway rats, certain species of cockroaches, and even white-tailed deer, have benefited from the conversion of wildlands to urban, industrial, or agricultural landscapes. Because the habitat requirements of various wildlife species are so diverse, there is no question that some species will benefit and some species will not by wilderness designation. This, of course, makes the case for identifying wilderness-dependent and wilderness-associated wildlife for purposes of planning wilderness management for wildlife.

In short, the key to diverse wildlife is diversity of habitat. The U.S. wilderness system offers the possibility of preserving most of the nation's 261 ecosystems, as defined by the

Bailey-Küchler method, which reflects physical and biological factors. However, traditional emphasis on areas with dramatic high-country appeal at the expense of less scenic lowland areas will need to be changed if the wilderness system is to fulfill its potential in providing that diversity. By 1987, only 157 of the 261 ecosystems were represented in designated wilderness (Davis 1989).

Wilderness Wildlife as an Environmental Baseline and Laboratory

Wilderness areas have three crucial roles as permanent and natural laboratories: (1) as *invaluable reservoirs of genetic material* that preserve resource options and serve as storehouses of potential new products; (2) as *biological standards of comparison* with ecological communities more heavily affected by human activities; and (3) as sites for *integrated studies of the structure and function of natural ecosystems* (Cutler 1980).

Aldo Leopold promoted the theme in his writings that wilderness provides a standard against which the alteration of developed lands can be measured—that wild places reveal what the land was, what it is, and what it ought to be. He emphasized the importance of areas where evolution operates without hindrance from humans, thereby providing standards against which to measure the effects of development. Each biotic province, said Leopold (1949, 196), "needs its own wilderness for comparative studies of used and unused land."

The value of wilderness to understanding ecosystem relationships, including the regulation of wildlife populations, predator-prey relationships, diseases and hosts, and relations between herbivores and their physical and vegetal environments, has been amply demonstrated in many classical studies. For example, the intact flora and fauna of the Serengeti and Yellowstone wildlands have been priceless to science (Clark et al. 1999; Sinclair and Arcese 1995). The wildlands of boreal and Arctic regions have been similarly important to describing the simpler web of relations among caribou, moose, wolves, bears, and their biophysical environment (e.g., Gasaway et al. 1983), or simpler even yet, the classic relations among wolves, moose, and moose browse on Isle Royale (D. L. Allen 1979; McLaren and Peterson 1994). Wilderness areas also have offered the opportunity for numerous in-depth studies of individual species in comparatively intact ecosystems, including cougar in central Idaho (Hornocker 1970), black bears in deciduous forests of Great Smoky Mountains National Park (Pelton and van Manen 1996), chimpanzees and mountain gorillas in central and eastern Africa (Goodall 1986; Schaller 1963), and leopards in Kruger National Park, South Africa (Bailey 1993). Clearly, many insights into biological systems have depended and will continue to depend on setting aside and fully protecting wilderness areas and continuing ongoing research or initiating additional long-term integrated studies in relatively undisturbed wilderness laboratories.

Recreational, Aesthetic, Economic, and Political Values

The recreational and aesthetic uses of wilderness have increased rapidly during the past five decades, and the presence of wildlife is surely part of this wilderness lure. Many people come to view or photograph native species in natural settings. In season, others come to hunt or fish under primitive conditions. The fact that many other people are not deliberately seeking contact with wildlife in wilderness makes little difference; the incidental contact—the chance observation under natural conditions—can immeasurably enrich a wilderness experience. For some, even *danger* can be a positive aspect of a wilderness experience—knowing that the possibility exists for confrontation with dangerous wildlife. Close encounters with grizzlies, lions, rhinos, or other dangerous animals, or even hearing or sighting them, make for impressive recollections (Fig. 12.4).

Millions of people also enjoy wilderness wildlife vicariously by appreciating it through friends, stories, photos, and art; through television and movies; or simply by knowing that it is there. Even those who have never personally experienced wilderness have some vision of its enchantment due to the rich profusion of these media celebrating its beauty in images, words, and poetry that appeal to a strong human instinct for maintaining some connection to a more primitive life and time.

Quantifying the economic value of wilderness wildlife may be even more difficult than putting a dollar value on wilderness itself, but efforts to date show that both are extremely valuable. Wildlife is an important component for many visitors who use outfitters and guides to access wilderness—and supports a multimillion-dollar-a-year industry. Interestingly, in the Frank Church–River of No Return Wilderness where a recovering population of wolves has been reestablishing itself, at least part of the appeal is a chance of hearing wolves and seeing wolf sign. Internationally, wilderness wildlife is the appeal for a big business in both hunting and camera safaris, presenting wilderness wildlife in its native habitat, in films, videos, and television.

In developing countries, wilderness wildlife can have a more pragmatic (as well as economic) value to humans. The fish, birds, and mammals that live in wildlands are an

FIGURE 12.4 Grizzly bears are one of several species of wildlife requiring large tracts of undisturbed habitat. Most wilderness areas are too small or boundaries do not coincide with the natural movements of such animals, thus complicating effective management. Grizzly bears are currently confined to a few isolated and comparatively small wilderness ranges in the contiguous United States, primarily because humans are intolerant of the risks they pose to human safety and to livestock. Photo courtesy U.S. Fish and Wildlife Service.

important source of food and potentially a reason to value wilderness in areas where there are few other reasons to do so. The value of game meat is so high in developing countries that wildlife in most protected areas experience substantial harvest, mostly by market hunters. Whether legal or not, this activity provides a livelihood for many people living nearby and is important to local economies. Unfortunately, harvests of larger species are rarely sustainable in developing countries, yet it is the value placed on such animals as food that potentially motivates humans to change their practices to preserve wildlands and sustain the wildlife within.

Kellert's studies (1985) of Americans' attitudes toward wildlife indicate strong public support—especially by younger, more highly educated, and urban-dwelling Americans—for protecting many wilderness wildlife species such as wolves, eagles, and grizzly bears. Kellert argues that wildlife makes direct and vicarious contributions to the American quality of life, especially where contact with and awareness of such wildlife are inextricably woven into the fabric of a region's heritage. For example, findings from his study of Minnesota residents' attitudes toward timber wolves revealed that some people had *seen* and *heard* a timber wolf in the wild, and a majority reported having read an article about wolves within the previous year. The study revealed a strong "existence value" for wolves, and a majority of all publics except farmers agreed that (1) it was important to them to "know that wolves exist in Minnesota"; (2) "timber wolves are essential to maintaining the balance of nature"; and (3) they "would very much like to see a timber wolf in the wild" and "hear a wolf howl." A majority of all respondents except farmers and trappers opposed the idea of "giving preference to people who derive a living from the land over protection of the timber wolf." However, relatively few residents were willing to protect timber wolves if it meant excluding people from northern areas of Minnesota (Kellert 1985).

Other findings complicate the wolf picture. A national survey (Kellert 1985) indicated that wolves were among the least liked animals. Only 42 percent nationwide said they did not dislike wolves, but that figure jumped to 74 percent among Alaskans. Again, younger age, higher education, and especially knowledge of animals were correlated with a favorable view of wolves. Kellert observes that affection for wolves has increased, and that this trend can be expected to continue, especially among residents of regions containing wolves that are not potentially impacted by them.

Politically, wildlife is the most powerful focus for environmental protection in the world and some of the largest citizen environmental organizations in the United States are wildlife oriented (Hendee and Pitstick 1993). *Although wildlife is a common thread of these organizations, their interests go beyond wildlife and certainly include wilderness protection. They are a strong political force.*

In the participatory society of the United States, wilderness preservation could not have come to pass without a viable political constituency. Although this constituency has been composed of many groups, for one hundred years, wildlife supporters, including hunters and anglers, have played an important role, spurred by their recognition of wilderness as a fish and wildlife refuge. For example, the major 132-million-acre expansion of U.S. national forests came during the 1901 to 1909 presidency of Teddy Roosevelt, a founder of the Boone and Crockett Club—a big-game hunting club—who believed forest conservation and game conservation to be synonymous. It was the Boone and Crockett Club that urged the passage of the Yellowstone Game Protection Act of May 7, 1894, which was designed in part to protect Yellowstone wildlife from poachers and thus established the precedent of the national parks as game preserves (Trefethen 1975). When a wilderness bill was first introduced in the U.S. Congress in 1956, among its charter supporters were the National Wildlife Federation and the Wildlife Management Institute (Mercure and Ross 1970).

To be sure, hunters and anglers are not always supportive of a particular wilderness. They exhibit considerable concern about the ultimate levels of hunting and fishing to be allowed in wilderness, and there is concern that designation of wilderness, and in certain wilderness areas restoring wolves and grizzly bears, may limit harvests of big game. These concerns demonstrate the need for a basic philosophy incorporating wildlife into a wilderness framework—a philosophy emphasizing wildlife uses including appreciation, hunting, and fishing carried out in a manner consistent with wilderness values.

WILDLIFE-RELATED PROBLEMS IN WILDERNESS STEWARDSHIP

Wildlife issues are frequently a source of controversy in wilderness management. One survey of agency managers in the southwestern United States reported controversial wildlife-management practices in 53 of 273 wilderness areas (Czech and Krausman 1999). Water provision for wildlife (Fig. 12.5), trout stocking, feral burro control, and big-game surveys using motorized vehicles accounted for over half the controversies. The other half spanned diverse issues, including research with motorized access, mechanical and chemical vegetation control, reintroductions, mechanical animal capture, and restrictions on hunting and human access. Wilderness management involving wildlife is loaded with potential conflict and difficult problems—not just in the United States but internationally. Following are ten categories of wildlife-related problems in wilderness management with examples drawn from the United States and around the world.

MULTIPLE AGENCIES AND MISSIONS

A designated wilderness in the United States is almost always part of a larger federal agency jurisdiction, and the four federal land-managing agencies have wildlife-management missions and traditions predating the 1964 Wilderness Act. The situation is further complicated by the legal tradition dating back to the Magna Carta under which fish and wildlife devolved to state (not federal) authority, a status quo not changed by the Wilderness Act. For practical purposes, this means that the state manages the wildlife and the federal agency manages the habitat everywhere except in certain national parks. Obviously, careful coordination is needed, because populations and habitats are interdependent. There are differences of opinion about how far a state's legal authority extends to federal lands, and cutbacks in federal budgets can weaken the ability of federal agencies to maintain their traditional strength in some federal-state controversies over wilderness wildlife policies, for example, control of wild horse and burro populations to protect habitat on federal lands,

FIGURE 12.5 Supplementing natural water catchments for wildlife in wilderness areas raises management questions. In some national wildlife refuge wildernesses, legislation establishing the refuge or game range (and superseding wilderness designation within it) may call for management measures to ensure the survival of particular species. A rock tank catchment (left) in the Cabeza Prieta Wilderness, Arizona, helps ensure the survival of desert bighorn. A "guzzler" (center) and a water flue (right), both in the Desert National Wildlife Range, Nevada, provide vital water for wildlife. Photo courtesy U.S. Fish and Wildlife Service.

reintroduction of wolves and grizzlies, and fish stocking in high mountain lakes.

Management of wilderness wildlife is affected by a variety of constraints that have the potential to conflict with the goals of naturalness and solitude set forth in the Wilderness Act. For example, hunting and fishing are established activities in all national forest and BLM wildernesses, and in some national wildlife refuges; and fishing is allowed in some places in national parks. The Endangered Species Act (ESA), passed in 1973 (U.S. Public Law 93-205), requires preserving certain threatened or endangered species and critical habitat for those species in wilderness. This requirement may even call for using artificial devices and alterations if necessary, thereby conflicting with wilderness goals. Furthermore, ecological trends precipitated by previous management, such as unnatural conditions resulting from excluding fire, may be reflected dramatically in current wilderness wildlife populations.

Although we do not specifically review policies of U.S. agencies for wildlife in wilderness, we want to make the point that each agency responds to a different set of interest groups. Each has different philosophical traditions, a particular management legacy, and a different balance of disciplines among its professional personnel. These are important influences that directly and subtly shape management policy for wilderness (and its wildlife) within each agency. Funding for wilderness and wildlife management in wilderness is subordinate to other pressing concerns in all agencies.

LEGISLATIVE AND ADMINISTRATIVE DIRECTION

Wilderness management operates under a number of legislative and administrative constraints, including some directly related to wildlife. In the United States, the hierarchy of wilderness management direction includes the Wilderness Act of 1964 (U.S. Public Law 88-577); the *Code of Federal Regulations* that interpret and clarify that act; national wilderness management policies of the administering agencies as set forth in their manuals and handbooks; management plans for jurisdictions (such as parks, refuges, national forests, or public land in a state or region) that include wilderness; and individual wilderness management operating plans that specify how national direction will be applied on the ground. Other laws, such as the Endangered Species Act of 1973 (U.S. Public Law 93-205) and the Clean Air Act Amendments of 1977 (U.S. Public Law 95-95), further constrain management. In some cases, wilderness designation acts—laws designating individual areas as wilderness—specify how particularly controversial wilderness management issues will be handled (see Chapters 4 and 5), and these issues often affect fish and wildlife; for example, in wilderness designated by the 1994 California Desert Protection Act (U.S. Public Law 103-433), motorized access for fish and wildlife management purposes is allowed.

Grazing by horses, goats, sheep, and cattle is allowed by the Wilderness Act in some national forest and BLM wildernesses, with often substantial impacts on wildlife. The Colorado Wilderness Act of 1980 (U.S. Public Law 96-560) mandated guidelines for grazing of livestock in new wilderness areas in that state that have been adopted elsewhere (see Chapter 5). Grazing also brings pressures for predator control to the affected areas, additional human activity, pressure for developing fences and water sources, and consumption of forage, all with obvious local impacts on fish and wildlife.

Endangered species protection in the United States illustrates the problem of conflicting legislative direction. For example, the Wilderness Act limits managerial freedom to manipulate vegetative cover to perpetuate, improve, or alter an area's value for any particular purpose, including wildlife. On the other hand, the Endangered Species Act directs agencies to make sure no actions are taken that would "jeopardize the continued existence of any endangered species or threatened species or result in the destruction or adverse modification of critical habitat" (U.S. Public Law 93-205, 7a), including in wilderness. Thus, conforming to the Endangered Species Act may require management intervention (e.g., vegetative manipulation, predator control, and mechanized access) that would otherwise be restricted in wilderness. For example, in wilderness on the Cape Romain National Wildlife Refuge off the coast of South Carolina, raccoons prey on the eggs of the endangered sea turtle. A raccoon-trapping program to protect sea turtles, by itself an unnatural alteration of raccoon numbers, also requires periodic motorized access to tend traps.

Endangered species protection in wilderness may also require (under the Endangered Species Act) that managers intervene in natural processes. For example, taking condor eggs to be hatched and raised in captivity is not a natural solution, but it has been applied as a necessary approach to save the condor (Campbell 1984). In 1986, there were only six California condors remaining in the wild (many in wilderness) when adult condor number eight was captured. One condor produced twelve offspring in captivity before being released in April 2000 to join other condors flying free in California and Arizona.

The Historic Preservation Act of 1966 (U.S. Public Law 89-665) and related executive orders also result in activities that may be contrary to the intent of the NWPS—such as preserving or refurbishing historic structures that would otherwise be removed or allowed to deteriorate. Examples are cabins in some western wildernesses and lighthouses in wilderness on East Coast wildlife refuge islands and capes. Because such attractions draw excessive recreational use, wilderness wildlife may be locally impacted.

Wilderness in Alaska represents a special case because of provisions in ANILCA, the Alaska National Interest Lands Conservation Act of 1980 (U.S. Public Law 96-487). Following years of debate on ANILCA, Congress established wilderness areas in Alaska that are different in some important ways from those in the lower forty-eight states. In general, these units are open to most "customary" uses unless specifically closed following public hearings. Motorized access by plane, motorboat, or snowmobile is guaranteed to private inholdings such as homesteads, mining claims, trade centers, management cabins, and native lands. Subsistence hunting, trapping, and fishing may continue in wilderness, even in some designated national park units, and temporary facilities, equipment, and sometimes motorized access directly related to such activities are allowed. Subsistence taking of fish and wildlife provides important food, material, and livelihood for native and other communities (Muth 1990). However, subsistence uses continue to be controversial, with debate over who should qualify as a subsistence taker, and how the *need* for subsistence taking should be established (U.S. Department of Agriculture, Forest Service 2005).

DISEASE

In places, wilderness animals may be reservoirs of diseases that threaten humans or domesticated livestock. The dynamics of the brucellosis bacterium and the rinderpest virus, harbored by bison in Yellowstone and African antelope in the Serengeti, respectively, are among the best studied in this regard (Meagher and Meyer 1994; Dobson 1995). Brucellosis can induce abortions whereas rinderpest causes death outright in domesticated cattle. Concerns among livestock producers in areas as far apart as Wood Buffalo National Park, Canada; Yellowstone National Park, Wyoming; and Serengeti National Park, Tanzania, have led to heated debates about the merits and means of limiting potential for the exchange of these diseases between wild and domesticated bovids. Ironically, both brucellosis and rinderpest were introduced to native ungulates in these wild areas from infected cattle. More extreme yet in its effects is trypanosomiasis, otherwise known as sleeping sickness (Murray and Njogu 1989). This disease-causing parasite is endemic among native African herbivores and is typically transmitted to humans and domesticated livestock by the tsetse fly. Not only are the tsetse fly and trypanosomiasis associated with wild areas, this lethal combination of organisms has been implicated in maintaining vast tracts of de facto wilderness, especially in central Africa.

Although wild animals can harbor diseases harmful to domesticated animals, the reverse also is true. Wild canids are probably most affected by diseases transmitted by domesticated animals—primarily dogs. Rabies threatens the endangered Ethiopian wolf and wild dogs in Africa; distemper destroys bat-eared foxes in the Serengeti; leishmaniasis affects crab-eating foxes in Brazil; and mange imperils the extremely rare Mednyi Island arctic fox (Fanshawe et al. 1991; Goltsman et al. 1996; Gottelli and Sillero-Zubiri 1992).

Control of disease transmittal in wildlands or in transition zones between wilderness and agricultural lands is difficult and complicated. The effects of mange on arctic foxes and the effects of rinderpest on cattle and wild bovids have been curbed in the short term by administering vaccines. However, the long-term efficacy of these vaccination programs remains in doubt (Dobson 1995); proposals to control brucellosis have ranged from eradicating infected wildlife, to vaccinating wild and domesticated animals, to creating buffer zones. In all cases, development of solutions has been hindered either by lack of information, lack of resources, or conflict among management agencies and affected groups. In most cases, resolution of wilderness-related disease problems requires broad-scale integrated programs involving many wilderness and non-wilderness stakeholders, including wildlife- and land-managing agencies and diverse political interests (Wobeser 1994).

DEPREDATION

Wilderness areas are often a source of native animals that prey on domesticated animals or consume or trample crops on nearby agricultural lands. These problems range from depredation of oats and beehives by bears in northern regions, to predation on domesticated dogs in rural and urban areas by wolves and cougars, to trampling and depredation of grain by Asiatic elephants, to predation on cattle, goats, sheep, and yaks by large cats wherever they occur. In many instances, the depredating animals are injured, young, and inexperienced, or old and in decline. Quite often, they are dispersing juveniles, pushed out of prime wilderness habitats by older individuals. For example, most of the black bears involved in conflicts over beehives in the Peace River region of Alberta, Canada, were juvenile males, presumably dispersing from wildlands to establish their own adult ranges (Mattson 1990). In most cases, these sorts of problems are exacerbated by or are the direct outcome of human actions. Predation by cougars on dogs in California is associated with urban areas expanding into historical cougar range (Torres et al. 1996). Depredation of livestock is often the result of humans removing native prey and replacing it with cattle or other domesticated herbivores. This has been observed for wolves and bears in the southwestern United States (Brown 1983, 1985); for tigers in southeastern Tibet and snow leopards in adjacent southwest China (Qiu 1996; Schaller et al. 1988); for leopards, lions, and tigers in India (Oza 1974; Sekhar 1998); and for wolves in Italy (Meriggi and Lovari 1996).

Wild herbivores can cause changes in the vegetation of wilderness areas that are incompatible with management goals or the values of some people. In the case of introduced nonnative ungulates, this can take the form of unprecedented change in plant communities not adapted to large-scale herbivores. Such was the case with red deer and Himalayan thar introduced into New Zealand by European settlers wanting hunting opportunities. These species caused widespread damage to native vegetation and were then subjected to massive, sustained animal-control programs, but only after damage was well advanced (Lever 1985). An interesting twist to this scenario is provided by the history of mountain goats in Olympic National Park, Washington. Here, a species native to the continent was introduced outside of its recent historical range with damaging effects to sensitive native vegetation (Houston et al. 1994) and accompanying concerns about maintaining the integrity of the wilderness ecosystem. Among native ungulates, there are well-documented effects by white-tailed deer on woody vegetation in the eastern United States where populations of this species have surged to historically unprecedented densities (Porter 1996). More controversial are contentions that dense populations of wapiti in North America have caused accelerated erosion and the decline of aspen, willows, and other browse in wilderness parks ranging from Banff in Canada to Yellowstone and Rocky Mountain National Parks in the United States (Wagner et al. 1995). For both species, extirpation of native predators and declining harvest by humans are implicated as ultimate causes of "unnatural" densities of these native herbivores.

Nonnative fish have threatened the integrity of aquatic systems in wilderness areas in many parts of the world. In the western United States, lake trout introduced by humans, legally or otherwise, have decimated native cutthroat trout populations. Most recently, lake trout were discovered in Yellowstone Lake in Yellowstone National Park, where they threaten to reduce cutthroat trout populations by as much as 70 percent, with dramatic negative effects on the thirty species of vertebrates that prey on cutthroat, but not lake trout (Kaeding et al. 1996). Native and nonnative fish stocked in mountain lakes formerly without fish predators because of physical barriers have similarly devastated resident populations of native amphibians. For example, a study of 2,200 lakes in the John Muir Wilderness and Kings Canyon National Park backcountry in California concluded that fish stocking had an adverse effect on native biota, especially the mountain yellow-legged frog (Knapp and Matthews 2000; Matthews and Knapp 1999).

Although resolving depredation problems can be technically simple, it is often politically controversial. Where depredation on livestock or crops is a problem because native prey has been replaced by livestock, or because livestock have depleted native forage on which wilderness wildlife depend, removal of the livestock or restoration of native herbivores is an obvious solution. However, in most instances such actions are not acceptable because they threaten people's livelihoods and traditional lifeways. Monetary compensation for losses and selective removal of depredating animals have often been successful in both preventing future losses and recruiting the acceptance of native predators by livestock owners. Proponents of wolf reintroduction and recovery in U.S. wilderness have proposed such accommodation, such as the Defenders of Wildlife's Wolf Compensation Fund in Yellowstone National Park and the northern Rocky Mountains (Defenders of Wildlife 2008). Where there are perceived problems with overgrazing or overbrowsing by native ungulates, reintroduction of native predators and reinstitution of human harvest are obvious solutions. Ironically, a reduction in predator controls might solve the problem. However, violent prejudices against predators or hunting often preclude such responses and contribute to protracted debates about whether wild areas are overpopulated and the nature of the solution. Where nonnative species are destroying native species, the obvious wilderness solution is removal of the nonnatives. However, in most instances, there is no means for their complete removal; thus, the often-costly control programs are an uncertain, long-term solution. As with the solution of disease problems, resolving depredation problems usually requires the development of broad integrated programs, responsive to the outlooks and concerns of affected stakeholders and the interests and sensibilities of the involved management agencies.

HUMAN SAFETY

A host of animals, ranging from invertebrates, such as scorpions and tarantulas, to megaherbivores, such as bison and Cape buffalo, pose a threat to human safety in wilderness. Annually worldwide, up to three thousand people are seized, mutilated, and, in most cases, eaten by crocodiles (Alderton 1991). In Africa, 40 of 444 Nile crocodiles (nearly 10 percent) had human remains in their stomachs at the time of their death. Elephants in Southeast Asia regularly kill people and cause such damage in Sumatra that a "school" was set up in the 1980s to modify the behavior of individual elephants thought to be amenable to reform (McNeeley and Wachtel 1988). However, in most cases the perception of risk from wilderness wildlife is at variance with reality. Although cougars are thought by many to pose a grave threat to humans in the United States and Canada, only ten people were killed by cougars in these countries from 1890 to 1990 (Beier 1991). Even where an animal species causes significant harm to humans, most injury or death occurs at boundaries of settled and wild lands or is associated with incursions by humans involved with economic activities such as honey or firewood gathering in tiger reserves in India. In places, risk to humans within wilderness areas has been distinguished from risk incurred at boundaries. For example, in North America, about 1 out of 320,000 visitors is injured by black or grizzly bears in the wilderness settings of Kluane and Denali National Parks (Herrero and Fleck 1990). Wherever such rates have been compiled, rates of injury caused by animals are almost always trivial compared to injury by other causes.

Although some risk is incurred by the mere presence of large carnivores and herbivores or poisonous reptiles and invertebrates, in most cases there are identifiable factors, including human behavior, that strongly affect that risk. As with cases of depredation, the mammalian predators most prone to injuring humans are young, sick, or enfeebled. Among bears, risk of injury is aggravated where individual animals have either lost their fear of humans through constant exposure (i.e., habituated) or sought out foods made available by humans (Herrero and Fleck 1990). In general, habituation and unsanitary conditions around human facilities aggravate the risks posed by most large carnivores and herbivores. Ignorant behavior by humans greatly increases any risk. *Most injuries to humans caused by large herbivores, such as bison in U.S. national parks, are precipitated by humans unknowingly trespassing on an animal's personal space, or unwittingly posing a threat to a female's young.*

Given the factors that precondition risk of human injury by animals in wilderness settings, potential solutions are often straightforward. In virtually all cases, research and education that identify risks and preventative behavior are important (Sanyal et al. 2006). *Keeping wildlife wild (and thus shy of people) is a key. Sanitizing human facilities, such as campsites, can alleviate many conflicts with carnivores. Restriction of humans to certain places and times also can minimize volatile encounters and allow animals to avoid humans.* These kinds of restrictions are the basis for the outstanding safety record of the bear-viewing program at McNeil River Falls, Alaska (Aumiller and Matt 1994). *As a final recourse, removal of animals known to injure humans may be necessary, especially if this behavior has become habitual or was an obvious act of predation.* Enforcement of policies to reduce risks is a necessary adjunct to any of these measures.

REINTRODUCTIONS

Wilderness areas are often prime sites for reintroducing native species extirpated or endangered by historical human practices. This general principle is well illustrated by efforts to reintroduce or otherwise restore wolves and grizzly bears in the Rocky Mountains of the United States (see "Case Discussion" at the end of the chapter). The essential feature of habitat suitable for both species is extensive areas free of roads and human activity, which largely corresponds with designated or candidate wilderness areas. Other features critical to recovery of these carnivores, such as abundant native prey, also are associated with wildlands. Wilderness is a similarly desirable setting for reintroducing and restoring a host of other species, including ungulates such as the Arabian oryx, North American desert bighorn sheep, and Przewalski's horse. For these species, wilderness conditions help ensure freedom from human disturbance, competition from livestock for forage, and transmittal of disease from domesticated animals. However, most restoration projects are fraught with controversy because of feared curtailment of traditional human prerogatives and economic activities in reintroduction areas. Ignorance of the risks or impositions posed by species contemplated for reintroduction typically aggravates fears. As in many other areas, disagreement over goals for reintroduction of grizzly bears into central Idaho was further exacerbated by disagreement over goals between state and federal governments (MacCracken et al. 1994).

Reintroduction programs have been subject to numerous appraisals designed to elicit general lessons and principles. These principles apply as much in wilderness settings as in any other. *In general, the release of more animals over a longer period in good to excellent habitat within the core of historical range greatly enhances the likelihood of success* (Griffith et al. 1989; Wolf et al. 1996). *Omnivores and mammals are more likely to successfully establish compared to animals with specialized dietary needs and birds.* However, successful programs are built on more than biological principles. Reintroduction success is more likely where local residents are heavily involved or where economic, political, and cultural factors are otherwise given great consideration (Fig. 12.6), and where organizational arrangements are made to enhance learning, flexibility, and attention to local and regional power structures (Reading et al. 1997). As with the management of disease and depredation problems, successful reintroduction of animals into wilderness areas must be built on broad-based integrated programs that are attentive to biological, economic, political, social, and cultural perspectives.

FIGURE 12.6 Fishing can be an important part of an overall wilderness experience. An angler tries his luck at Johnson Lake in the Anaconda-Pintler Wilderness, Montana. Photo courtesy W. E. Steuerwald, U.S. Forest Service.

LIVESTOCK GRAZING

Worldwide, livestock use many of the areas that are designated wilderness by law, policy, or custom, and where they do not graze inside such areas, they may use those that are adjacent. Livestock can create problems when they become the prey of native carnivores. However, in most instances, depredation becomes an issue only when wildlands have been overstocked with livestock and native prey have declined. Although poaching and hunting by the pastoralists who accompany livestock contribute to native prey declines, competition for forage is often the most important factor. Such has been the case for Tibetan antelope (chirus), wild asses (kiang), wild yaks, and Tibetan gazelle in competition with increasing numbers of domesticated sheep, goats, and yak on the Tibetan plateau (Schaller 1998); Mongolian and goitered gazelles in competition with 33 million head of livestock in Mongolia (Lhagvasuren et al. 1999); wild reindeer in competition with domesticated reindeer in Siberia (Heptner et al. 1989); and Nilgiri tahr in competition with cattle and goats in south India (Mishra and Johnsingh 1998).

In the United States, livestock grazing was provided for in wilderness on the national forests and BLM lands where it was already established. Concerns about effects on native animals, historic and ongoing, have been greatest in the more arid southwestern regions of the country. Livestock have affected native animals by altering hydrologic regimes, depleting forage, displacing other herbivores, and providing a rationale for predator "control." Historically, livestock aggravated conflicts with bears, wolves, and cougars in the United States, leading to aggressive control programs that resulted in the extirpation of these animals over large parts of their range (Brown 1983, 1985). The negative effects of livestock on wilderness animals are well documented and numerous.

The simplest solution to the negative effects of livestock on wilderness animals is to remove the livestock. However, this solution is rarely politically viable because such action can harm the economic well-being of livestock owners and threaten traditional lifeways. Livestock owners can be politically powerful. Moreover, livestock owners can cause harm if they resort to increased poaching and other illegal practices in retaliation for policies that they perceive as harmful to their interests. In developing countries especially, alleviating grazing damage caused by domestic animals requires provision of alternate livelihoods or compensation to ranchers and increasing the levels of education and economic well-being in rural areas (Mattson 1997).

HUNTING AND POACHING

Humans kill wilderness animals for many reasons, including economic gain, prestige from trophies, acquiring food, resolution of conflict, and alleviating danger. Legal harvest of wildlife under established regulations distinguishes hunting from poaching. However, regardless of the legality, where the motives to harvest outstrip political will or resources for enforcement, wilderness animals subject to overharvest can decline in number, even to the point of extirpation. Problems of this nature are most acute in developing nations, where the acquisition of protein or cash or the preservation of property and other sources of livelihood can literally be a matter of life and death. In developed countries, it is easier to enforce sustainable harvests of wilderness animals.

Hunting to provide protein and to generate income for market hunters is commonplace in tropical areas of Africa, Asia, and South America, with often dire consequences for the affected wilderness animals. Larger species and those animals that live within nine miles of human settlements or protected area boundaries are often subject to unsustainable harvest, especially where there are gaps in enforcement of existing regulations (e.g., Campbell and Hofer 1995; Wilkie et al. 1998). Primates, antelopes, tapirs, pachyderms, and birds are all affected. For those animals that are unfortunate enough to sport highly valued body parts, such as pelts, horns, and medicinal bones and organs, space offers little protection. Poachers travel to the most remote regions because of often-substantial monetary rewards and this affects such animals as elephants, rhinos, bears, leopards, and tigers (e.g., Douglas-Hamilton and Douglas-Hamilton 1992; Penny 1988). However, in addition to regulations and enforcement, enduring solutions lie in tackling root causes such as poverty, malnutrition, human population growth, and the demand for animal body parts fostered in affluent countries (Heinen 1996; Poffenberger 1990; Shaw 1989). Although not addressing demand, some of the greatest successes in conserving tigers and elephants have been achieved by banning international trade in body parts of these species (Chadwick 1992).

In many parts of Europe and North America, conservation programs have been so successful that the debate in places may be whether there are too many rather than too few animals. For example, the alteration of vegetation communities associated with extraordinarily high densities of elk and deer in many U.S. and Canadian national parks is a consequence of eliminating native predators and terminating hunting or other herd-reduction programs (Wagner et al. 1995). In places, low densities of moose (called elk in Eurasia) have been ascribed to high densities of predators such as wolves and bears and a change in predator-prey dynamics induced by past human harvests of both predators and prey (Ballard and van Ballenberghe 1997). Whether it is desirable or not to harvest elk and deer in U.S. and North American parks, or to reduce densities of predators to increase moose populations in boreal wilderness areas, is a matter of goals and philosophies. To date, most parks have opted to minimize direct intervention with natural predators whereas provincial and state game-management agencies have killed predators under a variety of programs designed to boost moose numbers, with varying success.

Hunting can cause changes in prey behavior and the selective pressures acting on prey species. Whether such changes are philosophically good or bad is a function of values, but in wilderness the ecological goal is to preserve "natural" behaviors and selective pressures.

A greater concern about hunting is how harvest biased toward trophy males might affect the vigor and evolutionary trajectory of hunted species. Trophy hunting by highly technologically advanced and well-equipped humans, supported by guides and trackers, does not resemble natural predation under primeval conditions where predators, human or otherwise, usually took sick, young, or old animals and were relatively unsuccessful at capturing the most vigorous. However, in modern trophy hunting, males in their prime are the principal targets. Selective removal of such animals reduces the likelihood that their genes associated with vigor, strength, and breeding success will survive. Over the long term, this could produce species that are less resilient to environmental stressors, with less evolutionary potential. Although this issue is conceptually well founded and there is evidence for these effects, trophy hunting remains commonplace.

HUMAN-CAUSED DISTURBANCE

Human-caused disturbances or human-induced changes in wildlife behavior are of greatest concern when they affect the survival or reproduction of wildlife species or lead to changes in behavior, such as conditioning to humans or human-related foods, that increase chances of human injury or death. Although short-term responses by mammals to disturbances by humans have been observed for many species in wilderness settings, few effects on survival or reproduction have been documented (Anderson 1995) (Fig. 12.7). In fact, with frequent exposure, flight or other signs of alarm and stress abate for many mammals as they habituate to the presence of humans. Although this minimizes direct harm to responding animals, it reduces wildness and many habituated rodents and carnivores turn to exploiting human-related foods. When such food conditioning or habituation involves large mammals, chances of

FIGURE 12.7 University of Arizona wildlife researcher Paul Krausman studies desert bighorn sheep in the Pusch Ridge Wilderness near Tucson. Most wilderness wildlife is studied using simple means and methods, such as observation, that are compatible with wilderness conditions. Photo courtesy John C. Hendee.

human injury escalate (Herrero and Fleck 1990), and habituated animals threatening humans are likely to be shot.

In contrast to mammals, there is conclusive evidence of widespread effects of human intrusion on the reproductive success and survival of birds, especially waterfowl (Anderson 1995; Anthony et al. 1995; Bowles 1995; Knight and Gutzwiller 1995). Humans can easily displace breeding birds from their nests, resulting in increased chances of nest failure. Moreover, when humans visit nests, they often attract predators that eat the eggs. Under extreme circumstances, these kinds of human-related effects have been blamed for 70 percent to 90 percent declines in hatching rates among ground-nesting waterfowl. Wildlife photographers cause great harm because they often closely approach birds and their nests. For the same reasons, photographers have been singled out for disturbing mammals and contributing to undesirable changes in their behavior. In concept, the means for minimizing harm to nesting birds are straightforward. If the location and timing of nesting are known, public education is key and such areas can be placed off-limits to humans for the duration of the nesting season, as with bald eagle fishing sites in Glacier National Park, Montana.

Whether wildlife flee humans or not in wilderness settings raises interesting questions about the presumed "natural" behavior of the wild animals. Many wilderness visitors cherish the opportunity to view wildlife and at close quarters. However, functionally, where frequent encounters with humans result in rewards for wildlife, such as food, most animals lose what fear they have and tolerate or even seek out the presence of humans. Conversely, where encounters with humans typically lead to distress, injury, or death, most animals come to fear humans and flee our presence. The latter response seems more closely related to the wilderness goal of keeping wildlife wild.

LONG-TERM, LONG-DISTANCE EFFECTS

Increasingly, wilderness areas are islands in an ocean of humans and human-influenced environments. At the very least, most wilderness areas are small in comparison to surrounding lands managed for other goals. Moreover, wilderness boundaries are highly permeable—animals migrate in and out and humans and their effects intrude. As wilderness size declines, the proportional effects of "edge" increase to the point where no core area remains immune to outside influences (Noss et al. 1999). For example, most populations of northern birds and most populations of larger animals cross the boundaries of wilderness areas that often constitute the core of their ranges to use nearby agricultural or multiple-use lands outside. Whereas these animals might be protected from harassment and hunting inside wilderness boundaries, they are often disturbed, pursued, and killed by humans outside, with the effects on behavior, population size, and population structure transmitted back. Similarly, and especially in developing countries, humans often intrude from the periphery with undesirable effects that include poaching, grazing, deforestation, and other alterations of vegetation. This leaves only the largest wilderness areas with portions free of these smaller-scale transboundary effects, and such wilderness areas are now rare outside of harsh Arctic, boreal, or tropical environments.

Human impacts are also transmitted at broader scales. Oceanic harvest of anadromous fish such as salmonids or disruption by dams of their river travel routes impacts distant terrestrial spawning grounds. Loss of salmon in the upper Columbia River basin in the Pacific Northwest of the United States has been implicated in the declines and extirpations of several salmon predators, including grizzly bears (Merrill et al. 1999). In fact, grizzly bears occupying the Yellowstone ecosystem in Wyoming are a stellar example of a wilderness species affected by long-distance human impacts. In the eastern part of the ecosystem, grizzly bears obtain much of their energy from adult army cutworm moths that aggregate in alpine talus to feed on tundra nectar during crepuscular and nighttime hours (French et al. 1994). However, moth populations are heavily affected by distant agricultural practices where the moths complete their life cycle in croplands. Yellowstone's bears are similarly threatened by human introduction of threats to two of their critically important foods, such as spawning cutthroat trout and seeds of the whitebark pine tree (Reinhart et al. 2001). Introduction of the nonnative cutthroat predator, the lake trout, and the highly lethal fungal pathogen, white pine blister rust, threatens to substantially reduce numbers of cutthroat trout and white bark pine, respectively. Thus, even in a wilderness ecosystem of more than 10,000 square miles, wilderness animals are affected by human activities far outside the area.

At the regional and global scales, no wilderness area is immune to changes in air quality and climate regimes due to human activities that have changed and will continue to change the global climate. Prognoses for changes in regional precipitation are highly uncertain. However, climatologists are gaining confidence in their predictions for the warming of high latitudes, along with the melting of polar ice, the northern advance of warmer water in the Pacific, and possible retreat south of the Gulf Stream in the Atlantic. Animals of the most remote wilderness regions of the world—walruses, belugas, narwhals, ringed seals, and polar bears—stand to be devastated by such changes. The receding Arctic sea ice has reduced the hunting range of polar bears seeking seals as food among ice floes (Regehr et al. 2007) and caused polar bears to compete with brown bears along shorelines. Polar bears have been known to swim more than fifty miles out to sea in search of floes as platforms for hunting seals.

The northern advance of tree line stands to rob caribou and wild reindeer of much of their prime summertime barren ground forage. Salmon, already buffeted by loss of spawning grounds in inland streams, would lose much of their oceanic food in the Pacific. In small wilderness islands farther south, populations of wilderness-dependent and wilderness-associated species will likely be lost as suitable habitat declines in their current ranges (Dobson et al. 1999).

Transboundary effects yield straightforward management implications. If wilderness managers want to achieve their wildlife-related goals, their efforts must be integrated with those of others managing adjacent lands, even across international boundaries, the snow leopard being an Asian example (Singh and Jackson 1999) and the mountain caribou a North American example. Whether the above are part of such integrated management or not, wilderness managers must at least be aware of broader-scale transboundary influences and, where there is little option to control such effects at their source, strive to manage to mitigate them within wilderness boundaries. This kind of broad-scale thinking and integration is arguably one of the greatest challenges to land managers, including those responsible for wildlife in wilderness.

WILDERNESS MANAGEMENT OBJECTIVES AND GUIDELINES FOR WILDLIFE

What do all the foregoing examples and reasoning lead to in terms of what we should be trying to achieve (objectives) for wildlife in wilderness, and how we should go about doing that (guidelines)? This section recommends objectives for wildlife in the U.S. NWPS and suggests guidelines for wilderness management to meet those objectives.

The objectives for wilderness wildlife describe conditions to be sought with respect to wildlife in wilderness; they provide criteria against which to evaluate management policies and actions—that is, will such policies/actions move things toward such conditions? The objectives are deliberately broad to apply to all types of wilderness, regardless of agency jurisdiction. Again, although these objectives are aimed at the U.S. wilderness system, we believe there is considerable relevance to wildlife in wilderness systems around the world:

1. Seek natural distribution, numbers, population composition, and interaction of Indigenous species of wildlife.
2. Allow natural processes, as far as possible, to control wilderness ecosystems and their wildlife.
3. Keep wildlife wild, with its behavior altered as little as possible by human influence.
4. Permit viewing, hunting, and fishing where such activities are (a) biologically sound, (b) legal, and (c) carried out in the spirit of a wilderness experience.
5. Favor the protection and restoration of threatened and endangered species and wildlife dependent on or associated with wilderness conditions.
6. Minimize degradation of wilderness qualities—naturalness, solitude, and absence of permanent, visible evidence of human activity—while managing wildlife in wilderness.

These objectives for wildlife do not displace the book's overriding theme that a wilderness philosophy and ethic, based on an appreciation of all affected values, should guide all wilderness management actions. *Only necessary actions to achieve objectives are justified, and they must employ the minimum methods (tools, regulations, and force) that are required.* Here are guidelines for three key elements of wilderness wildlife in the U.S. NWPS: the role of (1) natural processes, (2) fishing and hunting, and (3) wildlife research in wilderness.

Allow Natural Processes to Shape Wilderness Habitat

Designated wilderness is ideally a place where nature rolls the dice, and the resulting naturalness, whatever its characteristics, defines that wilderness. Artificial-habitat manipulations that may be desirable for fish and wildlife management on non-wilderness lands and waters are not consistent with either the Wilderness Act or the wilderness ethic that should guide its implementation. When there are legal exceptions to the natural-process criterion, *the most natural practices and tools should be used in a manner exerting the minimum impact on wilderness naturalness and solitude.*

Wilderness means *natural.* Vegetation should reflect natural conditions, substantially unaltered by people and their influence. Lakes and streams should reflect undisturbed watersheds and channels. Animal life should approach natural numbers of native species; at some times and places this will mean reduction or loss of some wildlife populations. The principal task facing wilderness managers is to promote conditions that will allow natural processes to operate as freely as possible. Where it is clear that human actions have compromised natural processes, managers may have to mimic or simulate natural processes, for example by eliminating non-native species or introducing fire where it has been unnaturally suppressed. However, this should be done as briefly and naturally as possible. The idea that wilderness should be a "vignette of primitive America" should not suggest a static picture of a pre-European settlement landscape maintained in suspended animation by whatever techniques are needed. *Wilderness, rather, should be a vignette of natural dynamic forces characteristic of early America, operating as freely as possible.* Ecological change will be inevitable and constant, its velocity dependent on the ecosystem involved and global climate change. This means that there may be times in its ecological succession when a wilderness or portions thereof will not be a particularly good habitat for some wildlife species or particularly appealing to visitors.

Sometimes it will be possible even to restore aspects of a wilderness that was previously impacted by human influence, and natural recovery may occur. Native animals that have disappeared may be reintroduced or recover naturally. The effects of minor grazing or forest cutting can be erased, over time, by plant succession. Fire and other natural disturbances can initiate new cycles of plant and animal life, as they did before the coming of modern humans.

A natural-processes criterion argues for artificial control of native predatory animals, insects, or plant diseases only when resources are threatened outside the wilderness. We believe that in wilderness, outbreaks of insects or disease should simply run their course as a contribution to a constantly changing natural ecosystem. Of course, massive infestations of nonnative organisms originating outside wilderness can be a major threat, as with the introduced gypsy moth in Shenandoah National Park, Virginia; the balsam woolly aphid in declining stands of native Fraser fir in Great Smoky Mountains National Park; and the southern pine beetle and hemlock woolly adelgid in the southeastern United States. White bark pine is declining because of infestations by the nonnative white pine blister rust in the western U.S.

Where predators are present, they should be permitted their natural role. Where the predators are absent, however, prey populations may irrupt and unnaturally threaten the continued existence of a natural environment. In such a case, native predatory species might be reintroduced, their prey cropped, or both, the goal being to return the wilderness to natural conditions.

Moreover, again we must be cautious in our definition of "natural." In the United States, an Adirondack forest in the East, by virtue of natural plant succession, may have recovered from logging a century ago, and in some ways be more natural today than a western mountain forest never logged but held for the past seventy-five years in unnatural plant succession by efficient fire-prevention and fire-control programs. Because wilderness fire may be the most effective and appropriate habitat-management tool, fire management and wilderness wildlife management must go hand in hand.

Encourage Angling and Hunting Styles Compatible with Wilderness Experiences

Fishing is a traditional recreation activity in many wilderness areas and is allowed by the U.S. Wilderness Act under the direction of state fish and wildlife agencies. Fishing can achieve its finest quality in wilderness, can be a scarce wilderness-dependent experience, and can be a means to realize wilderness values. Wilderness is a place where one can partially experience or imagine a primitive experience of living off the land. Catching a few fish and eating them in camp with family or friends can contribute to the overall quality of a wilderness experience for many users. Such fishing is best as *part* of a wilderness experience, rather than the *reason* for a wilderness visit. Thus, *fishing regulations for wilderness should encourage a focus on the overall wilderness experience rather than on the size of the catch.*

Stocking of certain high lakes in many western states was begun years ago to transport to suitable waters those species prevented from entering by natural barriers. Some of these lakes are now in designated wilderness, and their stocking and the method used (aerial stocking) are among the most controversial wilderness management issues. There is no question that artificial stocking compromises the naturalness of wilderness, particularly the specific aquatic ecosystems that are affected (Knapp and Matthews 2000; Matthews and Knapp 1999). *Stocked fisheries are advisable in designated wilderness only where the practice is clearly established and carried out in a manner to minimize its impact on wilderness ecosystems, qualities, and the experiences of visitors.* The long-range policy question is whether any artificial stocking in designated wilderness is too harsh a conflict with naturalness. Although fishing is an attraction for some visitors to high mountain lakes, hikers and others were just as numerous and spent just as much time at the lakeshores (Hendee et al. 1977).

Ideally, wilderness hunting should be scientifically managed to ensure that the natural ecosystem is not impaired; just intensive and selective enough (with regard to age and sex) to protect the natural behavior and dynamics of game populations; mimicking as closely as possible the pattern of harvest by natural predators; with very little or no hunting of predators near the top of the food chain, such as wolves and grizzlies; and with no hunting where either a lack of knowledge or lack of regulatory staff makes the consequences on naturalness uncertain. In reality, *lack of control and knowledge about the effects of predation makes it impossible to predict with certainty how hunting affects wildlife naturalness. The need for a conservative approach is thus imperative.*

Where hunting is allowed in wilderness, it should feature those aspects of the sport that are uniquely wilderness-dependent—in keeping with wilderness criteria of naturalness, solitude, and contrast with civilization. It is not realistic to expect that all wilderness hunters would voluntarily restrict themselves to muzzle loaders or bow and arrows. However, conforming to a wilderness ethic implies restrictions on extravagant base camps or extensive communication equipment that alters the remoteness and solitude of the wilderness experience. Wilderness is the one place hunters can confront game on its terms, and this opportunity should be encouraged for those who can appreciate the unique values of such an experience. *In wilderness, hunting, like fishing, should be part of a wilderness experience and not the sole reason for it.*

Focus Research on Essential Priorities Using Appropriate Methods

In using wilderness as a wildlife research laboratory, scientists should be under the same constraints as everyone else not to degrade its naturalness or solitude. Although some instrumentation may be necessary to adequately characterize the biological systems of wilderness, physical structures as a general rule are not in keeping with wilderness character, nor are such visible techniques as painting large markings on live-trapped animals for later observation from an obtrusive helicopter.

We must recognize, however, that radiotelemetry and aircraft, used with discretion, are invaluable means—sometimes the only means—of obtaining data on the natural distributions and behaviors of large wilderness wildlife species. However, temperance is needed. For example, radio collars are an intrusion on wilderness and may constitute a form of harassment that threatened species such as grizzlies do not need and that possibly distorts natural behavior (Krausman and Hewert 1983). At the very least, radio collars can be color matched to the animal to minimize visual impact. Marking with paint or dye, colored tags, and streamers should be discouraged.

Many wildlife research opportunities are essentially wilderness-dependent; they require vast natural areas that may ultimately exist only in wilderness. Nonetheless, *the scientific community should be sure that investigations in wilderness seek essential data, that wilderness is not open to casual inquiry with intruding mechanization, and that the scholar above all is governed by a wilderness ethic. Wilderness wildlife research should, in short, be restricted by overriding principles of wilderness management: do only what is necessary and use the minimum methods, approaches, or tools required.* Wilderness wildlife research proposals need to be strongly justified as sources of information about natural processes that are necessary to wilderness wildlife protection, to improved understanding of global human impacts on nature, and thus to the continued well-being of humans on the planet. If sophisticated mechanization is allowed in wilderness for scientific purposes that are not so justified, it could set a precedent, opening the door to excessive development for other purposes, such as radar installations for monitoring air traffic, telecommunication installations, instrumentation of headwaters for water-yield information for commercial irrigation, climatological data gathering for weather forecasting, and so forth. Currently, agency policies permit sophisticated instrumentation and mechanization for wilderness wildlife research, usually on a case-by-case basis.

Long-standing questions about what factors control populations of ungulates and larger carnivores can only be studied in large, undisturbed areas (Peek 1989). Research to test competing hypotheses about how ungulate populations are naturally regulated (habitat vs. predation), in addition to other important hypotheses, will require relatively intact ecosystems found only in large wilderness areas. This work has important ramifications for management of native ungulates outside and inside wilderness (Fig. 12.8). In the United States, several benchmark studies of predator-prey and habitat relationships have been conducted in relatively undisturbed wilderness settings. For example, studies such as

FIGURE 12.8 Bison range in Yellowstone National Park, habitat for one of the largest remaining free-roaming bison herds in the world. Photo courtesy David Mattson, USGS Forest and Rangeland Ecosystem Science Center.

those of moose-wolf interaction in Isle Royale National Park (Allen 1979; McLaren and Peterson 1994), cougar-ungulate interaction in the Frank Church–River of No Return Wilderness (Hornocker 1970), and desert bighorn sheep habitat relationships in Arizona (Etchberger et al. 1989; Krausman et al. 1989) should be extended and replicated in other systems and with other species. Such research requires large wilderness areas.

The greatest scientific values of wilderness are based on the opportunity to study, over long periods of time, large ecosystems that are relatively undisturbed. Wilderness wildlife is an integral part of these ecosystems, sometimes dependent on and reflective of their degree of naturalness and solitude. Study of wilderness wildlife ecosystems can reveal how natural processes work, thereby providing baseline information for assessing human influence on the area, and from which management guidelines can be derived. Such information is increasingly important as resource management becomes more complex and human influences expand. At the same time, wildlife research methods in wilderness must be balanced against the need to maintain wilderness values.

SUMMARY

We recommend applying, as necessary, the foregoing objectives and guidelines for wilderness wildlife management. These guidelines are intended to supplement, not substitute for, wilderness management principles suggested in Chapter 7 and elsewhere in the book. However, it is unrealistic to seek one monolithic approach to wilderness wildlife management. Wilderness management needs to be tailored to the unavoidable constraints of the individual situations. For this reason, *a wilderness ethic is always needed as a guiding policy.* This chapter has sought to set the stage for such considerations. It has also sought to remind readers of the interdependency of wilderness and wildlife.

Wildlife is only one component of the wilderness resource and cannot be managed separately. Wildlife uses, such as recreational fishing and hunting, must be governed not just by the capacity to produce "surplus" populations of fish and game but by the impact of such uses on the naturalness of relationships among all components of the wilderness resource. A fish population in a particular lake may have the capacity to support a certain level of fishing pressure, but the acceptable level must also be measured against the aggregate impact of anglers and other users on the vegetation and soil around the lake or along the stream, and on the solitude of visitors' experiences.

Finally, we wish to emphasize the importance of wildlife in the wilderness web. Wilderness lets us view the natural processes by which the land and the living things on it have achieved their characteristic forms and by which they maintain their existence. It makes us aware of the incredible intricacies of plant and animal communities, their intrinsic beauty, and of contrasts with creeping degradation in many non-wilderness settings. These things add up to the great lesson of human-environment interdependency: that air, water, soils, trees, plants, insects, birds, fishes, mammals, and people are all part of the same scheme—an intricately woven landscape fabric. Snip one thread and the entire cloth begins to unravel; stitch up one tear and you begin to repair the whole.

Exposure to wilderness and wildlife, even indirectly, can be a doorway to the ecological understanding of the complete human interdependence with our environment and with life everywhere (Fig. 12.9). The ultimate wilderness and wildlife value may be their contribution to the development of a culture that will secure the future of an environment fit for life and fit for living, and to an appreciation of all those amenities that are inexorably linked to the prosperity of the human spirit.

FIGURE 12.9 Sighting a bald eagle near Anchorage, Alaska, a national symbol of the U.S. and a charismatic bird, is a long-remembered event by visitors in and around wilderness. Photo courtesy Carol Dawson.

STUDY QUESTIONS

1. What are four categories of wilderness wildlife? How important is wilderness to each?
2. What are some ways in which wilderness contributes to the preservation of wildlife?
3. How should hunting and fishing in wilderness be different from those activities outside of wilderness?
4. What are some wilderness uses allowed under the Wilderness Act, and some mandates under other legislation, that have the potential to complicate wilderness wildlife management?
5. What are some effects on wildlife of human activities in wilderness, on adjacent non-wilderness lands, and elsewhere?
6. What are some broad objectives for wilderness wildlife management that may be used as criteria for choosing and judging management guidelines (policies and actions)?
7. What are some of the problems wilderness managers face as they attempt to allow "natural processes" to shape wilderness wildlife habitat?
8. How can wilderness wildlife research benefit wildlife and humankind? How can wilderness managers and researchers preserve other wilderness values while conducting necessary research?

ACKNOWLEDGMENTS

This chapter was written for the first and second editions by John C. Hendee and Clay Schoenfeld. David Mattson of the U.S. Geological Survey's Forest and Rangeland Ecosystem Science Center, Northern Arizona University, Flagstaff, joined John C. Hendee in coauthoring a substantial revision for the third edition. The fourth and fifth editions include minor editing by John C. Hendee and Chad P. Dawson.

CASE DISCUSSION: WOLF REINTRODUCTION CONTROVERSY

In 1995, fourteen wolves were released in Yellowstone National Park from a trap-and-transfer agreement with Alberta, Canada. In 1996, another twenty wolves from British Columbia were released in Yellowstone and Montana. Wolves were confirmed in various adjoining states within two years, and the population of wolves recovered in the greater Northern Rockies (Idaho, Montana and Wyoming) to an estimated seventy-one breeding pairs and more than a thousand wolves by 2005 (U.S. Department of the Interior 2008, Defenders of Wildlife 2008). The wolf recovery projects were strongly supported by numerous individuals during various hearings regarding proposed plans and management actions and by the Defenders of Wildlife, National Wildlife Federation, and Friends of Animals (Defenders of Wildlife 2008).

The gray wolf was protected under the Endangered Species Act (U.S. Public Law 93-205) and was listed as endangered and under the management authority of the U.S. Fish and Wildlife Service until February 27, 2008 (U.S. Department of the Interior 2008). In spite of this federal protection, several wolves were killed by hunters and ranchers in the area after the reintroduction, who sometimes claimed they thought they were shooting coyotes or dogs running in the wild.

In an effort to compensate ranchers and livestock owners for losses due to wolf predation on livestock, the Defenders of Wildlife established a Wolf Compensation Fund from donations to help offset any verified economic losses by ranchers. Between 1987 and 1999 the fund paid $80,000 to 85 ranchers. The number of compensation payments increased as the wolf recovery grew, until by 2005 and 2006, the fund paid out more than $100,000 each year to 70 or 80 ranchers (Defenders of Wildlife 2008).

During 1995 to 2007, several serious legal actions were taken to block the wolf reintroduction by the American Farm Bureau, Montana Stock Growers, Wyoming Farm Bureau, and others (Defenders of Wildlife 2008). In addition, Idaho, Montana, and Wyoming proposed to the FWS in 2003 through 2007 that they be allowed to manage wolves under state wolf management plans they developed, following the delisting of wolves in the region from protection under the ESA (U.S. Department of the Interior 2008). In 2006, the Idaho Department of Fish and Game announced plans to radio-collar one wolf in each pack, including those in the Frank Church–River–of–No–Return Wilderness and the Selway-Bitterroot Wilderness.

In 2007, the FWS proposed removing the gray wolf from the endangered listing under the ESA and allowing states to manage wolf populations according to their state management plans. In 2008, the FWS completed the process to remove the gray wolf from ESA protection and returned wolf management back to state agencies (U.S. Department of the Interior 2008). Returning wolf management back to the state fish and wildlife agencies means that public hunting and other wolf control programs would be implemented, such as allowing states to kill wolves in areas where deer and elk populations are declining and allowing landowners to kill wolves caught attacking dogs or livestock in Idaho, Montana, and Wyoming. However, in 2022, a federal court decision was made to protect wolves in all of the lower 48 states, except the Northern Rocky Mountain region, and that court decision is being challenged in federal courts with the final outcome not yet known.

CASE DISCUSSION QUESTIONS

1. Are gray wolves in the northern Rockies and Yellowstone National Park wilderness-dependent or wilderness–associated wildlife?
2. Which types of groups of people are for and against wolf reintroduction and why? What are their arguments for and against wolves?
3. How does this wolf reintroduction case match up with the six wildlife-management objectives proposed in this chapter?
4. Some recent research (U.S. Department of the Interior 2008) suggests that wolves are moving elk out of meadow areas that had been overgrazed by elk and allowing for recovery of the natural vegetation and other wildlife species, thereby reducing elk populations. Others claim that elk are declining due to direct wolf predation and that wolves have become too abundant. How do you focus research on these issues to decide what needs to be done for naturalness?

REFERENCES

Alderton, David. 1991. *Crocodiles and Alligators of the World.* New York: Facts on File.

Allen, D. L. 1979. *Wolves of Minong.* Boston, MA: Houghton Mifflin.

Allen, Durward. 1966. The preservation of endangered habitats and vertebrates of North America. In: Darling, F. Fraser; Milton, John, eds. *Future environments of North America.* Garden City, NY: The Natural History Press, pp. 22–37.

Anderson, Stanley. 1995. Recreational disturbances and wildlife populations. In: Knight, Richard L.; Gutzwiller, Kevin J., eds. *Wildlife and Recreationists.* Washington, DC: Island Press, pp. 157–181.

Anthony, Robert G.; Steidl, Robert J.; McGarigal, Kevin. 1995. Recreation and bald eagles in the Pacific Northwest. In: Knight, Richard L.; Gutzwiller, Kevin J., eds. *Wildlife and Recreationists.* Washington, DC: Island Press, pp. 223–241.

Aumiller, Larry D.; Matt, Colleen A. 1994. Management of McNeil River State Game Sanctuary for viewing of brown bears. *International Conference on Bear Research and Management.* 9: 51–61.

Bailey, Theodore N. 1993. *The African Leopard: Ecology and Behavior of a Solitary Felid.* New York: Columbia University Press.

Ballard, Warren B.; van Ballenberghe, Victor. 1997. Predator/prey relationships. In: Franzmann, Albert W.; Schwartz, Charles C., eds. *Ecology and Management of the North American Moose.* Washington, DC: Smithsonian Institution Press, pp. 247–273.

Barnes, R. F. W.; Lahm, S. A. 1997. An ecological perspective on human densities in the central African forests. *Journal of Applied Ecology.* 34: 245–260.

Beier, Paul. 1991. Cougar attacks on humans in the United States and Canada. *Wildlife Society Bulletin.* 19: 403–412.

Berger, Joel; Cunningham, Carol. 1994. Active intervention and conservation: Africa's pachyderm problem. *Science.* 263: 1241–1242.

Bowles, Ann E. 1995. Responses of wildlife to noise. In: Knight, Richard L.; Gutzwiller, Kevin J., eds. *Wildlife and Recreationists.* Washington, DC: Island Press, pp. 109–156.

Brown, David E., ed. 1983. *The Wolf in the Southwest: The Making of an Endangered Species.* Tucson: University of Arizona Press.

———.1985. *The Grizzly in the Southwest.* Norman: University of Oklahoma Press.

Campbell, Ken; Hofer, Heribert. 1995. People and wildlife: Spatial dynamics and zones of interaction. In: Sinclair, A. R. E.; Arcese, Peter, eds. *Serengeti II: Dynamics, Management, and Conservation of an Ecosystem.* Chicago: University of Chicago Press, pp. 534–570.

Campbell, Sheldon. 1984. Restocking the wilderness with captive-bred animals: The California condor story. In: Martin, Vance; Inglis, Mary, eds. *Wilderness: The Way Ahead.* Middleton, WI: Lorian Press, pp. 108–113.

Caro, Tim M.; O'Doherty, Gillian. 1999. On the use of surrogate species in conservation biology. *Conservation Biology.* 13: 805–814.

Carr, Archie. 1973. *The Everglades.* New York: Time-Life Books. 184p.

Chadwick, Douglas H. 1992. *The Fate of the Elephant.* San Francisco: Sierra Club Books.

Clark, Tim W.; Curlee, A. Peyton; Minta, Steven C.; Kareiva, Peter M., eds. 1999. *Carnivores in Ecosystems: The Yellowstone Experience.* New Haven, CT: Yale University Press.

Cronin, Matthew A.; Ballard, Warren B.; Bryan, James D.; Pierson, Barbara J.; McKendrick, Jay D. 1998. Northern Alaska oil fields and caribou: A commentary. *Biological Conservation.* 83: 195–208.

Cutler, M. Rupert. 1980. Wilderness decisions: Values and challenges to science. *Journal of Forestry.* 78(2): 74–77.

Czech, Brian; Krausman, Paul R. 1999. Controversial wildlife management issues in southwestern U.S. wilderness. *International Journal of Wilderness.* 5(3): 22–28.

Dasmann, Raymond F. 1966. *Wildlife Biology.* New York: John Wiley.

Davis, George D. 1989. Preservation of natural diversity: The role of ecosystem representation within wilderness. In: Freilich, Helen R., comp. *Wilderness Benchmark 1988: Proceedings of the National Wilderness Colloquium;* January 13–14, 1988; Tampa, FL. Asheville, NC: U.S. Department of Agriculture, Forest Service, Southeastern Forest Experiment Station, pp. 76–82.

Defenders of Wildlife. 2008. A chronology of wolf recovery in the northern Rockies. www.defenders.org/programs_and_policy/wildlife_conservation/imperiled_species/wolves; accessed January 31, 2008.

Dobson, Andy. 1995. The ecology and epidemiology of rinderpest virus in Serengeti and Ngorongoro Conservation Area. In: Sinclair, A. R. E.; Arcese, Peter, eds. *Serengeti II: Dynamics, Management, and Conservation of an Ecosystem.* Chicago: University of Chicago Press, pp. 485–505.

Dobson, Andy; Ralls, Katherine; Foster, Mercedes; Soulé, Michael E.; Simberloff, Daniel; Doak, Dan; Estes, James A.; Mills, L. Scott; Mattson, David; Dirzo, Rodolfo; et al. 1999. Corridors: Reconnecting fragmented landscapes. In: Soulé, Michael E.; Terborgh, John, eds. *Continental Conservation.* Washington, DC: Island Press, pp. 129–170.

Douglas-Hamilton, Ian; Douglas-Hamilton, Oria. 1992. *Battle for the Elephants.* New York: Viking.

Dublin, Holly T. 1995. Vegetation dynamics in the Serengeti-Mara ecosystem: The role of elephants, fire, and other factors. In: Sinclair, A. R. E.; Arcese, Peter, eds. *Serengeti II: Dynamics, Management, and Conservation of an Ecosystem.* Chicago: University of Chicago Press, pp. 71–90.

Etchberger, Richard C.; Krausman, Paul R.; Mazaika, Rosemary. 1989. Mountain sheep habitat characteristics in the Pusch Ridge Wilderness, Arizona. *Journal of Wildlife Management.* 53(4): 902–907.

Fanshawe, John H.; Frame, Lory H.; Ginsburg, Joshua R. 1991. The wild dog—Africa's vanishing carnivore. *Oryx.* 25: 137–146.

French, Steven P.; French, Marilynn G.; Knight, Richard R. 1994. Grizzly bear use of army cutworm moths in the Yellowstone ecosystem. *International Conference on Bear Research and Management.* 9: 389–399.

Gasaway, William C.; Stephenson, Robert O.; Davis, James L.; Sheperd, Peter E. K.; Burris, Oliver E. 1983. Interrelationships of wolves, prey, and man in interior Alaska. *Wildlife Monographs.* 84: 1–50.

Goltsman, M.; Kruchenkova, E. P.; Macdonald, D. W. 1996. The Mednyi Arctic foxes: Treating a population imperilled by disease. *Oryx.* 30: 251–258.

Goodall, Jane. 1986. *The Chimpanzees of Gombee: Patterns of Behavior.* Cambridge, MA: Belknap Press of Harvard University Press.

Gottelli, Dada; Sillero-Zubiri, Claudio. 1992. The Ethiopian wolf—an endangered endemic canid. *Oryx.* 26: 205–214.

Griffith, Brad; Scott, J. Michael; Carpenter, John W.; Reed, Christine. 1989. Translocation as a species conservation tool: Status and strategy. *Science.* 245: 477–480.

Heinen, Joel T. 1996. Human behavior, incentives, and protected area management. *Conservation Biology.* 10: 681–684.

Hendee, John C.; Clark, Roger N.; Dailey, Thomas E. 1977. Fishing and other recreation behavior at roadless high lakes: Some management implications. Research Note PNW-304. Portland, OR: U.S. Department of Agriculture, Forest Service, Pacific Northwest Forest and Range Experiment Station.

Hendee, John C.; Pitstick, Randal. 1993. The growth of environmental and conservation-related organizations: 1980–1991: *Renewable Resources Journal.* 10(2): 6–11.

Heptner, V. G.; Nasimovich, A. A.; Bannikov, A. G. 1989. *Mammals of the Soviet Union.* Vol. I: *Ungulates.* New York: E. J. Brill.

Herrero, Stephen; Fleck, Susan. 1990. Injury to people inflicted by black, grizzly, or polar bears: Recent trends and new insights. *International Conference on Bear Research and Management.* 8: 25–32.

Hochbaum, H. Albert. 1970. Wilderness wildlife in Canada. In: McCloskey, Maxine E., ed. *Wilderness: The Edge of Knowledge.* San Francisco: Sierra Club, pp. 23–33.

Hornocker, Maurice. 1970. An analysis of mountain lion predation on mule deer and elk in the Idaho Primitive Area. *Wildlife Monographs.* 21.

Houston, Douglas B.; Schreiner, Edward G.; Moorhead, Bruce B. 1994. *Mountain Goats in Olympic National Park: Biology and Management of an Introduced Species.* NPS/NROLYM/NRSM-94/25. Washington, DC: National Park Service.

Jachmann, H.; Berry, P. S. M.; Imae, H. 1995. Tusklessness in African elephants: A future trend. *African Journal of Ecology.* 33: 230–235.

Kaeding, Lynn R.; Boltz, Gary L.; Carty, Daniel G. 1996. Lake trout discovered in Yellowstone Lake threaten native cutthroat trout. *Fisheries.* 21: 16–20.

Kellert, Stephen R. 1985. Public perceptions of predators, particularly the wolf and coyote. *Biological Conservation.* 31: 167–189.

Knapp, Roland; Matthews, Kathleen. 2000. Nonnative fish introductions and the decline of mountain yellow-legged frog from within protected areas. *Conservation Biology.* 14: 428–438.

Knight, Richard L.; Gutzwiller, Kevin J., eds. 1995. *Wildlife and Recreationists.* Washington, DC: Island Press.

Krausman, Paul R.; Hervert, John J. 1983. Mountain sheep responses to aerial surveys. *Wildlife Society Bulletin.* 11(4): 372–375.

Krausman, Paul R.; Leopold, Bruce D.; Seegmiller, Rick F.; Torres, Steven G. 1989. Relationships between desert bighorn sheep and habitat in western Arizona. *Wildlife Monographs.* 102.

Leopold, A. Starker. 1966. Adaptability of animals to habitat change. In: Darling, F. Fraser; Milton, John, eds. *Future Environments of North America.* Garden City, NY: The Natural History Press, pp. 66–75.

Leopold, Aldo. 1933. *Wildlife Management.* New York: C. Scribner's Sons.

———. *A Sand County Almanac, and Sketches Here and There.* New York: Oxford University Press.

Lever, Christopher. 1985. *Naturalized Mammals of the World.* New York: Longman.

Lhagvasuren, B.; Dulamtseren, S.; Amgalan, L.; Mallon, D.; Schaller, G.; Reading, R.; Mix, H. 1999. Status and conservation of antelopes in Mongolia. *Scientific Research of the Institute of Biological Sciences.* 1: 96–108.

MacCracken, James G.; Goble, Dale; O'Laughlin, Jay. 1994. Grizzly bear recovery in Idaho. Idaho Forest, Wildlife and Range Policy Analysis Group, Report No. 12. Moscow: University of Idaho.

Matthews, Kathleen R.; Knapp, Roland A. 1999. A study of high mountain lake fish stocking in the U.S. Sierra Nevada Wilderness. *International Journal of Wilderness.* 5(1): 24–26.

Mattson, David J. 1990. Human impacts on bear habitat use. *International Conference on Bear Research and Management.* 8: 33–56.

———. 1997. Wilderness-dependent wildlife: The large and the carnivorous. *International Journal of Wilderness.* 3(4): 34–38.

McLaren, B. E.; Peterson, R. O. 1994. Wolves, moose, and tree rings on Isle Royale. *Science.* 266: 1555–1558.

McNeeley, Jeffrey A.; Wachtel, Paul S. 1988. *Soul of the Tiger.* New York: Doubleday.

Meagher, Mary; Meyer, M. E. 1994. On the origin of brucellosis in bison of Yellowstone National Park: A review. *Conservation Biology.* 8: 645–653.

Mercure, Delbert V., Jr.; Ross, William M. 1970. The Wilderness Act: A product of congressional compromise. In: Cooley, Richard; Wandesforde-Smith, Geoffry, eds. *Congress and the Environment.* Seattle: University of Washington Press, pp. 47–64.

Meriggi, Alberto; Lovari, Sandro. 1996. A review of wolf predation in southern Europe: Does the wolf prefer wild prey or livestock? *Journal of Applied Ecology.* 33: 1561–1571.

Merrill, Troy; Mattson, David J.; Wright, R. Gerald; Quigley, Howard B. 1999. Defining landscapes suitable for restoration of grizzly bears *Ursus arctos* in Idaho. *Biological Conservation.* 87: 231–248.

Milton, John P. 1972. The web of wilderness. *Living Wilderness.* 35(16): 14–19.

Mishra, Charudutt; Johnsingh, A. J. T. 1998. Population and conservation status of the Nilgiri tahr *Hemitragus hylocirus* in Anamalai Hills, south India. *Biological Conservation.* 86: 199–206.

Muchaal, Pia K.; Ngandjui, Germain. 1999. Impact of village hunting on wildlife populations in the western Dja Reserve, Cameroon. *Conservation Biology.* 13: 385–396.

Murie, Margaret E.; Murie, Olaus. 1966. *Wapiti Wilderness.* New York: Knopf.

Murray, Max; Njogu, A. R. 1989. African trypanosomiasis in wild and domestic ungulates: The problem and its control. *Symposium of the Zoological Society of London.* 61: 217–240.

Muth, Robert M. 1990. Community stability as social structure: The role of subsistence uses of natural resources in southeast Alaska. In: Lee, Robert G.; Birch, Donald R., Jr., eds. *Community and Forestry.* Boulder, CO: Westview Press.

Nash, Roderick. 1970. Wild-deor-ness. In: McCloskey, Maxine E., ed. *Wilderness: The Edge of Knowledge.* San Francisco: Sierra Club, 36.

Noss, Reed F.; Dinerstein, Eric; Gilbert, Barrie; Gilpin, Michael; Miller, Brian J.; Terborgh, John; Trombulak, Steve. 1999. Core areas: Where nature reigns. In: Soulé, Michael E.; Terborgh, John, eds. *Continental Conservation.* Washington, DC: Island Press, pp. 99–128.

Olson, Sigurd. 1963. *Listening Point.* New York: Knopf.

Oza, G. M. 1974. Conservation of the Asiatic lion: Now limited to Gujarat State, India. *Biological Conservation.* 6: 225–227.

Peek, James M. 1989. Natural regulation of ungulates: What constitutes a real wilderness? *Wildlife Society Bulletin.* 8(3): 217–227.

Pelton, Michael R.; van Manen, Frank T. 1996. Benefits and pitfalls of long-term research: A case study of black bears in Great Smoky Mountains National Park. *Wildlife Society Bulletin.* 24: 443–450.

Penny, Malcolm. 1988. *Rhinos: Endangered Species.* New York: Facts on File.

Poffenberger, Mark, ed. 1990. *Keepers of the Forest: Land Management Alternatives in Southeast Asia.* West Hartford, CT: Kumarian Press.

Porter, William F. 1996. Management of overabundant species in protected areas: The white-tailed deer as a case example. In: Wright, R. Gerald, ed. *National Parks and Protected Areas: Their Role in Environmental Protection.* Cambridge, MA: Blackwell Science, pp. 223–248.

Qui, Ming J. 1996. Tiger-human conflict in southeastern Tibet. *Oryx.* 30: 5–6.

Reading, Richard P.; Clark, Timothy W.; Griffith, Brad. 1997. The influence of valuational and organizational considerations on the success of rare species translocations. *Biological Conservation.* 79: 217–225.

Regehr, E. V.; Lunn, N. J.; Amstrup, S. C.; Stirling, I. 2007. Effects of earlier sea ice breakup on survival and population size of polar bears in western Hudson Bay. *Journal of Wildlife Management.* 71(8): 2673–2683.

Reinhart, Daniel P.; Haroldson, Mark; Mattson, David J.; Gunther, Kerry A. 2001. Effects of exotic species on Yellowstone's grizzly bears. *Western North American Naturalist.* 61(3): 277–288.

Russell, Andy. 1971. *Grizzly Country.* New York: Knopf.

Sanyal, N.; Krumpe, E. E.; Vanormer, C. 2006. Modeling encounters between backcountry recreationists and grizzly bears in Glacier National Park. *International Journal of Wilderness.* 12(3): 24–29, 48.

Schaller, George B. 1963. *The Mountain Gorilla: Ecology and Behavior.* Chicago: University of Chicago Press.

———. 1998. *Wildlife of the Tibetan Steppe.* Chicago: University of Chicago Press.

Schaller, George B.; Junrang, Ren; Mingjiang, Qiu. 1988. Status of the snow leopard *Panthera uncia* in Quinghai and Gansu Provinces, China. *Biological Conservation.* 45: 179–194.

Sekhar, Nagothu U. 1998. Crop and livestock depredation caused by wild animals in protected areas: The case of the Sariska Tiger Reserve, Rajasthan, India. *Environmental Conservation.* 25: 160–171.

Shaw, R. Paul. 1989. Rapid population growth and environmental degradation: Ultimate *versus* proximate factors. *Environmental Conservation.* 16: 199–208.

Simberloff, Daniel. 1998. Flagships, umbrellas, and keystones: Is single-species management passé in the landscape era? *Biological Conservation.* 83: 247–257.

Sinclair, A. R. E.; Arcese, Peter, eds. 1995. *Serengeti II: Dynamics, Management, and Conservation of an Ecosystem.* Chicago: University of Chicago Press.

Singer, Francis J. 1984. Effects of wild pig rooting in a deciduous forest. *Journal of Wildlife Management.* 48(2): 466–473.

Singh, Jaidev "Jay"; Jackson, Rodney. 1999. Transfrontier conservation areas: Creating opportunities for conservation, peace and the snow leopard in Central Asia. *International Journal of Wilderness.* 5(3): 7–12.

Tardiff, Sandy, E.; Stanford, Jack A. 1998. Grizzly bear digging: Effects on subalpine meadow plants in relation to mineral nitrogen availability. *Ecology.* 79: 2219–2228.

Terborgh, John; Estes, James A.; Paquet, P.; Ralls, K.; Boyd-Heger, Diane; Miller, Brian J.; Noss, Reed F. 1999. The role of top carnivores in regulating terrestrial ecosystems. In: Soulé, Michael E.; Terborgh, John, eds. *Continental Conservation.* Washington, DC: Island Press, pp. 39–64.

Torres, Steven G.; Mansfield, Terry M.; Foley, Janet E.; Lupo, Thomas; Brinkhaus, Amy. 1996. Mountain lion and human activity in California: Testing speculations. *Wildlife Society Bulletin.* 24: 451–460.

Trefethen, James B. 1975. *An American Crusade for Wildlife.* New York: Boone and Crockett Club.

U.S. Department of Agriculture, Forest Service. 2005. What can I do in wilderness? Alaska National Interest Lands Conservation Act (ANILCA) and wilderness in national forests. Ketchikan, AK: U.S. Department of Agriculture, Forest Service.

U.S. Department of the Interior. 2008. Endangered and threatened wildlife and plants. Final rule designating the northern Rocky Mountain population of gray wolf as a distinct population segment and removing this distinct population segment from the federal list of endangered and threatened wildlife. 50 *Code of Federal Regulations,* Part 17: February 27, 2008; *Federal Register.* 70(39): 10,514–10,560.

U.S. Public Law 88-577. The Wilderness Act of September 3, 1964. 78 Stat. 890.

U.S. Public Law 89-665. National Historic Preservation Act of October 15, 1966. 80 Stat. 915.

U.S. Public Law 93-205. Endangered Species Act of December 28, 1973. 87 Stat. 884.

U.S. Public Law 95-95. Clean Air Act of August 7, 1977, as amended. 42 Stat. 7401.

U.S. Public Law 96-487. Alaska National Interest Lands Conservation Act of December 2, 1980. 96 Stat. 487.

U.S. Public Law 96-560. Colorado Wilderness Act of December 22, 1980. 94 Stat. 3265.

U.S. Public Law 103-433. California Desert Protection Act of October 31, 1994. 108 Stat. 4473.

Wagner, Frederic H.; Foresta, Ronald; Gill, R. Bruce; McCullough, Dale R.; Pelton, Michael R.; Porter, William F.; Salwasser, Hal. 1995. *Wildlife Policies in the U.S. National Parks.* Washington DC: Island Press.

Wilkie, David S.; Curran, Bryan; Tshombe, Richard; Morelli, Gilda A. 1998. Modeling the sustainability of subsistence farming and hunting in the Ituri Forest of Zaire. *Conservation Biology.* 12: 137–147.

Wilson, Edward O. 1984. Million-year histories: Species diversity as an ethical goal. *Wilderness.* 48(165): 12–17.

Wobeser, Gary A. 1994. *Investigation and Management of Disease in Wild Animals.* New York: Plenum Press.

Wolf, C. Magdalena; Griffith, Brad; Reed, Christine; Temple, Stanley A. 1996. Avian and mammalian translocations: Update and reanalysis of 1987 survey data. *Conservation Biology.* 10: 1142–1154.

CHAPTER

13

Potential Threats to Wilderness Resources and Values

INTRODUCTION

Formal protection of wilderness conditions and natural processes starts with wilderness designation and then relies on good planning and management to achieve the purposes of the Wilderness Act of 1964 (U.S. Public Law 88-577), to "secure for the American people of present and future generations the benefits of an enduring resource of wilderness." These National Wilderness Preservation System areas, individually and collectively, exist in a global context and the nation's larger, natural landscape and public-land heritage. *Wilderness is affected both internally and externally by many environmental and social conditions, influences, and changes that may threaten wilderness resources and values, now or in the future.*

Since 1964, the NWPS has grown from 9 million acres to more than 111.8 million acres. These areas encompass more than 4 percent of the land area of the United States and more than 2 percent of the land area of the contiguous forty-eight states, with additional designations expected in the future. *The intended resources and values of the Wilderness Act—which we interpret in this book as naturalness, opportunities for solitude, and wildness—will become even more scarce in the future because the vast remainder of the land will have been urbanized, cultivated, roaded, harvested, mined, drilled, scraped, heavily managed, or otherwise developed for human uses.* Thus, the 4.5 percent of the landscape set aside to remain "untrammeled" as wilderness will also represent remnants of many ecosystems, wild conditions, and natural landscapes that have either disappeared or been altered. Given the growth and urbanization of our population in the United States and the world, wilderness may represent the remaining wild conditions and natural landscapes in the future, plus one of the only remaining opportunities for solitude and some forms of unconfined and primitive recreation. Wilderness, like parks—another wildland resource—is jeopardized worldwide by at least forty-eight specific threats to ecological and social values, as noted by Machlis and Tichnell (1985) in their survey of threats to the world's parks identified by park managers.

The loss of naturalness and wildness in wilderness may be incremental, but ultimately, in the long term, it is dramatic. The exponential development and use of our natural-resource base over the last one and one-half centuries are having profound direct and indirect impacts on all other resources of the planet. Consider the profound changes to the natural ecosystems and landscape in the brief human history on the planet. Drive the Lewis and Clark Trail today from St. Louis to the Pacific Northwest, and read their diaries and narratives of discovery from only two hundred years ago. Or ask anyone who is fifty to eighty years old or a lifelong wilderness traveler what changes they have seen in the landscape or wilderness in their lifetime.

Of course, population growth and economic development are driving the worldwide use of resources and contribute directly and indirectly to the loss of naturalness and wildness. Environmentally concerned individuals and wilderness visitors, while enjoying the resources, also contribute incrementally to losing naturalness and solitude (Fig. 13.1). The dramatic changes of the past and present on our global environment and the unnatural influences manifested by them will have a profound effect on wild landscapes and wilderness areas of the world.

FIGURE 13.1 Visitors hike a popular trail on Mount Washington, New Hampshire. They have views of the Presidential Range–Dry River Wilderness on the White Mountain National Forest. High-use recreation access is one threat to wilderness resource conditions and to visitor wilderness experiences. Photo courtesy Chad P. Dawson.

The incremental loss of naturalness is subtle and harder to detect than more dramatic changes, especially by the majority of people who live and work in urbanized environments and do not have the experience or training to observe and comprehend the longer-term and more gradual changes that are occurring within wilderness. We rely on scientists, managers, and wilderness advocates who study and monitor changes to report them; yet the information may be overlooked by the public and be regarded as less relevant than more urgent day-to-day matters. Moreover, the ubiquitous focus on today's urgent issues makes it difficult to appreciate the important long-term impacts and changes that are occurring in wilderness areas over time. Only through long-term resource monitoring in wilderness and other natural, protected areas can we document and understand the global changes that are taking place. This is essential not just for wilderness protection today and in the future, but to safeguard the human condition.

Many public controversies about wilderness arise when a particular user group or special-interest group contends that their interests are not being served in some way by agency management practices in a particular wilderness. It is well documented that recreational use of wilderness areas is growing, and the public continues to support and be committed to the concept of wilderness and the NWPS. However, there is concern among the federal agencies, such as the U.S. Forest Service, that the resources committed to protect and manage the wilderness have not kept pace with their needs (Dombeck 1999). Some wilderness managers believe that we should be planning wilderness for a century or longer time frame and not the ten-year plans typical of most federal and state agencies. A longer planning horizon might help avoid short-term contentiousness between vested interests and managers who focus primarily on today's needs and wants, often to the detriment of long-term naturalness and "untrammeled" conditions. *Managing the long-term threats to wilderness resources and values is an important challenge, if we are to achieve the purposes of the Wilderness Act of 1964—to preserve wilderness now and for future generations.*

WILDERNESS CONDITIONS AND VALUES AFFECTED BY THREATS

We define threats to wilderness as a general concept and focus on change agents or processes that negatively or adversely impact wilderness resource conditions and values. Although we present some examples of impacts on wilderness resource conditions and values, our focus is mostly on the *change agents—what causes the impacts. These change agents or causes of impacts on natural conditions and values come directly or indirectly from human activities, because natural disturbances (e.g., lightning-caused fires, volcanoes, and hurricanes) are considered natural processes.* For example, increasing visitor use of wilderness areas (i.e., a change agent) can impact opportunities to experience wilderness values due to crowding, visitor conflicts, loss of solitude, and from impacts on wilderness resources such as loss of vegetative ground cover at campsites and trail erosion.

Our intent is to summarize threats (i.e., causes of impacts) to wilderness and the NWPS. We do not intend *to exhaustively list the impacts* to wilderness conditions (i.e., ecosystem effects) nor the impacts to wilderness values (i.e., human-dimension effects). In this chapter, we draw on the recent, excellent work by scientists at the Aldo Leopold Wilderness Research Center in Missoula, Montana (Cole 1994; Cole 2001b; Cole and Landres 1996; Landres et al. 1998a; Landres, Morgan et al. 1999; Landres, Brunson et al. 2000; Cole, McCool et al. 2000).

Changes that affect wilderness can be considered threats when they are not part of the natural conditions and processes that shaped, or were historically part of, the wilderness. The Wilderness Act of 1964 (U.S. Public Law 88-577) describes wilderness as "an area where the earth and community of life are untrammeled by man . . . retaining its primeval character and influence . . . protected and managed so as to preserve its natural conditions . . . affected primarily by the forces of nature." The natural conditions and attributes of wilderness character that are of most concern have been previously summarized as including air quality, aquatic systems, landforms, soils, vegetation, fish and wildlife, ecosystems and landscapes, and wilderness experiences (Cole 1994; Cole and Landres 1996). These threats will be expanded in this chapter, based on available information (Dawson et al. 2016).

In defining natural conditions and naturalness, we think of the ecological concept of historical range of variation (HRV), which recognizes that natural conditions are not static, but in constant change (Aplet and Keeton 1999; see Chapter 10). For example, during the two-hundred- to three-hundred-year period since white settlement began in the United States, a given wilderness site might have appeared quite different from time to time, although remaining within its HRV, due to ecological succession and perturbations of fire, wind, and natural forces.

Naturalness is a relative concept and is often applied to wilderness, but it is difficult to relate to threats or degradation to wilderness conditions and functions. Landres, White et al. (1998b) have defined *naturalness* as the compositions, structures, and functions within an area that are unaffected by contemporary (roughly since the time of European settlement) anthropogenic (human) influences. This definition of naturalness requires explicit spatial bounds, temporal bounds, and value judgments. Natural variability applies too because none of the wilderness conditions remains the same over time or at other locations. *Natural variability* is the natural change in the compositions, structures, and functions of an ecological system over time (temporal variability) and from one place to another (spatial variability), and this variability can be described statistically (Landres, White et al. 1998b).

In wilderness, the management guideline is to allow ecosystems to function within their HRV, without being manipulated. This does not mean returning ecosystems to any preexisting condition (e.g., presettlement conditions) or maintaining them in any particular set of conditions, but rather to allow natural processes and variability to continue (Landres, Morgan et al. 1999). As the impacts of global climate change have become more evident, there is increased discussion of how managers can address the changes in HRV (see Chapter 10).

Wildness refers to the absence of human controls and manipulation on the naturalness and natural variability within an area (Landres, Brunson et al. 2000). While the biological aspects of *naturalness* are a requisite component of wilderness, the concept of *wildness* carries with it the connotation of social and biological values that benefit society because the area is unmodified by civilization and human manipulation.

Threats to wilderness conditions, attributes, and values are complex, interactive, and both internal and external to wilderness. *Internal threats to wilderness may come from remnants of previous activities, as well as the so-called permitted but nonconforming uses under the Wilderness Act,* like grazing, mineral exploration, and mining. For example, previous human activities, such as logging, homesteading, or road building, can affect natural conditions even in the present-day wilderness. Grazing is an obvious impact, even though it may be legal. These impacted conditions may recover to some degree once natural processes are allowed to operate freely, but such impacts are not likely to be completely removed. *External threats* may come from air pollution, acid rain, pesticide drift from agriculture, or forest-management practices on nearby lands. External threats may originate some distance from a particular wilderness and still alter its biota, like hunting migratory wildlife species, invasions of nonnative plant species, or some larger influences like global climate change.

Actual or potential impacts from changing conditions will vary depending on a myriad of factors ranging from social and political trends to global climate shifts. *Further complicating any prediction about the impacts is that threats to wilderness conditions can be multiple and interactive so that the cumulative changes and cascading effects are more extensive or far-reaching than any single threat.* Given the complexity of ecosystems and the natural environment, and our limited knowledge of historic conditions and processes, predicting impacts from changing conditions is problematic (Fig. 13.2). Research and professional experience, however, suggest that there are some high-priority threats, and these are discussed in the following sections.

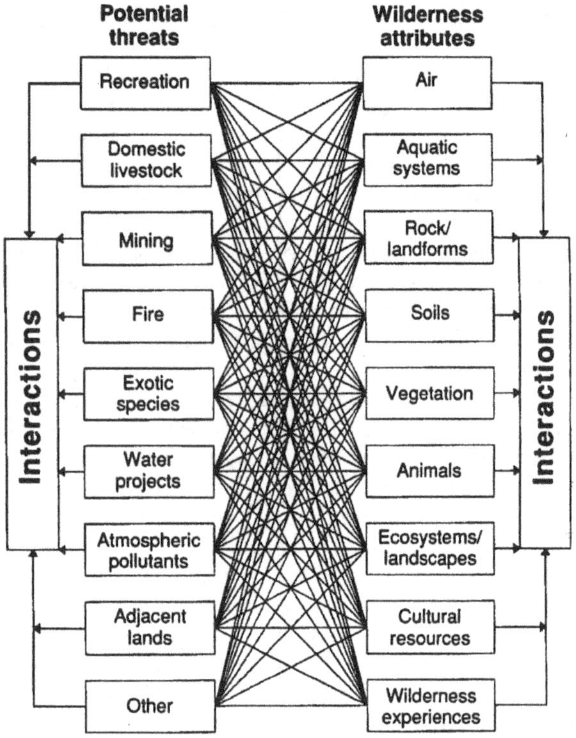

FIGURE 13.2 Cole and Landres (1996, 169) charted the complex linkages and interactions among potential threats to wilderness and wilderness attributes.

TABLE 13.1 The Top Ten Potential Threats to the NWPS, for 2014 to 2034, Indicated by Planners and Managers as High to Very High Threats (Dawson et al. 2016)

Potential Threats	Percentage
Lack of political and financial support for wilderness protection and management	74
Invasive species	56
Disconnected urban audiences	53
Adjacent land management and use	44
Legislation designating wilderness with compromised wilderness conditions or special provisions for management	41
Wildland fire suppression and management	39
Motorized and mechanical equipment trespass and illegal use	38
Fragmentation and isolation of wilderness as ecological islands	38
Aircraft noise and airspace reservations	37
Increasing and changing noncommercial recreation	35

POTENTIAL THREATS TO WILDERNESS

A 2014 survey of NWPS planners and managers asked managers to rate twenty-four potential threats to wilderness (Dawson et al. 2016). The top ten threats rated as high or very high by planners and managers, for the period 2014 to 2034, are shown in Table 13.1. Threats to wilderness and their impacts are so numerous that it would be too exhaustive to present all of them here (Cole, McCool et al. 2000). For example, grazing (a potential threat) can have different impacts on vegetation, soil, erosion, water quality, and native animal and fish populations. Because our knowledge about these threats and how to manage them in wilderness is limited, we organize and limit our discussion to them.

A study of threats to wilderness in the U.S. Forest Service's Northern Region (Cole 1994) provides a prototype matrix of how to organize potential threats and information about them. For example, comparing the importance of impacts to wilderness, such as those caused by nonnative species of flora and fauna (noxious weeds, nonnative fish and wildlife) and the state of knowledge about those impacts suggests that more knowledge is needed to fully understand ecosystem impacts. When the knowledge of potential threats is combined with the importance of each for wilderness management, a more comprehensive pattern can emerge for scientists and managers. Cole (1994) compiled such a table for the U.S. Forest Service's Northern Region and concluded that a better understanding of the effects of fire on the wilderness attributes of soils, vegetation, and ecosystems was one of the most important information needs for wilderness managers.

The following information focuses on the categories of threats to wilderness that are mentioned in the relevant literature (Dawson et al. 2016) and that have come up in our discussions about wilderness threats with wilderness managers and researchers. From this information, we identified nineteen threats to wilderness conditions and values, including *fragmentation and isolation of wilderness as ecological islands; threatened and endangered species; increasing commercial and public recreation use; livestock grazing; nonnative species; administrative access, facilities, and intrusive management; adjacent-land management and use; inholdings; mining; wildland fire suppression; air quality; water projects and water quality; advanced technology; motorized and mechanical equipment trespass and legal use; aircraft noise and airspace reservations; urbanization and encroaching development; global climate change; legislation that designates new wilderness areas with compromised wilderness conditions; and lack of political and financial support for wilderness protection and management.*

These internal and external threats are acting on wilderness in the United States with considerable variation among geographic regions and between remote and more urban-proximate wilderness. Some of these threats are the focus of management actions to minimize or mitigate them, and these threats may diminish over time even as new threats and controversies appear.

FRAGMENTATION AND ISOLATION OF WILDERNESS AS ECOLOGICAL ISLANDS

The threat here is that wholly functioning wilderness ecosystems cannot be adequately protected if they are not large enough and/or are disconnected from surrounding natural habitat. The more than 800 units within the NWPS tend toward smaller rather than larger units. The NWPS does include some very large areas, such as the four wildernesses in Alaska that include more than 5 million acres, but beyond those there are only 19 wildernesses that are more than 1 million acres in size within the NWPS (Landres and Meyer 2000).

The most common (42 percent) wilderness size is from 10,000 to 50,000 acres (Landres and Meyer 2000). Thus, because of their relatively small size, most wilderness areas are ecological "islands," and more vulnerable to external forces than larger areas, especially those of 1 million acres or more. Due to the small size of most wilderness areas and the lack of connecting corridors of wildland between them, wildlife naturalness is strained because entire home ranges of some species are not provided. Furthermore, beyond the necessary movement for seasonal changes, a mixing of species populations is also essential to genetic diversity upon which their long-term health and survival depend. *One of the greatest impacts of fragmentation may be on the species that require large undisturbed home ranges, such as wolves and grizzly bears.* Further aggravating fragmentation in other countries may be transboundary effects where international borders may increase isolation of wilderness from other habitats, with major impacts on key wildlife species (Singh and Jackson 1999) (see Chapter 12).

The lack of large or "corridor-connected" wilderness units is most pronounced in the eastern United States. Landres and Meyer (2000, 9) report that "western states have an average of 37 wildernesses per state, compared to an average of five wildernesses per state in the east. Western states have an average of 5.4 percent of their land area in wilderness, compared to the eastern states average of 0.5 percent land area in wilderness. The average size of a wilderness in the western states is 101,610 acres, compared to 23,657 acres in the eastern states." Thus, fragmentation threats to wilderness ecosystems and landscapes is greater in the East and further complicated because creating corridors to connect ecological islands is more difficult and expensive.

So how can the undesirable impacts of fragmentation be minimized (i.e., loss of genetic diversity and protection of wholly functioning ecosystems)? We think the following actions would help. *Pursue additional wilderness designations and protection of other park and wildland areas, with connecting corridors that create larger blocks of natural functioning ecosystems.* The campaign to establish a mosaic of protected lands from Yellowstone to the Yukon (Y2Y) with connecting corridors is an example (Posewitz 1998). Furthermore, *manage wildlands that are related to or adjacent to wilderness with policies that support or restore ecosystem integrity for the combined area* (Landres, White et al. 1998b). Proposals calling for restoring wild conditions on lands adjacent to wilderness, such as in the northeastern U.S. (Klyza 2001), require enhanced communications and coordination among different land-management agencies and units. The biosphere-reserve concept, which may have wilderness in a central core with gradations of decreasing wildness and coordinated management radiating from the central area, is a good strategy that has been applied in the United States (e.g., Yellowstone National Park) and many other countries (Singh and Jackson 1999).

THREATENED AND ENDANGERED SPECIES

Losing threatened and endangered species, or sometimes intrusive actions to save them, can threaten wilderness naturalness and solitude. Some threatened and endangered species are wilderness-dependent, requiring naturalness and solitude, whereas others merely find favorable, protected conditions in wilderness. The impacts of adjoining-land-management activities and global changes add to the pressures on threatened and endangered species in wilderness, and the smaller or more fragmented the area, the greater the pressure.

Protecting endangered species may require (and this is legally allowed in U.S. wilderness) special efforts, including manipulating the wilderness environment to favor the continuation of such species. At first, this may seem like the most logical course of events to "restore" that which is threatened and endangered. However, there is concern that such well-intentioned activities may cause a secondary and tertiary wave of perturbations within the ecosystem that we cannot predict due to our limited knowledge of interactions and interdependencies. For example, some people may question whether efforts to protect bighorn sheep in the Cabeza Prieta Wilderness, Arizona, by providing water are necessary or whether such action sustains bighorn sheep populations at unnaturally high levels (or, for that matter, other species benefiting from the water). Furthermore, is the invasion of naturalness and solitude by truck traffic to supply the water worth the impacts? The answers to such questions are not easy and pose a two-edged sword. Losing an endangered species would impact naturalness but so might intrusive efforts to save it. (Fig. 13.3)

The many external threats and global changes compound pressures on threatened and endangered species in wilderness such that their probability for survival may be reduced, in spite of what wilderness and wildlife managers might do to save them. In addition, with any intervention to solve one problem, there may be undesirable side effects. Therefore, managers should be cautious and humbled by the

FIGURE 13.3 Wilderness managers must balance what some may perceive as intrusive water facilities and using mechanized vehicles against sustaining an endangered species, in this case desert bighorn. This four-part facility, a catchment, underground storage, water diversion, and walk-in drink tanks, in Arizona's North Maricopa Mountains Wilderness (BLM) eliminates the use of a tank truck and helps the sheep that are cut off from their natural water supply by agricultural and residential development. Photo courtesy Chad P. Dawson.

complexity of ecosystems and not assume that we can surgically or technologically solve all the problems that occur (see Chapter 10).

INCREASING COMMERCIAL AND PUBLIC RECREATIONAL USE

Wilderness has visitation levels, at least in places, that cause unacceptable biological and ecological impacts. *As commercial and public recreation use of wilderness increases, impacts accumulate, and managers then increasingly regulate and control use* (Cole 1996a; also see Chapters 14, 15, 16). Recreation use and visitor management are already intense in some high-use locations in the more popular wilderness areas and in some places, some times, and some seasons, even in many large and remote areas. *Therefore, the impacts of commercial and recreation use and efforts to control them may threaten wilderness resources and values.* The choice described by David Cole (2000) was natural, wild, uncrowded (meaning regulated), or free (unregulated): which of these should our wilderness be? Clearly, if we want to preserve wilderness resources it will take considerable regulation of visitors in many places. However, such regulation takes away the freedom and spontaneity that many visitors associate with wilderness experiences.

The social and ecological impacts of users and their activities on the wilderness flora and fauna, especially at heavily used sites, have been the subject of concern by managers for decades (see Chapter 15; Hammitt and Cole 1998; Liddle 1997; Manning 1997; Marion et al. 1993). For example, the biophysical impacts from recreational activities were studied in six wilderness areas in Oregon and Washington. The impacts were found to be considerable in specific sites (Cole, Watson et al. 1997), but may represent only a small percentage of the total wilderness area. However, these impacts are highly noticeable to users, especially at heavily visited campsite areas or vistas along lakeshores and become a high priority for management restrictions or closure of such areas to allow for recovery.

The need to balance wilderness recreation use among user groups and against environmental protection standards is increasing because of rising demand and because commercial and public recreation users often have divergent motivations, expectations, and management preferences (Hunger et al. 1999). Although commercial outfitters provide access to members of the public who might not otherwise have a wilderness or wild river experience, they do charge fees and operate to make a profit. The cost of acquiring a commercial outfitter and guiding permit can be very high, tens or hundreds of thousands of dollars, which requires steep user fees to offset and still make a business profit. This profit motive and competition between outfitters often encourages a "catering to customer comforts and convenience" orientation, which may further aggravate impacts on naturalness and solitude. This also creates pressures on the wilderness managing agencies against tightening environmental standards for uses taking place under permits for which fees are charged, especially as they become more dependent on recreation revenues.

We believe the guiding principles for allocating wilderness uses should be to protect wilderness conditions and processes, as well as opportunities for solitude and primitive recreation. Already allocation decisions between commercial and public recreation use have proven difficult administratively, and conflicts between these two categories of use have stalled completion of some Wilderness Management Plans. The Forest Service has not enforced social standards for wilderness visitation in some high-use locations and fears that limiting visits there would only displace use to pristine locations, with an overall loss of wilderness values (Oye 2001). However, others are critical of the Forest Service for *not* enforcing social and environmental standards in its plans and demanding a rollback in use, a strict nondegradation policy, and use limits wherever ecological conditions or solitude are threatened (Worf 2001).

Management response to the express needs of special-interest recreation groups will be difficult for a variety of reasons. Many special-interest groups in wilderness are also some of the strongest supporters of wilderness, and their continued support is essential. Some accommodation may be necessary because wilderness management is also political and becoming more so (Mackey 1998). Besides, the need to maintain and build partnerships with established users, just to get wilderness work done, makes them special stakeholders. Furthermore, special-interest groups can often rationalize their requests as appropriate in wilderness because they may have been doing that activity or something similar for years or decades, with no negative feedback (e.g., no specific regulations against the activity or no enforcement if there was such a regulation). For example, placing fixed anchors on rock-climbing routes in wilderness may seem reasonable to climbers (e.g., safe anchors for ascents, rappelling, or rescue), and where fixed anchors have been tolerated for years, enforcement now may be difficult.

LIVESTOCK GRAZING

Livestock grazing by cattle, sheep, and recreational pack stock is allowable in wilderness where it existed before wilderness designation, but it is a definite threat to naturalness and social values. The use of public lands for grazing is a long-standing one, but increasingly controversial, and is more so in wilderness because of its impacts on naturalness of the ecosystem. Permitting grazing in wilderness was a compromise that was politically necessary to achieve passage of the Wilderness Act, but it conflicts with the wilderness goals of maintaining primeval character, preserving natural conditions, and leaving wilderness affected primarily by the forces of nature (McLaran 1990). Thus, livestock grazing, like mining, is called a "nonconforming but allowed" use. Fortunately, because of the economic costs of operating in remote wilderness locations, many grazing permits are no longer actively used.

Besides the impacts on soil and water (Derlet et al. 2008), consuming and trampling vegetation for forage may be in direct competition with wildlife, or cause changes in the composition of the forage base, which may also affect wildlife (Murray 1997). Livestock grazing is selective of plant species and influences plant composition on wilderness meadows and range. Such influences do not necessarily favor the wildlife that must compete for the same forage. Moreover, where livestock graze there are accompanying pressures for predator control. A related concern is that livestock and associated management activities and facilities may negatively impact visitors' experiences. Another negative impact on social values is the public perception of the loss of naturalness and wildness through the direct ecological impacts of grazing and associated management activities on wilderness conditions.

The legal basis and various issues surrounding grazing as a permitted use in wilderness are discussed further in Chapter 14. Grazing is an allowable (even though nonconforming) use of wilderness, and this was further confirmed in the Colorado Wilderness Act (U.S. Public Law 96-560) in 1980 and its accompanying committee report (see Chapter 5). In addition, grazing is strongly protected by the politically powerful livestock industry. *However, there is a continuing struggle between that industry and wilderness supporters to keep grazing and its accompanying facilities within limits that will sustain wilderness resources and values. Wilderness managers need public support for a strong, pro-wilderness stand on this issue.*

NONNATIVE SPECIES

The invasion of wilderness ecosystems by nonnative species is an increasing, direct threat to wilderness naturalness. Nonnative and nonindigenous species are increasing in places in wilderness at an alarming rate and can trigger ecosystem changes and displace native species. For example, noxious weeds like knapweed, star thistle, cheatgrass, leafy spurge, and others have outcompeted native species and are rapidly spreading (Asher and Harmon 1995). Secondary impacts may then result from control efforts, which may include biological, chemical, and physical control mechanisms. Although the goal of such action is to maintain or restore natural conditions and natural functioning ecosystems, some management actions can cause other problems because often previous conditions cannot be restored without other perturbations to the ecosystem (Cole 1996a; Worf 1997). Manipulation for restoration may be appropriate if it is the minimum tool to restore natural conditions, but there is often a lack of baseline information about natural and wild conditions and processes to guide such intrusive management (see Chapter 10). The loss of wildness is one of the largest concerns because management actions and impacts may not be justified even if they can restore naturalness.

How do nonnative species get introduced to wilderness? Some have been introduced inadvertently and others deliberately. For example, noxious weeds may be introduced through pack-stock feces, even though "clean" feed may be packed for the duration of the pack trip in wilderness. Burros and wild horses may also migrate in and out of wilderness, importing seeds. Some noxious-weed seeds come in on the feet and clothes of hikers, or in the droppings of seed-eating birds. It is hard to prevent these inadvertent introductions. However, some nonnative species have been deliberately introduced, such as eastern brook trout stocked in western wilderness lakes. Even fish stocking beyond natural population levels, or where fish did not naturally occur, may impact other natural, native species like yellow-legged frogs displaced by fish stocking in wilderness of the High Sierra of California (Matthews and Knapp 1999). The translocation of mountain goats to areas where they were not indigenous is another example of deliberate introduction of a nonindigenous species (Carter 1997).

Furthermore, attempts to remove nonnative species can cause additional changes beyond restoring the preexisting "natural" conditions. Wilderness is not isolated from the global climate changes that are occurring and which may favor nonindigenous species, such as noxious weeds whose spread is encouraged in some places by drought and warming of the climate. Some nonnative pests are so firmly established that they have defied serious efforts to control them in nonwilderness; for example, blister rust invading white bark pine in high-elevation wilderness sites in the West and chestnut blight in the East. In such cases, as with infestation of native pests, allowing natural ecological response and succession to the place is probably the best approach. *It seems wilderness managers, in trying to preserve naturalness, must often decide which is a more serious intrusion—the problem or the cure. The cure, in the end, may not work and could wreak havoc of its own in a natural ecosystem and diminish wildness.*

ADMINISTRATIVE ACCESS, FACILITIES, AND INTRUSIVE MANAGEMENT

Wilderness management can threaten naturalness with too-frequent mechanized access, excessive facilities, and methods that are too intrusive. Mechanized access to wilderness by managers is legal under the Wilderness Act when it is the minimum method to accomplish a legitimate wilderness or endangered species purpose, such as facility construction and maintenance (e.g., fencing in an approved grazing allotment), forest-fire suppression, search-and-rescue operations, accessing inholdings, and so forth. Such management activities may be in support of any legitimate wilderness purpose, such as visitor use, grazing, mining, commercial outfitting, maintaining historic structures, and trail construction. The test is whether such access is the minimum necessary method to accomplish the purposes of the Wilderness Act.

Mechanized access can be a threat to wilderness naturalness and solitude, in spite of a manager's best intentions. So, rigorous interpretation of what is the minimum method to accomplish the purposes of the Wilderness Act will go a long way toward protecting the wilderness from mechanized access, excessive facilities, and intrusive management. Such a minimum-tool requirement becomes even more important with passage of wilderness laws like the California Desert Protection Act (U.S. Public Law 103-433), which provides for mechanized access to support fish and wildlife management (not just endangered species) and law enforcement in the 69 BLM wilderness areas it established.

Some examples of administrative management impacts on wilderness are evident in wildlife and fisheries management. For example, such management can include the removal of nonindigenous species, such as mechanical capture of feral burros, aerial application of chemicals to kill "undesirable" fish species before reintroducing native species, fish stocking, removal of nonindigenous species by physical or chemical applications, water developments (guzzlers) for wildlife and their servicing in arid environments, and mechanized wildlife surveys or research methods. Czech and Krausman (1999) reported that 53 of 273 wilderness areas in the southwestern United States acknowledged using controversial wildlife-management activities, a great number of which involved mechanized access. Often the purpose of the activity is questionable, such as stocking to maintain unnaturally productive fisheries (Duff 1995); translocation or maintenance of nonindigenous (but huntable) populations of mountain goats; or construction, reconstruction, maintenance, and servicing of water developments for desert bighorn sheep.

For today's wilderness managers, who may not be familiar with primitive tools used in earlier days, such as crosscut saws, pack strings, hand tools, and pioneering skills, it may be especially tempting to use mechanized access and tools, or to establish and maintain supporting facilities. However, a lax attitude toward minimum tools by managers will threaten wilderness naturalness and solitude by making approval of mechanized intrusions too easy.

ADJACENT-LAND MANAGEMENT AND USE

Wilderness can be impacted by management activities and use on adjacent lands. The influence of external factors on wilderness management is a concern for managers because they often have little or no control over what happens on lands adjacent to wilderness areas (Stokes 1996; Kelson and Lilieholm 1997; Dawson et al. 2016). For example, timber harvests and grazing disrupt naturalness in adjacent ecosystems with ripple effects on wilderness. These are probably best illustrated by big-game wildlife examples but may be as important to other species. Suppression of insects, disease, and fire, and predator control all impact adjacent ecosystems. Roads serving management and use on adjacent lands may make access to wilderness easier or harder, thereby affecting the amount or location of recreation use. Everything is connected within and adjacent to wilderness. *Wilderness planners and managers need to expand their awareness and communication beyond wilderness boundaries and seek to coordinate adjacent-land management activities to minimize their impacts on wilderness* (Landres, White et al. 1998b) (Fig. 13.4).

INHOLDINGS

Inholdings of private or public lands within wilderness areas can create impacts because inholders have a right to reasonable access and use of their property. The impacts can be similar to the dilemmas of adjacent-land management. Some inholdings are dramatic, such as old mining claims, homesteads, and ranches, which may provide their modern owners prime access to wilderness surroundings. In recent years, there have been proposals to build resorts on private inholdings, such as in Montana's Absaroka-Beartooth Wilderness. Sometimes motorized access is granted to inholders on primitive roads, by aircraft, or by boat. Inholdings may be used by commercial outfitters to provide sites for support facilities and services (e.g., cabins, stock facilities, aircraft landing fields), access (e.g., interior private roads), and visitor facilities (e.g., hiker huts, outfitter camps, private dwellings). The most extensive examples of this type of access are under ANILCA (U.S. Public Law 96-487), which provides for various types of access to private or public land inholdings within Alaskan wilderness areas (Tanner 2004).

Federal agencies try to buy out wilderness inholdings, but they are constrained by their overall acquisition priorities and available funds. In the eastern United States, agencies can even condemn lands under authority of the so-called Eastern Wilderness Act (U.S. Public Law 93-622)—although to date it has not been done. A more recent threat to wilderness has been the subdivision and development (or proposal for same) of wilderness inholding lands by developers or speculators who perceive the opportunity to make a profit from the development or may be more likely to inflate the value of the lands and pressure a federal agency to buy them out at a high price.

FIGURE 13.4 Keeping vehicles out of wilderness is a difficult management task requiring road closures and signs listing prohibited uses (left), wilderness signage to identify entry to wilderness (center), and boundary marking (right). Jim Mahoney, formerly of the BLM Phoenix office, inspects a shotgun-blasted wilderness boundary sign. Installing, maintaining, and monitoring wilderness boundary signs are an important management challenge. Photo courtesy Marilyn Hendee.

MINING

Mineral exploration has been phased out under the Wilderness Act, but mining and extraction from established claims are allowed and are a threat to wilderness resources and values. For example, oil development is being considered in Alaska and silver mining on existing claims near the Cabinet Mountains Wilderness in Montana. Another potential threat may be on wilderness lands purchased by the federal government without subsurface rights to minerals and oil and gas, such as in wilderness areas designated in the eastern United States. Thousands of mining claims exist in BLM wilderness study areas and may impact wilderness in the future if mineral prices rise, leading to pressures for mineral claim development and subsequent mining operations at these locations. Mines can produce acid drainage and siltation that lead to reduced water quality in streams and water bodies. The negative social and ecological impacts of mining on wilderness conditions—naturalness and wildness—are extensive. Even mines that have played out may continue to impact wilderness because of old buildings, many of them eyesores, junk heaps, and roads that continue to erode and which are invitations to vehicle-trespass into wilderness.

WILDLAND FIRE SUPPRESSION

Fire prevention and suppression, adjacent to and in wilderness, change ecosystems that are fire-dependent by reducing the natural-fire frequency, leading to changes in ecosystem structure and composition (see Chapter 11). Allowing natural processes like fire to continue to function as part of wilderness ecosystems and landscapes is now recognized by wilderness managers as important to providing diversity and natural variation. The tendency has been for federal agencies to suppress most fires, partially because they fear that the fires would spread to inholdings of other landowners or to adjacent, non-wilderness lands and, in part, because of political pressure to suppress them.

For most of the last century, fires were aggressively attacked, and public vigilance was enlisted in a very successful fire-prevention campaign. This was before it was understood how fires renew ecosystems and provide variation and diversity to fire-dependent ecosystems. The massive stand-replacing fires that occurred in Yellowstone National Park in 1988 and the forest's subsequent regeneration helped educate the general public and wilderness users about the value of fires, although for many people, including some powerful legislators, the fires only confirmed rigid views that fire was bad. A similar process was at work in the intense fire year of 2000, when more than 6 million acres burned in more than 80,000 wildland fires, according to the National Interagency Fire Center (2008). Past fire suppression has allowed tremendous fuel loads to build up, and the loads are contributing to the larger and more intense wildfires. The complexity of fire as a natural disturbance process and the natural ecological functions it serves—especially in fire-dependent ecosystems—are becoming more completely understood (see Chapter 11). Although greater understanding is evolving, additional research and scientific inquiry are needed (Parsons 2000).

Although fire suppression, leading to the loss of natural fire regimes and inhibiting fire as a natural process in wilderness, is identified as a problem (Stokes 1996), reversing the trend will be difficult. *Due to the political volatility of any potential "let burn" policies for fire, the managed approach to allow fire in specific management prescriptions and controls will continue even in wilderness. Nature, however, will continue to restore fire whenever conditions permit.*

AIR QUALITY

Polluted air is a threat to wilderness naturalness because of its physical and biological impacts. It reduces visibility that may impact wilderness experiences. Good visibility within a wilderness naturally enhances the enjoyment of that environment and vice versa. For example, the San Gorgonio Wilderness near Los Angeles has severely diminished visibility due to urban air pollution. Bell et al. (1985) have reported that visual impairment from pollution can influence visitors to natural areas (e.g., Grand Canyon) to change trip schedules or to choose another location that has better visibility.

Air quality creates impacts on wilderness because of visibility and other indirect biological impacts. In the eastern United States, acid rain from industrial and urban emissions can be especially harmful to high-elevation ecosystems. For example, atmospheric pollutants, sometimes deposited in acid rain or acid fog, can cause changes in vegetation and water quality, and thus affect fish and amphibian populations. Forest fires contribute to air pollution, and managers deciding to allow wildfires to burn in wilderness ecosystems should consider such effects. Even in prescribing wilderness burns, agencies must consider atmospheric effects, but by selecting conditions that minimize impacts on air quality, such managed fires may differ in time, scale, and postfire effects from the natural fire they are supposed to simulate.

The Clean Air Act of 1977 (U.S. Public Law 95-95) provides a complex legal framework that further protects wilderness air resources. In particular, the amendments to this legislation in 1997 required the U.S. Forest Service to monitor 88 (Class I) wilderness areas for visibility and mandated action if air quality deteriorated. All other wilderness areas are under Class II protection, which charges wilderness managers with responsibility to protect air-quality-related values like visibility. In 1996, the USFS notified the State of Washington that visibility in the Alpine Lakes and Goat Rocks Wilderness areas was adversely impacted by a coal-fueled power plant in Centralia, Washington, and subsequently a mediated settlement for air-quality improvement was completed (Stokes 1999). *Thus, wilderness areas, and the air quality in them, serve as an important indicator of overall ambient air quality in the United States.*

WATER PROJECTS AND WATER QUALITY

Reconstructing and maintaining dams and reservoirs in wilderness for water storage threaten wilderness solitude and the physical landscape; the demand for water produced in wilderness threatens in-stream flows; and wilderness water quality is threatened by some nonconforming wilderness activities, like grazing and mining.

Water storage facilities built before the Wilderness Act was passed continue to be used and maintained. New facilities are allowed under the Wilderness Act if under an executive order from the president. Water storage facilities require maintenance or reconstruction that is very controversial because of the mechanized intrusions that are required. For example, maintenance of dams in the Emigrant Wilderness of California and in the Selway-Bitterroot Wilderness of Montana and Idaho is a contested public issue that has both proponents and opponents. These water storage facilities provide badly needed water for irrigation of agriculture in valleys below the wilderness, especially during late summer months when water is scarce. However, opponents point out that reservoirs behind dams are an unnatural influence in wilderness, and as these dated facilities wear out they should not be repaired, restored, or replaced—nature should be allowed to restore the landscape to its natural condition.

There is also interest and competition in the western United States for water flowing from remote forest and mountain wilderness areas by those wanting more irrigation water and conservation interests wanting to maintain fisheries, aquatic biota, and wildlife. The need to maintain minimum in-stream flows to protect endangered salmon runs in the Northwest is an example.

The quality of water in some wilderness streams may be affected by runoff from grazed areas that can include animal feces and sediments at higher than natural levels. A few streams and rivers that flow through wilderness areas have their origins outside wilderness and may bring water with pollutants and unnatural soil sedimentation levels to these wilderness rivers. Water quality is also affected by recreation visitors and their activities, such as disposing of food scraps and human waste. *Protection of natural water regimes and water quality in wilderness is an increasingly important challenge.*

ADVANCED TECHNOLOGY

High-tech clothing and camping equipment are widely used in wilderness, and some fear this protective technology may be partially cutting visitors off from the wilderness experience they seek (Knapp 2000). Certainly, modern gear makes wilderness visits easier and may also account for some of the increased visitor use.

The use of GPS equipment, cell phones, radios, beepers, pagers, and other electronic technology by search-and-rescue personnel is generally accepted, much like the use of other high-tech climbing gear. However, the use of electronic technology by wilderness visitors raises questions over whether they are having a primitive-recreation experience with the naturalness and solitude for which wilderness exists (Nash 1996; Shultis 2000).

Using aircraft in search-and-rescue operations is somewhat accepted when human life seems at stake, but it is controversial unless really needed because it intrudes on the experiences of other visitors. Aircraft, even where legal landing strips exist in wilderness, impacts the experience and solitude of wilderness visitors who see or hear them (Meyer 1999).

Modern electronic communication and navigation devices and evolving mechanical transport equipment may give visitors a false sense of security and contribute to irresponsible behaviors based on assumptions that rescue is only a cell phone call and helicopter flight away. Similarly, visitors may tend to rely on technological devices to navigate and not be able to read maps or orient themselves in a wilderness environment, thus robbing themselves of the benefits of wilderness discovery and self-reliance, while also increasing the probability of the need for search-and-rescue activities. The intrusions of advanced technology, at the very least, remind visitors of the society and busy work world that they may be trying to take a break from while in the wilderness.

MOTORIZED AND MECHANICAL EQUIPMENT TRESPASS AND LEGAL USE

Motorized equipment or mechanical vehicles trespass, such as all-terrain vehicles (ATVs), snowmobiles, or mountain bikes, can dilute wilderness solitude and damage resources. For example, operators of snowmobiles and ATVs can travel cross-country and enter a wilderness area inadvertently or deliberately for convenience, recreation, and exploration. Such trespassing usually goes undiscovered in remote locations. In some areas, vehicles are intentionally taken into or across wilderness areas as a shortcut or used by adjoining landowners or users who do not view their use of a remote area as an intrusion on the wilderness.

Management use of motorized vehicles and mechanical equipment is legal in wilderness where it is the minimum method for accomplishing a legitimate wilderness purpose, and in some areas for certain activities, such as wildlife management and law enforcement. Thus, depending on the area or purpose, visitors may see mechanical equipment used by management even in remote wilderness areas, such as helicopters for search and rescue, ATVs for beach patrols, four-wheel-drive vehicles for wildlife-management activities, and chain saws for trail construction or maintenance.

The extent to which motorized and mechanical equipment is used by managers affects a visitor's frame of reference as to what is acceptable in wilderness. *Our view is that motorized and mechanical equipment is justified in wilderness management only when it is absolutely essential as the minimum necessary tool to accomplish the purposes of the Wilderness Act. Very often primitive skills can do the work just as well, economically, and always with less impact on the wilderness environment and experience of wilderness visitors.*

AIRCRAFT NOISE AND AIRSPACE RESERVATIONS

Aircraft overflights of wilderness by commercial and military aircraft cause noise and air pollution, can disturb wildlife and visitors through visual and noise intrusion, and dilute solitude with a reminder of society and modern technology (Tarrant et al. 1995). There are private and public airfields on federal lands that are still in operation within wilderness areas in Montana and Idaho and which are used by private visitors, as well as flights by outfitters moving supplies and customers in and out (Meyer 1999). The Frank Church–River of No Return Wilderness (FC-RONRW) has thirty-one operating airstrips on federal and private lands within the wilderness boundary, and the twelve located on federal land handle about 5,500 landings per year (U.S. Department of Agriculture, Forest Service 1998). The Forest Service has little discretion in managing these airstrips because the Central Idaho Wilderness Act (Public Law 96-312) establishing the FC-RONRW stated that these uses "shall" be permitted, which is different from the Wilderness Act, which stated such uses "may" continue subject to secretary of agriculture regulations. Wilderness areas in Alaska are another example where preexisting use of aircraft, especially floatplanes, is allowed to continue under their wilderness designation legislation, such as ANILCA (U.S. Public Law 96-487).

Intrusions on wilderness airspace, especially by low-flying military aircraft on training flights in the West or by commercial sightseeing flights in places such as the Grand Canyon, affect wildlife and may offend people who experience them (Tarrant et al. 1995). The California Desert Protection Act contains wording that the designation of 69 wilderness areas by that act would not limit military overflights (U.S. Public Law 103-433). Only in the BWCAW has an airspace reservation been established, limiting overflights in the affected area to above 5,000 feet by presidential order. *Managers with little authority to restrict airspace activity may need to encourage the public to speak out and object to excessive airspace intrusions.*

URBANIZATION AND ENCROACHING DEVELOPMENT

*Urban sprawl and development encroach on wilderness boundaries and dilute wilderness with sights, sounds, and diminished remoteness, while increasingly urbanized wilderness visitors are suspected of being more tolerant of crowding and less wild conditions (*Fig. 13.5*).*

Sprawling urban development has dramatically affected wilderness conditions with smog, encroaching roads and ease of access, noise, and casual day use in urban-proximate wilderness areas like San Gorgonio outside Los Angeles. The Pusch Ridge Wilderness near Tucson, Arizona, has suburban backyards abutting its wilderness boundary. Even a remote area such as the Mission Mountains Wilderness in Montana is being affected by development in surrounding valleys, because it is isolating a remnant population of grizzly bears and threatens the bears' viability by interfering with the genetic interchange with other breeding populations (Stokes 1999).

Unfortunately, Congress ruled against buffer zones to insulate wilderness from adjacent non-wilderness activities—first in the Endangered American Wilderness Act (U.S. Public Law 95-237), which encouraged wilderness designation near cities to provide urban populations with much-needed recreation, and then the New Mexico Wilderness Act (U.S. Public Law 96-550), which specifically prohibited buffer zones to insulate wilderness from adjacent activities, as did several subsequent wilderness designation acts.

FIGURE 13.5 As urban and resort developments advance into wildlands, they may directly adjoin wilderness boundaries. Here a resort development and blacktop road crowd the boundary of Red Rock–Secret Mountain Wilderness (USFS) near Sedona, Arizona. Photo courtesy John C. Hendee.

The increasingly urbanized and diverse population in the United States may bring different values to wilderness in the future (Hammitt and Schuster 2000; Watson and Landres 1999) that will raise questions for managers, such as are today's urban visitors more likely to tolerate crowding in high-use wilderness areas? The fear is that increasingly urbanized visitors to wilderness may be satisfied with trips to crowded and heavily impacted wilderness due to their lack of previous experience in a more pristine area and a frame of reference more tolerant of crowding and oblivious to impacts. The Forest Service, and presumably the other agencies, have been reluctant to restrict use to achieve solitude standards. They are reluctant because of political pressure and concern that turning some visitors away from high-use areas would displace some of them to other less used and more pristine wilderness sites, thereby spreading impacts to those areas. Unfortunately, some visitors who go to these high-use destinations may assume that whatever crowded and impacted conditions they experience are what is intended for wilderness by managers. *Helping visitors to understand what wilderness is, what it is meant to provide, and why management is necessary is an ongoing educational challenge for managers* (Stokes 1999). Creating and implementing management strategies to address increasing wilderness use and impacts will continue to be difficult and controversial because there are many views on what approach is best (Cole 2001a; Oye 2001; Spring 2001; Worf 2001).

GLOBAL CLIMATE CHANGE

The impacts of global climate change and changes in weather regimes are having impacts on wilderness conditions and processes beyond the HRV. National and international panels of experts have acknowledged that the scientific evidence for global climate change is well documented, even though the causes for change are still being debated, such as human-caused versus natural variation (Intergovernmental Panel on Climate Change 2007). The need for some management response to these ongoing climate changes and the potential for subsequent impacts to biological and physical resources were summarized in a report by the U.S. Government Accounting Office (2007) that strongly pointed out the responsibility of federal land-management agencies to give guidance and direction to agency managers on how to address the effects of climate change on federal land and resources. Examples of the effects of climate change are numerous, especially in northern latitudes, where evidence has mounted to show later freeze-up of water bodies, earlier thaw dates, declining glacial ice, changes in permafrost, changes in forest vegetation, more extensive forest insect outbreaks, changes in bird and animal ranges, and other evidence of environmental change (Hinzman et al. 2005). Some situations have caught the attention of the public and the media about how rapidly environmental change is occurring, such as the loss of sea ice in northern latitudes, which has reduced the primary habitat for polar bears and has led to a decline in their abundance. Consequently, the U.S. Department of the Interior proposed listing the polar bear as a threatened species under the U.S. Endangered Species Act of 1973 (U.S. Public Law 93-205) (U.S. Department of the Interior 2007).

LEGISLATION DESIGNATING AREAS WITH COMPROMISED WILDERNESS CONDITIONS

There has been a long-standing concern about the tendency to politically compromise with non-wilderness interests and uses when designating wilderness areas, even going back to 1935 when, upon founding the Wilderness Society, Bob Marshall commented, "in the past far too much good wilderness has been lost by those whose first instinct is to compromise." The Wilderness Act required more than fifty versions before passage in 1964 due to the need to reach a political compromise with what we now refer to as permitted but nonconforming uses like grazing and mining (Fig. 13.6). Many subsequent laws designating wilderness areas have included special provisions and management direction along with permitted and nonconforming uses like aircraft or motorboat access to some areas.

Managing special provisions within wilderness requires a thorough understanding of the Wilderness Act and the intent of an area's designating legislation (Meyer 1999), along with some information about the values of wilderness held by those for whom a special provision was included (Watson et al. 2004). Even with substantial information and legislative background to guide managers, special provisions can conflict with other visitor and user interests and the overall purpose of the Wilderness Act (Nickas and Proescholdt 2005).

The history of the Wilderness Act and subsequent designation laws suggests that some compromise is necessary in our political system to achieve designation of new areas, and there are still many candidate areas to be considered (Scott 2004). However, the concern is that special provisions and nonconforming uses diminish wilderness character and cause unintended negative consequences for the quality of wilderness in the NWPS. For example, grazing guidelines first put in a Colorado wilderness bill in 1980 set a precedent to include special grazing provisions in subsequent wilderness legislation as well as to allow more motorized access for grazing interests (Nickas and Proescholdt 2005).

FIGURE 13.6 Old mining and homestead sites, such as this one in the Old Woman Mountains Wilderness (BLM) in California, concentrate ruins, trash, and buildings that detract from wilderness conditions and, in addition, attract curious visitors to unsafe situations. Photo courtesy Marilyn Hendee.

LACK OF POLITICAL AND FINANCIAL SUPPORT FOR WILDERNESS PROTECTION AND MANAGEMENT

The ideal expressed in the Wilderness Act, to preserve an enduring resource of wilderness, is threatened by a lack of political support sufficient to generate the finances needed for its protection and management. Concern is expressed among federal agencies, such as by former U.S. Forest Service chief Mike Dombeck (1994, 4), "that the resources committed to protect and manage wilderness have not kept pace with our needs," particularly for fieldwork budgets and staff. The evidence of such neglect is wide and deep. Wilderness plans are still in progress, have not been started, or are in need of revision and updating for many areas. Numerous wilderness study areas and roadless areas are still being evaluated to determine if they should be added to the NWPS, and many areas recommended for wilderness years ago have not been acted on by Congress. Without the efforts of volunteers and partnerships with environmental organizations and cooperators, much more wilderness work would be undone.

Brown (2000) observed that there was need for more wilderness research, planning, and management training to support current wilderness field staff. Management concerns also included fragmenting administrative responsibilities for wilderness management and multiple jurisdictions for some federal wilderness areas. For example, the Bob Marshall Wilderness administratively includes four national forests and five ranger districts, and the Frank Church–River of No Return Wilderness covers parts of six national forests (but is managed by four of them and six ranger districts).

The threat to wilderness from the lack of adequate resources for protection and management of the NWPS is caused by the lack of political support for this purpose among the larger public—as expressed by Congress. This may be the most serious threat to wilderness in the long run because it is at the root of establishing and maintaining high standards of wilderness naturalness and solitude and funding the people and programs require to implement them. The fact that projections by agencies like the Forest Service show recreation as a larger source of revenue than commodity programs, like timber, should help (Mackey 1998), but there is also competition among different kinds of recreation, with wilderness being classified as a small program compared to motorized recreation including car camping and off-road vehicle (ORV) use.

Strong support for wilderness by visitors and environmental-protection groups is key, but even among wilderness users there are divisive forces among commercial and public visitors, educational groups, and organized wilderness programs for personal growth and therapy—not to mention grazing, hunting, and endangered-species interests. *Support by wilderness constituencies for legislative appropriations to the agencies and administrative decisions to allocate more funding and staff resources to wilderness management by the agencies are key to future wilderness protection (Stokes 1999).* However, this too is problematic because the question remains: will visitors support use restrictions on popular wilderness sites so that a higher standard of naturalness and solitude can be achieved? *Expanded public information and education programs are essential components for the future success of wilderness management, and these education programs must reach beyond visitors to the broader public.*

MONITORING THREATS TO WILDERNESS

Growing wilderness visitation and use and the need for continued and additional research, planning, and monitoring to produce the information necessary to guide management of wilderness resources and values have been reported by numerous authors (Cole 1996b; Cole and Landres 1996; Cole, Watson et al. 1997; Manning and Lime 2000; Cole and Wright 2003). The Forest Service developed a national framework for monitoring wilderness character based on key words and phrases in the wilderness definition of the 1964 Wilderness Act (Landres et al. 2005).

The threats to wilderness documented in this chapter underscore the need for monitoring wilderness conditions and processes so that baseline conditions can be documented and the changes from internal and external factors can be quantified. The objectives and standards identified in planning and visitor management (Chapters 8 and 9) illustrate the quantifiable and objective measures that managers must use for comparison to decide if management actions are working or if new actions are required. Unfortunately, standards outlined in wilderness plans (such as for high-use wilderness sites) have not always been enforced (Oye 2001; Worf 2001). Only with well-established and long-term monitoring can we hope to document and understand natural conditions and processes in wilderness to track the degree to which wilderness standards are being maintained or degraded.

A final point, but very important to protecting wilderness against threats, is that U.S. wilderness primarily falls under the jurisdiction of four federal agencies: the U.S. Forest Service in the Department of Agriculture and in the U.S. Department of Interior, the National Park Service, the U.S. Fish and Wildlife Service, and the Bureau of Land Management. Wilderness is managed by these agencies as part of their larger land-management system, and they all struggle with threats to wilderness and shortage of finances and people (Henry 1996; Jerome 1996; Jarvis 1996; Stokes 1996; Brown 2000). All the agencies managing wilderness need more people and more resources to meet their wilderness management responsibilities.

SUMMARY

One of the greatest values of wilderness to future generations will be "naturalness" because 95 percent of the land will have been heavily managed, manipulated, and developed for human uses. The more than 4.5 percent of the landscape that remains "untrammeled" as wilderness in the United States is a scarce resource and may soon represent the only natural remnants of many ecosystems and landscapes with "wildness." *Threats to wilderness are generally defined as change agents that cause impacts on wilderness resource conditions and values—what causes the impacts—not the impacts themselves.*

We described nineteen short-term and long-term, direct and indirect threats to wilderness resources and values that must be addressed to achieve the purposes of the Wilderness Act for present and future generations. Changes that affect wilderness can be considered threats when they are not part of (or weaken) the natural conditions, processes, and variability that were historically part of a wilderness area. Threats to wilderness conditions, attributes, and values are complex, interactive, and both internal and external to wilderness. We provide many examples of both threats and their impacts.

These internal and external threats are both actual and potential threats to wilderness now, with considerable variation among geographic areas of the United States, and between remote and more urban-proximate wilderness areas. Some of these threats and examples are current controversies that may diminish over time, but new controversies will appear. However, this discussion highlights the need for continued planning, management, and monitoring of wilderness conditions and processes if we are to continue to enjoy the values of wilderness now and in the future.

STUDY QUESTIONS

1. Define two threats to wilderness.
2. When are wilderness conditions and values affected by threats?
3. Why is understanding wilderness threats important for wilderness management?
4. What are examples of threats that are internal to wilderness?
5. What are examples of threats that are external to wilderness?
6. How does monitoring contribute to wilderness planning and management?
7. Why are nonindigenous and nonnative species a threat to wilderness naturalness?
8. Why is there a shortage of people and financial resources to manage wilderness?

ACKNOWLEDGMENTS

Chad P. Dawson and John C. Hendee were lead authors of this new chapter for the third edition and changes and additions in the fourth and fifth editions.

For contributions and review comments on drafts of the third edition of this chapter, we thank the following colleagues. We have benefited from their review comments and input, but we alone bear responsibility for any errors, mistakes, and opinions reported herein: David N. Cole, research biologist, Aldo Leopold Wilderness Research Institute, Missoula, Montana; Ed Krumpe, professor of resource recreation and tourism, University of Idaho, Moscow, Idaho; and Peter Landres, biologist, Aldo Leopold Wilderness Research Institute, Missoula, Montana.

CASE DISCUSSION: INTERACTIONS BETWEEN POTENTIAL THREATS TO WILDERNESS

Most of the potential internal and external threats to wilderness are interactive and complex in how they affect wilderness character and attributes. Threats like global climate change are more difficult for managers to address because they were not part of the natural conditions in wilderness and were not included in the HRV for an area. Thus, there is a great deal of uncertainty on how to manage the effects and impacts of this threat.

The U.S. Government Accounting Office conducted a study in 2006 and 2007 to understand how federal land-management agencies were addressing the global climate change threat to federal lands. The report stated (2007, 44–45):

> Climate change has already begun to adversely affect federal resources in a variety of ways. Most experts with whom we spoke believe that these effects will continue—and likely intensify—over the coming decades. Some federal resources, depending on a variety of factors, may be more vulnerable than others. Because this issue is long term, global, and may affect federal resources in a number of ways, it will require foresight on the part of federal agencies to prepare for and minimize the adverse effects of climate change. However, federal resource management agencies have not yet made climate change a high priority. BLM, FS, FWS, NOAA, and NPS are generally authorized, but not specifically required, to address changes in resource conditions resulting from climate change in either their resource management actions or planning efforts. However, none of these agencies have specific guidance in place advising their managers how to address the effects of climate change in either their resource management actions or planning efforts. The resource managers with whom we spoke stated that in the absence of such guidance, they are unsure whether or how to take the effects of climate change into account when carrying out their responsibilities. Such uncertainty may, as unanticipated circumstances arise, force resource managers to set their own priorities, which may be inconsistent with those of the agencies' management and may result in misdirected efforts and wasted resources. Because there is growing evidence that climate change is likely to have wide-ranging consequences for the nation's land and water resources, elevating the importance of the issue in their respective strategies and plans would enable BLM, FS, FWS, NOAA, and NPS to provide effective long-term stewardship of the resources under their purview.
>
> To better enable federal resource management agencies to take into account the existing and potential future effects of climate change on federal resources, we recommend that the Secretaries of Agriculture, Commerce, and the Interior—in consultation with the Director of FS; the Administrator of NOAA; and the Directors of BLM, FWS, and NPS, respectively—develop clear, written communication to resource managers that explains how managers are expected to address the effects of climate change, identifies how managers are to obtain any site-specific information that may be necessary, and reflects best practices shared among the relevant agencies, while also recognizing the unique missions, objectives, and responsibilities of each agency.

CASE DISCUSSION QUESTIONS

1. Give three examples *of internal* threats to wilderness that interact with the impacts of global climate change and create situations that threaten wilderness conditions and processes.
2. Give three examples *of external* threats to wilderness that interact with the impacts of global climate change and create situations that threaten wilderness conditions and processes.
3. Why is it necessary to understand the interaction between different threats to wilderness to provide better stewardship for wilderness?
4. When the federal agencies that manage wilderness develop "best management practices" to address global climate change, should those practices include monitoring activities or not? Why or why not?

REFERENCES

Aplet, G. H.; Keeton, W. S. 1999. Application of historical range of variability concepts to the conservation of biodiversity. In: Baydack, R. K.; Campa, H., III; Haufler, J. B., eds. 1999. *Practical Approaches to the Conservation of Biological Diversity.* Covelo, CA: Island Press, pp. 71–86.

Asher, J. E.; Harmon, D. W. 1995. Invasive exotic plants are destroying the naturalness of U.S. wilderness areas. *International Journal of Wilderness.* 1(2): 35–37.

Bell, P. A.; Malm, W.; Loomis, R. J.; McGlothin, G. E. 1985. Impact of impaired visibility on visitor enjoyment of the Grand Canyon. *Environment and Behavior.* 17(4): 459–474.

Brown, P. 2000. Issues in the quality of U.S. wilderness management. *International Journal of Wilderness.* 6(2): 3–4.

Carter, D. 1997. Maintaining wildlife naturalness in wilderness. *International Journal of Wilderness.* 3(3): 17–21.

Cole, David, N. 1994. The wilderness threats matrix: A framework for assessing impacts. Research Paper INT-RP-475. Ogden, UT: U.S. Department of Agriculture, Forest Service, Intermountain Research Station.

———. 1996a. Ecological manipulation in wilderness: An emerging management dilemma. *International Journal of Wilderness.* 2(1): 15–18.

———. 1996b. Wilderness recreation in the United States—trends in use, users, and impacts. *International Journal of Wilderness.* 2(3): 14–18.

———. 1997. Recreation management priorities are misplaced—allocate more resources to low-use wilderness. *International Journal of Wilderness.* 3(4): 4–8.

———. 2000. Soul of the wilderness: Natural, wild, uncrowded, or free? Which of these should wilderness be? *International Journal of Wilderness.* 6(2): 5–8.

———. 2001a. Balancing freedom and protection in wilderness recreation use. *International Journal of Wilderness.* 7(1): 12–13.

———. 2001b. Management dilemmas that will shape wilderness in the 21st century. *Journal of Forestry.* 99(1): 4–8.

Cole, David N.; Landres, Peter B. 1996. Threats to wilderness ecosystems: Impacts and research needs. *Ecological Applications.* 6(1): 168–184.

Cole, David N.; McCool, S. F.; Borrie, William T.; O'Loughlin, Jennifer. 2000. *Wilderness science in a time of change conference*, Vol. 5: *Wilderness Ecosystems, Threats, and Management;* May 23–27, 1999; Missoula, MT. RMRS-P-15-Vol-5. Ogden, UT: U.S. Department of Agriculture, Forest Service, Rocky Mountain Research Station.

Cole, David N.; Watson, A. E.; Hall, T. E.; Spildie, D. R. 1997. High use destinations in wilderness: Social and biophysical impacts, visitor responses, and management options. Research Paper INT-RP-496. Ogden, UT: U.S. Department of Agriculture, Forest Service, Intermountain Research Station.

Cole, David N.; Wright, V. 2003. Wilderness visitors and recreation impacts: Baseline data available for twentieth century conditions. General Technical Report RMRS-GTR-117. Ogden, UT: U.S. Department of Agriculture, Forest Service, Rocky Mountain Research Station.

Czech, B.; Krausman, P. R. 1999. Controversial wildlife management issues in southwestern U.S. wilderness. *International Journal of Wilderness.* 5(3): 22–28.

Dawson, Chad P., Ken Cordell, Alan E. Watson, Ramesh Ghimire, and Gary T. Green. 2016. The US wilderness managers survey: Charting a path for the future. Journal of Forestry 114(3):298-304.

Derlet, R. W.; Carlson, J. R.; and Richards, J. R. 2008. Risk factors for coliform bacteria in Sierra Nevada Mountain wilderness lakes and streams. *International Journal of Wilderness.* 14(1): 28–31.

Dombeck, Mike. 1999. A wilderness agenda and legacy for the U.S. Forest Service. *International Journal of Wilderness.* 5(3): 4–6.

Duff, D. A. 1995. Fish stocking in U.S. federal agency areas: Challenges and opportunities. *International Journal of Wilderness.* 1(1): 17–19.

Hammitt, W. E.; Cole, D. N. 1998. *Wildland Recreation: Ecology and Management.* New York: John Wiley.

Hammitt, W. E.; Schuster, R. M. 2000. Wilderness use in the next 100 years. *International Journal of Wilderness.* 6(2): 12–13.

Henry, W. 1996. Status and prospects for wilderness in the U.S. National Park Service. *International Journal of Wilderness.* 2(3): 19, 47.

Hinzman, L. D.; Bettez, N. D.; Bolton, W. R.; Chapin, F. S.; et al. 2005. Evidence and implications of recent climate change in northern Alaska and other Arctic regions. *Climate Change.* 72: 251–298.

Hunger, D. H.; Christensen, N. A.; Becker, K. G. 1999. Commercial and private boat use on the Salmon River in the Frank Church–River of No Return Wilderness, United States. *International Journal of Wilderness.* 5(2): 31–36.

Intergovernmental Panel on Climate Change. 2007. *Climate Change 2007: The Physical Science Basis; Summary for Policy Makers.* Switzerland, Geneva: Intergovernmental Panel on Climate Change.

Jarvis, J. 1996. Status and prospects for wilderness in the U.S. Bureau of Land Management. *International Journal of Wilderness.* 2(3): 23.

Jerome, P. 1996. Status and prospects for wilderness in the U.S. National Wildlife Refuge System. *International Journal of Wilderness.* 2(3): 22, 47.

Kelson, A. R.; Lilieholm, R. J. 1997. The influence of adjacent land activities on wilderness resources: U.S. wilderness manager perceptions. *International Journal of Wilderness.* 3(1): 25–28.

Klyza, C. M., ed. 2001. *Wilderness Comes Home: Rewilding the Northeast.* Middlebury Bicentennial Series in Environmental Studies. Hanover, NH: Middlebury College Press and University Press of New England. 320p.

Knapp, D. 2000. Activities and technology: Technology and wilderness in the 21st century. *International Journal of Wilderness.* 6(2): 20.

Landres, Peter B.; Boutcher, S.; Merigliano, L.; Barns, C.; Davis, D.; Hall, T.; Henry, S.; Hunter, B.; Janiga, P.; Laker, M.; McPherson, A.; Powell, D.; Rowan, M.; Sater, S. 2005. Monitoring selected conditions related to wilderness character: A national framework. General Technical Report RMRS-GTR-151. Fort Collins, CO: U.S. Department of Agriculture, Forest Service, Rocky Mountain Research Station.

Landres, Peter B.; Brunson, M. W.; Merigliano, L.; Sydoriak, C.; Morton, S. 2000. Naturalness and wildness: The dilemma and irony of managing wilderness. In: Cole, D. N.; McCool, S. F.; Borrie, William T.; O'Loughlin, Jennifer, eds. *Wilderness science in a time of change conference*, Vol. 5: *Wilderness Ecosystems, Threats, and Management;* May 23–27, 1999; Missoula, MT. RMRS-P-15-Vol-5. Ogden, UT: U.S. Department of Agriculture, Forest Service, Rocky Mountain Research Station, pp. 377–381.

Landres, Peter B.; Marsh, S.; Merigliano, L.; Ritter, D.; Norman, A. 1998a. Boundary effects on wilderness and other natural areas. In: Knight, R. L.; Landres, P. B., eds. *Stewardship Across Boundaries.* Covelo, CA: Island Press, pp. 117–139.

Landres, Peter B.; Meyer, S. 2000. National wilderness preservation system database: Key attributes and trends, 1964 through 1999. General Technical Report RMRS-GTR-18-Rev. ed. Ogden, UT: U.S. Department of Agriculture, Forest Service, Rocky Mountain Research Station.

Landres, Peter B.; Morgan, P.; Swanson, F. J. 1999. Overview of the use of natural variability concepts in managing ecological systems. *Ecological Applications.* 9(4): 1179–1188.

Landres, Peter B.; White, P. S.; Aplet, G.; Zimmermann, A. 1998b. Naturalness and natural variability: Definitions, concepts, and strategies for wilderness management. In: Kulhavy, D. L.; Legg, M. H., eds. *Wilderness and Natural Areas in Eastern United States: Research, Management and Planning.* Nacogdoches, TX: Stephen F. Austin State University, Arthur Temple College of Forestry, Center for Applied Studies, pp. 41–50.

Liddle, M. J. 1997. *Recreation Ecology: The Ecological Impact of Outdoor Recreation and Ecotourism.* New York: Chapman and Hill.

Machlis, G. E.; Tichnell, D. L. 1985. *The State of the World's Parks.* Boulder, CO: Westview Press. 131p.

Mackey, C. 1998. U.S. wilderness management in the 21st century—politics, policy, and partnerships. *International Journal of Wilderness.* 4(3): 6–11.

Manning, R. E.; Lime, D. W. 2000. Defining and managing the quality of wilderness recreation experiences. In: Cole, D. N.; McCool, S. F.; Borrie, William T.; O'Loughlin, Jennifer, eds. *Proceedings: Wilderness Science in a Time of Change,* Vol. 4: *Wilderness visitors, experiences, and visitor management;* May 23–27, 1999; Missoula, MT. RMRS-P-15. Ogden, UT: U.S. Department of Agriculture, Forest Service, Rocky Mountain Research Station, pp. 13–52.

Manning, Robert E. 1997. Social carrying capacity of parks and outdoor recreation areas. *Parks and Recreation.* 32: 32–38.

Marion, Jeffrey L.; Roggenbuck, Joseph W.; Manning, Robert E. 1993. Problems and practices in backcountry recreation management: A survey of National Park Service Managers. NPS/NRVT/NRR-93/12. Denver, CO: U.S. Department of the Interior, National Park Service. 64p.

Matthews, K. R.; Knapp, R. A. 1999. A study of high mountain lake fish stocking effects in the U.S. Sierra Nevada wilderness. *International Journal of Wilderness.* 5(1): 24–26.

McLaran, M. P. 1990. Livestock in wilderness: A review and forecast. *Environmental Law.* 20(4): 857–889.

Meyer, S. S. 1999. The role of legislative history in agency decision-making: A case study of wilderness airstrip management in the United States. *International Journal of Wilderness.* 5(2): 9–12.

Murray, M. P. 1997. High elevation meadows and grazing: Common past effects and future improvements. *International Journal of Wilderness.* 3(4): 24–28.

Nash, Roderick. 1996. A wilderness ethic for the age of cyberspace. *International Journal of Wilderness.* 2(3): 4–5.

National Interagency Fire Center. 2008. Historical wildland fire information. www.nifc.gov/fire_info/fire_stats.htm; accessed February 1, 2008.

Nickas, G.; Proescholdt, K. 2005. Keeping the wild in wilderness: Minimizing nonconforming uses in the National Wilderness Preservation System. *International Journal of Wilderness.* 11(3): 13–18.

Oye, G. 2001. A new wilderness recreation strategy for National Forest Wilderness. *International Journal of Wilderness.* 7(1): 13–15.

Parsons, David J. 2000. Restoration of natural fire to United States wilderness areas. In: Watson, A. E.; Aplet, G. H.; Hendee, J. C., comps. *Personal, Societal, and Ecological Values of Wilderness: 6th World Wilderness Congress Proceedings on Research, Management, and Allocation,* vol. II; October 24–29, 1998; Bangalore, India. RMRS-P-14. Ogden, UT: U.S. Department of Agriculture, Forest Service, Rocky Mountain Research Station, pp. 42–47.

Posewitz, Jim. 1998. Yellowstone to the Yukon (Y2Y): Enhancing prospects for a conservation initiative. *International Journal of Wilderness.* 4(2): 25–27.

Scott, D. 2004. *The Enduring Wilderness: Protecting our Natural Heritage Through the Wilderness Act.* Golden, CO: Fulcrum.

Shultis, J. 2000. Activities and technology: Gearheads and golems—technology and wilderness recreation in the 21st century. *International Journal of Wilderness.* 6(2): 17–18.

Singh, Jaidev; Jackson, Rodney. 1999. Transfrontier conservation areas: Creating opportunities for conservation, peace and the snow leopard in Central Asia. *International Journal of Wilderness.* 5(3): 7–13.

Spring, Ira. 2001. If we lock people out, who will fight to save wilderness? *International Journal of Wilderness.* 7(1): 17–19.

Stokes, J. 1996. Status and prospects for wilderness in the U.S. Forest Service. *International Journal of Wilderness.* 2(3): 20–21.

———. 1999. Wilderness management priorities in a changing political environment. *International Journal of Wilderness.* 5(1): 4–8.

Tanner, R. J. 2004. Subsistence, inholdings, and ANILCA: The complexity of wilderness stewardship in Alaska. *International Journal of Wilderness.* 10(2): 18–22.

Tarrant, M. A.; Haas, G. E.; Manfredo, M. J. 1995. Factors affecting visitor evaluations of aircraft overflights of wilderness areas. *Society and Natural Resources.* 8: 351–360.

U.S. Department of Agriculture, Forest Service. 1998. Frank Church–River of No Return Wilderness programmatic and operational management plans, Vols. I and II, Draft Environmental Impact Statement. U.S. Department of Agriculture, Forest Service, Intermountain and Northern Regions.

U.S. Department of the Interior. 2007. Endangered and threatened wildlife and plants: 12-month petition finding and proposed rule to list the polar bear (*Ursus maritimus*) as threatened throughout its range. *Federal Register*. 72(5): 1064–1099. Washington, DC: U.S. Department of the Interior, U.S. Fish and Wildlife Service, 50 *Code of Federal Regulations*, Part 17.

U.S. Government Accounting Office. 2007. Climate change: Agencies should develop guidance for addressing the effects on federal land and water resources. Washington, DC: U.S. Government Accounting Office.

U.S. Public Law 88-577. The Wilderness Act of September 3, 1964. 78 Stat. 890.

U.S. Public Law 93-205. The Endangered Species Act of December 28, 1973. 87 Stat. 884.

U.S. Public Law 93-622. (Eastern Wilderness) Act of January 3, 1975. 88 Stat. 2096.

U.S. Public Law 95-95. Clean Air Act of August 7, 1977. 42 Stat. 7401.

U.S. Public Law 95-237. Endangered American Wilderness Act of February 24, 1978. 92 Stat. 40.

U.S. Public Law 96-312. Central Idaho Wilderness Act of July 23, 1980. 94 Stat. 948.

U.S. Public Law 96-487. Alaska National Interest Lands Conservation Act (ANILCA) of December 2, 1980. 94 Stat. 2371.

U.S. Public Law 96-550. (New Mexico Wilderness) Act of December 19, 1980. 94 Stat. 3221.

U.S. Public Law 96-560. (Colorado Wilderness) Act of December 22, 1980. 94 Stat. 3266.

U.S. Public Law 103-433. California Desert Protection Act of October 31, 1994. 108 Stat. 4473.

Watson, A.; Landres, P. 1999. Changing wilderness values. In: Cordell, H. Ken, ed. *Outdoor Recreation in American Life: A National Assessment of Demand and Supply Trends.* Champaign, IL: Sagamore Publishing, pp. 384–388.

Watson, A. E.; Patterson, M.; Christensen, N.; Puttkammer, A.; Meyer, S. 2004. Legislative intent, science and special provisions in wilderness: A process for navigating statutory compromises. *International Journal of Wilderness.* 10(1): 22–26.

Worf, Bill. 1997. Response to "Ecological manipulations in wilderness" by Dr. David N. Cole. *International Journal of Wilderness.* 3(2): 30–31.

———. 2001. The new Forest Service wilderness recreation strategy spells doom for the National Wilderness Preservation System. *International Journal of Wilderness.* 7(1): 15–17.

CHAPTER

14

Wilderness Use and User Trends

INTRODUCTION

Wilderness is preserved because society believes it has values that justify its protection. Many of these values are based on some kind of direct or indirect human use. This is clear in the Wilderness Act of 1964 (U.S. Public Law 88-577), which states that wilderness is to be preserved for "use and enjoyment" by "the American people of present and future generations." The Wilderness Act defined wilderness as having "outstanding opportunities for solitude or a primitive and unconfined type of recreation . . . [and] may also contain ecological, geological, or other features of scientific, educational, scenic, or historical value." As use of wilderness has increased and the human effects have been documented by studies, we have learned that primitive recreation and experiencing the naturalness and solitude of wilderness yield important inspirational, physical health, personal growth, and psychological values.

Understanding wilderness uses and users is essential to wilderness management. Diverse wilderness uses reflect a range of wilderness values and these uses may overlap or threaten other values, thereby creating conflicts. Most wilderness management is use management, trying to protect the wilderness from use impacts and seeking to minimize conflicts among different wilderness recreation users (e.g., hikers, horseback riders, river rafters, mountain climbers) and among recreation and other uses where allowed (e.g., grazing, commercial outfitting, scientific study) (Fig. 14.1).

There are many kinds of wilderness use and they reflect a variety of user values (Driver et al. 1987; Alan E. Watson et al. 1999). Broad-based support exists for the concept of wilderness among the general U.S. population, mostly based on ecological and environmental quality and various values and uses that occur both off-site and on-site (Cordell, Tarrant et al. 1998; Cordell, Murphy et al. 2005). In this chapter, we first identify several different types of wilderness use and discuss the known or estimated amount of each use. Then, because information is limited for all uses except recreation and because recreation is an important use of most wildernesses, the remainder of the chapter reviews the amount, character, and distribution of wilderness recreational use and the characteristics of users. Trends in use and speculation about future use then complete the chapter.

FIGURE 14.1 Educating visitors about appropriate behavior and permitted activities in wilderness areas is becoming increasingly important as more people discover wilderness and decide to have a day or overnight experience there. Here a U.S. Forest Service volunteer wilderness information specialist informs visitors about Oregon's Three Sisters Wilderness. Photo courtesy Barbara Merlin.

WILDERNESS-DEPENDENT USES

Wilderness uses vary in the degree to which they depend on wilderness biological or physical conditions. Based on wording in the Wilderness Act and established knowledge of the concept, when we refer to wilderness conditions we mean the qualities of "naturalness and solitude." Some uses depend on these wilderness conditions but other uses do not. Some activities do not depend on wilderness conditions but can be enhanced by them. Thus, *there is a wilderness-dependence continuum of activities ranging from those dependent on wilderness, to those not dependent but enhanced by wilderness conditions, and to those activities not dependent at all on wilderness conditions.*

For example, commodity uses, such as grazing or mining, take place in some wildernesses but do not depend on or require wilderness conditions—they can take place just as well outside of wilderness. Thus, where they are allowed in wilderness they may be called "nonconforming" uses. Certain recreational uses also do not depend on wilderness conditions. Some recreational activities in wilderness—for example, campers playing catch in a wilderness meadow—just take place in a wilderness without depending on the naturalness and solitude qualities of the wilderness. Some activities are enhanced for some people by wilderness conditions, even if the activity is not dependent on naturalness and solitude. For example, observing wildlife or fly-fishing in solitude on a wilderness stream may be enhanced by the wilderness setting. Other activities—for example, observing the results of natural ecological processes on the landscape, experiencing solitude and isolation, and facing the challenges of traveling and living in a large, roadless, undeveloped area—clearly depend on wilderness settings.

Thus, the wilderness-dependence continuum is another concept for planning and management of wilderness activities. Wilderness dependence considers the importance of the setting to the recreation experience, as do the recreation opportunity spectrum (ROS) and wilderness opportunity spectrum (WOS).

The concept of wilderness dependency is basic to managing wilderness use. Chapter 7 sets forth a management principle that wilderness-dependent activities should be favored over those that can be carried on outside wilderness. It is easier to understand this principle if you envision a wilderness area that is approaching the limit of use that it can sustain without unacceptable loss of naturalness and solitude. In such situations, wilderness managers must often make a decision among competing uses, and which of them are more dependent on wilderness conditions. These decisions are always subjective and open to interpretation, but we believe that managers can and must make such judgments in their planning and management processes (see Chapters 8 and 9).

WILDERNESS USE: AN OVERVIEW

Wilderness users and uses are inherently complex (Fig. 14.2). Knowledge of user characteristics and the type and extent of use is essential, especially if management seeks to emphasize freedom and minimize visitor regimentation and rely instead on more indirect and light-handed approaches such as information and education to manage use and users.

PUBLIC RECREATIONAL USE OF WILDERNESS

Recreation is the most obvious and most often reported wilderness use. In 2005, wilderness recreation totaled 10.7 million visits per year to the NWPS and 16.3 on-site days (Bowker et al. 2007). Wilderness use and recreational visitation are often treated as synonyms; however, wilderness use is a broader term that includes *nonrecreational* use, such as grazing. On-site recreational use certainly involves the largest numbers of direct wilderness users, has important localized impacts, and poses major management challenges.

The research information on recreational use and biophysical impacts from recreational use to campsites and trails is limited. Cole and Wright (2003) reported that only 56 percent of the 625 wilderness units in the NWPS in 2000 had any research information on visitors and their impacts. The types of information available in wilderness units were mostly for campsite impacts (51 percent), visitor use (24 percent), and trail conditions (9 percent).

Recreational use of wilderness includes many diverse activities, such as hiking, hunting, fishing, rock climbing, photography, cross-country skiing, snowshoeing, and viewing wildlife. Wilderness visitors take short walks, day hikes, long backpacking trips, and everything in between. Some visitors ride horses on day trips, or on long trips with pack animals to carry gear and supplies. Other visitors walk, leading pack animals like horses, burros, mules, or llamas. Some wilderness users float rivers in boats, canoes, kayaks, or rafts. *In some places,* where legislation and regulations permit, some visitors use motorboats, snowmobiles, cruise ships, or airplanes (for example in Alaska and Idaho) for access. People travel in all types and sizes of groups—mostly in small groups with family or friends—and some travel alone. Other visitors travel with commercially outfitted groups composed of friends and strangers, or in groups sponsored by youth, church, environmental, adventure, or educational organizations. Increasingly in recent years, some visit wilderness with organized wilderness programs for personal growth, therapy, and healing.

COMMERCIAL RECREATIONAL USE IN WILDERNESS

In addition to public recreation use, the Wilderness Act authorizes commercial recreational use of wilderness. *There are two main kinds of commercial-recreation influences on wilderness:* from *outfitting and guiding,* and from *equipment rental.* Outfitters and guides accompany some visitors,

FIGURE 14.2 The 1964 Wilderness Act allows many uses of wilderness, in addition to individual enjoyment. A commercial outfitter has tents for clients, left; scientists at work, second from left; a dam in Colorado's Mount Zirkel Wilderness, third from left, is an example of water development; and a band of sheep in a meadow on the Bridger Wilderness in Wyoming, right, is an example of livestock grazing. Photos (left) courtesy Richard Walker and others. U.S. Forest Service.

adding to and facilitating the use of wilderness. This use usually involves travel by horse or boat, although outfitted and guided backpacking trips are a growing commercial use. Other outfitting businesses involve only equipment rental—the visitors travel without guides using rental equipment, such as canoes, rafts, backpacking gear, cross-country skis, and tents. Rental equipment is available in most major urban areas and cities and towns near wilderness. In a few places, one can rent pack animals like burros or llamas, but few outfitting businesses rent pack stock for do-it-yourself wilderness travel, out of concern for liability issues and proper care of the animals.

Commercial outfitting is a traditional use of wilderness permitted by the Wilderness Act and encouraged by wilderness managers in some areas where use capacity is available and the proposed activities are compatible with wilderness conditions. Outfitters provide camping and food services, but the key services are specialized transportation (horses, rafts, skis, etc.) and specialized equipment (lightweight camping gear, etc.). Outfitting in some areas has drawn criticism for stressing comfort, convenience, and excessive facilities and technology that conflict with wilderness values. Nationwide, the wilderness outfitting business is responding to the need for wilderness use that is light on the land, leaving no trace of use, and responsive to wilderness values and other users. Some outfitters and some wilderness areas reflect this trend more than others. Furthermore, outfitters can serve as educators of wilderness users to enhance visitor experiences and to promote wilderness stewardship.

Commercial outfitters and guides must obtain special-use permits to take groups into wilderness, and this gives managers a chance to influence their operations toward wilderness-compatible activities and the timing and location of use. In many wilderness areas, there is great competition for permits and, as established carrying capacities are reached, commercial outfitter use may be limited in favor of public recreational use. This is a contentious issue, but the underlying problem is the overall increased use of wilderness (Mackey 1998).

Outfitter use varies from none in some wildernesses, such as the smaller wilderness areas in the eastern and western United States, to a large amount of use in others, particularly on whitewater rivers such as the Colorado, Middle Fork of the Salmon, and Selway where specialized equipment and skill are necessary for safe travel. In general, outfitters and guides play a more important role in larger wildernesses where horse travel is common or on wilderness rivers where specialized watercraft and skill are required.

Institutional outfitting, which is outfitting and guiding by schools or by nonprofit educational and religious organizations, is a large and growing source of wilderness use. For example, colleges, universities, and schools take classes to wilderness and operate outdoor recreation, education, and adventure programs, and provide equipment-rental services to support such activities. Institutional use of wilderness is becoming a contentious issue because of the size of groups, the overall extent of use, and the fact that permits are not required in many wilderness locations. Commercial outfitters complain that, as wilderness-use limits are reached and exceeded, their permits are being restricted while institutional outfitters are not required to obtain permits and are less restricted.

INDIRECT USE OF WILDERNESS

Besides direct, on-site recreational use of wilderness, there are several important types of indirect use (Haas et al. 1986; Cordell, Bergstrom, and Bowker 2005). *Millions of people who never set foot in wilderness may nevertheless derive satisfaction from experiencing wilderness indirectly.* This indirect use comes about vicariously through reading; viewing television, photos, and films; and listening to lectures and accounts by others. Indirect use also comes about when people stay at resorts or recreation areas near wilderness where they can view wilderness scenery and experience the excitement in wilderness gateway communities, such as observing trailhead use, outfitter activities, and marketing materials about visiting wilderness areas.

One study documented greater population growth in western counties having wilderness than in counties without wilderness (Ridzitis 1996). The attraction of small towns and rural areas near wilderness and wildlands for retirees and persons with portable jobs is increasingly noted (Powers 1996). Indirect and vicarious enjoyment is a major use of wilderness. Without wilderness, the values and benefits from these experiences would be diminished. Historical and past accounts of wilderness encounters would remain but would provide only a partial substitute because they lack the possibility that one could actually visit the area. Indirect use produces and is produced by five types of values: option, existence, bequest, pre- and post-trip values, and ecological services.

1. *Option value: Many people value the possibility that they will someday visit wilderness* (Schuster et al. 2005), so keeping open the option to visit wilderness is important to them and, in that sense, they are indirect users. Whether or not they ever actually visit a wilderness, it is worth something to them to know they could. Another option value revolves around the idea that wilderness and other protected areas provide a storehouse of resources that could be used in other ways, such as for commodity production, in the future if needed by society.
2. *Existence value: For many other people, the simple fact that wilderness exists has value even though they may not actually visit it* (Fisher and Krutilla 1972; Schuster et al. 2005).
3. *Bequest value: Many people want to leave wilderness as an inheritance to later generations.* Schuster et al. (2005) report that a larger percentage of the general public in the United States held bequest values, existence values, and option values than the estimated percentage of the population who directly visited wilderness for recreational use (Bowker et al. 2007).
4. *Pre- and post-trip values: Recreation experiences have not only an on-site phase but also four off-site phases: anticipation, travel to, travel from, and recollection.* Compared to most other types of recreation, wilderness visits have particularly long and well-developed off-site anticipation and recollection phases. Planning frequently starts well in advance, and experiences are often relived afterward so that even direct on-site recreational use involves substantial indirect use.
5. *Ecological services: By far the largest wilderness value in estimates of economic values of wilderness is ecological services*, such as providing clean air and water, biodiversity, and a place where natural processes and conditions

exist relatively unmodified (Loomis and Richardson 2001). Similarly, these ecological service values are the ones reported most frequently by the majority of the U.S. population as important wilderness benefits (Cordell, Murphy et al. 2005).

Some wildernesses or portions thereof are closed to all direct, on-site recreational use, such as some FWS wilderness areas (e.g., small islands). In addition to the indirect or vicarious use these areas provide, they provide breeding grounds that produce many birds and wildlife that are observed and enjoyed outside the wilderness.

Any estimate of the amount of indirect wilderness use would be very difficult to make, and none is available. Sales of wilderness-related books and videotapes are substantial, and films with nature-wilderness themes draw crowds. Numerous popular television programs have been based on nature and wilderness. The clearly positive influence of wilderness on classic North American artists, philosophers, and writers is also well recognized (Mortensen 1997, 1999), and wilderness continues to inspire poetry, prose, and thoughts from less prominent artists and authors. We speculate that the number of indirect users is probably much greater than the number of actual visitors, and this may be part of what is reflected in the broad public support for wilderness.

SCIENTIFIC USE OF WILDERNESS

One of the major values of wilderness is its potential for scientific use. *Wilderness serves as a living laboratory, particularly for ecological studies looking at interrelationships (e.g., wildlife dynamics and fire-dependent communities) in relatively natural, unmodified conditions and large areas.* Aldo Leopold (1941) wrote of wilderness as a land laboratory or environmental baseline, valuable for learning how healthy ecosystems function and how humans affect the ecosystem.

As the rest of the world becomes more developed and modified by humans, the contrast between wilderness and non-wilderness increases, and the value of wilderness as a natural laboratory will increase. Thus, *scientific use is a substantial wilderness-dependent activity because of the scarcity of unmodified natural ecosystems.* Not only is wilderness valuable for the scientific opportunities it provides (wilderness for science), but science is proving a valuable source of new information to support wilderness and wildland conservation (science for wilderness) (Parsons 2000).

Certain types of environments, such as high mountain and alpine environments and more recently semiarid areas and deserts, are especially well represented in designated wilderness, so studies of them are often located in wilderness. Studies of glaciers are a notable example. Such scientific study is not directly dependent on wilderness; the subject just happens to be there. Other scientific use also depends on objects that happen to be located in wilderness, such as cultural and archeological sites (e.g., prehistoric campsites, dwellings, and artifacts) and paleontological resources (e.g., dinosaur fossils). Scientific activities at such sites often pose management dilemmas because research often requires excavation and removal of fossils and artifacts and protection of exposed sites. These resources are also very vulnerable to removal by visitors. Seeing fossils and artifacts in their natural wilderness setting is also valuable, and the experience is diminished by their removal. Protecting cultural and archeological resources is an important issue, especially in some wilderness areas in the southwestern United States.

Wilderness provides a good setting for some kinds of social and psychological research (Manning 1989). *Isolating small groups of people and their close interdependence in the face of the challenges of wilderness experiences provides valuable behavioral research opportunities.* Numerous research studies on wilderness user experiences, "opportunities for solitude," and tolerance for crowding have been conducted in wilderness areas, wildlands, and wild and scenic river areas (Absher and Lee 1981; Cole, Watson, and Roggenbuck 1995; Graefe et al. 1984; Gramann 1982; Manning 1985). This research on users has evolved from descriptive to more comprehensive studies seeking to better understand the benefits from user experiences in wilderness conditions (Driver et al. 1987; Ewert and MacEvoy 2000). Some studies looked at the reactions of visitors to contacts with other visitors (encounter preferences), the normative behavior of users (Hall and Shelby 1996; Alan Watson et al. 1996; Williams et al. 1991), and the importance and meaning of privacy and solitude (Dawson and Hammitt 1996; Hammitt and Madden 1989). Such research has important implications for wilderness management based on understanding what are reasonable standards and the expectations of visitors. The results of such studies and numerous others have changed wilderness visitor planning and management (see Chapters 8 and 9).

One of the important scientific values of wilderness is its function as a natural gene pool. In wilderness, the natural genetic diversity of native plants and animals is protected from direct manipulation. Outside wilderness, this diversity is often reduced unnaturally as some species become extinct and as domestic species are selectively bred for greater uniformity, increased yield, and other characteristics. Although the wilderness system represents some types of environments and ecosystems very well, such as alpine areas and deserts, other environments, such as grasslands and low-elevation forests, are not well represented (Davis 1989; Cordell, Murphy et al. 2005). The scientific value of wilderness increases as these yet unrepresented environments and ecosystems are added to the NWPS.

Scientific use of wilderness is pervasive but difficult to measure, and there is some difference of opinion as to its extent (Franklin 1987; Butler and Roberts 1986; Reed et al. 1989; Carr and Hendee 1993). The four federal management agencies tend to focus on what is present in the wilderness resource and what is changing, via inventory and monitoring studies. The USFS, NPS, and colleges and universities focus on studies to better understand ecological relationships in wilderness. The USFS, colleges, and universities also conduct studies focused on assisting managers in their wilderness management efforts.

Science was one of the stated purposes of wilderness from its original definition in the Wilderness Act in 1964. However, how much research is needed to fulfill the mandate of the Wilderness Act that science is one specific purpose of wilderness? The concept of wilderness includes the value of undeveloped lands with minimal human influence that serve as a baseline of environmental integrity and where scientific research can measure natural processes. Because the authors of the Wilderness Act could not have anticipated the current size of the NWPS, the need for scientific information to support management of this system is greater than they originally expected. Furthermore, global environmental conditions have changed at an unexpectedly rapid rate, and this highlights the need for basic research and monitoring within the NWPS to describe changes in natural processes and conditions (Dawson et al. 2016; Cordell and Young 1993; Hendee and Ewert 1993; Hendee 1995). As other lands become more impacted by human influences, the value of wilderness as a natural, protected laboratory increases.

A panel that reviewed the USFS research program in wilderness outlined three complementary categories of research (Parsons 2007, 35):

1. Science for wilderness: science that informs effective stewardship and management of wilderness, including the status and trends of ecological conditions
2. Wilderness for landscape sustainability: science that improves understanding of the contributions of wilderness to the ecological processes, services, and integrity of larger landscapes
3. Wilderness for science: science that uses wilderness and similarly managed lands as laboratories to understand the causes and consequences of environmental change, minimally confounded by other influences

These three categories reflect the legislative intent of the Wilderness Act and the place of wilderness on the extreme end of the environmental modification spectrum.

Sharing and disseminating wilderness research findings have taken on several forms to accommodate a worldwide interest in such information. For example, The World Wilderness Congresses, such as the eighth in Alaska in 2005 (Alan Watson, Sproull et al. 2007), have provided opportunities for compiling, comparing, and exchanging information among scientists from several countries through research symposia and published proceedings. The *International Journal of Wilderness* (www.ijw.org) was launched in 1995 and provides a forum for findings of wilderness studies. National conferences on Wilderness provide scientists the opportunity to share studies, perspectives, and dialogue about wilderness research and scientific issues (examples: Cole, McCool, Borrie et al. 2000a, 2000b; Cole, McCool, Freimund et al. 2000; McCool et al. 2000a, 2000b).

EDUCATIONAL USE OF WILDERNESS

Wilderness is used for educational purposes by providing sites for field trips, study areas for student research, and a source of instructional examples. Reed et al. (1989) reported that 39 percent of NWPS wilderness managers claimed that environmental and conservation education programs were being conducted in their wilderness in 1987. Some educational uses based on lessons from large-scale, long-term ecological processes may be dependent on wilderness, but for many topics, other lands are available and may be more suitable and accessible.

Some educational uses are more akin to recreational use: wilderness as a setting for teaching travel and survival skills. Many colleges and universities have field trips that teach wilderness skills. Some youth organizations, such as the Boy Scouts, teach outdoor-living skills, with their application in wilderness as a pinnacle of achievement. Whether such use is really dependent on wilderness is questionable, but trips to wilderness as a test and demonstration of scouting skills are classic. What they really need is large, unroaded areas with wilderness or primitive conditions. Other courses also teach wilderness values as well as low-impact use techniques, and this kind of education would be considered more wilderness-dependent.

Educational use of wilderness is not directly measured, but some indication of its magnitude can be gained from data on wilderness-related courses in colleges and universities and wilderness experience programs. Dawson and Hendee (2004) reported that in 2002 more than 1,500 students took primary wilderness-related courses at colleges and universities in the United States, which was 30 percent lower than what was reported in a 1982–1983 study (Hendee and Roggenbuck 1984). Most courses focused on wilderness appreciation and use, legislation and policy, protection and management, and history. Some university courses use wilderness experiences to teach other subjects, such as environmental education and natural ecosystems. These activities and educational programs can be highly dependent on wilderness.

Studies in the mid-1990s document more than two hundred education-oriented wilderness experience programs (excluding higher education institutions and youth organizations), ranging from several large programs serving hundreds of students (e.g., Teton Science School in Wyoming) to many small programs serving fewer than one hundred per year (Dawson, Tangen-Foster et al. 1998; Friese, Hendee et al. 1998).

Today, schools and higher education institutions may be exempt in some locations from having to secure special-use permits to use wilderness, which means there are no systematic data about the extent of this so-called institutional outfitting. Commercial outfitters would like to see permits and fees for educational use to help share the financial and use quota burden imposed on them (Mackey 1998).

WILDERNESS EXPERIENCE PROGRAMS FOR EDUCATION, PERSONAL GROWTH, THERAPY, AND HEALING

Wilderness experience programs (WEPs) take paying clients into wilderness or comparable lands to develop their human potential through education, personal growth (including leadership and organizational development) or therapy, and healing (Friese, Hendee et al. 1998). One nationwide survey of wilderness managers found that, among those reporting WEP use in wilderness they administered, two-thirds said such use was increasing, and more than one-third expressed concern over environmental impacts of use by such groups or conflicts with other users (Gager et al. 1998). Because of this user's size and prospects for continued growth, permitting and dealing with the use of wilderness for education, personal growth, therapy, and healing are a growing management challenge, yet a growing opportunity for enhanced social benefits from wilderness. Many policy and management issues surrounding access to the wilderness affect these organized uses (Ewert et al. 1999).

Wilderness personal growth, therapy, and healing programs use various combinations of challenge, adventure, and reflective activities to help participants get in touch with themselves. Wilderness personal-growth programs seek empowerment; they stretch participants' abilities and determination in wilderness activities to teach them that their capabilities exceed what they imagined and, therefore, they may be self-limiting their performance back home in their daily lives. Wilderness therapy and healing programs seek healing and restored normal functioning; they use the natural consequences of wilderness living and primitive skills to cleanse participants of unwanted stresses and restore normal functioning, with individual and group therapy and solo experiences to help connect participants with their inner selves as the ultimate source of their afflictions—and their recovery. Although wilderness personal growth, therapy, and healing programs serve all age classes, adolescents and young adults are the most frequent participants. Enhanced self-esteem and other variations in empowerment are the most consistently reported outcomes for individuals in studies of wilderness experience programs (Friese, Pitmann et al. 1996).

Surveys in the 1990s (Russell et al. 2000; Dawson, Tangen-Foster et al. 1998; Friese, Hendee et al. 1998) identified more than 230 personal-growth programs, ranging from several large programs serving thousands of clients annually—Outward Bound, by far the largest, served about 30,000 clients in 1998—to many small programs serving fewer than 100 clients annually.

Studies of WEPs and their clients document the attraction of wilderness for personal-growth and healing uses and their partial or complete dependence on wilderness to deliver program outcomes and achieve important social benefits. *Is the use of wilderness for the education, personal growth, therapy, and healing of young people as important as—or more important than—commercial or public recreational use? Increasingly, wilderness managers will face such questions in policy and actions that favor one kind of use over another when limiting use to protect wilderness naturalness and solitude.* Moreover, questions of what kinds of groups have to obtain permits and how much they must pay to use wilderness will be more contentious and more frequently debated.

COMMODITY USES

Several on-site commodity uses take place in wilderness, including mining, logging, water storage and use, and grazing. These uses, sometimes referred to as "nonconforming but allowed" uses, are generally limited to where they were established before designation of the area as wilderness.

Mining for Minerals, Oil, and Gas

The 1964 Wilderness Act permits mining under certain prescribed situations but under agency regulations designed to minimize adverse impacts on the wilderness environment (see Chapters 4 and 5 for legislative details). However, active mining in wilderness is minimal and prospecting for new claims has largely ended (Reed et al. 1989). The bigger threat may be on wilderness lands purchased by the federal government without subsurface rights to minerals, oil, and gas, such as areas designated in the eastern United States. However, even in these circumstances, development in or near wilderness would be constrained to help protect the wilderness environment.

Almost all mining claims in wilderness areas are located on national forest and in BLM wildernesses (Browning et al. 1988). Most national parks are closed to all mining, but six were established with special provisions permitting mining. Three of these parks contain wilderness and two have areas proposed for wilderness designation. In 1976, Congress passed the Mining in the Parks Act (U.S. Public Law 94-429), closing these six national parks to new mining claims and placing all existing claims under strict new regulations. The 1964 Wilderness Act permitted continued staking of claims in national forest wilderness only and only until the end of 1983. However, because claims could be staked only on public-domain lands, prospecting was limited to wildernesses in the western United States, which are part of the public domain, and was excluded from wildernesses in the East. In the East, almost all the land in wildernesses, mostly national forests and national parks, was private land before it was acquired by the federal government. However, in many cases the mineral rights on these lands were not acquired by the government, and therefore mining is a possible use but still subject to restrictions to protect the wilderness. These private mineral rights in wilderness areas in the East could be acquired from willing sellers, or even by condemnation (although politically it would be difficult), if funds were available to compensate owners.

Mining in wilderness on the national forests is limited. Mining claims are numerous—just how numerous is hard to determine because historically there was no requirement to notify the USFS when claims were filed on national forest-land. Claims were recorded in county courthouses, along with many other claims on other lands. In 1985, it was estimated that there were at least 10,000 mining claims on national forests (Wilkinson and Anderson 1985). In 1984, the Cabinet Mountains Wilderness in Montana on the Idaho border reported more than 700 claims, and in 1998, a proposal was made by a mining company to extract silver from under the Cabinet Mountains Wilderness using a tunnel system operating from just outside the wilderness area. However, reported mineral extraction from any wilderness is limited to nonexistent. This low level of mining is consistent with mineral surveys of wilderness and primitive areas that have failed to turn up major mineral deposits that are economically viable. Restrictions on mining activity to protect the wilderness make mining more difficult and costly and probably have discouraged development (Reed et al. 1989).

Oil and gas production is under lease inside wilderness in several locations, one example being Indian Mounds Wilderness, managed by the Forest Service in east Texas (Evans 1986). Such use, which operates under different laws than hard-rock mining, is a possibility in many places and is very controversial.

The BLM policy for their primitive areas generally prohibited mining and mineral leasing except for valid existing rights established before the area was administratively designated as a primitive area. Under provisions of the Federal Land Policy and Management Act of 1976 (FLPMA) (U.S. Public Law 94-579), mining claims can continue to be filed on lands included in wilderness study areas, provided that the claims or subsequent mining operations do not impair the suitability of such areas for subsequent designation as wilderness. Mining operations in effect on the date of FLPMA can continue, provided they do not cause unnecessary or undue degradation.

When BLM lands are legislatively designated as wilderness, they are withdrawn from all new mining and mineral leasing. BLM regulations for designated wilderness require holders of existing mining claims to file a plan of operations outlining how mining is to be undertaken. Before such plans are approved, the BLM must determine if the mining claim is valid. If it is not, actions are taken to invalidate the claim. Thousands of mining claims exist in BLM WSAs. In future years, these claims may pose a wilderness management problem for the BLM if mineral prices rise, leading to pressures for mineral claim development and subsequent mining operations at these locations. However, although expanding mining seems to be a declining threat, many wilderness areas contain old mining sites littered with old equipment and junk and accessed by abandoned roads that are in visual contrast to the natural landscape and an attractive nuisance for potential vehicle trespass.

Logging

Except for limited timber cutting necessary for mine timbers, logging has been allowed in just one wilderness, the Boundary Waters Canoe Area Wilderness (BWCAW) in Minnesota—an area that has been an exception within the NWPS in a number of ways (Duncan and Proescholdt 1999; Gladden 1990). The Boundary Waters Canoe Area Wilderness Act of 1978 (U.S. Public Law 95-495) was the first amendment to the Wilderness Act, and its purpose was to end logging inside the area. Before logging was ruled out in 1978, the BWCAW locations in which it was permitted had been progressively reduced over the years, with about two-thirds still available for timber harvest before the 1964 Wilderness Act and about 40 percent thereafter.

Another possible exception is cutting trees to control insects or disease, especially if an infestation threatens areas outside the wilderness. This was done in several national forest wildernesses in Texas to control southern pine beetles (Billings 1986), but it was very controversial.

Water Storage and Use

Facilities for water storage, such as dams and reservoirs, are permitted by the 1964 Wilderness Act. New water storage projects require presidential approval, but many small reservoirs built before enactment of the Wilderness Act continue to be used and maintained. It is very controversial when heavy maintenance or reconstruction using mechanical equipment is proposed for these aging facilities, as in the Selway-Bitterroot Wilderness.

Not all wilderness water storage and use depend on dams inside wilderness. Many wildernesses store vast amounts of water in the form of snow and provide a large proportion of the stream flow in the West. *Measuring snowpack to forecast stream flow poses a conflict with wilderness values when snow is measured with the aid of mechanized access, such as helicopters, snowmobiles, or electronic snow-weight recording devices connected to radio transmitting stations.* Wilderness managing agencies have responded somewhat differently to this conflict. In wilderness on the national forests, except where helicopter use was well established before the Wilderness Act, snow measurements are made without mechanized equipment, usually by crews traveling on cross-country skis or snowshoes. Electronic devices that measure and transmit snow data from wilderness locations are not permitted. The U.S. Department of the Interior policy for national park wilderness is to permit existing water-resource-monitoring devices to be retained in wilderness. New devices will be placed in wilderness only if the secretary of the interior decides that essential information cannot be obtained from locations outside the wilderness, and that the proposed device is the minimum necessary to successfully and safely accomplish the objective.

Most irrigation water impoundments in wilderness were built long before the Wilderness Act. Typically, the dams are low and constructed of local rock and timber, and shoreline vegetation has had many years to adjust to the higher water level. Thus, many of the reservoirs are not conspicuously unnatural unless water levels fluctuate greatly. *In 1987, about one hundred dams existed in national forest wilderness, with many others in national park wilderness and backcountry, but reservoirs are concentrated in a few areas, so most wildernesses do not have such developments.* A few dams in the BWCAW, built to facilitate log drives, are still in place and maintain lake levels; many more dams have been abandoned. Some wildernesses in the Sierra Nevada of California have small dams to control minimum stream flows so streams will not go dry in late summer with a resulting loss of fish. There are also a few high, concrete hydroelectric dams in the California Sierras, built before the Wilderness Act, with the last such dam constructed on the Rubicon River in the Desolation Wilderness in 1963, only one year before the Wilderness Act became law. Again, as these water storage and control facilities need heavy maintenance, reconstruction, or replacement, this becomes controversial because of perceived conflicts with wilderness values.

Grazing

Grazing by livestock—sheep, cattle, and a few horse herds—is also allowed by the Wilderness Act where it existed before wilderness designation. Wilderness in the national forests and on BLM lands accommodates virtually all of the livestock grazing in wilderness. Commercial grazing (as opposed to grazing of pack stock used for wilderness travel) is less common now as a commercial activity than in past decades. Changes in the livestock business generally have reduced the economic attractiveness of extensive grazing in wilderness because of the higher costs of grazing there compared to less remote pastures. Transporting livestock to and from wilderness, maintaining minimal facilities, hiring herders, and alleged losses to predators are some categories of increased costs related to grazing in wilderness compared to other lands. An emphasis on intensive production of forage on more suitable lands and more use of feedlots in the livestock industry further accentuates the cost differences. The fact that fees for grazing public land are low compared to these for private lands probably contributes to maintaining the wilderness grazing demand at current levels.

Nevertheless, grazing takes place on about one-third of the wilderness lands in the NWPS excluding Alaska (Reed et al. 1989). In 1986, about 14 percent of all cattle and sheep grazing in the National Forest System grazed in wilderness at least part of the year. Because grazing seasons are short in many wildernesses, the percentage of Animal Unit Months (AUMs) in wilderness is lower. (One AUM equals one cow or five sheep for one month.) Figures on grazing in BLM primitive areas and potential wilderness are unavailable, but grazing on these lands is substantial; more than 80 percent of BLM land in the western states (excluding Alaska) is subject to grazing (Foster 1976). Grazing often generates pressures for mechanized access to maintain facilities, such as water tanks and for predator control. These issues pose a difficult policy challenge for wilderness managers.

Grazing in wilderness will continue where it was an established use of public lands before the Wilderness Act passed in 1964 or in areas designated later. Public-land grazing has a strong and active political constituency in the livestock industry. Without provision for grazing to continue where it was already established, the Wilderness Act could not have passed—so grazing was included as one of the so-called allowed but nonconforming uses, along with mining.

Since 1964, objections to grazing in wilderness and on all public lands have increased, and in response, livestock interests have supported legislation and management direction to protect (and possibly strengthen) their right to graze and manage livestock grazing in wilderness, as the following brief review based on Browning et al. (1988) and McClaran (1990) documents. See also Chapter 5 on grazing.

Livestock grazing is a well-established use of wilderness on national forest and BLM lands, with rights embodied in the Wilderness Act and supported (some would say expanded) in subsequent legislation. These rights must be respected, but we share the concern of other wilderness supporters over the impacts of excessive grazing on biodiversity and naturalness of wilderness ecosystems, and on the aesthetic experience of visitors. Mitigation of these impacts involves good grazing management; grazing levels that leave necessary forage for wildlife; better monitoring; and wider recognition by wilderness managers, grazers, and visitors of potential ecological changes triggered by grazing before they become obvious physical impacts. Murray (1997) described ecological changes and impacts of grazing in high-elevation meadows, the underlying idea being that the more we understand about such impacts, the more vigilant we can be about imminent impacts, and the more likely they will be minimized. *We need more educational information about grazing levels and potential impacts, so wilderness visitors better understand the issues and can work with wilderness managers and the livestock industry to keep grazing within acceptable limits.* Fragile sites, such as high elevations and arid lands, are especially vulnerable and require vigilance by all interests.

WILDERNESS RECREATIONAL USE

Much more is known about recreational use than the other wilderness uses. As an important use, recreation is the focus of a large part of all wilderness management and thus warrants a detailed discussion. However, *wilderness is not just a primitive-recreation area. The Wilderness Act makes it clear that wilderness is established for many purposes, with "primitive and unconfined recreation" only one purpose* (see Chapter 4).

The discussion of wilderness recreational use covers three main points: (1) methods for estimating wilderness recreational use, (2) use characteristics, and (3) distribution of use. *An understanding of the amount, character, and distribution of recreational use is essential to wilderness management because such use is the cause of many impacts, the source of many wilderness values and potential funding* (Fig. 14.3). The amount of wilderness use an area receives is one basis for allocating agency funds among areas for wilderness management, and, as use increases faster than agency funding, user fees may be a larger source of funding support for wilderness management (Mackey 1998; Alan Watson 1998). Therefore, some review of the methods used to measure wilderness use, and their advantages and shortcomings, is necessary background to understand how and where they might be used to determine the extent of use in a wilderness area.

METHODS FOR ESTIMATING WILDERNESS RECREATIONAL USE

Wilderness recreational use is difficult to measure. The typical wilderness has many remote access points with some larger ones, such as the BWCAW and the Selway-Bitterroot Wilderness, having seventy or eighty entries. Even some smaller wilderness areas have more than ten access points. Many access sites are distant from ranger stations or contact points and are, thus, difficult and expensive to monitor.

Compared to use of developed sites, wilderness recreational use is light and variable, and often of low density. This makes it prohibitively expensive to observe all entry points—some would have no use at all to observe on some days. Use is dispersed over such a wide area that it is nearly impossible to make any sort of direct head count, as can be done in developed recreation areas. Therefore, a variety of direct and indirect methods for counting and estimating wilderness use have been devised, including observations, electronic counters, automatic cameras, trail registers, mandatory permits, remote sensing, and guessing. These different methodologies each have their strengths and weaknesses and should be used with some knowledge of the sampling requirements and reliability in estimating total use, information provided

FIGURE 14.3 A campsite near Bald Knob Lake in the Absaroka-Beartooth Wilderness, Montana, bears out data that campers prefer to be near lakes and streams. Data on the amount, character, and distribution of recreational use are important information for wilderness managers. Photo courtesy U.S. Forest Service.

in more detail in sources such as the following (Hollenhorst et al. 1992; Hornback and Eagles 1999; Watson et al. 2000; Yuan et al. 1995).

Direct Counting Methods

Four methods of direct counting are generally used to help estimate use and monitor use patterns: (1) direct observation, (2) field interviews, (3) voluntary self-registration at trailheads, and (4) mandatory permits. Direct observation and field interviews are generally used to determine registration compliance to adjust self-registered use data as a basis to estimate total use, or to obtain more detailed information about users, such as use patterns or user characteristics. Self-registration and mandatory permits are often used because of their greater accuracy in estimating use and lower costs than other direct counting methods.

1. *Direct observation*—Systematically observing visitors or interviewing them (on-site interviews or requesting names and addresses for follow-up mail surveys) at a sample of trailheads on sample days has been used for decades, but the highly variable use pattern makes reliable sampling difficult and costly. Large samples are difficult to obtain. For example, the 74,000-acre Mission Mountains Wilderness in Montana has nineteen access points. If the main use season is about 100 days, this results in 1,900 date/place combinations from which to sample. One person working full-time could sample about 4 percent of these, and then only for about six or seven hours per day because of travel time. Stratified sampling that would concentrate on more often used times and places could increase the efficiency of sampling but would require some reasonably accurate information on use patterns to plan the stratification of the sample.
2. *Field interviews*—Formal or informal interviews may be used to help managers understand user interests, preferences, and needs. Formal interviews are used in scientific studies to estimate visitor use and visitor characteristics, such as when gathering information in a study of visitor-to-visitor encounters and conflicts, or to gather names and addresses for a follow-up mail survey. Sometimes, managers informally interview visitors to get their impressions, learn how many other visitors they have seen, and gather information to help them understand wilderness visitors.
3. *Voluntary self-registration*—Many estimates of wilderness use are based on voluntary self-registration at trail registers. Party size, method of travel, date of entry, length of stay, some data on destination (or itinerary), and visitor residence are usually obtained. The problem, of course, is that some or even many visitors do not register (Lucas 1983; Lucas and Kovalicky 1981; Petersen 1985). As these studies document, some kinds of visitors—especially equestrians, hunters, people making very short visits, and lone individuals—are less likely to register than others. Thus, the resulting raw registration data not only underestimates use but may also provide biased estimates of its composition. Efforts have been made to develop systems for adjusting trail-register data with adjustment factors applied to raw data from the trail-register cards to compensate for nonregistration. It is necessary to observe a sample of registration behavior to develop the adjustment factors because studies indicate that registration rates may be somewhat to highly variable among users, between wildernesses, and over time. It is also essential to periodically check actual registration rates and recalibrate the adjustment factors employed as a basis for use estimates. Currently, use estimates based on trail registers have a large, but usually unknown, margin of error. If registration rates could be raised, trail registers would be a better basis for use estimates. Low registration rates require large expansion factors and yield undependable estimates.
4. *Mandatory wilderness permits—The most accurate wilderness use data probably come from mandatory visitor-permit systems* (Hendee and Lucas 1973; Hollenhorst et al. 1992; Yuan et al. 1995). In 1980, about 70 percent of NPS wildernesses and about 50 percent of USFS wildernesses required visitors to obtain permits, a practice also common in Canadian wilderness-type areas (Washburne and Cole 1983). Today, in almost all cases, permits must be obtained from the managing agency, but in some wildernesses, permits are issued by cooperators, such as at stores by resort employees, or are self-issued by visitors at trailheads. Permits provide all the information obtained from trail registers, in addition to greater detail on planned routes of travel and the chance for managers to encourage environmentally and socially sensitive behavior, plus answer user questions if the permits are issued in person.

Some types of visitors (day users, local users, visitors from nearby resorts) are less likely to get permits, just as some visitors do not register at trail registers. Permit compliance varies, although it is usually higher than for trail self-registers and ranges from two-thirds to more than 80 percent of users. Compliance is high in most national parks, in the BWCAW, and in some other areas where the permit system has been in effect for more than twenty years and visitor awareness is high, but compliance might be lower in areas with newer permit requirements. Availability of permits at times and places convenient for visitors will increase compliance, and public information programs and enforcement are necessary. *We strongly recommend an adaptive and comprehensive approach to develop the best permit system for a given area and especially its application only when and where needed as the minimum approach to managing wilderness use.*

Commercial outfitters and guides and nonprofit organizations that bring paying clients to wilderness areas are required to apply for a special-use permit and pay a fee based on numbers of paying clients and revenues from them. This provides a good measure for commercial use of wilderness, although it is a small proportion of total use. Institutional outfitters (e.g., schools, universities, colleges, and churches) usually are not required to obtain permits, yet they may account for substantial use, and there are pressures for policy changes to require them to register their use in some way and pay fees (Mackey 1998).

Indirect Counting Methods

Three general methods of indirect counting are sometimes used to estimate use and monitor use patterns: (1) digital photography, (2) electronic trail-traffic counters, and (3) educated estimations. The third approach, making educated guesses and estimates about visitor use, is more commonly used than is appropriate and produces precise-sounding numbers that can be deceiving, if warnings about the limitations of educated guessing are not made explicit. Other indirect methods, like pressure-plate counters, have been tried on trails but are not widely used due to inaccuracies and maintenance problems.

1. *Digital photography*—Automatic photography has evolved rapidly in the last fifty years. For example, digital cameras set to expose one frame at preset intervals have been used along with a date and time stamp of each observation. More recent photographic systems used in some wildernesses employ a digital camera automatically triggered by an electronic traffic counter to film a few frames or images of the visitors. Group size, direction, and method of travel are evident, and day users can be distinguished from overnight backpackers. For protection of visitors' privacy, no identification of individuals is made, and only public areas through which visitors pass are filmed, such as trails, but not campsites or swimming areas. This system has worked well for managers and researchers, with similar results to direct observation but at a much lower cost (Hollenhorst et al. 1992; Lucas and Kovalicky 1981; Petersen 1985; Yuan et al. 1995).
2. *Electronic trail-traffic counters*—The use of automatic electronic trail-traffic counters has been increasing because of newer technology and improved operation. *Active infrared systems* emit an infrared beam onto a receiver or reflector and register a count whenever there is an interruption by something moving through the beam. Another type of operating unit is recommended for specialized situations, but reduced accuracy is to be expected because of how the system operates. *Passive infrared systems* emit an infrared beam that directly reflects off a passerby and registers a count whenever the reflected light is detected. However, due to lower accuracy with this technology, they should be reserved for areas that require lightweight units that can be quickly installed and removed. At best, automatic electronic trail-traffic counters indicate the number of large moving objects passing during a time interval or date. They cannot identify whether the objects were hikers, packhorses, elk, or cows, or how they clustered into parties or groups. Neither direction (entry or exit) nor route of travel can be obtained from the counters. Nevertheless, automatic counters are being used in a number of wildernesses and provide useful information, especially to estimate trends in use and to estimate compliance with voluntary self-registration at trailheads and mandatory permits (Hollenhorst et al. 1992; O'Rourke 1994; Yuan et al. 1995).
3. *Educated guesses and estimates*—Wilderness use can be estimated by means of an educated guess, hopefully based on previous years' reports and an assumed rate of change. Estimates might be modified according to various observations—how many cars were parked at a particular access this year compared to previous years (but possibly at different times of the season and on different days of the week), changes in the number of guests reported by outfitters, wilderness rangers' impressions, ranger observation, casual field interviews, unusual weather patterns, and so on. Some wildernesses show wide fluctuations in use from year to year. Often this comes from the use of guessing (more educated or less), which may be more sensitive to changes in who does the guessing, or in whose observations are used from year to year, than actual visitation. Educated guessing may be the most commonly used method for estimating wilderness use, especially for smaller or less heavily used areas, or to adjust estimates based on partial data from one of the other use measurement methods. *We recommend that educated guessing to estimate wilderness use be based on as much data and experience as possible.*

Individual wildernesses use a wide variety of techniques to estimate use, and most areas report using a combination of techniques. Accuracy varies from good to very poor. Unfortunately, the methods used and the expected accuracy levels are not reported in annual use estimates. Adjustments for noncompliance on permits or trail registers, if made at all, are too often essentially guesswork based on unsystematic spot checks.

Thus, estimates of wilderness recreational use are based on a mixture of methods, and the resulting level of accuracy is uncertain and probably only fair. Whether errors compensate more than they accumulate is unknown. Comparisons of visitor use over long time periods are particularly unreliable. Older estimates were mainly guesses, and as more accurate methods were adopted, large changes in reported use often developed. Therefore, historical records of trends in use must be treated cautiously.

Certainly, wilderness recreation use data need not be perfect to be useful, but any improvement adds to the management value of the information. The need for accuracy in use data is greater in wilderness areas for which new management plans are being developed or in areas that are monitoring to evaluate the achievement of planning and management standards.

Units of Measurement

Several units of measurement are used to report recreational use, and the units vary between the Forest Service and NPS. The BLM and FWS do not routinely collect visitor data, which leads to further confusion in comparing data between agencies. The most common units of measure are listed in Table 14.1. Unfortunately, the standard units used by the NPS, the overnight stay, and the USFS, recreation visitor day (RVD; twelve-hour period), are not directly comparable. For an individual, overnight stays could be converted approximately into RVDs by multiplying by two and one-half on the assumption that a one-night stay usually is made up of a large part of the first day and the last day; for example, 8 AM to 8 PM, 8 PM to 8 AM, and 8 AM to noon.

This array of units for quantifying and expressing recreation use is confusing. Which of the units is best? As is so often the case, the answer is "that depends." Specifically, it depends on the purpose for which the data are to be used. Often, several different units are relevant to a particular purpose.

For indicating potential impact on camping areas and camper congestion, the overnight stay or the group night provides the most relevant information. With respect to solitude, the visitor day, which reflects how many people are present over time, is generally useful, and people at one time (PAOT) is particularly appropriate. For total use, the number of visitors or RVDs is

TABLE 14.1 Units of Measure for Wilderness Recreational Use

Unit	Definition	Use and Comments
Visit (or occasion)	The entry of 1 person into an area, regardless of length of stay (repeat visits counted)	NPS; USFS before late 1960s
Visitor day (recreation visitor day or RVD)	1 person present for 12 hours, or equivalent (2 for 6 hours, etc.)	USFS standard unit since 1965
Person day	1 person for 1 calendar day (7 hours or more; 5–7 hours = 0.75; 3 to 5 hours = 0.5; 15 minutes to 3 hours = 0.25; less than 15 minutes = 0)	USFS standard before 1965 (noncomparable to visitor days)
People at one time (PAOT)	Total number of people present at one time, usually a day	Some USFS reports, some research
Overnight stay	1 person passing 1 night within an area (could also be called a visitor night)	NPS standard unit
Group (or party) visits	The entry of 1 group (party) into an area	Some research

most useful. Total visits are a partial measure of visitors, are much easier to obtain, and give a general idea of total use tied to certain types of managerial workloads, such as making visitor contacts and issuing permits. However, the great variation in the definitions of use labeled "visits" makes this a difficult unit of measure to interpret. A brief visit of an hour or so counts the same as a two-week stay. Visits can be used with visitor days, if both are reported, to gain an idea of length of stay.

Different wilderness management agencies do not use comparable terms to report use. The USFS is the only agency to report recreational use separately for each wilderness area, and it has used the RVD as the unit of measure since 1965. The NPS reports backcountry camping use, which is defined as camping in minimally developed areas not reached by roads, and has used the overnight stay as a unit of measure since 1971. Thus, aggregate NPS "backcountry overnight stay" includes stays in wildernesses and are the best available information used to represent NPS wilderness camping use, but it does not include wilderness day use, which is often substantial in national parks.

Because wilderness recreation management often competes with other kinds of outdoor recreation for public funding, and use is one indicator of needed management support, more complete and accurate reports of wilderness use are needed. We recommend that at least visits and visitor days be reported annually for all wildernesses in the NWPS. Overnight stay would be a useful supplemental unit, although calling it a visitor night would clarify its meaning. We recommend use of one or more of the use-measurement methods and less reliance on educated guessing to report use estimates. Sometimes educated guessing to estimate recreational use is better than nothing, but we recommend that it be supported by as much actual data and manager experience as possible.

Trends in Recreational Use

In the following discussion, we summarize and interpret recreation use trends from numerous user studies conducted over the last fifty years (Bowker et al. 2007; Cole 1996; Loomis et al. 1999; Lucas 1980; Lucas and Stankey 1988; Roggenbuck and Lucas 1987; Roggenbuck and Watson 1989).

Wilderness recreation use has increased greatly since 1964. Although use was widely reported to have leveled off and even declined in some areas by 1986, it grew substantially in the 1990s despite predictions that it would not (Cordell, Bergstrom, Hartmann et al. 1990; Cordell and Teasley 1998). Cole (1996) points out that *these "level off or decline" visitor use predictions were incorrect and were detrimental because of their unsupported implications that it might be possible to ease management regulations and promote more wilderness use, and that recreational demand was no longer an appropriate argument for designating additional wilderness areas.*

Overall use of the NWPS was estimated by Cole (1996) as about 17 million RVDs in 1994. This was a sixfold increase between 1965 and 1994, an average annual increase of 6.3 percent. He reported that the majority of this increase was due to additional units and acreage being designated as part of the NWPS, which grew from the original 54 USFS wildernesses and more than 9 million acres in 1964 to 630 areas and more than 103 million acres by the mid-1990s. Cole further reported that *recreational use in the original 54 USFS wilderness areas increased 86 percent from about 3 million RVDs in 1965 to 5.5 million RVDs in 1994; the remaining increase of 11.5 million RVDs was attributed to the addition of 576 wilderness areas designated as part of the NWPS since 1964* (Cole 1996).

Estimates of wilderness use derived from a National Visitor Use Monitoring Project were 10.7 million trips and 16.3 million on-site days (i.e., visits) in 2002 (Bowker et al. 2007). According to Cole (1996) these estimates of wilderness use are difficult to compare with those by other researchers using agency reports and RVD units. The apparently higher numerical estimate by Cole (1996) of 17 million visitor days in 1994 is partially explained by the fact that RVDs are twelve-hour time periods and on-site days are visits from one to twenty-four-hours in length. Thus, the 2002 estimate of use by Bowker et al. (2007) is an actual increase over 1994 even though conversion from one unit of measure to another is, at best, a crude estimation.

The intensity of wilderness use (RVDs per acre) is also an important issue for managers. This type of use-intensity analysis is valuable for assessing concentrations of use across specific areas. Because it can be misleading when the distribution of the use is not known and because use is usually highly concentrated in certain popular locations, it may appear acceptable as an average (RVDs per acre) for an entire wilderness area. For example, across the NWPS one would expect wilderness management needs (and needs for funding) to be greatest for areas with the highest use per acre. *Now that growth in acreage of the NWPS is slowing because most large areas are already designated and use is projected to continue to increase, it is reasonable to expect that use intensities will increase as well—and so will the need for funding of wilderness management.*

FIGURE 14.4 Riders and pack stock travel through a mountain meadow; such trips are a traditional way of visiting a few large western wilderness areas. Hiking is a much more common method of travel. Photo courtesy U.S. Forest Service.

USE CHARACTERISTICS

In the following discussion, we summarize and interpret recreation characteristics from numerous user studies conducted over the last fifty years (Cole, Watson, and Roggenbuck 1995; Lucas and Stankey 1988; Roggenbuck and Lucas 1987; Roggenbuck and Watson 1989; Washburne and Cole 1983; Watson et al. 1999).

Length of Stay

Most wilderness visits are for short stays, day use, or only one- to two-night stays. Many small or medium-sized wildernesses are predominantly day-use areas. Studies report that the amount of visits that are day use ranges from 7 to 68 percent, but surveys are available for only a few areas. Among the areas for which visitor surveys are available, only the Bob Marshall Wilderness, the Weminuche and Rawah in Colorado, and some wildernesses in Wyoming have little day use. These areas are large, with the wilderness boundary miles beyond the trailhead in many places, and this limits day use. Most large wilderness areas are also far from large population centers, making them more often destinations for vacations and trips of several days, rather than weekend or day hiking trips. The limited information on trends in use suggests that the percentage of day users may be increasing in some areas.

Long wilderness trips of a week or more account for less than one-tenth of all visits, even in the very large areas such as the BWCAW, the Bob Marshall Wilderness, and the Selway-Bitterroot Wilderness (Fig. 14.4). Visits to many areas, especially small to medium-sized wildernesses, averaged two to three days, with wildernesses in the eastern and western United States having similar average lengths of stay (two to three days) and percentage day use (25 to 70 percent). Lengths of stay in most areas having had more than one survey have become a little shorter in more recent years. *Thus, wilderness use trends toward shorter wilderness trips and more day use seem consistent with other sociodemographic trends, such as individuals and families having shorter blocks of leisure time.*

Group Size

Parties of wilderness visitors are generally small; from one-half to three-fourths of the parties at all areas for which data have been collected range *from two to four persons. Few persons visit wilderness alone (less than 10 percent of all visitor groups), and parties of more than ten people account for only about 5 percent of all groups.*

Party size is declining in areas where long-term studies have been conducted. More wilderness areas have restricted party size, and even organized groups have become smaller because managers and organization leaders have become concerned about the impact of large groups on the wilderness environment and on other visitors.

Method of Wilderness Travel

The most common method of wilderness travel in almost all areas is, with few exceptions, hiking. Most visits to the BWCAW are made in paddled canoes. Some BWCAW visitors use limited horsepower, outboard-motor-powered canoes or boats in specifically designated zones, and a few hike, snowshoe, or ski.

The BWCAW is one of the few areas in the lower forty-eight states where mechanized travel is permitted within a wilderness. Other exceptions include a few large wilderness areas in the West that have wilderness landing fields or airstrips that predate the Wilderness Act or wilderness designation. Such airstrips have remained open to airplanes, such as in the Great Bear Wilderness in Montana and the Selway-Bitterroot Wilderness and Frank Church–River of No Return Wilderness in Montana and Idaho. In these large areas, airstrips are often the starting point of outfitted horseback or float trips. The Frank Church–River of No Return Wilderness includes twelve airstrips on federal land that have about 5,500 total aircraft landings per year, and the USFS is very limited in its ability to close the airstrips based on the statutory history of the wilderness's designation and legislative intent and language, but it is allowed to restrict use levels (Meyer 1999). The BWCAW has a policy not found elsewhere that not only prohibits air access but bans low-altitude flights as a result of Executive Order 10092, signed by President Truman in 1949.

In Alaska, floatplanes and "flight seeing" remain an important method of physical and visual access to vast wilderness

FIGURE 14.5 Flight seeing over Denali Wilderness, a National Park Service–managed wilderness, with a commercial outfitter. Federal legislation allows such business in some wilderness area airspace. Photo courtesy Chad P. Dawson.

areas under ANILCA legislation (Fig. 14.5). Similarly, places like Glacier Bay Wilderness are visited by large cruise ships whose passengers view tidal glaciers and passively experience the rugged mountain ranges in some coastal wilderness areas of Alaska (Fig. 14.6).

Horse use is important in some large western wilderness areas but still accounts for a minority of total use in the NWPS today. Managers of seven wilderness areas in the western United States estimated that more than half of the visitors to these areas rode horses in 1978 (Washburne and Cole 1983), and later studies reported that half of the visitors to the Bob Marshall Wilderness rode horses. Horse use is common in some southwestern wilderness areas, such as the Superstition Wilderness, where most of the visitors riding horses are day users. Horses and pack stock are few or uncommon in wildernesses in the eastern United States and are a small part of the use in most western states.

Recreational Activities

Wilderness visitors participate in a variety of recreational activities in addition to hiking. Surveys of visitors indicate that, on the average, respondents participated in *three major activities: fishing, photography, and nature study.* Studies suggest that in western areas, about one-half or more of the visitors

FIGURE 14.6 A commercial cruise ship enables passengers to visit Glacier Bay Wilderness, a National Park Service–managed area in Alaska. Photo courtesy Chad P. Dawson.

fish, but for many the fishing is somewhat incidental rather than central to their experience. We suggest that the increased wilderness use since these studies were conducted includes a declining percentage of anglers compared to other recreational activities.

Photography and nature study are also major activities in most places, and swimming is common in many southern areas. These are all low-impact, nonconsumptive uses that are compatible with wilderness values. Hunting varies from a minor to fairly common use in national forest wildernesses, but it is generally prohibited in national park wildernesses. Mountain climbing is an infrequent use in most areas but common in a few areas, as are other activities that involve challenge and adventure experiences (Ewert and Hollenhorst 1997). The frequency of these activities varies by area, but overall, the percentage of visitors who hunt or fish has declined while mountain climbers have increased (Cole et al. 1995).

Season and Timing of Use

Summer is the highest use season in virtually all areas, and that trend continues. A few large western U.S. areas have considerable fall big-game hunting, but in two areas studied (the Bob Marshall and the Selway-Bitterroot), summer visitors still substantially outnumber fall hunters. Many areas in the South, Southwest, and California deserts have substantial winter and spring use. In the North, and at higher elevations in the mountains, winter use is light but much more common than in previous decades and is growing (Fig. 14.7). Wilderness experience programs aimed at personal growth, education, or therapy operate year-round to achieve program objectives and to take advantage of client demand, weather conditions, and opportunities for solitude.

Many areas experience weekend peaks in use, especially the smaller, more accessible wildernesses near urban population centers, such as the San Gorgonio and San Jacinto Wildernesses near Los Angeles, Pusch Ridge Wilderness near Tucson, Superstition Mountains Wilderness near Phoenix, Alpine Lakes Wilderness near Seattle, and others. Some small wilderness areas in the East do not have extreme weekend peaks in use but do report the majority of annual use occurring during weekends.

Type of Group

In almost all wilderness areas, the majority of the use takes place in small groups of family and close friends. Wilderness

FIGURE 14.7 Increasingly, visitors use wilderness in winter, such as ski tourers in the Selway-Bitterroot Wilderness, Idaho. Photo courtesy Richard Walker.

use by organization-sponsored groups accounts for a small percentage of use such as conservation groups, outdoor recreation clubs, youth groups such as Boy and Girl Scouts, youth camps, and church groups. Several studies indicate an increase in use by wilderness experience programs aimed at personal growth, education, or therapy and healing (Dawson et al. 1998; Friese et al. 1998, 1999; Russell et al. 2000; Gager et al. 1998).

Visitor Residence

Some nationally and internationally known wilderness areas draw visitors from all over the nation and other countries. Some internationally attractive wildernesses are the Bob Marshall and Selway-Bitterroot in Montana; John Muir Wilderness, Yosemite, and Sequoia–Kings Canyon National Parks in California; the Great Smoky Mountains in the East; and the BWCAW in Minnesota. Wilderness and backcountry visitors come from many states and foreign countries, but even the majority of visitors to these wildernesses are from the state or region near the wilderness. The discretionary time and income necessary to travel to a distant wilderness area may limit some wilderness visitation. Numerous studies indicate that, although wilderness is a national resource, *wilderness recreational use typically comes from the home state and region or nearby urban population centers.* This may reflect the constraints of time, expense, and distance as well as residents' preferences for local opportunities. Some implications are that *wilderness is needed fairly close to population centers with nationwide distribution to make wilderness readily accessible to the nation's population.*

DISTRIBUTION OF USE

The geographical distribution of wilderness recreational use is very uneven—there are many people in a few places and only a few in many other wilderness locations. This unevenness of use is a pattern among wildernesses as well as within individual wildernesses. For a variety of reasons, *some wilderness areas, some trailheads, and some popular destinations attract more use than others.* This geographical concentration of use is compounded by the tendency for use to further concentrate during the summer and high-use weekends (example: Dawson et al. 2001).

These concentrations of wilderness use pose management challenges and opportunities. Wilderness areas vary in their capacity to absorb the social and environmental impacts of use, and some of the most popular locations (such as high alpine lakes) may be sensitive to physical impacts from use and intrusions on solitude. The tendency of use to concentrate in time and space enables management to focus efforts and resources on these locations and time periods, whether by redirecting, dispersing, or limiting use.

It is clear that an *even distribution—the same amount of use on every acre, or of every mile of the trail system, and of every campsite—is neither possible nor desirable.* It is impossible—barring total regimentation—for several reasons. Trips vary in length; thus, the interior is not as heavily visited as the periphery. In addition, most use is on trails or water routes, which typically branch and diverge, so the main trunks inevitably carry more people than the branches. Furthermore, evenly dispersed use is undesirable for two reasons. *First*, different parts of a wilderness vary in their response to visitor use and, in their capacity, to sustain use while maintaining the existing environmental conditions. *Second*, with evenly distributed use, persons who really prize solitude and are willing to travel or hike a long way to find it would be frustrated. At the same time, persons not strongly motivated to seek solitude would encounter fewer visitors than they would willingly accept and, in a sense, the greater solitude would not be appreciated by them.

We suspect that in some areas present use is too unevenly distributed and that some redistribution to reduce the extremes is desirable. Badly overused wilderness recreation sites are a common problem. The issue of optimum-use distributions, and how to encourage them, is considered in many sections of this book—in discussions of principles of management (Chapter 7), planning (Chapter 8), carrying capacity (Chapter 9), impacts of recreation activities and use (Chapter 15), and visitor use management (Chapter 16).

Distribution of Use Between Areas

Reported visitor use shows substantial variation between individual wildernesses in the national parks and national forests. The reasons for such different drawing power are poorly understood, but the most heavily used national forest and national park wildernesses are almost all located relatively close to large population concentrations. California, Minnesota, southern Appalachian, and New England wildernesses receive the most intense recreational-use pressure, but other wildernesses in these same regions are lightly used.

Location near many people is associated with heavy use, but a reputation as an attractive area is also necessary if heavy use is to actually occur. Publicity in national magazines and guidebooks has contributed to the popularity of some wildernesses and some particular trails in specific wildernesses. Some people have speculated that merely designating an area as wilderness attracts extra use, that is, assigning an area a name and identifying it as wilderness may stimulate use. This is often called the "designation effect." However, there is scant evidence that designation results in increased use.

By themselves, visitor totals indicate little about pressure on the resource. For one thing, areas vary in size. For example, although total use in the Collegiate Peaks Wilderness in Colorado and the Bob Marshall Wilderness in Montana was about the same in 1986, the Bob Marshall is about six times as large as the Collegiate Peaks. To aid in comparing areas and use pressure, acreage and use figures can be expressed in terms of visitor days per acre. However, such comparisons do not fully reveal the degree of visitor congestion and the resulting pressure on the soil, vegetation, water, and wildlife because they ignore the length of the use season. For wilderness in the northern Rocky Mountains or Cascade Range of the Pacific Northwest, the main use season might be only two or three months long (because of trails blocked by snow or streams too high to ford). In milder regions, the use season is much longer. Therefore, adding a length-of-season adjustment to annual-use-per-acre figures would show even higher use levels per acre for areas with primarily summer use. For instance, the more than 1 million visitor days of use in the BWCAW were not distributed evenly over an entire year—about 90 percent of the visitor days took place from late May through September.

Moreover, *the proportion of usable or effective acreage is not the same in all wildernesses.* Although all wilderness acreage is available in the sense that it provides at least a backdrop and space for isolation (and, of course, is a vital part of the natural ecosystem), only a portion is used directly by visitors. The amount of land available for use is affected by steepness of slope; type of vegetation; and the extent of area in lakes, streams, and wet, boggy soils.

Usable acreage is also influenced by the degree of access development, number of trails, and travel routes. Most wilderness use is restricted to travel on the existing trail systems, areas directly adjacent to trails, and overnight camping areas. Some areas have an extensive trail network whereas other areas have only a few trails. Therefore, *for more detailed indices of accessibility and use of an area, the entry points per thousand acres and miles of trail per thousand acres can be calculated for comparisons between different wilderness areas.* Miles of trail per thousand acres and number of entry points per thousand acres relate more closely to use capacity than does gross area, and they vary considerably among wilderness areas.

Distribution of Use Within Areas

Use distribution within any particular wilderness can vary as much as use between wilderness areas. For instance, in the BWCAW in 1986, about 61 percent of the user groups entered through only ten of the area's eighty-seven entry points; in fact, two entry points accounted for almost one-fourth of all groups entering the area. The 1999 use pattern in the Great Gulf Wilderness in New Hampshire is representative of the uneven spatial distribution of use in many wildernesses. The use data for the Great Gulf Wilderness indicate a pattern of uneven hiking use on some trail segments, particularly those associated with the Appalachian Trail and the summits of the Presidential Range (Dawson, Simon et al. 2001).

Heavy concentration of visitor use along only a few trail miles is common. For example, in the entire Spanish Peaks (CO) trail system the 10 percent of trail miles with the heaviest use accounted for about one-half of all visitor miles in 1970 (Lucas 1980). In 1999, the Great Gulf Wilderness had a concentration of 70 percent of visitor miles of use were on only 40 percent of the available miles of trail, and that is a moderately high concentration index for a small wilderness (Dawson, Simon et al. 2001).

Horse use tends to be more concentrated than hiking use. Horse parties go farther, but they tend to stay on main trails. As a result, areas with heavy horse use generally have more concentrated total use patterns, although horse use accounts for a small portion of use in most areas. The Bob Marshall, the only area studied where a majority of visitors traveled by horse, also showed the most concentrated use.

Short trips, both in distance traveled and duration, are characteristic for most wildernesses. One study of eight Montana and Idaho wildernesses and backcountry areas revealed that only about 14 percent of the visitors traveled more than twenty miles round-trip in the wilderness—a statistic that refutes the notion that wilderness trips are typically long.

Distribution of visitor use in many of the arid wildernesses of the southwestern United States is heavily influenced by water availability (Fig. 14.8). Visitor use is concentrated around primary water sources. In the Superstition Wilderness, Arizona, visitors are educated to camp in designated sites that have one maintained fire ring and allow access to water sources, but that are far enough away from those sources to reduce social and biological impacts. All other campsites are rehabilitated and fire rings are removed whenever they occur to keep the majority of the area wild and to provide opportunities for solitude.

An effective way of describing wilderness congestion resulting from use concentration is to measure the number of other parties a group could be expected to meet (encounters) per hour or day in a given portion of the wilderness. Six wilderness areas in the Alpine Lakes, Washington; Mount Jefferson, Oregon; and Three Sisters Wildernesses, Oregon, reportedly had extremely high encounter rates in 1991–1992 that exceeded the level preferred by most users (Cole et al. 1997). For example, weekend visitors to the Alpine Lakes Wilderness typically encountered thirty to forty other groups, whereas weekday visitors encountered ten to fifteen other groups. In

FIGURE 14.8 Lakes and river attract wilderness visitors, but camping on shorelines and riverbanks is often prohibited to reduce visual and biophysical impacts of visitors. Photo courtesy U.S. Forest Service.

the two Oregon wilderness areas, weekend visitors typically encountered twelve to eighteen other groups, whereas weekday visitors encountered five to eleven other groups. Most visitors reported that they expected these levels of encounters and were not overly bothered by them, and only 10 to 23 percent supported management actions to reduce use levels.

Use patterns in most wildernesses are strongly trail-related. The only exceptions are areas where water access predominates, as in coastal southeastern Alaska and the BWCAW. Studies of national forest wilderness areas estimate that fewer than 20 percent of the visitors to most areas travel cross-country and an even smaller percentage of the total distance covered in the wilderness is off-trail travel. Cross-country travel depends on the motives, interests, and experience of the visitors, as well as on the terrain and vegetation. Wilderness visitors are almost locked into trails in steep, heavily forested country, especially in areas with many downed trees, such as forest zones in the northern Rockies, much of the Cascades, and the East. Alpine, plateau-like areas and desert terrain have more cross-country opportunities.

Six wilderness areas in the Alpine Lakes, Mount Jefferson, and Three Sisters Wildernesses had substantial recreational use impacts in 1991–1992, showing on system trails, social trails, campsites, and lakeshores (Cole, Watson et al. 1997). Most visitors reported that they noticed these impacts, that the impacts detracted from their wilderness experience, and that they were supportive of management programs to close trails and campsites or to reestablish vegetation in some areas.

Campsite use tends to be uneven. Extrapolating from research studies and professional experience, the most popular campsites share characteristics of (1) proximity to water, possibly with fishing opportunities; (2) scenic and/or water views; (3) closeness to a trail (several hundred feet); (4) availability of enough level land to camp (4 percent or less slope); and (5) availability of firewood (not as important as it used to be thanks to portable stoves and Leave No Trace awareness). Where wilderness users do not camp so close to water, the impacts on the riparian zone are minimized, and lakeshore access is available to more users. Sites close to water are highly attractive, and getting people to change their selection of campsites may not be easy.

A study of campsites over twelve to sixteen years in portions of Montana's Selway-Bitterroot and Lee Metcalf Wildernesses and Oregon's Eagle Cap Wilderness, showed an increase in the number of campsites by 53 percent, 84 percent, and 123 percent, respectively (Cole 1993). The ecological impacts of locating and managing campsites are discussed in Chapter 15, but the key point here is that campsite impacts increase substantially with campsite proliferation, expansion, and, to a lesser extent, existing campsite deterioration.

USER CHARACTERISTICS

A common stereotype of wilderness visitors pictures them as young, athletic, wealthy, leisured, and urban. This stereotype is commonly reflected in congressional testimony on wilderness designation proposals and in public meetings on proposals in small towns near wilderness. Although there are some distinguishing sociodemographic characteristics among wilderness users, we must be careful to recognize that statistical averages represent some portion of the users but that there are many different types of users with a variety of interests, needs, and expectations. The diversity of experiences sought by visitors is one of the reasons that the WOS is used by wilderness planners to compare the wilderness experience opportunities within each wilderness area and among the wilderness areas of a region. Both stereotypes and diversity of users have been documented in studies over fifty years (Dovark et al. 2012; Cole and Williams 2012; Cole and Hall 2008; Hall and Cole 2007; Cook and Borrie 1995; Cole et al. 1995; Watson et al. 1992; Roggenbuck and Lucas 1987; Roggenbuck and Watson 1989; Stankey 1971). These studies are among the most complete and are the basis for much that follows—we will not cite them repeatedly.

Characteristics of wilderness users are from many studies, and these characteristics are similar from area to area, even in different parts of the nation. In general, wilderness users, compared to the general population, tend to be young but with all age groups represented, predominantly male but with increasing numbers of women, from urban areas but largely near the wilderness area visited, above average in income, well-educated and in professional or technical occupations or students.

Only a few important changes seem to have occurred in the sociodemographic characteristics of users over the last fifty years—higher average education levels, older average age, and a greater percentage of women. Even though there are reported changes in these sociodemographic characteristics, there are no reported shifts in the kinds of trips that users take or their preferences for conditions they encounter.

Given the high educational level of users, managers can continue to inform and educate them about wilderness management issues and the need for management actions and programs. Understanding who the wilderness visitors really are is important for both wilderness policy makers and wilderness managers—wilderness visitors are important clients. Policy makers need to know who receives the benefits gained from financial and natural-resource allocations to wilderness designation and management. Managers also need to understand who their clients are so they can communicate with them through information and education programs (e.g., Internet communications for off-site information and education before visitors arrive at the wilderness area).

Age: *Wilderness visitors tend to be younger on average than the general population, yet all age groups are fairly well represented.* Data from about thirty wilderness user studies show that a substantially higher percentage of sixteen- to twenty-five-year-olds and twenty-six- to thirty-five-year-olds are present among wilderness visitors than are in the general population. The thirty-six to forty-five age group is often more highly represented than the general population, and the forty-six to fifty-five age group is present in similar proportion to the general population. The fifty-six and older age groups are underrepresented among wilderness visitors as they are in all types of outdoor recreation. Children under sixteen years of age are substantially underrepresented in most wilderness studies, not because they do not use wilderness, but rather because most studies typically do not survey users less than sixteen years of age. All these age figures refer to individual visitors, but a number of studies have been based on party leaders or persons registering for the entire party, and these persons tend to be older.

Gender: *Wilderness use has sometimes been viewed as a male activity* because studies from 1960 through 1983 reported that 70 to 85 percent of users were male. The larger, horse-oriented wildernesses average less female visitation; the smaller hiking areas average a little more. More recent studies indicate an increasing percentage of women visitors, although their participation rates are still lower than those of male visitors.

Residence: *Most wilderness visitors are from large cities and urban areas, as are most Americans.* However, because visitors do not typically travel long distances to visit wilderness, the proportion from urban places depends largely on the degree of urbanization in that region. *Most wilderness users are from the state or region in which the wilderness area is located.* In general, the percentage of city and urban visitors to wilderness is in proportion to the population of the state or region. Although current residence of wilderness visitors has been reported in most studies as overwhelmingly urban, early studies (Hendee et al. 1968; Lucas 1980; Outdoor Recreation Resources Review Commission [ORRRC] 1962) found more visitors with rural backgrounds. Due to the urbanization of American society in the past fifty years, more adults may be visiting wilderness now because they were introduced to the activity by their urban parents, friends, organizational outfitters (such as youth and college outdoor programs), and wilderness experience programs.

Income: *Wilderness visitors are moderately above average in income, as are most types of outdoor recreationists, but all income levels are represented among wilderness visitors.* The above-average income levels reflect the high education levels and high proportion of users with professional occupations among wilderness visitors, and lower-income visitors may actually be student participants. High incomes are not necessary to visit wilderness, although a certain level of investment is required for equipment and transportation to access the wilderness. Thus, wilderness users need some modest level of disposable income. However, because most users visit a wilderness in the state or region near their residence, typical expenditures for wilderness visits are low compared to many other kinds of recreation away from home.

Education: *One of the most distinguishing characteristics of wilderness visitors is high educational levels.* Studies show that with few exceptions, 60 to 85 percent of the visitors to most wildernesses have attended college, and 20 to 40 percent have done graduate study. This higher level of education among wilderness users is related to their above-average income and professional and technical occupations. Why this association between educational attainment and wilderness use? Higher education may sensitize people to the natural world and global concerns about environmental issues such as wilderness. Students at institutions of higher education are more likely to have friends, acquaintances, and organizations that visit wilderness and who may introduce them to such activities.

Occupation: *Studies show that persons in professional or technical occupations and students form the majority of visitors to most wildernesses* and are more prevalent among wilderness visitors than in the general population. Generally, from 20 to 40 percent of the visitors of working age are in professional or technical work, with about one-fourth of the adults and young adults being students. This is related to the high education of wilderness visitors.

Racial and Ethnic Minority Participation: *It is a matter of concern that racial and ethnic minorities are greatly underrepresented among wilderness visitors, although their participation is increasing.* Some researchers have pointed out that the increasing racial and ethnic diversity of the U.S. population is having a small impact on increased wilderness use, but that shifts in user values may occur as more ethnic minorities participate. Nevertheless, the future of wilderness depends on public support, and wilderness needs the support of the increasingly diverse U.S. population.

FACTORS AFFECTING WILDERNESS USE TRENDS AND MANAGEMENT

Wilderness recreation use has increased greatly over the last fifty years, and overall use of the NWPS is expected to continue. In some areas, visitor use has leveled off and even declined, whereas in other areas there are steady increases. There are numerous factors that affect outdoor recreation and wilderness use (Cordell 1999; Cordell, Tarrant et al. 1998; O'Leary et al. 1989; Bowker et al. 2007), such as sociodemographic changes in the population and other factors (Watson 2000). In the following discussion, we consider several of these factors and their potential impacts on recreational use of wilderness in the future—leading toward a projection of future wilderness use.

CHANGING AGE AND POPULATION STRUCTURE

Many studies of outdoor-recreation participation have pointed to age as one important predictive measure of future participation, but *the gradual aging of the U.S. population has not led to a decline in wilderness participation as predicted by some researchers.* English and Cordell (1985) found that participation rates among different age cohorts have risen steadily since 1960, a fact that suggests that the expected dampening effects of increased age on recreation participation might be even smaller in the future than predicted in the past. One of the most fundamental changes in U.S. society today is the increasing age of the population. The median age of the U.S. population was 30 in 1980 and was projected to reach 36 by the year 2000 (Dwyer 1994) (U.S. Bureau of the census reported it was 35.3 in 2001).

Although the population is aging, short-term changes are not dramatic in those age classes who visit wilderness the most. The percentage of people in the eighteen-to-twenty-four age class has declined slightly since 1980; they were 10.4 percent of the population in 1990 but will account for only 8.3 percent by the year 2025 (Dwyer 1994). The percentage of the population that is sixty-five years of age or older is projected to increase from 12.6 percent in 1990 to 21 percent by the year 2025. Decreases in youth under eighteen and increases in the percentage of sixty-five and over should have little effect on wilderness use because neither age class has previously been a major wilderness user. The twenty-five-to-sixty-four-year-old age class, the most inclined to visit wilderness of any age class, will continue to increase as the short-term change is toward a "middle-aged" population. *Thus, the changing age and population structure in the United States should have neither a major dampening nor increasing influence on wilderness use.*

POPULATION DISTRIBUTION AND DIVERSITY

Changes in the U.S. population distribution might seem to have the potential to impact wilderness use trends in specific regions, but such changes to date are not evident. The U.S. population is projected to grow at a rate of 0.5 percent per year and the majority of that growth (50 million people between 1990 and 2025) is expected to be among racial and ethnic minority groups (Dwyer 1994). Wilderness recreation use is slowly increasing among nontraditional ethnic and racial groups, and this change is most often reported in wilderness areas near urban centers (Cook and Borrie 1995). This trend toward more racial and ethnic diversity is reflected in the U.S. population and recreation participation in general (Dwyer 1994). *Despite the slow increases toward more diverse racial, ethnic, and gender participation in wilderness recreation, this remains a major issue* for wilderness. Whatever the reasons for the underrepresentation of racial and ethnic minorities in wilderness recreation (cultural, economic, or preference), *we need to remove the barriers to participation so the benefits and values of wilderness can be more widely shared—and so wilderness protection will be a relevant goal for our increasingly diverse public.*

LEISURE TIME

One to several days are necessary to visit most wilderness areas—time for travel to and from the site and the experience. Such blocks of time seem harder to come by for many people in our fast-paced society, despite data that suggest that we have more leisure time. Many factors, such as age, education, income, race, residence, and marital status, affect the amount of leisure time available to individuals (Robinson and Godbey 1997). For example, today in a typical family with children, both spouses may hold jobs and these dual-income households are on the rise. Although the discretionary income of such households increases, it may be difficult to coordinate household duties, vacations, and free time between spouses so that "getaway" weekends or longer periods are possible. Thus, trends reducing blocks of discretionary time may have a dampening effect on wilderness visitation and encourage shorter trips and use of nearby wilderness areas. However, total recreation use of wilderness is increasing despite a pattern toward shorter trips to nearby wilderness areas.

EFFECTS OF EXPANDING THE WILDERNESS SYSTEM

One explanation for much of the increased total use of wilderness is the increase in the size of the NWPS. However, expanding the NWPS with new areas has probably had little effect—besides reclassifying the roadless-area-recreation visitors from that category to the wilderness-visitation category. In some cases, the visits were already being counted as wilderness. For example, some of the growth in acres and number of areas in the NWPS resulted from shifts of national forest primitive areas to wilderness or official designation of portions of a national park as wilderness. In both cases, the recreational use was already being counted. These areas have long been specially designated and widely perceived as wilderness, even though not technically designated.

Whatever effect major expansion of the wilderness system may have had on past use of older areas, its future effect probably will be less. The large expansion of the system (Landres and Meyer 1998) in the 1980s (more than 300 areas and 71 million acres added between 1980 and 1985) and the 1990s (69 wilderness areas and more than 4 million acres in the California Desert Protection Act—U.S. Public Law 103-433) is not likely to be duplicated in years to come (J. Watson and Brink 1996). *We anticipate a gradual increase in the NWPS some of which may include designation of already-proposed roadless areas in several large national parks.*

INCREASING EDUCATIONAL LEVELS AND ENVIRONMENTAL AWARENESS

The most distinguishing socioeconomic characteristic of wilderness users is their high educational levels, and this portion of the population is increasing. The percentage of the population with a college degree or some postsecondary education continued to increase through the 1990s. The introduction of wilderness activities and interests to students through college courses, field trips, outdoor clubs, and other students will likely continue as a factor contributing to future wilderness use.

Some aspects of higher education stimulate environmental concerns in both science-oriented and other students, as global environmental concerns have social, economic, historic, and political influences that are part of a liberal education program. Wilderness is related to these concerns because it is a benchmark and symbol of environmental quality. Therefore, wilderness continues to have broad public recognition as a positive environmental concern and support for wilderness is part of the agenda for numerous environmental and conservation groups.

CHANGING TRANSPORTATION COST AND SUPPLY

The impact of gasoline prices on travel for wilderness recreation does not appear to have been great in the past. Availability of gasoline has been a problem from time to time, but typically this has been a temporary situation that took place without a long-term impact on wilderness use. *Most wilderness visitors live relatively close to the areas they visit, so it seems unlikely that fuel cost and availability would affect overall wilderness recreation use to any large extent. However, substantial increases in the price of fuel (i.e., increases above the rate of inflation) may decrease wilderness visitation.*

CHANGING RECREATIONAL PREFERENCES AND ORGANIZED USE

Population surveys of recreation participation suggest that *in recent decades the amount of U.S. citizens participating in wilderness-related activities, such as hiking and backpacking, has remained relatively close to 5 percent* (Cordell and Teasley 1998). Numerous changes in the preferences of users for wilderness-dependent activities have increased use among existing and new clientele. Wilderness experience programs (WEPs) offered by organized groups for purposes of personal growth or education have evolved into a growing source of wilderness visitation, often introducing new clientele to the benefits of wilderness use (Fig. 14.9). Adventure-recreation activities and programs that combine risk and recreation in a wilderness setting are also growing, and challenge both users and managers alike to respond to this growing phenomenon (Ewert and Hollenhorst 1997; Ewert et al. 1999). These and other developments in recreation, conservation, environmental group membership, and public information campaigns continually change and shape the public perception of wilderness as a concept and as a place to recreate.

CONSTRAINTS TO WILDERNESS VISITATION

Green et al. (2007) report that there are three categories of wilderness constraints that are impediments or barriers to participation for some people: personal (e.g., don't have enough time because of long work hours or long school hours), structural (e.g., wilderness areas are crowded), and psychological (e.g., concerned for my personal safety). Their study supported the concerns by managers that minorities, women, low-income, elderly, and less educated people had higher probabilities of feeling constrained from participating in recreational activities in wilderness areas than other user groups.

CHANGING WILDERNESS CONDITIONS AND REGULATIONS

Despite some managerial concerns about declining quality of wilderness conditions and experiences, there is little evidence that user dissatisfaction is negatively influencing wilderness use levels. Most studies of wilderness users report high levels of satisfaction with their experience (Lucas 1985; Van Horne et al. 1985). An analysis of studies comparing satisfaction levels at three wilderness areas with earlier studies indicates that the majority of users were very satisfied with their trip or rated its quality as being very good (Cole et al. 1995).

FIGURE 14.9 Hundreds of organizations use wilderness as places for personal growth, therapy, education, and leadership development. The Anasazi Foundation, Phoenix, Arizona, takes at-risk youth on personal growth and healing trips in wilderness and deserts of the Southwest. Photo courtesy Jenny Dixon, Anasazi Foundation.

However, increased wilderness use, new equipment, increased education and information, and more intense management may combine to diminish the free and easy wilderness access of an earlier era when going on a wilderness trip was less complicated and less regulated. Wilderness conditions do affect visitor use, such as following fires when visitor use declines for a period of time immediately after the fire (Borrie and McCool 2007; Brown et al. 2008). Furthermore, there may be some users who were displaced by some of these changes or who no longer go to wilderness (i.e., dissatisfied users may not be returning to wilderness; see Chapter 16), but the total extent of such temporal and spatial shifts in use or nonparticipation is not known.

Most wilderness areas now have more regulations related to visitor safety and behavior. Fires may be prohibited in some areas and stoves required. Giardia infection is a risk in most wilderness waters and makes boiling water, filters, or purification tablets essential. Today, visitors might receive a weather report on their cell phone, thereby reducing uncertainty about what to expect. Today, they are also more likely to encounter more people with whom to share and obtain information. Is wilderness now a different experience?

How will visitors respond to such changes? Although these changes in the wilderness experience may not appeal to some who had a different experience in wilderness decades ago, today's wilderness experiences have great appeal to most current users, as witnessed by their high satisfaction ratings reported in studies of wilderness users and growing wilderness use.

WILDERNESS RECREATION USE PROJECTIONS

Projections of wilderness recreational use have been limited by basic-use data and by incomplete knowledge of the relationship of wilderness use to causal factors. Probably the earliest projection of wilderness use was made in 1961 as part of the USFS National Forest Recreation Survey project (U.S. Department of Agriculture, Forest Service 1961). That unpublished projection predicted a tripling of wilderness use days by 1976 and a more than eightfold growth by the year 2000—about a 5 percent average annual rate of increase. The actual reported figure for 1976 turned out to be 7,105,600 twelve-hour visitor days, compared to the projection of 5,804,000 use days—a different unit of measure. One cannot convert from one unit to the other with precision, but when adjusted, the projected 1976 use amounts to roughly 8.5 million visitor days. Thus, in 1976, projected use was about 20 percent higher than reported use. The projection technique was based on simple assumptions with no supporting research (there was almost no recreation research at the time) and based on projections of population, income, leisure time, and travel variables that are almost as hard to project as recreation use itself.

At almost the same time, wilderness use was projected as part of the Outdoor Recreation Resources Review Commission (1962) studies, which worked only with USFS data (no wildernesses were managed by any other agency before 1964). Their projected rates of increase were similar to those of the USFS study (about 5 percent per year), which is interesting because their projection procedures were quite different.

Almost twenty years elapsed before other wilderness use projections were developed in response to requirements in the Forest and Rangeland Renewable Resource Planning Act of 1974 (Public Law 93-378). Jungst and Countryman (1982) developed several models, yielding projected average annual rates of increase of 2.6 percent to 7.2 percent to the year 2020. This wide range reflects the uncertainty of projections. We know now that the mid to low range of predicted increase in use was closer to what has happened.

Cordell et al. (1990) estimated that wilderness demand would be greater than supply by the year 2000 and would continue to increase through the year 2040, based on projection of activities that could occur in wilderness. For example, demand for day hiking was expected to increase beyond 1987 at a rate of 1 percent per year, whereas primitive camping was projected to increase at a rate of 0.5 percent per year (Cordell et al. 1990). Although this does not provide real numbers for comparison, it reflects the notion of steadily increasing use of a wilderness system that could eventually have demand exceeding supply at acceptable levels of quality.

Overall use of the NWPS was estimated by Cole (1996) as about 17 million RVDs in 1994. This is a sixfold increase between 1965 and 1994 and an average annual increase of 6.3 percent.

Loomis et al. (1999) forecast wilderness use to increase at 0.5 percent annually in northeastern and Rocky Mountain wilderness areas and 1 percent annually in the southeastern and Pacific coastal states wilderness areas over the next several decades. This is less than the rate of increase (1.4 percent annually) reported by Cole (1996) for national forest areas from 1989 through 1994.

Bowker et al. (2007) estimated NWPS use was 10.7 million trips and 16.3 million on-site days (i.e., visits) in 2002, based on a National Visitor Use Monitoring Project. Their estimates of wilderness use are difficult to compare with other researchers using RVD units (Cole 1996). The visitor use projections by Bowker et al. (2007) are based on population projections and other factors. They project use to increase from 16.9 million visits in 2010 to 17.3 million visits in 2020 and 17.9 million visits in 2030. While the total number of participants and total number of visits will increase, the per capita participation will decline over that period.

Wilderness recreation use will remain an important use of the national forests and national parks; it is about 7 percent of total national parks' overnight use and about 5 percent of total recreational use of the national forests. Wilderness managed by the BLM is lightly used but growing as designated wildernesses and wilderness study areas are being discovered. Use of U.S. Fish and Wildlife refuge wilderness is limited in many areas, but accessible areas will be discovered by wilderness visitors seeking less crowded naturalness and solitude.

We have limited research information and capabilities to project wilderness use. A handful of studies agree that wilderness use will increase, but they do not agree on the projected rates of increase. All studies have predicted the steady, slow to modest increases seen in the last twenty to fifty years. The greatest difficulty in projecting future wilderness use is due to the limited current and past wilderness use information, as past use is the basis for estimating future use.

Wilderness and recreation use trends are complex (Clawson 1985), but *our overall conclusion is that wilderness recreation use will continue to grow steadily at a moderate rate of about 2 percent per year*. This is supported by a combination of factors, previously reviewed in this chapter, including the social and demographic structure of the U.S. population, recreational preferences, and tastes of an increasingly environmentally aware and more highly educated and informed public.

SUMMARY

Wilderness stewardship and management depend on an understanding of all the varied uses wilderness receives, because most wilderness management is management of use of one kind or another, primarily recreation use. Some uses depend on wilderness, but others just take place there and are less dependent on legally mandated wilderness qualities of naturalness and solitude. Recreation is the most obvious wilderness use, but other uses include commercial recreation; indirect, off-site uses like ecological services and benefits from wilderness; scientific uses; educational uses; personal growth, therapeutic, and healing uses; and a variety of direct commodity uses such as mining, water storage, and grazing.

Wilderness recreation use provides quality opportunities and benefits to users, and challenges to wilderness managers (Fig. 14.10). More is known about recreational use than most other wilderness uses, because it has been more widely studied and measured. Many direct- and indirect-use measurement methods are employed, ranging from direct observation and use counts at a sample of access points and days to self-registration and visitor permits. Recreational use is reported in many different statistical units, but visitor days (USFS) and overnight stays (NPS) are most common; unfortunately, they are not easily comparable.

FIGURE 14.10 The Aldo Leopold Wilderness Research Institute in Missoula, Montana, administered by the U.S. Forest Service, specializes only in wilderness research. Other federal agencies participating in the institute are the Bureau of Land Management, U.S. Geological Survey, U.S. Fish and Wildlife Service, and National Park Service. Photo courtesy Chad P. Dawson.

Most wilderness visits are short, with day use predominating in many—maybe most—areas. Parties visiting wilderness usually are small, most often two to four people, representing families or small groups of friends. The most common method of travel is hiking, with a few exceptions. Activities vary, with fishing, photography, nature study, and swimming reported as popular wilderness pursuits. Summer is the main use season, with exceptions mainly in the South and Southwest and at low elevations. Winter use in the North and at high elevations is growing. Many areas have sharp weekend peaks in use. Some visitors to many wildernesses come from all over the nation, but most wilderness visitors live relatively close to the area visited. Wilderness recreational use is distributed very unevenly; some wildernesses, entry points, trails, and campsites are heavily used, while many others have little use. A completely even distribution is not possible and not desirable, but, in many areas, use probably is too uneven and concentrated at attractive locations.

Wilderness visitors are younger than the general population, but all age groups are represented. About one-fourth of wilderness visitors are female, but this proportion has increased over time. Visitors are only slightly more likely to be urban residents and are moderately above average in income, as are most types of outdoor recreationists. The single most distinguishing socioeconomic characteristic of wilderness visitors is a high level of education, and this is related to their better-than-average incomes and typically professional and technical occupations.

Trends in wilderness recreational use have changed greatly over the years. In the 1940s, 1950s, and 1960s, use increased rapidly, but in recent decades, the rate of growth, though still high, has gradually slowed. In the 1980s and early 1990s, use fluctuated in many wilderness areas but total NWPS use has shown an overall increase. Projections of future use vary widely. Although annual percentage growth is slow, total use in the NWPS has grown dramatically due to the growth in the number of wilderness areas as acreage added to the wilderness system over the last forty years. Understanding recreational use and users helps managers to protect the valuable wilderness resource, manage wilderness conditions, and influence the wilderness user's experience (Fig. 14.11).

STUDY QUESTIONS

1. Why is understanding wilderness use important for wilderness management?
2. What are the important nonrecreational uses of wilderness? Which ones are substantially wilderness-dependent?
3. What are the different methods to measure recreation use in wilderness? What strengths and weaknesses does each method have?
4. In what way is wilderness use unevenly distributed? What are the major management implications of that use distribution pattern?
5. What are the major characteristics of wilderness recreation use?
6. What are the major characteristics of wilderness visitors?
7. What are the trends in recreational use of wilderness? What are the major management implications of those trends?

FIGURE 14.11 As visitor use increased in the popular Three Sisters Wilderness of central Oregon, Les Joslin, above, began a wilderness information specialist program while he was a USFS employee, to advance wilderness education and train volunteers to inform and educate visitors about low-impact hiking and camping, and to assist rangers with gathering visitor statistics and maintaining trailhead facilities. Photo courtesy Pat Joslin.

ACKNOWLEDGMENTS

Chad P. Dawson completed minor editing of this chapter for the fourth and fifth editions. The third edition of this chapter was a major revision by Chad P. Dawson and John C. Hendee from the second edition chapter on wilderness use; we recognize and thank our colleague, Robert Lucas, for his seminal efforts on the topic in the first two editions. We thank numerous colleagues for reviews of the third edition, as we have benefited from their review comments, but we alone bear responsibility for any errors and opinions reported herein.

CASE DISCUSSION: CONSTRAINTS TO WILDERNESS PARTICIPATION

The benefits and values of wilderness are available to all segments of society; however, not all segments of the U.S. population visit wilderness or, at least, not at the same participation rates. Green et al. (2007) studied wilderness participation to understand whether different segments of the population did not want to participate or if they did want to participate and perceived some type of constraints that kept them from visiting wilderness. Their study supported the concerns by managers that minorities, women, low-income, elderly, and less educated people had higher probabilities of feeling constrained from participating in recreational activities in wilderness areas than other user groups. Additionally, immigrants felt they were more constrained from participating in recreational activities in wilderness than people born in the United States. The NWPS protects wilderness areas for the use, enjoyment, and benefit of all society, but not all want to or perceive they can take advantage of these direct wilderness experience opportunities. Managers need to be aware that various segments of the public face one or more constraints to their participation in wilderness—and sometimes information and education programs or other management activities can reduce those real or perceived constraints.

CASE DISCUSSION QUESTIONS

1. If managers can find ways to reduce the constraints felt by minorities, women, low-income, elderly, and less educated people, how do you think more participation by each of these segments will change the use characteristics (e.g., methods of wilderness travel, season, and timing of use) reported in wilderness visitor studies in the future?
2. How will the factors affecting wilderness use trends and management (reported in this chapter) positively or negatively impact those segments of society—minorities, women, low-income, elderly, and less educated people—that are already underrepresented in wilderness visitation?
3. Why is it important for managers to attempt to reduce or remove the real and perceived constraints to participation in recreation activities in wilderness for minorities, women, low-income, elderly, and less educated people?

REFERENCES

Absher, James D.; Lee, Robert G. 1981. Density as an incomplete cause of crowding in backcountry settings. *Leisure Sciences.* 4(3): 231–247.

Billings, Ronald F. 1986. Coping with forest insect pests in southern wilderness areas, with emphasis on the southern pine beetle. In: Kulhavy, David L.; Conner, Richard N., eds. *Wilderness and Natural Areas in the Eastern United States: A Management Challenge.* Nacogdoches, TX: Stephen F. Austin University, School of Forestry, Center for Applied Studies, pp. 120–125.

Borrie, William T.; McCool, S. F. 2007. Describing change in visitors and visits to the "Bob." *International Journal of Wilderness.* 13(3): 28–33.

Bowker, J. Michael; Murphy, D.; Cordell, H. Ken; English, D. B. K.; Bergstrom, J. C.; Starbuck, C. M.; Betz, C. J.; Green, G. T.; Reed, P. 2007. Wilderness recreation participation: Projections for the next half century. In: Watson, A.; Sproul, J.; Dean, L. *Science and stewardship to protect and sustain wilderness values: Eighth World Wilderness Congress Symposium*; September 30–October 6, 2005; Anchorage, AK. Proceedings RMRS-P-49. Fort Collins, CO: U.S. Department of Agriculture, Forest Service, Rocky Mountain Research Station, pp. 367–373.

Brown, R. N. K.; Rosenberger, R. S.; Kline, J. D.; Hall, T. E.; Needham, M. D. 2008. Visitor preferences for managing wilderness recreation after wildfire. *Journal of Forestry.* 106(1): 9–16.

Browning, J. A.; Hendee, J. C.; Roggenbuck, J. W. 1988. 103 *Wilderness Laws: Milestones and Management Direction in Wilderness Legislation, 1964–1987.* Bulletin 51. Moscow: University of Idaho, College of Forestry, Wildlife and Range Sciences.

Butler, Lisa Mathis; Roberts, Rebecca S. 1986. Use of wilderness areas for research. In: Lucas, Robert C., comp. *Proceedings: National Wilderness Research Conference*: Current research; July 23–26, 1985; Fort Collins, CO. General Technical Report INT-212. Ogden, UT: U.S. Department of Agriculture, Forest Service, Intermountain Research Station, pp. 398–405.

Carlson, T. 2000. The Eagle Cap Wilderness permit system: A visitor education tool. *International Journal of Wilderness.* 6(2): 27–28.

Carpenter, M. Ralph; Bowhis, Donald R. 1976. Attitudes toward fishing and fisheries management of users in Desolation Wilderness, California. *California Fish and Game.* 62(3): 168–178.

Carr, D.; Hendee, J. C. 1993. An assessment of current research in the National Wilderness Preservation System. In Hendee, J. C.; Martin, V. G., eds. *Proceedings of the 5th World Wilderness Congress Symposium: International wilderness allocation, management and research*; September 24–October 2, 1993; Tromso, Norway. Ojai, CA: The WILD Foundation, pp. 297–303.

Clawson, Marion. 1985. Outdoor recreation: Twenty-five years of history, twenty-five years of projection. *Leisure Sciences.* 7(1): 73–100.

Cole, David N. 1993. Campsites in three western wilderness: Proliferation and change in condition over 12 to 16 years. Research Paper INT-463. Ogden, UT: United States Department of Agriculture, Forest Service, Intermountain Research Station.

———. 1996. Wilderness recreation use trends, 1965 through 1994. Research Paper INT-RP-488. Ogden, UT: U.S. Department of Agriculture, Forest Service, Intermountain Research Station.

Cole, David N. and Daniel R. Williams. 2012. Wilderness Visitor Experiences: Lessons from 50 Years of Research. In: Cole, David N., comp. 2012. Wilderness visitor experiences: Progress in research and management. Proc. RMRS-P-66. Fort Collins, CO: U.S. Department of Agriculture, Forest Service, Rocky Mountain Research Station: 3–20.

Cole, David N.; Hall, Troy E. 2008. Wilderness visitors, experiences, and management preferences: How they vary with use level and length of stay. Res. Pap. RMRS-RP-71. Fort Collins, CO: U.S. Department of Agriculture, Forest Service, Rocky Mountain Research Station. 61 p.

Cole, David N.; McCool, S. F.; Borrie, William T.; O'Loughlin, Jennifer. 2000a. *Wilderness science in a time of change conference*, Vol. 4: *Wilderness Visitors, Experiences, and Visitor Management*; May 23–27, 1999; Missoula, MT. RMRS-P-15. Ogden, UT: U.S. Department of Agriculture, Forest Service, Rocky Mountain Research Station.

———. 2000b. *Wilderness science in a time of change conference*, Vol. 5: *Wilderness Ecosystems, Threats, and Management*; May 23–27, 1999; Missoula, MT. Proceedings RMRS-P-15. Ogden, UT: U.S. Department of Agriculture, Forest Service, Rocky Mountain Research Station.

Cole, David N.; McCool, S. F.; Freimund, W.; O'Loughlin, Jennifer. 2000. *Wilderness science in a time of change conference*, Vol. 1: *Changing Perspectives and Future Directions*; May 23–27, 1999; Missoula, MT. RMRS-P-15. Ogden, UT: U.S. Department of Agriculture, Forest Service, Rocky Mountain Research Station.

Cole, David N.; Watson, Alan E.; Hall, T. E.; Spildie, D. R. 1997. High use destinations in wilderness: Social and biophysical impacts, visitor responses, and management options. Research Paper INT-RP-496. Ogden, UT: U.S. Department of Agriculture, Forest Service, Intermountain Research Station.

Cole, David N.; Watson, Alan E.; Roggenbuck, Joseph W. 1995. Trends in wilderness visitors and visits: Boundary Waters Canoe Area, Shining Rock, and Desolation Wilderness. Research Paper INT-RP-483. Ogden, UT: U.S. Department of Agriculture, Forest Service, Intermountain Research Station.

Cole, David N.; Wright, V. 2003. Wilderness visitors and recreation impacts: Baseline data available for twentieth century conditions. General Technical Report RMRS-GTR-117, Ogden, UT: U.S. Department of Agriculture, Forest Service, Rocky Mountain Research Station.

Cook, Barbara; Borrie, William T. 1995. Trends in recreation use and management of wilderness. *International Journal of Wilderness.* 1(2): 30–34.

Cordell, H. Ken 1999. *Outdoor Recreation in American Life: A National Assessment of Demand and Supply Trends.* Champaign, IL: Sagamore Press.

Cordell, H. Ken; Bergstrom, J. C.; Bowker, J. Michael. 2005. *The Multiple Values of Wilderness.* State College, PA: Venture Publishing.

Cordell, H. Ken; Bergstrom, John C.; Hartmann, Lawrence A.; English, Donald B. K. 1990. An analysis of the outdoor recreation and wilderness situation in the United States: 1989–2040. A Technical Document Supporting the 1989 USDA Forest Service RPA Assessment. General Technical Report RM-189. Fort Collins, CO: U.S. Department of Agriculture, Forest Service, Rocky Mountain Forest and Range Experiment Station.

Cordell, H. Ken; Murphy, D.; Riitters, K.; Harvard, J. E., III. 2005. The natural ecological value of wilderness. In: Cordell, H. Ken; Bergstrom, J. C.; Bowker, J. Michael. *The Multiple Values of Wilderness.* State College, PA: Venture Publishing, pp. 205–249.

Cordell, H. Ken; Tarrant, M. A.; McDonald, B. L.; Bergstrom, J. C. 1998. How the public views wilderness: More results from the USA survey on recreation and the environment. *International Journal of Wilderness.* 4(3): 28–31.

Cordell, H. Ken; Teasley, Jeff. 1998. Recreational trips to wilderness: Results from the USA national survey on recreation and the environment. *International Journal of Wilderness.* 4(1): 23–27.

Cordell, H. Ken; Young, M. 1993. An evaluation of wilderness research: State of knowledge and research needs for the 90's. Proceedings of the 1992 *Annual Meeting of the Society of American Foresters*; October 25–28, 1992; Richmond, VA: Washington, DC: Society of American Foresters, pp. 196–201.

Davis, George D. 1989. Preservation of natural diversity: The role of ecosystem representation within wilderness. In: Freilich, H. R., comp. *Wilderness benchmark 1988: Proceedings of the National Wilderness Colloquium*; January 13–14, 1988; Tampa, FL. General Technical Report SE-51. Asheville, NC: U.S. Department of Agriculture, Forest Service, Southeastern Forest Experiment Station, pp. 76–82.

Dawson, Chad P.; Cordell, Ken; Watson, Alan E.; Ghimire, Ramesh; Green, Gary T. 2016. The U.S. Wilderness Managers Survey: Charting a Path for the Future. *Journal of Forestry.* 114(3): 298–304.

Dawson, Chad P.; Hammitt, William E. 1996. Dimensions of wilderness privacy for Adirondack Forest Preserve hikers. *International Journal of Wilderness.* 2(1): 37–41.

Dawson, Chad P.; Hendee, J. C. 2004. Wilderness-related courses in natural resources programs at U.S. colleges and universities. *International Journal of Wilderness.* 10(1): 33–36.

Dawson, Chad P.; Simon, M.; Oreskes, R.; Davis, G. 2001. Great Gulf Wilderness use estimation: Comparisons from 1976, 1979, and 1999. In: Kyle, G., ed. *Proceedings of the 2000 Northeastern Recreation Research Symposium*; April 2–4, 2000; Bolton Landing, NY. General Technical Report NE-276. Newton Square, PA: U.S. Department of Agriculture, Forest Service, Northeastern Research Station, pp. 283–288.

Dawson, Chad P.; Tangen-Foster, Jim; Friese, Gregory T.; Carpenter, Josh. 1998. Defining characteristics of U.S.A. wilderness experience programs. *International Journal of Wilderness.* 4(3): 22–27.

Driver, B. L.; Nash, Roderick; Haas, Glenn. 1987. Wilderness benefits: A state-of-knowledge review. In: Lucas, Robert C., comp. *Proceedings: National Wilderness Research Conference: Issues, State-of-Knowledge, Future Directions*; July 23–26, 1985; Fort Collins, CO. General Technical Report INT-220. Ogden, UT: U.S. Department of Agriculture, Forest Service, Intermountain Research Station, pp. 294–319.

Duncan, Richard A.; Proescholdt, K. 1999. Protecting the Boundary Waters Canoe Area Wilderness: Litigation and legislation. *Denver University Law Review.* 76(2): 621–658.

Dwyer, John F. 1994. Customer diversity and the future demand for outdoor recreation. General Technical Report RM-252. Fort Collins, CO: U.S. Department of Agriculture, Forest Service, Rocky Mountain Forest and Range Experiment Station.

Dvorak, Robert G.; Watson, Alan E.; Christensen, Neal; Borrie, William T.; Schwaller, Ann. 2012. The Boundary Waters Canoe Area Wilderness: Examining changes in use, users, and management challenges. Res. Pap. RMRS-RP-91. Fort Collins, CO: U.S. Department of Agriculture, Forest Service, Rocky Mountain Research Station. 46 p.

English, Donald B. K.; Cordell, H. Ken. 1985. A cohort-centric analysis of outdoor recreation participation changes. In: Watson, Alan E., ed. *Proceedings: Southeastern Recreation Research Conference*; February 28–March 1, 1985; Myrtle Beach, SC. Statesboro, GA: Georgia Southern College, Department of Recreation and Leisure Services, pp. 93–110.

Evans, Kent E. 1986. Indian Mounds Wilderness Area: Perceived wilderness qualities and impacts of oil and gas development. In: Kulhavy, David L.; Conner, Richard N., eds. *Wilderness and Natural Areas in the Eastern United States: A Management Challenge. Nacogdoches*, TX: Stephen F. Austin University, School of Forestry, Center for Applied Studies, pp. 156–165.

Ewert, A.; MacEvoy, L. 2000. The effects of wilderness settings on organized groups: A state-of-knowledge paper. In: McCool, S. F.; Cole, D. N.; Borrie, William T.; O'Loughlin, Jennifer, comps. *Wilderness science in a time of change conference*, Vol. 3: *Wilderness as a Place of Scientific Inquiry*; May 23–27, 1999; Missoula, MT. RMRS-P-15. Ogden, UT: U.S. Department of Agriculture, Forest Service, Rocky Mountain Research Station, pp. 13–26.

Ewert, Alan W.; Hendee, John C.; Davidson, Sam; Brame, Richard; Mackey, Craig. 1999. Wilderness access issues for education, personal growth and therapeutic use. *International Journal of Wilderness.* 5(3): 13–18.

Ewert, Alan W.; Hollenhorst, Steven J. 1997. Adventure recreation and its implications for wilderness. *International Journal of Wilderness.* 3(2): 21–26.

Fisher, Anthony; Krutilla, John V. 1972. Determination of optimal capacity of resource-based recreation facilities. *Natural Resources Journal.* 12(3): 417–444.

Foster, John D. 1976. Bureau of Land Management primitive areas—are they counterfeit wilderness? *Natural Resources Journal.* 16(3): 621–663.

Franklin, Jerry F. 1987. Wilderness ecosystem research—a scientific perspective. In: Lucas, Robert C., comp. *Proceedings: National Wilderness Research Conference*: Issues, State-of-Knowledge, Future Directions; July 23–26, 1985; Fort Collins, CO. General Technical Report INT-220. Ogden, UT: U.S. Department of Agriculture, Forest Service, Intermountain Research Station, pp. 42–46.

Friese, Gregory T.; Hendee, John C.; Kinziger, Michael L. 1998. Wilderness experience program industry in the United States: Characteristics and dynamics. *Journal Experiential Education.* 21(1): 40–45.

Friese, Gregory T.; Kinziger, Michael L.; Hendee, John C. 1999. History and status of use of wilderness for personal growth. In: Cordell, K., ed. *Outdoor Recreation in American Life: A National Assessment of Demand and Supply Trends.* Champaign, IL: Sagamore Press, pp. 380–384.

Friese, Gregory T.; Pittman, T.; Hendee, John C. 1996. *Studies of the Use of Wilderness for Personal Growth, Therapy, Education and Leadership Development: An Annotation and Evaluation.* Moscow: University of Idaho Wilderness Research Center.

Gager, Dan; Hendee, J. C.; Kinziger, M.; Krumpe, E. 1998. What managers are saying and doing about Wilderness Experience Programs. *Journal of Forestry.* 96(8): 33–37.

Gladden, James N. 1990. *The Boundary Waters Canoe Area: Wilderness Values and Motorized Recreation.* Ames: Iowa State University Press.

Graefe, Alan R.; Donnelly, Maureen P.; Vaske, Jerry J. 1986. *Crowding and specialization: A reexamination of the crowding model: Proceedings of the National Wilderness Research Conference*; July 23–26, 1985; Fort Collins, Colorado. Ogden, UT: U.S. Department of Agriculture, Forest Service, Intermountain Research Station, pp. 333–338.

Gramann, James H. 1982. Toward a behavioral theory of crowding in outdoor recreation: An evaluation and synthesis of research. *Leisure Sciences.* 5(2): 109–126.

Green, G. T.; Bowker, J. Michael; Johnson, C. Y.; Cordell, H. Ken; Wang, X. 2007. An examination of constraints to wilderness visitation. *International Journal of Wilderness.* 13(2): 26–36.

Haas, Glenn E.; Herman, Eric; Walsh, Richard. 1986. Wilderness values. *Natural Areas Journal.* 6(2): 37–43.

Hall, Troy E.; Cole, David N. 2007. Changes in the motivations, perceptions, and behaviors of recreation users: Displacement and coping in wilderness. Res. Pap. RMRS-RP-63. Fort Collins, CO: U.S. Department of Agriculture, Forest Service, Rocky Mountain Research Station. 37 p.

Hall, Troy; Shelby, Bo. 1996. Who cares about encounters? Differences between those with and without norms. *Leisure Sciences.* 18: 7–22.

Hammitt, William E.; Madden, Mark A. 1989. Cognitive dimensions of wilderness privacy: A field test and further explanation. *Leisure Sciences.* 11: 293–301.

Hendee, John C. 1995. Universities must play a larger role in wilderness. *Trends.* 32(1): 22–27.

Hendee, John C.; Catton, William R., Jr.; Marlow, Larry D.; Brockman, C. Frank. 1968. Wilderness users in the Pacific Northwest—their characteristics, values, and management preferences. Research Paper PNW-61. Portland, OR: U.S. Department of Agriculture, Forest Service, Pacific Northwest Forest and Range Experiment Station.

Hendee, John C.; Clark, Roger N.; Dailey, Thomas E. 1977. Fishing and other recreation behavior at high-mountain lakes in Washington State.

Research Note. PNW-304. Portland, OR: U.S. Department of Agriculture, Forest Service, Pacific Northwest Forest and Range Experiment Station. 27p.

Hendee, John C.; Ewert, Alan. 1993. Wilderness research: Future needs and directions. *Journal of Forestry*. 91(2): 18–21.

Hendee, John C.; Lucas, Robert C. 1973. Mandatory wilderness permits: A necessary management tool. *Journal of Forestry*. 71(4): 206–209.

Hendee, John C.; Roggenbuck, Joseph W. 1984. Wilderness-related education as a factor increasing demand for wilderness. In: *Proceedings of the International Forest Congress 1984: Forest Resources Management—The Influence of Policy and Law*; August 6–7, 1984; Quebec City, Quebec, Canada. Bethesda, MD: Society of American Foresters, pp. 273–278.

Hof, John G.; Kaiser, H. Fred. 1983. Projections of future forest recreation use. Resources Bulletin W0-2. Washington, DC: U.S. Department of Agriculture, Forest Service.

Hollenhorst, Steven J.; Whisman, Steven A.; Ewert, Alan W. 1992. Monitoring Visitor Use in Backcountry and Wilderness: A Review of Methods. General Technical Report PSW-134. Albany, CA: U.S. Department of Agriculture, Forest Service, Pacific Southwest Research Station.

Hornback, K. E.; Eagles, P. F. J. 1999. *Guidelines for Public Use Measurement and Reporting at Parks and Protected Areas*. Gland, Switzerland, and Cambridge, UK: International Union for Conservation of Nature and Natural Resources (IUCN). 90p.

Johnson, Janet L.; Franklin, Jerry F.; Krebill, Richard G., coords. 1984. *Research natural areas: Baseline monitoring and management*: Proceedings of a symposium; March 21, 1984; Missoula, MT. General Technical Report INT-173. Ogden, UT: U.S. Department of Agriculture, Forest Service, Intermountain Forest and Range Experiment Station.

Jungst, Steven E.; Countryman, David W. 1982. Two regression models for projecting future wilderness use. *Iowa State Journal of Research*. 57(1): 33–41.

Landres, Peter; Meyer, S. 1998. National Wilderness Preservation System Database: Key attributes and trends, 1964 through 1998. Research Paper RMRS-GTR-18. Ogden, UT: U.S. Department of Agriculture, Forest Service, Rocky Mountain Research Station.

Leopold, Aldo. 1941. Wilderness as a land laboratory. *The Living Wilderness*. 6(3): 3.

Loomis, J.; Bonetti, K.; Echohawk, C. 1999. Demand for and supply of wilderness. In: Cordell, K., ed. *Outdoor Recreation in American Life: A National Assessment of Demand and Supply Trends*. Champaign, IL: Sagamore Publishing, pp. 351–375.

Loomis, J.; Richardson, R. 2001. Economic values of the U.S. wilderness system: Research evidence to date and questions for the future. *International Journal of Wilderness*. 7(1): 31–34.

Lucas, Robert C. 1980. Use patterns and visitor characteristics, attitudes, and preferences in nine wilderness and other roadless areas. Research Paper INT- 253. Ogden, UT: United States Department of Agriculture, Forest Service, Intermountain Research Station. 89p.

———. 1983. Low and variable visitor compliance rates at voluntary trail registers. Research Note INT-326. Ogden, UT: U.S. Department of Agriculture, Forest Service, Intermountain Forest and Range Experiment Station.

———. 1985. Visitor characteristics, attitudes, and use patterns in the Bob Marshall Wilderness Complex, 1970–82. Research Paper INT-345. Ogden, UT: U.S. Department of Agriculture, Forest Service, Intermountain Research Station.

Lucas, Robert C.; Kovalicky, Thomas J. 1981. Self-issued wilderness permits as a use measurement system. Research Paper INT-270. Ogden, UT: U.S. Department of Agriculture, Forest Service, Intermountain Forest and Range Experiment Station.

Lucas, Robert C.; Stankey, George H. 1988. *Shifting trends in wilderness recreational use: Proceedings of the National Outdoor Recreation Forum*; January 13–14, 1988; Tampa, FL. General Technical Report SE-52. Asheville, NC: U.S. Department of Agriculture, Forest Service, Southeastern Forest Experiment Station, pp. 357–367.

Mackey, C. 1998. U.S. wilderness management in the 21st century—politics, policy, and partnerships. *International Journal of Wilderness*. 4(3): 6–11.

Manning, R. E. 1989. Social research in wilderness: Man in nature. Wilderness Benchmark Conference. In: Freilich, Helen R., comp. *Proceedings of the National Wilderness Colloquium*; January 13–14, 1988; Tampa, FL. General Technical Report SE-51. Asheville, NC: U.S. Department of Agriculture, Forest Service, Southeastern Forest Experiment Station, pp. 120–132.

Manning, Robert E. 1985. Crowding norms in backcountry settings: A review and synthesis. *Journal of Leisure Research*. 17(2): 75–89.

Marnell, Leo F. 1977. Methods for counting river recreation users. In: *Proceedings: River Recreation Management and Research Symposium*; January 24–27, 1977; Minneapolis, MN. General Technical Report NC-28. St. Paul, MN: U.S. Department of Agriculture, Forest Service, North Central Forest Experiment Station, pp. 77–82.

McClaran, M. P. 1990. Livestock in wilderness: A review and forecast. *Environmental Law*. 20(4): 857–889.

McCool, S. F.; Cole, D. N.; Borrie, William T.; O'Loughlin, Jennifer. 2000a. *Wilderness Science in a Time of Change Conference*, Vol. 2: *Wilderness Within the Context of Larger Systems*; May 23–27, 1999; Missoula, MT. RMRS-P-15. Ogden, UT: U.S. Department of Agriculture, Forest Service, Rocky Mountain Research Station.

———. 2000b. *Wilderness Science in a Time of Change Conference*, Vol. 3: *Wilderness as a Place of Scientific Inquiry*; May 23–27, 1999; Missoula, MT. RMRS-P-15. Ogden, UT: U.S. Department of Agriculture, Forest Service, Rocky Mountain Research Station.

Meyer, S. S. 1999. The role of legislative history in agency decision-making: A case study of wilderness airstrip management in the United States. *International Journal of Wilderness*. 5(2): 9–12.

Mortensen, Charles O. 1997. Henry David Thoreau: A lecture and a wilderness legacy. *International Journal of Wilderness*. 3(4): 13–16.

———. 1999. Environmental perception: The influence of wilderness on United States artists, writers and their legacy. In: Watson, A. E.; Aplet, G. H.; Hendee, J. C., comps. *Personal, societal, and ecological values of wilderness: 6th World Wilderness Congress Proceedings on Research, Management, and Allocation*, Vol. II; October 24–29, 1998; Bangalore, India. RMRS-P-14. Ogden, UT: U.S. Department of Agriculture, Forest Service, Rocky Mountain Research Station, pp. 120–122.

Murray, Michael P. 1997. High elevation meadows and grazing: Common past effects and future improvements. *International Journal of Wilderness*. 3(4): 24–28.

O'Leary, Joseph T.; Dottavio, F. D.; McGuire, F. 1989. Social factors in recreation participation and demand: Implications from national surveys; January 13–14, 1988; Tampa, FL. General Technical Report SE-52. Asheville, NC: U.S. Department of Agriculture, Forest Service, Southeastern Forest Experiment Station, pp. 275–289.

O'Rourke, Deb. 1994. *Trail traffic counters for Forest Service trail monitoring*. Missoula, MT: U.S. Department of Agriculture, Forest Service, Technology and Development Program.

Outdoor Recreation Resources Review Commission (ORRRC). 1962. *Wilderness Recreation—A Report on Resources, Values, and Problems*. ORRRC Study Report 3. Washington, DC: U.S. Government Printing Office. 352p.

Oye, Garry. 2001. A new wilderness recreation strategy for national forest wilderness. *International Journal of Wilderness*. 7(1): 13–14.

Parsons, D. J. 2000. Wilderness science and the Aldo Leopold Wilderness Research Institute. International Journal of Wilderness. 6(1): 44.

———. 2007. An outside assessment of wilderness research in the Forest Service. *International Journal of Wilderness*. 13(3): 34–35, 39.

Petersen, Margaret E. 1985. Improving voluntary registration through location and design of trail registration stations. Research Paper INT-336. Ogden, UT: U.S. Department of Agriculture, Forest Service, Intermountain Forest and Range Experiment Station.

Powers, Thomas M. 1996. Wilderness economics must look at wilderness thru the windshield not the rearview mirror. *International Journal of Wilderness*. 2(1): 5–9.

Reed, Patrick; Haas, Glenn; Beum, Frank; Sherrick, Lois. 1989. Nonrecreational uses of the National Wilderness Preservation System. A 1988 telephone survey. In: Freilich, Helen R., comp. *Proceedings of the National Wilderness Colloquium*; January 13–14, 1988; Tampa, FL. General Technical Report SE-51. Asheville, NC: U.S. Department of Agriculture, Forest Service, Southeastern Forest Experiment Station, pp. 220–228.

Ridzitis, Gundars. 1996. *Wilderness and the Changing American West*. New York: John Wiley.

Robinson, John P.; Godbey, Geoffrey. 1997. *Time for Life: The Surprising Ways Americans Use Their Time*. University Park: Pennsylvania State University Press.

Roggenbuck, Joseph W.; Lucas, Robert C. 1987. Wilderness use and user characteristics: A state-of-knowledge review. In: Lucas, Robert C., comp. *Proceedings: National Wilderness Research Conference: Issues, state-of-knowledge, future directions*; July 23–26, 1985; Fort Collins, CO. General Technical Report INT-220. Ogden, UT: U.S. Department of Agriculture, Forest Service, Intermountain Research Station, pp. 204–245.

Roggenbuck, Joseph W.; Watson, Alan E. 1989. Wilderness recreation use: The current situation. In: Watson, Alan E., comp. *Outdoor recreation benchmark 1988: Proceedings of the National Outdoor Recreation Forum*; January 13–14, 1988; Tampa, FL. General Technical Report SE-52. Asheville, NC: United States Department of Agriculture, Forest Service, Southeastern Forest Experiment Station, pp. 346–356.

Russell, Keith; Hendee, John C.; Phillips-Miller, D. 2000. How wilderness therapy works: An examination of the wilderness therapy process to treat adolescents with behavioral problems and addictions. In: McCool, S. F.; Cole, D. N.; Borrie, William T.; O'Loughlin, Jennifer, comps. *Wilderness Science in a Time of Change Conference*, Vol. 3: *Wilderness as a Place of Scientific Inquiry*; May 23–27, 1999; Missoula, MT. RMRS-P-15-Vol-3. Ogden, UT: U.S. Department of Agriculture, Forest Service, Rocky Mountain Research Station, pp. 207–217.

Schuster, R. M.; Tarrant, M.; Watson, A. 2005. The social value of wilderness. In: Cordell, H. Ken; Bergstrom, J. C.; Bowker, J. Michael. *The multiple values of wilderness*. State College, PA: Venture Publishing, pp. 113–142.

Stankey, George H. 1971. Myths in wilderness decision making. *Journal of Soil and Water Conservation*. 25(5): 183–188.

U.S. Department of Agriculture, Forest Service. 1961. Table G, Summary of National Forest Recreation Survey, Form No. 10, Appendix 18, National Forest Recreation Survey (NFRS). Unpublished report on file at U.S. Department of Agriculture, Forest Service, Intermountain Research Station, Forestry Sciences Laboratory, Missoula, MT.

U.S. House of Representatives. 1979. House Report 96-617. 96th Congress, 1st Session. Washington, DC: U.S. Government Printing Office. 14p.

U.S. Public Law 88-577. The Wilderness Act of September 3, 1964. 78 Stat. 890.

U.S. Public Law 93-378. The Forest and Rangeland Renewable Resources Planning Act (RPA) of 1974. 88 Stat. 476.

U.S. Public Law 94-429. Mining in the Parks Act of September 28, 1976. 90 Stat. 1342.

U.S. Public Law 94-579. (Wilderness Study) Federal Lands Policy and Management Act (FLPMA) of 1976. 90 Stat. 2743.

U.S. Public Law 95-495. (Boundary Waters Canoe Area Wilderness) Act of October 21, 1978. 92 Stat. 1649.

U.S. Public Law 96-560. (Colorado Wilderness) Act of December 2, 1980. 94 Stat. 2371.

U.S. Public Law 96-550. (New Mexico Wilderness) Act of December 19, 1980. 94 Stat. 3221.

U.S. Public Law 98-406. Arizona Wilderness Act of August 28, 1984. 98 Stat. 1485.

U.S. Public Law 98-428. Utah Wilderness Act of September 28, 1984. 98 Stat. 1657.

U.S. Public Law 98-550. Wyoming Wilderness Act of October 30, 1984. 98 Stat. 2807.

U.S. Public Law 99-504. Nebraska Wilderness Act of October 20, 1986. 100 Stat. 1802.

U.S. Public Law 100-225. (El Malpais and Cebolla Wilderness) Act of December 31, 1987. 101 Stat. 1539.

U.S. Public Law 103-433. California Desert Protection Act of October 31, 1994. 108 Stat. 4473.

Van Horne, Merle J.; Szwak, Laura B.; Randall, Sharon A. 1985. Outdoor recreation activity trends—insights from the 1982–83 nationwide recreation survey. In: Wood, James D., Jr., ed. *Proceedings: National Outdoor Recreation Trends Symposium II*; February 25–27, 1985; Myrtle Beach, SC. Atlanta, GA: U.S. Department of the Interior, National Park Service, Southeast Regional Office, pp. 109–130.

Washburne, Randel F.; Cole, David N. 1983. Problems and practices in wilderness management: A survey of managers. Research Paper INT-304. Ogden, UT: U.S. Department of Agriculture, Forest Service, Intermountain Forest and Range Experiment Station.

Watson, A., Sproull, J.; Dean, L. 2007. *Science and stewardship to protect and sustain wilderness values: Eighth World Wilderness Congress Symposium*. September 30–October 6, 2005; Anchorage, AK. Proceedings RMRS-P-49. Fort Collins, CO: U.S. Department of Agriculture, Forest Service, Rocky Mountain Research Station.

Watson, Alan E. 1998. Sustainable financing of parks and protected areas—are user fees the answer? *International Journal of Wilderness*. 4(3): 5.

Watson, Alan E. 2000. Wilderness use in the year 2000: Societal changes that influence human relationships with wilderness. In: Cole, D. N.; McCool, S. F.; Borrie, William T.; O'Loughlin, Jennifer, comps. *Wilderness Science in a Time of Change Conference*, Vol. 4: *Wilderness Visitors, Experiences, and Visitor Management*; May 23–27, 1999; Missoula, MT. RMRS-P-15. Ogden, UT: U.S. Department of Agriculture, Forest Service, Rocky Mountain Research Station, pp. 53–60.

Watson, Alan E.; Cole, D. N.; Friese, G. T.; et al. 1999. Wilderness uses, users, values and management. In: Cordell, K., ed. *Outdoor Recreation in American Life: A National Assessment of Demand and Supply Trends*. Champaign, IL: Sagamore Publishing, pp. 377–401.

Watson, Alan E.; Cole, D. N.; Roggenbuck, J. W. 1995. Trends in wilderness use characteristics. In: Thompson, J. L.; Lime, D.; Gartner, B.; Samer, W., comps. *Proceedings of the Fourth International Outdoor Recreation and Tourism Trends Symposium*; May 14–17, 1995; St. Paul, MN. St. Paul: University of Minnesota, College of Natural Resources, pp. 68–71.

Watson, Alan E.; Cole, D. N.; Turner, D. L.; Reynolds, P. S. 2000. Wilderness Recreation Use Estimation: A Handbook of Methods and Systems. Research Paper RMRS-GTR-56. Ogden, UT: U.S. Department of Agriculture, Forest Service, Rocky Mountain Research Station.

Watson, Alan E.; Hendee, John C.; Zaglauer, Hans P. 1996. Human values and codes of behavior: Changes in Oregon's Eagle Cap Wilderness visitors and their attitudes. *Natural Areas Journal*. 16(2): 89–93.

Watson, Alan E.; Williams, Daniel R.; Roggenbuck, Joseph W.; Daigle, John J. 1992. Visitor characteristics and preferences for three national forest wildernesses in the South. Research Paper INT-455. Ogden, UT: U.S. Department of Agriculture, Forest Service, Intermountain Research Station.

Watson, J.; Brink, P. 1996. The California Desert Protection Act—a time for desert parks and wilderness. *International Journal of Wilderness*. 2(2): 14–17.

Wilkinson, Charles F.; Anderson, H. Michael. 1985. Wilderness. *Oregon Law Review*. 64(1, 2): 334–370.

Williams, Daniel R.; Roggenbuck, Joseph W.; Bange, Steve. 1991. The effect of norm-encounter compatibility on crowding perceptions, experience, and behavior in river recreation settings. *Journal of Leisure Research*. 23(2): 154–172.

Worf, Bill. 2001. The new Forest Service wilderness recreation strategy spells doom for the National Wilderness Preservation System. *International Journal of Wilderness*. 7(1): 15–17.

Yuan, Susan; Maiorano, Brian; Yuan, Michael; Kocis, Susan M.; Hoshide, Gary T. 1995. Techniques and equipment for gathering visitor use data on recreation sites. 2300-Recreation, 9523-2838-MTDC. Missoula, MT: U.S. Department of Agriculture, Forest Service, Technology and Development Program.

CHAPTER

15

Ecological Impacts of Wilderness Recreation and Their Management

by Jeffrey L. Marion

INTRODUCTION

Four parallel trails gouged into a wildflower-dotted alpine meadow, dense clusters of denuded campsites with exposed tree roots, numerous trees scarred by tethered livestock or felled for firewood. Such recreational impacts are all too common in wilderness. Managing resources to prevent and restore such impacts is a challenge, along with related challenges for managing quality wilderness experiences.

This chapter begins with a discussion of the significance of recreational impacts, its purpose being to bring recreation impacts into perspective with other wilderness management problems. This is followed by a description of important types of recreational impacts, those principally caused by hikers, backpackers, and pack and saddle stock use. The bulk of this chapter is devoted to visitor impacts on ecological conditions and processes as informed by recreation ecology science and management experience. Effective visitor impact adaptive management strategies, actions, and monitoring for trails and campsites are described to support the goal of sustainably accommodating visitation that protects the quality of wilderness resources and experiences.

SIGNIFICANCE OF WILDERNESS RECREATION IMPACTS

Understanding the goals of wilderness management allows evaluations of the significance of recreation impacts in wilderness areas. The relevant phrases in the Wilderness Act state that wilderness is an area "which is protected and managed so as to *preserve its natural conditions* and which *generally* appears to have been affected *primarily* by the forces of nature, with the imprint of man's work *substantially* unnoticeable" (emphasis added). Clearly some recreation impacts will be tolerated, but the integrity of relatively undisturbed wilderness ecosystems should not be substantially compromised, and the evidence of impacts should not be conspicuous (Marion 2016). The first of these objectives is concerned with protecting wilderness ecosystems; the second, with protecting the quality of wilderness visitor's experiences. Both are important.

What constitutes a substantial compromise of ecosystem integrity? There is no clear answer to this question. The most important impacts are those that seriously disrupt ecosystem composition, structure, and function and that are evident at large spatial scales. Impacts that alter a large proportion of a relatively rare ecosystem are particularly detrimental, as are long-term changes like soil loss. But where impacts occur repeatedly, even easily reversible changes are important.

However, ecological impacts of recreation that are spatially extensive, severe, or involve long-term changes are commonly viewed by managers as unacceptable. (Fig. 15.1). For example, the creation of thousands of unnecessary low use campsites, each with nearly complete loss of vegetation cover and varying degrees of soil loss. *At the site scale, recreation impacts can be as profound as any human threat to wilderness. However, these impacts are generally localized to trail and camping infrastructure elements* (Marion 2016). Managers can sustainably design and manage these infrastructures to minimize the areal extent and severity of impact on the wilderness landscape, with the intensity of management mitigating the intensity of use—up

FIGURE 15.1 Generally, recreation impacts that are either spatially extensive, such as this half of an extremely large campsite in the Boundary Waters Canoe Area Wilderness, Minnesota, or are particularly severe and/or long-lasting, such as the substantial soil loss on this hiking and stock trail in the Haleakala National Park Wilderness, Hawaii, represent "unacceptable" impacts that wilderness managers may seek to mediate.

to the point of maintaining natural conditions and being substantially unnoticeable as required by the Wilderness Act. The majority of trail and camping traffic is spatially concentrated and contained along a few major trails and at a few popular destinations. This leaves the majority of most wilderness areas essentially unvisited and virtually undisturbed by recreation use.

In summary, wilderness recreation management goals commonly seek to "avoid or minimize" recreation impacts, particularly those that are spatially extensive (e.g., excessively large campsites, campsite proliferation, or wide trails), particularly severe (e.g., extensive vegetation loss or soil exposure, large numbers of damaged or felled trees), or that are long-lasting (e.g., excessive soil loss, broad dispersal of invasive flora/fauna) (Cole 1989a, 1992; Marion et al. 2016). The Wilderness Act statement that "the imprint of man's work is substantially unnoticeable" suggests that the "footprint" of recreation impact should be constrained, by "avoiding" unnecessary impacts like trash, tree damage, unburied human waste, and wildlife obtaining human food, and "minimizing" the necessary footprint of campsite and trail infrastructures and the severity of various impacts.

Generally, then, recreation use, although causing locally severe impacts, does not substantially compromise wilderness landscapes. At large scales, from the perspective of landscapes, watersheds, and regions—recreation impacts are generally not of great spatial extent. At these scales, far more potent threats to the integrity of wilderness ecosystems exist, including largely external threats from global climate change, air and water pollution, and the increasing frequency and severity of forest fires (Cole and Landres 1996; Landres et al. 2001). Internal threats can also be consequential, including the impacts of fire suppression, livestock grazing, or fish stocking expressed over vast wilderness acreages.

There are several important exceptions to this generalization about large ecological impacts related to recreation use and management, however. *The first exception is the introduction of fish, often exotic species, into lakes and streams without fish or predator species and the subsequent removal of fish through angling.* In such cases, adding a new component to the food chain has impacted entire aquatic ecosystems (Pister 2001). It has reduced, displaced, or eliminated competitors or prey organisms and has stimulated the invasion or expansion of predatory populations. Fish stocking has been implicated as one of the reasons for the decline in amphibian populations in Sierra Nevada wilderness areas (Knapp 2005). Discontinuing the artificial stocking of fish may allow natural processes to regulate or eliminate introduced fish populations.

A second recreation impact that can be important at large spatial scales is wildlife disturbance from hunting, and recreation-related unintentional wildlife disturbance and feeding. Hunting is generally allowed in wildernesses outside of national parks. Although long-lasting effects are difficult to document, unintentional disturbance and wildlife feeding that encourages food attraction behaviors has undoubtedly altered the distribution, population structure, and behavior of many wildlife species (Marion 2019).

Third, the introduction of invasive and non-native wildlife and plant species into an ecosystem is disruptive and, in conjunction with climate change perturbations, causes substantial changes at the landscape level (Marion 2019).

Finally, less common but also important, is physical alteration or pollution of uncommon ecosystems or sites inhabited by rare plants or animals. Disturbance of such places can lead to disturbance or loss of rare species or degradation of certain rare or sensitive ecosystem types. Water pollution, intensive trampling in rare species habitats, and camping impacts along desert riparian corridors or in cryptobiotic soils are examples of such impacts that occur in certain wildernesses (Marion 2016).

The effect of recreation impacts on visitors, in contrast, is most important when it is apparent at the site scale—the scale most relevant to human interaction with the natural environment, where people spend the most time (D'Antonio et al. 2012). Consequently, recreation use—because it may severely impact ecosystems at the site scale—commonly compromises the goal of avoiding conspicuous evidence of human impact. The importance of such impacts depends on the visitor's sense of aesthetics and sensitivity to change, whether or not the change is perceived as desirable, and the magnitude of the change (D'Antonio et al. 2012).

Few studies of visitor perceptions of impacts show much relationship between visitor satisfaction and amount of impact. Visitors consistently report that recreation resource impacts would adversely affect their wilderness experience (White et al. 2008). However, some visitors do not notice ecological impacts that have occurred and of those who do notice impacts, many do not perceive common impacts as particularly undesirable (Cole and Williams 2010). For example, Farrell et al. (2001) found that 50 to 75 percent of visitors noticed visually obvious impacts like vegetation loss and soil exposure but 70 percent felt they were positive and provided functional benefits like places to tent or cook.

In contrast, wilderness managers are more aware than visitors of the deleterious nature of recreation impacts, such as unnecessarily large or numerous campsites, or the effects of soil erosion and sedimentation on waterways (Farrell et al. 2001; Marion et al. 1993). Managers are charged by legal mandates to "preserve natural conditions," regardless of visitor's awareness of their impacts. Thus, most of the discussion that follows pertains to management objectives that focus on avoiding and minimizing recreation impacts to appropriate and acceptable levels, as opposed to maintaining ecosystem integrity. The Visitor Use Management (VUM) framework described in Chapter 9 has emerged as a key planning and decision-making process that more explicitly specifies thresholds (standards) of acceptable change, and an adaptive management process for avoiding and minimizing unnecessary or unacceptable resource and social impacts.

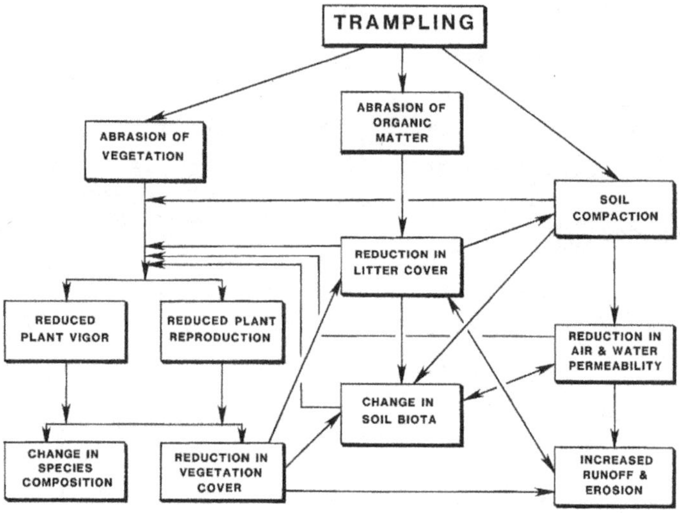

FIGURE 15.2 A conceptual model of trampling effects (partially based on Liddle 1979). Note the numerous reciprocal and cyclic relationships between soil and vegetational impacts.

RECREATIONAL ACTIVITIES AND ASSOCIATED IMPACTS

Now that the importance of recreation impacts considering wilderness management goals has been reviewed, let's take a more detailed look at the effects of various recreation activities on the wilderness resource.

TRAMPLING

The adage to "take nothing but pictures and leave nothing but footprints" is outdated. In many places, too many footprints have left an unwanted legacy. The effects of human trampling have been investigated for almost a century, and we now understand many of the general effects of trampling. These can be conveniently displayed in a conceptual model (Fig. 15.2). Initial and light trampling causes a reduction in plant height and loss of vigor and reproductive capacity, while pulverizing underlying organic materials (Hammitt et al. 2015; Marion et al. 2016). Further trampling reduces vegetation and organic litter cover and begins to compact underlying soils. Over time, recurring traffic causes the replacement of trampling-susceptible woody plants and tall broad-leafed forbs with trampling-resistant grasses, sedges, and prostrate forbs. Persistent and intensive trampling generally causes the loss of most plants, organic litter and the organic soil horizon, exposing and compacting mineral soils, which become vulnerable to displacement or erosion from water and wind (Hammitt et al. 2015; Marion et al. 2016).

Various physiological and morphological changes occur when vegetation is trampled. These include reductions in plant height, stem length, and leaf area, as well as in number of plants that flower, number of flower heads per plant, and seed production (Liddle 1997). In Glacier National Park, Montana, Hartley (1999) found reduced carbohydrate reserves in the roots of trampled glacier lilies, presumably a response to a reduction in the ability to photosynthesize after trampling. All of these changes are manifested in reduced vigor and reproduction, leading to less plant biomass and cover (Pickering 2010). Thus, *most recreation sites exhibit a gradient in which the cover of vegetation decreases from undisturbed vegetation to places where trampling stress is concentrated, and no vegetative cover survives. The sharpness of this gradient depends on both the fragility of the vegetation and whether the trampling is concentrated or dispersed.* For example, the gradient from no vegetation to normal cover levels is very narrow along trails, where trampling is concentrated, particularly trails through fragile vegetation or steep terrain, and quite wide on campsites and flat terrain, where trampling is dispersed, particularly campsites in resistant vegetation (Cole and Monz 2004).

Plants vary in their ability to tolerate trampling. Some plants are even favored by trampling—not as a direct response to abrasion, but in response to reduced competition with other plants and favorable changes in microclimate that result from trampling (Cole 1995b; Pickering and Growcock 2009). For example, trampling increases light levels—changes that favor certain species. Many, if not most, of these favored species are introduced plants that are brought into the wilderness by humans and pack stock (Pickering and Mount 2010).

The vegetation that grows in areas of moderate disturbance, then, is very different in composition from undisturbed vegetation. It consists of trampling-tolerant survivors as well as either native or nonnative invading species capable of dispersing to the site and growing in disturbed environments. In Oregon's Eagle Cap Wilderness, dandelion, a Eurasian weed, is the most common plant on lower-elevation campsites. However, it cannot tolerate conditions on campsites in subalpine forests, where the most common plants are native rushes and sedges that are resistant to trampling, although at reduced densities (Cole 1982b). Nonnative species are more invasive on low-elevation trails and recreation sites. The tendency for the original species on a site to decrease in density, creating more favorable conditions for new invading species, explains the observation that species richness—the number of different species occupying a site—often increases with low levels of trampling before declining to zero as trampling intensifies (Liddle 1997).

Knowledge of which species and types of plants are most tolerant of trampling can be useful in locating sites within resistant vegetation types. Generally, broad-leaved herbs, ferns, low shrubs, and tree seedlings have little tolerance and are mostly eliminated on recreation sites (Hammitt et al. 2015; Marion et al. 2016). Tree and shrub reproduction is often eliminated on forested recreation sites to create space for tenting and cooking, portending major, undesirable changes on these sites when the overstory dies. Larger trees and shrubs may be unaffected by trampling but can be subjected to deliberate mutilation—carved initials and ax scars and felling for firewood. They may also have reduced growth rates, increased water stress, damage to root systems, and increased windthrow due to soil erosion (Marion 2019).

Cole (1995b) has shown that the ability of groundcover plants to resist trampling *(resistance)* decreases with height and that broad-leaved herbs are less resistant than flexible grass-like plants and shrubs. Herbs growing in shade are particularly intolerant of trampling because *adaptations to shading*—possession of large, thin leaves and tall stems—make these plants vulnerable when trampled. This explains the common finding that trampling of forested sites generally results in more rapid loss of ground layer vegetation than trampling of open woodlands, shrublands, and meadows with groundcover dominated by narrow and flexible leafed grasses and sedges (Cole 1993b). Low shrubs, such as heather, are relatively resistant to trampling stress, but once damaged they recover slowly. Their *resilience* (recoverability) is low. Grasses and sedges have high resistance and resilience and are highly tolerant of repeated cycles of trampling and recovery over time, frequently replacing less tolerant species (Cole 1995b).

Normally, less than one-half of a given volume of soil is solid matter; the rest is pore space that contains air and water (Hammitt et al. 2015). *Trampling compacts the soil—compressing soil particles and pore spaces.* Larger pores—those that permit rapid percolation of water after precipitation and are normally occupied by air—are severely reduced by trampling. This can indirectly affect vegetation and soil microbiota, causing oxygen shortages and reductions in water availability. Along with the greater difficulty plant roots have penetrating compacted soils, these changes generally reduce plant vigor and retard reproduction and establishment of new seedlings. However, plants vary in their response to compaction, with adverse impacts to some species while others respond positively to low levels of compaction (Godefroid and Koedam 2004).

In addition to effects on above ground vegetation, compaction affects soil biota with both negative and positive effects (Beylich et al. 2010). Larger organisms, such as earthworms, that help rejuvenate the soil and recycle nutrients find it more difficult to penetrate dense soils (Wellnitz et al. 2020). Microbial communities develop in response to plant exudates and organic matter. In turn, they contribute to ecosystem functioning by metabolizing nutrients, transforming soil organic matter, producing phytohormones, and contributing to soil food webs. Microbiota are likely to be adversely affected by lower oxygen concentrations (Zabinski 1997). Zabinski et al. (2002) found evidence of dramatic changes in soil biology on Oregon's Eagle Cap Wilderness campsites, including reductions in the biomass and respiration of soil microbes, their functional diversity, and the quantity of soil nitrogen available to most plants (potentially available nitrogen). Other field studies have shown that lower levels of soil compaction can also increase microbial biomass (Beylich et al. 2010). Adverse effects on mycorrhizal fungi, which improve nutrient uptake and water absorption in plants, could be a limiting factor in revegetating disturbed areas.

Perhaps of greater importance, compaction drastically reduces the rate at which water filters into the soil. The rate of water infiltration into campsite soils was reduced by two-thirds in the Bob Marshall Wilderness (Cole 1983b), and three-fourths in Grand Canyon National Park (Cole 1986b). Water that does not permeate soil runs off across the surface, increasing erosion potential and decreasing soil moisture. Although this reduced water supply is unlikely to be a problem in areas with plentiful rainfall, it is likely to increase water stress for plants in arid environments or during dry periods of the year.

Erosion is likely to be a more common and important problem. Deeply eroded trails are unsightly and difficult to use. Erosion on campsites and recreation sites can expose roots and rocks and contribute to site expansion (Eagleston and Marion 2017b). Moreover, the formation of soil is such a slow process that erosion can be considered an irreversible process with negative long-term consequences.

Abrasion and loss of organic matter exacerbate many of these same problems. Thick organic litter and an underlying organic soil horizon protect deeper mineral soils from much of the direct impact of trampling and decrease surface water runoff (Hammitt et al. 2015; Marion et al. 2016). Intensive trampling pulverizes and accelerates the loss of these organic layers, resulting in increased mineral soil exposure, compaction, runoff, and soil loss. This leads to further loss of organic litter carried off by surface water runoff.

CAMPFIRES

Collecting and burning wood in campfires have their own unique impacts. As with trampling, these impacts are *aesthetic (fire rings, blackened rocks, and charcoal) and ecological (felled trees, deadwood removal, and sterilized soils).* Ecological impacts result from the removal of wood from large areas around campsites and burning the wood in campfires.

Trampling from firewood collection can greatly enlarge the area affected by camping activities. In Great Smoky Mountains National Park in Tennessee and North Carolina, the area disturbed by firewood collection was typically more than nine times the size of devegetated areas on campsites. Within these larger areas the number of live and dead trees and woody fuels was reduced. Pieces of wood one to three inches in diameter were barely one-third as abundant as in undisturbed areas (Bratton et al. 1982). In Oregon, Hall and Farrell (2001) reported a 40 percent reduction of woody material around wilderness campsites, and in Yellowstone National Park, Taylor (1997) found reduced tree sapling density more than one hundred feet from the center of campsites.

Loss of large woody debris is likely to adversely affect the populations of invertebrates, small mammals, and birds that use the wood as a food source or living place (Bull et al. 1997). The elimination of large (more than three-inch-diameter) woody debris plays an important role in the ecosystem (Maser et al. 1988). It has unusually high water-holding capacity; accumulates nitrogen, phosphorus, and sometimes calcium and magnesium; and is an important site for nitrogen-fixing microorganisms. Of particular importance, *ectomycorrhizal fungi*—organisms that develop a symbiotic association with the roots of most higher plants, improving the plants' ability to extract water, nitrogen, and phosphate from less fertile soils—are also concentrated in decayed wood. Eliminating large woody debris is likely to reduce site productivity, particularly on droughty and infertile soils. *Clearly, loss of large woody debris due to firewood collection is a serious impact where a sizable area is affected. Collection of wood that can be broken by hand is likely to have fewer negative effects.*

A wood campfire severely affects the area burned. Fenn et al. (1976) found that a single intense campfire burned 90 percent of the organic matter in the upper inch of soil. Fires also cause pronounced changes in soil chemistry (Arocena et al. 2006). Reported fire effects include the loss of nitrogen, sulfur, and phosphorus; increases in pH, copper, and zinc; and reductions in the moisture-holding capacity, infiltration rates, and microbiotic populations of soil. Some visitors burn trash and food that attract wildlife to campsites. Davies (2004) found that many non-wood products leave increased levels of toxic materials in campfire ashes and that noxious gas emissions pose a threat to human health. The effects of repeated campfires substantially alter and sterilize soils, rendering them less hospitable to plants and soil fauna and requiring more than a decade to recover.

TRAILS

Wilderness managers recognize that formal trail systems are necessary, particularly in moderate to high use areas of wilderness, to accommodate safe, low impact travel for hikers and horse riders. Without formal trails, visitors would develop much larger duplicative networks of informal (visitor-created) trails (ITs) that are less sustainable, resulting in greater aggregate area of impact and greater impact severity (Spernbauer et al. 2023; Wimpey and Marion 2011). Wilderness managers often have adopted a "Containment" strategy that condones the design, construction, and maintenance of sustainable formal trail networks—as appropriate to wilderness conditions and settings—to provide access to popular geographic areas, features, and destinations. Resource impacts associated with trail construction and maintenance work are viewed as "acceptable," a necessary trade-off to avoid and minimize recreation impacts that would be far greater if not provided. Likewise, visitors are strongly encouraged to remain on formal trails, though managers increasingly recognize the need for some ITs to provide lower use access to campsites, water, or climbing and fishing spots. In more remote and low use areas a "Dispersal" strategy that condones off-trail cross-country traffic is often permissible, provided lasting resource impacts are largely avoided (Marion 2016).

Unfortunately, many formal trail networks consist largely of old non-recreational "legacy" alignments that were not sustainably designed (Marion 2023). With intensive use, such trails frequently develop unacceptable problems with widening, soil loss, or muddiness, requiring improved reroutes or substantial recurring maintenance (Fig. 15.3). Similarly, unsustainable ITs receiving even moderate use often develop

FIGURE 15.3 Trails in flat terrain like this one in the Popo Agie Wilderness, Wyoming (left), are susceptible to muddiness, widening, and informal trail proliferation. Excessively steep trails aligned with the fall line like this section of the Appalachian Trail in Maine (right), are susceptible to soil displacement and loss, and widening as visitors move laterally to find the easiest route through rocks and roots.

severe impacts that require development of new formal trails or recurring maintenance and restoration work. Nevertheless, there is general acceptance by wilderness managers of the need to employ some level of a Containment strategy that encourages the concentration of intensive traffic on a limited set of sustainable formal trails.

Trail construction and maintenance activities clearly remove vegetation cover and reconfigure soil profiles, though before-and-after studies have not quantified such impacts (Marion 2023). Trail construction can create new and altered habitats, often by hindering the flow of water or intercepting and transporting water down-trail, which artificially manipulate soil moisture and vegetation composition and cover. Sustainable trail design practices, discussed in a later section, seek to make trails "hydrologically" invisible, though some drainage practices require visually obvious "engineered" features, such as rock or wood water bars, whose efficacy also require frequent recurring maintenance (Marion and Wimpey 2017). Improved "no maintenance" practices include designing side-hill trails that inherently shed water by rolling slightly up and down along the contour or constructing appropriately spaced grade reversals built into the tread (Hesselbarth et al. 2007; Marion 2023).

Post-construction trail impact studies document subsequent trail corridor impacts from opening tree and shrub canopies and creating a barren, compacted trail tread that may alter water drainage patterns. Common trail clearing widths and heights range from 4 by 8 feet for hiking trails to 8 by 10 feet for stock trails. In forested areas this clearance can increase light intensity and alter the composition of trailside vegetation (Adkison and Jackson 1996; Pickering and Hill 2007). Vegetation composition along well-established trails also shifts in response to increased trampling and grazing, increased nutrients from horse manure and urine, altered soil moisture, and delayed snowmelt. Trail users also introduce and disperse non-native plant species (Pickering and Hill 2007), as has been reported for the Boundary Waters Canoe Area Wilderness, Minnesota (Dickens et al. 2005) and Glacier National Park, Montana (Tyser and Worley 1992).

While trail treads are generally appreciated and accepted by recreationists, they do notice visually obvious trail impacts, particularly those that increase the difficulty of travel (Moore et al. 2012). Small amounts of soil loss and limited trail widening may not be noticed, but deep ruts and exposed roots and rocks negatively affect both the quality of visitor experiences and hiking difficulty. Excessive muddiness and trail widening due to trails in wet soils or used during wet seasons can also create challenging problems, leading visitors to create new ITs around large mudholes. While most trail impacts are confined to a corridor usually no more than eight feet wide, compositional changes can occur more than twenty feet away (Adkison and Jackson 1996; Pickering and Hill 2007). This is particularly true for some non-native invasive plants able to replace native species in undisturbed settings.

PACK AND SADDLE STOCK

Pack and saddle stock trample vegetation and soil along trails and on campsites, as hikers do, though these changes are more pronounced than those noted in the sections on trampling and trail impacts. In this section the impacts of recreational stock use are summarized, along with the changes occurring on meadows grazed by recreational stock. It's important to note that horse and pack stock use have declined markedly since their heyday in the 1960s and 70s, though their use remains prevalent in some wildernesses (Cole 2013a). For example, between 1980 and 2013 groups with pack stock in Sequoia Kings Canyon National Park have declined almost 30 percent and private stock use has declined by almost 60 percent.

Research has repeatedly revealed that recreational stock contributes substantially greater resource impact to vegetation, soil, and water quality than hiking or backpacking (Newsome et al. 2004; Cole and Spildie 1998). *Studies show that llamas cause substantially less impact than horses both to existing trails* (DeLuca et al. 1998) *and to vegetation* (Cole and Spildie 1998). These findings suggest that trail proliferation and impact can occur more rapidly with stock use than with hiker use. The trails created will be wider, deeper, more compacted, and less vegetated.

At campsites, the magnitude of impact caused by stock groups compared to hiking groups is even more pronounced than on trails. Loss of organic soil horizons, increased compaction, and decreased infiltration rates are particularly pronounced on stock sites. Exotic species accounted for just 5 percent of the vegetation cover on backpacker sites, compared with 43 percent of the cover on stock sites (Cole 1983b). *Seeds of exotic plants contained in horse feed readily germinate and grow on disturbed campsites and trailsides* (Pickering and Mount 2010).

Grazing areas are affected primarily by trampling and grazing (defoliation), although defecation may also cause minor changes. Grazing, by removing plant leaves, disrupts the ability of plants to manufacture food. Excessive and repetitive defoliation depletes food reserves, reducing plant vigor and reproductive capacity. Numerous studies have illustrated that grazing can reduce current growth; stem, leaf, and seed stalk heights; reproductive activity; basal and foliar cover; and root growth (McClaran and Cole 1993; Vernon et al. 2022). Loss of vigor, in turn, makes vegetation more susceptible to trampling damage, particularly penetration of the vegetative mat by stock hooves, and results in a reduction in vegetative cover.

Trampling causes changes in vegetation and soil conditions that are more severe due to the substantially greater weight and ground pressures exerted by horses and pack stock (Newsome et al. 2004). Of particular concern in grazing areas is disturbance of wet meadows (Vernon et al. 2022). Wet soils, thick organic deposits, and vegetation mats are all susceptible to deformation and disintegration when trampled. Heavy trampling of wet riparian meadows can lead to a surface of broken sod and hummocks, increased erosion, lowering of water tables, streambank erosion, and degraded water quality (Derlet et al. 2008; Vernon et al. 2022).

Plants differ in their susceptibility to grazing, much as they differ in susceptibility to trampling. Those capable of growing from buds close to or under the ground are more likely to survive close grazing than those with buds located where they can be removed by grazing. Of greater importance are plant compositional changes due to preferential grazing, with the least palatable species surviving and replacing the preferred grazed species (Vernon et al. 2022).

WILDLIFE DISTURBANCE

Recreation activities can impact wildlife in four key ways (Knight and Cole 1995b; Marion 2019). Animals can be indirectly affected through *habitat modification* or through *pollution and feeding* (particularly via access to visitor trash and food). They can be directly affected through *exploitation—hunting, fishing, trapping, or collecting*. Finally, wildlife can be *directly disturbed*—either intentionally or unintentionally (Larson et al. 2016). The short-term responses of individuals to disturbance, including flight and elevated heart rates, have been well-investigated. Less studied are longer lasting effects, or how entire populations or communities of animals are affected (see also Chapter 12).

Campsite and trail creation and expansion, along with firewood collection, can affect small mammal and bird populations by altering food sources and living places and eliminating protected sites (Cole and Landres 1995; Hammitt et al. 2015; Marion et al. 2016). Such habitat modifications are also likely to affect reptile and amphibian populations. Trash, food scraps, unstored food, and smellables (e.g., lotions, toothpaste, etc.) around campsites also attract animals, ranging from invertebrates to certain birds; and small to large mammals such as skunks and bears (Marion 2019).

Although habitat modification and improper disposal of food and trash are the major source of impact for smaller wildlife species, most of these changes are highly localized and, except for the attraction of "pest" species to campsites, not evident to most users (Cole and Landres 1995). Wildlife impacts become more important when a species' entire habitat is disturbed, when the affected species are rare or threatened, or when visitor safety is at risk (Gutzwiller and Cole 2005). For example, improper food storage and disposal have frequently been implicated as the causal factor contributing to decisions to capture and move or euthanize bears. Consequently, effective low impact food-storage practices and/or facilities are necessary and have been implemented in many areas (Fig. 15.4) (Marion 2019).

FIGURE 15.4 The Leave No Trace "Respect Wildlife" principle directs visitors to safely store all food, trash, and smellables to prevent their access from wildlife. Failure to do this leads to unsafe food attraction behaviors that too frequently result in wilderness staff having to trap or kill bears.

Perhaps of more widespread importance to the achievement of wilderness management goals are the impacts resulting from stocking fish, angling, hunting, and unintentional disturbance of wildlife. As noted earlier, hunting outside of national parks and angling are legal, accepted practices that are likely to continue in wilderness, despite their considerable alteration of natural conditions (Anderson 1995). Fish stocking, which has been shown to adversely affect amphibian populations (Knapp 2005), is a common practice in wildernesses with mountain lakes (Landres et al. 2001). Recently, it has been discontinued in a few wildernesses.

Unintentional disturbance of wildlife, particularly of birds and large mammals, has undoubtedly altered the distribution, structure, and behavior of animal populations (Knight and Cole 1995b). Where harassment affects an entire population or affects most of a species' habitat, this disturbance may be highly disruptive. Disturbance of wildlife by recreationists produces excitement or stress in animals (Hammitt et al. 2015; Marion et al. 2016). This may lead to panic, exertion, disruption of essential functions such as breeding or nesting, temporal or spatial displacement from preferred habitats, and sometimes death (Marion 2019). Animals that are healthy and have ample food and places to escape to are better able to withstand human disturbance than animals that are underfed, highly parasitized, experiencing severe weather, giving birth or nesting, or lacking secure areas for escape (Cole and Landres 1995). Negative impacts to animals—in terms of increased energy expenditures or radical changes in behavior or distribution—also increases as disturbance becomes more frequent and more unpredictable (Knight and Cole 1995a).

Generalizing about human disturbance is difficult because of the considerable variability among and within species (Gutzwiller and Cole 2005). Effects on wolves, which are relatively intolerant of disturbance, are much more serious than effects on coyotes. Similarly, effects on eagles, which may not return to feeding sites for several hours after being disturbed, are more serious than effects on jays. Within species, prior experience with humans strongly tempers responses. Some individuals can learn to tolerate benign, consistent, and predictable disturbances (Knight and Temple 1995). However, differences between hunted and nonhunted populations can be profound, with hunted animals eliciting much stronger and negative reactions to human disturbance. Individuals giving birth or with young are also more prone to disturbance.

Wildlife disturbance can be exacerbated by attempts to disperse visitors from popular parts of the wilderness to less visited places. Such dispersal occurs when use levels are limited in popular places, forcing visitors to seek out less popular places. It can also result from educational messages asking visitors to avoid popular places or to seek out places where opportunities for solitude are greater. Where increased dispersal occurs, the frequency of wildlife harassment may increase and the size of secure areas where harassed animals can escape may decrease. Consideration of the impacts of altered visitor use distributions on wildlife may allow managers to reduce negative consequences to wildlife.

Winter use of wilderness can also be particularly problematic. Cross-country skiing and snowshoeing can stress populations at a time of year when they are least able to tolerate it. Cassirer et al. (1992) documented disturbance of elk by cross-country skiers in Yellowstone National Park. Ferguson and Keith (1982) have documented a tendency for elk and moose to move away from trails being used by skiers.

Other important problems occur when use is concentrated on limited critical habitat. Recreation use around desert waterholes and salt licks can cause more substantial problems than one would expect from the same amount of dispersed use (Marion 2019). The presence of visitors can displace wildlife from critical access to water or salt. Nor does the use need to coincide with the presence of animals; summer and fall grazing by recreational stock can reduce available food sources critical to native grazers in the winter. Such impacts can be minimized by identifying habitats critical to wildlife at various seasons and developing plans that protect them.

INTRODUCTION/DISPERSAL OF NON-NATIVE SPECIES

Wilderness visitors can also intentionally or unintentionally introduce non-native flora and fauna and disperse them along trails or at campsites (Eagleston and Marion 2017a; Randal 2000). Some species are highly invasive, able to outcompete and replace native species with widespread ecosystem alterations that can compromise ecological integrity (Randal 2000; Pickering and Mount 2010). Examples include buffelgrass, cheatgrass, Japanese stiltgrass (Fig. 15.5), zebra and quagga mussels, Asian carp, and earthworms. Eagleston and Marion (2017a) found 21 non-native plant species on campsites in the Boundary Waters Canoe Area Wilderness, Minnesota in 1982 and 2014, but with increases in the mean number of non-native plants per site (1.8 to 4.0) and the number of campsites they occupied (55 to 74). A study on the spread of invasive earthworms in this wilderness found earthworm biomass and density to be consistently more abundant on islands with campsites than those without, due to their release by fishermen (Wellnitz et al. 2020). The Leave No Trace "*Leave What you Find*" principle advises visitors to "Avoid introducing or transporting non-native species," but a recent study of Appalachian Trail thru-hikers found that their knowledge of non-native plant problems and low impact practices was limited, though most thru-hikers when made aware of the problem stated a willingness to learn and apply practices to limit the introduction and spread of non-native species (Dolman and Marion 2022).

FIGURE 15.5 The highly invasive Japanese stiltgrass (Microstegium vimineum) has replaced native trailside vegetation along this segment of the Appalachian Trail in Pennsylvania. It is shade-tolerant, unlike most grasses and nonnative plants, and able to replace native vegetation in undisturbed areas.

WATER POLLUTION AND DISPOSAL OF HUMAN WASTE

Common types of water pollution include erosional run-off from trails and campsites and areas grazed by pack stock. In a Kings Canyon National Park study, health hazards were minimal, but recreation use was associated with a number of changes in the basic ecology of lakes (Silverman and Erman 1979). The most heavily used lakes had less nitrate, more iron, and more aquatic plants than other lakes (Taylor and Erman 1979). The authors suggest that recreation use—through erosion of trails and campsites, improper waste disposal, destruction of vegetation, and campfires—may cause an increase in trace elements, such as iron, the absence of which formerly limited plant growth. Stimulated plant growth results in increased nitrogen uptake and, therefore, decreased nitrate levels. Insects, aquatic worms, and small clams were more abundant on the bottom of more heavily visited lakes (Taylor and Erman 1980).

Lake eutrophication effects are felt throughout the food chain and such changes are long-lasting (Dynowski et al. 2019). In the Kings Canyon National Park study mentioned earlier, changes were still prevalent at Bullfrog Lake, a formerly heavily used lake, sixteen years after it had been closed to camping and grazing. Such changes are important because they are likely to affect the functioning of entire lake ecosystems, particularly for heavily used lakes. Wilderness aquatic ecosystems may be particularly sensitive to disruption by recreation use, suggesting studies on shifting recreational trails, campsites, and grazing away from streams and lakeshores could provide information for improved management strategies. Marion et al. (2018) provide a review of this topic, relevant regulations, ecological and social considerations, and guidance for decision-making. While camping setbacks of 100–200 feet are generally sufficient to allow vegetation and/or intact organic litter to filter drainage prior to reaching water bodies, camping in rock/gravel and bedrock settings along shorelines or in flatter shoreline areas and sandbars kept barren by flooding is often equally acceptable, particularly in canyons where more distant camping locations are rare to non-existent.

Most management concerns with water pollution have centered on the potential for transmitting disease by organisms present in water. Organisms capable of causing illness in humans include viruses, bacteria, and protozoans (Cilimburg et al. 2000). *Three prominent sources of water contamination are (1) visitors, their dogs, and pack stock; (2) domestic livestock; and (3) wildlife.* Even when human-related contamination is absent, potentially harmful bacteria and pathogens are naturally occurring in soils, water, and sediments (Silsbee and Larson 1982).

Water-quality studies in mountainous wilderness in the West have generally found very low levels of bacterial contamination, even in areas of concentrated use. For example, a study in Sequoia and Kings Canyon National Park, California, found minimal contamination by fecal contamination and other sources throughout the park (Clow et al. 2013). However, concentrations of *Escherichia coli* and total coliform bacteria were significantly different when sampled upstream and downstream from mixed-use stock and backpacking sites but were not different for backpacking sites. Basin hydrological characteristics and wildlife were also cited as influential.

A similar earlier study in this park and Yosemite National Park also attributed greater coliform concentrations to areas used by pack stock (Derlet et al. 2008). Along the Colorado River in Grand Canyon National Park, Arizona, water was unfit for drinking, but coliform levels were generally low except during and after flood events (Brickler et al. 1983). Here the primary source of contamination appeared to be domestic livestock or wildlife. Springs and streams in Great Smoky Mountains National Park exceeded maximum permissible levels of coliform bacteria, but contamination did not appear related to recreation use (Silsbee and Larson 1982).

Even where contamination is not evident, transmission of disease can occur. Surface waters in many wildernesses are contaminated with the protozoan *Giardia lamblia* (Cilimburg et al. 2000). As with bacterial contamination, many experts believe that humans, domestic animals, and wildlife all act as hosts capable of spreading the organism. Beaver have most frequently been implicated as a major source of giardia contamination.

Proper control of waste, from both humans and recreational stock, is still important, although this will not eliminate health hazards. Toilets may be appropriate at sites that receive heavy, consistent use throughout the season. However, knowledge is insufficient to provide specific guidance about when toilets are or are not necessary. Toilets are a reminder of civilization and may not be consistent with the spirit of wilderness, so their sensitive use and location are important considerations. Maintaining stock more than 200 feet from water may lessen contamination by their wastes (Marion 2014).

MANAGING CAMPSITES

In wilderness areas the most common and observable impacts of recreation are generally those associated with campsites and trails.

CAMPSITE IMPACTS

Most overnight wilderness visitors spend the majority of their time at campsites. Unfortunately, this increases the opportunity for substantial resource impacts to occur at campsites, including negative aesthetic impacts that may diminish the quality of their experiences. Although natural conditions are desirable, some clearing of vegetation is necessary to provide space for tents, cooking, and social activities. This causes many campers to select sites with some vegetation loss and view small barren areas of impact for tenting and cooking as "positive," because they recognize and accept campsites as an essential infrastructure element necessary for camping to occur (Farrell et al. 2001). Problems arise when campsites grow substantially in size, proliferate within attractive destination areas, or develop particularly severe or avoidable impacts, such as large numbers of damaged or felled trees. However, recreation ecology studies can aid managers by identifying influential factors that affect the areal extent and severity of camping impacts, knowledge that can be applied to shift camping to more sustainable campsites. A sustainable campsite is one that *can accommodate the intended type and amount of use over time without unacceptable levels of expansion, degradation, maintenance, and social crowding or conflict* (Marion et al. 2018).

Temporal Patterns of Impact

Many studies have examined the temporal changes in vegetation and soils that occur on wilderness campsites. Vegetation cover and organic litter is trampled and lost on new sites, exposing soils that compact, reduce water infiltration, and increase water runoff. Recreation ecology studies reveal that vegetative impacts occur quickly, with initial and low levels of use (less than 12 nights/year) responsible for 60 to 95 percent of total impact occurring on high use sites (greater than 50 nights/year) (Marion et al. 2016). Impacts to soils are more distributed over time, with the pulverization and limited loss of organic litter occurring with light traffic, the exposure and loss of underlying organic soil with intermediate traffic, and the exposure, compaction, and loss of underlying mineral soil occurring with heavy traffic. Over time well-established campsites generally develop exposed mineral soil in core use areas, pulverized organic litter in secondary use areas, and trampling-resistant plants colonizing sunny peripheral areas.

Plant compositional changes occur on campsites over time, with a reduction in species diversity as large numbers of trampling-susceptible plants are lost, though small numbers of resistant native and non-native species often become a component of campsite flora. While the removal of smaller woody vegetation often occurs with initial use, mature trees are more resistant to recreational impact. However, mature trees eventually die from disease, wildfires, cumulative visitor damage, girdling from tethered stock, felling for firewood, or are cut because they become hazardous (Eagleston and Marion 2017b). Furthermore, their replacement on campsites is rare to non-existent; on forested campsites canopies inevitably open, creating unnatural clearings in forested landscapes (Fig. 15.6). On the beneficial side, the substantially increased sunlight permits shade intolerant but trampling resistant grasses and sedges to colonize the less trafficked peripheral use areas, retaining soil and improving site aesthetics (Marion et al., 2023).

Recreation ecology studies reveal that campsites have a typical "*life history*" marked by rapid and substantial impacts occurring in the initial "*development*" phase (Cole and Hall 1992). On newly established campsites at Pennsylvania's Delaware Water Gap National Recreation Area, the majority of resource impact that occurred over the six years following campsite opening occurred during the first season of use (Marion and Cole 1996). In years two through six, impacts increased incrementally at a diminishing rate. Site and vegetative durability can influence the initial rates of impact: a single night of camping in undisturbed forest in the Popo Agie Wilderness, Wyoming, eliminated about 60 percent of the vegetation, while four nights eliminated over 80 percent (Cole and Monz 2003). On neighboring grassy meadow sites, four nights of camping had no effect on vegetation cover during the first year of use, less than 20 percent after the second year, with no further loss during the third year.

Well-established campsites exist in a "*dynamic equilibrium*" phase where resource conditions remain largely static, with smaller changes occurring in response to variations in annual cycles of visitation and recovery, as influenced by

FIGURE 15.6 Photo comparison of tree loss over thirty-two years (1982 to 2014) on a Boundary Waters wilderness campsite (from Eagleston and Marion 2017b).

numerous use-related, environmental, and managerial factors. For example, on long-established campsites in the Eagle Cap Wilderness, mean vegetation cover was 15 percent in 1979, 12 percent in 1984, and 19 percent in 1990 (Cole and Hall 1992). Other impacts tend to deteriorate more steadily because recovery processes are so slow. For example, mineral soil exposure on Eagle Cap campsites increased from 33 percent in 1979 to 41 percent in 1984 and 44 percent in 1990 (Cole and Hall 1992). In these western coniferous forests, the accumulation of new organic litter cannot keep pace with the rate of litter loss from campsite use. In eastern forests, however, the amount of mineral soil might be more stable because litter production is greater. However, tree damage tends to be cumulative, worsening over time due to slow recovery rates; campsites lose trees over time as felled trees are rarely ever replaced by onsite regeneration.

A "*recovery*" phase is initiated when campsites are closed to use, with studies consistently revealing a slow process easily halted by recurring use. Once impacted, a high degree of campsite impact can be sustained by relatively low levels of ongoing use. This is particularly true at higher elevations where recovery rates are low (Cole and Spildie 2007). Once campsites are closed to camping, recovery rates are highly variable, depending on the extent of impact and environmental conditions. Substantial recovery in a few years can occur on eastern low elevation campsites in resilient vegetation types, particularly in sunny settings (Cole and Ferguson 2009; Marion and Cole 1996; Reid and Marion 2004). Recovery is low in drier high-elevation western sites, to intermediate in lower-elevation sites and moist high-elevation sites (Spildie et al. 2000; Cole and Spildie 2007).

Regardless, *recovery rates are always slower than deterioration rates, and recovery cannot occur unless nearly all site use is curtailed.* Wilderness managers are often reluctant to erect the ropes and signs that can be effective in keeping day and overnight visitors off closed campsites. Closing and rerouting trails to avoid flat areas that will forever be prone to campsite proliferation and expansion is the most effective practice, though popular destination areas that exert a strong permanent attraction to visitors can be an exception (Cole 1982a; Cole and Ferguson 2009). Effective closure in these areas generally benefit from a well-communicated and enforced designated site camping policy. *Much can be accomplished by carefully integrating the planning of trail and campsite infrastructures that meet the needs of visitors while minimizing the areal extent and severity of resource impacts.*

In summary, most camping impacts occur rapidly with initial and low use before reaching a more stable "dynamic equilibrium" between cycles of annual use and recovery. The key to minimizing aggregate impact therefore lies in limiting campsite numbers, concentrating use on small sustainable campsites, and recognizing that once created, campsites are difficult to close and restore to natural conditions.

Spatial Patterns of Impact

Dispersed "unconfined" camping policies have failed to effectively address the chronic "twin problems" of campsite expansion and proliferation. "Avoidable" impact occurs when campsites grow to excessively large sizes and when policies permit excessive campsite proliferation, generally in popular flat destination areas. Surveys reveal that visitors commonly select and develop campsites in flat terrain that offers no resistance to campsite expansion and proliferation, which is pronounced during periods of peak use, e.g., a "musical chairs" dilemma of too many campers and too few campsites (Marion et al. 2023). Dispersed camping policies, referenced in this chapter as *Unconfined* camping, but with variations also called at-large and unregulated camping, have failed to address campsite expansion, proliferation, or to substantially disperse visitors away from crowded attraction features, or anywhere close to the extremely low levels of traffic necessary to preclude lasting resource impacts and creation of new campsites (Cole 1982a).

Campsite expansion is a significant chronic problem in most popular areas with flat terrain. Studies demonstrate that campsites vary substantially in size between and within wilderness areas. For example, in the flatter valleys of the Bob Marshall Wilderness, Montana, mean campsite size is much larger (3,000 sq ft) than in the Grand Canyon National Park (500 sq ft), where sloping terrain and woody shrubs restrain site expansion (Cole and Hall 1992). In a Pacific Crest Trail campsite study (Marion et al. 2023) reported mean campsite sizes ranging from 1,760 sq ft in the flat valley bottom settings common to North

Cascades National Park, Washington, to 560 sq ft in the sloping and rocky terrain typical of California's Desolation Wilderness or Mt. Whitney's Outpost and Trail Camps (550 sq ft), John Muir Wilderness.

These surveys document maximum campsite sizes ranging from 2,000 to 16,000 sq ft. For example, Marion et al. (2020a) in their representative study of Appalachian Trail recreation sites, documented that campsites exceeding 2,000 sq ft, which they named "mega-sites," were responsible for a remarkable 13 percent of the aggregate areal extent of impact. It's important to consider that the current infrastructure of campsites in nearly all wilderness areas were originally selected and created by visitors, including in areas with designated site camping. Studies reveal that visitors frequently select campsites in large flat areas near water, attractive locations that should be considered "unsustainable" due to their inherent susceptibility to the twin problems of campsite expansion and proliferation (Marion et al. 2020b, 2023).

The extremely wide range in campsite mean sizes documented in wilderness surveys reveal that many campsites are substantially larger than is likely needed, representing unnecessary or "avoidable" impact. Campsites ranging from 600 to 1000 sq ft should be capable of accommodating all but the largest horse groups and could be considered an "acceptable range," with larger sites responsible for "avoidable impact" that unacceptably increases the area over which vegetation is trampled and lost, soils are exposed to compaction and erosion, and trees have roots exposed, are damaged, or are felled. Minimizing the size of campsites can minimize both the aggregate areal extent of camping impact and effectively constrain impact severity, including the area of vegetation loss, soil exposure, and number of damaged or felled trees (Arredondo et al., 2021).

Campsite proliferation is the second significant chronic problem, both in popular destination areas and along higher use trail corridors. Dispersed/unconfined camping policies allow visitors the freedom to create new campsites in nearly any location, though use of existing "well-established" campsites and setbacks from water and trails are increasingly encouraged. Campsite studies consistently find that this camping policy measurably exacerbates campsite proliferation problems, creating mostly non-sustainable campsites, often with high site densities that manifest the additional social/experiential problems of crowding, conflicts, and noise (Marion et al. 2018; Marion et al. 2020a, 2023). For example, the large Appalachian Trail study documented increasing numbers of "mega-clusters" of campsites with aggregate area of impact exceeding 10,000 sq ft, causing substantially degraded ecological and social/experiential conditions (Daniels & Marion 2006; Marion et al. 2020a); the ten largest mega-clusters accounted for 30 percent of the aggregate areal extent of impact from all overnight sites (163,112 sq ft of 539,939 sq ft).

Cole noted that "the most frequently used sites were not disturbed much more than sites used no more than once a week" (Cole 1982b), indicating little "cost" from substantially increasing use to a subset of campsites. A subsequent survey revealed steadily worsening campsite proliferation—the number of campsites at seven high-use lakes increased by 134 percent 15 years, with campsite density in 1990 exceeding two sites per 2.5 acres and "many clusters of sites so dense that it is difficult to tell where one site ends, and another begins" (Cole 1993a). This study also documented an 84 percent increase in campsite numbers within Montana's Lee Metcalf Wilderness from 1972 to 1988, attributed to "an increase in site-pioneering behaviors" and "passive campsite management programs."

In a study investigating the management of high-use destinations in Oregon and Washington wilderness areas, Cole et al. (1997) describe the challenges and costs faced by managers and the need to base decisions on a thorough understanding of changing resource and social conditions, effects on visitors, and how visitors feel about alternative management options. Similar campsite proliferation problems were documented in this study in the Three Sister's Wilderness, Oregon, also managed under a dispersed/unconfined camping policy. In the popular Sunshine-Obsidian Falls area they found 269 campsites (796 sq ft mean size) and the highest campsite density across all six high-use study areas, with 398,265 sq ft aggregate areal extent of "infrastructure" impact, including campsites (214,223 sq ft) and formal/informal trails (Fig. 15.7).

Dispersed/unconfined camping polices, intended to resolve unacceptable resource and social impacts in popular areas by dispersing use, often contributes to significant campsite expansion, proliferation, and social/experiential problems. As described by Cole et al. (1997), reducing use in popular areas inevitably shifts that use and impact to nearby lower use areas, where visitors select and create many new unsustainable campsites and expand aggregate impact. Recreation ecology studies demonstrate that impacts occur rapidly with initial and low levels of use, while recovery rates are quite low (Cole and Monz 2003; Cole and Spildie 2007). Furthermore, social

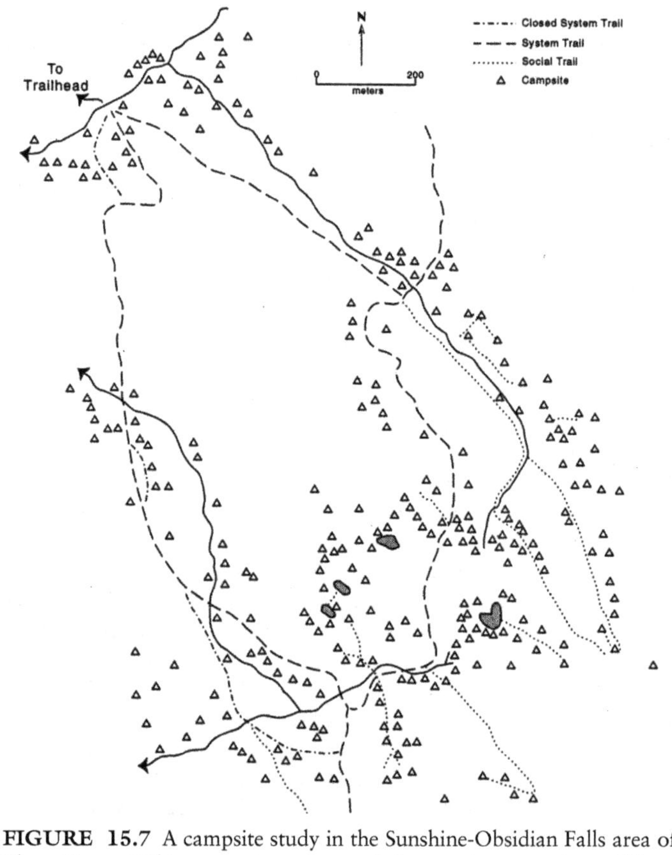

FIGURE 15.7 A campsite study in the Sunshine-Obsidian Falls area of Three Sisters Wilderness, Oregon, revealed extensive campsite proliferation, with 269 campsites averaging only 4.5 nights/year of use and 214,223 sq ft of aggregate campsite impact (796 sq ft/site) (Cole et al. 1997).

science studies of wilderness visitors consistently document their preference and support for educational and designated site camping options over limitations on use to achieve improved resource and/or social conditions (Hall et al. 2010).

In their study of high use wilderness destinations, Cole et al. (1997) concluded that efforts to reduce use likely have more negative than positive consequences: "neither encounter levels nor physical impacts would be reduced to any meaningful extent; many people would be denied access, and adjacent areas of the wilderness would likely experience increased encounter levels and levels of physical impact." The authors concluded that: (1) the primary physical impact problem associated with camping is the large aggregate areal extent of camping impact, which is largely linked to campsite proliferation; (2) education by itself is insufficient to concentrate use; and (3) direct management in the form of an intensive program of site management is needed to shift use to a subset of existing campsites (Cole et al. 1997).

FACTORS THAT INFLUENCE AMOUNT OF IMPACT

Aggregate campsite impact is a product of both the areal extent of impacted places and impact intensity. *The core factors influencing how much change occurs on an individual site are: (1) the amount and frequency of use the site receives, (2) the environmental conditions and sustainability of the site, and (3) the type and behavior of visitor use.* Management tools exist to effectively manipulate all three of these factors to avoid and minimize visitor impact.

Amount of Use

An historic assumption had been that the amount of use a trail or campsite receives is most important in determining its impact, but recreation ecology studies demonstrate that other factors are often more influential. To illustrate, consider that campsites in the Eagle Cap Wilderness used only a few nights per year have been severely altered (Table 15.1) (Cole 1982b). Most of the vegetation cover and tree seedlings have been lost, soil has been compacted, and trees have been damaged or felled. Moderately used campsites, occupied about one night per week during the main use season, differed in the following ways: site size and devegetated area were usually much larger, exposure of tree roots was greater, organic horizons were thinner, and changes in the composition of vegetation groundcover (floristic dissimilarity) were more pronounced. Heavily used campsites, occupied several nights per week, differed from moderate use sites only in that organic horizons were thinner. Similarly, in an experimental study, one night of camping on previously unused sites caused significant vegetation loss in all four vegetation types studied (Cole 1995a), while four nights of camping caused less than twice the impact of one night.

TABLE 15.1 Relationship between Selected Campsite Impact Indicators and the Amount of Use a Site Receives (Cole 1982b)

	Amount of Use		
Impact Indicator	**Light (n=6)**	**Moderate (n=6)**	**Heavy (n=10)**
Campsite Size (sq ft)	517	2,411	2,207
Devegetated Area (sq ft)	205	1,313	1,001
Vegetation Cover (%)	9	6	4
Bare Soil (%)	14	20	35
Floristic Dissimilarity (%)	31	60	64
Bulk Density Increase (%)	16	11	16
Infiltration Rate Decrease (%)	8	57	12
Organic Horizon Decrease (%)	3	21	68
Organic Matter Increase (%)	19	26	20
pH Increase (%)	3	5	11
Damaged Trees (%)	74	85	97
Trees with Exposed Roots (%)	3	33	39
Felled Trees (%)	43	12	34
Tree Seedlings (#/acre)	430	738	827

These and numerous other recreation ecology studies describe *a general sigmoidal relationship between use and impact* (Cole 2021). Above a relatively low threshold of trampling pressure, impacts occur rapidly until an inflection point is reached where per capita impacts begin to substantially diminish with increasing use (Marion et al. 2016). Research in the United States indicates this inflection point occurs around 15 nights of camping or 250 passes along a trail, at which point additional use causes substantially less per capita impact (Cole 1995a,b; Marion 2016; Monz et al. 2010). However, these values vary with different environmental settings and vegetation resistance and resilience.

Environmental Conditions and Sustainability of the Site

Topography, substrates, rockiness, and vegetation significantly affect campsite sustainability. A recent study that modeled a large and diverse array of field-assessed and GIS-derived indicators to determine their influence on campsite size and area of vegetation loss identified topography as the most influential factor (Arredondo et al. 2021). The steepness of the terrain surrounding Appalachian Trail campsites was evaluated through GIS calculations within a 33-ft wide doughnut-shaped buffer. The model predicted that a campsite located in flat terrain would be 1,367 sq ft, a campsite where 50 percent of the buffer had slopes greater than 15 percent grade would be 840 sq ft, and a campsite where 100 percent of the buffer had slopes above 15 percent would be 506 sq ft, holding other variables constant.

Another influential factor revealed by the campsite size modeling study was off-site rugosity, generally rockiness but also substantially uneven terrain, which similarly acts to deter campsite expansion and proliferation and spatially concentrate visitor activities (Arredondo et al. 2021). The influence of these factors was highly evident throughout the Sierran range in the Pacific Crest Trail study (Marion et al. 2023). In California's Mt. Whitney area in the John Muir Wilderness 20,000 visitors/year attempt to hike Mt. Whitney, with many camping overnight at Trail Camp (also known as High Camp due to its 12,000 ft elevation). Rockiness in this area has effectively minimized campsite expansion, limiting the sizes of the 59 campsites (87 separate use areas) to a mean size of 490 sq ft (269 sq ft median size) (Fig. 15.8). The aggregate area of camping disturbance was also remarkably low (28,891 sq ft) given the exceptionally high use this camp receives. This was the only area surveyed in the PCT-wide campsite study that had optimal scores for two measures of campsite expansion potential.

FIGURE 15.8 Rugosity, including rockiness and uneven terrain, effectively limit campsite expansion in areas like Mt. Whitney's Trail Camp (left) and most Desolation Wilderness lake basins (right).

Similarly, in northern California's popular Desolation Wilderness west of Lake Tahoe, rockiness was cited as the principal factor limiting campsite mean sizes to only 259 to 527 sq ft in six lake basins where rockiness was common and campsite expansion was rated extremely low (Figure 15.8). Despite these small campsite sizes, a dispersed/unconfined camping policy with heavy visitation has led to significant problems with site proliferation causing substantial aggregate area of impact (109,283 sq ft), with many of the sites appearing to receive low use.

A final noteworthy finding from the campsite size modeling study was a highly significant positive relationship between increasing overstory tree cover and area of vegetation loss (Arredondo et al. 2021). Sunny campsites with less than 5 percent tree cover had a mean area of vegetation loss of 75 sq ft and mean vegetation cover of 58 percent, compared to intermediate 25 to 50 percent tree cover with mean area of vegetation loss of 1,055 sq ft and mean vegetation cover of 32 percent. The substantial trampling resistance and resilience of sun-loving grasses and sedges greatly exceed that of shade-tolerant broad-leaved herbs, a finding repeatedly documented in experimental trampling studies (Cole 1993b, 1995b). For example, Cole (1993b) found that as few as 20 hikers can trample and remove 50 percent of the vegetation cover in the least resistant vegetation type, forested herbs and ferns, while in the most resistant type, meadow sedges and grasses, 600 hikers are needed to remove 50 percent vegetation cover. This suggests that meadow campsites can absorb more than 30 times as much use as the forested campsites, with the same level of vegetation damage. Clearly, where people camp is critical in determining both the percentage and areal extent of vegetation loss.

Similarly, in the Boundary Waters Canoe Area Wilderness, tree loss from 1982 to 2014 on well-used designated campsites caused a 29 percent reduction in canopy cover, with the additional sunlight allowing a 16 percent increase in campsite vegetation cover (Eagleston & Marion 2017b). More specifically, grass/sedge cover on campsites increased to 42 percent in 2014, while just 3 percent in nearby forested "control" sites; in contrast, herb cover was 7 percent on campsites but 29 percent in controls. Grasses were found to colonize all but the more intensively trafficked campsite areas in sunny locations, effectively limiting the area of vegetation loss and filtering soil from water runoff. In the Ansel Adams Wilderness, California, most campsites around the heavily visited Thousand Island Lake are in gently sloping terrain with open grassy meadows, interspersed with bedrock (Marion et al. 2023). While campsite proliferation has long been a problem, the PCT study found campsites to be widely distributed with an exceptionally small mean size (358 sq ft) and area of vegetation loss (138 sq ft) (Marion et al. 2023). In the southern PCT near the border, the large annual "bubble" of thru-hikers is accommodated in federal and commercial campgrounds predominantly on grassy campsites with little to no bare soil; campground managers reported no resource protection problems.

In summary, the influential relationships described here indicate that the areal extent and severity of camping impact can be effectively minimized by concentrating use on the most durable substrates and shifting camping from flat, smooth terrain to sloping/rocky/uneven terrain, and from forests with high herbaceous groundcover to sunny dry meadows with grass/sedge groundcover.

Type and Behavior of Site Users

Certain types of impact on campsites are determined almost entirely by the behavior of campers, including damage to trees, feeding of wildlife, and "pollution" of the site with campfire ashes, charcoal, food wastes, greywater, and trash.

Many studies have documented the efficacy of the Leave No Trace (LNT) national educational program. Manning (2003) emphasizes that such educational programs provide managers with an indirect, light-handed visitor management tool that has been strongly favored by wilderness visitors. The use and efficacy of LNT messaging to reduce impacts that are strongly related to visitor behaviors are addressed in Chapter 16. Three other characteristics of user groups also influence campsite impact—group size, length of stay, and if they use pack stock.

Group Size

Organized groups, frequently comprised of younger or older participants, are responsible for introducing large numbers of novice visitors to protected areas. Such groups often serve racially and economically diverse urban populations that frequently lack opportunities to learn about or visit wilderness. *Most importantly, organized groups provide the necessary training in outdoor practices that shape the life-long outdoor skills and ethics of their participants, which with agency collaboration, can include the best available low impact practices and enlist their participants in stewarding and protecting our public lands* (Daniels and Marion 2005).

The effect of group size on campsite resource conditions has not been formally studied. Larger groups ostensibly require more tenting sites, though shifting from 1- to 2-person tents to 3+ person tents or hammocks can substantially offset the areal extent of impact. Reducing group sizes could reduce campsite sizes, though any benefits are likely to be fully offset by an increased number of groups requiring an increased number of campsites. There is little reason to believe that group size should affect any other characteristic of established campsites. An exception has been "avoidable" campfire-related impacts often associated with youth groups, particularly tree damage and felling, though these are effectively addressed by discouraging or prohibiting the use of axes, hatchets, and saws. Large groups are most problematic when they create substantial new impact by using a too-small campsites and pioneering new tenting areas; such impact is avoided when they split their group to camp on two or more campsites or find and use a larger campsite (ATC 2015). Some agencies are creating group use campsites that effectively concentrate resource impacts while avoiding social impacts through increased separation from other campsites. Large groups can also avoid impact by camping on particularly durable substrates (snow, bedrock, gravel, naturally barren shoreline sandbars).

Length of Stay

Limits on length of stay have been placed both on the maximum number of nights allowed in the wilderness and at individual campsites. Imposing area-wide length-of-stay limits is unlikely to have any effect on site impacts (Cole et al. 1987). The major benefit of such a regulation is to allow more people to use an area in which total use is limited. Length-of-stay limits for campsites are likely to do little to improve campsite conditions in popular parts of wildernesses or on popular campsites because new groups are likely to occupy sites shortly after they are vacated. *Generally, the most valuable use of length-of-stay limits is minimizing time spent at little-used sites.* This goal is most effectively accomplished through education, particularly when such limits are not imposed on heavily used sites. Such limits could prevent serious deterioration of low use sites. However, in these cases *the common fourteen-day limit is too long. To avoid damaging little-used areas, sites should not be used more than a night or two except on durable surfaces* (Marion 2016).

Stock Use

Studies have consistently found that groups with pack stock disturb a larger area than hikers, in part due to the need for stock confinement areas. These camps also often have greater soil disturbance, a result of trampling by heavy shod animals, and more tree damage, a result of tying horses to trees for extended periods (Cole 1983b). This topic, along with sustainable management practices, is discussed in more detail later.

CAMPSITE MANAGEMENT STRATEGIES

As previously reviewed, the traditional dispersed/unconfined camping policy sometimes seeks to disperse and reduce camping in high use areas and shift it to new or lower use campsites in nearby areas (Table 15.2). While the intent is to reduce resource and social impacts, campsite studies have frequently

TABLE 15.2 Camping Management Strategies, Options, and Guidance (Marion et al. 2018)

Camping Strategy	Use Level	Description and Guidance
UNCONFINED (Dispersed Camping)	All	Visitors have the freedom to select or create a campsite in the location of their choice, sometimes modified by guidance to avoid locations close to water or formal trails or to use an existing site. Campsite proliferation, expansion, excessive aggregate resource impact, and high-density camping are frequently significant problems in popular areas.
DISPERSAL STRATEGY Natural Site Camping	Low-Mod	Visitors are asked to camp on durable *previously undisturbed* surfaces at low use levels that avoid lasting impact. Requires visitors to learn and consistently apply low impact natural site camping practices.
CONTAINMENT STRATEGY Established Site Camping	All	Visitors are *encouraged to camp* on a subset of sustainable Established campsites selected by managers to promote resource protection and desired social qualities. Supporting actions include accurate maps to aid visitors in finding Established campsites and a program to close and restore non-selected campsites.
Designated Site Camping	Mod - High	Visitors are *required to camp only* on a subset of sustainable designated campsites selected by managers to promote resource protection and desired social qualities. Managers generally mark designated campsites on maps, have signs and/or some facilities, and remove hazardous trees. Designated sites are managed as "first-come first-served" or more rarely through a reservation system.

documented an expansion in aggregate measures of camping impact (Marion et al. 2018; Marion et al. 2023). Studies have revealed little measurable recovery occurring in higher use areas under this policy, particularly on attractive sites close to water bodies and formal trails (Cole 2013b; Cole et al. 1997). Cole (2013a) concluded that these policies often result in a "profound proliferation of campsites" and that studies documenting improved campsite conditions often "reflects successful implementation of a use concentration or containment strategy."

The sigmoidal use-impact relationship (Fig. 15.9) reveals several very important implications for camping management (Marion 2016). Consider a hypothetical example of three "typical" wilderness campsites that each receive about fifteen nights of camping/year, near the inflection point. If that aggregate use were *fully dispersed* to forty-five previously unused sites, research reveals that no lasting impact would occur, provided they were not reused for the balance of that year (Cole, 1995a; Marion 2016). This finding supports a pure *Dispersal* strategy (Table 15.2), with *Natural* site camping one to several nights/year, levels that preclude lasting impact. Alternately, the least sustainable two campsites could be closed, and all use shifted to the third, with aggregate impact substantially reduced from three sites with an "a" level of impact to a single site with a "b" level of impact. This outcome supports a *Containment* strategy (Table 15.2), with two options: *Established* site camping where visitors are encouraged to use a subset of the most sustainable campsites selected to optimize visitor and management preferences, or *Designated* site camping where visitors are required to use a preferred reduced set of sustainable campsites. Finally, the sigmoidal use-impact relationship illustrates that limiting visitor use will do little to reduce resource impacts on campsites and trails unless use is restricted to levels below the inflection point, where use and impact are strongly related (Cole 1995a,b).

The Dispersal Strategy

Camping impacts could be avoided by the effective implementation of a pure *Dispersal* strategy and *Natural* site camping (Marion et al. 2018). *The objective of a Dispersal strategy is to avoid any resource impact that persists for more than one year.* Even in forests with trampling susceptible forbs and ferns the impact from one night of camping/year would recover to fully natural conditions within a year (Cole 1995a). However, several nights/year could be accommodated in dense forests with little to no ground vegetation, or in sunny dry meadows with complete grass/sedge cover (Cole and Monz 2003). Gravely/sandy shorelines and bedrock can support substantially more nights of camping without showing lasting impact. Dispersed Natural site camping is frequently the most applicable and appropriate strategy in lower use areas, though experienced visitors could successfully apply this practice in moderate use areas.

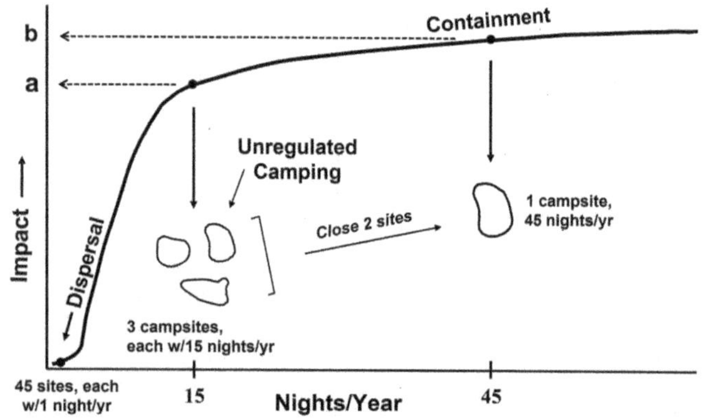

FIGURE 15.9 The general relationship between amount of use and amount of impact is asymptotic. This relationship provides an important empirical basis for effective dispersal and containment impact management strategies. (Marion 2016).

The effective application of the Dispersed Natural site camping option requires that all group members employ the following key low impact practices:

1. In a natural area beyond sight or greater than 100 yds from any existing trails/campsites, select the most durable surface available: snow, bedrock, mostly non-vegetated gravel, sand, or organic litter, or grasses/sedges in sunny dry meadows.
2. Minimize vegetation disturbance by focusing camping activities on the most durable surfaces and dispersing all traffic, particularly to water and trails. Camp one to a few nights but move camp if visible vegetation or soil disturbance begins to appear (no campfires that leave lasting impacts).
3. Naturalize the area to disguise it by placing sticks, rocks, or organic materials to conceal traces of your stay. It must be *highly unlikely* for other visitors to find and reuse your spot.

A significant risk with this camping option is that visitors who fail to adhere to this guidance may create new campsites, so it is most appropriate for conscientious and experienced visitors (Marion 2018). Campsite maintenance practices for this strategy involve having agency staff or volunteers follow informal side-trails and check likely off-trail camping areas to locate and restore any visible traces of camping. This would include the thorough removal and restoration of any fire sites, scattering of collected firewood and rocks used to anchor tents, and the return of rocks, sticks, and organic litter to any visible tenting or cooking areas (Marion et al. 2020b). All work should seek to appear "natural" so as not to attract further attention and possible reuse. Temporary "No Camping" or "Closed for Recovery" signs may prove necessary for sites that receive recurring use, adding symbolic fencing with nylon cord to encircle campsites with persistent use that prove difficult to close.

The Containment Strategy

The objective of a *Containment* impact management strategy is to limit aggregate impact by concentrating high levels of use on a minimum number of preferred sustainable campsites (Marion 2016). Preferred sites are those with attributes desired by visitors (e.g., close to a water source with smooth, well-drained tent sites), and attributes preferred by managers (e.g., greater than 200 ft from water, trails, other sites). Sustainable sites are those surrounded by sloping, rocky, or uneven terrain that constrain site proliferation and expansion. This strategy is most appropriate in moderate to high use areas, where managers limit campsite numbers to achieve high use while actively discouraging the creation of new campsites (Marion 2016).

Containing use is a well-developed resource protection principle, though some consider it inappropriate in wilderness—labeling it the "sacrifice" policy. This label is inaccurate because it implies that managers have the option of avoiding impacted campsites. While that may be possible with Dispersed Natural site camping, it is not possible in moderate to high use areas, where the choice lies between: (1) large numbers of visitor-selected campsites prone to site expansion that receive low to moderate use, or (2) small numbers of management-selected campsites resistant to site expansion that receive heavy use. The second option most effectively minimizes aggregate camping impact (Arredondo et al. 2021; Marion 2016).

The traditional construction of appropriately located and sustainably designed formal trails recognize and condone the need for a containment strategy, which effectively constrains intensive traffic to narrow treads able to accommodate visitation while limiting the areal extent and severity of impact. This same strategy is equally applicable to camping management with intensive camping constrained to a limited number of appropriately located sustainable campsites through voluntary use (Established sites) or required use (Designated sites). Exhorting visitors to stay on formal trails when possible is an example of appropriate containment, just as encouraging visitors to use a subset of carefully selected campsites and avoid expanding or creating new campsites is. The traditional application of a containment strategy to trails can and should be fully extended to campsites.

Wilderness can accommodate a range of opportunities; having some popular areas with limited numbers of impacted campsites managed through a Containment strategy with Established or Designated site camping, while most of the remote and low use areas remain largely undisturbed by camping through a pure Dispersal strategy. Concentrating use on a limited subset of sites will not result in ever-increasing deterioration if preferred sustainable sites are selected and low impact visitor behaviors are encouraged.

In summary, for popular destination areas two options are a limited number of deteriorated sites—the result of containment—and many deteriorated sites—the result of traditional Dispersed/Unconfined camping policies. The *Containment* strategy provides managers with more numerous and effective tools for achieving and balancing recreation provision and resource protection objectives. Established site camping promotes visitor freedom while retaining substantial management influence, while Designated site camping can be efficiently implemented if necessary for exceptionally popular areas where greater control is required. Recreation ecology studies consistently demonstrate the efficacy of a *Containment* strategy (Eagleston and Marion 2017b; Marion and Farrell 2002; Marion et al. 2018, 2020b, 2023; Reid and Marion 2004; Spildie et al. 2000). For example, Spildie et al. (2000) found that a containment strategy applied to a subalpine lake basin in Idaho's Selway-Bitterroot Wilderness reduced campsite area by 37 percent and bare area by 43 percent in five years; a containment strategy with Established site camping adopted at Virginia's Shenandoah National Park reduced campsite area by 50 percent in just three years (Reid and Marion 2004).

As illustrated in Table 15.2, *Established* site camping is a voluntary option while *Designated* site camping is a regulatory option—both options restrict use to a management-selected subset of heavily used campsites that effectively minimize aggregate resource and social impact in areas of moderate to high use. Both options confer several beneficial advantages. They allow managers to identify campsites that ensure the desired social/experiential conditions sought for a zone by specifying distances between campsites and setbacks from trails. Resource protection is enhanced by selecting setbacks from water, cultural/historic features, vistas, or sensitive flora/fauna, and by avoiding flatter terrain in favor of sloping, uneven, or rocky terrain that inherently compels the spatial concentration of visitor activity to smaller campsites. These are attributes of *sustainable* campsites derived from recreation ecology research.

Once again, the effective implementation of the *Containment* strategy requires visitors to adopt a few key low-impact practices:

1. Find and use only the sustainable well-established or designated campsites selected by managers, preferably distant from other groups. Select a sufficiently large campsite or split your group to two sites.
2. Concentrate camping activities on the most durable surfaces available: rock or mostly non-vegetated soil or organic litter. Never enlarge a campsite. If a campfire is permitted, build it only in an existing fire ring, use dead and down wood breakable by hand, lit only when truly needed and keep it small to conserve wood supplies. Burn wood and coals to ash and extinguish everything to cold touch.
3. Clean the campsite and fire site of *all* trash and food scraps, scatter cold crushed charcoal and ashes away from camp.

Combined ground-based and GIS campsite inventories can apply improved site selection and sustainability indicators (described below) to select subsets of existing or new campsites better able to accommodate heavy use while minimizing expansion and impact severity (Marion et al. 2023). Campsites can be either clustered or dispersed; their spatial distribution and density is selected by managers to ensure desired social and experiential conditions.

Campsites are selected by managers to meet both visitor's and manager's preferences, and to minimize the aggregate areal extent of impact and impact severity through the selection and heavy use of sustainable campsites. Studies of visitor campsite preferences indicate they want sites with sufficient smooth gently sloped areas for tenting and proximity to water (Brunson and Shelby 1990; Cole and Hall 2009; White et al. 2001). Isolation from other campsites through distance or vegetative screening is preferred, along with a good cooking area, scenic beauty, bug-free, and firewood availability. Managers prefer campsites located more than 100–200 feet from formal trails, water, and other campsites, areas with little to no tall tree cover (to avoid hazardous trees), and high potential for the ability to dig effective cat-holes for human waste disposal. Additional considerations for campsite selection include avoiding proximity to sensitive rare and endangered flora and fauna, cultural, and historic resources.

Important considerations when deciding to employ a *Containment* strategy include: (1) assurance that use is being moved from less sustainable to more sustainable campsites, (2) the ability to effectively direct visitors to the preferred sites and to close/restore the unnecessary campsites, (3) the ability to select or create enough sustainable campsites in the locations where they are needed and promote high levels of site use, and (4) the ability to maintain the campsites to minimize their size, impact severity, and safety. Volunteer stewards can assist with these tasks, but they do represent a significant staffing challenge.

Established site camping is most appropriate in moderate use areas, where visitors are asked to voluntarily select and use only well-established campsites, avoiding the creation of new campsites or use of sites that managers have closed because they were unnecessary, less sustainable, or in less desirable locations. Established campsites should be used more than about 25 nights/year to obtain the level of containment that minimizes the aggregate areal extent of camping impact. Managers can also promote their use by showing their locations on paper or phone app maps, and/or by paint blazing their access trails and locations. Camping can also be prohibited within specified setbacks (e.g., greater than 200 ft) from water or formal trails. Such setbacks can be shown on paper or phone app maps with shaded areas, clearly communicating areas off-limits to camping and facilitating educational and enforcement actions.

Designated site camping is most appropriate in high use areas, where visitors are required to select and use only designated campsites. This affords complete control to select preferred sustainable campsites, focusing heavy use on sites selected to minimize site expansion, resource impacts, and intersite distances needed to promote or ensure desired levels of social interaction or solitude. Designated site camping improves managers' ability to balance competing recreation provision and resource protection demands by concentrating intensive use, preferably greater than 50 nights/year, to achieve optimal levels of containment and reductions in the areal extent of impact. However, visitors may lose the freedom to camp in a preferred location and enforcement is often necessary to ensure adequate compliance.

THE ADAPTIVE MANAGEMENT PROCESS

Manipulating factors that influence the amount of recreational impact offer the most effective management "toolbox" strategies and actions for reducing resource and social impacts. A core set of these strategies and actions are presented in Table 15.3 and discussed in this section for both trails and campsites (Marion 2016). Additional relevant guidance is provided in Cole et al. (1987) and Anderson et al. (1998). These toolbox strategies and actions are most effectively applied within an *adaptive management process* used to thoroughly evaluate problems in comparison to prescriptive management objectives as reflected by indicators and thresholds (standards) and used to formulate and apply effective impact-reduction strategies and actions, with periodic monitoring to evaluate efficacy and the need for further corrective actions (Marion 2016; Marion et al, 2020b). This may be conducted as part of a formal decision-making framework like VUM (see Chapter 9) but can also be conducted more informally. When possible, consult a team of knowledgeable and experienced individuals with expertise in recreation resources management, visitor management, social science, recreation site and trail management, natural resource management, and interpretation. Visit the site(s) where the natural resource or social problems are

TABLE 15.3 Core "Toolbox" Management Strategies and Actions for Avoiding or Minimizing Resource and Social Impacts (Adapted from Cole et al., 1987; Marion 2003)

Core Strategies	Core Management Actions
Manage Use Levels	Redistribute, discourage, or limit use (e.g., set access point or travel zone quotas). Redistribute or reduce use during times of peak use, in high use locations, or when impact potential is high.
Modify the Location of Use	Concentrate use on sustainable expansion-resistant trails and campsites to limit the aggregate area of impact. Disperse use on durable substrates at levels that prevent formation of trails and campsites. Encourage or require visitors to camp out-of-sight or a minimum distance from trails and campsites. Restrict certain types of use to specific locations (e.g., restrict horses to trails and campsites designed for their use).
Increase Resource Resistance	Construct, reconstruct, or maintain impact-resistant trails and campsites (e.g., construct side-hill trails and campsites). Install resource protection facilities that attract and spatially concentrate visitor activity (e.g., tent pads, fire rings, stove rocks, food storage facilities)
Modify Visitor Behavior	Persuasive communication, interpretation or education—encourage Leave No Trace practices when traveling and camping. Regulation and enforcement—prohibit or require certain practices and equipment when traveling and camping (e.g., feeding wildlife, woods tools, safe food/trash storage).
Close/Rehabilitate Resources	Close and rehabilitate unnecessary or less sustainable trail segments and campsites.

occurring and guide discussions to apply the following two-step problem analysis process (Marion et al. 2020b):

1. **Identify and Evaluate the Problem**
   a. *Describe area and use(s)*—provide background information about the area, facilities, and history of visitor use.
   b. *Describe problem(s)*—briefly describe the facility, resource, and social impact problems that are occurring.
   c. *Problem significance*—consider if and why the impacts are significant or unacceptable to managers and visitors based on area objectives and desired conditions.
   d. *Previous management actions*—describe the history of the problems and previous actions; discuss the effectiveness of these actions and why they did or didn't work.
   e. *Causes and influential factors*—discuss the underlying causes for the impacts and the role of non-causal but influential factors that may intensify impacts. Consider use-related factors (type and amount of visitor use, visitor behavior, use density), environmental factors (soil and vegetation type, environmental sensitivity, topography), and managerial factors (siting, design, construction, and maintenance of facilities, visitor and impact management strategies). Consult recreation ecology or social science research and managerial expertise to better understand the role of causal and influential factors.
2. **Identify and Evaluate Strategies and Actions**
   a. *List potential strategies and actions*—create a comprehensive list of appropriate and potentially effective management strategies and actions. Strategies are broad approaches (e.g., modify visitor behavior, manage sites and facilities) and actions are the specific means used to implement a strategy (e.g., educate visitors, relocate campsites).
   b. *Evaluate strategies and actions*—discuss and evaluate the following attributes for each strategy and action: potential effectiveness, management feasibility (cost, staffing, long-term maintenance), advantages/disadvantages (e.g., costs to visitor freedom), expected visitor compliance, etc.
   c. *Formulate recommendations* through group discussion, develop and write recommendations that reflect the group's consensus views. Describe the recommended action or group of actions to implement first and what might be tried next if these are ineffective.

Manage Use Levels

As previously described and illustrated in Fig. 15.9, a strategy of reducing use to limit resource impacts is ineffective unless camping is restricted to below the use-impact curve inflection point (about 15 nights on campsites or 250 passes on trails in many U.S. areas) (Marion 2016). Below the inflection point, such actions are effective and above it reducing use is unlikely to promote measurable recovery. Restricting use in areas where a pure dispersal strategy with Natural site camping could be helpful, though more influential factors include the availability of resistant settings and visitors who know and apply dispersed Natural site camping and travel practices.

In places where use levels are high—the usual case where rationing has been implemented—use limits are likely to be more effective in limiting the *number* of campsites than the *severity* of impacts on individual sites. *To limit the number of campsites, however, it is critical to simultaneously employ the containment strategy to constrain camping to limited numbers of heavily used Established (greater than 25 nights/year) or Designated (greater than 50 nights/year) campsites, rather than retaining large inventories of more lightly used campsites.* In general, then, use limits are likely to be effective *only when* supporting pure dispersal or containment strategies. For example, limiting the number of thru-hikers starting their hikes each day on popular long-distance trails can flatten the annual moving "bubble" of peak use that contribute to unacceptable levels of campsite expansion and proliferation (Marion et al. 2020b, 2023).

Modify the Location of Use

The dispersed Natural site camping option shifts camping to a multitude of impact-resistant locations while also providing visitors with the greatest freedom to camp out of sight and sound of other visitors (Cole 1995a; Marion et al. 2018; Marion et al. 2020b). If it can be effectively implemented, this option preserves the highest quality resource conditions and visitor experiences, perpetuating the "ideal" wilderness situation with no lasting evidence of camping impact. The containment strategy shifts camping to a limited number of intensively used resistant locations that limit aggregate resource impact (Marion 2016; Marion et al. 2018; Marion et al. 2020b). This strategy also allows managers to establish any preferred distances between selected campsites, trails, and shorelines, a necessary level of control that effectively balances the protection of desired resource and social/experiential conditions and visitor freedom in areas of moderate to high use.

When effectively implemented, both strategies protect resource and social conditions while maximizing visitor access, minimizing the need for direct regulatory use limits that visitors find most objectionable. The dispersal and containment strategies can effectively distribute camping locations to maintain wilderness solitude and natural quiet and avoid the crowding and conflicts that commonly occur when visitors have complete freedom to select and create campsites. However, a key challenge is facilitating visitor navigation in finding/using Established or Designated campsites. While campsite paint blazes on these campsites and their access trails can help, the increasing use of navigational phone apps offers the greatest promise when preferred sustainable campsites are included on maps.

As previously described, attributes of campsite sustainability can be incorporated into campsite selection indicators (Marion et al. 2020b, 2023). Visitor/manager preference and sustainability indicators have been developed and incorporated into a new "Recreation Site Library of Monitoring Protocols" (Marion 2021). As reviewed in previous sections, campsites located in sloping, rocky, or uneven terrain effectively minimize the aggregate areal extent of impact when visitors simply respond to inhospitable tenting conditions in offsite areas, an improvement over reliance on educational or regulatory interventions.

Marion et al. (2023) developed and applied GIS protocols in the Desolation Wilderness to identify large numbers of possible *"naturally occurring" side-hill campsite locations* in the 200–1000 sq ft size range that were predominantly surrounded by sloping terrain (greater than 15 percent) or rockiness and uneven ground, though the analytical procedures required sub-meter LiDAR-derived Digital Elevation Model elevation data. Ground-truthing surveys during field work revealed that many of these locations were indeed sufficiently flat and optimal for campsites: some had even been found and used by visitors, while others would need to be cleared of low shrubs or trees and receive limited substrate manipulation to create ideal tenting spots.

Another option is for managers or volunteers to construct side-hill campsites in sloping terrain that is sufficiently steep (greater than 15 percent) to prevent campsite expansion and proliferation problems (Arredondo et al. 2021). GIS analyses can easily produce color-coded slope maps for trail corridors showing the optimal locations for constructing side-hill campsites, preferably near water sources (Marion et al. 2020b). This concept was discovered in a wilderness campsite study at Isle Royale National Park, which had remarkably small campsites and the lowest disturbed area per overnight stay (3.7 sq ft) of any prior study for which this measure could be computed (Marion and Farrell 2002). Just like a constructed side-hill trail, park staff had created these campsites in sloping terrain by cut-and-fill work to form sustainable campsites that remained small and in good condition under intensive long-term use (greater than 45 nights/year). Surrounding steeply sloped terrain compelled the spatial concentration of camping activities to these small "footprint" sites more naturally and effectively than could educational or regulatory practices.

To evaluate the potential of side-hill campsites in other settings they were experimentally constructed at two areas identified as having the greatest areal extent and severity of camping impact on the Appalachian Trail in 2001 (Daniels and Marion 2006; Marion 2003). Their management efficacy over time was investigated at Annapolis Rocks in Maryland before and after closing a large dense cluster of nineteen campsites and shifting use to fourteen new well-distributed side-hill sites (Fig. 15.10). The aggregate areal extent of camping impact was reduced 86 percent, from 43,099 sq ft in 2002 to 6,243 sq ft in 2003 following the first year of use. Study of recovery rates and impacts on the side-hill campsites were continued to 2019, when all original campsites were judged fully recovered.

Travel-related impacts are also effectively managed through the dispersal and containment strategies. In areas of low use or highly durable substrates sufficient dispersal of traffic can effectively prevent the formation of lasting trails (Cole 1995a). Leave No Trace practices promote "meadow walking," with visitors traveling side-by-side rather than in-line to avoid lasting damage to vegetation (Marion 2016). Travel on gravel, rock, snow, or barren shorelines and dry washes can occur at substantially higher levels of use with no lasting impact. In areas of moderate to high use the most effective practice is the containment of traffic to sustainably designed and managed formal trail networks. Traffic is constrained to well-used durable treads where the sigmoidal use impact relationship (refer to Fig. 15.9) reveals that little additional per capita impact will occur. A permanent program of trail maintenance can keep trails adequately drained, hardened where needed, and avoids use rationing except in response to crowding.

Increase Resource Resistance

As previously described, research has documented considerable differences in vegetative resistance and resilience (Cole 1993b, 1995b). Thus, vegetative impact can be minimized by encouraging hiking or camping in resistant vegetation types or discouraging it from fragile types (e.g., desert cryptogamic soil crusts of cyanobacteria, mosses, and lichens). Unfortunately, camping in meadows has been discouraged or prohibited in some western wilderness areas, often attributed to problems in the 1960s and 70s when large stock groups created unacceptable damage to wet meadows through intensive grazing and trampling (Marion et al. 2023). However, substantial recreation ecology research over the last three decades has consistently revealed that sunny meadow vegetation is exceptionally resistant to travel and camping, and resilient in riparian zones with adequate soil moisture (Cole and Monz 2003). Therefore, visitors could be encouraged to camp on dry, sunny meadows and grasslands, particularly when interspersed with woody vegetation that limits visibility from trails and other campsites and provides some shade for visitors.

Areas managed for pure dispersal include cross-country zones that lack trail or campsite infrastructures, though trails may be needed to facilitate safe, efficient, and low-impact travel. Tread maintenance and hardening may be appropriate

FIGURE 15.10 Annapolis Rocks side-hill campsite photo and cross-sectional diagram (Marion et al. 2020a).

for trails, but campsites would generally be discouraged in favor of Natural site camping. Areas managed for containment should always benefit from sustainably designed and managed trail and campsite infrastructures able to accommodate intensive long-term use while retaining acceptable resource and social conditions.

When monitoring reveals resource impacts trending or crossing over to unacceptable conditions a problem assessment process can be employed to suggest optimal remedies, which may include trail or campsite relocations to more sustainable locations, improved drainage, site hardening work with rock, such as stone pitching or steps on trails, use of naturally-occurring or constructed side-hill campsites, or installation of site facilities like anchored rock or steel campfire facilities, improved tent pads, toilets, or food storage facilities.

A widely acknowledged recreation management maxim is: "Intensive use requires intensive management" (Marion et al. 2020b). *Campsite maintenance actions and facilities may be needed and justified to enhance the efficacy of the containment strategy and improve the spatial concentration of visitors to reduce the areal extent of impact.* Campsite maintenance for established or designated campsites consist of actions that attract and "pull" visitors to use these preferred campsites, and "push" actions that dissuade visitors from camping elsewhere (Marion et al. 2023, 2020b). Agency staff and volunteers can promote visitor use and satisfaction with campsites through limited maintenance work to keep them clean, attractive, and small (Marion and Sober 1987). *Pull* actions also include ensuring an adequate number of ideal tenting areas to accommodate the range of visitor group sizes, with limited maintenance to ensure tent spots are smooth and well-drained to attract and concentrate use (Marion et al. 2023).

Push actions include having agency staff and volunteers locate and restore any campsites or visible traces of camping not on the established or designated site list. Staff are directed to follow all informal side-trails and check likely off-trail camping areas for these sites, particularly after peak periods of use. Practices, efficacy monitoring, and adaptive management decision-making should generally follow that described under the Dispersal strategy for all non-selected campsites — the objective is to shift all camping to the preferred and sustainable established or designated campsites. Newly created and lower use campsites are generally easier to close and restore but some campsites in attractive locations may prove difficult to close despite repeated efforts.

When appropriate, existing, or new preferred and sustainable campsites could be added to the established/designated site list to accommodate increasing camping demand and enhance the efficacy of closures. This could include incorporation of naturally occurring or constructed side-hill campsites in sloping or rocky terrain. If unacceptable levels of site proliferation and/or continued high use of non-selected campsites occur and are not resolved through increased educational or site restoration actions in areas of established site camping, then the affected areas could be switched to the designated site camping option. If such problems persist in areas of designated site camping, then enhanced education and restoration can be followed by increased law enforcement, and ultimately by rationing use (Marion et al. 2023).

Cole's hypothetical campsite impact modeling work (1992) demonstrates that the degree to which activities are spatially concentrated is the factor that most influences the amount of vegetation loss on campsites. At high levels of spatial concentration, that occurs on campsites surrounded by sloping, uneven, or rocky terrain, neither the amount of use a site receives nor the fragility of the vegetation have much influence on the amount of vegetation loss (Arredondo et al. 2021). Activity concentration is the factor explaining why campsite expansion pressure declines with increasing levels of site use—there is a natural tendency to concentrate activity on the already denuded parts of large, well-impacted campsites. The opposite is true when dispersed natural site camping occurs on vegetation, where vegetation loss is strongly influenced by amount of use, vegetation fragility, and the extent to which camping activities are dispersed. For this camping option, when camping cannot be focused on durable non-vegetated surfaces, dispersing activities on resistant and resilient grasses results in substantially less vegetative loss than on fragile forested broadleaf forbs.

The implications of these findings are that under a Containment strategy, any actions that managers can take to increase the spatial concentration of visitor activities will effectively reduce the areal extent of camping impact and vegetation loss. Installation of site facilities provides an effective option to pull visitors to core campsite areas and constrain site expansion pressures, particularly in flat terrain where site expansion is a chronic problem (Marion et al. 2020b; Marion et al. 2023). Managers can spatially concentrate cooking, eating, and social activities to a single location by providing small permanent fire rings with ice-berged (implanted) rocks and/or "kitchen rocks" for stoves (Fig. 15.11); use of large

FIGURE 15.11 In the Daniel Boone National Forest, Kentucky, volunteers in a Leave No Trace Master Educator course created a kitchen rock (left) and a small fire ring with ice-berged border rocks (center) to attract and concentrate cooking and social activities on campsites, with an alternate design shown on the right.

and embedded rocks will minimize their movement to other locations by visitors and identify the "preferred" single location for these primitive facilities to visitors, agency staff, and volunteers (Reid and Marion 2005). Any additional fire sites should be removed and restored to focus use and impact at the preferred location—allowing these locations to migrate around over time unnecessarily enlarges campsites. When these actions are ineffective, anchored steel fire rings have been used to denote a designated campsite and/or campfires can be restricted to steel rings to protect against wildfires (Fig. 15.12a). Both are true in the Boundary Waters (Fig. 15.12b), while in other areas, like the Chisos Mountains of Big Bend National Park, Texas, wooden camping posts designate campsites and attract and spatially concentrate use to their vicinity.

The installation of such facilities to protect wilderness resources and for visitor safety can be justified when lesser actions have been ineffective, though not for visitor comfort and convenience. Trail networks exemplify an almost universally accepted infrastructure "facility" that serves to attract and concentrate impact to a narrow highly resistant tread. Similarly, an infrastructure of carefully selected campsites with optimal tenting sites and anchored fire or cooking sites serve the identical function of attracting and concentrating intensive use to minimize resource and social impact.

Marion (1995) documents the effects of management actions taken to increase the concentration of use at two spatial scales on canoe campsites at Delaware Water Gap over a five-year period. Intersite concentration of use was increased by reducing the number of designated sites 25 percent. Managers increased the intrasite concentration of use by installing anchored fire grates on each site. This facility attracted and concentrated activities to its vicinity more than in previous years when visitors built multiple campfire sites that moved around campsites over time. As a result, mean campsite size decreased more than 30 percent. These combined actions reduced the aggregate area of campsite disturbance 50 percent—from 6.9 acres in 1986 to 3.5 acres in 1991. Although these actions increased the intensity of use on individual campsites, there was no resultant increase in intensity of impact on individual sites (Marion and Cole 1996), consistent with the asymptotic use-impact relationship (refer to Fig. 15.9).

Other beneficial facilities include pit or composting toilets, which become more essential as site use levels increase, food storage facilities and designated eating areas to separate eating and tenting areas and avoid dangerous wildlife food attraction behaviors, and visually obvious tent pads to constrain site sizes when intensive camping must occur in flat terrain (Fig. 15.12) (Marion et al. 2023, 2020b). Managers can use monitoring data and adaptive management decision-making to determine if such facilities are needed or justified as a minimum tool to effectively limit resource or wildlife impacts. Evaluation of trade-offs may support such decision-making, for example, providing food storage or toilet facilities versus requiring visitors to carry bear-resistant food canisters or to pack out their waste.

Modify Visitor Behavior

Visitor behavior can influence the type and extent of both resource and social impacts, particularly related to uninformed, unskilled, or careless behaviors (Manning 2003). Persuasive

FIGURE 15.12 Campfires can be restricted to anchored steel fire rings, which can also be used to denote a designated site (a [top left], b [top center]). These facilities attract and spatially concentrate cooking, eating, and social activities to preferred durable locations, reducing campsite sizes. Seating logs (c [top right]) and stove rocks (b) also help attract and concentrate use. Toilets like this durable fiberglass model (d [bottom left]) are often needed on high use designated campsites, and steel food storage boxes (e [bottom center]) can effectively reduce wildlife food attraction behaviors, trampling disturbance, and tree damage. Visitor and wildlife safety can be enhanced by providing designated food cooking/eating areas, and tent pads (f [bottom right]) can effectively constrain campsite sizes when camping must be accommodated in flat terrain.

communication is a preferred and effective indirect management tool for both informing visitors about the consequences of their actions and of low impact outdoor practices and ethics that can avoid or minimize resource and social impacts (Kidd et al. 2015; Lawhon et al. 2019; Marion 2014, 2016).

Managers can alternately employ direct regulatory measures to require low impact practices or equipment (e.g., hanging bear-bags or carrying canisters to protect food, trash, and smellables), limit certain types or amounts of use (e.g., trailhead quotas or restrictions on horse use), or prohibit high impact practices or equipment (e.g., littering, felling woody vegetation, carrying woods tools). Regulations directly curtail visitor freedoms, though they generally require a high level of enforcement to ensure their efficacy. Regulatory options may be needed to address illegal behaviors (e.g., theft of archaeological artifacts, hunting out of season) or critical high impact behaviors for which voluntary educational interventions were insufficiently effective (e.g., building a campfire during periods of high wildfire danger). For example, three decades of employing a strong Leave No Trace educational program at the Boundary Waters Canoe Area Wilderness has failed to prevent the felling of trees for campfire wood, as evidenced by an average loss of 160 trees/acre on 81 campsites assessed in 1982 and 2014, and 1,438 tree stumps located on or near these campsites (18/site) (Eagleston and Marion 2017b). These findings prompted the authors to suggest prohibiting woods tools (axes, hatchets, saws), as this might more effectively curtail the loss of large wood and the scarring and felling of trees. A meta-analysis of campfire-related impacts and their management from seven protected areas found that restrictive campfire policies and bans did not effectively reduce campfire-related tree damage (Reid and Marion 2005). Regardless, regulations have the potential to antagonize visitors rather than enlist their voluntary support and enforcement is often costly and difficult to sustain.

Disposal of human waste in individual "cat-holes" is recommended where soil is available and use levels are not too high (Marion 2014). Research in Montana's Bridger Range has shown that significant numbers of intestinal pathogens in feces survived an entire year of burial (Temple et al. 1982). Depth of burial (two to eight inches) did not influence pathogen survival; neither did it matter whether disposal occurred at high or low elevations, in forest or meadow. The possibility of disease transmission is long-lasting, emphasizing the need for deep burial away from trails, campsites, and water.

Cat-holes are ineffective in settings with particularly low rates of decomposition, including deserts, wetlands, and areas that lack soil or have prolonged cold temperatures. Such areas are more appropriate for toilets or "carry-out" policies using waste aggregation and gelling (WAG) bags, portable toilets (rafters/horse packers), or other means (Marion 2014). The types of toilets found in wilderness include composting, pit, and above ground moldering systems common along the Appalachian Trail. For a comprehensive review of options see ATC (2014) and Civil and McNamara (2000).

Close and Rehabilitate the Resource

A final strategy in the management toolbox is the closure and recovery to natural conditions of campsites or trails that are unnecessary, poorly located (e.g., in non-sustainable locations), or have unacceptable conditions. However, before seeking to close and restore campsites and trails managers may recognize and consider two critical challenges: (1) closure should generally not be considered unless use is shifted from impact-susceptible to impact-resistant locations, although social considerations (crowding, conflict, or visitor safety) may also provide justification, and (2) impact rates greatly exceed recovery rates; recurring use can halt resource recovery unless the unwanted campsites are effectively closed (Cole & Ranz 1983; Therrell et al. 2006).

It's important to emphasize that temporary closures as part of a campsite rest-rotation system are unlikely to be successful because several years of slow recovery can be quickly reversed by relatively low levels of subsequent use. The effectiveness of a temporary campsite-closure program was monitored around Big Creek Lake in the Selway-Bitterroot Wilderness, Montana, where seven out of fifteen campsites were closed to allow recovery. Eight years after closure, vegetation cover on closed sites was still only a third that of offsite controls, and mineral soil exposure was 25 percent compared with only 0.1 percent on controls (Cole and Ranz 1983). The most profound change after the closures was that visitors created seven new campsites close to the closed sites, on which conditions rapidly deteriorated. Within eight years of their creation, loss of vegetation and soil exposure were as high on new sites as on long-established sites.

The effective implementation of a dispersal strategy is greatly enhanced by the capability to quickly identify, close, and restore emerging campsites before they become well-established. This is also important for a containment strategy, which shifts use from campsites that are unnecessary or located in poor or impact-susceptible locations, to more optimal and sustainable locations. Sometimes large numbers of existing non-selected campsites need to be effectively closed and restored. Thus, both strategies rely on staff or volunteer capabilities to close and restore emerging and well-established campsites.

The closure and recovery of new or lightly impacted campsites in areas managed for dispersed Natural site camping is comparatively easy compared to closures of well-established campsites in high use travel zones. Patrols are most effective when conducted soon after peak use periods to locate and naturalize newly pioneered sites so they will not be found and reused. Relocating formal trails to avoid accessing particularly attractive flat areas with large numbers of campsites that require closure is a "forever" solution that successfully resolved unacceptable campsite expansion and proliferation impacts at Slaughter Gap in Georgia's Blood Mountain Wilderness Area. Alternative camping was provided at several new side-hill campsite locations, allowing complete recovery of the original camping areas (Marion et al. 2020b). Cole and Ferguson (2009) describe how trail relocation, education, campsite closure, and restoration in the Caney Creek Wilderness of Arkansas successfully reduced campsite numbers 40 percent, from 91 in 1994 to 54 in 2007.

Substantial restoration work on recently closed sites is best avoided until staff can confirm that closure actions have effectively halted further visitor use; continued trampling can negate staff-intensive restoration work and expensive plant stock (Therrell et al. 2006). Applying a phased process is the most pragmatic approach. ***Phase 1*** activity includes efficient naturalization work requiring a few staff 10–20 minutes, such as dragging large woody debris and spreading organic litter across the site (Reid & Marion 2004). When possible, a few larger rocks and partially rotted logs placed across tenting areas or felling/moving large dead trees across the site are even more helpful. The objective is to naturalize and hide the site, and to obstruct the best tenting areas with logs or rocks that are not easily moved. Monitoring the efficacy of this work, particularly after busy or peak use weekends is helpful, with reapplication to quickly restore sites that are reused. A phone app can be used to efficiently track the locations, actions, and timing of such activity and their efficacy over time.

If repeated Phase 1 work is ineffective, then more intensive ***Phase 2*** work may be necessary and justified, including actions such as encircling the site with nylon cord as symbolic fencing to deter entry, posting temporary site closure signs, and site ruination and restoration practices such as ice-berging large rocks and logs or digging shallow depressions and mounding soil in flat tenting areas. Felling or moving large trees onto the site is very effective. These practices focus on deterring continued visitor use but still rely on the slow process of unassisted vegetative recovery. Recovery rates can be accelerated with a variety of assisted recovery practices, including scarification (loosening compacted soils), organic mulching, transplanting, seeding, and water/fertilization (Cole 2013b; Cole et al. 2012; Cole & Spildie 2007; Willard et al. 2007; Zabinski et al. 2002). Although commonly applied to restore closed campsites, some of these techniques can also be used on open campsites to close unnecessary portions and reduce their size (Marion & Sober 1987). Communicating alternative locations to visitors and providing educational messages that convey compelling reasons for the closures can also be helpful. Enhanced visitor patrols, contact, and enforcement activity of site closures may also be necessary for particularly challenging areas, though moving formal trails away from the problem area is generally the most effective long-term solution.

If sites are effectively closed by the activities described above, then managers may elect to allow natural unassisted recovery to occur over time (Therrell et al. 2006). Recovery rates are dependent on many factors, including length of growing season, soil texture, fertility, moisture, sunlight penetration, elevation, and size of the disturbed area and severity of disturbance (Reid & Marion 2004; Cole 2013b). For example, recovery rates on large highly impacted campsites can require decades in subalpine and alpine ecosystems because of short growing seasons, poor substrates, and low rates of plant establishment and growth (Cole 2013b; Moritsch and Muir 1993; Scherrer & Pickering 2006; Willard et al. 2007; Zabinski et al. 2002). In contrast, Marion and Cole (1996) found substantial unassisted vegetative recovery of well-established campsites in just five years in a Pennsylvania riparian floodplain.

Site rehabilitation is an appropriate means for correcting past abuses, but it should be used judiciously. In addition to being costly, it interjects horticulture and landscaping into the wilderness. *As a general principle, rehabilitation should be used to restore wilderness campsites that can be well-protected through a new management program. Clearly, it is much better to avoid damage than to try to fix it after it occurs.*

MONITORING CAMPSITE CONDITIONS

Effective campsite management requires data on the number, location, sustainability, and condition of campsites and trends over time. Periodic assessments to monitor changing campsite numbers and conditions are essential to adaptive management programs that seek to balance the provision of recreation use with the negative consequences of resource and social impacts. These same protocols can also be applied to day use recreation sites, or sites closed to recovery, if desired. Well defined and implemented monitoring programs provide baseline data and periodic updates over time for assessing the efficacy of impact management strategies and actions to assess progress in accomplishing management objectives. Monitoring data also permit comparison of current conditions to indicator thresholds/standards of quality to objectively evaluate the acceptability of changes in resource and social conditions within VUM frameworks. Finally, monitoring data may also be analyzed to identify the most influential factors that managers can manipulate to accommodate increasing visitation while minimizing negative impacts (Marion 2025).

A variety of different campsite monitoring protocols have been developed. Early systems (Cole 1989b; Marion 1991) have been replaced with newer protocols and with data efficiently recorded in the field using smartphone data-entry apps (Marion 2021, 2025). Techniques vary in the quantity, quality, and cost/efficiency of information produced. Techniques that produce data that are *accurate* (close to the "true" value) and *precise* (repeated separate measures that are similar) for many impact indicators generally are less *efficient* (require more time) than those that produce summary assessments or less precise and accurate data, requiring trade-offs between these three attributes. Four standard options include photographs, condition classes, ratings, and measurements, along with combinations of these.

Photographs

Digital photos are a common monitoring option that can help verify campsite identity between monitoring cycles, visually document site conditions to supplement data, and illustrate site attractiveness, features, and environmental/topographic context. Unfortunately, photo quality can vary based on the camera, lighting, and field staff skill. However, substantial improvements in phone cameras now provide high resolution digital photos to be linked to monitoring data and transferred and stored efficiently. Experimentation with UAV or pole-mounted cameras and structure-from-motion or LiDAR digital elevation model data collection is possible and may offer benefits (Wang and Watanabe 2022), but such methods remain beyond the core monitoring capabilities of many land management agencies.

Condition Classes

Subjectively assessed descriptive campsite condition class ratings have long been a traditional option used to track changes in general site conditions (Fig. 15.13). No information on campsite size or other types of impact are provided. Such ratings are efficiently applied but yield ordinal scale ratings that can be ranked, but the differences between values are not the same and ratings should not be averaged. For example, a site rated class 4 should not be considered *twice* as degraded as a class 2 site, and a 1500 sq ft class 3 campsite is generally considered to represent greater impact than a 500 sq ft class 4 campsite. This method also overestimates impact in areas with little natural vegetative cover or organic litter.

Rock 0	Site is predominantly on rock surfaces; the effects of trampling are difficult to see or assess.
Class 1	Site barely distinguishable; slight loss of vegetation cover and/or minimal disturbance of organic litter.
Class 2	Site obvious; vegetation cover lost and/or organic litter pulverized in primary use areas.
Class 3	Vegetation cover lost and/or organic litter pulverized on much of the site, some bare soil exposed in primary use areas.
Class 4	Nearly complete or total loss of vegetation cover and organic litter, bare soil widespread.
Class 5	Soil erosion obvious, as indicated by exposed tree roots and rocks and/or gullying.

FIGURE 15.13 Descriptive condition class ratings based on increasing levels of change in vegetation, organic litter, and soil.

Ratings

This option generally allows for three to five ratings for each of six to twelve indicators of campsite condition, with rating descriptors based on increasing levels of impact for each indicator. Common indicators include site size, loss of vegetation cover, increased soil exposure, damaged trees, felled trees, trees with root exposure, site development, informal trails, and litter. A typical indicator like site size with three ratings illustrates this option but reveals several deficiencies.

Site Size: (1) Site area less than 400 sq ft, (2) Site area = 401–800 sq ft, (3) Site area greater than 800 sq ft.

A trade-off exists between having fewer ratings, making the rating process easy but yielding less accurate data, or having more rating categories, increasing the difficulty of selecting accurate ratings unless site measurements are made, and reducing assessment efficiency. Experience also reveals that managers often do not calibrate rating categories to the distribution of values for each indicator (Marion 1991). For example, if 70 percent of all campsites fall into a single rating category then it is defined so broadly as to be insensitive to alert managers to changing conditions. Even if rating categories are initially calibrated to include equal percentages of campsites, change over time will cause imbalances, but recalibration would prevent comparisons to baseline data. Another deficiency is that the final rating category must encompass an unlimited upper range, e.g., site sizes greater than 800 sq ft, a significant deficiency as a 3x or 5x increase in site size would not be documented and conveyed to decision-makers. Finally, ratings are also ordinal data with the same analytical deficiencies described for condition classes.

Measurements

When possible, a more optimal approach is to employ counts and measurements for each indicator using a ratio scale with equal distance between values and a true zero point. This option generally increases assessment accuracy (closeness to the "true" value) and precision (how close repeated separate measures are to each other) but decreases efficiency (indicator assessment time). However, experience indicates that travel time to and from campsites greatly exceeds assessment times at each location, suggesting that increased assessment time for counts and measurements is generally inconsequential compared to the higher utility of measurements. Walking around a campsite to evaluate the level of tree damage to yield a count instead of a rating requires little additional time; however, measuring campsite sizes is an important but necessary exception. Areal measures of the extent of camping impact are considered the most salient of indicators as a core resource protection goal is to minimize the spatial "footprint" of visitor impact (Cole 1989a; Marion 1995, 2016; Marion et al. 2018).

The most efficient and accurate collection of campsite size and location is to walk site boundaries with a submeter GPS unit using differential correction post processing (Marion 2024). In areas with narrow or high canyon walls or dense tree cover a preferred option is the variable radial transect method, though the most efficient option is the geometric figure method (Marion 1991, 1995). Regardless, all measurements contain an element of measurement error, though protocols can be applied to estimate error and determine a "minimum detectable change" for indicators (Cole 1989b; Williams and Marion 1995). Developing and applying measurement-based protocols in a comprehensive monitoring manual with careful staff training and quality assurance protocols can maximize the ability to measure actual changes in resource conditions and minimize measurement error.

Marion (2025) provides monitoring guidance and a library of "best practice" protocols that include common inventory, sustainability assessment, site preference, and impact indicators for wilderness area recreation sites and campsites, developed for VUM adaptive management decision-making. This book supplements the Interagency VUM Council's Monitoring Guidebook (IVUMC 2019); staff can select and include only those indicators they deem necessary. Inventory indicators include attributes like site location, assessment date, unique site number, GPS coordinates, site designation, and use type. Campsite sustainability attributes include a site's expansion potential as constrained by sloping terrain, rugosity (rockiness, uneven terrain), and wet substrates; and tree canopy cover (the best sites are very sunny or shady). Site preference indicators include locational attributes best derived from GIS procedures (distance to formal trail, water, and nearest campsite), intersite visibility, usable tent pads, and various campsite functionality attributes (e.g., ease of water access, site developments, dead/hazard trees, and scenic attractiveness). These attributes can be used to select campsites that will be sustainable and highly preferred by visitors and managers for high use established or designated campsites. Impact indicators include digital photos, condition class, and numerous resource condition indicators (site size, on- and off-site vegetation cover and exposed soil, tree damage and root exposure, stumps, access trails, fire sites, and human waste, including computational guidance to calculate seven additional indicators).

Data entry smartphone apps are a highly efficient means for collecting and storing site inventory and monitoring data, with automatic transfer to "cloud storage" when there is cellular or Wi-Fi connectivity (Marion 2025). Digital field data entry forms with full guidance "lookups" can be developed that prevent missing data, error check data values, allow dictated comments, take and link digital photos, and collect accurate GPS points through access to five GNSS constellations of GPS satellites. Carrying back-up batteries to recharge phones at night allows monitoring on multi-day trips. The additional use of sub-meter accuracy high-end GPS units improves locational accuracies and efficiencies in collecting site size data. Reference data and photos from prior monitoring cycles can be referenced on smartphones to aid in recording different values when there is "compelling" evidence of change (Marion 2025).

Regardless of the technique employed, wilderness campsites should be inventoried periodically, perhaps every five years or when changes become evident. This will show whether the number of sites is increasing or decreasing and whether conditions on individual sites are improving or deteriorating. Standards/thresholds are employed as part of an adaptive management VUM program to identify unacceptable change and initiate a problem analysis process to select and implement corrective actions, with success evaluated through future monitoring.

MANAGING TRAILS

Impacts on and along trails result from the trampling of hikers and pack stock and the effects of rain, freeze/thaw, and wind as mediated by trail design, construction, and maintenance. As previously discussed, *these impacts include progressive changes in the loss of vegetation and organic litter cover and soil, plant composition, exposure of mineral soil, soil loss from compaction, displacement, and water/wind erosion, and changes in trail corridor habitats, primarily from increased sunlight and altered drainage.*

Sustainable trails are designed, constructed, and managed to accommodate their intended types, amounts, and seasons of use to provide high quality visitor experiences while protecting the trail infrastructure and adjacent natural resources (Marion and Wimpey 2017). The design, construction, and maintenance of sustainable trails minimizes most trail impacts and accommodates intended types and amounts of use with acceptable levels of trail impact. Unfortunately, many wilderness trails were originally constructed for non-recreational purposes and these "legacy" trails can fail to sustain intensive recreational traffic (Marion 2023). Some legacy trails date back to indigenous peoples and settlement periods, while others include logging, firefighting, administrative, agriculture, and visitor-created alignments. Severe trail resource impacts are frequently attributable to sustainability deficiencies than to recreational use. Consider any longer wilderness trail that receives the same type, amount, and season of use— while most of the tread is likely in good condition, short sections may be excessively wide, eroded, or muddy (Marion 2023). As in camping management, the most effective long-term solutions generally involve improving infrastructure location and sustainability and concentrating visitation on it—a containment strategy. And similarly, limiting use is an ineffective practice, though shifting heavy stock traffic to separate and more sustainably developed and hardened trails is an effective practice to accommodate their considerably higher ground pressure and alleviate possible social conflicts.

TRAIL IMPACTS

Trail erosion and muddiness make trails difficult to use while widening increases the infrastructure "footprint" of impact; all three are most effectively resolved through site management actions such as reroutes and tread maintenance, only tread widening is significantly related to amount or type of use. Informal (visitor-created) trails result from visitors "voting with their feet" to circumvent tread problems or reach desired locations not accessed by formal trails, such as campsites and waterbodies. Severe trail impacts are objectionable because they degrade the transportation and recreational utility of trails, increase hiking difficulty, and threaten visitor safety and resource protection objectives while reducing aesthetic qualities, visitor satisfaction, and the socio-economic benefits of trails and wilderness (Leung and Marion 1996; Marion 2023).

Erosion

Soil loss measurements on trails often largely document soil erosion from water drainage, but also soil displacement, soil compaction, and wind erosion in dry climates and seasons when traffic abrades and loosens soil particles. Soil loss on trails is often considered the most significant long-term form of recreational impact because: (1) soil creation requires many thousands of years, (2) secondary impacts to vegetation, water quality, and aquatic flora and fauna can occur from eroded soil, (3) eroded treads increase hiking difficulty and reduce hiking enjoyment and safety, and (4) eroded treads are more challenging to drain and repair (Marion and Wimpey 2017; Meadema et al. 2020). Eroded treads often create increased rugosity and tread unevenness in the form of deep erosional ruts and exposed roots and rocks that can contribute to trail widening and informal trail creation as visitors seek to circumvent all three tread impacts (Leung and Marion 1996).

Cole's longitudinal study of trails in the Selway-Bitterroot Wilderness, Montana, revealed the occurrence of modest erosion but soil deposition along trails was more common (Cole 1991). Soil eroded from treads or onto trails from higher side-slopes is often deposited elsewhere on the trail, with soil loss occurring only where water drains along and off the trail in larger volumes. Trail stewards work to keep trails "hydrologically invisible" through various tread drainage actions that minimize the amount of water captured and diverted down trails. Soil loss occurs most frequently in less sustainably fall-aligned or steep segments that make maintenance difficult, ineffective, or costly.

Data from a large representative study of the Appalachian Trail (AT) with measurements taken at 3,150 transects illustrate the nature and severity of soil loss (Marion et al. 2020a). Soil loss is most efficiently assessed with maximum incision measures as the largest perpendicular distance between a transect tape positioned to reflect the post-construction, pre-use tread surface and the current tread surface. Soil loss is more accurately assessed with the cross-sectional area indicator that calculates aggregate planar soil loss within tread boundaries from the transect tape to the tread surface, generally from perpendicular measures taken at a fixed interval of two to five inches.

Muddiness

Tread muddiness is not a typical problem, though it can be common in wet seasons, during snowmelt, in areas with high rainfall, and in low flat areas with wet and organic soils (Meadema et al. 2019). *Muddiness is particularly disturbing to hikers and pack stock users, and its presence frequently causes trail users to widen trails when they circumvent deep mud and standing water.* It's generally the trail steward's responsibility to provide visitors with usable trails; both research and management experience reveal that the best solution to muddiness is avoidance of low areas with wet soils, or incorporation of sustainable tread design, maintenance, and tread drainage practices to traverse wet areas while keeping treads dry (Hesselbarth et al. 2007; Marion and Leung 2004; Meadema et al. 2020).

Trail Widening

Trail widening is a salient impact indicator because managers seek to minimize the "footprint" of visitor impact; keeping trails no wider than necessary to accommodate intended trail use minimizes the aggregate areal extent of trail impact. Trails with smooth, dry, "usable" treads, absent steep grades, and steps generally remain narrow in width; eroded trails with ruts, exposed roots and rocks, and muddiness frequently become excessively wide as visitors move laterally to select the easiest route of travel (Cole 1991; Tomczyk et al. 2016; Wimpey and Marion 2010). Trail widening can also be caused by visitors moving laterally to pass others or to allow side-by-side travel (Leung and Marion 1996, Marion 2023). Lateral movement is also easy on fall-line trails and those in flat terrain, both are inherently susceptible to trail widening (Meadema et al. 2020; Tomczyk et al. 2016, 2017).

Tread width is generally assessed as the width of trail transects between stakes placed at the outer boundaries of visually obvious disturbance to vegetation, organic litter, and substrates from trail traffic. In the AT study, median trail width was 22.2 in with a range of 0 to 197 in; 59 percent of the AT is less than 2 feet in width while 7 percent exceeds 4 feet. The widest segments were associated with the use of woods roads, eroded steep fall-line segments, or flat terrain and high use (Marion et al. 2020).

Informal Trails

Informal trails (ITs) are defined as visitor-created trails created to access locations generally not serviced by formal trails (e.g., campsites, water), though some ITs are created as undesirable shortcuts or are duplicative and parallel to other trails. ITs result when visitors "vote with their feet" and leave the formal trail system out of a desire to reach a non-accessed destination or dissatisfaction with a trail's conditions or route. While at least some ITs are a necessary component of trail infrastructure systems, they also degrade and remove vegetation, displace wildlife, alter hydrology and habitats, disperse non-native plants, and fragment landscapes (Barros et al. 2013; Monz et al. 2010; Wimpey and Marion 2011). Research has demonstrated that ITs are less sustainably designed than formal trails and most are not maintained to facilitate use or avoid degradation. For example, many ITs are created as the shortest route to a desired destination and include fall-line alignments, often on steep slopes and following the "path of least resistance" to avoid dense vegetation or challenging terrain. They can be created from low levels of off-trail traffic and managers are frequently challenged to deter their use or restore them to natural conditions (Cole et al. 1997; Hockett et al. 2017; Park et al. 2022).

SUSTAINABLE TRAIL DESIGN AND MANAGEMENT

While many trail science studies document trail conditions and impacts, some conduct relational analyses to identify the influence of various environmental, managerial, and use-related factors. Knowledge of the most influential factors is used to develop best management practices for designing, constructing, and managing "sustainable trails," defined as trails able to accommodate intended types, amounts, and seasons of use to provide high quality visitor experiences while protecting the trail infrastructure and adjacent natural resources (Marion and Wimpey 2017).

As an example, relational analyses of soil loss on trails have identified several key factors, including a trail's grade and alignment angle relative to prevailing slopes (Leung and Marion 1996; Marion and Wimpey 2017; Marion 2023). *Trail soil loss increases exponentially with increasing trail grade, while trails aligned with the "fall line," perpendicular to contour lines, can be nearly impossible to drain water from.* Climatic factors like amount of rainfall and snowmelt, trail substrates (particularly the amount of rock), density and efficacy of tread drainage features, and type, amount, and season of recreational traffic are also influential factors (Marion and Wimpey 2017; Meadema et al. 2020).

These findings indicate that soil loss can be minimized by avoiding steep trail grades, generally those exceeding 10 percent, unless trail substrates have substantial amounts of rock, rock steps, and/or exceptionally effective drainage (Meadema et al. 2020; Marion 2023) (Fig. 15.14). *Trail stewards find it easiest to drain water from side-hill trails angled across steeper terrain, avoiding alignments close to fall-lines that are difficult to drain, and contour lines, where water can pool on flat treads.* A trail's alignment angle to the prevailing landform grade has been termed its trail slope alignment, the smallest angle between the azimuths of the trail and landform fall-line, measured in degrees ranging from 0 (fall-line) to 90 degrees (contour line) (Leung and Marion 1996; Marion and Wimpey 2017). The most sustainable trails have side-hill slope alignments of about 23 to 80 degrees (Fig. 15.14). Trail practitioners prefer to use trail slope ratio, calculated by dividing the landform grade by trail grade and ranging from 0 (contour line) to 1 (fall-line), and applying the "Half Rule" guidance to avoid alignments closer to the fall-line with slope ratio values greater than 0.50 (IMBA, 2004; Meadema et al. 2020). The influence of trail slope alignment or slope ratio on soil loss is low below landform grades of 10 percent but increases substantially with increasing trail and landform grades (Olive and Marion 2009; Marion and Wimpey 2017; Meadema et al. 2020). Even on sustainably designed side-hill trails constructed with out-sloped treads, tread compaction and berm formation below treads in the first few years of use often allow water to be intercepted and channeled down treads, hence the perennial need for effective tread drainage.

Trail widening is also influenced by these trail alignment attributes; as landform grades become steeper, fall-line trails become significantly wider while side-hill trails become narrower (Wimpey and Marion 2010). *Shifting trails to side-hill alignments in landform grades exceeding 10 percent is always the most optimal permanent action to constrain trail widening.* Trails in flatter terrain are far less sustainable for two core reasons: (1) the loss of organic materials and soil leaves their treads below the ground surface making them prone to muddiness and rutting, and (2) the ease of lateral traffic movement promotes trail widening and the formation of secondary parallel trails. Use-related factors that widen trails include heavy use, particularly when visitors frequently pass each other or choose to travel side-by-side, or type of use, ranging from narrow hiking trails to wider horse and stock trails (Wimpey and Marion 2010). Tread condition factors that widen trails include muddiness, erosion that causes deep ruts or exposed

FIGURE 15.14 Constructing or relocating trails to side-hill alignments in steeper terrain (a [upper left]) is the most sustainable long-term solution to most trail impacts. If steeper trail grades are required, then rock steps are necessary to prevent soil displacement and erosion (b [upper center]). Trails supporting heavy pack stock traffic must often be armored with rock (c [upper right], d [lower left]). Fixes for trail widening and multiple trail development often employ rock, wood, and plantings to define preferred treads and encourage vegetative restoration (e [lower center], f [lower right]).

rocks and roots, steep trail grades and fall-line alignments, and uneven or naturally rocky treads.

Visitors generally select the easiest and smoothest walking surfaces, moving laterally around mud or exposed ruts, rocks, and roots. Trail stewards have developed numerous trail management practices to avoid this tendency and focus traffic on narrow treads (Wimpey and Marion 2010). The "push" and "pull" actions described for resolving campsite impacts can also be applied to trails. Pull actions attract and concentrate traffic on formal treads by making them the most appealing alignment. Stepping stones, boardwalks, or puncheons facilitate a dry traverse of seasonally or permanently wet soils. Trail relocations avoid sections of unsustainable or severely eroded treads that cannot be easily corrected. Alternately, trail stewards can add rocks and mineral soil to fill ruts and/or cover exposed roots/rocks to attract visitors back to a smooth, usable tread; on side-hill trails these materials are most easily obtained from upslope trailsides (Marion 2023).

Push actions deter off-trail traffic by making any inviting trailsides less desirable than hiking on the tread (Wimpey and Marion 2010). If steps and water bars are installed the adjacent off-site areas must be made less desirable for walking with large rocks. Traffic can be "recentered" on narrow treads by occasionally ice-berging large trailside rocks or cutting out narrow sections of trees that fall across trails, or even moving logs into perpendicular trailside positions (Fig. 15.14e). Jumbled rocks in trailside areas or rock scree walls have been effective in protecting sensitive subalpine and alpine vegetation; low symbolic fences of nylon cord are even more effective, though generally appropriate only as a temporary solution in wilderness (Park et al. 2022).

Additional push actions include the careful trimming of woody trailside trees and shrubs to create a dense "wall" of vegetation that centers traffic; in open forests and meadows, grasses and sedges resistant and resilient to trampling will colonize trailsides and resist trail widening pressures (Marion 2023). Trail borders of parallel logs are effective but only if they can be installed without impeding tread water drainage. Reliance on logs or woody branches is also less desirable as they rot or burn in wildfires, whereas rocks are permanent. Educational messaging can emphasize the need for remaining on formal trail treads to minimize trail widening and impacts to sensitive off-trail vegetation or fragile cryptogamic soil crusts (Park et al. 2022).

Use-related factors, the type, amount, behavior, and season of visitor use represent a final group of influential factors that managers may consider when balancing the provision of visitor use with resource protection mandates. *Amount and type of use are generally less important in causing erosion than trail location and design features.* Several studies have found that trails were not substantially deeper where use levels were higher (Cole 1991; Marion and Leung 2001). On a sustainable moderate use trail the impact of doubling traffic would result in little additional resource impact—at this point the sustainability of a trail's alignment and maintenance are substantially more important determinants of degradation than amount of use (Marion 2016, 2023).

Whether trail use is by hikers or by groups with pack and saddle stock is an important indicator of potential erosion. A small bearing surface carrying heavy weight, a horse's hoof can generate pressures of up to 1,500 pounds per square inch. These pressures, along with sharp metal shoes, cause stock to deform, break up, and displace tread substrates while compacting subsurface soils, making them less permeable to water. Detached soil is more easily eroded and makes trails dustier when dry and muddier when wet. In an experimental study, DeLuca et al. (1998) found that horse traffic resulted in much higher sediment yields (an indicator of erosion potential) from established trails than either hikers or llamas. Use of stock on well-used trails designed and maintained for their use is generally sustainable (Fig. 15.14c,d). However, *on little-used trails that are steep, pass through wet meadows, and are seldom maintained, stock use can be much more damaging than hiker use.*

Season of use and visitor behavior can also be influential, asking visitors to use the most sustainable trails during rainy or snowmelt seasons. In general, it's best not to relocate a trail unless (1) the new section is in a more resistant environment or is better designed, and (2) hikers can be kept off the old section of trail. Unless both these conditions are met, relocation may only compound problems by disturbing new sites.

In summary, sustainable trails employ side-hill alignments in sloping terrain with low (3 to 10 percent) grades, trail slope alignments of 23 to 80 degrees, appropriate densities of grade reversals or effectively maintained tread drainage features, and other features (e.g., board walks, rock steps, armored rockwork) needed to protect treads from soil displacement, erosion, widening, and muddiness. Wilderness trails intentionally managed as primitive and challenging should also be sustainably managed, their difficulty should be tied to alignments through steep but rocky terrain rather than degraded tread conditions. Additionally, a greater emphasis on sustainable trail alignments and maintenance practices are necessary to address climate change, including substantially higher or lower precipitation and large rainfall and flooding events, and an increased frequency and intensity of forest fires, hurricanes, tornadoes, and derecho events. Limiting amount of trail use is less effective than restricting a trail's type or season of use.

Trail Maintenance

Sustainable trail design and construction substantially reduces long-term trail degradation and the intensity and frequency of tread maintenance. For example, trails with lower grades lessen both soil erosion and displacement and side-hill trail alignments always facilitate tread water drainage to lower slopes, while steeper trail side-slopes permanently deter trail widening. Thus, less sustainable alignments close to the fall-line require either relocation, more frequent and intensive tread maintenance, or acceptance of inevitable tread degradation in the form of tread soil loss and widening. Intensive tread maintenance in the form of a constructed rock staircase may still require frequent maintenance if built in the same eroded fall-line tread; water, particularly weathering from freeze/thaw cycles, will continue to degrade the rockwork. Similarly, trails in flat terrain are less sustainable and require persistent long-term maintenance to address tread incision and rutting that attracts water, creates muddiness, and/or causes widening. Most areas are not completely flat and GIS slope maps can assist in optimally routing trails to whatever side-hill terrain does exist.

Nevertheless, trail maintenance is an essential but sometimes neglected component of trail infrastructure management. Core trail maintenance practices are described in Hesselbarth et al. (2007), Birkby (2006), and many other guidance books and agency guidance documents (see Marion 2023). Surprisingly, the current guidance is experientially based—relevant research has rarely been conducted. Trail maintenance consists of two somewhat related components: vegetation management and tread management. Our focus here is predominantly on tread management to deter trail degradation. Vegetation management includes the removal of trees and limbs that fall within trail corridors, and the trimming of woody and taller herbaceous vegetation that encroach on trailways to impede traffic, generally with clearing widths and heights of 6 × 8 feet for hiking trails and 8 × 10 feet for pack stock trails. Exceptions may include removing narrower sections of fallen trees to "center" traffic and allowing closer growth on lower trailsides to discourage traffic and tread failures on lower, weaker trailsides. Trees and shrubs growing within the specified trailway dimensions are generally removed, including seedlings, as they require more intensive annual trimming.

Tread maintenance is primarily about avoiding water from flowing or standing on trail treads, though tread armoring with rock, particularly on steep trail grades, may also be essential. Tread drainage begins by ensuring that all permanent and intermittent water drainages intersected by trails are effectively passed across trails with open wide deep ditching or swales, armored with rock when needed. Surface water can still flow onto trails, so for trails in sloping terrain the most sustainable tread drainage features are short complete reversals of trail grade, forcing all water off the trail and requiring no annual maintenance. *Tread grade reversals*, also referred to as Coweeta or terrain dips, are designed and built into new trails so that as a trail descends the tread briefly swings uphill for ten or more feet before resuming a descending grade again (Hesselbarth et al. 2007; Birkby 2006). Any water flowing down the trail is forced off by gravity and no recurring maintenance should be necessary to maintain its efficacy. These features are best constructed by curving the trail around large rocks or trees to "anchor" the tread and guide traffic.

A variety of similar tread drainage features can be installed through maintenance following trail construction and use. Again, nomenclature varies, but *rolling grade dips* are like grade reversals, where maintainers excavate a descending section of tread to create a wide lower grade dip that drains water in a sheet flow, adding the excavated soil to a downhill section of tread, elevating it to ensure all water leaves the tread (Hesselbarth et al. 2007). Effective features are more than 10 feet in length with wide deep dips that will not clog and require annual maintenance. A similar but less effective *grade dip* involves excavation of the dip portion without elevating the downslope tread. A comparable feature known as a *knick* can be constructed adjacent to lower terrain by excavating a wide semicircular area about 10 feet long to ensure sheet flow runoff of all tread water to the lower ground (Hesselbarth et al. 2007). This feature essentially removes a wide section of the lower trailside berm and creates a more steeply outsloped tread, greater than 10 percent, ensuring that all water drains from the trail even if some leaves or needles collect in the area.

Reliance on *outsloped treads* to drain water is rarely successful unless treads have substantial rock to resist cupping and prolong the outsloped surface (Marion 2023). The formation of berms along lower trailsides are caused by accumulating organic debris and soils displaced or eroded downslope from the tread—these can be scraped or dug further downhill to reestablish the tread outslope and drainage, but this often involves substantial work and removes trailside vegetation cover. *Rock or wood water bars* are no longer recommended as these features require more time to acquire and move the material, physically obstruct the trail, and are often knocked out of place by trail traffic or cause hikers to move around them (Marion 2023). Long experience has revealed that water bar ditches are frequently too narrow and easily clogged by organic debris and eroded sediments, too short, and have frequently been constructed at angles that slow water drainage sufficiently for sediments to be dropped within the ditching, reducing their efficacy. Regardless of the specific drainage feature employed, clogging by deposited organic litter and eroded soil is the enemy of all—efficacy is increased by creating exceptionally wide dips and knicks, and annual maintenance to keep them cleaned of accumulated debris (Marion 2023). Trail maintainers can substantially improve their tread drainage knowledge and efficacy by hiking their trails to observe tread drainage during heavy rain.

Tread drainage features must be spaced closely enough so that water cannot build up excessive volume, speed, and erosive power. The density or spacing of tread drainage features varies with trail grade, substrate, and other factors, ranging from about 30 feet for steep trails (greater than 15 percent grade), 75 feet for trails with 11 to 15 percent grades, 150 feet for trails with 7 to 10 percent grades, and 300 feet for trails with 3 to 6 percent grades. Spacing should be reduced if tread substrates are coarse-grained and/or lack rock and are easily displaced/eroded; spacing can be increased if tread substrates have larger amounts of clay, gravel, and rock, particularly with stone pitching, steps, or bedrock. This general guidance was assembled from the widely variable experientially based literature; comprehensive experimentally designed studies to derive this guidance for natural-surfaced trails has not been published. There is evidence from both trail stewards and research suggesting that the amount of rock in tread substrates is more influential in deterring soil loss and displacement than soil texture (percent sand, silt, and clay) (Marion 2023; Marion et al. 2020).

Stream crossings require special attention to select the best "control points" for bridges, with narrow deep channels and high, stable abutment locations, or fords, with wide, safe, shallow, slow-moving waters, and stable rock substrates. Trails should employ side-hill descents to crossings in a downstream orientation and outsloped treads maintained to remove all water with sheet flow, allowing vegetation and organic materials to filter out sediments before reaching waterbodies (Marion 2023; Marion and Leung 2004).

Informal Trail Management

Since informal trails result from visitor behaviors, their management may be improved by considering and, in some instances, accommodating visitor desires. Managers pragmatically recognize that formal trails simply cannot be routed to every destination that visitors desire to access, so networks of ITs are both unavoidable and necessary. Examples of such destinations include: (1) camping infrastructures with thousands of campsites and access trails to water, and (2) day-use sites at vistas, swimming holes, fishing/hunting spots, climbs, waterbodies/shorelines, and historic, cultural, and geologic features of interest. The chief management challenge for these "access" ITs is that they are generally not sustainably designed or maintained and often have duplicative and braided trails to the same destination. For example, ITs that directly descend a steep slope to water can become severely degraded with high use, or visitors in riparian canyon areas form extensive braided IT networks due to use at different levels of flooding. In popular areas, managers can consider developing new formal sustainable trails based on existing use patterns to replace the most heavily used nonsustainable ITs. This strategy can substantially reduce the aggregate areal extent and severity of visitor impact and avoid routings through sensitive environments. Monitoring data describing existing spatial networks of ITs and their condition can be a critical aid in decision-making to identify new formal trails that will effectively supplant the greatest amount of IT impact.

Multiple trails result from visitors avoiding muddiness, deep ruts, or exposed roots and rocks. Alternate parallel trails are then formed by hikers and horses seeking easier travel (Fig. 15.14f). While education may help, a more effective long-term solution may be to relocate the degraded formal trail to a more sustainable side-hill alignment that avoids those problems, unless it's easier to correct the degraded conditions. In Yosemite National Park it was difficult to discourage use of deeply rutted meadow trails, often with several parallel trails developing. Moving trails out of flat meadows to side-hill alignments is the most effective long-term practice, but engineering solutions such as turnpike construction have also been effective (Marion 2023). Although engineered treads may be preferable to severe impacts, it should not be the first option considered in wilderness. Armoring treads, widening and filling in deep ruts, and using trailside rocks, logs, and plantings to compel visitors to remain on treads can also be helpful (Fig. 15.14e, f) (Marion 2023).

Shortcutting switchbacks can be addressed with education and trail design or maintenance. Visitors should be asked to not cut switchbacks. However, improved trail designs and maintenance practices can more effectively avoid providing visitors with those opportunities. Signs in the wilderness are also an undesirable intrusion and should be avoided unless necessary. Closely stacked switchbacks should be avoided and the use of wide turns that incorporate natural rock, carefully placed sections of downed trees or logs, shrubby vegetation, or scree walls can effectively deter their creation and use.

Lightly used ITs, particularly in more remote low use areas, represent "avoidable" impact that could be restored to natural conditions if visitors practiced full dispersal of foot traffic. This involves educating visitors to recognize when shifting to a dispersed "most durable surface" mode of travel is appropriate. Managers can assist by working to close and restore any existing or newly created trails, though in some instances establishing a sustainable formal trail may be necessary. If dispersal practices are not successful, use limitation may be an effective impact reduction strategy if use levels are within the lower range where a significant use-impact relationship exists (refer back to Fig. 15.9). Regardless, use should not be purposefully dispersed to trailless areas from heavily used destinations.

MONITORING TRAIL CONDITIONS

The effective management of trail infrastructures requires data on the number, location, sustainability, and condition of trails and trends over time. Periodic assessments provide data for adaptive management programs, such as evaluating the efficacy of trail management actions and guiding trail use, creation/closure, location, and maintenance decisions. The advent of accurate GPS devices, GIS analytical capabilities, and the efficient collection of field data with phone apps now greatly facilitates the acquisition and use of such data. Such efforts should begin with a comprehensive inventory of all formal and informal trails using the most accurate submeter GPS units available. Positional accuracy is critical to future GIS analyses with new high-resolution LiDAR-derived topographic data that can assist managers in evaluating trail sustainability, trail relocations, and spatial analyses involving other important spatial datasets (e.g., administrative boundaries, hydrological features, vegetation/soil maps, rare species, sensitive resources, campsites, historic/cultural sites) (Marion et al. 2011b).

A variety of trail inventory and condition assessment protocols have been developed and applied, with the most comprehensive and early review provided by Cole (1983a). He described three types of trail surveys: replicable measurements of a sample of trail segments, rapid surveys of sampled trails, and complete censuses of trail problems. *Replicable measurements* can be applied at permanent sampling points to provide the most accurate monitoring of change, though time requirements generally limit sample sizes. *Rapid surveys* are the most common type, with field staff employing a fixed sampling interval with a random start and larger numbers of rapidly assessed sample points. Leung and Marion (1999b) assessed the influence of sampling interval on the accuracy of estimates of the lineal extent of trail impacts at Great Smoky Mountain National Park, finding that a sampling interval of less than 328 feet (100 m) is best. *Trail problem census surveys* are best at identifying all instances of specific trail "problems," defined in terms of impact severity like soil loss exceeding one foot in depth, trail width exceeding four feet, and excessive muddiness (Leung and Marion 1999a).

More recent reviews of trail survey methodologies are provided by Marion et al. (2011b) and Marion (2025), describing trail attribute inventories, condition assessments, and prescriptive management assessments. *Attribute inventories* can include census data on trail types, numbers, lineal extent, locations, hiking difficulty/accessibility, maintenance features (e.g., bridges, signs, drainage features, steps), and sustainability attributes (trail/landform grades, trail slope alignment, substrates), all of which are spatially referenced with accurate GPS units for use in GIS analyses. *Trail condition assessments* employ point sampling, problem assessment, or condition class surveys to characterize the types, severity, and locations of various trail impacts. Trail *prescriptive management* assessments are conducted by knowledgeable trail stewards to evaluate and document maintenance needs, trail sustainability, and relocation options. The results of these surveys can be used to model and estimate the labor and materials needed to manage or maintain trails, including their associated facilities. Examples of data from these types of surveys and their application to management are included in Marion and others (2011b) and a comprehensive library of indicators and field protocols in Marion (2025).

TRAIL REHABILITATION

Much of the information about rehabilitating campsites also applies to trails, particularly the need to eliminate all use, and the techniques for reestablishing vegetation. The major difference is the common need to halt further erosion and replace lost soil. This requires identifying the sources of water intercepted by the trail and improving drainage straight across the trail. If it's a sidehill trail, additional soil can be pulled up or down to create a substantial outslope that deters further use and approximates the original pre-trail terrain. Locating sources of soil and rocks in flatter terrain can be more challenging, but includes scree slopes, excavated material from a new adjacent trail, streambed gravel or sandbars, and hidden, camouflaged borrow pits. Otherwise check dams and plantings can reduce water flow and allow deposition to fill in ruts with sediments over time (Rochefort and Gibbons 1992). Revegetation is most effective if the trail ruts are filled to grade with soil.

Multiple trails have been successfully restored in Yosemite National Park by cutting off the sod ridges between multiple trails at the level of the trail tread and stacking them in the shade (Eagen et al. 2000). The soil beneath the trails and the ridges was dug up to eliminate compaction, and soil was added to bring the trail up to the level of the surrounding meadow. Finally, the sod was divided into transplant plugs and planted. Bay and Ebersole (2006) documented successful transplanting of alpine turf onto closed trails in the Sangre de Cristo Wilderness, Colorado. Rock walls were built across trails, backfilled with rock from talus slopes and raw soil, and planted with pieces of turf cut for the new trail.

Trail closure and restoration work should not be done until staff can develop and open a more sustainable relocated trail segment or direct use to an alternate formal or informal trail to provide visitors with a clear alternative route to where they want to go. The final step is to block and naturalize with organic material, brush, logs, and rocks all connecting or visible parts of the closed trail segment to disguise it and discourage further use. Temporary "Closed for Recovery" signs are also helpful.

MANAGING PACK AND SADDLE STOCK

Despite a strong tradition of stock use in some wilderness areas, it's clear that social, environmental, and administrative costs associated with stock use are much more pronounced than with a comparable amount of hiker use (Newsome et al. 2004). As an example, trails that support stock use must be more sustainably designed, constructed, and maintained with armored rock treads in places, special puncheon construction to cross wet soils, greater maintenance of tread substrates and drainage, wider and taller vegetative clearing, and more rapid removal of obstructing trees, and much larger and more expensive bridges than required for foot travel (Marion 2023). These challenges, along with the continuing decline of stock use, greater scientific documentation of their substantial impacts, and conflicts and complaints from other wilderness visitors has prompted managers to increase regulations restricting grazing, off-trail traffic, and the trails and campsites that are open to stock use.

As with campsite and trail impact, many factors influence the severity of stock impacts. The most important are frequency and amount of use, group size and behavior, time of use, and location of use. These inform and define four primary management strategies described below.

LIMITING USE

Stock impacts on trails and campsites occur rapidly at low use levels; generally, all stock use must be restricted, there are few situations where reducing its use will improve conditions (Newsome et al. 2004). Reducing use can help minimize social and aesthetic impacts (e.g., horse user–hiker conflicts or manure). An important exception is limits on stock use in specific meadows to avoid overgrazing. This has been done in Sequoia–Kings Canyon National Parks, with monitoring use and meadow conditions to determine when closures are needed (DeBenedetti and Parsons 1983). Some meadows have length-of-stay limits so that more users can share the limited resource.

ENCOURAGING LESS DAMAGING BEHAVIOR

Low-impact stock practices can do much to reduce impacts, particularly in and around campsites. Most tree damage—a particularly severe problem on stock sites—results from felling trees for firewood or tent poles and tying horses to trees. Available techniques and equipment (Aaland 1993; Brame and Cole 2011; Leave No Trace 1996) can largely eliminate these practices and their resultant impacts. For example, stock can be moved to small confinement areas as soon as they are unloaded to reduce campsite impacts.

Limiting group size is currently the most common pack-stock management practice in wilderness, ranging from five to thirty-five animals per group, with twenty-five being the most common limit (Monz et al. 2000). *Available research indicates that such high limits will have very little beneficial effect.*

Stock impacts can also be reduced by requiring groups to carry stock feed, thus eliminating grazing damage (McClaran and Cole 1993). Many wildernesses with stock use have taken this action, with others recommending packing in feed. In many places, groups bring enough feed to supplement grazing, though there is a trade-off as additional animals are needed to pack the feed. Certified weed-free hay is now required in many wilderness areas.

MANAGING THE TIMING OF USE

Although only a few wildernesses prohibit stock use during certain seasons of the year (McClaran and Cole 1993), the impact of both grazing and trampling is highly dependent on seasonal variables, particularly the phenology or stage in the annual cycle of plant development and soil moisture content. *Generally, plants and soil are most vulnerable to disturbance during spring and early summer when plants are using stored nutrients for growth and when soils are water saturated.* This has led Sequoia–Kings Canyon National Parks—the backcountry area with the longest history of research and management of pack-stock impact—to make their primary management tool a system of opening dates that are determined from information on the type of hydrologic year (wet, normal, and dry). A research program examined the composition and distribution of the parks' major forage areas as well as their susceptibility to differing intensities, frequencies, and times of stock use (DeBenedetti and Parsons 1983).

MANAGING THE LOCATION OF USE

Among the most effective ways to manage the impact of stock is to control where stock use is allowed. In most NPS wilderness, stock use is only permitted on established trails (McClaran and Cole 1993) and some parks—like Glacier National Park—only allow overnight stock use in selected areas judged to be resistant to stock impact. One approach is to permit stock use on trails and campsites specifically selected and managed for horse use. Limiting stock use to a subset of trails can substantially reduce trail maintenance costs and provides places for hikers to go where they will not see stock groups. Another common action is to keep stock away from stream banks, lakeshores, and wet meadows to reduce severe trampling damage and the risk of water pollution.

Confinement of stock impacts (limiting where they can go) offers a reasonable compromise between providing horse riders with access to wilderness and confining the heavy impacts typically associated with stock use (Spildie et al. 2000).

SUMMARY

This chapter has identified the various environmental impacts of wilderness recreation use and described many management strategies and practices for avoiding or minimizing them. Several general points summarize this material.

First, impact is inevitable wherever wilderness recreation use occurs. Therefore, consistent with the goal of providing recreation opportunities, management can only limit impact, not prevent it. Nevertheless, to limit impact from increasing incrementally, with little ability to constrain cumulative impacts, setting specific prescriptive objectives and thresholds (standards) that effectively define and limit unacceptable levels impacts may, through monitoring of conditions, allow managers to identify when specific impacts have become so pronounced as to require management attention.

Second, there is a large "toolbox" of strategies and actions available to help managers constrain each type of impact. Managers can perform a problem analysis to select actions that effectively manipulate the most influential factors revealed by research and monitoring. The Visitor Use Management framework provides an effective and defensible adaptive management process for evaluating problems against management objectives and thresholds, selecting and implementing effective strategies and practices, and evaluating their efficacy and need for additional actions through monitoring.

Preoccupation with "too much use" as the major cause of problems and use limitation as the principal solution can prevent consideration of more effective or preferred actions. For example, relying on use rationing in high use areas, when research reveals that more sustainable trail and camping management actions are more effective and better received by the public (Marion et al. 2023). Amount of use is just one of many factors influencing amount of impact; often it is one of the less influential factors. Likewise, limiting use is only one of many alternative management techniques, and it has often unnecessarily restricted visitor access while failing to resolve substantial problems with campsite proliferation, expansion, or trail impact. When use is limited, other actions—such as confinement of impact—are often necessary if use limits are to be beneficial. However, there clearly are situations where use must be limited. *More consideration should be given to limiting use and promoting dispersed travel and camping practices in lightly used wilderness areas—to ensure the protection of natural conditions.*

Finally, wilderness management may benefit from equal concern for both the quality of wilderness visitors' experiences (see Chapter 16) and environmental impacts. The two are inextricably bound together; actions that affect one will affect the other—in positive and negative ways. Therefore, clearly defined resource and social problems, influential factors, and how alternative actions will affect the quality of both resource conditions and visitor experiences can promote effective management. Consideration of how an action to correct a problem in one specific place will affect other places and other wilderness conditions may result in improved management outcomes. Effective integration of ecological and visitor experience concerns, both in research and management, may lead to more holistic approach that will make wilderness management more effective.

STUDY QUESTIONS

1. Describe how recreation *ecology* as a field of study may provide the information needed to describe the interrelationships of various use-related, environmental, and managerial factors that influence recreation impacts.
2. Provide definitions of vegetation resistance and resilience, describe the types of plants that have and lack these traits, and how such knowledge can be applied to accommodate recreational use while minimizing vegetation loss.
3. Draw and label a diagram of the use-impact relationship for camping impact and discuss its implications for camping management in low use and high use areas.
4. Describe why dispersed (unconfined) camping and use limitation can be ineffective policies in high use wilderness areas. Describe and discuss alternative camping policies for both high use and low use areas. If you had to select a single indicator to monitor camping impact, what would you choose and why?
5. Compare and contrast the magnitude and nature of impacts caused by hiker groups and stock groups. Describe where impacts are similar and where they differ. Describe how low impact practices could aid in minimizing recreation impacts.
6. Discuss the three most common types of trail impacts and describe at least two influential factors for each impact. To what extent would reducing trail use improve trail conditions?

ACKNOWLEDGMENTS

This chapter is a major revision by Jeffrey L. Marion, U.S. Geological Survey and Virginia Tech, Field Station Leader, for the fifth edition. The fourth edition was authored by David Cole, the author's principal mentor and long-time colleague. I express my thanks to David, to numerous graduate students who became my closest colleagues, and to all federal land management agency staff, volunteer stewards, and Leave No Trace educators for their professionalism in preserving our incredible wealth of protected natural areas. Any use of trade, firm, or product names is for descriptive purposes only and does not imply endorsement by the U.S. government.

CASE DISCUSSION: BASELINE WILDERNESS DATA

Research information on recreational use and biophysical impacts from recreational use on campsites and trails may be limited. Cole and Wright (2003) reported that only 56 percent of the 625 wilderness units in the NWPS in 2000 had any research information on visitors and their impacts. The types of baseline information available in wilderness units were mostly for campsite impacts (51 percent), visitor use (24 percent), and trail conditions (9 percent). Monitoring campsite and trail conditions supports understanding what management strategies and actions may effectively constrain impacts and support an effective adaptive management process.

CASE DISCUSSION QUESTIONS

1. What specific information would you seek in *current resource conditions* and baseline studies of campsites regarding area of disturbance, vegetation and soil changes, spatial patterns of impact, and campsite proliferation in order to make decisions about management strategies and actions needed to constrain visitor impacts on *campsites*?
2. What information about amount, type, season, and location of *visitor use* do you need to decide how to manage use and control impacts on *campsites*?
3. What specific information would you seek in *current resource conditions* and baseline studies of trails regarding soil loss, trail widening, muddiness, and informal trails to make decisions about management strategies and techniques needed to control visitor impacts on *trails*?
4. What information about amount, type, season, and location of *visitor use* do you need to decide how to manage use and control visitor impacts on *trails*?

REFERENCES

Aaland, Dan. 1993. *Treading Lightly with Pack Animals.* Missoula, MT: Mountain Press.

Adkison, Gregory P.; Jackson, Marion T. 1996. Changes in ground-layer vegetation near trails in midwestern U.S. forests. *Natural Areas Journal.* 16: 14–23.

Anderson, Dorothy H.; Lime, David W.; Wang, Theresa L. 1998. *Maintaining the Quality of Park Resources and Visitor Experiences: A Handbook for Managers.* St. Paul: University of Minnesota, Department of Forest Resources, Cooperative Park Studies Unit.

Anderson, S. H. 1995. Recreational disturbance and wildlife populations. In: Knight, Richard L.; Gutzwiller, Kevin J., eds. *Wildlife and Recreationists: Coexistence Through Management and Research.* Washington, DC: Island Press, pp. 157–168.

Arocena, Joselito M.; Nepal, Sanjay K.; Rutherford, Michael. 2006. Visitor-induced changes in the chemical composition of soils in backcountry areas of Mt Robson Provincial Park, British Columbia, Canada. *Journal of Environmental Management.* 79: 10–19.

Arredondo, Johanna R.; Marion, Jeffrey L.; Meadema, Fletcher P.; Wimpey, Jeremy F. 2021. Modeling areal measures of campsite impacts on the Appalachian National Scenic Trail to enhance ecological sustainability. *Journal of Environmental Management.* 279, 111693.

ATC. 2014. Backcountry Sanitation Manual, 2nd ed. Appalachian Trail Conservancy, Green Mountain Club, USDA Forest Service, National Park Service.

ATC. 2015. Organized Group Management Manual. Appalachian Trail Conservancy. Harpers Ferry, WV.

Barros, Agustina; Gonnet, Jorge; Pickering, Catherine. 2013. Impacts of informal trails on vegetation and soils in the highest protected area in the Southern Hemisphere. *Journal of Environmental Management.* 127: 50–60.

Bay, Robin F.; Ebersole, James J. 2006. Success of turf transplants in restoring alpine trails, Colorado, US. *Arctic, Antarctic, and Alpine Research.* 38: 173–178.

Beylich, Anneke; Oberholzer, Hans-Rudolf; Schrader, Stefan; Hoper, Heinrich; Wilke, Berndt-Michael. 2010. Evaluation of soil compaction effects on soil biota and soil biological processes in soils. *Soil and Tillage Research.* 109: 133–143.

Birkby, Robert C. 2006. *Lightly on the Land: The SCA Trail-Building and Maintenance Manual.* 2nd ed. Student Conservation Association, Inc. Seattle, WA: The Mountaineers.

Brame, Rich; Cole, David. 2011. NOLS Soft Paths: Enjoying the Wilderness without Harming it. 4th ed., National Outdoor Leadership School. Stackpole Books, Mechanicsburg, PA.

Bratton, Susan P.; Stromberg, Linda L.; Harmon, Mark E. 1982. Firewood-gathering impacts in backcountry campsites in Great Smoky Mountains National Park. *Environmental Management.* 6: 63–71.

Brickler, Stan; Tunnicliff, Brock; Utter, Jack. 1983. Use and quality of wildland water: The case of the Colorado River corridor in the Grand Canyon. *Western Wildlands.* 9(2): 20–25.

Bridle, Kerry L.; Kirkpatrick, J. B. 2005. An analysis of the breakdown of paper products (toilet paper, tissues and tampons) in natural environments, Tasmania, Australia. *Journal of Environmental Management.* 74: 21–30.

Bromley, Maria; Marion, Jeffrey L.; Hall, Troy. 2013. Training to teach Leave No Trace: Efficacy of Master Educator courses. *Journal of Park and Recreation Administration.* 31: 62–78.

Brunson, M.; Shelby, B. 1990. A hierarchy of campsite attributes in dispersed recreation settings. *Leisure Sciences.* 12: 197–209.

Bull, E. L.; Parks, C. G.; Torgersen, T. R. 1997. Trees and logs important to wildlife in the Interior Columbia River Basin. General Technical Report PNW-GTR-391. Portland, OR: USDA, Forest Service, Pacific Northwest Research Station.

Cassirer, E. Frances; Freddy, David J.; Ables, Ernest D. 1992. Elk responses to disturbance by cross-country skiers in Yellowstone National Park. *Wildlife Society Bulletin.* 20: 375–381.

Cilimburg, Amy; Monz, Christopher; Kehoe, Sharon. 2000. Wildland recreation and human waste: A review of problems, practices, and concerns. *Environmental Management.* 25: 587–598.

Civil, K.; McNamara, B. 2000. Best Practice Human Waste Management Workshop. Workshop Proceedings and Papers, Canberra and Jindabyne, Australian Alps Liaison Committee, Australian Alps National Parks.

Clow, David W.; Forrester, Harrison; Miller, Benjamin; Roop, Heidi; Sickman, James O.; Ryu, Hodon; Domingo, Jorge S. 2013. Effects of stock use and backpackers on water quality in wilderness in Sequoia and Kings Canyon National Parks, USA.

Cole, David N. 1982a. Controlling the spread of campsites at popular wilderness destinations. *Journal of Soil and Water Conservation.* 37: 291–295.

———. 1982b. Wilderness campsite impacts: Effect of amount of use. Research Paper INT-284. Ogden, UT: USDA, Forest Service, Intermountain Forest and Range Experiment Station.

———. 1983a. Assessing and monitoring backcountry trail conditions. Research Paper INT-303. Ogden, UT: USDA, Forest Service, Intermountain Forest and Range Experiment Station.

———. 1983b. Campsite conditions in the Bob Marshall Wilderness, Montana. Research Paper INT-312. Ogden, UT: USDA, Forest Service, Intermountain Forest and Range Experiment Station.

———. 1986a. Ecological changes on campsites in the Eagle Cap Wilderness, 1979–1984. Research Paper INT-368. Ogden, UT: USDA, Forest Service, Intermountain Research Station.

———. 1986b. Recreational impacts on backcountry campsites in Grand Canyon National Park, Arizona, USA. *Environmental Management.* 10: 651–659.

———. 1989a. Area of vegetation loss: A new index of campsite impact. Research Note INT-389. Ogden, UT: USDA, Forest Service, Intermountain Research Station.

———. 1989b. Wilderness campsite monitoring methods: A sourcebook. General Technical Report INT-259. Ogden, UT: USDA, Forest Service, Intermountain Research Station.

———. 1991. Changes on trails in the Selway-Bitterroot Wilderness, Montana, 1978–89. Research Paper INT-450. Ogden, UT: USDA, Forest Service, Intermountain Research Station.

———. 1992. Modeling wilderness campsites: Factors that influence amount of impact. *Environmental Management.* 16: 255–264.

———. 1993a. Campsites in three western wildernesses: Proliferation and changes in condition over 12 to 16 years. Research Paper INT-463. Ogden, UT: USDA, Forest Service, Intermountain Research Station.

———. 1993b. Trampling effects on mountain vegetation in Washington, Colorado, New Hampshire, and North Carolina. Research Paper INT-464. Ogden, UT: USDA, Forest Service, Intermountain Research Station.

———. 1995a. Disturbance of natural vegetation by camping; experimental applications of low- level stress. *Environmental Management.* 19: 405–416.

———. 1995b. Experimental trampling of vegetation, II: Predictors of resistance and resilience. *Journal of Applied Ecology.* 32: 215–224.

———. 2013a. Changing conditions on wilderness campsites: Seven case studies of trends over 13 to 32 years. Gen. Tech. Rep. RMRS-GTR-300. Fort Collins, CO: USDA, Forest Service, Rocky Mountain Research Station. 99 p.

———. 2013b. Long-term effectiveness of restoration treatments on closed wilderness campsites. *Environmental Management.* 51: 642–650.

———. 2021. Reflections on the early history of recreation ecology. *Parks Stewardship Forum* 37: 379–400.

Cole, David N.; Ferguson, David N. 2009. A relatively nonrestrictive approach to reducing campsite impact Caney Creek Wilderness, Arkansas. *International Journal of Wilderness.* 15: 20–25.

Cole, David N.; Hall, Troy E. 1992. Trends in campsite condition: Eagle Cap Wilderness, Bob Marshall Wilderness and Grand Canyon National Park. Research Paper INT-453. Ogden, UT: USDA, Forest Service, Intermountain Research Station.

Cole, David N.; Hall Troy E. 2009. Perceived effects of setting attributes on visitor experiences in wilderness: Variation with situational context and visitor characteristics. *Environmental Management.* 44: 24–36.

Cole, David N.; Hammond, Timothy P.; McCool, Stephen F. 1997. Information quantity and communication effectiveness: Low-impact messages on wilderness trailside bulletin boards. *Leisure Sciences*. 19: 59–72.

Cole, David N.; Landres, Peter B. 1995. Indirect effects of recreationists on wildlife. In: Knight, R. L.; Gutzwiller, K. J., eds. *Wildlife and Recreationists: Coexistence through Management and Research*. Washington, DC: Island Press, pp. 183–202.

———. 1996. Threats to wilderness ecosystems: Impacts and research needs. *Ecological Applications*. 6: 168–184.

Cole, David N.; Monz, Christopher A. 2003. Impacts of camping on vegetation: Response and recovery following acute and chronic disturbance. *Environmental Management*. 32: 693–705.

———. 2004. Spatial patterns of recreational impact on experimental campsites. *Journal of Environmental Management*. 70: 73–84.

Cole, David N.; Parsons, David J. 2012. Campsite impact in the wilderness of Sequoia and Kings Canyon National Parks: 30 years of change. Final Research Report. Ogden, UT: USDA, Forest Service, Rocky Mountain Research Station, Aldo Leopold Wilderness Research Institute.

Cole, David N.; Petersen, Margaret E.; Lucas, Robert C. 1987. Managing wilderness recreation use: Common problems and potential solutions. General Technical Report INT-259. Ogden, UT: USDA, Forest Service, Intermountain Research Station.

Cole, David N.; Ranz, Beth. 1983. Temporary campsite closures in the Selway-Bitterroot Wilderness. *Journal of Forestry*. 81: 729–732.

Cole, David N.; Spildie, David R. 1998. Hiker, horse, and llama trampling effects on native vegetation in Montana, USA. *Journal of Environmental Management*. 53: 61–71.

———. 2007. Vegetation and soil restoration on highly impacted campsites in the Eagle Cap Wilderness, Oregon. General Technical Report RMRS-GTR-185. Fort Collins, CO: USDA, Forest Service, Rocky Mountain Research Station.

Cole, David N.; van Wagtendonk, Jan W.; McClaran, Mitchel P.; Moore, Peggy E.; McDougald, Neil K. 2004. Response of mountain meadows to grazing by recreational pack stock. *Journal of Range Management*. 57: 153–160.

Cole, David N.; Watson, Alan E.; Hall, Troy E.; Spildie, David R. 1997. High-use destinations in wilderness: Social and biophysical impacts, visitor responses, and management options. Research Paper INT-RP-496. USDA, Forest Service.

Cole, David N.; Williams, Daniel R. 2010. Wilderness Visitor Experiences: A Review of 50 Years of Research. In: Cole, David N., comp. 2012. Wilderness visitor experiences: Progress in research and management; 2011 April 4–7; Missoula, MT. Proc. RMRS-P-66. Fort Collins, CO: USDA, Forest Service, Rocky Mountain Research Station. 219 p.

Cole, David N.; Wright, Vita. 2003. Wilderness visitors and recreation impacts: Baseline data available for twentieth century conditions. RMRS-GTR-117. Ogden, UT: USDA, Forest Service, Rocky Mountain Research Station.

Daniels, Melissa L.; Marion, Jeffrey L. 2005. Communicating Leave No Trace ethics and practices: Efficacy of two-day trainer courses. *Journal of Park and Recreation Administration*. 23: 1–19.

Daniels, Melissa L.; Marion, Jeffrey L. 2006. Visitor evaluations of management actions at a highly impacted Appalachian Trail camping area. *Environmental Management*. 38(6): 1006–1019.

D"Antonio, Ashley; Monz, Christopher; Newman, Peter; Lawson, Steve; Taff, Derrick. 2012. The effects of local ecological knowledge, minimum-impact knowledge, and prior experience on visitor perceptions of the ecological impacts of backcountry recreation. *Environmental Management*. 50(4): 542–554.

Davies, M. A. 2004. What's burning in your campfire: Garbage in, toxics out. Tech Tip 0423-2327-MTDC. Missoula, MT: USDA, Forest Service, Missoula Technology and Development Center, 8 p.

DeBenedetti, Steven H.; Parsons, David J. 1983. Protecting mountain meadows: A grazing management plan. *Parks*. 8(3): 11–13.

DeLuca, T. H.; Patterson, W. A., IV; Freimund, W. A.; Cole, D. N. 1998. Influence of llamas, horses, and hikers on soil erosion from established recreation trails in western Montana, USA. *Environmental Management*. 22: 255–262.

Derlet, Robert W.; Ger, K. Ali; Richards, John R.; Carlson, James R. 2008. Risk factors for coliform bacteria in backcountry lakes and streams in the Sierra Nevada Mountains: A 5-year study. *Wilderness and Environmental Medicine*. 19: 82–90.

Dickens, Sara Jo M.; Gerhardt, Fritz; Collinge, Sharon K. 2005. Recreational portage trails as corridors facilitating non-native plant invasions of the Boundary Waters Canoe Area Wilderness, USA. *Conservation Biology*. 19: 1553–1657.

Dolman, Megan R.; Marion, Jeffrey L. 2022. Invasive plant hitchhikers: Appalachian Trail thru-hiker knowledge and attitudes of invasive plants and Leave No Trace practices. *Journal of Outdoor Recreation and Tourism*. 40: 100581.

Dynowski, Piotr; Senetra, Adam; Zrobek-Sokolnik, Anna; Kozlowski, Jacek. 2019. The impact of recreational activities on aquatic vegetation in alpine lakes. *Water*. 11: 173.

Eagen, Sean; Newman, Peter; Fritzke, Susan; Johnson, Louise. 2000. Restoration of multiple-rut trails in the Tuolumne Meadows of Yosemite National Park. In: Cole, D. N.; McCool, S. F.; Borrie, William T.; O'Loughlin, Jennifer, comps. *Wilderness Science in a Time of Change*. Vol. 5: *Wilderness Ecosystems, Threats, and Management*; May 23–27, 1999; Missoula, MT. RMRS-P-15. Ogden, UT: USDA, Forest Service, Rocky Mountain Research Station, pp. 188–192.

Eagleston, Holly; Marion, Jeffrey L. 2017a. "Naturalness" in designated wilderness: Long-term changes in non-native plant dynamics on campsites, Boundary Waters, Minnesota. *Forest Science*. 64: 50–56.

Eagleston, Holly; Marion, Jeffrey L. 2017b. Sustainable campsite management in protected areas: A study of long-term ecological changes on campsites in the Boundary Waters Canoe Area Wilderness, Minnesota, USA. *Journal of Nature Conservation*. 37: 73–82.

Farrell, Tracy A.; Hall, Troy E.; White, Dave D. 2001. Wilderness campers' perception and evaluation of campsite impacts. *Journal of Leisure Research*. 33: 229–250.

Farrell, Tracy A.; Marion, Jeffrey L. 2001. Identifying and assessing ecotourism visitor impacts at eight protected areas in Costa Rica and Belize. Environmental Conservation. 28: 215–225.

Fenn, Dennis B.; Gogue, G. Jay; Burge, Raymond E. 1976. Effects of campfires on soil properties. Ecological Service Bulletin 5. Washington, DC: DOI, National Park Service.

Ferguson, Michael A. D.; Keith, Lloyd B. 1982. Influence of Nordic skiing on distribution of moose and elk in Elk Island National Park, Alberta. *Canadian Field-Naturalist*. 96: 69–78.

Godefroid, S.; Koedam, N. 2004. Interspecific variation in soil compaction sensitivity among forest floor species. *Biological Conservation*. 119: 207–217.

Gutzwiller, Kevin J.; Cole, David N. 2005. Assessment and management of wildland recreation disturbance. In: Braun, C. E., ed. *Techniques for wildlife investigations and management,* 6th ed. Bethesda, MD: The Wildlife Society, pp. 779– 796.

Hall, Troy E.; Farrell, Tracy A. 2001. Fuelwood depletion at wilderness campsites: Extent and potential ecological significance. *Environmental Conservation*. 28: 241–247.

Hall, Troy E.; Seekamp, Erin; Cole, David N. 2010. Do recreation motivations and wilderness involvement relate to support for wilderness management? A segmentation analysis. *Leisure Sciences*. 32: 109–124.

Hammitt, William E.; Cole, David N.; Monz, Christopher A. 2015. Wildland recreation: Ecology and management, 3rd ed., John Wiley & Sons.

Hampton, Bruce; Cole, David. 2003. *Soft Paths,* 3rd ed. Mechanicsburg, PA: Stackpole Books. 225p.

Hartley, Ernest A. 1999. Visitor impact at Logan Pass, Glacier National Park: A thirty-year vegetation study. In: Harmon, D., ed. *On the Frontiers of Conservation: Proceedings of the 10th George Wright Society Biennial Conference on Research and Resource Management in Parks and Public Lands;* March 22–26, 1999; Asheville, NC. Hancock, MI: The George Wright Society, pp. 297–305.

Hesselbarth, Woody; Vachowski, Brian; Davies, Mary Ann. 2007. Trail Construction and Maintenance Notebook: 2007. 0723-2806-MTDC. Missoula, MT: USDA, Forest Service, Missoula Technology and Development Center.

Hockett, Karen S.; Marion, Jeffrey L.; Leung, Yu-Fai. 2017. The efficacy of combined educational and site management actions in reducing off-trail hiking in an urban-proximate protected area. *Journal of Environmental Management.* 203: 17–28.

Holmquist, Jeffrey G.; Schmidt-Gengenbach, Jutta; Roche, James. 2015. Stream macroinvertebrates and habitat below and above two wilderness fords used by mules, horses, and hikers in Yosemite National Park. *Western North American Naturalist.* 75: 311–324.

IVUMC (Interagency Visitor Use Management Council). 2019. Monitoring Guidebook: Evaluating Effectiveness of Visitor Use Management. Denver, CO. https://visitorusemanagement.nps.gov/VUM/Framework.

Kidd, Abigail M.; Monz, Christopher; D'Antonio, Ashley; Manning, Robert E.; Reigner, Nathan; Goonan, Kelly A.; Jacobi, Charles. 2015. The effect of minimum impact education on visitor spatial behavior in parks and protected areas: An experimental investigation using GPS-based tracking. *Journal of Environmental Management.* 162: 53–62.

Knapp, Roland A. 2005. Effects of non-native fish and habitat characteristics on lentic herpetofauna in Yosemite National Park, USA. *Biological Conservation.* 121: 265–279.

Knight, Richard L.; Cole, David N. 1995a. Factors that influence wildlife responses to recreationists. In: Knight, R. L.; Gutzwiller, K. J., eds. *Wildlife and Recreationists: Coexistence through Management and Research.* Washington, DC: Island Press, pp. 71–80.

———. 1995b. Wildlife responses to recreationists. In: Knight, Richard L.; Gutzwiller, Kevin J., eds. *Wildlife and Recreationists: Coexistence through Management and Research.* Washington, DC: Island Press, pp. 51–70.

Knight, Richard L.; Temple, Stanley A. 1995. Origin of wildlife responses to recreationists. In: Knight, R. L.; Gutzwiller, K. J., eds. *Wildlife and Recreationists: Coexistence through Management and Research.* Washington, DC: Island Press, pp. 81–91.

Landres, Peter; Meyer, Shannon; Matthews, Sue. 2001. The Wilderness Act and fish stocking: An overview of legislation, judicial interpretation, and agency implementation. *Ecosystems.* 4: 287–295.

Larson, Courtney L.; Reed, Sarah E.; Merenlender, Adina M.; Crooks, Kevin R. 2016. Effects of recreation on animals revealed as widespread through a global systematic review. *PLoS One.* 11(12): e0167259.

Lawhon, Ben; Taff, B. Derrick; Newman, Peter; Vagias, Wade M.; Miller, Zachary D. 2019. Understanding attitudes and support for Leave No Trace: Informing communication strategies with frontcountry State Park visitors. *Journal of Outdoor Recreation, Education, and Leadership.* 11: 37–52.

Leave No Trace. 1996. Horse Use Leave No Trace: Outdoor Skills and Ethics. Leave No Trace, Boulder, CO.

Leung, Yu-Fai; Marion, Jeffrey L. 1996. Trail degradation as influenced by environmental factors: A state-of-the-knowledge review. *Journal of Soil and Water Conservation.* 51: 130–136.

———. 1999a. Assessing trail conditions in protected areas: An application of a problem-assessment method in Great Smoky Mountains National Park, USA. *Environmental Conservation.* 26: 270–279.

———. 1999b. The influence of sampling interval on the accuracy of trail impact assessment. *Landscape and Urban Planning.* 43: 167–179.

Liddle, Michael J. 1997. *Recreation Ecology: The Ecological Impact of Outdoor Recreation and Ecotourism.* London: Chapman and Hall.

MacArthur, Robert A.; Geist, Valerius; Johnston, Ronald H. 1982. Cardiac and behavioral responses of mountain sheep to human disturbance. *Journal of Wildlife Management.* 46: 351–358.

Manning, Robert E. 2003. Emerging principles for using information/education in wilderness management. *International Journal of Wilderness.* 9: 20–7.

Marion, Jeffrey L. 1991. Developing a natural resource inventory and monitoring program for visitor impacts on recreation sites: A procedural manual. NPS/NRVT/NRR-91/06. DOI, National Park Service.

———. 1995. Capabilities and management utility of recreation impact monitoring programs. *Environmental Management.* 19: 763–771.

———. 2003. Camping impact management on the Appalachian National Scenic Trail. Appalachian Trail Conference, Harper's Ferry, WV. 109 p.

———. 2014. Leave No Trace in the outdoors. Stackpole Books, Mechanicsburg, PA.

———. 2016. A review and synthesis of recreation ecology research supporting carrying capacity and visitor use management decision making. *Journal of Forestry.* 114: 339–351.

———. 2019. Impacts to wildlife: Managing visitors and resources to protect wildlife. U.S. Interagency Visitor Use Management Council, Contributed Paper. https://visitorusemanagement.nps.gov.

———. 2021. Guidance on developing recreation impact monitoring programs for wilderness. Report prepared for the Arthur Carhart National Wilderness Training Center, Missoula, MT.

———. 2023. Trail sustainability: A state-of-knowledge review of trail impacts, influential factors, sustainability ratings, and planning and management guidance. *Journal of Environmental Management.* 340; 117868.

———. 2025. Recreation Ecology Guidance for Campsite and Trail Monitoring and Visitor Use Management. Draft Open-Access Book. Virginia Tech University, Blacksburg, VA. Available at https://www.researchgate.net

Marion, Jeffrey L.; Cole, David N. 1996. Spatial and temporal variation in soil and vegetation impacts on campsites. *Ecological Applications.* 6: 520–530.

Marion, Jeffrey L.; Farrell, Tracy A. 2002. Management practices that concentrate visitor activities: Camping impact management at Isle Royale National Park, USA. *Journal of Environmental Management.* 66: 201–212.

Marion, Jeffrey L.; Leung, Yu-Fai. 1997. An assessment of campsite conditions in Great Smoky Mountains National Park Research/Resources Management Report. Atlanta, GA: DOI, National Park Service, Southeast Regional Office.

———. 2004. Environmentally sustainable trail management. In: Buckley, R., ed. *Environmental Impact of Tourism.* CABI Publications, Cambridge, MA, pp. 229–244.

———. 2011. Indicators and protocols for monitoring impacts of formal and informal trails in protected areas. *Journal of Tourism and Leisure Studies.* 17: 215–236.

———. 2001. Trail resource impacts and an examination of alternative assessment techniques. *Journal of Park and Recreation Administration.* 19: 17–37.

Marion, Jeffrey L.; Merriam, L. C. 1985. Recreational impacts on well-established campsites in the Boundary Waters Canoe Area Wilderness Bulletin. AD-SB-2502. St. Paul: University of Minnesota, Agricultural Experiment Station. 16p.

Marion, Jeffrey L.; Arredondo, Johanna; Wimpey, Jeremy; Meadema, Fletcher. 2018. Applying recreation ecology science to sustainably manage camping impacts: A classification of camping management strategies. *International Journal of Wilderness.* 24: 84–100.

Marion, Jeffrey L.; Leung, Yu-Fai; Eagleston, Holly; Burroughs, Kaitlin. 2016. A review and synthesis of recreation ecology research findings on visitor impacts to wilderness and protected natural areas. *Journal of Forestry.* 114: 352–362.

Marion, Jeffrey L.; Leung, Yu-Fai; Nepal, Sanjay K. 2006. Monitoring trail conditions: New methodological considerations. *The George Wright Forum.* 23(2): 36–49.

Marion, Jeffrey L.; Reid, Scott E. 2007. Minimising visitor impacts to protected areas: The efficacy of low impact education programmes. *Journal of Sustainable Tourism.* 15: 5–27.

Marion, Jeffrey L.; Roggenbuck, Joseph W.; Manning, Robert E. 1993. Problems and practices in backcountry recreation management: A survey of National Park Service managers. NPS/NRVT/NRR-93/12. Denver, CO: DOI, National Park Service.

Marion, Jeffrey L., Sober, Toivo. 1987. Environmental impact management in the Boundary Waters Canoe Area Wilderness. *North. J. Appl. For.* 4: 7–10.

Marion, Jeffrey L., Wimpey, Jeremy. 2011. Informal Trail Monitoring Protocols: Denali National Park and Preserve. Final Report to the DOI, National Park Service, Denali National Park. Denali Park, AK.

Marion, Jeffrey L., Wimpey, Jeremy. 2017. Assessing the influence of sustainable trail design and maintenance on soil loss. *Journal of Environmental Management.* 189: 46–57.

Marion, Jeffrey L.; Wimpey, Jeremy; Arredondo, Johanna; Meadema, Fletcher. 2020a. Improving the Sustainability of the Appalachian Trail: Trail and Recreation Site Conditions and Management. Final Report to the DOI, National Park Service, Appalachian Trail Park Office and the Appalachian Trail Conservancy, Harpers Ferry, WV.

———. 2020b. Sustainable Camping "Best Management Practices." DOI U.S. Geological Survey, Virginia Tech Field Unit. Final Research Report to the DOI, National Park Service, Appalachian Trail Park Office, and the Appalachian Trail Conservancy, Harpers Ferry, WV. 59 p.

———. 2023. Improving the Sustainability of Camping Management on the Pacific Crest National Scenic Trail. Final Report to the USDA, Forest Service, Pacific SW Region, Vallejo, CA and the Pacific Crest Trail Association, Sacramento, CA. 119 p.

Marion, Jeffrey L.; Wimpey, Jeremy. 2017. Assessing the influence of sustainable trail design and maintenance on soil loss. *Journal of Environmental Management.* 189: 46–57.

Marion, Jeffrey L.; Wimpey, Jeremy; Lawhon, Ben. 2018. Conflicting messages about camping near waterbodies in wilderness: A review of the scientific basis and need for flexibility. *International Journal of Wilderness.* 24: 68–81.

Marion, Jeffrey L.; Wimpey, Jeremy; Park, Logan O. 2011a. Informal and Formal Trail Monitoring Protocols and Baseline Conditions: Acadia National Park. Final Report to DOI, National Park Service, Acadia National Park. Bar Harbor, ME.

———. 2011b. The science of trail surveys: Recreation ecology provides new tools for managing wilderness trails. *Park Science.* 28: 60–65.

Maser, C.; Tarrant, R. F.; Trappe, J. M.; Franklin, J. F. 1988. From the forest to the sea: A story of fallen trees. General Technical Report PNW-GTR-229. Portland, OR: USDA, Forest Service, Pacific Northwest Research Station.

McClaran, Mitchel P.; Cole, David N. 1993. Pack stock in wilderness: Use, impacts, monitoring, and management. General Technical Report INT-301. Ogden, UT: USDA, Forest Service, Intermountain Research Station.

Meadema, Fletcher.; Marion, Jeffrey L.; Arredondo, Johanna; Wimpey, Jeremy. 2020. The influence of layout on Appalachian Trail soil loss, widening, and muddiness: Implications for sustainable trail design and management. *Journal of Environmental Management.* 257 p.

Miller, Scott G.; Knight, Richard L.; Miller, Clinton K. 2001. Wildlife responses to pedestrians and dogs. *Wildlife Society Bulletin.* 29: 124–132.

Monz, Chris A.; Cole, David N.; Leung, Yu-fai; Marion, Jeffrey L. 2010. Sustaining visitor use in protected areas: Future opportunities in recreation ecology research based on the USA experience. *Environmental Management.* 45: 551–562.

Monz, Chris A., Marion, Jeffrey L., Goonan, Kelly A., Manning, Robert E., Wimpey, Jeremy, Carr, Chris. 2010. Assessment and monitoring of recreation impacts and resource conditions on mountain summits: Examples from the Northern Forest, USA. *Mountain Research and Development.* 30: 332–343.

Monz, Christopher; Roggenbuck, Joseph; Cole, David; Brame, Richard; Yoder, Andrew. 2000. Wilderness party size regulations: Implications for management and a decision-making framework. In: Cole, D. N.; McCool, S. F.; Borrie, William T.; O'Loughlin, Jennifer, comps. *Wilderness Science in a Time of Change*, Vol. 4: *Wilderness Visitors, Experiences, and Visitor Management;* May 23–27, 1999; Missoula, MT. RMRS-P-15. Ogden, UT: USDA, Forest Service, Rocky Mountain Research Station, pp. 265–273.

Moore, Roger L.; Leung, Yu-Fai; Matisoff, Craig; Dorwart, Catherine; Parker, Alan. 2012. Understanding users' perceptions of trail resource impacts and how they affect experiences: An integrated approach. *Landscape and Urban Planning* 107: 343–350.

Moritsch, Barbara J.; Muir, Patricia S. 1993. Subalpine revegetation in Yosemite National Park, California: Changes in vegetation after three years. *Natural Areas Journal.* 13: 155–163.

Newsome, David; Cole, David N.; Marion, Jeffrey L. 2004. Environmental Impacts Associated with Recreational Horse-riding. In: Buckley, Ralf, ed., Environmental Impact of Tourism, Cambridge, MA: CABI Publishing. pp. 61–82.

Olive, Nate D.; Marion, Jeffrey L. 2009. The influence of use-related, environmental, and managerial factors on soil loss from recreational trails. *Journal of Environmental Management.* 90: 1483–1493.

Park, Logan O.; Marion, Jeffrey L.; Wimpey, Jeremy F. 2022. Efficacy of combining education and site management in reducing off-trail travel in a fragile biotic community, Acadia National Park. *Journal of Interpretation Research.* 28(1).

Pickering, Catherine M. 2010. Ten factors that affect the severity of environmental impacts of visitors in protected areas. *Ambio.* 39: 70–77.

Pickering, Catherine M.; Growcock, Andrew J. 2009. Impacts of experimental trampling on tall alpine herbfields and subalpine grasslands in the Australian Alps. *Journal of Environmental Management.* 91: 532–540.

Pickering, Catherine M.; Hill, Wendy. 2007. Impacts of recreation and tourism on plant biodiversity and vegetation in protected areas in Australia. *Journal of Environmental Management.* 85: 791–800.

Pickering, Catherine M.; Mount, Ann. 2010. Do tourists disperse weed seed? A global review of unintentional human-mediated terrestrial seed dispersal on clothing, vehicles and horses. *Journal of Sustainable Tourism.* 18: 239–256.

Pister, Edwin P. 2001. Wilderness fish stocking: History and perspective. *Ecosystems.* 4: 279–286.

Randall, John M. 2000. Improving management of nonnative invasive plants in wilderness and other natural areas. In: Cole, D. N.; McCool, S. F.; Borrie, William T.; O'Loughlin, Jennifer, comps. *Wilderness Science in a Time of Change*, Vol. 5: *Wilderness Ecosystems, Threats, and Management;* May 23–27, 1999, Missoula, MT. USDA Forest Service, Proceedings RMRS-P-15-Vol-5, Rocky Mountain Research Station, Ogden, UT, pp. 64.

Reid, Scott E.; Marion, Jeffrey L. 2004. Effectiveness of a confinement strategy for reducing campsite impacts in Shenandoah National Park. *Environmental Conservation.* 31: 274–282.

———. 2005. A comparison of campfire impacts and policies in seven protected areas. *Environmental Management.* 36: 48–58.

Rochefort, Regina M.; Gibbons, Stephen T. 1992. Mending the meadow: High-altitude meadow restoration in Mount Rainier National Park. *Restoration & Management Notes.* 10(2): 120–126.

Scherrer, P.; Pickering C. M. 2006. Recovery of alpine herbfield on a closed walking track in the Kosciuszko Alpine Zone, Australia. *Arctic Antarctic Alpine Res.* 38: 239–248.

Settina, Nita; Marion, Jeffrey L.; Schwartz, Forrest. 2020. Leave No Trace communication: Effectiveness based on assessments of resource conditions. *Journal of Interpretation Research.* 25: 5–25.

Silsbee, David G.; Larson, Gary L. 1982. Bacterial water quality: Springs and streams in the Great Smoky Mountains National Park. *Environmental Management.* 6: 353–359.

Silverman, G.; Erman, D. C. 1979. Alpine lakes in Kings Canyon National Park, California: Baseline conditions and possible effects of visitor use. *Journal of Environmental Management.* 8: 73–87.

Spildie, David R.; Cole, David N.; Walker, Sarah C. 2000. Effectiveness of a confinement strategy in reducing pack stock impacts at campsites in the Selway-Bitterroot Wilderness, Idaho. In: Cole, D. N.; McCool, S. F.; Borrie, William T.; O'Loughlin, Jennifer, comps. *Wilderness Science in a Time of Change*, Vol. 5: Wilderness ecosystems, threats, and management; May 23–27, 1999; Missoula, MT. RMRS-P-15. Ogden, UT: USDA, Forest Service, Rocky Mountain Research Station, pp. 199–208.

Spernbauer, Bettina S.; Monz, Christopher; D'Antonio, Ashley; Smith, Jordan W. 2023. Factors influencing informal trail conditions: Implications for management and research in Urban-Proximate parks and protected areas. *Landscape and Urban Planning* 231: 104661.

Stewart, D. P. C.; Cameron, K. C. 1992. Effect of trampling on the soils of the St. James Walkway, New Zealand. *Soil Use and Management.* 8: 30–36.

Stewart, William P. 1989. Fixed itinerary systems in backcountry management. *Journal of Environmental Management.* 29: 163–171.

Taylor, J. 1997. Leave only footprints? How backcountry campsite use affects forest structure. *Yellowstone Science.* 5(1): 14–17.

Taylor, T. P.; Erman, D. C. 1979. The response of benthic plants to past levels of human use in high mountain lakes in Kings Canyon National Park, California, USA. *Journal of Environmental Management.* 9: 271–278.

Taylor, T. P.; Erman, Don C. 1980. The littoral bottom flora of high elevation lakes in Kings Canyon National Park. *California Fish and Game.* 66: 112–119.

Temple, Kenneth L.; Camper, Anne K.; Lucas, Robert C. 1982. Potential health hazard from human waste in wilderness. *Journal of Soil and Water Conservation.* 37: 357–359.

Therrell, Lisa; Cole, David; Claassen, Victor; Ryan, Chris; Davies, Mary Ann. 2006. *Wilderness and Backcountry Site Restoration Guide.* 0623 2815. Missoula, MT: USDA, Forest Service, Missoula Technology and Development Center.

Tomczyk, A. M.; White, P. C. L.; Ewertowski, M. W. 2016. Effects of extreme natural events on the provision of ecosystem services in a mountain environment: The importance of trail design in delivering system resilience and ecosystem service co-benefits. *Journal of Environmental Management.* 166: 156–167.

Tomczyk, A. M.; Ewertowski, M. W.; White, P. C. L.; Kasprzak, L. 2017. A new framework for prioritising decisions on recreational trail management. *Landscape and Urban Planning.* 157: 1–13.

Tyser, R. W.; Worley, C. A. 1992. Alien flora in grasslands adjacent to road and trail corridors in Glacier National Park, Montana, USA. *Conservation Biology.* 6: 253–262.

Vernon, Marian E.; Campos, Brent R.; Burnett, Ryan D. 2022. Effects of livestock grazing on the ecology of Sierra meadows: A review of the current state of scientific knowledge to inform meadow restoration and management. *Environmental Management.* 69: 1118–1136.

Wang, Ting; Watanabe, Teiji. 2022. Monitoring campsite soil erosion by structure-from-motion photogrammetry: A case study of Kuro-Dake campsites in Daisetsuzan National Park, Japan. *Journal of Environmental Management.* 314: 115106.

Wellnitz, Todd; Barlow, Jenna L.; Dick, Cory M.; Shaurette, Terrance R.; Johnson, Brian M.; Wesley, Troy; Weiher, Evan. 2020. Campsites, forest fires, and entry point distance affect earthworm abundance in the Boundary Waters Canoe Area Wilderness. *PeerJ. 8:e8656.*

White, David D.; Hall, Troy E.; Farrell, Tracey A. 2001. Influence of ecological impacts and other campsite characteristics on wilderness visitors' campsite choices. *Journal of Park & Recreation Administration.* 19(2): 83–97.

White, Dave D.; Virden, Randy J.; van Riper, Carena J. 2008. Effects of place identity, place dependence, and experience-use history on perceptions of recreation impacts in a natural setting. *Environmental Management.* 42: 647–657.

Whittaker, Doug; Knight, Richard L. 1998. Understanding wildlife responses to humans. *Wildlife Society Bulletin.* 26: 312–317.

Willard, B. E.; Cooper, D. J.; Forbes, B. C. 2007. Natural regeneration of alpine tundra vegetation after human trampling: A 42-year data set from Rocky Mountain National Park, Colorado, USA. *Arctic Antarctic Alpine Res.* 39: 177–183.

Wimpey, Jeremy F.; Marion, Jeffrey L. 2010. The influence of use, environmental and managerial factors on the width of recreational trails. *Journal of Environmental Management.* 91: 2028–2037.

Wimpey, Jeremy; Marion, Jeffrey L. 2011. A spatial exploration of informal trail networks within Great Falls Park, VA. *Journal of Environmental Management.* 92: 1012–1022.

Zabinski, Catherine A.; DeLuca, Thomas H.; Cole, David N.; Moynahan, October S. 2002. Restoration of highly impacted subalpine campsites in the Eagle Cap Wilderness, Oregon. *Restoration Ecology.* 10: 275–281.

Zabinski, Catherine A.; Gannon, J. E. 1997. Effects of recreational impacts on soil microbial communities. *Environmental Management.* 21: 233–238.

CHAPTER

16

Wilderness Visitor Management: Stewardship for Quality Experiences

by Chad P. Dawson and Paul A. Dawson

INTRODUCTION

Wilderness visitor management is guided by two objectives of the Wilderness Act (U.S. Public Law 88-577): *(1) maintaining the natural setting and (2) providing "outstanding opportunities for solitude or a primitive and unconfined type of recreation."* Wilderness recreation use can impact a relatively pristine natural setting, and too many visitor contacts diminish solitude, cause conflicts, and affect naturalness. *The objectives of wilderness management—maintaining naturalness and solitude—must be achieved without diminishing the wilderness character of an area.* Fortunately, most benefits reported in studies of wilderness use—benefits that can be categorized as relating to development of self, development of community or group social connections, and spiritual development—are related to wilderness naturalness and solitude.

The Wilderness Act generally, and the spirit of the wilderness idea, restrict developing facilities for visitors, such as paving trails and providing campsite facilities, because that would reduce naturalness and interfere with primitive and unconfined experiences. Wilderness managers must rely primarily on means other than physical site development to control unwanted impacts and to manage visitor use. Furthermore, because the public holds many different values for wilderness (Cordell et al. 1998) and these values are sometimes in conflict with one another or wilderness (Manning and Valliere 1996), it is necessary for managers to specify the standards of naturalness and solitude for each wilderness area and the desired conditions for visitor use management (Chapter 9) (Fig. 16.1).

The Wilderness Act (with a few specific exceptions) prohibits structures or installations, commercial enterprises, permanent or temporary roads, motorboats, motorized equipment, aircraft, or mechanical transport (such as bicycles). Established nonconforming uses, such as mechanized access to operate and maintain existing dams or airplane access, have been authorized by legislation in certain areas. The Wilderness Act also provides for some specific, limited exceptions to these constraints *for some wilderness areas*, including existing private rights; emergencies involving the health and safety of people within a wilderness; fire, insect, and disease control; water control structures; access to private inholdings, valid

FIGURE 16.1 The quality of a visitor's wilderness experience is influenced by many biological, social, and managerial conditions. Managers can, to some extent, control many of these conditions to maximize a visitor's opportunities for a high-quality experience. Photo courtesy U.S. Forest Service.

mining claims, and occupancies; commercial operations by outfitters and guides; continuing motorboat and aircraft use established before an area was designated wilderness; and livestock grazing. Many of these exceptions do not apply to wilderness in the national parks where, conceptually, park-resource protection is more comprehensive due to other NPS legislative mandates.

Some wilderness designation laws expand some of the exceptions to the Wilderness Act, compromises that were necessary to their passage. For example, the Alaska National Interest Lands Conservation Act of 1980 (ANILCA, U.S. Public Law 96-487) added 56 million acres to the wilderness system and specifically incorporates the original Wilderness Act's direction, "except as otherwise expressly provided for." However, because of special conditions in Alaska, ANILCA provides for some activities and facilities not permitted in wilderness elsewhere, including continued traditional access by airplanes (mostly floatplanes), motorboats, and snowmobiles; continued use of public and private cabins and possible construction of new public cabins; continued subsistence uses of fish, wildlife, and plants; and provision for fish production facilities such as fish ladders and hatcheries.

Critically important for wilderness visitor management, the Wilderness Act grants exceptions to most of the prohibitions (except commercial enterprises and permanent roads) for administrative activities, but limited to management action "*necessary to meet* minimum requirements [emphasis added] *for the administration of the area for the purpose of this Act.*" This does not give managers a free hand. Management actions must pass stringent tests: the action must be the minimum necessary to meet the requirements for managing the area as wilderness under the Wilderness Act. This minimum requirement concept gives rise to management directives in the Forest Service that managers use the "minimum tools" or "minimum requirements" necessary to achieve specific wilderness management objectives. In general, the act's purposes are to "assure that an increasing population, accompanied by expanding settlement and growing mechanization, does not occupy and modify all areas within the United States . . . to secure . . . an enduring resource of wilderness" and to "preserve its wilderness character" while devoting areas to recreational, scenic, scientific, educational, conservation, and historical use consistent with wilderness.

Management actions cannot be justified because they are convenient or economical or because they achieve non-wilderness goals, such as increasing stream flows, increasing fish and wildlife populations, or controlling predators or other natural processes such as fire, unless they threaten resources outside the wilderness. Determining whether a decision or proposed action is consistent with the act or qualifies for an exception is an important challenge for wilderness managers, and many have agonized over decisions as to whether specific actions are defensible under the act or qualify for the exception.

MANAGING FOR WILDERNESS EXPERIENCES

Wilderness experiences—experiences dependent on wilderness conditions—can be affected by many factors, some of which managers can control or influence. The Wilderness Act's statement of policy is that wilderness areas "shall be administered for the use and enjoyment of the American people in such manner as will leave them unimpaired for future use and enjoyment as wilderness." Wilderness experiences can be addressed using the four qualities that are identified in the Act and listed in the Interagency Strategy to Monitor Trends in Wilderness Character (Landres et al. 2005):

1. Untrammeled—an area free from human control and manipulation. The Act directs the federal agencies to use restraint in its actions and minimize impacts that control or interfere with natural processes and conditions within the wilderness ecosystem. Even managerial attempts to restore natural conditions or eliminate invasive species is considered trammeling. Wilderness visitors and managers are encouraged to use restraint and humility when entering into a wilderness area and not interfere with natural processes and conditions, manipulate the landscape, or the community of life.
2. Natural—an area where wilderness ecosystem conditions and processes are free from the effects of human civilization. Wilderness naturalness is diminished and/or enhanced by three factors:
   a. *Recreational impacts can diminish naturalness through evidence of recreation use*, such as devegetated campsites, eroded shortcuts across trail switchbacks, social trails between campsites, depleted firewood and littering, and impact on wildlife. Recreational pack-stock use can also diminish naturalness from trampling and overgrazing in points of confinement such as picket lines or tether trees. It is the "apparent naturalness" perceived by visitors that is important under the Wilderness Act and also to visitors' experience, as most are not professionally trained to perceive actual naturalness. Some of the more severe

types of impacts (e.g., trail erosion) probably detract from visitor experiences, but more common impacts (e.g., loss of ground-cover vegetation at campsites or lower branches on trees near campsites broken off for firewood) likely have less of an effect, positive or negative, on visitors' experiences (Shelby and Harris 1985, 1986).

b. *Nonconforming but allowed uses have great potential for diminishing the apparent naturalness of wilderness.* Cattle and sheep grazing presents an agricultural rather than wild appearance, and livestock may also consume and trample areas of ground vegetation and create muddy conditions at watering holes and points of concentration. Old mining roads and mine tailings and debris are an obvious intrusion on naturalness, just as are buildings from "inholding" landowners. Even vehicle tracks from administrative access by wilderness managers, where allowed, will diminish apparent naturalness, and managers must set a positive example by using such access only for emergencies.

c. *Natural ecological processes like wildfire, insect infestations, wind, avalanche danger, and fish and wildlife are apparent signs of naturalness that visitors can identify.* The types of wildlife observed, the frequency of observation, and the behavior of animals observed are all expressions of the degree of naturalness. Truly natural conditions may not facilitate the degree of wildlife or scenic visibility most visitor's desire. For example, timber harvest areas may improve visibility and concentrate wildlife more than dense mature forests. Furthermore, wilderness wildlife includes many animal species besides deer, elk, moose, eagles, and other charismatic fauna that most people particularly enjoy seeing. In certain wildernesses, some natural species may be considered dangerous to visitors, especially grizzly bears but also black bears, bison, moose, and poisonous snakes and scorpions. Visitors vary in their appreciation of dangerous animals and in their willingness to share the wilderness with them. Natural wildlife populations often will not be of an optimum size for visitor preferences. For example, deer and elk will usually not be as numerous under wilderness conditions as they might be if the area were managed to maximize their numbers. Some wild animals, such as wolves, wolverines, and cougars, are secretive and difficult to observe.

Wildfire is a natural process that can dramatically affect "apparent naturalness" and, in addition, insects, diseases, climate change, windstorms, floods, landslides, volcanic eruptions, and avalanches can alter vegetation and the landscape over large areas. The type of vegetation, its patterns, and wildlife populations are strongly influenced by fire in most wildernesses. Most visitors will not recognize unnatural successional changes, such as meadows invaded by trees or aspen being replaced by spruce, as the result of wildfire prevention and control.

Visitation generally declines for several years immediately following a fire (Borrie and McCool 2007; R. Brown et al. 2008). However, in many cases, visitors prefer conditions that occur several years after fire has played its natural role, when the landscape is more open and lower vegetation is conducive to larger populations of deer, elk, grouse, songbirds, and other wildlife. On the other hand, when a wilderness fire is actually burning, it can be a threat to visitors, temporarily close some trails, and generate smoke that burns one's eyes and throat and impairs visibility. However, many knowledgeable wilderness visitors support policies to allow fires to burn in fire-dependent ecological communities, and support for fire's natural role has grown over time as a result of education and experience (McCool and Stankey 1986; Taylor and Mutch 1986; Watson et al. 1996).

Natural and wild conditions are not automatically recognized and appreciated by visitors during their experiences. However, "apparent naturalness," meaning "generally appears to have been affected primarily by the forces of nature," is required by the Wilderness Act. *One challenge for managers is to educate visitors more fully to understand natural processes so they can better appreciate experiencing naturalness in wilderness.* Besides, we believe the wildness of wilderness is closely linked to the benefits of experiencing wilderness where natural conditions help people come into balance as human beings, and this is one reason for increasing use of wilderness for personal growth, therapy, and healing (White and Hendee 2000).

3. Undeveloped—an area retaining its primeval character, without permanent improvements, and without human habitation. Section 4(c) explicitly states that "there shall be no temporary road, no use of motor vehicles, motorized equipment or motorboats, no landing of aircraft, no other form of mechanical transport, and no structure or installation." Many wilderness areas have had some of these prohibited activities and structures "grandfathered" into the enabling area designation through special provisions in the legislation for structures or activities that existed prior to designation. Examples of special provisions include: grazing rights, fences, administrative cabins or access roads, airfields, radio towers, motorized access, mechanical transport, water impoundments for agricultural or human consumption, and many other special provisions. While these special provisions make these activities or structures legal, they remain a non-conforming wilderness use and can negatively affect the wilderness experience of visitors.

4. Solitude or Primitive and Unconfined—an area "has outstanding opportunities for solitude or a primitive and unconfined type of recreation." *Visitors' experiences are strongly affected by other visitors and their actions; generally, social conditions affect experiences more than natural conditions.* Outstanding opportunities for solitude or a primitive and unconfined type of recreation are affected by the presence of other visitors, conflicts with other visitors, and visitor behavior.

Crowding as a management concern stems from the Wilderness Act's provisions for "outstanding opportunities for solitude," and also from common conceptions about wilderness as a place away from people. Research on wilderness visitors' reasons for visiting wilderness supports the importance of uncrowded conditions and solitude to many visitors. "Solitude" is something of a misnomer, as a common definition of solitude is "the state of being alone." Although some individuals go alone into wilderness, most go with a small group and think of solitude as their group being separate from other groups. *Solitude in wilderness generally refers to a group of visitors meeting relatively few other groups of visitors.* Meeting no other visitors at all is not as desirable to many visitors as meeting just a few other parties (Cole et al. 1995). Solitude is certainly not the only appeal of wilderness, and for many visitors it is not the most important, but it is a required condition and largely subject to control by managers who may establish acceptable use levels to protect solitude and naturalness.

Wilderness solitude is a relative term. What is considered crowding or solitude depends on crowding norms held by visitors and those they encounter and is influenced by the situation or location in which encounters occur (Hammitt and Patterson 1991; Manning 1985, 1986; Patterson and Hammitt 1990; Cole et al. 1995; Stankey 1973; Watson 1995a). However, there is some concern that newer visitors to wilderness are experiencing more crowded conditions than historically were present, and thus, some speculate that newer visitors may be more tolerant of crowding, as that is how they first experienced wilderness.

Solitude is also impacted by encounters with nonrecreational uses, such as grazing and uses outside wilderness that create noise, air pollution, or visual distractions, including aircraft overflights. Encounters with different types of visitors (e.g., large groups), with different styles of use (e.g., horse users), or for different purposes (e.g., wilderness therapy or personal growth) may affect visitors' experiences more than sheer numbers of other parties. Such unwanted encounters often are one-sided impacts to one type of visitor (Marion et al. 1993). Where motorized use is allowed (e.g., outboard motor use in parts of the BWCAW), visitors not using motorized means of travel prefer not to meet those who are (Cole et al. 1995), but the motorized visitors do not object to the nonmotorized visitors (Adelman et al. 1982). Horse-hiker conflict is usually not as severe, but it also tends to be one-sided, with the hikers complaining more. Large parties are objectionable to persons in the typical smaller groups. Visitors with outfitters and those accessing the wilderness on their own may object to each other, and so may hunters and nonhunters. Reducing such visitor conflicts through wilderness management helps protect solitude.

How managers operate and behave, especially their attitudes and presence, can greatly affect wilderness visitors' experience. Experiences can be damaged by management actions intended to solve other problems, poorly designed actions, or implementation that has not been well thought out. Direct types of management have more potential for adverse effects than indirect approaches. Excessive regulations, in particular, can detract from the sense of "primitive and unconfined recreation" called for in the Wilderness Act. Managerial presence, if too prominent or regulatory in attitude, can diminish the sense of experiencing wilderness and facing challenges on one's own. Managerial contact with visitors is usually preferable outside rather than inside the wilderness, although wilderness rangers usually report a pleasant reception to low-key or less formal encounters with visitors in the wilderness.

DESIRED CONDITIONS FOR WILDERNESS EXPERIENCES

The Visitor Use Management (VUM) framework presented in Chapter 9 outlines how to manage for appropriate wilderness conditions and visitor experiences. One of the steps in the VUM planning process is to define the desired conditions in a wilderness area that are appropriate within the Act, agency policies, and the designation legislation for that area. Desired conditions are aspirational and may not coincide with current existing conditions. Desired conditions describe the visitor opportunities and experiences, resource conditions, and recreational facilities and services that an agency plans to provide and maintain in a given wilderness area.

All conditions and influences on wilderness experiences will vary from wilderness to wilderness and within each wilderness, based on variation in use levels as affected by differences in ease of access, attractions, and natural conditions. Such variation is advisable, as long as all desired conditions meet basic wilderness definitions and agency policies. Total quality in wilderness experiences will be greater by providing a diverse set of opportunities so that people, who vary in their desires and abilities, can find what they want (i.e., a spectrum of opportunities among wilderness areas). Different portions of each wilderness also vary in capacity, durability, and suitability for various types of use; some areas are reached by easy trails close to a trailhead; and some places are very remote with no trails. Thus, there is a wilderness opportunity spectrum (WOS) among most wilderness areas and even within an individual wilderness area (e.g., trailless zones or no-campfire zones).

VISITOR MANAGEMENT CONSIDERATIONS

Wilderness visitor management is one of the most compelling issues facing wilderness managers today. *Two main considerations of wilderness visitor management are (1) to provide visitors opportunities for quality wilderness experiences and (2) to limit impacts on resources caused by visitor use.* These two considerations are interrelated. Aspects of them are sometimes subtle and easily overlooked, but they are becoming increasingly important for managers who must implement the seemingly conflicting goals of providing quality experiences while limiting site impacts (Anderson et al. 1998; Manning and Lime 2000; Dawson et al. 2016). For example, many resource impacts (such as heavily compacted soil in campsites) also affect visitor experiences, so reducing impacts might improve visitor experiences. However, management to control impacts (such as hardening camping sites) could adversely affect visitors' wilderness experiences by reducing naturalness. Often a management solution to one problem has side effects creating another problem, so evaluating alternatives and consequences is critical. A careful assessment of likely management impacts on both visitors and wilderness resources is always required (Fig. 16.2).

FIVE TYPES OF VISITOR IMPACTS

Wilderness managers must consider five types of undesirable visitor actions and their associated effects and

FIGURE 16.2 Most wilderness stewards regard vegetation damage at campsites, such as trampling and loss of ground vegetation, as seen here in Oregon's Eagle Cap Wilderness, as a serious problem. Many visitors, however, are bothered more by crowding and conflicts with other visitors. Photo courtesy U.S. Forest Service.

appropriate responses. Table 16.1 outlines the basis of classifying visitors' awareness and motivation, and how they relate to laws and regulations. The five types of visitor actions to be managed are (1) illegal actions, (2) careless actions, (3) unskilled actions, (4) uninformed actions, and (5) unavoidable impacts.

Type 1. Illegal Actions

Illegal actions violate agency regulations established to implement the Wilderness Act or subsequent wilderness designation legislation. Examples are the illegal use of chainsaws or mechanical and motorized vehicles in wilderness, such as mountain bikes, motorbikes, ATVs, or snowmobiles, that disrupt other visitors' experiences and damage wildlife, soil, and vegetation. The managerial response would be law enforcement but also to communicate legal restrictions so users will know the rules. Managers understand that some illegal actions may result from ignorance and can be treated more like Type 3 (unskilled) or Type 4 (uninformed) behaviors. Mechanized violations, for example, can be addressed by informing visitors of appropriate non-wilderness areas for semiprimitive, motorized recreation.

TABLE 16.1 Five Types of Problem Visitor Actions, Examples, and General Management Responses

Type of Visitor Action	Example	Management Response
1. Illegal actions	Mechanized trespass (e.g., motorbikes or snowmobiles)	Direct regulations supported by law enforcement
2. Careless actions	Littering, nuisance activity (e.g., shouting)	Persuasion, education about impacts, rule enforcement
3. Unskilled actions	Ditching around a tent or burning garbage	Primarily education about low-impact practices, some rule enforcement
4. Uninformed actions	Concentrated visitor use	Education-information programs
5. Unavoidable impacts	Physical impacts from careful use	Reduction of use levels to limit unavoidable impacts; relocation of use to more durable site

Type 2. Careless or Thoughtless Violations of Regulations

Examples of such behavior are littering, parties larger than specified in area restrictions, camping in closed areas, and building wood fires where prohibited. Managers can try to motivate visitors to alter such behavior by persuasion, by making it easier to behave appropriately, and by discouraging inappropriate behavior. For example, managers might provide litterbags or encourage litter pickup through direct appeals by rangers, or physically block old campsites during vegetation recovery. Providing information through effective communication will usually prevent careless actions or thoughtless violations. We must assume that visitors want to do the right action—if they know the rules and the reasons for the rules.

Type 3. Unskilled Actions Result from a Lack of Skills or Knowledge

These actions include such things as digging a drainage ditch around a tent or making a bed or shelter of boughs cut from live trees. Educating visitors about desirable practices can reduce impacts from unskilled actions, and, when necessary, enforcing rules against such actions—but always trying to educate visitors about why such rules are necessary. Wilderness education is making a difference, such as documented positive changes in wilderness visitor norms in Oregon's Eagle Cap Wilderness over a thirty-year period (Carlson 2000; Watson 2000; Watson et al. 1996). The Leave No Trace (LNT) education effort by public recreation agencies and many private organizations is an attempt to combat unskilled actions (Swain 1996).

Type 4. Uninformed Actions

These actions reflect behavior that might have been different if certain information had been communicated to visitors. Uninformed behavior can intensify some types of use impacts, such as by large numbers of visitors who enter a wilderness at a few well-known access points when they might have used other access points or non-wilderness lands if they had known about use patterns and alternative places. Wilderness recreation use is often poorly distributed and concentrated in a few places, which leads to impacts no matter how well visitors try to behave. Use wilderness education to provide visitors with prior information about alternatives available, or as a last resort, limit access at popular entry points.

Type 5. Unavoidable Impacts

These impacts reflect the reality that every party visiting a wilderness, by their very presence, causes some minimum, unavoidable impacts to other users (e.g., visitor crowding and conflicts) and to the resources (e.g., human and pack-stock impacts on trails, shoreline, campsites, and water conditions). For example, a party arrives on-site, another party is there and notices them, and the solitude of both parties is reduced. Vegetation under a tent or sleeping bag is unavoidably compressed and damaged to some extent. Trails will be impacted by travel, especially in wet conditions. LNT and minimum-impact educational programs have helped reduce visitor impacts that previously were thought to be unavoidable by teaching about appropriate behaviors and equipment (Swain 1996). However, even with all the skill and knowledge possible, there remains a minimum, unavoidable level of impact. If, after all actions to reduce visitor impacts have been taken (e.g., visitor education, trail relocation to more durable sites), the accumulation of unavoidable impacts is excessive, managers can still limit use after other options have been tried. Rolling back use levels may be politically difficult to impossible to achieve—hence, we suggest that use limits should be evaluated along with other options and not held back as the last alternative.

FIGURE 16.3 Most wilderness visitors tolerate fire rings and campfire remains, but many managers find such visual and physical evidence of use unacceptable. Photo courtesy U.S. Forest Service.

MANAGERS' AND VISITORS' PERCEPTIONS OF PROBLEMS

Visitors often see wilderness problems differently than managers, and managers must understand this when defining problems and seeking solutions. Visitors are looking for an experience, and they may not be aware of their impacts, whether careless, unskilled, or uninformed. Trying to change visitor behavior is difficult at best, but it becomes almost impossible if visitors do not perceive their actions and impacts as undesirable, no matter how inappropriate managers may believe it to be. Thus, *education must be a key component of visitor management, and to be good and effective wilderness educators, managers must try to see situations from a visitor's point of view.*

Washburne and Cole (1983) reported that managers of different wilderness areas usually indicated similar visitor and environmental problems and usually saw these problems as localized rather than widespread. *Managers considered human impacts on campsite vegetation and trail conditions to be the most serious problem* (Manning et al. 1996; Marion et al. 1993; Washburne and Cole 1983). The managers surveyed perceived ecological impacts as a problem more often than visitor crowding or conflicts among users. However, Manning and Lime (2000, 41) compared four studies of National Park Service wilderness managers from 1979 to 1993 and concluded that *crowding and visitor conflicts are increasingly reported by managers and that carrying capacity "has become a pervasive but largely unresolved issue."*

Most surveys of wilderness visitors indicate more concern with social conditions, such as crowding, conflict between visitor groups, and littering, than with resource conditions, such as campsite and trail impacts (Cole et al. 1995; Shelby and Harris 1985) (Fig. 16.3).

Generally, managers are more concerned about protecting wilderness resources from physical and biological impacts, whereas visitors are more concerned about their experience, including access and freedom from intrusions by other users (Lucas 1985; Cole et al. 1995). These perspectives are understandable as managers are responsible for protecting the resource and visitors want to experience it. A survey of agency-training needs revealed that managers are aware that they need to be better trained and educated to effectively and efficiently meet the present and future demands of visitor management in wilderness (Conrad 1997; Dawson et al. 2016).

Educating visitors is a key management strategy; introducing such concepts as LNT helps visitors minimize their impacts. Other wilderness ethics programs are equally important to foster stewardship and protection of wilderness resources. Hansen (1990) cautioned wilderness managers that information and education programs must be consistent from the front desk at visitor centers and ranger stations to the information dispersed at trailheads and in the backcountry.

Because the supply of wilderness is ultimately limited, we urge visitor management approaches (e.g., information and education) that make it easier—not harder—for visitors to experience wilderness, including redirecting visitor recreation use to a more appropriate activity site or wilderness with less use, if necessary. Extensive evidence shows that wilderness settings lead to wilderness-related experiences that are good for people because they provide numerous human benefits (Watson et al. 2016).

Balancing wilderness visitor use against maintaining naturalness and solitude is the essence of the wilderness stewardship challenge before managers today and into the future, especially in view of the need to restore ecological conditions and processes in some wilderness areas. Thus, wilderness management should focus on providing opportunities for visitor experiences and encouraging appropriate wilderness use within the sustainable capacity of the resource—such ideals must be the hallmark of stewardship in the next century (Watson et al. 2016; McCool and Freimund 2016).

VISITOR IMPACTS ON SOCIAL CONDITIONS

The condition of the natural environment in wilderness influences the social setting and the overall experience of visitors while on their trip in wilderness. The social setting is additionally influenced by each visitor's behavior, by visitor interactions within their group, and by between-group interactions during their various activities on a wilderness trip. Visitors arrive in wilderness settings having made choices and have expectations of what they will see and experience. Ultimately, visitors seek some level of satisfaction with their trip and will cope with various environmental and social conditions to arrive at overall trip satisfaction.

In addition to directly accounting for social settings, managers must be aware of how actions and decisions that modify management and resource settings affect the social setting. Managers need to develop a perspective on the natural and social settings important to visitor experiences when making decisions. This section briefly presents basic elements of the social experience and how visitors handle social interactions with each other during on-site experiences and provides a selected summary of recreation management research as it applies to wilderness.

Satisfactions and Benefits

Visitor satisfaction with recreational experiences has been measured using overall trip satisfaction as well as breaking the experience into component types of satisfactions (so-called multiple satisfaction approach). At the most basic level, trip satisfaction measures ask the visitor to indicate satisfaction with the overall experience, and, while this is helpful, it may hide some dissatisfactions that visitors have with certain aspects of the trip that would be helpful for managers to know. Overall trip satisfaction measures do not easily translate to management applications. Multiple satisfaction measures are more complex and target specific attributes associated with the setting and experience that are important to visitors and management. The multiple satisfaction approach can target aspects of the resource that management can control (e.g., trail condition, campsite condition, educational materials, etc.) and the social experiences, such as solitude, encounters, and crowding. Managers can use this approach to determine if the on-the-ground condition matches the objectives and standards defined in the management plan or the desired conditions in visitor use management.

Driver and Brown (1983) used expectancy theory to develop the concept that people engage in particular activities and choose particular resource and social settings in order to fulfill personal needs and outcomes from recreational experiences. Needs were defined as a preexisting condition that developed prior to the recreational activity. The visitor chooses the activity, resource, and social setting with the expectation that the combination will satisfy their needs. The visitor may have a need to escape industrial areas, for family bonding, excitement, adventure, social interaction, learning, stress reduction, and so on. The desire to fulfill a need is what motivates people to engage in wilderness recreation.

Satisfactions and benefits are the outcomes of the recreational experience in wilderness. An individual's satisfaction leads to personal benefits such as improved health, reduced stress, and improved capacity to perform his or her work. The combined benefits to individuals form the overall social benefits to society, such as preserving cultural heritage, family cohesion, reduced need for medical services, and others. Societal benefits are the important outcomes from recreational use of such resource and social settings as found in the NWPS (Watson et al. 2016; McCool et al. 2007; Driver et al. 1987).

Normative Behavior

In outdoor recreation management, social norms are defined as shared beliefs about how people ought to behave in certain settings (Heywood 1996) and about standards for what is acceptable and not in certain settings (Hall and Shelby 1996). Normative beliefs guide human behavior during the recreation experience. In wilderness settings normative beliefs concerning appropriate camping practices guide how backpackers establish campsites and interact within the camping group. Norms also govern interactions among groups.

Social norms evolve over time and vary among settings. Normative beliefs concerning standards describe the ability of wilderness visitors to accept on-site conditions, including physical impacts, management regimentation, and crowding. Impacts or crowding beyond the level accepted by visitors adversely affect their wilderness experience. Managers need to be aware that social norms may exist among the people using a wilderness. Social norms can be used to judge the acceptability of management decisions affecting the resource as well as to assess the impact of management decisions on the wilderness experience.

Visitors influence each other's behavior. Information and education programs have a multiplier effect because what is learned and shared directly or indirectly (for example, observations by novice visitors of the behavior of more experienced visitors) and can either reinforce or undermine management actions and programs. Managers can influence normative behavior by modeling, informing, and educating visitors about appropriate behaviors in various wilderness settings (for example, small wood fires where allowed, LNT practices for camping, and selecting campsites within approved regulations).

Visitor Density and Perceived Crowding

Density is an objective measure that counts the number of people in the system using methods such as number of backcountry permits issued, trail counters, or parking lot vehicle counts. Management can control density by limiting the number of permits issued, restricting parking, or charging an access fee. The question in terms of the visitor experience becomes, when is visitor density so high that the experience is undermined? (Freimund and Cole 2001)

Crowding is defined as a negative and subjective evaluation of visitor density that is made by the visitor (Kuss et al. 1990). Solitude is an important component of the wilderness experience, and some limited social interactions are often expected in wilderness. Thus, solitude, crowding, and high or low density are all relative terms. Whether a wilderness visitor experiences solitude or crowding in a given situation is relative to the individual's perception of that setting.

Stress and Coping with Dissatisfying Conditions

Crowding, solitude, and density are relative terms and management application of them is complex. Contributing to the complexity of these terms is the fact that visitors often initiate a coping process in response to dissatisfying situations (Schuster et al. 2006). The coping process is when an individual appraises a social or resource situation as problematic and then chooses coping mechanisms that are thought to be able to help solve the problem. These coping mechanisms may be problem-focused and directed at modifying the problem environment or emotion-focused and directed at reducing the internal and personal effect of the problem.

Common coping mechanisms employed in responding to social interactions include rationalization, product shift, displacement, avoidance, confrontation, and problem-solving (Anderson and Brown 1984; Schuster et al. 2006). Rationalization occurs when the individual justifies the difference between on-site conditions and expectations to maintain cognitive consistency. Product shift is a change in the definition of the recreation opportunity after the on-site experience does not match expected conditions. Displacement is categorized as intersite displacement (leaving the area altogether), intrasite displacement (participating in the same activity at a different location on the same area), and temporal displacement (participating in the same activity at the same location at a different time of day, week, or year). Avoidance is when a hiker uses subtle behaviors to avoid contact with other wilderness users (e.g., change hiking pace, leave a view spot early, or avoid conversation). Confrontation occurs when the individual actually interacts with the party responsible for causing the problems. Finally, problem-solving occurs when the individual makes a plan of action that is designed to modify the environment or reduce and remove the source of the problem.

These issues are relevant to wilderness management because people cope with situations that are not perfect and it is difficult for management to assess experience quality from the visitors' perspective. For example, a wilderness visitor may experience a problem, cope with it, come to a personal resolution for the issue, and not accurately report to management that there was a problem requiring attention. Thus, increased visitor density does not necessarily lead to increased perception of crowding or to conflict, and visitor satisfaction measures as indicators of experience quality may be misleading if only overall trip satisfaction is measured. Most management plans require that social conditions be maintained within certain standards for encounter levels and perceptions of crowding. A report of high trip satisfaction or low perceptions of crowding are considered inaccurate by themselves, if the wilderness visitor had to employ coping mechanisms to achieve the perception of high trip satisfaction.

Sense of Place and Relationships with Place

Attachment to a place is the extent to which an individual values or identifies with a specific natural or cultural setting (Brooks et al. 2006; Farnum et al. 2005). Place dependence is relative to the importance of a resource or location in providing the social and natural setting necessary to participate in a given activity. The wilderness includes a collection of attributes that make it functional for specific purposes. Value is seen by the visitor as the degree to which a place provides satisfying hiking or kayaking locations, opportunities for solitude, or other amenities. Place identity refers to the dimensions of the individual's personality that are defined by the wilderness area, and is a subjective and cognitive combination of the ideas, beliefs, experiences, values, goals, and preferences that the individual holds in relation to that place.

The type and strength of attachment to the place influence the visitor's perception of acceptable resource, social, and management conditions. Managers should be aware that management plans and actions will be viewed differently depending on the place attachment of visitors. Place attachment also has been associated with the local public's reaction to local management actions and broad policy changes.

Visitor Conflict

Recreation conflict has been defined as goal interference attributable to another's behavior (Jacob and Schreyer 1980). Recreation conflict can be categorized into four types: (1) intra-activity conflicts, such as between hikers—maybe two different types of hikers; (2) inter-activity conflicts, such as between backpacking hikers and horse users; (3) conflict between recreation visitors and other permitted users of wilderness, such as between hikers and livestock grazing; and (4) conflict between recreational visitors and management activities and approaches, such as campsite or trail closures in a wilderness with limited alternatives (Schreyer 1990). Graefe and Thapa (2004) note that the true nature of conflict is more complicated than the four categories allow.

Conflict between visitors is often a greater problem in the visitors' opinions than sheer numbers of other visitors or crowding (Manning 1985; Stankey 1973). Conflict is usually between different types of visitors (e.g., horse users and backpackers, large and small groups), but sometimes it is caused by objectionable behavior by similar visitors. Large parties are a source of dissatisfaction to most visitors (Cole et al. 1995). Some conflict stems from different methods of travel, particularly where motorized use is permitted as an exception; for example, outboard motors in parts of the BWCAW (Cole et al. 1995). Horse/hiker conflicts tend to affect hikers more than horse users (Lucas 1985).

Conflict is often defined as asymmetrical, meaning that one group experiences conflict while the other does not. For example, wilderness horse riders have expressed concern over llama pack-stock users while llama packers do not perceive problems with horse riders (Blahna et al. 1995). Asymmetry should not be assumed; a manager should fully investigate conflicts as they arise. Conflict is just as likely to arise within groups as between groups.

Defining conflict as a discrepancy in social values provides another approach to understanding and anticipating conflict. Vaske et al. (1995) found that even though hikers on Mount Evans in Colorado did not see hunters, they expressed conflict with the group based on differing social values. When asked a hypothetical question concerning potential hiker-hunter interactions in wilderness, hikers indicated that they would consider seeing hunters as problematic (Schuster, Hammitt, Moore, and Schneider 2006) (Fig. 16.4).

Activity style or technology has been used as a categorization variable for conflict. For example, hikers have been found to perceive conflict with mountain bikers due to a perceived difference in social values (Carothers et al. 2001), but the conflict was asymmetric; mountain bikers and dual-sport participants were less likely to perceive conflict with hikers.

FIGURE 16.4 For many visitors, seeing wildlife is an important part of their wilderness experience. Photo courtesy U.S. Forest Service.

Conflict may occur within an individual. Problems may arise during the trip that have no bearing other than on the individual or blame cannot be placed on another. For example, wilderness hikers have been found to experience problems due to getting lost, poor planning, or undertaking an endeavor that is beyond their physical ability. While a manager may not have control over these situations, being aware of them is useful in the larger process of assessing visitor experiences and managing the overall social setting.

ELEMENTS OF VISITOR USE SUBJECT TO MANAGEMENT

Management cannot control or shape all the things that influence wilderness visitors' experience. Two aspects of visitor behavior potentially influenced by management are (1) effects on the resource and (2) effects on the experiences of others. How and to what extent can these different aspects of use and their impacts be modified? How can management of each aspect reduce ecological impacts and improve the quality of the visitor's experience? We address these questions in considering seven use characteristics that are subject to management: amount and distribution of use, redistribution of visitor use, method of travel, party size, length of stay, visitor behavior, and visitor effects on the resource.

AMOUNT AND DISTRIBUTION OF USE

Managers can directly restrict the numbers of visitors, if that is the minimum requirement (or minimum-tool decision) necessary for keeping impacts within the physical or social carrying capacity of the area. Restricting use would be a last resort after other approaches have failed and impacts cannot be reduced sufficiently in other ways. For example, use can be managed *indirectly*, through education, by closing roads, by lengthening trails, or by changing signs; or *directly*, by limiting the number of use permits issued or redirecting use with a ranger on-site. However, direct management of use should be a last resort after educational and indirect approaches have been exhausted. One must weigh the results of limiting the number of visitors entering an area against tighter regulation of use inside the wilderness, which may reduce the quality of visitor experiences. A balance is needed, and the optimum solution may vary from area to area.

Wilderness use distributions are typically uneven both spatially and temporally, partly due to uninformed actions by users who possess limited knowledge of alternative opportunities. However, sometimes concentrated use reflects easily accessible trails, differences in the attractiveness of locations, or proximity to population centers. A manager's first tool for altering the distribution of use is providing information and education to users. If that does not solve problems, limits might be imposed at entry points, or in certain travel zones, but always with information on alternatives made available to visitors.

REDISTRIBUTING VISITOR USE

Efforts to redistribute recreational use need to be linked to well-defined management objectives to avoid simply spreading problems more widely. Wilderness managers may have five objectives for seeking to shift some use:

1. *Redistribute some use from wilderness to non-wilderness locations*—Many wilderness visitors probably would be as well or better satisfied in areas having more facilities and managed more intensively for dispersed recreation than wilderness can be. These users are seeking relatively primitive, roadless recreation experiences but do not require or necessarily want the completely undeveloped conditions provided in designated wilderness. Managers, by providing information about roadless, non-wilderness locations, might redirect users to desired experiences, thereby relieving pressures on wilderness while at the same time better meeting some visitors' desires. Similarly, information about wilderness needs to stress its special character and de-emphasize recreation opportunities that are not wilderness dependent.
2. *Redistribute some use from heavily used to less used wilderness areas, or non-wilderness and roadless areas as appropriate*—Use can vary greatly from wilderness to wilderness, and sometimes a lightly used wilderness, capable of absorbing more use, can handle redistributed use from a heavily used area. Through information and education, managers can encourage those visitors who are particularly concerned with solitude and relatively undisturbed conditions to visit more lightly used areas, thereby diverting visitation from heavily used areas. If this approach does not shift use as much as desired, it might need to be supplemented by more direct controls. *Shifting visitors among wilderness areas needs to be carefully planned,* because it could result in merely exporting problems from one area to another or reducing diversity within the wilderness system—the concept of a wilderness opportunity spectrum (WOS) suggests intensity-use variation from unused or lightly used areas to more heavily used areas, but still within desired wilderness conditions.
3. *Redistribute use within a wilderness to less used entry points, trails, and campsites*—Several access points, a few miles of trail, and a few campsites usually account for a large proportion of all use in any wilderness area. Providing information about less used locations might

help guide visitors to them. If that doesn't work, more direct redistribution of use could be imposed through wilderness permits or on-site dispersal by wilderness rangers. Again, these direct measures should be imposed as last-resort actions, with care not to just redistribute the impacts.

4. *Redistribute some use to more appropriate locations or time periods better able to accommodate the use*—For example, horse users might be encouraged to go to places with more abundant forage and durable sites to reduce physical impacts, and to places where hikers do not usually go to reduce hiker–horse user conflicts. In another example, visitors might be diverted from specific locations like the shoreline of alpine lakes because of sensitive vegetation or wildlife use (Fig. 16.5). Most wilderness areas have short use seasons and many have sharp weekend and holiday peaks. Managers might shift use from peak times to low-use periods, first by providing information about peak and off-peak use levels, such as weekdays versus weekends and summer versus fall seasons. Finally, the minimum tool to accomplish the redistribution may be to restrict peak use, while at the same time providing advance information about anticipated peak-use situations and alternative areas.
5. *Concentrate heavy use in more popular locations that are durable and already impacted*—For example, in some wilderness areas recreation visitors might be encouraged to go to already popular places with durable sites to reduce physical impacts to other areas in the wilderness and allow some visitors who seek solitude to have that experience in the rest of the wilderness area. If visitors are appropriately informed about the different experiences possible in other areas, it may improve their satisfaction and reduce user conflicts. However, this strategy can have two serious drawbacks: some visitors may become accustomed to the higher-impacted sites and crowded conditions in those areas and assume that these conditions are generally allowable in wilderness, and the impacts on these high-use recreation sites may become unacceptable in wilderness. This strategy may require a rest and rotation of sites to allow heavily impacted sites to recover, if research shows that such an approach is ecologically appropriate in a reasonable time frame. If visitor impacts become unacceptable, it may require managers to limit use, especially during peak-use times.

FIGURE 16.5 Camping near lakes and streams, as in the Selway-Bitterroot Wilderness in Idaho, is often prohibited. Nevertheless, campers often claim they are unaware of the regulation or that they were unable to find a campsite farther from shores and banks. Photo courtesy Richard Behan.

METHOD OF TRAVEL

In most wildernesses, the majority of visitors hike, and horse travel is a distant second. In several areas, boats and canoes are used (some with motors in the BWCAW); kayaks and rafts are common on wilderness rivers; and floatplanes and bush planes are used to access certain areas in Alaska, Idaho, and Montana. The method of travel may provoke conflicts and increase both environmental and social impacts. Managers can use information and education to influence behavior of certain types of users to reduce impacts and conflicts (e.g., whitewater outfitters, horse users, and backcountry pilots). Some types of travel may be limited to certain portions of a wilderness and make it possible to separate different travel methods by giving each type of travel its own use zone or certain time of use. Commercial operators of cruise ships in coastal wilderness and flight seeing in Alaskan wilderness areas such as Glacier Bay and Denali require special considerations and management under ANILCA.

PARTY SIZE

Large parties are not common in most wildernesses, but when encountered they diminish other visitors' experiences (Cole et al. 1995) *and, almost inevitably, impact a larger area than smaller parties* (Cole 1986). Managers already limit party size in many places, but this technique needs to be supplemented with education and information. Frequently, large parties are sponsored by organizations that are easier to contact than independent users, and they may be willing to cooperate with wilderness managers.

The question of what limits to place on party size is controversial and becoming more so all the time. Should group size be limited to six, nine, twelve, or fifteen? There is no perfect and completely defensible answer. However, commercial outfitters complicate the question, as do wilderness experience programs that require some reasonable group size to remain cost-effective and economically viable as a business or program (Ewert et al. 1999). Good information on visitor use, user impacts, and carrying capacities, plus some common sense, are needed to make these decisions.

LENGTH OF STAY

Length of stay can be regulated, but generally it does not contribute much to overuse because few parties stay in the wilderness very long (Cole et al. 1995). Nevertheless, when a party stays for a long time at a popular campsite, it unfairly monopolizes the site and increases impacts. For these reasons, some limit on length of stay at any one campsite seems desirable. At locations where long stays are likely to occur, some reasonable limit on length of stay may permit more people to visit the area. Here again, good information and some common sense are needed to make decisions on limiting the length of stay.

VISITOR BEHAVIOR

Visitor behavior includes many actions: staying on trails or shortcutting them, choosing a spot to camp, whether a campfire is built and how big, how human wastes are handled, how noisy a group is, consideration of other visitors, littering, and so on. *Visitor management to influence behavior of wilderness users is key, and it can reduce or eliminate the need to regulate and control visitor use.* All undesirable behavior is not inevitable—it can be influenced through information and education programs. People can adopt new practices and, in fact, research has shown that visitors' understanding of appropriate behavior has changed and improved over time (Lucas 1985; Watson et al. 1996).

VISITOR EFFECTS ON THE RESOURCE

Resource impacts can be dramatically increased or reduced, depending on visitor behavior. "Illegal" and "careless" behaviors (Types 1 and 2) may be more serious than "inappropriate behavior by well-meaning but unskilled visitors" (Type 3), but unskilled behavior probably is most common.

Low-impact camping involves using appropriate equipment and learning camping and traveling skills that can reduce resource impacts. For example, the LNT program teaches visitors to use campfires only where wood is abundant, keep them small, and leave little evidence of the fire.

Horse use requires special skill and knowledge because of its potential for damaging soils and vegetation. Most horses weigh more than a thousand pounds, and their weight is supported on small hooves with iron shoes. A horse can eat twenty, thirty, or more pounds of grass and other forage each day in a wilderness. Excellent educational materials are available on how to minimize impacts while using horses. In addition, new, light equipment reduces the number of pack animals needed to transport camp equipment, thereby further reducing impacts and probably reducing visitor conflicts as well. In some areas, visitors are using llamas as pack animals, and they cause fewer impacts than horses or mules because they are lighter and have small, unshod hooves (Watson et al. 1998).

WILDERNESS VISITOR MANAGEMENT APPROACHES

Several important issues to consider in wilderness visitor management include direct versus indirect management, the role and use of regulations, location of management, and multiple combinations of management methods.

DIRECT VERSUS INDIRECT VISITOR MANAGEMENT

Direct management emphasizes regulating behavior by restricting individual choice as managers exert control over visitors (e.g., limiting use, permit requirements). Indirect management emphasizes influencing or modifying behavior by managing factors that influence visitors' decisions, such as information and education or making access difficult by limiting parking space or road maintenance, and individual visitors retain freedom to choose, as managers try to influence visitors while allowing more variation in use and behavior. Direct and indirect visitor management approaches are discussed in more detail later in this chapter (Manning and Lime 2000).

Usually, indirect management should be the first choice, with direct management used only when indirect means cannot achieve management objectives or desired conditions under VUM (Chapter 9). Most wilderness users prefer indirect management, and the concept of wilderness as an undeveloped, open, and unconfined place accentuates the desirability of a management philosophy that is as indirect, unobtrusive, and subtle as possible. Trying indirect management first agrees with the concept of minimum requirements and the minimum-tool decision that calls for the minimum necessary regulation, action, or force to solve a problem or meet a management objective or desired condition (Fig. 16.6). Education is the key to most indirect management, and it is the main managerial response to careless, unskilled, and uninformed actions (Types 2, 3, and 4 in Table 16.1).

Direct controls are sometimes necessary to address specific problems, but they should be applied in ways that allow as much visitor freedom as possible. Respecting visitor freedom is a high priority in wilderness visitor management. Visitor regulation is a common and important tool for managing both ecological impacts and visitor experiences, but it should be applied with restraint and only when and where it is the minimum necessary approach to solve the problem.

There is some controversy about whether indirect management approaches are as effective as direct management (Shindler and Shelby 1993; Manning and Lime 2000). Some wilderness researchers suggest that indirect and direct management approaches should not be thought of as alternatives to each other, but rather that they form a continuum of management options (McCool and Christiansen 1996) and many of these techniques can be used effectively together (Cole, Watson, Hall et al. 1997).

Employing several approaches and techniques in a balanced, coordinated manner is needed in wilderness visitor management. Relying on only one approach is less effective than an integrated combination of approaches—neither regulations, education, design, nor wilderness rangers alone are the complete answer. *No single approach is so powerful, versatile, or acceptable in terms of costs to visitors or managers that it constitutes a panacea.*

REGULATION—COSTS AND BENEFITS

Regulations can help solve some problems, but solutions are not free. For managers, the costs of regulation are in terms of employees' time needed to explain and enforce them, plus potential political costs from angry letter writers and e-mailers, phone callers, or even campaigns led by people offended by government regulations. For visitors, the costs are in terms of hassles, oppressive reminders of civilization and government, lost opportunities, and diminished enjoyment from wilderness experiences that are not "primitive" or "unconfined," as called for in the Wilderness Act.

Costs to visitors stem in part from differences between the basic definitions of "recreation" and "regulation." Recreation is usually defined as a type of human experience that

FIGURE 16.6 Managers use trail design to indirectly influence a visitor's experience. Fording a stream (upper left) affects hikers differently than when crossing on a bridge (upper right). Removing trails and roads is an important management option as well. The Bureau of Land Management in California planned to remove hundreds of miles of old four-wheel-drive roads (lower left) following the 1994 passage of the California Desert Protection Act. In partnership with the Student Conservation Association (SCA), restoration began (lower right). Photos courtesy (upper left, upper right) U.S. Forest Service and (lower right, lower left) Dave and Katie Wash, BLM.

results from self-rewarding physical or mental engagements, based on personal free choice during leisure time. This definition stresses internal control and free choice for personal reasons that vary as much as do people. Regulation, in contrast, is defined as a means to govern and to produce order or uniformity. Thus, instead of the self-direction and diversity inherent in recreation, regulation invokes external direction and uniformity.

Tension between recreation and regulation intensifies in wilderness management because the Wilderness Act defines wilderness as an area that provides "outstanding opportunities for solitude or a primitive and *unconfined* [emphasis added] type of recreation." However, the act also requires managers to protect and manage wilderness "so as to preserve its natural conditions"—an almost impossible mandate if uncontrolled recreational use is allowed.

The appropriateness of wilderness recreation regulations depends largely on the benefits and costs of implementing a specific regulation, compared to nonregulatory alternatives for solving a particular problem. Furthermore, the effectiveness of wilderness recreation regulations depends on information and education programs combined with enforcement to support the regulations.

Research suggests that regulations can diminish the quality of recreational experiences. The notions of autonomy, solitude, and privacy have been found to be highly valued in studies of wilderness visitors (Dawson and Hammitt 1996; Hammitt and Madden 1989; Roggenbuck 1980; Brown and Haas 1980).

Regulation can be beneficial, and, of course, recreation in many sports and games requires rules. Eliminating some freedoms in outdoor recreation can create other, perhaps more valuable, freedoms. Safety is a case in point. For example, removing power boaters' freedom to operate in swimming areas greatly increases swimmers' freedom to swim safely. Even the benefits of safety can have costs if safety regulations detract from high adventure, such as preventing whitewater kayaking in difficult rapids. However, where safety is not an issue, regulations can also protect experiences for some by limiting the behavior of others—for example, regulations requiring quiet in campgrounds after 10 pm.

A key, positive role for regulation is allocating recreational opportunities. Thus, managers can separate conflicting types of recreation, such as snowmobiling and skiing. *Research suggests that wilderness visitors often accept regulations fairly well, particularly if the necessity is explained, and once they become accustomed to them.* For example, 82 percent of the applicants for the limited number of permits to visit California's San Gorgonio and San Jacinto Wildernesses supported restricted use of the area; even those turned away without a permit supported rationing through permits by three to one (Stankey 1979). Overall, recreational visitors and managers share a sense of wilderness stewardship to work as allies for wilderness protection—even when user regulations are necessary (Cole et al. 1995; McCool and Utter 1982; Lucas 1980).

ACCEPTANCE DOES NOT EQUAL PREFERENCE

Accepting restrictive regulations, as described earlier, should not be misinterpreted as indicating that visitors like such regulations. Surveys of visitor attitudes toward regulations show a strong preference for less authoritarian styles of wilderness management (Cole et al. 1995). In many cases, conscientious visitors accept restrictions because they see them as necessary to sustain desired wilderness conditions (Hendee and Lucas 1973). On the other hand, some people do object to regulations (Behan 1974). The debate over how much regulation should be used in wilderness management has been going on for several decades, with indirect—less regulatory and information-based—approaches finding more acceptance by managers as well as visitors.

Visitor acceptance can be a false indicator that all is well if people avoid an area because of what they view as restrictive regulations. These displaced visitors are succeeded or replaced by those who accept the regulation. This can give managers a false sense of success. Research on this invasion and displacement of one kind of user with another due to regulation avoidance is difficult and scarce.

USE OF REGULATIONS BY MANAGERS

Several surveys of wilderness management practices show that regulations are common. When asked what they considered the most effective management technique, the majority of managers listed educational contacts; however, a large minority also listed regulations, such as designated campsites and lakeshore camp restrictions, as the most effective techniques.

National Park Service managers reported that they used backcountry overnight permits (68 percent) and group-size restrictions (62 percent) most frequently and then a wide combination of direct and indirect measures to manage or influence user behavior, such as a minimum-impact backcountry-use education program (Manning et al. 1996; Manning and Lime 2000; Marion et al. 1993). Similarly, most state wilderness areas restrict overnight use by requiring that only designated campsites be used, and they use a broad array of direct and indirect measures to manage or influence user behavior.

GUIDELINES FOR REGULATION

Some general guidelines seem reasonable for managers trying to fit regulation into its appropriate place in recreation management, whether related to managing impacts or visitor experiences, or both.

1. *Do not regulate if effective nonregulatory alternatives exist*—Establishing a regulation, by itself, achieves nothing, although it may provide a sense of satisfaction that something is being done. This may be an illusion, as *the potential effectiveness of a regulation can be tested with three questions.*

   First, is it possible to inform visitors of each regulation? Managers have a responsibility to inform visitors before citing them for a violation. The old principle "ignorance of the law is no excuse" seems out of place in wilderness recreation management. Limited knowledge of recreation regulations is common. For example, the main reason given by violators of a regulation prohibiting camping too close to water in Forest Service–administered wilderness in Colorado was they were unaware of the regulation (Swain 1986).

   Second, if visitors are informed about a regulation and understand it, will their behavior change enough to solve the problem? The main underlying question here is whether visitors' motives, knowledge, attitudes, and behavior are understood well enough to estimate their response to a regulation. Visitor behavior can be changed by regulations, but drastic changes in behavior are usually difficult to achieve. For example, Swain (1986) found the main reason Colorado wilderness visitors violated a regulation requiring dogs to be kept on a leash was that they rejected the need for such a regulation, and many said they intended to continue to violate the regulation, even if it meant paying a fine.

   Third, if the regulation changes visitor behavior as intended, will it really help achieve management objectives? For example, if visitors do not camp at three existing campsites within 200 feet of an alpine lake, three new campsites may be established beyond 200 feet. Furthermore, parties eating lunch or just resting at the old campsites may trample the vegetation enough to prevent substantial recovery, resulting in six impacted sites where once there were only three.
2. *Try to develop effective nonregulatory (indirect) visitor management*—Some wilderness managers are too quick to assume that regulations are the only or the most effective way to achieve objectives, and thereby sell indirect, nonregulatory approaches short. For example, a wilderness ranger reported that about one-half of the campers in his area were camping within 200 feet of shorelines even after several years of intensive efforts to enforce a regulation prohibiting camping within 200 feet of lakes and streams. An information program that described the problem and explained the impacts of campsite location on the environment, on other visitors, and on sanitation might have changed specific campsite selection by visitors, at least as much as the regulation with its poor compliance. Wilderness visitors seem particularly well suited for management programs based on education and information, as one of the most distinguishing socioeconomic characteristics of wilderness visitors is a high level of education. Furthermore, wilderness visitors usually attach high personal importance to wilderness, which suggests many would respond to information about how to protect these areas.
3. *Explain why regulations are necessary*—An explanation of necessary regulations should improve compliance, especially in ways that relate to the most important aspects of the problem. Good explanations could reduce the costs to visitors by reducing perceptions of regulations as arbitrary and capricious and help make them seem more reasonable and necessary. Consistently explaining why regulations are necessary, from public information at the front desk to ranger contacts at the wilderness trailhead, helps support the rationale linking the regulation to a management problem.
4. *Provide the minimum regulation needed to solve the problem*—Regulations span a continuum from severe to relatively mild. Avoid regulations that are more strict or sweeping than needed or that restrict visitor behavior that is not part of the problem. Target the specific behavior as precisely as possible. For example, if campfires are causing unacceptable impacts at some locations, prohibiting fires would deal with the problem more precisely than banning camping in that area.

5. *Regulate outside wilderness rather than regulate activity inside wilderness*—In wilderness and backcountry recreation, more freedom and spontaneity can be preserved if most regulations are applied outside the area; those admitted to an area would be substantially free to travel and camp with little regulation. Limits on number of visitors admitted, party size, and method of travel are examples of "outside" regulations; assigned campsites or prescheduled travel itineraries are examples of "inside" regulations.
6. *Monitor the effects of the management actions on target problems*—Monitoring of some sort is essential because most management actions are taken with only a limited understanding of their likely consequences. Monitoring is the only way to determine what the management actions really accomplish. Managers must monitor effectively so they can use the information to minimize regulations if possible, to make necessary regulations more effective through progressive fine-tuning, and to develop other approaches to solving problems.
7. *When considering recreation regulations, remember that visitor use and enjoyment are one of the reasons why wilderness exists*—The Wilderness Act in section 2(a) established that wilderness was designated primarily to preserve natural ecosystems while also offering opportunities for wilderness experiences. However, support for wilderness depends also on the support of people who benefit from experiencing it. The visitors are interested in wilderness and should be viewed as being on the same team of wilderness supporters and advocates as the managers. *The goal of wilderness visitor management should be to allow and facilitate as much use and enjoyment of wilderness as is consistent with protecting wilderness naturalness and solitude—not to keep people out of wilderness.* The only precautionary note is that the first priority has to be preserving wilderness character, conditions, and processes.

LOCATION OF MANAGEMENT

Applying regulations outside of wilderness instead of inside is part of a more general wilderness visitor management issue. Contact with visitors, in person, through signs, and with printed materials, to present information about suggested practices, regulations, closures, or alternative places to visit is usually best delivered outside the wilderness. To do otherwise could detract from certain dimensions of the wilderness experience: the sense of isolation, solitude, freedom, and being in an undeveloped setting. Pre-trip contact usually is effective in achieving the desired change in behavior because the information comes early enough in the decision process to be used. Most visitors react positively to meeting rangers inside the wilderness, but there, wilderness rangers are concentrating on tasks that cannot be done outside the wilderness: monitoring conditions, dealing with violations of regulations, maybe checking permits, educating visitors engaging in inappropriate behavior, and correcting specific on-site problems such as litter cleanup or maintaining trails. Of course, many wilderness management tasks are performed outside wilderness, such as educational contacts, planning, and data analysis.

INDIRECT AND DIRECT MANAGEMENT TECHNIQUES

A variety of potential management tools and techniques are available to deal with specific recreation use situations, including indirect techniques and direct methods (Manning and Lime 2000). This section explains indirect and then direct methods, evaluates their effectiveness, discusses visitor attitudes about each, and the use of each method by wilderness managers.

INDIRECT MANAGEMENT TECHNIQUES

Indirect management techniques address factors that influence visitors, rather than directly controlling their behavior. Visitors are still free to choose. The main methods and techniques of indirect management include (1) physical design and alterations, (2) information and education programs, and (3) entry and eligibility requirements.

Physical Design and Alterations

Design of access, trails, and facilities can indirectly influence visitor behavior. All such designed works need to be developed sensitively to avoid conflicting with the objective of wilderness as an essentially undeveloped area that appears to be substantially natural.

Modifying access to wilderness entry points can alter the amount and type of use by making it harder or easier to visit an area. Because it takes place outside wilderness, many legal constraints on development are lifted. Closing the last section of road and thus making the trail longer to the wilderness boundary would reduce the total amount of use entering the wilderness. Such access modification would tend to reduce day use more than overnight use and reduce hiking more than horse use in areas with substantial levels of horse use. Reducing use can lessen impacts and increase opportunities for solitude for those who visit. Less day use inside wilderness may be desirable to protect naturalness and solitude, and indirect methods may be the minimum necessary action to achieve that objective.

Improving or extending roads closer to the wilderness boundary would likely have the opposite effect and would increase total use and raise the proportion of day users. This might direct some use away from other high-use areas where overuse problems exist, especially if combined with actions to make access more difficult to these problem areas. However, access should be improved only with great care. Improved access might attract new and more users, thereby creating impacts and displacing some visitors with few offsetting gains elsewhere.

Design and providing facilities at access points can increase or decrease particular types of use and total use. Horse facilities, such as unloading ramps, corrals, and hitch rails, encourage horse use. In water-oriented areas, such as the BWCAW, ramps for motorboats with trailers result in more boat use, and use by larger boats, than would a launch area requiring a carry-in portage from the parking area. The portage arrangement will screen out large boats and reduce boat use more than canoe use, with this effect more pronounced as the portage lengthens. The size of the parking area may have some effect on amount of use, although people determined to use that entry will usually park somewhere, like down the road, even if the parking lot is full. Conversely, parking should not be expanded where increased use is not desired, because it invites more use. Visitor acceptance of access modifications has been rarely studied over long time periods (Stankey 1973, 1980).

Trail design and maintenance are potentially powerful indirect management tools. Trails are an acceptable, unobtrusive way to influence use patterns. Some areas deliberately left without trails will be lightly used. Trails can be built or rebuilt to be either easy or steep and difficult; trail length from point A to B can often vary considerably; and bridges can be built to facilitate use, or trails can require fording streams, which will reduce or divert use and may stop all travel during high water. Some trails are too rough, rocky, or poorly cleared of blown-down trees for horses and pack animals, but hikers can scramble over and through them, thus separating these types of use.

Trail design can markedly influence the quality of the experience for the people who hike or ride horses. A trail can lead travelers to changing vistas or travel straight through dense forests. Trails can lead to varied vegetation, rock outcrops, glimpses of water, and perhaps increase chances of seeing wildlife. *Trail design offers a unique opportunity for managers to subtly diversify and enhance visitors' experiences. The sequence of visual experiences and level of challenge are strongly influenced by trail design.*

A trail can pass right by popular campsites, thus encouraging their use, or swing around through trees or behind a ridge to shift camping use elsewhere. This can also reduce feelings of crowding by limiting campers' observation of trail traffic, and vice versa. A winding, up-and-down trail cuts down on observation of other travelers just as long, straight, flat stretches increase sightings. Carefully thought-out trail loops or alternate routes can reduce visual and direct encounters among users. Trail design and maintenance also can facilitate or discourage off-trail travel and exploration, such as a rocky, steep trail section with a few strategic natural barriers left in place near the trailhead.

Despite the potential power of trail design and maintenance, its practical use is constrained. Most trails were built more than fifty years ago, before areas were designated as wilderness, primarily as an administrative transportation system, especially for fire control. In the West, early trails were intended mainly for horse travel, not hikers. Most trails were designed to get crews and supplies to fires quickly or to service fire lookouts. Neither environmental protection nor visitor experiences were important objectives. Budgets for trail work have been low for many years and maintenance of existing trails (often by volunteers working for expenses) consumes most of the available funds; some trails are not maintained; and most new trail construction is replacement of poorly located short sections of old trail.

Visitors generally agree that trails are appropriate and desirable (Stankey and Schreyer 1987; Lucas 1980). Visitors often complain about trail conditions and desire maintenance, not expansion, of trail systems and support for bridges, especially over large streams, is strong and widespread (Stankey and Schreyer 1987; Cole et al. 1995; Lucas 1985).

Signs within the wilderness can indirectly affect visitor experiences in several ways (Fig. 16.7). Directional signs,

FIGURE 16.7 A national park backcountry directional sign without mileage, such as this (left), is appropriate and welcomed by most wilderness visitors. A cairn, a pile of rocks, marks a trail project (right) in a BLM-managed desert wilderness near Phoenix; cairns can mark routes across sandy, rocky terrain where the tread may not be obvious. Such direct, low-key management also keeps visitors from trampling sensitive vegetation. Photo courtesy Chad P. Dawson.

usually at trail junctions, make travel easier and perhaps a bit safer. They also are an obvious work of people and reduce the visitor's sense of adventure and reliance on map-reading and orienteering skills. Most wildernesses have trail signs, but most managers strive for a balance, installing only limited, simple signs, particularly at confusing junctions. Wilderness signs might provide only one destination and not give distances or only give a directional arrow and a trail name. Interpretive signs, explaining natural features or historic events, are rarely used in wilderness, but such information is often shared via guidebooks and printed materials. Visitors generally support the concept of directional signs and some complain if they are scarce (Lucas 1985; Stankey and Schreyer 1987).

Trail systems and associated bridges and signs are the main recreational facilities in most wildernesses. Some wildernesses have other facilities at campsites, such as outhouses, corrals, and hitch rails, and a few areas have shelters, all to protect wilderness resources. *The use of facilities should be carefully limited in wilderness to avoid changing visitor experiences to something other than primitive. Facilities should not be provided for the comfort and convenience of visitors, but rather to protect wilderness resources and values, including wilderness experiences.* For example, bridges are often out of place in wilderness, and we suggest their limited use or removal unless they are a minimum tool to provide safety or have major historical significance.

Generally, visitors seem to support little or no development of campsites and little increase in numbers of campsites (Stankey and Schreyer 1987). Visitors tend to reject fireplaces but loose, rock fire rings are generally acceptable. Minimum-impact and LNT camping programs are influencing visitor behavior toward small campfires or no fires in environmentally sensitive locations and the use of camp stoves to reduce resource and visual impacts.

Outhouses vary from enclosed, roofed buildings such as the pit toilets found in developed campgrounds to simple boxes that rise only a few feet above the ground. Wilderness toilets are found in many parts of the country (Fig. 16.8). In heavily used areas, toilets concentrate human waste and may reduce the chance of visitor contact with it. In places with limited soil near campsites, as in the BWCAW, toilets must be carefully placed to reduce the chance of water pollution.

FIGURE 16.8 A variety of toilet designs address a basic human impact in wilderness. They range from a toilet made from wood in the BWCAW (left), to nonnatural material design (center), to a box toilet (right), commonly used at higher elevations in the Selway-Bitterroot Wilderness. Photos courtesy U.S. Forest Service and Chad P. Dawson.

Attitudes about toilets seem to vary in relation to the perceived need in each area and to local management practices where their use is established and they are accepted (Lucas 1985; Cole et al. 1995; Stankey and Schreyer 1987). *Toilets, and any obvious maintenance programs, are an obvious sign of humans that intrudes on wilderness conditions; they should be used only where they are clearly essential—the minimum tool to solve the sanitation problem of human waste disposal.*

Corrals for holding horses near camps are found in some western wildernesses, particularly at outfitter camps (Fig. 16.9). They confine the impacts of horses to a selected area, which is usually devoid of vegetation, and minimize impacts elsewhere. Corrals are probably a good example of a "sacrifice area" as well as a work of humans, but they may be the best solution to limiting impacts of parties with large numbers of stock camp, especially for long periods, such as during hunting seasons. Visitors' experiences are affected by the presence of a structure, associated soil and vegetation impacts, and, for horse users, a way to avoid the challenge of keeping stock nearby under primitive conditions. Horse users generally favor corrals; hikers oppose them, as demonstrated in several studies (Hendee et al. 1968; Lucas 1980, 1985).

FIGURE 16.9 Corrals, like hitch posts, somewhat mitigate the impact of horses at campsites in some western wilderness areas, such as this one at an outfitter camp in the Bob Marshall Wilderness. Photo courtesy U.S. Forest Service.

Shelters, usually open-fronted Adirondack types, are found in some areas, mostly in national forests in the Pacific coast states, the Northeast, and national parks. Enclosed public cabins are found in national forest wilderness in Alaska and are specifically authorized there by law. *Shelters concentrate impacts and, though they might have some advantages for that purpose, they also serve as an attractive nuisance and draw curious visitors.* Shelters reduce the primitive character of the visitor's experience and can reduce solitude because a number of parties often share a shelter overnight. Shelters function primarily as comfort and convenience facilities, with little or no wilderness protection role except in Alaska, where severe weather and bear danger confront visitors, and in Great Smoky Mountains National Park, where open shelters can be closed to keep out bears. *Shelters are out of place in wilderness, and we suggest their gradual elimination unless they are a minimum tool to provide safety or have major historical significance.*

Information and Education Programs

Informing and educating wilderness visitors and prospective visitors are a highly acceptable indirect management action—maybe the most important of all wilderness management techniques (Manning and Lime 2000). Information and education do not alter the wilderness resource directly, as do facilities, and they do not regulate or control visitors. Visitors retain freedom to choose, but their choices are made with more information. Good information and education programs enable visitors to have more rewarding experiences and help them build appreciation for appropriate wilderness behavior, experiences, and values; come closer to achieving the type of experience they are seeking; and develop a deeper appreciation for an area. Such programs also help managers achieve objectives by influencing where visitors go, what they do, and how they do it (Roggenbuck and Ham 1986; Roggenbuck 1992). *Effective programs require consistent messages and information from initial contacts with personnel at the front desk, trailhead message boards, and contacts with rangers at trailheads or on patrol in the wilderness.*

Wilderness visitors provide a particularly good audience for information and education programs because one distinguishing socioeconomic characteristic of wilderness visitors is high education levels. Most wilderness visitors also place a high personal value on wilderness, suggesting that most want to use it carefully. Information and education are now widely used by wilderness managers, and many are quite creative. Visitor acceptance of educational programs is rarely a concern, but effectiveness often is a question (Queen et al. 1998; Cole, Hammond et al. 1997; Roggenbuck and Manfredo 1990; Doucette and Cole 1993; Roggenbuck and Berrier 1982). More research is needed on the effectiveness of wilderness education on wilderness behavior. Some studies do document effects on the behavior of wilderness visitors, such as a study of the use of bulletin board messages at trailheads to increase visitor knowledge of low-impact wilderness travel practices (Cole, Hammond et al. 1997). Another study documented improvement in wilderness behavior norms over a twenty-eight-year period, perhaps reflecting that wilderness education efforts are working (Watson et al. 1996).

Manning and Lime (2000) developed eleven guidelines from a review of published literature for effective use of information and education programs in recreation management:

1. Personal contact by wilderness field staff can be an effective approach to delivering information and education programs.
2. Use of multiple techniques is more effective than single techniques.
3. Information and education programs are more effective with younger, less experienced, and less knowledgeable visitors.
4. Brochures, personal messages, and audiovisual programs are generally considered more effective than the same information on signs.
5. Messages may be more effective for visitors during their pre-trip planning or early in their trip.
6. Messages from highly credible sources are more effective.
7. Computer-based information dissemination can be an effective technique.
8. Volunteers, outfitters, and commercial guides are effective and efficient communication partners.
9. Information on the impacts, costs, and consequences of inappropriate visitor behavior can be an effective approach to reinforcing communication messages.
10. Role modeling of appropriate behavior by wilderness field staff can be an effective approach to reinforcing communication messages.
11. Messages should be targeted to specific visitor audiences when possible.

Wilderness information and education efforts are widespread, and managers can seek to shift use by making visitors more aware of alternative destinations and by providing descriptive information that makes some places more attractive and others less so (Washburne and Cole 1983; Lucas 1981; Krumpe and Brown 1982; Roggenbuck and Berrier 1982; Lime and Lucas 1977). Such use of information to influence use distributions can have desirable effects even if there is little change in overall distributions, because the information can help visitors better match their desires to the characteristics of alternative places. For example, it may make it possible for those who prefer lower levels of encounters to visit lightly used areas offering such opportunities, and those who better tolerate higher levels of contact with other visitors go to areas where more encounters are likely. Information can help visitors develop realistic expectations, and this will improve their satisfaction, at least when satisfaction is defined as the congruence between expectations and reality of the experience.

Information campaigns must be geared to management objectives. Managers must decide if their objectives call for bringing about a general redistribution (say, from heavy- to light-use areas) or site-specific redistribution, such as from an impacted area to another location (probably a more appropriate objective), or help visitors match their desires and experiences better, such as suggesting new trails and other wilderness or non-wilderness locations (a very appropriate objective and probably the easiest to achieve). Objectives should also guide the design and conduct of the information materials, media, and campaign strategy.

The information must be available to visitors in the planning stage of their trip. After people have arrived at an access point, it usually is too late to influence that trip, although later trips might be affected. If information does not reach a large proportion of visitors early, its potential for improving experiences is severely reduced.

Information and education are important tools for managing the distribution of wilderness use, but they are not a panacea for use distribution problems. In fact, unless used with care (e.g., providing consistent messages for all types of contacts), information could stimulate use and create problems that would not otherwise occur. Shifting use is a two-edged sword, and managers need to be aware and careful to monitor the degree to which they are solving overuse problems or merely redistributing such problems.

Leave No Trace (LNT)

The emphasis on minimum-impact education programs has increased and substantial staff and budget are devoted to them, such as "pack it in and pack it out," or Leave No Trace (LNT) (Swearingen and Johnson 1995; Roggenbuck 1992; Marion et al. 1993; Washburne and Cole 1983). *Minimum-impact education programs reach the public through a variety of contacts: at federal agency information centers and ranger stations, at agency websites, at schools and colleges, through organizations such as the Boy Scouts, at sporting goods and outdoor-recreation equipment stores, at wilderness access points, and on wilderness trails and campsites through wilderness rangers.* Some organizations have developed their own low-impact education programs, such as the Boy Scouts, Outward Bound, the National Outdoor Leadership School, and the Wilderness Education Association. The content of such programs has varied because some minimum-impact practices were based on judgment and assumptions and not all been based on field research (Cole 1989). The LNT program has support throughout the NWPS and is designed to develop outdoor skills and ethics among wilderness users and managers (Swain 1996). The LNT educational program focuses on camping and visitor behavior and stresses resource impacts and social impacts on other visitors' experiences. With the LNT program, consistency in low-impact education has emerged, even among the different federal land-managing agencies. This consistency is most apparent in *the following seven principles that are the crux of the LNT program* (Leave No Trace 2023):

1. *Plan ahead and prepare for your wilderness trip*—Take time to learn about the area you plan to visit so you know what to expect. Choose the area you visit based on an evaluation of your group's ability and interest in minimizing impact. Travel in small groups and take appropriate equipment. Stoves either eliminate or reduce firewood gathering and campfire impact. On backcountry trips, repackage food so that potential trash is reduced. Finally, if traveling in bear country, be prepared and knowledgeable.
2. *Travel and camp on durable surfaces*—This is a complex principle derived from research about the relationship between amount of use and amount of impact. Numerous studies show that a little use causes a lot of impact. Consequently, in popular high-use areas it may be best to concentrate use and impact. Stay on established trails and select a campsite that is already impacted. Select a site that is large enough to accommodate your group. Set up tents and the "kitchen" in places that have already been disturbed. In remote and relatively pristine places, disperse use and impact. Spread out while hiking and select a campsite with no evidence of previous use. Try to select travel routes and campsites that are durable. Disperse tents, activities, and traffic routes when camped in a pristine area and naturalize the site when you leave—so the next group that happens by will not recognize it as a campsite. Keep stays short in pristine areas.
3. *Dispose of waste properly*—Pack out litter and waste food. Be particularly careful in bear country. Dispose of human waste appropriately, in toilet facilities if they are provided, pack out solid human waste, or bury in cat holes at least 200 feet away from water, trails, and campsites. Do all washing away from camp and never directly in streams, lakes, or springs.
4. *Leave what you find*—Never blaze trees, leave flagging, or build cairns along trails. Never make trenches around tent sites or build campsite "improvements." Avoid damaging live trees and plants and leave natural objects and cultural artifacts.
5. *Minimize campfire impacts*—Cook on stoves and minimize the use of campfires. If having a fire in a high-use area, use an existing fire ring. Keep fires small. Use only dead and downed wood that can be broken by hand. Burn the fire until only ash or small coals are left. Be sure the fire is out and scatter ashes, leaving a clean and attractive fire ring. In remote areas, select a durable fire site. Use a fire pan, build the fire on a mound of mineral soil, or build it in a shallow pit in mineral soil. Do not line the fire with rocks. Naturalize the fire site when you leave.
6. *Respect wildlife*—Travel quietly and give animals the space they need to feel secure. Avoid approaching animals too closely, feeding them, or touching them. Learn about locations and times of the year when disturbance of animals is particularly problematic. Leave pets at home or keep them under control.
7. *Be considerate of other visitors*—Travel and camp quietly. Minimize undesired contact between groups by taking breaks and camping away from the trail, from public attractions (such as lakeshores), and from other groups.

The LNT program recognizes that educational programs need to be tailored to individual places and user groups. Behavior that may be appropriate in one place may be disastrous in another place. This is most obvious in Principle 2, "Travel and camp on durable surfaces." Recommended behaviors in popular high-use areas (where use concentration is appropriate) are precisely the opposite of recommended behaviors in remote areas (where use dispersal is appropriate). More subtle differences also exist. For this reason, the LNT program has developed different outdoor-skill booklets for different ecoregions in the United States, as well as for specialized recreational activities (such as river floating, horse use, rock climbing, snow camping, caving, and sea kayaking). Visitors must accumulate knowledge and the wisdom to consider the many variables that determine the most appropriate behavior for any given situation. Then they must use their own judgment.

The following six principles of low-impact education (not just LNT) have been compiled from both experience and research:

1. *Educational programs should be guided by specific objectives*—It is important to identify specific problems and the users that are the primary cause of the problems. Messages should be targeted to these specific users and problems. Certain behaviors and user groups will be more amenable to change than others. A two-goal approach to education is (a) to deal with specific problems caused by specific users and (b) to invest in the education of future users (such as through grade school programs).
2. *Messages should be clear, concise, and consistent*—Visitors may be overloaded with information and, consequently, miss the most important points if managers do not target their messages. Target a few specific behaviors. Short statements that demonstrate the desired behavior and why it is important are most likely to be effective. When more than one educational channel is used, or when different personnel are involved, the message should be consistent. Otherwise, visitors are likely to be confused. Unfortunately, the complexity of certain judgments essential for low-impact recreation (such as where to camp) often defies concise, consistent answers.
3. *Timing of educational messages is important*—To steer visitors away from heavily used areas, or to influence the equipment they take, messages must be sent to prospective visitors when they are planning their trip. Timing of messages should be varied for maximum effectiveness.
4. *A combination of techniques is likely to be most effective*—A combination of techniques allows messages to be repeated. It also makes it more likely that most visitors will be contacted. Different messages and media may be needed for different user groups. There is no reason, for example, to burden backpackers with all of the details of low-impact stock use. However, backpackers are more effectively contacted through messages on bulletin boards than stock users are.
5. *Messages should be presented in a professional manner*—Text and dialogue should be accurate and easy to understand. High-quality equipment and materials should be used for graphics and productions. The options for effective communication increase if celebrities or respected individuals deliver the messages.
6. *Personnel must be personable, well trained, and committed*—Personnel who contact the public should be cheerful, polite, outgoing, and have a positive attitude. The

ability to speak well, give accurate directions, and listen attentively to the visitors' experiences is helpful. Personnel need to be trained in communication skills and low-impact techniques. They should have personal knowledge of the area. Finally, they must believe in the value of minimizing recreation impact and the importance of education in achieving that goal.

Entry and Eligibility Requirements

To drive an auto, a person must demonstrate knowledge and skill to obtain a license. In most states, youngsters must pass a course in firearms safety to obtain a hunting license. Although human life is not so directly at stake as in driving or hunting, a test or schooling in wilderness skills might be required to certify knowledge and skills of wilderness visitors. The course might be a few hours long and be conducted in a classroom, using films or video and presenting general principles, ethics, and outdoor skills, or it could be longer and take place in the outdoors.

Some professional guides, but not the general public, must pass tests, except for mountain climbers or whitewater boaters in some places. A license requirement might be imposed only for heavily used areas, or only for those areas where use is actually rationed. In such places, wilderness permits might be issued only to visitors with licenses or to groups with a licensed leader, and others could gain experience by visiting other less used areas. Obviously, the details could be complicated and administrative costs could be high. Costs for visitors would vary with the length and demands of a course but could be substantial. Training would have to be readily available; otherwise, eligibility requirements could foster elitism—such as making some wilderness experiences only available to those with the time and money to take the training.

Although it may seem elitist to make wilderness entry dependent on certification of knowledge, there are some segments of wilderness visitors who are already going in that direction for trip leaders. For example, to be employed by most wilderness experience programs (WEPs), guides are usually required to have some level of first aid training, ranging from basic first aid to wilderness first-responder certification (Tangen-Foster and Dawson 1999). Different levels of training and certification are required to keep participants in the program safe and healthy; however, WEPs are required to obtain permits to use federal lands, and this requires proof of insurance. The insurance companies base the program's eligibility and rates on several factors, including staff training (usually requiring first aid) and experience.

Fees are another entry technique for potentially modifying use and are classified here as an indirect technique because they do not directly control what visitors do. Even if a cost is imposed for the privilege of visiting an area, the choice of paying or not is up to the visitor. To date, it is not clear that any wilderness fees have been charged at a sufficient level to reduce or shift use, but the potential exists to do that with increased entry fees or parking fees, and would likely have the following effects.

Use of wilderness for recreation is managed to maintain both the quality of the resource and the experience for the visitors. The related planning, management, and enforcement activities to support recreational use of wilderness are costly in terms of staff and funding. Recreation managers have been increasingly challenged to provide quality opportunities with limited or declining budgets.

Most wilderness visitors object to fees, and visitor studies usually show charging is one of the least popular methods of controlling use (Stankey and Schreyer 1987). Yet, *visitor surveys show that fees at any reasonable level would not be beyond the incomes of most visitors, and fee levels proposed to date are still a small fraction of the travel expense of a wilderness trip.* One study showed greater support for fees that would be used to protect and manage the area visited, rather than going into the general treasury (Martin 1986). Fees are increasingly used for river management where entry permits are required (Welsh 2000).

The resistance is partially related to past wilderness user experiences and whether or not such fees have been traditionally collected in a given wilderness area for public access (Richer and Christensen 1999; Williams et al. 1999; Winter et al. 1999). Because wilderness management needs to grow and limited budgets continue, the concept of wilderness user fees will seem more appropriate, and fees will play a larger role in wilderness management in the future.

A fee approach to gaining access to wilderness and supporting wilderness management is controversial for numerous reasons, such as the social equity of charging user fees to the public when some segments of the public may not have the same financial resources as others. Others object to charging the public for using "their public land." Probably one of the more troubling aspects of this approach is the tendency for recreational users who pay a use fee to be viewed as customers, and with that perspective comes the associated expectation to satisfy customer needs and wants, which may run contrary to the resource-protection purposes of the Wilderness Act.

DIRECT MANAGEMENT TECHNIQUES

Direct management techniques regulate and restrict individual choices, such as rationing use with permits. Management may be able to exercise a high degree of control based on its regulatory authority, but generally this has high administrative costs, and occasionally political costs from offended users who formally object. To some extent, direct management is a less sensitive approach and requires less knowledge about visitors, their behavior, and likely responses to management actions than does indirect management. Some knowledge of use is needed, but if use is to be directly controlled to fit management objectives, less real understanding of it is required. Like all wilderness management efforts, direct regulation should only be used when it is the minimum tool to meet the objective or solve the problem (Manning and Lime 2000).

This means that direct management may be somewhat easier than indirect management and perhaps demands less professional skill. Direct management does, however, have more potential for confrontations, conflict, controversy, and political backlash than indirect techniques, and it carries higher costs for visitors. Direct techniques should be applied only as a last resort, after indirect methods have been used as effectively as possible and have accomplished as much as they can to solve management problems, thereby leaving direct control as the minimum tool that will work.

Direct management can also include "strong suggestions" to discourage certain behaviors, as well as regulating and rationing use. Many of the specific techniques discussed next can take the form of either regulations or recommended practice. The distinction is basically whether or not noncompliance is considered a punishable violation. This distinction is probably significant to visitors. A recommendation or suggestion, no matter how strong, imposes less on the quality of visitors' experiences than an enforceable regulation. Unfortunately, suggestions are more of a burden for conscientious visitors than for indifferent or careless visitors, but most wilderness visitors do care about their impacts. A strong suggestion costs less for managers to implement than a formal regulation requiring field enforcement and follow-up. Managers should strive to encourage visitors to adopt desired behavior as a social norm, and to the extent this occurs, social pressure can become a substitute for agency regulations.

All regulations can either be uniform within a wilderness or vary from location to location. Several types of visitor behavior can be managed directly. Although there are numerous direct management actions that might be applied, only *eight techniques of direct management found to be effective are discussed here: (1) group or party size limits, (2) length-of-stay limits, (3) campfire restrictions, (4) prohibiting certain types of use, (5) camping setbacks, (6) rationing use, (7) restrictions on activities, and (8) litter control.*

Group-Size Limits

Setting a maximum group or party size has both social and environmental benefits. The social impacts are that *most visitors feel large parties are out of place in wilderness and quality of experience suffers* (Cole et al. 1995; Marion et al. 1993). Quality of experiences will be raised if contacts with large parties are reduced or eliminated. Environmental impacts are that *unusually large parties expand the area impacted at campsites and probably intensify impacts in the core of the campsite.* Slow recovery of soil and vegetation means that the effects of a large group persist for a long time, probably for years, continuing to degrade visitor experiences (Cole 1987).

Maximum group sizes are usually set based on typical group sizes and the estimated resource and social impacts of large groups. Maximum group size tends to vary with the character of an area. For example, large groups may be appropriate where large rafts are needed to run rough, whitewater rivers. Large groups and parties with large numbers of stock are not common, so most visitors are not inconvenienced. Research has found somewhat mixed attitudes about regulating group size, but a majority of visitors in most wildernesses studied supported a twelve-person limit (Lucas 1980, 1985) and strongly supported party size limits, particularly in more heavily used areas (Cole et al. 1995; Stankey and Schreyer 1987).

Monz et al. (2000) reported that the use of group-size limits has been relatively stable within the federal agencies over fifteen years, and by 1999 the U.S. Forest Service (73 percent) and National Park Service (68 percent) wilderness areas were more likely to use group-size restrictions than the Fish and Wildlife Service (22 percent) and the Bureau of Land Management (17 percent) wilderness areas. The main reasons for using group-size restrictions were to reduce environmental impacts of recreational use and conflicts between groups. *The most common group size for people is ten (with a median of twelve people) in federal wilderness areas that use such a restriction.* The most common group size for horses and pack stock is twenty-five (with a median of fifteen people) in federal wilderness areas that use such a restriction.

There seems to be a consensus that group-size limits, perhaps six to twelve persons, are reasonable in most places. However, party size limits for commercial outfitters and wilderness experience programs promise to be extremely controversial because these organizations must take enough clients at a time to be economically viable. Perhaps party size limits for public use could be smaller than for such organizations running wilderness trips under special-use permits that may also require other restrictions. Consistent application of regulations is best when and where it is possible and appropriate.

Length-of-Stay Limits

Although it can be regulated, length of stay usually does not contribute much to overuse. Few parties stay very long, and those that do usually remain in more remote, less used, less impacted areas and have well-experienced visitors able to minimize impacts. An exception may be commercial outfitter and big-game hunter camps, which may feature rustic convenience facilities to ease hardship for paying guests or make bad weather tolerable. However, an extended wilderness adventure to the most remote possible location seems particularly consistent with the purposes of the Wilderness Act and with the philosophy of wilderness. Publicizing length-of-stay limits may diminish feelings of wilderness, even for visitors planning short trips. Enforcement is difficult without a permit system.

On the other hand, when a party stays for a long time at one popular campsite, it is unfair to other visitors and hard on the site. Therefore, a reasonable limit on length of stay at any one campsite seems useful. Time spent at a particular location—not total time in the wilderness—seems to be the critical issue.

Limits on time allowed at any one wilderness campsite are not as common as general limits on visits to an entire area. This is true despite the fact that general stay limits are harder to enforce and, in our view, less useful than campsite limits. Seven to fourteen days are the most common campsite limits, especially in national forest wilderness. Hunting camps, usually set up in one place for a week or two in national forest wildernesses, may account for keeping some limits longer. National park limits usually range from one to three nights. Short limits, probably in the three- to five-day range, may be best for quality of experience and also may be desirable for minimizing ecological impacts.

Campfire Restrictions

Campfires can be prohibited or discouraged throughout a wilderness or at certain places, such as near timberline where trees are scarce and slow-growing and near water sources. Campfires can be restricted to officially designated sites, sometimes to designated fire rings or grates. Campfires have substantial environmental impacts (Cole and Dalle-Molle 1982) because fire destroys vegetation where it is built, and it sterilizes soil. Fuel-wood gathering can cause more widespread and often severe impacts. Campfires and horse use often produce the most conspicuous impacts in a wilderness.

Campfires are a traditional part of wilderness experiences and a treasured part of a camping trip for most people.

Educational efforts to encourage limited use of fires, small fires, minimum-impact fire-building and fuel-collection methods, and use of gas camp stoves for most cooking must be a major part of any wilderness management program. Timberline or subalpine areas may require special education efforts or, in some cases, prohibit campfires. Extreme fire danger may require temporary or seasonal prohibitions on campfires or cooking fires.

Campfire policies vary among wilderness managing agencies. Restrictions on the use of fires by wilderness visitors are common and widespread.

The presence of rock campfire rings also affects visitor experiences. In many wildernesses, especially in national forests, managers encourage visitors to remove all fire rings and "naturalize" the site to remove all evidence of fire. The objective is to eliminate the visual impact of campfires, but the ecological effects may remain. Wilderness rangers may then remove the fire rings visitors do not obliterate, often over and over again. This approach does not always make sense because of the cost of such work and the potential spread of ecological impacts to more areas rather than a few. A better approach is the common policy of leaving one or a few well-located, small fire rings per site, or removing rings only where fires or all camping is prohibited or discouraged. This seems in line with visitor desires, requires less work by wilderness rangers, and probably limits environmental impacts.

Prohibiting Certain Types of Use

Some types of recreational uses conflict with each other, and the impacts are often greater on one type of user than on others. Hikers tend to dislike encountering horse users and their impacts (Shew et al. 1986), and occasionally hikers bother horse users. Many people in small groups dislike meeting large parties. Canoeists usually resent motorboats in the few wildernesses where some motorized use is allowed. Prohibiting one type of use may improve the quality of experience for other visitors, but it obviously eliminates or diminishes experiences for those whose use is prohibited. Enforcement is relatively easy when prohibited uses are conspicuous, but confrontations may result.

Prohibiting certain uses in some sections of a wilderness is common, such as camping near or at some alpine lakes. Many areas have prohibitions on horses near lakes and streams or on some trails, and a few have seasonal prohibitions on horses. Much of the limited motorized use allowed under special exceptions is also zoned; for example, outboard-motor use in the Boundary Waters Canoe Area Wilderness (BWCAW) is allowed only on certain waterways.

Excluding some uses from certain sections of a wilderness can be a valuable management tool, but it should be done with restraint. Associated education efforts to reduce objectionable behavior and to increase mutual understanding and acceptance should be tried first, and if successful, may avoid the need for prohibitions or limit the extent of the area to which they need to be applied.

Camping Setbacks

In many wildernesses, camping within a certain distance from water, trails, and other campsites is discouraged or prohibited. A 100-foot or 300-foot no camping zone is most common. These setbacks are encouraged or required to minimize resource impacts and social impacts such as crowding and visual intrusions that may be reduced by encouraging people to select secluded sites that are screened from view by topography and vegetation, rather than imposing a specific setback.

Some water setback restrictions refer only to lakes, some to lakes and streams, occasionally only to named streams. The various reasons given for setbacks include reduced environmental impacts, but shorelines are not necessarily more ecologically fragile, and water pollution depends on where wastes are disposed, not where camp is located. Social reasons are just as important, and benefits to quality of experience include a reduced sense of crowding, keeping camps out of the foreground of the prime visual attraction, and keeping the lakeshore or streamside open to all visitors without disturbing the privacy of someone's camp.

Restricting camping near water goes against a natural preference of users. Water is a powerful attraction, and most people want to be close to it, mainly for aesthetic reasons. Locations of existing campsites, developed by visitors' free choices over many years, clearly reflect a preference for certain locations near water, trails, and other users. Besides reducing quality by forcing people out of preferred sites, the supply of potential places to camp is greatly reduced, and in steep, mountainous country, there may be no legal places to camp near many destinations because reasonably level ground is often only available near water.

Setback enforcement is difficult. In most places with setback regulations, the majority of violation notices are issued for camping too close to water (Swain 1986). Visitors feel harassed, which is unpleasant for them and for wilderness rangers. Furthermore, Robertson (1986) found that wilderness campers were poor judges of distance from shorelines.

Many wildernesses also prohibit camping within a certain distance of trails. Trail and water no camping zones often overlap, and this closes large areas to camping. If areas suitable for camping do exist, many new campsites will develop behind the setback, while the old ones closer to water typically persist, increasing total impacts considerably (Cole 1989).

Although setback regulations are useful, they may have negligible effects in reducing impacts and a negative effect on visitor experiences. Benefits are not clear and may not justify the full-scale enforcement campaign that is required. Educating visitors on where to camp and why to camp there is the first course of action. The advantages of not camping too close to water should be explained, with specific distances from water, trails, and other campers suggested as the minimum setback wherever possible. Reasons for picking a site screened from view should be explained. The relationship of impacts to campsite choices could be explained too so that visitors can make reasonably good choices under varying conditions.

Rationing Use

Rationing use can have major potential impacts on quality of experiences. Those permitted to enter enjoy heightened solitude and quality of experience, but the experience is eliminated for those whose wilderness entry is denied. For example, a study of the BWCAW in 1991 concluded that overnight visitors who reported having difficulty finding a campsite would benefit from reduced limits of visitor use (Watson 1995b).

How use is rationed can further impact experiences. For example, several techniques can be used to limit or redistribute use. A daily limit can be set for the number of parties or individuals at all or certain entry points, with the itinerary up to each party admitted; they can make it up as they go. Alternately, limits could also be placed on the number of persons staying overnight in the entire area, in specific zones, or at individual campsites—in this case, numbers admitted would depend on the itineraries and planned lengths of stay of parties already in the area. If use were at capacity, the numbers leaving would determine numbers to be admitted. Numerous approaches exist for allocating use within an area and among user groups, but not all approaches are equally acceptable to each user group (Cable and Watson 1998).

Limited entry permits are administered in a variety of ways. Some or all are usually available for advance reservations. Some proportions of the available permits often are held back for last-minute drop-ins on a first-come, first-served basis (sometimes called queuing, from the similarity to waiting in line to be served). A mixture of reservation and first-come, first-served permits seems to be fair. People who carefully plan far ahead, perhaps because of their need to schedule vacations and travel far from their home to the wilderness, need some advance assurance they will be able to visit the wilderness. Other people cannot or at least prefer not to plan so much, and operate in a more spontaneous, spur-of-the-moment fashion. If they live close to the wilderness—and most visits are from persons in the same state or region—they may want to react to last-minute, personal schedule or weather changes by going to the wilderness or not. A mixed system gives each type of person a chance at a permit.

Limited permits could also be issued through a lottery, as is done for some big-game hunting, such as bighorn sheep permits or some river-running permits like the Colorado River in the Grand Canyon. Handling such a lottery for entries on various dates is more cumbersome than assigning hunting permits for clearly defined hunting seasons, but it is done regularly for river-floating permits. However, few wilderness areas use a lottery approach, and visitors surveyed twenty-five years ago did not like the lottery concept. In fact, it was the least favored alternative for limiting use, except on rivers where visitors had become familiar with lotteries and accepted them. Welsh reports that lottery systems with call-ins for cancellations and unallocated launch dates were used in 2000 on fourteen rivers where use was rationed and applications were required by management (Welsh 2000).

All of these and other approaches to rationing use—reservations; lottery; first-come, first-served; fees; and skill/knowledge—have different combinations of advantages and disadvantages. Each system benefits certain types of visitors at the expense of others. Visitor acceptance varies from low to high, with some uncertainty. Difficulty of administration varies, but only queuing (first-come, first-served) seems to be easy.

The psychological impacts of rationing systems and the ethical burden on managers warrant comment. For example, whether a party is in the "wrong" campsite by choice or accident, confronting them is likely to be an uncomfortable experience for everyone. Apprehensions about whether they comply could nag visitors. Will a ranger come along? Will the ranger check their permit? Will the ranger reprimand them, make them move, issue a violation notice, or what? If the camp quota is not full anyway, will the ranger take this into account? If the ranger ignores the violation, is this fair to the party that reluctantly passed a lovely, empty campsite late in the day and pushed on with aching feet until sunset to reach the campsite listed on their permit? Rationing systems can cause a loss of the sense of wilderness freedom and exploration, sometimes for a fairly small gain, and place the wilderness rangers in a particularly uncomfortable enforcement role. *Rationing is needed in many wildernesses and wilderness locations, but we urge their implementation as a last resort,* when they truly are the minimum tool to solve overuse problems and their application is the most light-handed approach possible.

One way to lessen the impacts of controlled itineraries is to inform and educate visitors, before their arrival at the wilderness entry points, about the need for and use of rationing. Wilderness travel information can indicate the alternatives and opportunities, such as the likelihood that a particular campsite or entry through a particular trailhead will be available. Such information can be sent by mail or accessed by the visitor via the Internet to help facilitate travel planning (Queen et al. 1998). This is an example where the rapid expansion of electronic communication can facilitate wilderness information and education for better visitor management.

Restricting Activities

Many specific activities can be prohibited or discouraged to reduce or eliminate adverse impacts on the environment and visitor experiences. Whatever restrictions managers impose, they must monitor results to justify whether to continue the restrictions, drop them, or modify them. For example, target shooting might be restricted to protect solitude. Hunting and fishing regulations can also affect wilderness use and activities, can contribute or detract from experiences, and alter the wildlife and fish presence and behavior in the ecosystem. Naturalness, of course, is the goal. Fish and game management, however, is a state responsibility except in national parks and wildlife refuges. State fish and game departments may object to special regulations for wilderness hunting or fishing because they feel they unfairly discriminate against these activities.

Litter Control

Litter cleanup and efforts to further reduce littering, usually through the "pack-it-in, pack-it-out" regulation, are essential in wilderness (Fig. 16.10). *Research indicates that litter detracts seriously from wilderness experiences and is a common user complaint identified in visitor studies* (Cole et al. 1995). Litter is an intrusion from civilization on the wilderness experience. Many believe that a clean site is more likely to be kept clean by users; at least it improves their experience and provides no excuse to leave it littered because it was found that way.

Litter does not all have to be removed by agency employees. In field tests in developed campgrounds, rangers contacted families with children and solicited the help of children to clean the campground in return for fire-prevention and environmental-education-type rewards—badges, comic and coloring books, and so on. It is actually an "appeal system"

FIGURE 16.10 Controlling litter depends on cooperation between land managers and the public. While the "pack it in, pack it out" regulation generally works for most visitors to wilderness, occasionally it doesn't. Here, Sierra Club volunteers clean up litter at a campsite in the Selway-Bitterroot Wilderness. Photo courtesy Richard Walker.

FIGURE 16.11 Managing recreational use of wilderness requires combining a range of activities, such as designing and maintaining trails, providing wilderness rangers, conducting information and education programs, and writing regulations. Here a trail crew repairs a trail. Photo courtesy John C. Hendee.

because it was not feasible for wilderness rangers to carry a supply of incentive rewards, children are not as numerous, and many wilderness users objected to the notion of rewards—an appeal for visitors' help is all that is necessary. This is because the "pack-it, pack-it-out" norm has become accepted, and in some areas, visitors are not permitted to take bottles or cans with them (reusable containers are permitted). This regulation is reported to have greatly reduced littering.

WILDERNESS RANGERS

Wilderness rangers carry out many of the management actions discussed above. Most patrol the wilderness during its high-use season, usually summer or fall, but some are stationed at busy trailheads to contact entering visitors and are sometimes called wilderness information specialists. A few wilderness specialists are year-round employees who compile and analyze wilderness data and conduct public education programs during the off-season. Wilderness rangers can serve many functions, such as gathering field data on resource conditions, use, and visitor actions, provided they receive some of the training that a manager would receive (Conrad 1997). They can influence visitor behavior by suggesting, advising, and informing; enforcing regulations; modeling appropriate behavior; performing emergency trail repairs and minor maintenance; rehabilitating campsites; planning and directing or doing some cleanup; and giving emergency assistance. As field managers, wilderness rangers can help solve site-specific problems by providing visitors with information and advice, thereby reducing both the need for area-wide regulations and more direct controls.

Almost all wildernesses and backcountry areas have wilderness rangers, assistant rangers, or information specialists, although sometimes the large area patrolled by one person permits only superficial coverage. In some areas, great importance is placed on rangers as a key wilderness management tool. From all available evidence, wilderness rangers have proven to be very useful, and visitor surveys also indicate they are well accepted by the public. Because of their importance, rangers should be selected carefully for their interest and ability to work with people as well as their knowledge and concern for wilderness (Dawson 2007). Wilderness rangers are, in effect, people managers, and they are an agency's prime contact with the wilderness-using public (Fig. 16.11). Most are volunteers or seasonal employees, but this does not diminish the importance of their work and the need for their continued training.

MONITORING WILDERNESS EXPERIENCE QUALITY

Wilderness visitor management requires monitoring visitor experiences just as resource conditions must be monitored. Wilderness management affects visitor experiences, but unfortunately, monitoring methods for social conditions and experiences are not as well developed and tested as for resource conditions.

If the LAC or VUM system of managing visitors and their impacts (Chapter 9) is being used, monitoring of all established indicators is essential. Some of these indicators will relate to visitor experiences, such as the intensity of use and presence of impacts (Fig. 16.12). In any wilderness, whatever management system is being used, identifying problems requiring management attention comes best from systematic, objective monitoring. *The results of such monitoring indicate the effectiveness of the management actions and whether or not additional or different effort is needed.*

FIGURE 16.12 Maintaining naturalness and solitude even in the vast reaches of Alaska, here in the USFS-managed Stikine-LeConte Wilderness, is difficult because the Alaska National Interest Lands Conservation Act (1980) permits activities such as aircraft access usually prohibited elsewhere. Photo courtesy John C. Hendee.

To be most useful, information gathered by managers in the field needs to be collected systematically and accurately and carefully recorded. Three important aspects of wilderness experience quality to monitor are (1) encounter levels on trails and at campsites, (2) user-to-user and activity conflicts, and (3) visitor preferences and dislikes. The three aspects are reviewed next, but numerous other aspects can be monitored depending on the variables and standards specified in the LAC and VUM processes (see Chapter 9).

ENCOUNTER LEVELS ON TRAILS AND AT CAMPSITES

The number of parties met per day by a group while traveling on trails or waterways is a measure of the degree of solitude or crowding and is one measure of wilderness experience quality (Cole et al. 1995; Hammitt and Rutlin 1995). Encounter levels have been used as an LAC indicator. In one approach, wilderness rangers or volunteers tally how many individuals and groups they meet each day. To reasonably represent visitors' encounter experiences, the agency people must travel at about the same speed and for about the same distance as typical visitors. Tallies might be adjusted slightly to compensate for unusually fast, long travel, or for coverage of only part of a typical day's travel. It is essential to keep formal notes, preferably on a standard form, and to record the trail segments covered. Later these data can be compiled, compared, and evaluated for patterns and trends.

A second approach is for fieldworkers to ask visitors about their encounter experiences. "How many other individuals or groups did you meet yesterday?" (or "today" if late in the day). The visitors' route and answers would be noted. More encounter information can be obtained from visitors than in the first approach, and thus results would be more valid. Answers should be recorded systematically in a notebook with date and location of travel noted, along with weather, which might influence the data analysis.

Similarly, most visitors place particular value on a campsite isolated from the sight and sound of other campers. How successful they are in achieving isolation significantly influences their quality of experience. Therefore, managers should ask about and monitor campsite encounters, and LAC systems will usually include it as an indicator.

At least two approaches can be used to monitor campsite solitude: direct observation and asking visitors. Because of their comparability, both could be used and the results compared and combined to improve accuracy. In direct observation, a wilderness ranger records how many other occupied camps were observable from each occupied camp, and how much noise was audible. This must be done in the evening or early morning, when most people are in camp, which limits the number of observations possible.

If asking visitors about their experiences, when an individual or group is contacted in camp or on the trail, it would be common to ask: "Did you camp here last night?" or "Where did you camp last night?" Then, as appropriate: "Was anyone else camped nearby?" "How many groups?" "Was their noise or presence an intrusion for you?"

All the data need to be carefully recorded and the campsite identified as precisely as possible. Campsite environmental conditions should also be monitored, and a system of numbering campsites should be established to facilitate data recording and tabulation. Locating camps mentioned by groups on the trail may involve going over a map with them, and sometimes it will be imprecise, but a general area and good information can usually be recorded.

USER-TO-USER AND ACTIVITY CONFLICTS

The behavior of all visitors affects every user, to some extent, either positively (e.g., good social exchanges, trail information) or negatively (e.g., noise, dogs, litter, camping too close together). The degree to which visitors are impacted by others' behavior depends on their expectations and the degree to which the behavior is considered intrusive or not "normal" (Jacob and Schreyer 1980). Other users consider some activities intrusive because they diminish the solitude or naturalness experience they expect and seek in the wilderness. For example, campfire smoke may annoy other users, the sound of gunshots by hunters or shooters may not be appreciated by nonhunters or nonshooters, and pack-stock users may expect hikers to give them the right-of-way when they meet on the trail, which may not be acceptable to the hikers. These kinds of behaviors can become conflicts depending on the sensitivity of those who believe their experience has been intruded on or the expectations of visitors who believe such activities are unnecessary or inappropriate in wilderness (Watson 1994, 1995a).

PREFERENCES AND DISLIKES

Managers need some feedback from visitors on the things they prefer or dislike, whether as part of a formal monitoring program or not. Reactions to trail conditions, campsite conditions, litter, specific management policies or regulations, and information provided by the agency are all possible topics managers might want to monitor. There is no way to do this except by talking with or surveying visitors. The same questions or topics should be raised with all visitors. Some of these topics are commonly discussed with visitors anyway, especially trail conditions. In these cases, it is just a matter of making careful notes after the conversation. To be most useful, the information should be collected in a standardized way. Recording spontaneous complaints and compliments is fine, but mixing such responses with answers to specific questions such as "What do you think about our closure of camping at Summit Lake?" will produce a misleading, noncomparable mixture of information.

Visitors' preferences and dislikes (satisfactions and dissatisfactions) are a measure of the quality of the wilderness experience. Some factors affecting visitors' experiences include the number and type of user encounters (especially compared to prevailing visitor norms and expectations), perceptions of crowding, actual use levels, and trail and campground conditions (Manning 1999; Patterson and Hammitt 1990; Watson et al. 1997). These factors are complicated because visitors may adjust their expectations toward actual conditions, such as use levels after experiencing the area, or adopt other "coping" behaviors or rationales to achieve satisfaction (Hammitt and Patterson 1991).

ADAPTIVE MANAGEMENT AND BEST MANAGEMENT PRACTICES

The result of monitoring indicators under LAC or VUM helps inform achievement or not towards the desired conditions in wilderness and, in turn, suggest the continuation of current management or adjustments to management techniques or approaches (see Chapters 9 and 15). *Adaptive management is the systematic process of monitoring and management adjustments using LAC or VUM. The result of adaptive management over time is the evolution of best management practices (BMPs) that both provide quality visitor experiences and protect the wilderness resource as intended under the Wilderness Acts.* BMPs serve two purposes: (1) they are specific to the wilderness area being managed and improve management effectiveness and efficiency; and (2) they suggest management approaches and techniques that could be considered in other wilderness areas as potentially effective and efficient—in this way, they contribute over the years to the evolution of wilderness management.

SUMMARY

Visitor management is a vital part of wilderness stewardship. It is guided by two principal objectives: providing a natural setting (naturalness) and a special type of quality visitor experience (solitude and primitive and unconfined recreation). Both objectives are important, but sometimes achieving them results in conflict.

Five types of undesirable visitor actions exist, varying in severity, each with different kinds of appropriate management responses: illegal actions, careless actions, unskilled actions, uninformed actions, and unavoidable impacts. *Managers and visitors differ in their perceptions of problems;* managers tend to perceive resource impacts as more important than social problems, whereas visitors tend to take the opposite point of view. These divergent perceptions reduce compliance with regulations and acceptance of recommendations. Managers need to be aware of visitor perceptions and work to narrow gaps between visitors' views and their own perceptions of problems and conditions.

Many aspects of wilderness recreation use are subject to management: amount of use, distribution of use, timing of use, party size, length of stay, and visitor behavior. Professional management of these aspects of wilderness recreation to protect natural conditions and opportunities for appropriate wilderness experiences requires adequate data about visitors and resources, much of which currently are of poor quality or are completely lacking.

Some wilderness visitor management approaches are direct, imposing control on visitors' actions, whereas other approaches are indirect, altering factors that influence visitors' choices. In general, indirect approaches are preferable, being more consistent with wilderness concepts. Regulation is the prime example of a direct approach and should be used with restraint and only when it is the minimum tool to achieve the objective, and with recognition of the limitations on effectiveness of regulations. *Monitoring results of management actions is essential to improve future management, and much improvement is needed.* Management actions implemented from outside a wilderness are generally preferable to management inside, especially in regulations and educational contacts. No single management approach is effective enough or acceptable enough to be the sole choice; a blend of approaches is essential.

Many important values of wilderness depend on the quality of visitor experiences. Managers can influence the factors that affect experience quality: natural, social, and managerial conditions. Natural conditions depend on management of user impacts and allowing natural processes such as natural wildfires to shape wilderness ecosystems. Social conditions depend on management of solitude/crowding, user conflicts, and certain types of problem behavior. Management conditions influence visitor experiences through style and tone, as well as specific actions. Diversity in all three conditions is desirable.

Indirect management involves less restrictive visitor management techniques. Design can include access, trails, and facilities to modify the amount, type, and distribution of recreational use. Most specific facilities, other than trails and bridges, play only a limited role in wilderness, receive limited support from visitors, and need to be used with restraint. Information and education are essential tools for managing visitors and their experience quality. These tools can be used to redistribute use, to expand visitors' understanding of wilderness values, to reduce visitor conflict, and to increase visitors' skills in minimum-impact use (Fig. 16.13).

Direct management is a necessary tool in some situations. Regulations and use rationing are the most common direct management approaches, but strong recommendations are a less authoritarian alternative. Some types of regulations are more often appropriate than others. Group-size limits, length-of-stay limits at campsites, campfire restrictions in some situations, prohibiting some types of use in parts of areas, and sometimes rationing use all have their place in the manager's tool kit. Some management techniques are harder to justify, such as total prohibition of types of use allowed under the Wilderness Act and fixed itineraries under some permitting approaches. Rationing, when necessary, can be done in many ways, and each way has different advantages and disadvantages.

The importance of monitoring visitor experience quality and resource conditions cannot be overstressed. Monitoring visitor-to-visitor encounters on trails and at campsites, and

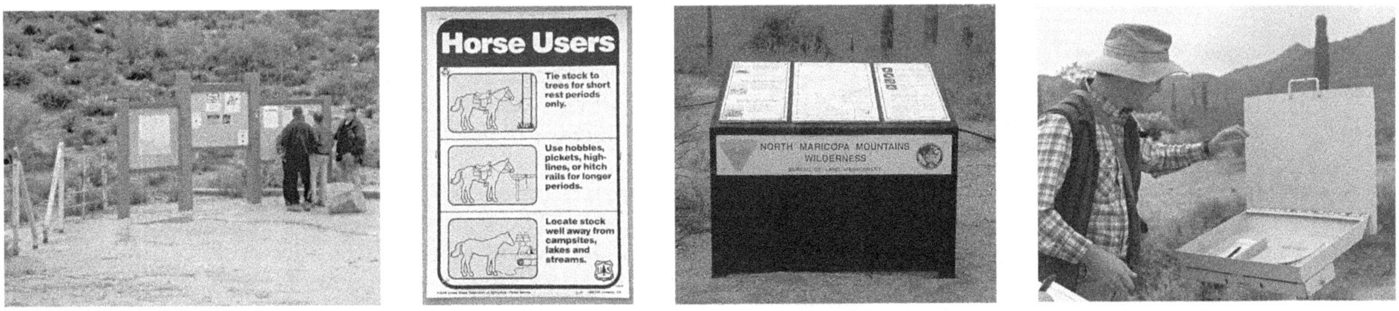

FIGURE 16.13 Designing entrances to wilderness parking and trailheads requires managers to balance between providing necessary facilities and information and being intrusive. Informational signs at trailheads (left) use brief, directed messages (left center), and more permanent displays educate visitors about the wilderness experience and low-impact recreation practices (right center). While permit systems can bring visitors and managers together for a face-to-face exchange at permit-issuing offices, even a self-issuing permit system (right) is an opportunity to educate visitors and collect information about them. Photos courtesy Marilyn Hendee and Chad P. Dawson.

satisfactions and dissatisfactions have been reported as some of the important visitor experience indicators.

Wilderness management should stress visitor education and information more, and regulation less, except where direct regulation is the minimum method to achieve planning standards and objectives. Visitor understanding of "wilderness ethics" and minimum impact use techniques is increasing. Benefits are apparent in many places: litter is less common, groups are smaller; and so on. Continuing emphasis on education can further improve visitor behavior and further diminish the need to rely on regulation. Managers need to evaluate their educational efforts and the results of education programs to guide their improvement.

Adaptive management is the systematic process of monitoring and management adjustments using LAC or VUM management approaches. The result of adaptive management over time is the evolution of best management practices (BMPs) that both provide quality visitor experiences and protect the wilderness resource as intended under the Wilderness Acts.

STUDY QUESTIONS

1. What two considerations are stressed in wilderness visitor management?
2. How can wilderness recreation use problems be classified? How does this relate to appropriate management responses to those problems?
3. How do managers' and visitors' perceptions of problems in wilderness compare?
4. What are some of the important concepts in understanding visitor impacts on social conditions in wilderness?
5. What aspects of wilderness recreation use can managers control or influence?
6. What is the nature of visitor experiences in wilderness? What are the main impacts on those experiences?
7. What characterizes direct and indirect visitor management? What are the strengths and weaknesses of each?
8. What are the main types of indirect visitor management? How would each affect experiences? What are the pros and cons of each?
9. What would a wilderness manager monitor to keep track of experience quality?
10. How could adaptive management and best management practices help provide quality wilderness experiences?

ACKNOWLEDGMENTS

This chapter was a major revision by Chad P. Dawson and Paul Dawson, USFS recreation specialist on the Coconino National Forest, in the fifth edition. The fourth edition included contributions from Dr. Rudy Schuster on the social aspects of visitors and their management and David N. Cole, former research biologist at the Aldo Leopold Wilderness Research Institute, on educational programs. We recognize and thank our colleague, Robert C. Lucas, for his seminal efforts on these topics in the first two editions.

CASE DISCUSSION: PREVENTIVE SEARCH AND RESCUE (PSAR)

Managers continually work at understanding visitors and the interest of various public groups in using wilderness directly or indirectly. Over time, wilderness managers have had to become students of sociology, recreation management, and other social sciences, in addition to the natural sciences and management, to be good stewards of the NWPS. Much of wilderness management work is people management—both direct and indirect management of people.

The Grand Canyon National Park employs staff to conduct a Preventive Search and Rescue (PSAR) program. The goal is to provide education about the hazards of hiking in the Grand Canyon, and the time and equipment necessary to successfully and safely complete a planned hike (Fig. 16.14). The PSAR program was started in 1997 to reduce the hundreds of heat-related illnesses park visitors were experiencing every summer. Some of those illnesses resulted in deaths that could have been avoided with better preparation and planning. PSAR Rangers patrol the upper portions of the main corridor trails, such as the Bright Angel and South Kaibab Trails, and ask hikers questions about their hiking plans. Although information about the trails is available at the Visitor Center and on signs posted at the trailheads, many hikers are still surprised when a PSAR Ranger talks to them about their planned hike. The PSAR Rangers who encounter hikers in the field are meant to help them have a safe and positive experience at the Grand Canyon. PSAR Rangers are also EMTs and are often the first park personnel on scene with an ill or injured hiker. PSAR Rangers carry basic medical gear

FIGURE 16.14 Hikers in Grand Canyon National Park face challenging hiking conditions to explore the canyon environment. Photo courtesy of Brian Dawson.

and can call for additional personnel if advanced medical or technical rescue skills are required (Grand Canyon National Park 2023).

The *10 Essentials* that PSAR Rangers talk about with visitors are organized into ten categories (Grand Canyon National Park 2023; The Mountaineers 2010):

1. *Navigation—Map, compass, and GPS system*
   Navigation systems are used when planning your route before your trip, and when you need help orienting yourself in your surroundings during your activity. Know how to use a topographical or relief map as well as your compass or GPS unit before going out.
2. *Sun Protection—Sunglasses, sunscreen, and hat*
   Sun protection is necessary to protect your skin and eyes against harsh UV rays that are responsible for sunburns and skin cancer. Consider using sunglasses, sunscreen, and hats. Sun-protection clothing such as pants and long sleeve shirts can also help minimize your exposure to the sun.
3. *Insulation—Jacket, hat, gloves, rain shell, and thermal underwear*
   Nature is unpredictable. Be prepared for sudden changes in weather conditions. Pack an extra layer of clothing that reflects the most extreme conditions you could encounter.
4. *Illumination—Flashlight, lanterns, and headlamp*
   Lighting is indispensable in the outdoors where no conventional light sources can be found. Items include flashlights, lanterns, and headlamps. Headlamps are the preferred light source because they are hands-free. Be sure to pack extra batteries.
5. *First-Aid Supplies*—First Aid Kit
   Be prepared for emergencies by packing first-aid supplies with you. Start with a pre-made kit and modify it to fit your trip and your medical needs. Check the expiration date on all items and replace them as needed. Consider including an emergency guide in case you are faced with an unfamiliar medical emergency.
6. *Fire*—Matches, lighter and fire starters
   Fire can be an emergency signal and a heat source for cooking and staying warm. Pack matches (preferably waterproof) and fire starters—items that catch fire quickly and sustain a flame (e.g., lighter). Familiarize yourself with the fire use regulations of your park before heading out. Learn more about campfires at https://www.nps.gov/articles/campfires.htm.
7. *Repair Kit and Tools*—Duct tape, knife, screwdriver, and scissors
   Carry a basic repair kit with you to help repair equipment. The kit should include items such as duct tape, a knife, and scissors. Consider packing a multi-tool, a compact version of many tools that can include a knife, screwdriver, can opener, etc. Be sure to bring any tools specific to your trip and your activity.
8. *Nutrition*—Food
   You should always be prepared for the possibility of changes to your trip plans. Pack an extra day's supply of food, preferably no-cook items that have good nutritional value in order to keep your energy high. Salty and easy to digest snacks (e.g., trail mix, nuts, and granola bars) work well for outdoor activities.
9. *Hydration*—Water and *water treatment supplies*
   Staying hydrated on your trip is of utmost importance! Physical activity increases your risk of dehydration (loss of water and salts from the body), which can lead to negative health consequences. *If you're active outdoors (hiking, biking, running, swimming, etc.), especially in hot weather, you should drink water often and before you feel thirsty.* Prepare your water before you need it and do not allow yourself to become dehydrated. Before heading out on your trip, be sure to identify if there are any bodies of water at your destination that you could collect water from and treat using your water treatment supplies.
10. *Emergency Shelter*—Tent, space blanket, tarp, and bivy
    Shelter is one of the most important elements during an emergency survival situation. It can protect you from severe weather conditions and exposure to the elements. A tent, tarp, bivy sack, or emergency space blanket are all lightweight options for emergency shelter.

CASE DISCUSSION QUESTIONS

1. What are some of the challenges of implementing this type of educational approach to supporting visitor experiences?
2. How does this educational approach help managers in their search and rescue efforts?
3. Most visitors will respond positively to PSAR, but how would we monitor the effectiveness of the program?

REFERENCES

Adelman, Bonnie J. E.; Heberlein, Thomas; Bonnicksen, Thomas M. 1982. Social psychological explanations for the persistence of a conflict between paddling canoeists and motor craft users in the Boundary Waters Canoe Area. *Leisure Sciences.* 5(1): 45–61.

Anderson, D. H.; Brown, P. J. 1984. The displacement process in recreation. *Journal of Leisure Research.* 16: 61–73.

Anderson, D. H.; Lime, D. W.; Wang, T. L. 1998. *Maintaining the Quality of Park Resources and Visitor Experiences: A Handbook for Managers.* St. Paul: University of Minnesota, Department of Forest Resources, Cooperative Park Studies Unit.

Behan, R. W. 1974. Police state wilderness: A comment on mandatory wilderness permits. *Journal of Forestry.* 72(2): 98–99.

Blahna, D. J.; Smith, K. S.; Anderson, J. A. 1995. Backcountry llama packing: Visitor perceptions of acceptability and conflict. *Leisure Sciences.* 17: 185–204.

Borrie, William T.; McCool, S. F. 2007. Describing change in visitors and visits to the "Bob." *International Journal of Wilderness.* 13(3): 28–33.

Brooks, J. J.; Wallace, G. N.; Williams, D. R. 2006. Place as relationship partner: An alternative metaphor for understanding the quality of visitor experience in a backcountry setting. *Leisure Sciences.* 28: 331–349.

Brown, Perry J.; Haas, Glenn E. 1980. Wilderness recreation experiences: The Rawah case. *Journal of Leisure Research.* 12(3): 229–241.

Brown, R. N. K.; Rosenberger, R. S.; Kline, J. D.; Hall, T. E.; Needham, M. D. 2008. Visitor preferences for managing wilderness recreation after wildfire. *Journal of Forestry.* 106(1): 9–16.

Cable, S.; Watson, A. E. 1998. Recreation use allocation: Alternative approaches for the Bob Marshall Wilderness Complex. RMRS-RN-1. Ogden, UT: U.S. Department of Agriculture, Forest Service, Rocky Mountain Research Station.

Carlson, T. 2000. The Eagle Cap Wilderness permit system: A visitor education tool. *International Journal of Wilderness.* 6(2): 27–28.

Carothers, P.; Vaske, J. J.; Donnelly, M. P. 2001. Social values versus interpersonal conflict among hikers and mountain bikers. *Leisure Sciences.* 23: 47–61.

Cole, David N. 1986. Ecological changes on campsites in the Eagle Cap Wilderness, 1979 to 1984. Research Paper INT-368. Ogden, UT: U.S. Department of Agriculture, Forest Service, Intermountain Research Station.

———. 1987. Research on soil and vegetation in wilderness: A state-of-knowledge review. In: Lucas, Robert C., comp. *Proceedings: National Wilderness Research Conference: Issues, state-of-knowledge, future directions*; July 23–26, 1985; Fort Collins, CO. General Technical Report INT-220. Ogden, UT: U.S. Department of Agriculture, Forest Service, Intermountain Research Station, pp. 135–177.

———. 1989. Low impact recreational practice for wilderness and backcountry. General Technical Report INT-265. Ogden, UT: U.S. Department of Agriculture, Forest Service, Intermountain Research Station.

Cole, David N.; Dalle-Molle, John. 1982. Managing campfire impacts in the backcountry. General Technical Report INT-135. Ogden, UT: U.S. Department of Agriculture, Forest Service, Intermountain Forest and Range Experiment Station.

Cole, David N.; Hammond, T. P.; McCool, S. F. 1997. Information quantity and communication effectiveness: Low-impact messages on wilderness trailside bulletin boards. *Leisure Sciences.* 19: 59–72.

Cole, David N.; Watson, Alan; Hall, Troy; Spildie, David. 1997. High-use destinations in wilderness: Social and biophysical impacts, visitor responses, and management options. Research Paper INT-RP-496. Ogden, UT: U.S. Department of Agriculture, Forest Service, Intermountain Research Station.

Cole, David N.; Watson, Alan E.; Roggenbuck, Joseph W. 1995. Trends in wilderness visitors and visits: Boundary Waters Canoe Area, Shining Rock, and Desolation Wilderness. Research Paper INT-RP-483. Ogden, UT: U.S. Department of Agriculture, Forest Service, Intermountain Research Station.

Conrad, Richard. 1997. National survey highlights agency training needs in the United States. *International Journal of Wilderness.* 3(4): 9–12.

Cordell, H. Ken; Tarrant, M. A.; McDonald, B. L.; Bergstrom, J. C. 1998. How the public views wilderness: More results from the USA survey on recreation and the environment. *International Journal of Wilderness.* 4(3): 28–31.

Cordell, Ken; Barns, Chris; Brownlie, David; Carlson, Tom; Dawson, Chad; Koch, William; Oye, Garry; Ryan, Chris. 2016. Wilderness stewardship in America today and what we can do to improve it. *Journal of Forestry.* 114(3): 292–297.

Dawson, Chad P. 2007. New opportunities for educating future wilderness and wildland managers in a changing technological world. *International Journal of Wilderness.* 13(3): 36–39.

Dawson, Chad P.; Hammitt, William E. 1996. Dimensions of wilderness privacy for Adirondack Forest Preserve hikers. *International Journal of Wilderness.* 2(1): 37–41.

Dawson, Chad; Cordell, Ken; Watson, Alan E.; Ghimire, Ramesh; Green, Gary T. 2016. The U.S. Wilderness Managers Survey: Charting a path for the future. *Journal of Forestry.* 114(3): 298–304.

Denburg, Ronald F. 1982. *Crowding and social displacement on the lower St. Croix and upper Mississippi Rivers.* Thesis. Madison: University of Wisconsin.

Doucette, J.; Cole, David N. 1993. Wilderness visitor education: Information about alternative techniques. Technical Report INT-295. Ogden, UT: U.S. Department of Agriculture, Forest Service, Intermountain Research Station.

Driver, B.; Brown, P. 1983. *Contributions of Behavioral Scientists to Recreation Resource Management.* New York: Plenum.

Driver, B.; Nash, R.; Haas, G. 1987. Wilderness benefits: A state of knowledge review. In: *National Wilderness Research Conference: Issues, State of Knowledge, Future Directions.* GTR-INT-220. Ogden, UT: U.S. Department of Agriculture, Forest Service Intermountain Research Station, pp. 294–319

Dvorak, R. G.; Borrie, William T. 2007. Changing relationships with wilderness: A new focus for research and stewardship. *International Journal of Wilderness.* 13(3): 12–15.

Ewert, A.; Hendee, J. C.; Davidson, S.; Brame, R.; and Mackey, C. 1999. Wilderness access issues for education, personal growth, and therapeutic use: A U.S. panel summary. *International Journal of Wilderness.* 5(3): 13–18.

Farnum, J.; Hall, T.; Kruger, L. E. 2005. Sense of place in natural resource recreation and tourism: An evaluation and assessment of research findings. General Technical Report PNW-GTR-660. Portland, OR: U.S. Department of Agriculture, Forest Service, Pacific Northwest Research Station.

Freimund, W. A.; Cole, D. N. 2001. Use density, visitor experience, and limiting recreational use in wilderness: Progress to date and research needs. In: Freimund, W. A.; Cole, D. N., eds. *Visitor Use Density and Wilderness Experience.* Missoula, MT: U.S. Department of Agriculture, Forest Service, Rocky Mountain Research Station, pp. 3–8.

Gager, Dan; Hendee, J. C.; Kinziger, M.; Krumpe, E. 1998. What managers are saying and doing about Wilderness Experience Programs. *Journal of Forestry.* 96(8): 33–37.

Gimblett, H. R.; Itami, R. M.; Richards, M. 2000. Simulating wildland recreation use and conflicting spatial interactions using rule-driven intelligent agents. In: Gimblett, H. R., ed. *Integrating GIS and Agent-Based Modeling Techniques for Understanding Social and Ecological Processes.* New York: Oxford University Press.

Graefe, A. R.; Thapa, B. 2004. Conflict in natural resource recreation. In: Manfredo, M. J.; Vaske, J. J.; Field, D.; Brown, P.; Bruyere, B., eds. *Society and Natural Resources: A Summary of Knowledge.* Jefferson City, MO: Modern Litho, pp. 209–224.

Grand Canyon National Park. 2023. Ten essentials. https://www.nps.gov/articles/10essentials.htm; accessed May 1, 2023.

Hall, T.; Shelby, B. 1996. Who cares about encounters? Differences between those with and without norms. *Leisure Sciences.* 18: 7–22.

Hammitt, William E.; Madden, Mark A. 1989. Cognitive dimensions of wilderness privacy: A field test and further explanation. *Leisure Sciences.* 11: 293–301.

Hammitt, William E.; Patterson, M. E. 1991. Coping behavior to avoid visitor encounters: Its relationship to wild land privacy. *Journal of Leisure Research.* 23(3): 225–237.

Hammitt, William E.; Rutlin, William M. 1995. Use encounter standards and curves for achieved privacy in wilderness. *Leisure Sciences.* 17(4): 245–262.

Hansen, Greg. 1990. Teaching a wilderness ethic—reaching beyond the forest. In: Lime, David, ed. *Managing America's Enduring Wilderness Resource.* St. Paul: University of Minnesota Extension Service, pp. 123–130.

Hendee, John C.; Catton, William R., Jr.; Marlow, Larry D.; Brockman, C. Frank. 1968. Wilderness users in the Pacific Northwest—their characteristics, values, and management preferences. Research Paper PNW-61. Portland, OR: U.S. Department of Agriculture, Forest Service, Pacific Northwest Forest and Range Experiment Station.

Hendee, John C.; Lucas, Robert C. 1973. Mandatory wilderness permits: A necessary management tool. *Journal of Forestry.* 71(4): 206–209.

Heywood, J. L. 1996. Social regularities in outdoor recreation. *Leisure Sciences.* 18: 23–37.

Jacob, Gerald R.; Schreyer, Richard. 1980. Conflict in outdoor recreation: A theoretical perspective. *Journal of Leisure Research.* 12(4): 368–380.

Krumpe, Edwin E.; Brown, Perry J. 1982. Redistributing backcountry use through information related to recreational experiences. *Journal of Forestry.* 80(6): 360–362.

Kuss, F. R.; Graefe, A.; Vaske, J. J. 1990. *Visitor Impact Management: A Review of Literature.* Washington, DC: National Parks and Conservation Association.

Leave No Trace Center for Outdoor Ethics. 2008. Leave No Trace Program Principles. Boulder, CO. www.lnt.org/programs/principles.php; accessed February 1, 2008.

Lime, David W.; Anderson, Dorothy H.; McCool, Stephen F. 1978. An application of the simulator to a river recreation setting. In: Shechter, Mordechai; Lucas, Robert C., eds. *Simulation of Recreational Use for Park and Wilderness Management.* Baltimore: Johns Hopkins University Press, pp. 153–174.

Lime, David W.; Lucas, Robert C. 1977. Good information improves the wilderness experience. *Naturalist.* 28(4): 18–20.

Lucas, Robert C. 1980. Use patterns and visitor characteristics, attitudes, and preferences in nine wilderness and other roadless areas. Research Paper INT-253. Ogden, UT: U.S. Department of Agriculture, Forest Service, Intermountain Forest and Range Experiment Station.

———. 1981. Redistributing wilderness use through information supplied to visitors. Research Paper INT-277. Ogden, UT: U.S. Department of Agriculture, Forest Service, Intermountain Forest and Range Experiment Station.

———. 1982. Recreation regulations—When are they needed? *Journal of Forestry.* 80(3): 148–151.

———. 1985. Visitor characteristics, attitudes, and use patterns in the Bob Marshall Wilderness Complex, 1970–1982. Research Paper INT-345. Ogden, UT: U.S. Department of Agriculture, Forest Service, Intermountain Forest and Range Experiment Station.

Lucas, Robert C.; Shechter, Mordechai. 1977. A recreational visitor travel simulation model as an aid to management planning. *Simulation and Games.* 8(3): 375–384.

Mackey, C. 1998. U.S. wilderness management in the 21st century—Politics, policy, and partnerships. *International Journal of Wilderness.* 4(3): 6–11.

Manning, Robert E. 1985. Crowding norms in backcountry settings: A review and synthesis. *Journal of Leisure Research.* 17(2): 75–89.

———. 1986. Density and crowding in wilderness: Search and research for satisfaction. In: Lucas, Robert C., comp. *Proceedings: National Wilderness Research Conference: Current Research;* July 23–26, 1985; Fort Collins, CO. General Technical Report INT-212. Ogden, UT: U.S. Department of Agriculture, Forest Service, Intermountain Research Station, pp. 440–448.

———. 1999. *Studies in Outdoor Recreation: Search and Research for Satisfaction,* 2nd ed. Corvallis: Oregon State University.

Manning, Robert E.; Ballinger, N.; Marion, Jeffrey L.; Roggenbuck, Joseph W. 1996. Recreation management in natural areas: Problems and practices, status and trends. *Natural Areas Journal.* 16: 142–146.

Manning, R. E.; Lime, D. W. 2000. Defining and managing the quality of wilderness recreation experiences. In: Cole, D. N.; McCool, S. F.; Borrie, William T.; O'Loughlin, Jennifer, comps. *Proceedings: Wilderness Science in a Time of Change.* Vol. 4: *Wilderness Visitors, Experiences, and Visitor Management;* May 23–27, 1999; Missoula, MT. RMRS-P-15. Ogden, UT: U.S. Department of Agriculture, Forest Service, Rocky Mountain Research Station, pp. 13–52.

Manning, Robert E.; Potter, Fletcher I. 1984. Computer simulation as a tool in teaching park and wilderness management. *Journal of Environmental Education.* 15(3): 3–9.

Manning, Robert E.; Valliere, William A. 1996. Environmental values, environmental ethics, and wilderness management—An empirical study. *International Journal of Wilderness.* 2(2): 27–32.

Marion, Jeffrey L.; Roggenbuck, Joseph W.; Manning, Robert E. 1993. Problems and practices in backcountry recreation management: A survey of National Park Service managers. NPS/NRVT/NRR-93/12. Denver: U.S. Department of the Interior, National Park Service.

Martin, Burnham H. 1986. Hikers' opinions about fees for backcountry recreation. In: Lucas, Robert C., comp. *Proceedings: National Wilderness Research Conference: Current Research;* July 23–26, 1985; Fort Collins, CO. General Technical Report INT-212. Ogden, UT: U.S. Department of Agriculture, Forest Service, Intermountain Research Station, pp. 483–488.

McCool, Stephen F.; Christensen, N. 1996. Alleviating congestion in parks and recreation areas through direct management of visitor behavior. In: *Crowding and Congestion in the National Park System: Guidelines for Management and Research.* St. Paul: University of Minnesota, Agricultural Experiment Station, pp. 67–83.

McCool, Stephen F.; Clark, R. N.; Stankey, G. H. 2007. An assessment of frameworks useful for public land recreation planning. General Technical Report PNW-GTR-705. Portland, OR: U.S. Department of Agriculture, Forest Service, Pacific Northwest Research Station. 125p.

McCool, Stephen F.; Stankey, George H. 1986. Visitor Attitudes Toward Wilderness Fire Management Policy: 1971–1984. Research Paper INT-357. Ogden, UT: U.S. Department of Agriculture, Forest Service, Intermountain Research Station.

McCool, Stephen F.; Utter, Jack. 1982. Recreation use lotteries: Outcomes and preferences. *Journal of Forestry.* 80(1): 10–11, 29.

Monz, C.; Roggenbuck, J.; Cole, D.; Brame, R.; Yoder, A. 2000. Wilderness party size regulations: Implications for management and a decision-making framework. In: Cole, D. N.; McCool, S. F.; Borrie, William T.; O'Loughlin, Jennifer, comps. *Proceedings: Wilderness Science in a Time of Change.* Vol. 4: *Wilderness Visitors, Experiences, and Visitor Management;* May 23–27, 1999; Missoula, MT. RMRS-P-15. Ogden, UT: U.S. Department of Agriculture, Forest Service, Rocky Mountain Research Station, pp. 265–273.

The Mountaineers. 2010. *Mountaineering: The Freedom of the Hills,* 8th ed. Mountaineer Books.

Muth, Robert M.; Clark, Roger N. 1978. Public participation in wilderness and backcountry litter control: A review of research and management experience. General Technical Report PNW-75. Portland, OR: U.S. Department of Agriculture, Forest Service, Pacific Northwest Forest and Range Experiment Station.

Oset, Robert. 1990. Administrative resuscitation: New life for wilderness education. In: Lime, David W., ed. *Managing America's Enduring Wilderness Resource: Proceedings of the Conference;* September 11–17, 1989; Minneapolis, MN. St. Paul: University of Minnesota, Extension Service, pp. 113–117.

Patterson, M.; Hammitt, W. E. 1990. Backcountry encounter norms, actual reported encounters, and their relationship to wilderness solitude. *Journal of Leisure Research.* 22(3): 259–275.

Peterson, M. R. 1996. Wilderness by state mandate: A survey of state-designated wilderness areas. *Natural Areas Journal.* 16(3): 192–197.

Queen, L.; Freimund, W.; Peel, S. 1998. Enhancing the potential for wilderness electronic communication. In: Kulhavy, D. L.; Legg, M. H., eds. *Wilderness and Natural Areas in Eastern North America: Research, Management and Planning.* Nacogdoches, TX: Stephen F. Austin State University, Arthur Temple College of Forestry, Center for Applied Studies, pp. 180–184.

Richer, Jerrell R.; Christensen, Neal A. 1999. Appropriate fees for wilderness day use: Pricing decisions for recreation on public lands. *Journal of Leisure Research.* 31(3): 269–280.

Robertson, Rachel D. 1986. Actual versus self-reported wilderness visitor behavior. In: Lucas, Robert C., comp. *Proceedings: National Wilderness Research Conference: Current Research*; July 23–26, 1985; Fort Collins, CO. General Technical Report INT-212. Ogden, UT: U.S. Department of Agriculture, Forest Service, Intermountain Research Station, pp. 326–332.

Roggenbuck, Joseph W. 1980. Wilderness user preferences: Eastern and western areas. In: *Proceedings: Wilderness Management Symposium*; November 13–15, 1980; Knoxville, TN. Atlanta, GA: U.S. Department of Agriculture, Forest Service, Southern Region, pp. 103–146.

———. 1992. Use of persuasion to reduce resource impacts and visitor conflicts. In: Manfredo, J. J., ed. *Influencing Human Behavior: Theory and Applications in Recreation, Tourism, and Natural Resources.* Champaign, IL: Sagamore Publishing, pp. 149–208.

Roggenbuck, Joseph W.; Berrier, Deborah L. 1982. A comparison of the effectiveness of two communication strategies in dispersing wilderness campers. *Journal of Leisure Research.* 14(1): 77–89.

Roggenbuck, Joseph W.; Ham, Sam. 1986. Use of information and education in recreation management. In: *A Literature Review: The President's Commission on Americans Outdoors.* Report M-59-M-71. Washington, DC: U.S. Government Printing Office.

Roggenbuck, Joseph W.; Manfredo, Michael J. 1990. Choosing the right route to wilderness education. In: Lime, David W., ed. *Managing America's Enduring Wilderness Resource;* September 11–17, 1989; St. Paul, MN. St. Paul: University of Minnesota, Extension Service, pp. 103–112.

Schreyer, R. 1990. Conflict in outdoor recreation: The scope of the challenge to resource planning and management. In: Vining, J., ed. *Social Science and Natural Resources Management.* Boulder, CO: Westview Press, pp. 13–31.

Schuster, R. M.; Cole, D.; Hall, T.; Baker, J.; Oreskes, R. 2006. Appraisal of and response to social conditions in the Great Gulf Wilderness: Relationships among perceived crowding, rationalization, product shift, satisfaction, and future behavioral intentions. In: Burns, R.; Robinson, K. *Proceedings of the 2006 Northeastern Recreation Research Symposium;* April 9–11, 2006; Bolton Landing, NY. General Technical Report NRS-P-14; Newton Square, PA: U.S. Department of Agriculture, Forest Service, Northern Research Station, pp. 488–496.

Schuster, R. M.; Hammitt, W. E.; Moore, D. 2006. Stress appraisal and coping response to hassles experienced in outdoor recreation settings. *Leisure Sciences.* 28: 97–113.

Schuster, R. M.; Hammitt, W. E.; Moore, D.; Schneider, I. E. 2006. Coping with stress resulting from social value conflict: Nonhunters' response to anticipated social interaction with hunters. *Human Dimensions of Wildlife Management.* 11: 101–113.

Shechter, Mordechai; Lucas, Robert C. 1978. *Simulation of Recreational Use for Park and Wilderness Management.* Baltimore: Johns Hopkins University Press.

Shelby, Bo; Harris, Richard. 1985. Comparing methods for determining visitor evaluations of ecological impacts: Site visits, photographs, and written descriptions. *Journal of Leisure Research.* 17(1): 57–67.

———. 1986. User standards for ecological impacts at wilderness campsites. In: Lucas, Robert C., comp. *Proceedings: National Wilderness Research Conference: Current Research*; July 23–26, 1985; Fort Collins, CO. General Technical Report INT-212. Ogden, UT: U.S. Department of Agriculture, Forest Service, Intermountain Research Station, pp. 166–171.

Shelby, Bo; Heberlein, Thomas A. 1984. A conceptual framework for carrying capacity determination. *Leisure Sciences.* 6(4): 433–451.

Shew, Richard L.; Saunders, Paul R.; Ford, Joseph D. 1986. Wilderness managers' perceptions of recreational horse use in the northwestern United States. In: Lucas, Robert C., comp. *Proceedings: National Wilderness Research Conference: Current Research*; July 23–26, 1985; Fort Collins, CO. General Technical Report INT-212. Ogden, UT: U.S. Department of Agriculture, Forest Service, Intermountain Research Station, pp. 320–325.

Shindler, B.; Shelby, Bo. 1993. Regulating wilderness use: An investigation of user group support. *Journal of Forestry.* 91: 41–44.

Smith, V. Kerry; Krutilla, John V. 1976. *Structure and Properties of a Wilderness Travel Simulator: An Application to the Spanish Peaks Area.* Baltimore: Johns Hopkins University Press.

Stankey, George H. 1973. Visitor perception of wilderness recreation carrying capacity. Research Paper INT-142. Ogden, UT: U.S. Department of Agriculture, Forest Service, Intermountain Forest and Range Experiment Station.

———. 1979. Use rationing in two southern California wildernesses. *Journal of Forestry.* 77(6): 347–349.

———. 1980. A comparison of carrying capacity perceptions among visitors to two wildernesses. Research Paper INT-242. Ogden, UT: U.S. Department of Agriculture, Forest Service, Intermountain Forest and Range Experiment Station.

Stankey, George H.; Baden, John. 1977. Rationing wilderness use. General Technical Report INT-198. Ogden, UT: U.S. Department of Agriculture, Forest Service, Intermountain Forest and Range Experiment Station.

Stankey, George H.; Schreyer, Richard. 1987. Attitudes toward wilderness and factors affecting visitor behavior: A state-of-knowledge review. In: Lucas, Robert C., comp. *Proceedings: National Wilderness Research Conference: Issues, State-of-Knowledge, Future Directions*; July 23–26, 1985; Fort Collins, CO. General Technical Report INT-220. Ogden, UT: U.S. Department of Agriculture, Forest Service, Intermountain Research Station, pp. 246–293.

Swain, Ralph W. 1986. *Colorado wilderness violators: Who they are and why they violate.* Thesis. Fort Collins: Colorado State University.

———. 1996. Leave No Trace (LNT)—Outdoor skills and ethics program. *International Journal of Wilderness.* 2(3): 24–26.

Swearingen, T. C.; Johnson, D. R. 1995. Visitors' responses to uniformed park employees. *Journal of Park and Recreation Administration.* 13: 73–85.

Tangen-Foster, J.; Dawson, C. P. 1999. Risk management programs in wilderness experience programs. *International Journal of Wilderness.* 5(3): 29–34.

Taylor, Jonathon G.; Mutch, Robert W. 1986. Fire in wilderness: Public knowledge, acceptance, and perceptions. In: Lucas, Robert C., comp. *Proceedings: National Wilderness Research Conference: Current Research;* July 23–26, 1985; Fort Collins, CO. General Technical Report INT-212. Ogden, UT: U.S. Department of Agriculture, Forest Service, Intermountain Research Station, pp. 49–59.

Underhill, A. Heaton; Xaba, A. Busa; Borkan, Ronald E. 1986. The wilderness use simulation model applied to Colorado River boating in Grand Canyon National Park, USA. *Environmental Management.* 10(3): 367–374.

U.S. Department of Agriculture, Forest Service. 1982. ROS user's guide. Washington, DC: U.S. Department of Agriculture, Forest Service.

U.S. Public Law 88-577. The Wilderness Act of September 3, 1964. 78 Stat. 890.

U.S. Public Law 96-487. Alaska National Interest Lands Conservation Act (ANILCA) of December 2, 1980. 94 Stat. 2371.

Vaske, J. J.; Donnelly, M. P.; Wittmann, K.; Laidlaw, S. 1995. Interpersonal versus social-values conflict. *Leisure Sciences.* 17: 205–222.

Wagar, J. V. K. 1940. Certified outdoorsmen. *American Forests.* 46(11): 490–492, 524–525.

Washburne, Randel F.; Cole, David N. 1983. Problems and Practices in Wilderness Management: A Survey of Managers. Research Paper INT-304. Ogden, UT: U.S. Department of Agriculture, Forest Service, Intermountain Forest and Range Experiment Station.

Watson, Alan E. 1994. The nature of conflict between hikers and recreational stock users in the John Muir Wilderness. *Journal of Leisure Research.* 26(4): 372–385.

———. 1995a. An analysis of recent progress in recreation conflict research and perceptions of future challenges and opportunities. *Leisure Science.* 17(3): 235–238.

———. 1995b. Opportunities for solitude in the Boundary Waters Canoe Area Wilderness. *Journal of Applied Forestry.* 12(1): 12–18.

———. 2000. Wilderness use in the year 2000: Societal changes that influence human relationships with wilderness. In: Cole, D. N.; McCool, S. F; Borrie, William T.; O'Loughlin, J, comps. *Proceedings: Wilderness Science in a Time of Change.* Vol. 4: *Wilderness Visitors, Experiences, and Visitor Management;* May 23–27, 1999; Missoula, MT. RMRS-P-15. Ogden, UT: U.S. Department of Agriculture, Forest Service, Rocky Mountain Research Station, pp. 53–60.

Watson, Alan E.; Asp, C.; Walsh, J.; Kulla, A. 1997. The contribution of research to managing conflict among national forest users. *Trends.* 34(3): 29–35.

Watson, Alan E.; Christensen, N. A.; Blahna, D. J.; Archibold, K. S. 1998. Comparing manager and visitor perceptions of llama use in wilderness. Research Paper RMRS-RP-10. Ogden, UT: U.S. Department of Agriculture, Forest Service, Rocky Mountain Research Station.

Watson, Alan E.; Hendee, John C.; Zaglauer, Hans P. 1996. Human values and codes of behavior: Changes in Oregon's Eagle Cap Wilderness visitors and their attitudes. *Natural Areas Journal.* 16(2): 89–93.

Welsh, R. T. 2000. A comparison of strategies for rationing and managing use on selected rivers in the United States in 1986 and 1998. In: Watson, A. E.; Aplet, G. H.; Hendee, J. C., comps. *Personal, Societal, and Ecological Values of Wilderness: Sixth World Wilderness Congress Proceedings on Research, Management, and Allocation.* Vol. II; October 24–29, 1998; Bangalore, India. Proceedings RMRS-P-14; Ogden, UT: U.S. Department of Agriculture, Forest Service, Rocky Mountain Research Station, pp. 25–34.

White, David; Hendee, J. C. 2000. Primal hypotheses: The relationship of naturalness and solitude to the wilderness experience benefits of self, development of community, and spiritual development. In: Cole, D. N.; McCool, S. F; Borrie, William T.; O'Loughlin, Jennifer, comps. *Proceedings: Wilderness Science in a Time of Change.* Vol. 3: *Wilderness Visitors, Experiences, and Visitor Management;* May 23–27, 1999; Missoula, MT. RMRS-P-15. Ogden, UT: U.S. Department of Agriculture, Forest Service, Rocky Mountain Research Station, pp. 223–227.

Williams, Daniel R.; Vogt, Christine A.; Vitterso, Joar. 1999. Structural equation modeling of users' response to wilderness recreation fees. *Journal of Leisure Research.* 31(3): 245–268.

Winter, Patricia L.; Palucki, Laura J.; Burkhardt, Rachel L. 1999. Anticipated responses to a fee program: The key is trust. *Journal of Leisure Research.* 31(3): 207–226.

Worf, William A. 1987. Introduction to wilderness research needs panel discussion. In: Lucas, Robert C., comp. *Proceedings: National Wilderness Research Conference: Issues, State-of-Knowledge, Future Directions;* July 23–26, 1985; Fort Collins, CO. General Technical Report INT-220. Ogden, UT: U.S. Department of Agriculture, Forest Service, Intermountain Research Station, pp. 351–352.

VI. The Future

CHAPTER

17

Future Issues and Challenges in Wilderness Stewardship

WILDERNESS—PAST, PRESENT, AND FUTURE

We have traced the development of the wilderness concept and discussed its various resources and values, threats to the resource, and management actions that might be employed to enhance and protect it. Wilderness, as a land-use concept, has evolved a great deal over the past century, particularly since passage of the Wilderness Act in 1964, as has our capacity to manage this valuable and dwindling resource (Fig. 17.1).

Aldo Leopold (1970, 264, 270) noted that "wilderness is the raw material out of which man has hammered the artifact called civilization"; he goes on to point out that only remnants of it remain, so that "*wilderness is a resource which can shrink but not grow* [emphasis added]." Invasions can be arrested or modified in a manner to keep an area usable either for recreation, or for science, or for wildlife, but the creation of new wilderness in the full sense of the word is impossible."

What about the future for wilderness? What we do today or fail to do today largely shapes the kind of wilderness future we will deal with tomorrow. The implications for thinking about wilderness as we begin the twenty-first century are clear. The kind of wilderness system we will have tomorrow—a generation or even a century from now—will depend on what we do today. This kind of thinking underlies *the concern of many for designating as much wilderness as possible now, to save it for tomorrow,* otherwise many options will be gone. In addition, the implications are just as clear for protecting wilderness already designated; if stewardship is lax, wilderness qualities and conditions will diminish or vanish.

In the U.S. experience with wilderness, defining, protecting, and managing it have derived from a particular set of environmental and sociopolitical conditions. To what extent will these conditions prevail in the future, and, to the extent that they change, how will the new conditions affect wilderness uses and values to society and needs for its management? Land at the primitive end of the environmental

FIGURE 17.1 Stewardship of the NWPS is receiving increased emphasis as more visitors and other users compete for access. Some of the primary objectives will be to maintain the qualities of natural ecosystems in settings such as the Spanish Peaks area of the Lee Metcalf Wilderness, Montana. Photo courtesy U.S. Forest Service.

modification spectrum has always had a tendency to move toward more modification, for many societal reasons. What Robert Marshall noted in 1937 (240) is still true today:

> The world is full of conflicts between genuine values. Often these conflicts are resolved entirely from the standpoint of one of the competing values, and thus whole categories of human enjoyment may be needlessly swept away. It is far more conducive to human happiness to attempt some rational balance that will make possible for the immensely different types of people the varied values they crave. Emphatically this is true of the conflict between the values created by the modification of Nature and the values of the primitive. . . The fate of unmodified Nature rests in the activity of its friends.

We contend that the quality of the National Wilderness Preservation System in the future depends substantially on the steps we take now to manage existing areas and on the high standards set for its stewardship. As we noted in the opening chapter, *failure to implement adequate and appropriate management and stewardship today will lead to diminished or diluted wilderness tomorrow.*

FIGURE 17.2 Wilderness in Alaska is relatively uncrowded and unaffected by human activities, but even here choices must be made between naturalness and wildness, lack of crowding and freedom. Photo courtesy David Cole.

A GROWING IMPORTANCE OF WILDERNESS

We see wilderness resources and values increasing in importance for many reasons—for direct recreational values; for therapy and healing, including respite for all users from the accelerating pace of today's world; for increased scientific understanding of natural systems and the world around us; as a source of material values to humankind, such as for medicines and genetic stock for agriculture and forestry; and as a protected source of ecological services, such as clean air, water, soil, and natural systems for climate regulation. Although economic estimates can place dollar amounts on some of these values, and the current economic estimates are impressive, they are but a partial indicator of the total value of wilderness. We see wilderness remaining as an important, even essential, component of a spectrum of land uses for a variety of important societal opportunities and values (Fig. 17.2). Nearly a half century ago, Howard Zahniser (1956, 41) eloquently stated some of those values:

> In the areas of wilderness that are still relatively unmodified by man it is, however, possible for a human being, adult or child, to sense and see his own humble, dependent relationship to all of life. In these areas, thus, are the opportunities for so important, so neglected a part of our education—gaining of the true understanding of our past, ourselves, and our world which will enable us to enjoy the conveniences and liberties of our urbanized, industrialized, mechanized civilization and yet not sacrifice an awareness of our human existence as spiritual creatures nurtured and sustained by and from the great community of life that comprises the wildness of the universe, of which we ourselves are a part.

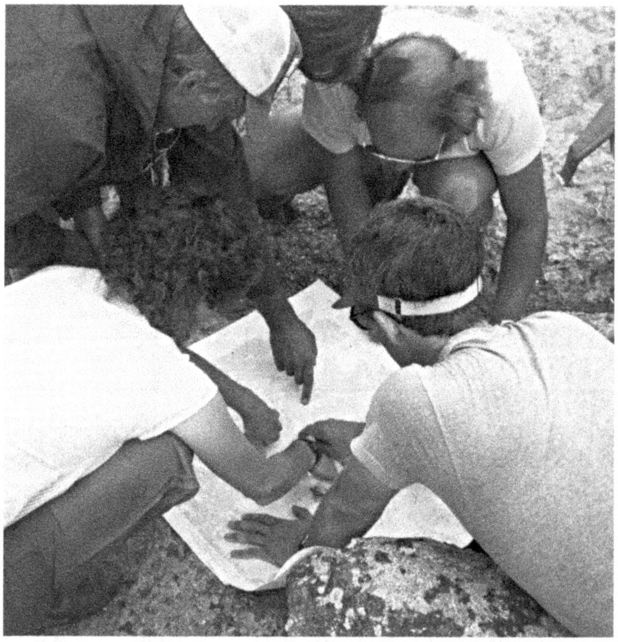

FIGURE 17.3 Wilderness provides a unique setting for educational, personal growth, and human development programs. A group supervised by the National Outdoor Leadership School plans a cross-country hike in the Popo Agie Wilderness, Wyoming. Photo courtesy National Outdoor Leadership School, Lander, Wyoming.

We are optimistic about the future of wilderness (Fig. 17.3). There is little evidence that public interest is waning, that wilderness is just a fad. To the contrary, wilderness has shown remarkable tenacity as an item on the public agenda, as evidenced by over 170 pieces of federal legislation strengthening and expanding the NWPS since 1964. We think public interest and advocacy will continue, empowered by modern communication methods such as the Internet. We also see growing international interest and support for wilderness as a measure of the imposing strength and resiliency of the wilderness idea.

Whatever problems exist with the wilderness program in the United States, it nevertheless stands as an achievement that many other nations envy and hold up as a model to which they aspire—namely a special category to protect lands with natural and wild values. For example, other nations now have wilderness systems by law or administrative mandate, and several others provide some protection for comparable areas (Chapter 3). Wilderness protection is on the international agenda as another means of protecting planet Earth, with a growing compilation of information and dialogue on international possibilities in proceedings of seven World Wilderness Congresses (Martin 2001).

Some wildernesses have experienced extremely high-use pressures, necessitating intensive management, including rationing use. Undoubtedly, such measures will continue to be needed, and in more areas. Nevertheless, we do not foresee a future in which people seeking a wilderness experience must always book a time and route months ahead, where rangers lurk behind every tree, and where every step is monitored by sensors and satellites. Many areas will be open and accessible without heavy-handed direct restrictions. Although the past decades have seen a generally steady increase in the level of recreational use of wilderness, the NWPS has also expanded. Unfortunately, the continuing growth and interest in wilderness leave managers constantly struggling to catch up (Dawson et al. 2016). This is unlikely to change because personnel and financial support are always issues, and management adjustments to increased visitation and impacts are too often made after the fact. Yet, the challenge and mandate for wilderness stewardship remain the same—to try to plan by looking ahead rather than continually reacting to the latest crisis, and to resist the temptation to come down hard on every issue, focusing instead on light-handed, indirect approaches to management, which most managers and users prefer.

Some authors over the last fifty years, like Howard Zahniser—chief architect of the Wilderness Act and former head of The Wilderness Society—warned that "we must not only protect the wilderness from exploitation. We must also see that we do not ourselves destroy its wilderness character in our own management programs. We must remember that the essential quality of the wilderness is its wildness" (Zahniser 1992, 52). The concern for management impacts on wilderness conditions is not a new issue. Cole (2000) puts forward the concept that wilderness can be managed for naturalness, wildness, uncrowded conditions (outstanding opportunities for solitude), or a minimum of management restrictions (unconfined type of recreation), but all four of these attributes of wilderness cannot be simultaneously maximized. Consequently, we support the idea of a wilderness opportunity spectrum whereby management and stewardship approaches seek to maintain diversity within and among wilderness areas—striving always to maintain minimum standards within the guidelines of the Wilderness Act and to maintain higher standards of wildness where they existed when an area was designated (Fig. 17.4).

FIGURE 17.4 When a road through the remote Izembeck National Wildlife Refuge on the Alaska Peninsula was proposed, the wilderness character of the area provided the strongest argument against development. Photo courtesy U.S. Fish and Wildlife Service.

TRENDS IN WILDERNESS USE

Wilderness recreation use provides quality experience opportunities and benefits to users and challenges to wilderness managers. More is known about recreational use than most other wilderness uses, because it has been more widely studied and measured. From the 1940s through the 1960s, recreation use increased rapidly, but in recent decades, the rate of growth, though still high, has gradually slowed. In the 1980s and early 1990s, recreation use fluctuated in many individual wilderness areas, but total NWPS recreation use increased.

Projections of future use vary widely. The annual percentage of growth in recreation use of wilderness has been small, but total use in the NWPS has grown dramatically due to the growth in the size of the NWPS as new areas have been added. Understanding wilderness recreation use and users is essential to protecting wilderness resources and values while managing conditions that influence wilderness users' experiences.

We expect recreational use of wilderness will continue to grow in the twenty-first century. We expect more demand for day use and easy access; more demand for information to support pre-trip planning and area selection; more demand for challenge, adventure, and risk-taking activities; and greater diversity among users (Chavez 2000; Hammitt and Schuster 2000). We are concerned about the trends toward the use of more high-tech equipment that may insulate visitors from truly experiencing wilderness conditions, and about the demand for more "packaged" and service-oriented commercial experiences, requiring less understanding of and connection with nature (Roggenbuck 2000; Shultis 2000).

Some social benefits are at stake in the kinds of use facilitated or stymied by wilderness management policies as we approach use limits in a growing number of areas. For example, will special permits for organized user groups favor commercial parties, science-education groups, organized youth groups, or wilderness therapy and personal growth groups? Increasingly, such difficult choices are directly influenced by wilderness planners and managers.

Although we are concerned about these pressures that may dilute the natural qualities and conditions of the wilderness experience and may have substantial negative impacts on wilderness, we hope that such pressures will be a force in helping define and strengthen a more biocentric approach to wilderness management and stewardship (Haas and Wells 2000). In the past, the case for more wilderness designation and management was often based on the need to set aside areas because of the increasing recreational demands placed on such natural areas. This argument will surely continue; however, *we believe it is time to expand the concern for wilderness, and the kind of wilderness system we need, by placing a greater emphasis on the full spectrum of environmental, scientific, and social values a NWPS can deliver* (Fig. 17.5). Examples of such values include expanding ecosystem representation; protecting genetic diversity; habitat for wilderness-dependent species, like grizzly bear; and the education, inspiration, and healing experiences that wilderness can provide.

FIGURE 17.5 Units of the National Wilderness Preservation System constitute about 4 percent of the land area of the United States, protect a range of ecosystems, and provide access to wild and varied landscapes. Everglades National Park Wilderness, Florida (upper left); Boundary Waters Canoe Area, Minnesota (upper right); Steens Mountain Wilderness, Oregon (lower left); and Aravaipa Canyon Wilderness, Arizona (lower right). Photos courtesy (upper left) National Park Service, (upper right) U.S. Forest Service, (lower left) Oregon Bureau of Land Management, (lower right) Arizona Bureau of Land Management.

During the twentieth century, increasing recreational use of wilderness provided an important and immediately obvious justification for its protection; it has also been the source of many of the most pressing management challenges. However, the case for wilderness is justified by much more than its recreation use, and more emphasis needs to be placed on its multiple values. While we, as wilderness stewards, are often preoccupied with direct use of wilderness, the public has broader interests in wilderness. Cordell and Stokes (2000) report that in national surveys the top five values of wilderness, reported by 75 percent or more of the populations, are protection for water quality, for wildlife habitat, for air quality, for endangered species, and for future generations.

Good wilderness management depends on an understanding of all the varied uses it receives, because most wilderness stewardship is management of use of one kind or another, but primarily recreational use. Some uses depend on wilderness, but others just take place there and are less dependent on wilderness attributes and conditions. Public recreation is the most obvious wilderness use, but other uses include commercial recreation; indirect, off-site uses like ecological services and benefits from wilderness; scientific uses; education uses; personal growth, therapeutic, and healing uses; and a variety of limited but legally allowable, direct commodity uses such as mining, water storage, and grazing.

MANAGEMENT CHALLENGES

The challenges facing wilderness managers are varied and numerous; however, the opportunities for success of and future of the NWPS are greater and worth the investment on behalf of present and future generations (Cordell et al. 2016; Dawson et al. 2016). For example, government financial constraints and demand for cost-effectiveness and efficient delivery of services will undoubtedly continue to increase, along with recognition that wilderness protection is possible only with adequate management and stewardship. We think this will require managers to continue to seek new ways of accomplishing their work, such as by forming alliances among different levels of government and among the public, private, and volunteer sectors of society. Many of these approaches are already well under way.

In recent years, many crucial wilderness management jobs—trail construction and maintenance, campsite rehabilitation, visitor contact—have been possible only because of volunteer or partnership participation. For example, trails have been built by conservation groups and local activity clubs (Wash and Wash 2001); volunteers have served as wilderness information specialists providing information to visitors and have collected basic resource inventory data (Joslin 2000); and interagency programs have trained scout leaders and leaders of specialized groups such as horse users to educate their constituents in LNT and minimum-impact skills (Miranda 2000). *Because partnerships and volunteers will be even more essential in the future, wilderness managers will need more and better new skills in managing and organizing people. Effective work by volunteers and other partners requires strong planning, facilitation, and guidance by professional wilderness managers. Volunteers and partnerships must supplement professional wilderness management, not substitute for it, or wilderness quality will diminish. At the same time, involving partners and volunteers in wilderness work yields important benefits to these workers, as well as to the wilderness they serve. We believe that the strength of support for and appreciation of wilderness in the future will depend on making wilderness experiences and wilderness work opportunities available to as wide a spectrum of people as possible.*

AN EXPANDING—BUT ULTIMATELY LIMITED—WILDERNESS SYSTEM

When the first edition of this book was published in 1978, fewer than 10 million acres of wilderness existed. Now more than 111 million acres exist, and the NWPS will likely grow beyond that. Few people, even among the most dedicated wilderness enthusiasts, foresaw such an outcome. It is further evidence, in our view, of the strength and breadth of support for wilderness in the United States.

Ironically, the rapid expansion of wilderness, coupled with the many laws Congress has passed to create such a system, has raised fears about a gradual deterioration in standards and perhaps a slow, insidious creep of more lenient, less demanding qualities for areas admitted to the NWPS. Yet, by and large, the demanding standards of wilderness management described by the act have been consistently endorsed by the words and actions of Congress. Moreover, although some areas admitted to the NWPS currently do not meet all standards for wilderness, they provide an opportunity for managers to begin the process of restoring wilderness qualities.

Threats to wilderness integrity and conditions are numerous and will require the vigilance of wilderness managers, rangers, users, volunteers, outfitters, and the general public. As Robert Marshall (1930, 52) commented more than nine decades ago: "There is just one hope of repulsing the tyrannical ambition of civilization to conquer every niche on the whole earth. That hope is the organization of spirited people who will fight for the freedom of the wilderness." Examples of threats to wilderness abound. Scores of wilderness plans are contentious and may only be resolved through compromise; unfortunately, the compromises will likely diminish wilderness qualities in some way.

With a growing national and international population, the need and demand for limited resources of all types will only increase in the United States and throughout the world. One of the greatest values of wilderness to future generations will be its relative "naturalness," because 96 percent of the land will have been heavily managed, manipulated, and developed for human uses. *Wilderness in the United States will be increasingly scarce, and the remnants of functioning ecosystems and natural landscapes that legislation protects will be stressed to provide places for escape from escalating pressures of modern life* (Fig. 17.6).

FIGURE 17.6 Backpackers gather at information panels at a wilderness trailhead (top); untrammeled wilderness spreads out in front of backpackers in Alaska (middle); a canoeist paddles in Minnesota's Boundary Waters (bottom). Photos courtesy (top) Alan Ewart, (middle) David Cole, (bottom) John Roggenbuck.

Protecting wilderness from threats to dilute or diminish its naturalness and solitude is the central mission of wilderness management and stewardship. The threats are many, and they can be internal or external to any given wilderness area. We discuss threats in Chapter 13, too many to summarize here. *However, anticipating and managing the short-term and long-term threats to wilderness resources and values are the most important challenges to achieving the purposes of the Wilderness Act for present and future generations—this is the challenge of wilderness management and stewardship* (Cordell et al. 2016; Dawson et al. 2016).

CONCLUSION

The future will further affirm the important symbolic values underlying preserving wilderness. In the last two hundred years, the United States has been settled and modified virtually from shore to shore and border to border. At more than 111 million acres, the wilderness system is, in one sense, vast. Yet, we must remember it constitutes slightly more than 4 percent of the nation's total land area and less than 2 percent of the lower forty-eight states. How much more will be added in future wilderness designations—perhaps another 1 percent, or about 25 million acres, of the United States? Thus, the wilderness values available to our society—historical, ecological, recreational, spiritual, inspirational—are ultimately finite. The availability of naturalness and solitude will depend on what natural land areas we have been able to preserve, and how well we have preserved them. The values of wilderness to present and future generations will change and shift with societal influences, but those values will not diminish. The challenge to provide adequate and appropriate management and stewardship of these protected natural land areas will increase dramatically in proportion to their scarcity.

The Wilderness Act is a historic and important landmark in U.S. history, but the concept of wilderness as a symbol of nature is not fixed in time or legislation—rather it is fixed in the human consciousness. As Olaus Murie (1961, n.p.) noted about the Arctic National Wildlife Range (now the Arctic National Wildlife Refuge): "Certainly a wilderness area, a little portion of our planet left alone, undeveloped, will furnish us with a number of very important uses. . . . We have only begun to understand the basic energies which through the ages have made this planet habitable. If we are wise, we will cherish what we have left of such places in our land." *The ultimate goal of this book is to help ensure that there are some wild places, some wildness, some naturalness, some solitude—some wilderness—for us to enjoy and for future generations to experience.*

REFERENCES

Chavez, Deborah J. 2000. Wilderness visitors in the 21st century: Diversity, day use, and preferences. *International Journal of Wilderness.* 6(2): 10–11.

Church, Frank. 1977. Wilderness in a balanced land use framework. Wilderness Resource Distinguished Lectureship No. 1. Moscow: University of Idaho Wilderness Research Center.

Clark, Jamie Rappaport. 2000. The value of wilderness to the U.S. National Refuge System. *International Journal of Wilderness.* 6(3): 7–11.

Cole, David N. 2000. Natural, wild, uncrowded, or free? Which of these should our wilderness be? *International Journal of Wilderness.* 6(2): 5–8.

Cordell, Ken; Stokes, Jerry. 2000. The social value of wilderness: A Forest Service perspective. *International Journal of Wilderness.* 6(2): 23–24.

Cordell, Ken; Barns, Chris; Brownlie, David; Carlson, Tom; Dawson, Chad; Koch, William; Oye, Garry; Ryan, Chris. 2016. Wilderness stewardship in America today and what we can do to improve it. *Journal of Forestry.* 114(3): 292–297.

Dawson, C. P. 2007. New opportunities for educating future wilderness and wildland managers in a changing technological world. *International Journal of Wilderness.* 13(3): 36–39.

Dawson, Chad; Cordell, Ken; Watson, Alan E.; Ghimire, Ramesh; Green, Gary T. 2016. The U.S. Wilderness Managers Survey: Charting a path for the future. *Journal of Forestry.* 114(3): 298–304.

Haas, Glenn; Wells, Marcella. 2000. A more pristine wilderness. *International Journal of Wilderness.* 6(2): 21–22.

Hammitt, William E.; Schuster, Rudy M. 2000. Wilderness use in the next 100 years. *International Journal of Wilderness.* 6(2): 12–13.

Joslin, Les. 2000. A community-based wilderness education partnership in central Oregon. *International Journal of Wilderness.* 6(3): 27–30.

Leopold, Aldo. 1970. *A Sand County Almanac with Essays on Conservation from Round River.* New York: Sierra Club and Ballantine Books. 295p.

Marshall, Robert. 1930. The problems of the wilderness. *The Scientific Monthly.* 43–52.

———. 1937. The universe of the wilderness is vanishing. *Nature Magazine.* 29(4): 235–240.

Martin, Vance. 2001. The World Wilderness Congress. *International Journal of Wilderness.* 7(1): 31–34.

Miranda, Barb. 2000. The central and southern Sierra wilderness education project: An outreach program that works. *International Journal of Wilderness.* 6(3): 35–37.

Murie, Olaus. 1961. Wilderness philosophy, science, and the Arctic National Wildlife Range. In: Dahlgren, G., Jr., ed. *Proceedings: Twelfth Alaskan Science Conference, American Association for the Advancement of Science*; August 28–September 1, 1961. Fairbanks: University of Alaska.

Roggenbuck, Joseph W. 2000. Meanings of wilderness experiences in the 21st century. *International Journal of Wilderness.* 6(2): 14–17.

Shultis, John. 2000. Gearheads and golums: Technology and wilderness recreation in the 21st century. *International Journal of Wilderness.* 6(2): 17–18.

Wash, Dave; Wash, Katie. 2001. The BLM partnership with the Student Conservation Association: Restoring wilderness in the California desert. *International Journal of Wilderness.* 7(1): 30.

Zahniser, Howard. 1956. The need for wilderness areas. *The Living Wilderness.* 21(59): 37–43.

Zahniser, Edward. 1992. *Where Wilderness Preservation Began: Adirondack Writings of Howard Zahniser.* Utica, NY: North Country Books.

Appendix A

The Wilderness Act

Act of September 3, 1964
(Public Law 88-577, 78 Stat. 890; 16 U.S. Code 1121, 1131-1136)
Sec 1. This Act may be cited as the "Wilderness Act" (16 U.S. Code 1121)

WILDERNESS SYSTEM ESTABLISHED STATEMENT OF POLICY

Sec. 2. (a) In order to assure that an increasing population, accompanied by expanding settlement and growing mechanization, does not occupy and modify all areas within the United States and its possessions, leaving no lands designated for preservation and protection in their natural condition, it is hereby declared to be the policy of the Congress to secure for the American people of present and future generations the benefits of an enduring resource of wilderness. For this purpose there is hereby established a National Wilderness Preservation System to be composed of federally owned areas designated by Congress as "wilderness areas," and these shall be administered for the use and enjoyment of the American people in such manner as will leave them unimpaired for future use and enjoyment as wilderness, and so as to provide for the protection of these areas, the preservation of their wilderness character, and for the gathering and dissemination of information regarding their use and enjoyment as wilderness; and no Federal lands shall be designated as "wilderness areas" except as provided for in this chapter or by a subsequent Act.

(b) The inclusion of an area in the National Wilderness Preservation System notwithstanding, the area shall continue to be managed by the Department and agency having jurisdiction thereover immediately before its inclusion in the National Wilderness Preservation System unless otherwise provided by Act of Congress. No appropriation shall be available for the payment of expenses or salaries for the administration of the National Wilderness Preservation System as a separate unit nor shall any appropriations be available for additional personnel stated as being required solely for the purpose of managing or administering areas solely because they are included within the National Wilderness Preservation System.

(c) A wilderness, in contrast with those areas where man and his own works dominate the landscape, is hereby recognized as an area where the earth and its community of life are untrammeled by man, where man himself is a visitor who does not remain. An area of wilderness is further defined to mean in this chapter an area of undeveloped Federal land retaining its primeval character and influence, without permanent improvements or human habitation, which is protected and managed so as to preserve its natural conditions and which (1) generally appears to have been affected primarily by the forces of nature, with the imprint of man's work substantially unnoticeable; (2) has outstanding opportunities for solitude or a primitive and unconfined type of recreation; (3) has at least five thousand acres of land or is of sufficient size as to make practicable its preservation and use in an unimpaired condition; and (4) may also contain ecological, geological, or other features of scientific, educational, scenic, or historical value.

PREVIOUSLY CLASSIFIED AREAS

Sec. 3. (a) All areas within the national forests classified at least 30 days before the effective date of this Act by the Secretary of Agriculture or the Chief of the Forest Service as "wilderness," "wild," or "canoe" are hereby designated as wilderness areas. The Secretary of Agriculture shall—

(1) Within one year after the effective date of this Act file a map and legal description of each wilderness area with the Interior and Insular Affairs Committees of the United States Senate and the House of Representatives, and such descriptions shall have the same force and effect as if included in this Act: Provided however, that, correction of clerical and typographical errors in such legal descriptions and maps may be made.

(2) Maintain, available to the public, records pertaining to said wilderness areas, including maps and legal descriptions, copies of regulations governing them, copies of public notices of, and reports submitted to Congress regarding pending additions, eliminations, or modifications. Maps, legal descriptions, and regulations pertaining to wilderness areas within their respective jurisdictions also shall be available to the public in the offices of regional foresters, national forest supervisors, and forest rangers.

(b) The Secretary of Agriculture shall, within ten years after the enactment of this Act, review, as to its suitability or nonsuitability for preservation as wilderness, each area in the national forests classified on the effective date of this Act by the Secretary of Agriculture or the Chief of the Forest Service as "primitive" and report his findings to the President. The President shall advise the United States Senate and House of Representatives of his recommendations with respect to the designation as "wilderness" or other reclassification of each area on which review has been completed, together with maps and a definition of boundaries. Such advice shall be given with respect to not less than one-third of all the areas now classified as "primitive" within three years after the enactment of this Act, not less than two-thirds within seven years after the enactment of this Act, and the remaining areas within ten years after the enactment of this Act. Each recommendation of the President for designation as "wilderness" shall become effective only if so provided by an Act of Congress. Areas classified as "primitive" on the enactment of this Act shall continue to be administered under the rules and regulations affecting such areas on the enactment of this Act until Congress has determined otherwise. Any such area may be increased in size by the President at the time he submits his recommendations to the Congress by not more than five thousand acres with no more than one thousand two hundred and eighty acres of such increase in any one compact unit; if it is proposed to increase the size of any such area by more than five thousand acres or by more than one thousand two hundred and eighty acres in any one compact unit the increase in size shall not become effective until acted upon by Congress. Nothing herein contained shall limit the President in proposing, as part of his recommendations to Congress, the alteration of existing boundaries of primitive areas or recommending the addition of any contiguous area of national forest lands predominantly of wilderness value. Notwithstanding any other provisions of this chapter, the Secretary of Agriculture may complete his review and delete such area as may be necessary, but not to exceed seven thousand acres, from the southern tip of the Gore Range–Eagles Nest Primitive Area, Colorado, if the Secretary determines that such action is in the public interest.

(c) Within ten years after the effective date of this Act the Secretary of the Interior shall review every roadless area of five thousand contiguous acres or more in the national parks, monuments and other units of the national park system and every such area of, and every roadless island within the national wildlife refuges and game ranges, under his jurisdiction on the effective date of this Act and shall report to the President his recommendation as to the suitability or nonsuitability of each such area or island for preservation as wilderness. The President shall advise the President of the Senate and the Speaker of the House of Representatives of his recommendation with respect to the designation as wilderness of each such area or island on which review has been completed, together with a map thereof and a definition of its boundaries. Such advice shall be given with respect to not less than one-third of the areas and islands to be reviewed under this subsection within three years after enactment of this Act, not less than two-thirds within seven years of enactment of this Act. A recommendation of the President for designation as wilderness shall become effective only if so provided by an Act of Congress. Nothing contained herein shall, by implication or otherwise, be construed to lessen the present statutory authority of the Secretary of the Interior with respect to the maintenance of roadless areas within units of the national park system.

(d) (1) The Secretary of Agriculture and the Secretary of the Interior shall, prior to submitting any recommendations to the President with respect to the suitability of any area for preservation as wilderness—

(A) give such public notice of the proposed action as they deem appropriate, including publication in the Federal Register and in a newspaper having general circulation in the area or areas in the vicinity of the affected land;

(B) hold a public hearing or hearings at a location or locations convenient to the area affected. The hearings shall be announced through such means as the respective Secretaries involved deem appropriate, including notices in the Federal Register and in newspapers of general circulation in the area: *Provided*, that if the lands involved are located in more than one State, at least one hearing shall be held in each State in which a portion of the land lies;

(C) at least thirty days before the date of a hearing advise the Governor of each State and the governing board of each county, or in Alaska the borough, in which the lands are located, and Federal departments and agencies concerned, and invite such officials and Federal agencies to submit their views on the proposed action at the hearing or by no later than thirty days following the date of the hearing.

(2) Any views submitted to the appropriate Secretary under the provisions of (1) of this subsection with respect to any area shall be included with any recommendations to the President and to Congress with respect to such area.

(e) Any modification or adjustment of boundaries of any wilderness area shall be recommended by the appropriate Secretary after public notice of such proposal and public hearing or hearings as provided in subsection (d) of this section. The proposed modification or adjustment shall then be recommended with map and description thereof to the President. The President shall advise the United States Senate and the House of Representatives of his recommendations with respect to such modification or adjustment and such recommendations shall become effective only in the same manner as provided for in subsections (b) and (c) of this section.

LIMITATION OF USE AND ACTIVITIES

Sec. 4. (a) The purposes of this chapter are hereby declared to be within and supplemental to the purposes for which national forests and units of the national park and national wildlife refuge systems are established and administered and—

(1) Nothing in this Act shall be deemed to be in interference with the purpose for which national forests are established as set forth in the Act of June 4, 1897 (30 Stat. 11), and the Multiple-Use Sustained-Yield Act of June 12, 1960 (74 Stat. 215).

(2) Nothing in this Act shall modify the restrictions and provisions of the Shipstead-Nolan Act (Public Law 539, Seventy-first Congress, July 10, 1930; 46 Stat. 1020), the Thye-Blatnik Act (Public Law 733, Eightieth Congress, June 22, 1948; 62 Stat. 568), and the Humphrey-Thye-Blatnik-Andresen Act (Public Law 607, Eighty-fourth Congress, June 22, 1956; 70 Stat. 326), as applying to the Superior National Forest or the regulations of the Secretary of Agriculture.

(3) Nothing in this Act shall modify the statutory authority under which units of the national park system are created. Further, the designation of any area of any park, monument, or other unit of the national park system as a wilderness area pursuant to this Act shall in no manner lower the standards evolved for the use and preservation of such park, monument, or other unit of the national park system in accordance with the Act of August 25, 1916, the statutory authority under which the area was created, or any other Act of Congress which might pertain to or affect such area, including, but not limited to, the Act of June 8, 1906 (34 Stat. 225; 16 U.S.C. 432 et seq.); section 3(2) of the Federal Power Act (16 U.S.C. 796[2]); and the Act of August 21, 1935 (49 Stat. 666; 16 U.S.C. 461 et seq.).

(b) Except as otherwise provided in this Act, each agency administering any area designated as wilderness shall be responsible for preserving the wilderness character of the area and shall so administer such area for such other purposes for which it may have been established as also to preserve its wilderness character. Except as otherwise provided in this Act, wilderness areas shall be devoted to the public purposes of recreational, scenic, scientific, educational, conservation, and historical use.

(c) Except as specifically provided for in this Act, and subject to existing private rights, there shall be no commercial enterprise and no permanent road within any wilderness area designated by this Act and, except as necessary to meet minimum requirements for the administration of the area for the purpose of this Act (including measures required in emergencies involving the health and safety of persons within the area), there shall be no temporary road, no use of motor vehicles, motorized equipment or motorboats, no landing of aircraft, no other form of mechanical transport, and no structure or installation within any such area.

(d) The following special provisions are hereby made:

(1) Within wilderness areas designated by this Act the use of aircraft or motorboats, where these uses have already become established, may be permitted to continue subject to such restrictions as the Secretary of Agriculture deems desirable. In addition, such measures may be taken as may be necessary in the control of fire, insects, and diseases, subject to such conditions as the Secretary deems desirable.

(2) Nothing in this Act shall prevent within national forest wilderness areas any activity, including prospecting, for the purpose of gathering information about mineral or other resources, if such activity is carried on in a manner compatible with the preservation of the wilderness environment. Furthermore, in accordance with such program as the Secretary of the Interior shall develop and conduct in consultation with the Secretary of Agriculture, such areas shall be surveyed on a planned, recurring basis consistent with the concept of wilderness preservation by the United States Geological Survey and the United States Bureau of Mines to determine the mineral values, if any, that may be present; and the results of such surveys shall be made available to the public and submitted to the President and Congress.

(3) Notwithstanding any other provisions of this Act, until midnight December 31, 1983, the United States mining laws and all laws pertaining to mineral leasing shall, to the same extent as applicable prior to the effective date of this Act, extend to those national forest lands designated by this Act as "wilderness areas"; subject, however, to such reasonable regulations governing ingress and egress as may be prescribed by the Secretary of Agriculture consistent with the use of the land for mineral location and development and exploration, drilling, and production, and use of land for transmission lines, waterlines, telephone lines, or facilities necessary in exploring, drilling, producing, mining, and processing operations, including where essential the use of mechanized ground or air equipment and restoration as near as practicable of the surface of the land disturbed in performing prospecting, location, and, in oil and gas leasing, discovery work, exploration, drilling, and production, as soon as they have served their purpose. Mining locations lying within the boundaries of said wilderness areas shall be held and used solely for mining or processing operations and uses reasonably incident thereto; and hereafter, subject to valid existing rights, all patents issued under the mining laws of the United States affecting national forest lands designated by this Act as wilderness areas shall convey title to the mineral deposits within the claim, together with the right to cut and use so much of the mature timber therefrom as may be needed in the extraction, removal, and beneficiation of the mineral deposits, if needed timber is not otherwise reasonably available, and if the timber is cut under sound principles of forest management as defined by the national forest rules and regulations, but each such patent shall reserve to the United States all title in or to the surface of the lands and products thereof, and no use of the surface of the claim or the resources therefrom not reasonably required for carrying on mining or prospecting shall be allowed except as otherwise expressly provided in this Act: Provided, that, unless hereafter specifically authorized, no patent within wilderness areas designated by this Act shall issue after December 31, 1983, except for the valid claims existing on or before December 31, 1983. Mining claims located after the effective date of this Act within the boundaries of wilderness areas designated by this Act shall create no rights in excess of those rights which may be patented under the provisions of this subsection. Mineral leases,

permits, and licenses covering lands within national forest wilderness areas designated by this Act shall contain such reasonable stipulations as may be prescribed by the Secretary of Agriculture for the protection of the wilderness character of the lands consistent with the use of the lands for the purposes for which they are leased, permitted, or licensed. Subject to valid rights then existing, effective January 1, 1984, the minerals in lands designated by this Act as wilderness areas are withdrawn from all forms of appropriation under the mining laws and from disposition under all laws pertaining to mineral leasing and all amendments thereto.

(4) Within wilderness areas in the national forests designated by this Act, (1) the President may, within a specific area and in accordance with such regulations as he may deem desirable, authorize prospecting for water resources, the establishment and maintenance of reservoirs, water-conservation works, power projects, transmission lines, and other facilities needed in the public interest, including the road construction and maintenance essential to development and use thereof, upon his determination that such use or uses in the specific area will better serve the interests of the United States and the people thereof than will its denial; and (2) the grazing of livestock, where established prior to the effective date of this Act shall be permitted to continue subject to such reasonable regulations as are deemed necessary by the Secretary of Agriculture.

(5) Other provisions of this Act to the contrary notwithstanding, the management of the Boundary Waters Canoe Area, formerly designated as the Superior, Little Indian Sioux and Caribou Roadless Areas, in the Superior National Forest, Minnesota, shall be in accordance with regulations established by the Secretary of Agriculture in accordance with the general purpose of maintaining, without unnecessary restrictions on other uses, including that of timber, the primitive character of the area, particularly in the vicinity of the lakes, streams, and portages: Provided, that nothing in this Act shall preclude the continuance within the area of any already established use of motorboats.

(6) Commercial services may be performed within the wilderness areas designated by this Act to the extent necessary for activities which are proper for realizing the recreational or other wilderness purposes of the areas.

(7) Nothing in this Act shall constitute an express or implied claim or denial on the part of the Federal Government as to exemption from State water laws.

(8) Nothing in this chapter shall be construed as affecting the jurisdiction or responsibilities of the several States with respect to wildlife and fish in the national forests (16 U.S.C. 1133).

RIGHTS OF NON-FOREST LANDS OWNERSHIP

Sec. 5. (a) In any case where State-owned or privately owned land is completely surrounded by national forest lands within areas designated by this Act as wilderness, such State or private owner shall be given such rights as may be necessary to assure adequate access to such State-owned or privately owned land by such State or private owner and their successors in interest, or the State-owned land or privately owned land shall be exchanged for federally owned land in the same State of approximately equal value under authorities available to the Secretary of Agriculture: *Provided, however*, that the United States shall not transfer to a State or private owner any mineral interests unless the State or private owner relinquishes or causes to be relinquished to the United States the mineral interest in the surrounded land.

(b) In any case where valid mining claims or other valid occupancies are wholly within a designated national forest wilderness area, the Secretary of Agriculture shall, by reasonable regulations consistent with the preservation of the area as wilderness, permit ingress and egress to such surrounded areas by means which have been or are being customarily enjoyed with respect to other such areas similarly situated.

(c) Subject to the appropriation of funds by Congress, the Secretary of Agriculture is authorized to acquire privately owned land within the perimeter of any area designated by this chapter as wilderness if (1) the owner concurs in such acquisition or (2) the acquisition is specifically authorized by Congress (16 U.S.C. 1134).

Sec. 6. (a) The Secretary of Agriculture may accept gifts or bequests of land within wilderness areas designated by this Act for preservation as wilderness. The Secretary of Agriculture may also accept gifts or bequests of land adjacent to wilderness areas designated by this Act for preservation as wilderness if he has given sixty days' advance notice thereof to the President of the Senate and the Speaker of the House of Representatives. Land accepted by the Secretary of Agriculture under this section shall become part of the wilderness area involved. Regulations with regard to any such land may be in accordance with such agreements, consistent with the policy of this Act, as are made at the time of such gift, or such conditions, consistent with such policy, as may be included in, and accepted with, such bequest.

(b) The Secretary of Agriculture or the Secretary of the Interior is authorized to accept private contributions and gifts to be used to further the purposes of this Act (16 U.S.C. 1135).

REPORT TO CONGRESS

Sec. 7. At the opening of each session of Congress, the Secretaries of Agriculture and Interior shall jointly report to the

President for transmission to Congress on the status of the wilderness system, including a list and descriptions of the areas in the system, regulations in effect, and other pertinent information, together with any recommendations they may care to make (16 U.S.C. 1136).

APPROVED SEPTEMBER 3, 1964

Legislative History:

House Reports: No. 1538 accompanying H.R. 9070 (Committee on Interior & Insular Affairs) and No. 1829 (Committee of Conference).

Senate Report: No. 109 (Committee on Interior & Insular Affairs). Congressional Record: Vol. 109 (1963):

April 4, 8, considered in Senate.

April 9, considered and passed in Senate.

Vol. 110 (1964): July 28, considered in House.

July 30, considered and passed in House, amended, in lieu of H.R. 9070.

August 20, House and Senate agreed to conference report.

Index

M

N

O

P

Q

R

Y

Z